COAL UTILISATION
Technology, Economics and Policy

COAL UTILISATION
Technology, Economics and Policy

L. Grainger and J. Gibson

Halsted Press

First published in 1981 by

HALSTED PRESS

A Division of JOHN WILEY & SONS, Inc., Publishers
605 Third Avenue, New York, N.Y. 10016

Library of Congress Cataloging in Publication Data

Grainger, Leslie
Coal utilisation.

1. Coal. I. Gibson, J. ‹Joseph. II. Title.
TP325.G732 1981 553.2'4 81-7249
ISBN 0-470-27272-4 AACR2

ISBN 0-470-27272-4

Printed in Great Britain by
King's English Bookprinters, Leeds

Filmset in Great Britain by
Mid-County Press, London

CONTENTS

List of Figures		xi
List of Tables		xvii
Foreword		xxi
SECTION I	INTRODUCTION TO COAL AND ITS UTILISATION	
Chapter 1	WHAT IS COAL?	3
	General Introduction	3
	History of Coal Usage	4
	Geological Origin	5
	Analysis and Testing of Coals	8
	Classification Systems	13
	Petrology and Macerals	14
	The Chemical Structure of Coal	20
	The Fine Physical Structure of Coal	30
	Chemical Reactions	30
Chapter 2	WHY IS COAL IMPORTANT?	31
	The Vital Importance of Energy	31
	The Role of Fossil Fuels	33
	The Inflexibility of Non-fossil Energy Sources	35
	Conclusions	36
Chapter 3	HOW MUCH COAL?	37
	Introduction	37
	Estimates of Coal Resources	38
	World Energy Conference 1977	40
	World Energy Conference 1980	44
	WAES and WOCOL	46
	International Energy Agency (IEA)	46
	Ford Foundation	47
	Estimates for the USA	47
	Conclusions	48

Chapter 4 HOW IS COAL WON? 49

Introduction 49
Underground Mining 49
Surface Mining 51
Coal Preparation 52
Transport and Storage 53
Future Prospects 55

Chapter 5 HOW CAN COAL BE USED? 56

Introduction 56
Combustion 56
Carbonisation 58
Gasification 59
Liquefaction 60
Chemicals 60
Conclusion 60

SECTION II TECHNOLOGY OF COAL UTILISATION

Chapter 6 COMBUSTION 65

Introduction 65
Some Basic Considerations 65
Pulverised Fuel Systems 68
Fixed Bed Combustion Devices 71
Fluidised Combustion 76
Flue Gas Treatment Technology 108
Coal–Oil Mixtures (COM) 111
Combustion of Low Grade Coal 115
Advanced Combustion Systems 117
Conclusions 119
Appendix 1: Fluid Bed Combustion R&D Projects 122

Chapter 7 THE CARBONISATION OF COAL 129

Introduction 129
Scientific Basis 130
Some Features of Coke Oven Design 136
Coke Quality 137
Coke Improvement: Blend Formulation 141
Other Carbonisation Processes 143
Formed Coke 147
Modern Developments in Slot Oven Practice 150
By-products 152
Activated Carbons 153
Environmental Aspects 157
General Conclusions on Carbonisation 160

Chapter 8 GASIFICATION 161

Introduction 161
Gasifiers 165
The Treatment of Gases Prior to Use 186
Status of Gasification Development Programmes 191
Gasification Systems 201
Some Common Development Areas 207
Conclusions 210

Chapter 9 LIQUEFACTION OF COAL 211

General Considerations and Background 211
Historical Background 218
Some Major Projects and Programmes 221
Research and Supporting Technology 251
Environmental Aspects 256
Conclusions 256

Chapter 10 CHEMICALS FROM COAL 257

Introduction 257
Historical Review 257
Gasification/Synthesis 260
Hydrogenation 272
Supercritical Gas Extraction 274
Pyrolysis and Hydropyrolysis 274
Acetylene 278
Carbon Products 279
Conclusions 280

Chapter 11 *IN-SITU* PROCESSES: UNDERGROUND COAL GASIFICATION 281

Introduction 281
Historical Background 281
General Principles 282
National Programmes 285
Environmental Impact 297
Utilisation of Gas 298
Instrumentation and Control 298
Other Possible *In-situ* Methods 300
Conclusions 300

SECTION III ECONOMICS OF COAL UTILISATION

Chapter 12 GENERAL CONSIDERATIONS OF COAL UTILISATION ECONOMICS 303

Introduction 303
Coal Conversion 303

	Availability and Cost of Coal	310
	Conclusion on Coal Supply	317
Chapter 13	ECONOMICS OF COMBUSTION AND POWER GENERATION	319
	Introduction	319
	The Economics of Coal Based Electricity Generation	320
	Industrial Combustion	332
	Domestic Combustion	334
	Coal–Oil Mixtures	335
	General Conclusions	336
Chapter 14	ECONOMICS OF CARBONISATION	337
	Introduction	337
	General Economic Factors	337
	Conclusions	344
Chapter 15	ECONOMICS OF COAL GASIFICATION	345
	Introduction	345
	PADB Studies	346
	Recent US Department of Energy Economic Assessments	351
	Recent German Estimates	355
	Other SNG Cost Assessments	356
	Low and Medium Btu Gas and Synthesis Gas	357
	Conclusions	359
Chapter 16	THE ECONOMICS OF LIQUEFACTION	362
	Introduction	362
	Sasol	362
	UK Assessment	363
	West German Assessments	366
	US Estimates	368
	Conclusions	372
Chapter 17	ECONOMICS OF CHEMICALS FROM COAL	376
	Introduction	376
	Synthesis Gas Routes to Chemicals	377
	Other Routes to Chemicals	385
	Conclusions	386
Chapter 18	ECONOMICS OF UNDERGROUND COAL GASIFICATION	388
	A General Framework	388
	Other Forecasts and Estimates	393
	Conclusions	396

Chapter 19 ENVIRONMENTAL IMPACT 398

Introduction 398
General Environmental Effects of Coal Utilisation 398
Utilisation: Combustion 403
Synthetic Fuels Manufacture 417
Conclusions 418

Chapter 20 MULTI-COMPONENT PLANTS: COALPLEXES 420

Introduction and General Concepts 420
Components of Coalplexes and their Relationships 421
Economics 431
The Establishment of Coalplexes 435
Conclusions 437

SECTION IV COAL IN ENERGY POLICIES

Chapter 21 DISTRIBUTION OF COAL IN RELATION TO ENERGY NETWORKS 441

Introduction 441
Distribution of Reserves 441
Transport 442
Economic Factors 443
Reflections and Conclusions 450

Chapter 22 RELATIONSHIP OF COAL TO NUCLEAR POWER AND OTHER ENERGY SOURCES 453

Introduction 453
Energy Demand/Supply Balance 454
Conventional Oil and Gas 455
Unconventional Oil and Gas 456
Limitations of Nuclear Power 457
Fusion Energy 459
Solar Energy 459
Geothermal Energy 461
Hydraulic Energy 462
Tidal Energy 462
Summary of Contribution from "Renewables" 463
Conservation 463
Other Energy Supply and Conversion Proposals 464
Conclusions 465

Chapter 23 COAL UTILISATION IN RELATION TO WORLD ENERGY STRATEGIES 466

Introduction 466
Existing Predictions and Policies for Coal 466
Research and Development 470

International Collaboration 479
Some Current Initiatives 482
The Strategic Importance of Coal and its Economic Impact 484
General Conclusions 487

Bibliography 489

Index 497

LIST OF FIGURES

1.1 Densities of coal macerals
1.2 World production of coal (semilogarithmic scale)
1.3 Diagram showing the process of decomposition, subsidence and pressure by which the progressive ranks of coal are formed, anthracite being the highest of all
1.4 Simplified form of Seyler's coal classification chart
1.5 Coal classification system used by the NCB (revision of 1964)
1.6 Molecular weight distribution of ethylated Pocahontas coal
1.7 X-ray diffraction curves of (a) vitrinites, (b) graphite
1.8 Variation of average layer diameter
1.9 Structural models of coal
1.10 Molar volume aromaticity correlation
1.11 Average aromatic ring size as a function of H_{ar}/C_{ar} and connectivity
1.12 Proposed molecular model of an 82% carbon vitrinite
2.1 World coal production trends
3.1 Percentage of the total cost found within each of the geological formations
3.2 World coal deposits. Survey of Energy Resources 1980 prepared by BGR, Federal Institute for Geosciences and Natural Resources, Hannover
6.1 Essential features of a pulverised-coal-fired water-tube boiler plant
6.2 Underfeed stoker
6.3 Vekos-Powermaster boiler with manual ash removal (courtesy Parkinson Cowan GWB Ltd)
6.4 Chain-grate stoker
6.5 Air heater combined cycle
6.6 Supercharged boiler combined cycle
6.7 "Ignifluid" furnace
6.8 BCURA vertical shell boiler
6.9 Babcock boiler converted to fluidised bed firing
6.10 Diagram of 4.5 MW vertical shell boiler
6.11 Cross-section of Vosper Thornycroft (UK) Ltd boiler
6.12 Diagram of 10 MW double-ended locomotive-type boiler
6.13 Diagram of Johnston Boiler Company's 3 MW boiler
6.14 Babcock Power Ltd's composite water tube/shell boiler
6.15 Diagram of fludised bed fired grass dryer
6.16 Arrangement of fluid bed combustor Mk VII
6.17 A section through the Grimethorpe combustor pressure shell
6.18 General layout of experimental fluidised-bed combustion plant at Grimethorpe
6.19 Fluidised-bed boiler, 30 MWe
6.20 Cross-section through the Georgetown AFB Boiler
6.21 Babcock and Wilcox Alliance Research Center fluidised bed combustion development facility, 3 MW
6.22 Pressurised fluidised bed combined-cycle pilot plant, 13 MW
6.23 Flue gas treatment techniques

6.24 Model anti-pollution coal-fired thermal power plant
6.25 Coal-oil mixture combustion
6.26 Open-cycle coal-fired MHD system
6.27 KDV-process of Steag AG
6.28 VEW coal conversion process
7.1 Volatile matter of coal in relation to softening and decomposition temperatures
7.2 Dilatometer curves of coals of different rank
7.3 Coefficient of contraction of coals of differing volatile matter content
7.4 Analysis of contraction rates
7.5 Variation of contraction rates with temperature
7.6 Variation of atomic spacing with atomic weight
7.7 Hypothetical condensation process leading to layer growth
7.8 Arrangement of crystals in (a) non-graphitising carbon: randomly oriented crystallites containing hexagon layer planes, cross-linked, large pores and (b) graphitising carbon: more compact structure, near parallel arrangement of crystallites, little cross-linking, small pores
7.9 Variation of stack layer number with carbonisation temperature
7.10 Flow of gases shown for coke oven gas firing
7.11 How coke is produced
7.12 Influence of nominal carbonisation temperature on contraction characteristics of coke
7.13 M40 indices of coke versus total dilation and volatile matter content of charge
7.14 Blend formulation
7.15 Flow diagram of the "Phurnacite" production process
7.16 Cross-section through Disticoke oven for making "Phurnacite"
7.17 Compounds taken from the by-products of coal carbonisation
7.18 Manufacture of activated carbon from coking coals: generalised flow diagram
7.19 Flow diagram for manufacturing of activated carbon from anthracite
7.20 Pore volume distributions by mercury porosimetry
8.1 Rotating grate gasifier
8.2 Process stages and uses of coal gasification
8.3 Steam decomposed and carbon dioxide produced in converting various feedstocks to SNG
8.4 Variation of oxygen demand with percentage methane produced directly in coal gasification processes
8.5 Classification of gasification processes (names in boxes are commercially available)
8.6 The Lurgi gasifier (a) dry ash; (b) slagging
8.7 Entrained dust gasification (Koppers-Totzek)
8.8 Shell-Koppers coal gasification process flowscheme, 150 t/d Hamburg pilot plant
8.9 Winkler gasifier
8.10 HYGAS pilot plant configuration
8.11 Bi-Gas process
8.12 Westinghouse fluidised bed gasification, two stage pressurised process
8.13 Exxon catalytic gasification process
8.14 Hydrogasification process
8.15 Cross-section of the nuclear heat supply system
8.16 Simplified process scheme for steam gasification of low volatile bituminous coal using nuclear process heat
8.17 Preliminary design of a technical scale gas generator suited for steam gasification of coal using nuclear heat
8.18 Flow sheet of the fixed bed pressurised coal gasification plant at Dorsten
8.19 Fluidised bed coal gasification for combined cycle power generation
8.20 Schematic diagram of CMRC gasification plant (PDU)
8.21 Fixed bed gasifier, slagging process

8.22 Low Btu industrial fuel gas process
8.23 Small scale industrial fuel gas process
8.24 Waste heat recovery combined cycle
8.25 Exhaust fired combined cycle
8.26 Supercharged combined cycle
8.27 Gasification for power generation. Plant Configuration of Lünen
8.28 Medium Btu gas slagging-gasifier route
8.29 Medium Btu gas Shell-Koppers route
8.30 Hydrogen from coal facility
8.31 Steam iron process
9.1 Tar yields by flash heating
9.2 Complex coal structures and hydrocracking
9.3 Extraction yields of British coals
9.4 The donor solvent effect in coal by hydrogenation
9.5 The composition of coals and coal extracts
9.6 Unit operations in the SGE process
9.7 Fischer-Tropsch synthesis on current South African processes
9.8 Sasol 2 feed and products
9.9 Block flow diagram of coal gasification and purification
9.10 Rectisol gas purification
9.11 Fischer-Tropsch fluid bed reactor (Synthol)
9.12 Block flow diagram of Fischer-Tropsch synthesis
9.13 Further processing of hydrogenation products
9.14 Pilot plant for coal hydrogenation of Saarbergwerke AG
9.15 Block diagram of the "coal oil" large scale testing plant (Ruhrkohle AG, Essen)
9.16 H-coal process
9.17 H-coal process reactor
9.18 Synthoil process
9.19 Zinc chloride catalyst process
9.20 SRC (solid) process–SRC1
9.21 SRC (liquid) process–SRC2
9.22 Simplified block diagram of the EDS process
9.23 Conversion of Monterey coal increases with increasing temperature and time
9.24 Conversion of Wysodak coal increases with increasing temperature and time
9.25 Liquid yield response differs for each coal
9.26 Entrained pyrolysis process (Garrett/Occidental)
9.27 Mobil-M process
9.28 4 b/d fixed bed pilot plant
9.29 4 b/d fluid bed pilot plant
9.30 Reaction path
9.31 NCB liquid solvent extraction process
9.32 NCB supercritical gas extraction process
9.33 Conceptual scheme based on liquid solvent extraction of coal
9.34 Conceptual scheme based on gas extraction of coal and SNG production from the residue
10.1 The petrochemical process
10.2 Building blocks
10.3 Steam cracking
10.4 Lurgi ammonia-from-coal process
10.5 Flow sheet and related data for a 1000 t/d ammonia plant (with Koppers-Totzek gasifiers)
10.6 Modified Lurgi gasification route, 1000 Mt/d
10.7 1000 Mt/d of NH_3; Koppers-Totzek gasification route
10.8 Lurgi low pressure methanol synthesis process

10.9 Veba-Chemie integrated ammonia/methanol process
10.10 Petrochemicals from synthesis gas; efficiency of each step shown
10.11 Proposed scheme for manufacture of liquid hydrocarbons from coal
10.12 Liquid solvent extraction material balance and product utilisation
11.1 The basic concept of underground gasification of coal
11.2 The vertical drilling method showing the three stages of development
11.3 Linked vertical wells process
11.4 Hoe Creek Number 3 test using directional drilling
11.5 Steeply dipping beds; Rawlins Number 1 experiment
11.6 The original longwall generator concept
11.7 Modified Morgantown layout using hydraulic fracturing
11.8 Typical generator at Lisicharsk, USSR
11.9 The development from drifts to boreholes
11.10 Drilling into a steeply inclined seam from the surface
11.11 Use of two air inlet boreholes in a steeply sloping seam
11.12 Early Russian achievements with directional drilling
11.13 Proposed method of development of Kholmogorsk, USSR
11.14 The blind borehole method
11.15 P5 layout at Newman Spinney
11.16 Upgrading process options
13.1 Electricity generating costs: central view
13.2 Economics of coal and nuclear (2000 MW stations) showing the effect of varying the main parameters (discount rate = 7%)
13.3 Electricity generating costs: sensitivity to lower discount rate and higher fuels costs
13.4 Breakeven load factor: sensitivity to discount rate and to fuel price escalation
13.5 Coal prices required for coal-firing to be competitive
13.6 Costs of coal and nuclear electricity
13.7 Percentage by which coal is cheaper than oil
15.1 Synthetic fuel prices for Lurgi gasification process
15.2 Synthetic fuel prices for Lurgi gasification process
15.3 Cost of gas from Koppers-Totzek process compared with SNG from Lurgi process
15.4 Computational costs of SNG
15.5 Electricity prices — Lurgi gasification versus solid fuel combustion
15.6 Electricity prices
15.7 US comparison of pipeline gas prices (1978 $)
15.8 Total capital requirements, million $ (C. F. Braun estimates 1 March 1976; 100% equity, 12% DCF; western sub-bituminous coal)
15.9 Shell-Koppers coal gasification costs
15.10 Cost of SNG + synthesis gas versus raw brown coal costs. Basis: W. Germany 1978
15.11 Koppers-Totzek gasifier — effect of lignite cost on gas cost
16.1 The economics of producing premium liquid fuels
16.2 Comparison of oil price rises with coal costs
16.3 Economics of coal liquefaction in the UK; July 1979
16.4 Economics of coal liquefaction in the UK; oil price doubled
16.5 Effect of coal costs on price of liquids
16.6 Gasoline from coal production
16.7 Product price against coal cost; "base" investment
16.8 Fuel costs for $1.5/GJ coal; "base" investment
16.9 Fuel costs for $1.5/GJ coal; "base" investment + 50%
16.10 Fuel costs for $0.8/GJ coal; "base" investment
17.1 Production costs of ammonia versus raw material costs (plant capacity 1000 t/d)
17.2 Production costs of coal-based ammonia versus plant capacity
17.3 Comparison of chemical plant synthesis gas costs made directly from coal and indirectly via synthetic gasoline produced in a very large gas + gasoline plant

18.1 Cost contours for the vertical drilling method
18.2 Comparative costs (thermal) of coal exploitation; surface and UCG
18.3 Comparative costs (electricity) of coal exploitation; surface and UCG
19.1 Trace elements from coal combustion and their position in periodic table
20.1 Coalplex incorporating gasification, synthesis and combustion
20.2 Coalplex incorporating liquefaction and fluidised combustion
20.3 Integration of power generation in a coal complex
20.4 Integrated plant for SNG and electric power
20.5 Coalplex incorporating gas extraction and combustion
20.6 Coalplex incorporating gas extraction, combustion and gasification
20.7 Coalplex incorporating gas extraction and hydrogasification
20.8 The Cogas process
20.9 Gas and chemicals coalplex
20.10 Simplified block flow diagram of the POGO complex
20.11 Profitability of the chemical-type Coalplex
20.12 Coal refinery and chemicals plant
20.13 Schematic diagram of the clean coke process
20.14 Cost advantage of a simple coalplex
20.15 Cost advantages of a flexible coalplex
21.1 Cost of energy transmission
21.2 Cost of electricity and SNG
21.3 Industrial fuel gas distribution cost
23.1 Development technologies and future markets for coal

LIST OF TABLES

1.1 Selected conventions categorised by number of properties used in classification, subdivided by parameters used
1.2 Hard coals. Classes of the International System compared with the Classes of the National Systems
1.3 Classification of coals by ASTM Ranking
1.4 ISO Coal Classification
1.5 Code numbers of brown coal types
1.6 Relation of macro- and micro-structures
1.7 American nomenclature
1.8 Correlation of Thiessen-Bureau of Mines Nomenclature (Thin Section Method) with Stopes-Heerlen Nomenclature (Polished Block Method) of Hard Coal
1.9 Data on the aromaticity of vitrinites
1.10 Size of aromatic clusters in vitrinites. Number of carbon atoms
1.11 Data on the hydrogen aromaticity (vitrinites). Fractional aromaticity f_a
1.12 Structural parameters for Iowa and Virginia vitrains
3.1 The distribution of world coal resources grouped by continents
3.2 Coal resources and reserves, major countries ($\times 10^9$ TCE)
3.3 Boundary values of the most important parameters when classifying coal according to rank
3.4 Total resources of solid fossil fuels *in situ*, and recoverable reserves and the calculated lifetime of these reserves based on the production level in 1978
3.5 Total resources of solid fossil fuels and recoverable reserves
3.6 World coal resources and reserves by major coal-producing countries
6.1 Coal-fired atmospheric fluidised-bed combustors in the United States
7.1 Cold briquetting processes
7.2 Hot briquetting processes
8.1 Pilot plants for coal gasification, in operation or under construction in West Germany
8.2 Coal gasification projects — W. Germany
9.1 Analysis of a typical gas extract, coal feed and residue
9.2 Pilot plants for hard coal hydrogenation
9.3 Concepts of industrial coal hydrogenation plants
9.4 Product yields and hydrogen consumption: H-coal
9.5 SRC2- Typical yields and operating conditions
9.6 Typical EDS yields for Illinois No. 6 Coal
9.7 Fixed bed MTG conditions and yields
10.1 Principal derivatives from benzole and coal tar
10.2 Primary intermediates and derivatives
10.3 Suitability of main coal conversion routes for chemicals production
10.4 Ammonia plants based on coal gasification
10.5 Typical raw gas compositions from coal gasifiers
10.6 Daily consumption of coal and oxygen to produce 1000 t/d ammonia

10.7 Inputs for 1000 t/d ammonia plant
10.8 Feed requirements per tonne of methanol
10.9 Characterization of distillates from pilot plant hydrocracking of coal extract solutions
10.10 Thermal cracking tests on coal oil distillates
10.11 Catalytic reforming of coal oil naphtha
10.12 Analysis of oil obtained by hydrotreatment of supercritical-gas extract of coal
10.13 Analyses of COED tars
12.1 Fuel conversions
12.2 Prices for primary fuels
12.3 Assumed energy prices in 2010 AD relative to 1978
12.4 Relative capital cost of conversion processes
12.5 Efficiencies and yields of conversion processes
12.6 Converted energy costs 1978
12.7 Energy costs 2010 ("cheap energy" scenario)
12.8 Energy costs 2010 ("expensive energy" scenario)
12.9 Projection of major exporters
12.10 World coal trade
12.11 Major exporters in 2000
12.12 World coal trade 1977
12.13 Coal trade as a proportion of total coal production
12.14 Range of possible coal export availabilities
12.15 Summary of steam coal import requirements
12.16 Delivered fuel prices: US Regions
12.17 Coal export price competitiveness
12.18 Estimated costs of imported steam coals delivered to Western Europe and Japan in the mid-1980s
12.19 Delivered cost of thermal coal in major consuming regions
13.1 Comparative generating costs: IEA estimates
13.2 IEA regional estimates
13.3 Costs of coal electricity
13.4 Percentage savings in electricity costs by fluidised combustion
13.5 Boiler costs
13.6 Annual home heating costs in UK
14.1 Cost breakdown for coke production
14.2 Coal by-product prices per ton
14.3 Nominal coking plant performance
14.4 Breakdown of capital cost estimates for coking plant processing 4200 t/d dry coal
15.1 Lurgi costs and efficiencies
15.2 Koppers Totzek costs and efficiencies
15.3 Cost breakdown of the Lurgi process
15.4 Cost and efficiencies of new SNG processes
15.5 Generation efficiencies
15.6 Comparative capital costs of power generation systems
15.7 US basic coal conversion economics
15.8 US gas cost estimates
15.9 Gasification of Eastern coals
15.10 Synthesis gas costs
15.11 Economic conditions for gas cost estimates
15.12 Summary of SNG costs
15.13 Capital and operating costs for SNG
15.14 Coal and oxygen use for MBG
16.1 UK estimates of refinery costs: oil and coal
16.2 Capital costs and efficiencies

16.3 UK petroleum product prices
16.4 Product yields from coal liquefaction
16.5 Mean product prices for coal liquefaction processes
16.6 Coal liquefaction costs
16.7 Parsons estimates of coal liquids costs
16.8 Generic data of representative routes to coal liquids
17.1 Estimated costs of energy sources
17.2 Cost data for ammonia production
17.3 Daily consumption and costs: 1000 t/d ammonia plants
17.4 Daily consumption and costs: 1000 t/d methanol plants
17.5 Estimates of investment and production costs: ammonia
18.1 Cost breakdown of the UCG system
18.2 Effect on cost of distance between boreholes
18.3 Effect on cost of overall efficiency
18.4 Effect on cost of operating pressure
18.5 Effect on cost of calorific value of gas
19.1 Mean urban sulphur dioxide and smoke concentrations
19.2 Possible annual sulphur pollution in the USA
19.3 Sulphur dioxide standards and guidelines adopted or proposed in various countries
20.1 Economic comparison of certain coalplexes
21.1 Long distance transport costs of coal
21.2 Costs for pipeline transport of coal
21.3 Long distance transport costs for coal
21.4 Transport costs for various forms of energy
21.5 Transport costs of various forms of energy derived from coal
21.6 US estimates of coal transport costs
22.1 World energy demand and potential supply
22.2 Potential world primary energy production
22.3 Total world conventional energy resources
22.4 Possible solar-derived energy contribution to energy needs in USA
22.5 Potential hydro-electric output
23.1 Funding of US Government R&D spending on coal
23.2 German Federal energy R&D
23.3 West German energy R&D: coal and other fossil sources of primary energy

FOREWORD

The authors who have collaborated in writing this book have also worked together for more than a decade in promoting Coal Utilisation R & D. They bear a substantial responsibility for the way the policy of the National Coal Board in this field has developed since 1966 and, more directly, for the programme of work at the Coal Research Establishment, Stoke Orchard, near Cheltenham.

After a period of relative neglect, R & D on Coal Utilisation has flourished in recent years, both in extent and the importance ascribed to it. A large amount of technical data has been obtained from the pioneering experimental work and this will form the foundation on which vast new industries can be based. The timing and organisation of the application of technical information into these new coal conversion industries represents, in the authors' view, the most important question in the whole field of energy, which is now widely recognised as a vital aspect of social and economic development. The scale of the new coal utilisation enterprises will be greater, and their success more critical, than that of any other development in the field of energy, including that of nuclear power or the renewable resources.

This book is, therefore, not directed specifically at technical experts in the field of coal utilisation, and in particular it is not intended to enlighten those who specialise in particular sections of this technology. It is, nevertheless, hoped that such experts may welcome an interpretation of their work in relation to the economic and policy background.

The book is intended mainly for those interested but not specialists in coal technology wanting to understand the possibilities which exist for the utilisation of coal and so that they may appreciate the significance of the decisions that need to be made. The book may appeal both to technical and non-technical people in the general field of energy, to those in politics, government, and the media who need to be informed about the prospects for coal and not least to those members of the general public who are concerned about energy futures. While some of the special terminology may have inevitably intruded, it is hoped that the language is mainly understandable to the general reader.

In order to assist the general reader who may wish to refer to the book for guidance on particular issues, it is constructed in a series of chapters, each of which is intended to be reasonably suitable for reading separately, although it is preferable that the reader should be conscious of the broad content of the rest of the book. The first five chapters are of an introductory and background nature. Then there follow six chapters dealing with the main branches of coal utilisation technology. The subsequent nine chapters deal with economics. There is one chapter on economics for each of the six main branches of technology and three chapters on general economic matters, including the impact of environmental constraints. The final three chapters in the book discuss the context within which political decisions about the role of coal will be taken; the authors' reactions to the existing policy situation are included in the last chapter.

The book is deliberately selective and subjective in character, reflecting the authors' views and judgements. The authors have drawn upon a number of documents of various kinds, not all of which are widely available. A selected bibliography is provided for the various topics covered in the book but the text does not generally contain specific references, since the authors' interpretation may not always coincide with the intentions of those providing the basic technical data.

The results of original work were originally reported in a variety of units. For the most part these have been translated into SI units, generally with a certain amount of rounding-off. Where appropriate, units used in the original publications are also included, alongside SI units.

The authors were fortunate to receive assistance from a large number of colleagues. Those at the Coal Research Establishment who were particularly involved, include:

Dr D. Merrick
Dr P. F. M. Paul
Dr W. G. Kaye
Mr B. Robson
Dr D. G. Madley
Mr J. Highley

Elsewhere in the National Coal Board, assistance was provided by Mr P. N. Thompson, Operational Research Executive, Mr T. L. Carr, Mining Research and Development Establishment, Mr R. W. C. Wheatley, Marketing Department and Mr R. J. Ormerod, Central Planning Unit. Miss M. D. Barnes, and staff of the Technical Intelligence Branch provided references for the bibliography, and Mr R. J. Jennings produced the index.

We would also like to express our thanks to those responsible for the compilation and typing of the book. These were headed by Nancy Reay, assisted by Betty Dungan, Sarah Wolsoncroft and Cherylanne Aveling.

The views represented are those of the authors and not those of any organisation with whom they have been associated.

SECTION I
Introduction to Coal and its Utilisation

Chapter 1

WHAT IS COAL?

GENERAL INTRODUCTION

Coal is a familiar substance and an important natural resource. It is also however a complex and diverse material. It has close links with peat, which can be considered a precursor, and also with the other fossil fuels, including petroleum and natural gas, which were formed by related processes. Oil shales and tar sands are other types of fossil fuels, with some similar background features. These fossil fuels are the most important readily available energy source and, within this group, coal represents by far the largest fraction of the resources and reserves.

Coal is not another form of carbon, like graphite or diamond. It consists of a complex mixture of organic chemical substances containing carbon, hydrogen and oxygen, together with smaller amounts of nitrogen, sulphur and some trace elements. Strictly speaking, coal is not a hydrocarbon; for the purposes of a book concentrating on coal utilisation, however, it may perhaps best be considered so. Coal is, in fact, a fossil, or an organic sedimentary rock, formed mainly by the action of temperature and pressure on plant debris and always has associated with it various amounts of moisture and minerals.

Coal is black or brownish-black in reflected light and its surface may be dull or bright, often in bands within the same deposit. The mode and ease of fracture vary widely, as does the hardness. These properties, and many others, are acquired during the processes of coal formation the details of which can also vary widely, giving rise to property differences.

Coal formation occurred in two distinct stages, one biochemical and the other geochemical. Differences in the plant material and the extent of its decomposition during the first stage largely account for the different petrographic components known as macerals (by analogy with the minerals in rocks); the subsequent action of differing degrees of pressure and heat over different periods of time during the geochemical stage, acting on the peat-like deposits, was responsible for the difference in coalification or maturity of the coals, referred to as the "rank" of the coal. The degree of coalification or rank increases throughout the series from lignite through low rank coal and high rank coal to anthracite. The carbon content increases with increase in coal rank, while the oxygen and hydrogen contents decrease, as does the reactivity of the coal.

Coal macerals are commonly grouped together under the names vitrinite, exinite and inertinite. These macerals can be identified by microscopy and concentrates can be prepared. Their properties are found to be different, although the differences become smaller as rank increases. This is illustrated in Fig. 1.1, which illustrates the convergence of the densities of different macerals at high rank.

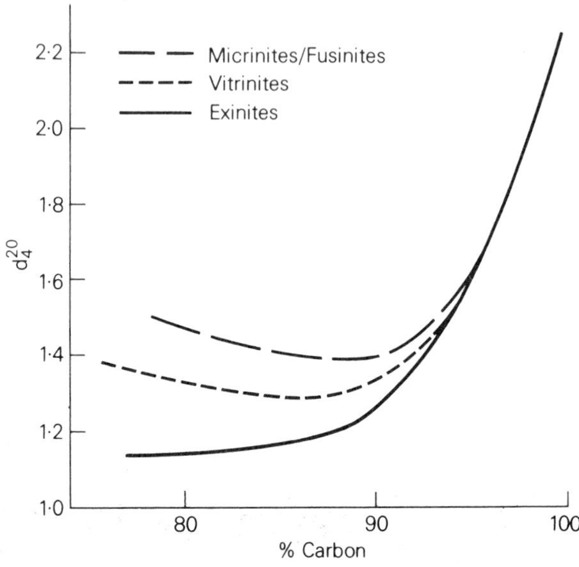

1.1 Densities of coal macerals

Coal, therefore, is not one but a whole family of substances covering a range of rank, each substance containing a number of macerals, moisture and mineral matter which differentiate it from the rest of the family. It is clearly important to be able to analyse and classify a given coal in such a way that its properties can be predicted. The constitution of coal is discussed later. It is worth mentioning at this point, however, that vitrinite (characteristically "bright coal") is commonly the major petrographic constituent of coals (in the UK and many other places) and the most consistent in its properties. Many scientific investigations of coal have therefore concentrated mainly on the vitrinite component, which can often be prepared in reasonable purity from large lumps of coal.

HISTORY OF COAL USAGE

The first knowledge and use of coal probably occurred several thousands of years ago. Coal seams outcrop frequently and, since coal has an unusual appearance and is combustible, it seems probable that it was known and used for ornaments, and possibly for heat, in prehistoric times. There are reports of Chinese knowledge of coal several thousand years ago. Coal mining may have become more common about 2000 years ago and the Romans certainly made use of coal during their occupation of Britain.

Coal was probably used in Germany in the 10th century. Traffic in coal in Britain was certainly established by the 13th century but the increasing depletion of wood in the 16th century gave the trade a tremendous boost. In Elizabethan times, London relied heavily on "Sea Coal", so called because it came by boat from the North East (its use at that time in primitive equipment also originated concern about environmental effects). The industrial revolution was fuelled on coal, both through

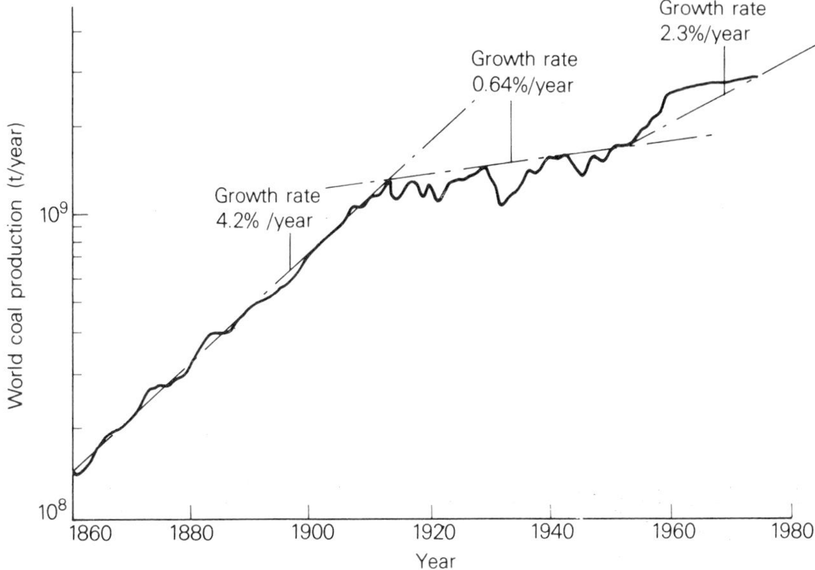

1.2 World production of coal (semilogarithmic scale).

the growth of the steam engine and by the development of iron-making using coal instead of charcoal (the supply of which was becoming limited). Coal was used in North America early in its colonisation and mines were important from the beginning of the 18th century in the industrialisation of the USA. For the last century or two, the world output of coal has increased inexorably, with only minor stutters for wars or recessions (Fig. 1.2). It could be concluded that coal output has been closely linked with economic or industrial growth. Originally, coal was probably used as a local heat source. In modern times the use of coal for electricity production is dominant, with the production of iron the second largest market. Until recently, the production of gas was also an important market for coal; this market may soon revive. Thus, coal has a long history, in which its role has varied but has always been an important one. This story is by no means nearing its end.

GEOLOGICAL ORIGIN

That coals were formed by the breakdown of plant remains is incontrovertible: the residues of plant life are readily seen in thin section which can be examined microscopically by transmitted light. More substantial remains of plants and trees can also be found in coal seams. The general scientific basis of coal formation is also well accepted, although the fine detail still provides some room for debate. This is not surprising in view of the fact that coals were mostly formed two to three hundred million years ago and that the general process was affected by important local differences. These differences, which could occur at all stages — the plant growth, the deposition and decomposition of vegetable matter, the deposition of non-vegetable matter and the time/temperature conditions of the overburdens — created the dissimilarities in the characteristics of the various coals. Coals are classified for both scientific and commercial purposes and the differences in formation history are related to these classifications.

The principal initiating requirements for coal formation are a swampy or marshy environment, climatic conditions favourable for rapid plant growth, with enough depth of water to exclude or severely restrict oxygen supply during the breakdown of the original plant material when it dies and falls into the water. The material is diverse and may include tree trunks, branches, twigs, leaves, bark and spores. The type of coal formed (including the distribution of macerals) depends on the type of plant material and the availability of oxygen. Excess oxygen, usually from fresh flowing streams, breaks the plant debris down, generally leaving no solid residue; some rare coals, of merely scientific interest, consist of the most stable elements, such as waxes, resins and spores. When conditions are highly anaerobic, putrefaction occurs, resulting in featureless sapropelic coals (cannel or boghead coals) which are relatively rare and of little commercial importance; they also have relatively high hydrogen and nitrogen contents. In between these limits, the humic coals, including almost all those of commercial value, are formed; the relative availability of oxygen affects the degree to which the hydrogen and nitrogen are removed.

It is also essential that the swampy area should continue to subside in order to allow further growth and accumulation. Such subsidence may also have effects on the deposition of fine mineral matter such as sand, mud and clay. These materials may be deposited to some extent during the periods of formation of the peat bogs or, if subsidence occurs erratically, there may be long periods when the water is too deep for plant life and only minerals are deposited, giving the typical layered structure of the coal measures. In the Northern Hemisphere, where warm moist conditions were prevalent, the coal seams themselves tend to be banded, with bright bands which may have arisen mainly from woody material. In the Southern Hemisphere, as for example in Southern Africa and India, plant growth often took place only during seasonal temperate conditions and there is little bright coal; in these circumstances fine clay particles may have been deposited with the plant debris and become involved in the coal substance.

In Britain, which is fairly typical of the Northern Hemisphere, coal deposits were laid down in the Carboniferous Period (225–300 million years ago). There are often up to 20–30 seams with thicknesses varying from a few centimetres to 3 metres (but exceptionally up to about 10 metres). The average thickness of seams worked is about 1.5 metres. Over the whole of the areas where coals are found, the sequence of sedimentary rocks is recognisable but the occurrence of individual coal seams is highly anomalous. This can be explained by the concept that the area which is now Britain was during that time a sedimentary basin, consisting generally of lagoons or estuaries in which silt was deposited. The various parts of the area were sinking at a variable rate; sometimes the silt deposits reduced the depth of water so that vegetated swamps formed and might persist for many centuries. These phases were terminated by periods of more rapid sinking, when silt deposits covered the humus or peat-like material. This sequence occurred frequently but, in Britain, not after the end of the Carboniferous Period. Subsequently, sedimentary rocks, often thousands of feet thick, were laid down and their influence brought about the coalification process. Sometimes the areas where coal was being formed were inundated by sea water, resulting in the contamination of the coals by various salts, of which sodium, potassium, calcium, magnesium, chlorine, fluorine and sulphur may be the most significant elements.

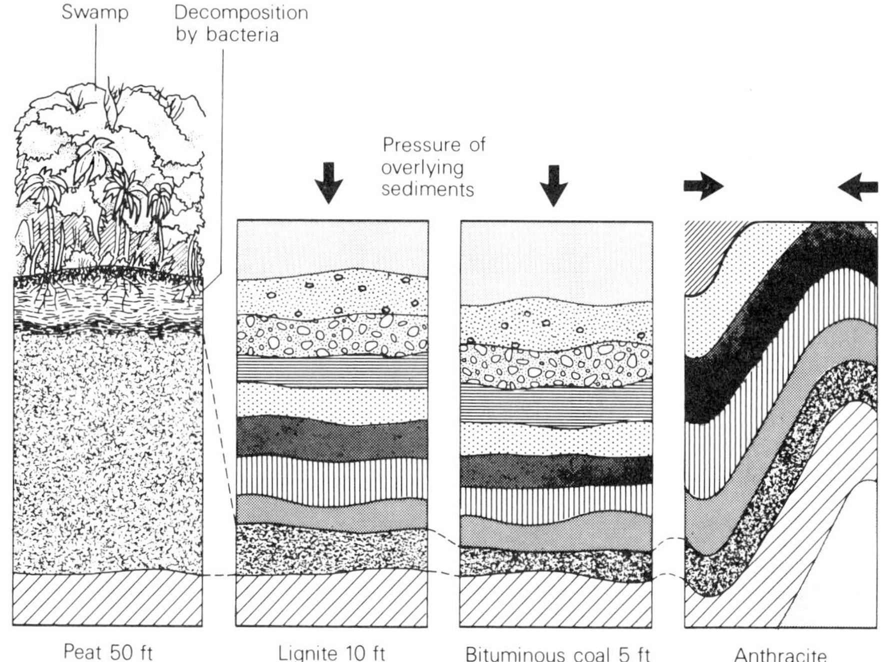

1.3 Diagram showing the process of decomposition, subsidence and pressure by which the progressive ranks of coal are formed, anthracite being the highest of all

In addition to the *in-situ* formation and deposition of plant remains, the drifting of debris from elsewhere into the collecting basins may have played some part, but more especially in cases where thick seams of relatively limited lateral extent were formed.

By the end of the biochemical stage, peat had been formed, a process which is still thought to be occurring in some places today. The characters of the distinctive macerals (described later) were by then already established. Present day peat deposits are less than one million years old. Coalification of the peat — the geochemical stage — took place over much longer periods of time and was accompanied by pressure and heating (generally modest) which were mainly the result of the overburden of more recent sedimentary rocks. This compressed the peat (Fig. 1.3) and also modified the chemical composition and properties of the seam, giving rise to a progressive increase in rank, the most important distinguishing feature of various coals. The coalification progression results in a continuous decrease in the oxygen and hydrogen contents (and corresponding increase in carbon). There is also a consequent decrease in the volatile matter (VM) which is driven off by the action of heat. The calorific value increases with rank and, as might be expected from the continuing effects of pressure, the coal generally becomes harder and stronger, with reduced porosity although the changes of hardness and strength with rank are complex overall, as discussed later.

The vitrinite maceral, being the most common and best known, is generally regarded as the marker for compositional changes related to rank, although similar changes occur also in the other macerals (except fusain, which owes its composition to selective chemical or biochemical activity). In addition to the normal pressure of the overburden, pressure arising from tectonic movements such as the bending and folding of strata can expedite the transformation; such additional pressures, and not merely time, are probably essential for progression to anthracite. Heat from igneous intrusions can also accelerate transformation and some anthracite-like coals, or even natural cokes, can be formed in this way. Some elegant studies have been made relating rank in a coal seam to distance from an igneous intrusion and it can be inferred that such effects are of relatively minor importance in coalification generally, compared with time and pressure. Time seems to be the most important factor, since most of the important bituminous coals were formed in the Carboniferous Period; a few coals are older and some somewhat younger but coals younger than about 60 million years are generally lignites or brown coal. There are, however, some Carboniferous coals which have not been transformed from the brown coal phase. presumably because they were not subjected to sufficient pressure.

ANALYSIS AND TESTING OF COALS

This subject is an important aid to the classification of coal. As might be expected for such a commercially important but heterogeneous and variable substance, there is a voluminous and expert literature on this subject. The methods of determining coal quality are mentioned here only to the extent necessary to appreciate the relationship between various coals and their uses. Coal quality may be approached from several aspects:

(a) Chemical;
(b) Physical nature and condition;
(c) Fundamental properties related to rank.

The classical analytical methods for coal are called "ultimate" and "proximate". The former determines precisely the proportion of the main chemical elements contained in the coal substance, i.e. carbon, hydrogen, nitrogen, sulphur, oxygen. The oxygen may be determined chemically but is usually estimated by difference. The results of the analysis may be presented on an as-received basis, a dry basis or, most commonly, a dry mineral matter free (dmmf) basis. In the latter case, an adjustment is required for changes in the ash which occur on heating (for example, carbonates evolve carbon dioxide).

The proximate analysis, so-called because it is less precise, is more generally useful; it consists of the determination of moisture, ash, volatile matter and (by difference) fixed carbon. Since results depend on experimental detail, prescribed, standard methods must be used; the actual volatile matter in particular is an arbitrary measure depending on the heating conditions. On a dry ash-free basis, the fixed carbon plus volatile matter equals 100% and either one may be used in classification systems. The moisture measured is the "inherent" moisture, as distinct from adventitious or added water; the inherent moisture is itself affected by the atmosphere so that care must be taken in the conditions of determination. Moisture is, however, an important indicator of rank and age (lower rank, younger coals having a higher moisture content) and of utilisation value. The estimation of calorific value is frequently carried out in parallel with the proximate analysis (from

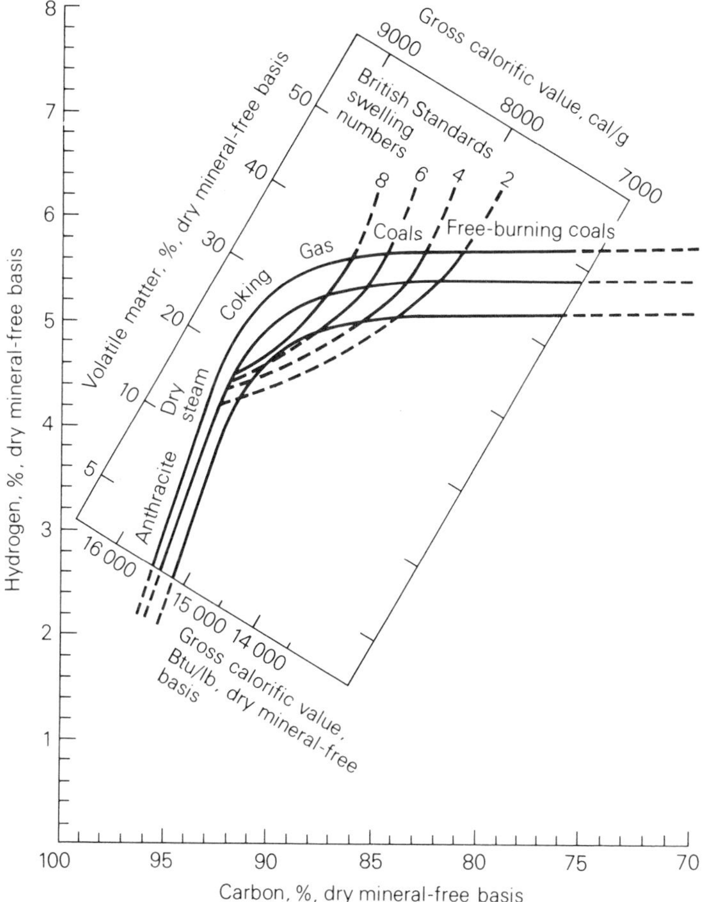

1.4 Simplified form of Seyler's coal classification chart

which it can be estimated approximately); again standard calorimetry procedures must be adhered to.

Typical values for the ultimate and proximate analyses of coals of different rank can be read from the Seyler chart and the UK National Coal Board (NCB) Classification system illustrated in Figs. 1.4 and 1.5. For example, the low rank high volatile free burning coals (volatile matter above 32%) have carbon contents below about 85%, oxygen up to about 15% and hydrogen above 5%. With increasing coal rank the oxygen content decreases first, with little change in the hydrogen content and consequently an increase in carbon content. In the later stages of coalification the hydrogen content decreases and anthracites with carbon and hydrogen contents in the region of 93 and 3%, respectively, are typical.

Other aspects relating to the physical properties or condition of the coal may be important for utilisation; these may include grade, sulphur content, size, free moisture, hardness and density. Grade, which is a commercial term of great importance, refers to the degree of contamination of the coal by non-plant

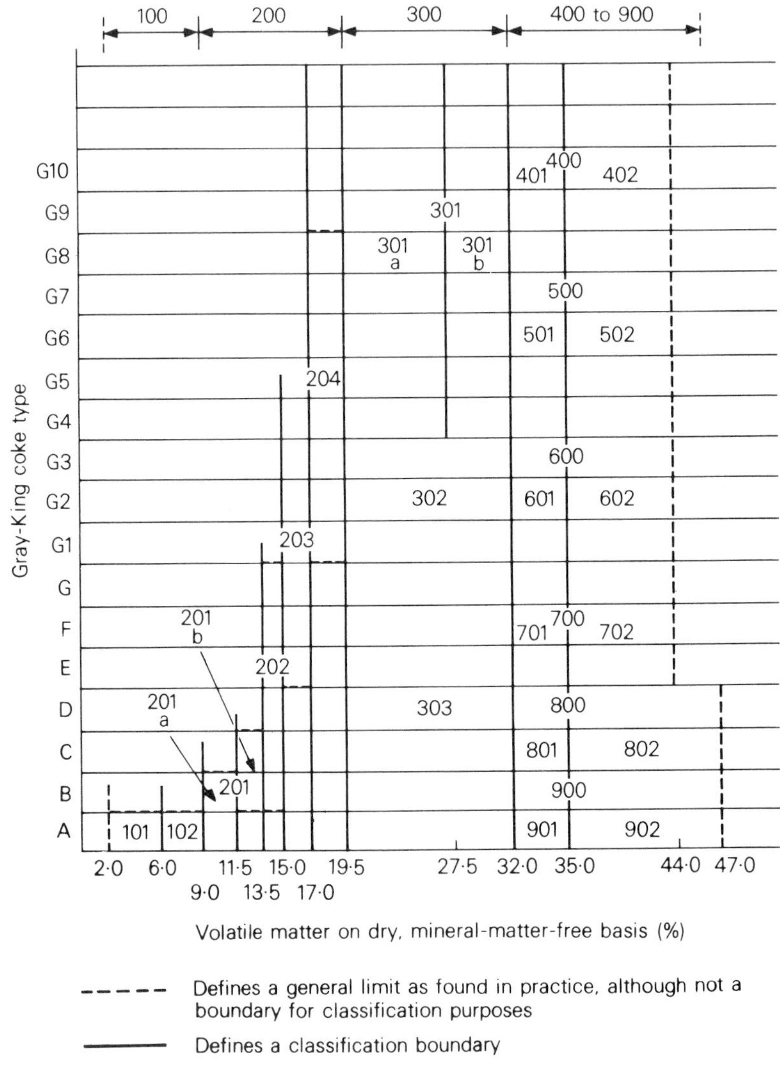

1.5 Coal classification system used by the NCB (revision of 1964)

substances. Contamination of significance may be by mineral matter, in total or as separate constituents, or by moisture. Some of these contaminants come from the original plant life or were introduced as air or water-borne mineral particles. They are then finely divided, intimately associated with the coal substance and are difficult to remove. Other material may come from thin bands of silt within the coal seam or by the roof and or floor being mined with the seam. In general, these latter materials are partly removed by coal preparation, so the amount of mineral matter in a coal as marketed is somewhat determined by expediency. Thus, while classification systems normally deal with the pure coal substance, grade is an essential supplement in commercial terms. Over recent years especially, sulphur has become an increasingly important aspect of coal quality. Generally, sulphur is said to be organic or pyritic: the former being related to the plant substance and the formation of the coal, the latter being discrete particles of iron pyrite, which can in theory be removed by coal preparation. However, the extent to which this separation is possible is affected by the particle size of the pyrite; if some is finely divided, the degree of grinding necessary to release it imposes practical or economic limitations.

The size, or size range, of a coal is determined partly by the inherent friability of the coal but also by the mining and handling methods; modern machines and practices are tending to produce finer coal. For some applications, fine coal is preferred; this is the case for pulverised firing of boilers or for coke ovens, since in both these cases further grinding of the coal as received is necessary. However, fine coals tend to adsorb and retain more moisture. This can cause difficulties, not only because ot the dilution of the heating value but also because severe handling problems can result, especially when fine clays are present with the coal. Sizes are determined by sieving through standard sieves (round or square holes must be specified); results are quoted by referring to the two sizes of holes over which and through which the coal has passed e.g. $\frac{1}{4}$–1 inch (or 6–25 mm). In some cases (e.g. -10 mm) no bottom size is specified and in others no top size (e.g. $+30$ mm) is given. Most countries give names to particular size ranges. In Britain, for example, "grains" can be $\frac{1}{8}$–$\frac{1}{4}$ inch (3–6 mm), "peas" $\frac{1}{4}$–$\frac{1}{2}$ inch (6–12 mm); singles are $\frac{1}{2}$–1 inch (12–25 mm) and so on up through doubles, trebles and cobbles.

The classical tests for mechanical properties, such as tensile strength, have been applied with some difficulty to coal for scientific purposes but the only property where practical guidance can be related to strength measurements is that of grindability. This is based on a standard grinding test determining a parameter called the Hardgrove Index. In the case of British coals, those with VM contents of about 15–30% are the easiest to grind, having Hardgrove Indices of about 100 (the higher the index the easier the grinding). Coals with a lower volatile matter are associated with the hard anthracites (Hardgrove Indices of about 50); coals having a higher VM content, up to about 40%, are also harder.

Density or specific gravity is also a property requiring close definition; this property should be tested and quoted in accordance with the purpose of the determination, using methods appropriate to that purpose. Requirements for density information cover a wide range of purposes and may include the estimation of coal in place in reserves, the bulk density of coals as a design parameter for storage or for coke oven charging equipment, the density of individual lumps or particles for gravity coal cleaning methods, or scientific investigations of coal structure. For reserve estimation — determining tonnage from *in-situ* volume measurements — although some allowance must be made for mineral matter, arbitrary figures are used with reasonable success in different countries, usually reflecting increasing density with rank. In the UK, for reserve estimation, bituminous coals and anthracites are taken as equivalent to 1.32 and 1.38 respectively in apparent specific gravity. In other countries, conventions range from about 1.3–1.4; some anthracite specific gravities over 1.7 have been quoted. Where discrete dirt bands are known to occur, they can be allowed for in these calculations at values of about 2.3. The relationship between the density of individual macerals and their carbon content has already been mentioned (Fig. 1.1). There is a shallow minimum of about 1.3 between 85–90% carbon, rising to over 1.8 at exceptionally high carbon levels. In the cleaning of coal, using jig washers or dense medium systems, the separation into coal and dirt is made on the basis of density. The density "cut point" chosen will depend on the type of coal being washed and the required product specification. The cut point will determine the yield of clean coal and its ash and sulphur content, as well as the proportion of "middlings" and dirt recovered from the process. A Baum jig will normally make a separation at a specific gravity between 1.60 and 1.80. Dense medium plant will normally prepare a clean coal of specific gravity less than 1.40, a middlings fraction of specific gravity in the range 1.40 to 1.80 and a refuse of specific gravity greater than 1.80. For the design of coal preparation facilities, coal particles are assigned a suitable density to suit the type of coal and the preparation requirements. In designing bunkers, the bulk density is the important factor and a value appropriate to the nature and size distribution of the particular coal has to be used. In addition to these factors, bulk density depends on the size of the container, the percentage of free moisture, the shape of the particles or lumps and the method of packing. Thus, for graded or small coal, bulk density can vary in the range 640–830 kg/m^3, while for fine coal (say 3 mm) the corresponding range would be 480–580 kg/m^3. In coke oven charging, a high "flow" bulk density is now considered desirable and may be improved by drying or pre-heating, typically increasing the density from 750 kg/m^3 to 825 kg/m^3; this treatment presumably works by assisting the particles to flow over one another more easily.

The other main properties of coals which are used for their classification or assessment are derived from petrology and from their behaviour on heating. Petrology is discussed later; its most practical relevance is that the reflectance of vitrinite varies with rank. This property can be used in classification systems and also as a routine on-line quality control check.

The behaviour of coal on heating is of particular importance in coking and is further discussed in the chapter on carbonisation; these properties do nonetheless have great importance in other uses, including combustion. Briefly, many coals soften on heating, starting from about 350°C. On further heating, particles of coal first become quite fluid, with some contraction as they fuse together; then they expand under the action of gas and vapour bubbles. Subsequently, they resolidify at about 500°C as a coherent cake which becomes coke on further heating. This sequence varies greatly in detail and extent depending on the coal; a strong coke, for example, requires a suitable degree of fluidity and gas generation to give cell walls which resist breakage. There are a number of tests, necessarily empirical in nature, which give indications of these complex processes. Four tests are most widely used and accepted internationally. In the first of these, "crucible swelling numbers" are determined by heating finely ground coal to 820°C and comparing the coke button with standard profiles; the numbers range between 0 (non-swelling) and 9 (most strongly swelling). In the Gray–King test, used extensively in the UK, the finely divided coal is heated in a retort to 600°C and the coke is again compared with standard photographs; the letters A to G are used progressively to designate coals that vary from non-coking to those that give a strong coke of the original volume. For coals of greater swelling power than G, subscripts (G_1, G_2, etc.) are used to denote the number of parts of electrode coke (inert) which must be added to one part of the coal to reduce the swelling to the standard G level. In the Audibert–Arnu Dilatometer Test changes in the length of a pencil of compacted coal particles on heating are followed by means of a loose piston in the cylinder containing the coal; the percentage contraction and the percentage dilatation are reported. The Roga test is used to determine the coking power of the coal more directly. A button of coke is produced and is then subjected to a repeated process of breaking in a revolving drum. From the sieve tests the Roga Index is calculated. Correlations between these four tests can be made, with limitations.

CLASSIFICATION SYSTEMS

Some form of classification is a prime necessity commercially because of the great diversity of coals. At the minimum this must give an indication of rank, as, for example: lignite, sub-bituminous coal, bituminous coal, anthracite. Generally more refined classification is necessary.

Perhaps the most famous classification is that of Seyler, first published in 1899. A simplified version is shown in Fig. 1.4. This chart enables coals to be classified by two pairs of parameters, either by carbon and hydrogen or by volatile matter and calorific value. Average values and likely ranges for types of coal are shown, together with nomenclatures and the area of coals with coking properties. This system has been of great value in codifying and classifying coal properties; it is however too complex for commercial use. In simplifying coal classification, volatile matter and calorific value have generally been used as the basis for differentiating rank in the main coal classification systems now in use.

In the classification of bituminous coals for use in coke making, the extent to which they swell and cake when heated is also of great importance and it has therefore proved necessary to introduce a second classification parameter to characterise coals for carbonization. Examples are the BS Swelling Number, the Gray–King coke type, the dilatation and the Roga Index discussed above. The classification system used in the UK (Fig. 1.5) is based on volatile matter and the Gray–King coke type; similar systems of greater or lesser complexity are employed in other countries. The analyses required to classify a coal in these systems are relatively easy and cheap to perform and the parameters measured are clearly related to the main commercial uses of coals. Table 1.1 lists some of the conventions and parameters used by various countries. A further comparison (after van Krevelen) is given in Table 1.2.

In the USA, the ASTM classification is most generally used (Table 1.3). The United Nations system (ISO) uses volatile matter supplemented by calorific value for high volatile matter coals (Table 1.4) and two methods of measurement related to coking (Table 1.5). The USSR classifies coal entirely on the reflectance of the vitrinite.

Advances in coal science and in the technology in the use of coal, combined with the prospect of increasing international coal trading, have led to the expression of views that a new classification system should be developed, and discussions are now taking place under the auspices of the United Nations. One difficulty is that existing systems use parameters measured on the whole coal, that is on the mixture of macerals present, without taking into account the relative proportions of different macerals. The resulting classification can then sometimes prove misleading, for example, when applied to coals with very high inertinite content from the Southern Hemisphere. Such problems might be overcome by the use of the reflectance of the vitrinite as a parameter of rank instead of volatile matter, coupled with some index of maceral composition (and presumably an index of swelling) but petrographic analyses are more difficult and expensive to carry out.

TABLE 1.1 Selected conventions categorised by number of properties used in classification, sub-divided by parameters used

Number of properties	*Parameter*	*Properties or tests used*	*Classification*
One	Rank	VM	Belgian: "Moniteur Belge-Belgisch Staatsblad" p. 9664
		Reflectance	USSR: GOST 21489
	Size		USSR: GOST 19242
Two	Rank	C and H *or* VM and CV	Seyler
		VM and Free Swell	FRG: Ruhrkohlen-Handbuch 1969
		VM and Gray-King	UK: NCB classification
Three	Rank	VM, CV and agglomerating test	USA: ASTM D 388
		VM, CV and Gray-King	India: ISI Bulletin, Vol. 19, pp 3504
	Rank and local usage	Tar yield, total moisture; local usage	ISO 2950
Four	Rank	VM, CV, Roga *or* Free Swell, Gray-King *or* dilatometer	UN: International Classification of Hard Coals by Type
		VM, CV, Free Swell and dilatometer	FRG: DIN 23 003
			France: NF M 10-001
			Italy: ENI/ENEL/FINSIDER draft
		VM, CV, Roga and dilatometer	Poland: PN 68/G-97002
Five	Rank and grade	VM, CV, Free Swell and Gray-King; ash	Australia: AS K 184

PETROLOGY AND MACERALS

As indicated previously, it is possible to distinguish different components (macerals) in coal by optical examination. Coal petrology is the study of these components and it is of importance in understanding the formation and the properties of coal. It is possible to prepare sections of coal sufficiently thin to be translucent and early microscopic studies were made using transmitted light which revealed the petrographic components by differences of colour and form.

The preparation of thin sections is, however, very laborious and can only be carried out on particles of coal above a certain size. Nowadays, therefore, coal samples are crushed, mounted in resin and polished for examination by reflected

TABLE 1.2 Hard Coals. Classes of the International System compared with the Classes of the National Systems. (After D. W. van Krevelen)

Classes of the international system			*Classes of national systems*								
	Parameters										
Class no.	*Volatile matter, %*	*Calorific value (calculated to standard moisture content), cal/g*	*Belgium*	*Germany*	*France*	*Italy*	*Netherlands*	*Poland*	*United Kingdom*	*United States*
0	0–3					antraciti speciali		meta-antracyt		meta-anthracite
1A	3–6.5		maigre	Anthrazit	anthracite	antraciti communi	anthraciet	antracyt	anthracite	anthracite
1B	6.5–10							polantracyt		
2	10–14		$\frac{1}{4}$ gras	Mager-kohle	maigre	carboni magri	mager	chudy	dry steam	semianthracite
3	14–20		$\frac{1}{2}$ $\frac{3}{4}$ gras	Esskohle	demigras	carboni semi-grassi	esskool	polkoksowy metakoksowy	coking steam	low-volatile bituminous
4	20–28			Fettkohle	gras a courte flamme	carboni grassi corta fiamma	vetkool	ortokoksowy	medium-volatile coking	medium-volatile bituminous
5	28–33		gras		gras propre-ment dit	carboni grassi media fiamma		gazowo-koksowy		high-volatile bituminous
6	>33 (33–40)	8450–7750		Gaskohle		carboni da gas	gaskool			
7	>33 (32–44)	7750–7200			flambant gras	carboni grassi da vapore	gasvlam-kool	gazowy	high-volatile	high-volatile bituminous
8	>33 (34–46)	7200–6100		Gas flamm-kohle	flambant sec			gazowo-plomienny		high-volatile bituminous
9	>33 (36–48)	<6100				carboni sechi	vlamkool	plomienny		subbituminous

TABLE 1.3 Classification of coals by ASTM Ranking

		Fixed carbon limits, % (dry, mineral-matter-free basis)		*Volatile matter limits, % (dry, mineral-matter-free basis)*		*Calorific value limits, Btu/lb. (moist, mineral-matter-free basis)*		
Class	*Group*	*Equal or greater than*	*Less than*	*Greater than*	*Equal or less than*	*Equal or greater than*	*Less than*	*Agglomerating character*
I Anthracite	1. Meta-anthracite	98	—	—	2	—	—	Non-agglomerating
	2. Anthracite	92	98	2	8	—	—	
	3. Semianthracite	86	92	8	14	—	—	
II Bituminous	1. Low volatile bituminous coal	78	86	14	22	—	—	
	2. Medium volatile bituminous coal	69	78	22	31	—	—	
	3. High volatile A bituminous coal	—	69	31	—	14 000	—	Commonly agglomerating
	4. High volatile B bituminous coal	—	—	—	—	13 000	14 000	
	5. High volatile C bituminous coal	—	—	—	—	11 500	13 000	
		—	—	—	—	10 500	11 500	Agglomerating
III Subbituminous	1. Subbituminous A coal	—	—	—	—	10 500	11 500	
	2. Subbituminous B coal	—	—	—	—	9 500	10 500	
	3. Subbituminous C coal	—	—	—	—	8 300	9 500	Non-agglomerating
IV Lignite	1. Lignite A	—	—	—	—	6 300	8 300	
	2. Lignite B	—	—	—	—	—	6 300	

Source: ASTM, Standard Specification D338

TABLE 1.4 ISO Coal Classification

GROUPS (determined by coking properties)			CODE NUMBERS										SUB-GROUPS (determined by coking properties)		
Group number	Alternative group parameters: Crucible-swelling number	Alternative group parameters: Roga index	The first figure of the code number indicates the class of the coal determined by volatile matter content up to 33% VM and by calorific parameter above 33% VM. The second figure indicates the group of coal, determined by caking properties. The third figure indicates the sub-group, determined by coking properties.										Sub-group number	Alternative sub-group parameters: Dilatometer	Alternative sub-group parameters: Gray-King
3	>4	>45					435	535	635 (V_C)				5	>140	>G3
						334 (V_A)	434 (V_B)	534	634				4	>50–140	G5–G8
						333	433	533	633	733 (V_D)			3	>0–50	G1–G4
						332a, 332b	432	532	632	732	832		2	≤0	E–G
2	$2\frac{1}{2}$–4	>20–45				323	423	523	623	723 (VI_A)	823		3	>0–50	G1–G4
						322 (IV)	422	522	622	722	822		2	≤0	E–G
						321	421	521	621	721 (VI_B)	821		1	Contraction only	B–D
1	1–2	>5–20			212	312	412	512	612	712	812		2	≤0	E–G
					211	311 (III)	411	511	611 (VII)	711	811		1	Contraction only	B–D
0	0–$\frac{1}{2}$	0–5		100 (I) A B	200 (II)	300	400	500	600	700	800	900	0	Non-softening	A
Class number			0	1	2	3	4	5	6	7	8	9			
Class parameters	Volatile matter (daf)		0–3	>3–10 (A: >3–6.5; B: >6.5–10)	>10–14	>14–20	>20–28	>28–33	>33	>33	>33	>33	As an indication, the following classes have an approximate volatile matter content of: Class 6 33–41%; 7 33–44%; 8 35–50%; 9 42–50%		
	Calorific parameter								>7750	>7200–7750	>6100–7200	>5700–6100			

(Determined by volatile matter up to 33% VM and by calorific parameter above 33% VM)

Note: (i) Where the ash content of coal is too high to allow classification according to the present systems, it must be reduced by laboratory float-and-sink method (or any other appropriate means). The specific gravity selected for flotation should allow a maximum yield of coal with 5 to 10% of ash.

(ii) 332a ... >14–16% VM
332b ... >16–20% VM

Gross calorific value on moist, ash-free basis (30 C, 96% humidity) kcal/kg.

TABLE 1.5 Code numbers of brown coal types

Group parameter: tar yield on the dry, ash-free basis, %	*Group number*	*Code numbers*					
>25	4	14	24	34	44	54	64
>20 to 25	3	13	23	33	43	53	63
>15 to 20	2	12	22	32	42	52	62
>10 to 15	1	11	21	31	41	51	61
⩽10	0	10	20	30	40	50	60
Class number		1	2	3	4	5	6
Class parameter: total moisture content of run-of-mine coal on the ash-free basis, %		⩽20	>20–30	>30–40	>40–50	>50–60	>60–70

light. The colour differences noted in transmitted light are then lost but the macerals can still be identified by differences in form and reflectance.

The system using reflected light was derived from the work of Dr Marie Stopes, starting in 1919 and developed into the Stopes–Heerlen System, which is widely used throughout the world. Initially she distinguished four rock types called lithotypes, in banded bituminous coal — vitrain, clarain, durain and fusain — simply by gross (and, it must be admitted, somewhat imprecise) observation, described as follows:

Vitrain: Bright, black, usually brittle, frequently with fissures.
Clarain: Semi-bright, black, very finely stratified.
Durain: Dull, black or grey-black, hard rough surface.
Fusain: Silky lustre, black, fibrous, soft, quite friable.

Subsequently, these distinctions were extended but the important step was the identification, by microscopic means using reflected light, of microlithotypes and individual macerals, each of the latter being distinguishable by shape or colours, or other observable properties. The relationship between lithotypes, microlithotypes and macerals is indicated in Table 1.6(a) and (b). It became clear that the character of the different microlithotypes depended on the proportions of the various macerals present. Somewhat different criteria were used in the USA and a different nomenclature adopted. A broad correlation is given in Table 1.7; Table 1.8 shows a more detailed breakdown.

For geological purposes, many macerals and sub-macerals are recognised and used in, for example, the identification of coal seams but for applications in coal utilisation it is often sufficient to group the macerals together under the headings vitrinite, exinite (or liptinite) and inertinite, the macerals in a given group in a coal having roughly similar properties. Thus the exinite has a lower density and higher volatile content, compared to the vitrinite, whereas the inertinite has a higher density and lower volatile content than the vitrinite. The properties of a particular coal will clearly vary if the proportions of the different maceral groups change, as can happen in practice for several reasons.

TABLE 1.6 Relation of macro- and micro-structures

(a)

Lithotype	*Microlithotype*
Vitrain	Vitrite Clarite (where V > E)
Durain	Clarite (where V < E) Durite Trimacerite
Fusain	Inertite (semi-fusinite and fusinite)
Clarain	may contain elements of any microlithotype or maceral group

(b) Microlithotype and maceral groups

	Microlithotype	*Maceral groups*
Monomaceral	Vitrite liptite inertite	Vitrinite/Humanite Exinite/liptinite Inertinite
Bimaceral	Clarite Durite Vitrinertite	V + E I + E V + I
Trimaceral	Duroclarite Clarodurite Vitrinertoliptite	V > I + E I > V + E E > I + V

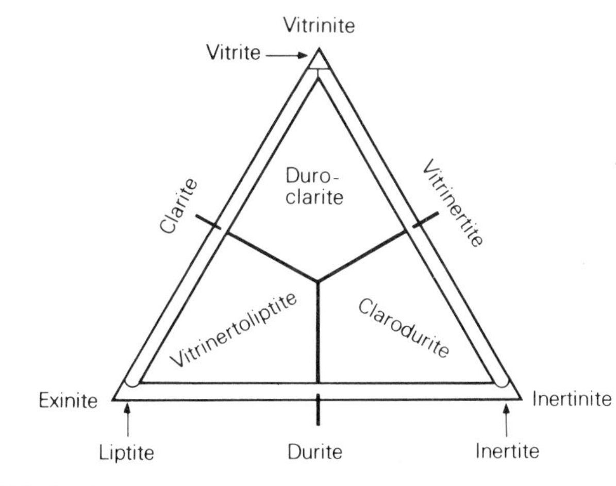

In addition to the determination of the proportions of the maceral groups in coal, the petrographer can provide a measure of the rank of the coal because it is found that the reflectance (of the vitrinite in particular) increases progressively with increasing rank. As discussed above in the section on classification systems, other parameters of rank in common use are based on measurement of a property of the

TABLE 1.7 American nomenclature (Thiessen-Spackman)

Anthraxylon	Equivalent to Stopes-Heerlen Vitrain
Attritus	Ground mass, a broad term referring to macerated plant debris from any source intimately mixed with mineral matter and coalified. It predominates in splint, cannel and boghead coals, and may be present in large amounts in bright coals.
Fusain	Same as Stopes-Heerlen Fusain.

whole coal and will therefore be affected by changes in maceral composition, but reflectance of the vitrinite is not so affected of course. The prime advantage of petrographic analysis lies in this ability to separate two basic variables, namely the rank and the maceral composition, which independently influence coal properties.

For characterisation purposes, the reflectance of vitrinite must be measured under strictly controlled conditions which are now in the process of being standardised internationally. With improved agreement between different analysts which can be expected in due course, commercial applications will undoubtedly increase. Among the present applications, the most important is the prediction of the strength of a metallurgical coke made from a blend of coals of different ranks. There are several methods, all of which depend essentially on two facts: (1) vitrinite and exinite are reactive (softening and swelling on heating), whereas inertinite and minerals are unreactive and a strong coke requires the optimum proportions; (2) the strength of the coke is affected by the proportions of coals of different ranks (reflectances).

Finally, it should be noted that mineral particles can be observed also in coal sections and form a major portion of the ash. These are most commonly detrital minerals deposited with the plant material, frequently clays or common minerals such as quartz or feldspar. Secondary minerals deposited from water percolation after the formation of the coal beds include kaolinite, calcite and pyrite. Many trace metals are also present but cannot be distinguished by petrology. Some may have come from the original plant life. These elements may have importance environmentally.

THE CHEMICAL STRUCTURE OF COAL

Knowledge of the carbon, hydrogen and oxygen content of a coal does not tell us much about its chemical make-up. Even in elementary organic chemistry, for example, a substance containing 52% carbon, 13% hydrogen and 35% oxygen might be ethanol or dimethyl ether and, without a knowledge of the molecular weight, the number of possible alternatives is boundless. Evidence of other sorts is necessary before it is possible to define the way in which the various atoms are assembled.

Many different methods of investigation have been used in conjunction to produce a fairly reliable description of the chemical structure of coal. This subject attracted considerable attention — particularly in the 1950s — and although notable advances have been made, it must be emphasised that not all of the facts fit and the argument continues.

In the classical chemical methods of determining structure, the empirical formula (relative numbers of combining chemical elements), the molecular weight and

TABLE 1.8 Correlation of Thiessen-Bureau of Mines Nomenclature (Thin Section Method) with Stopes-Heerlen Nomenclature (Polished Block Method) of Hard Coal[a]. Courtesy Internationale Kommission für Kohlenpetrologie

Transmitted light			*Reflected light*	
Thiessen-Bureau of Mines system			*Stopes-Heerlen system*	
Banded components	*Constituents of attritus*[b]		*Macerals*	*Groups of macerals*
anthraxylon (translucent)			vitrinite more than 14 μ in width	vitrinite
	translucent attritus	translucent humic matter	vitrinite less than 14 μ in width	vitrinite
		spores, pollen cuticles, algae	sporinite, cutinite, alginite	exinite
		resinous and waxy substance	resinite	exinite
attritus		brown matter (semitranslucent)	weakly reflecting semifusinite weakly reflecting massive micrinite weakly reflecting sclerotinite	inertinite
	opaque attritus	granular opaque matter	granular micrinite	inertinite
		amorphous (massive) opaque matter finely divided fusain, sclerotia	fusinite less than 37 μ in width strongly reflecting massive micrinite strongly reflecting sclerotinite	inertinite
fusain (opaque)			fusinite and semifusinite more than 37 μ in width	

Types of coal	*Quantitative statements* (*Transmitted light*)	*Microlithotypes* (*Reflected light*)	
Banded coals			
bright coal	more than 5% anthraxylon	vitrite	
	less than 20% opaque attritus	clarite	
semisplint coal	more than 5% anthraxylon	duroclarite	
	20–30% opaque attritus	vitrinertite	
splint coal	more than 5% anthraxylon	clarodurite	
	more than 30% opaque attritus		
Nonbanded coals			
cannel coal	less than 5% anthraxylon	durite[c]	cannel
boghead coal	less than 5% anthraxylon		boghead

[a] An exact correlation of the two systems is not possible.

[b] For the recognisable botanical entities (for instance "spores", "cuticles") the term "phyteral" may be employed.

[c] Durite is part of banded coals with less than 5% vitrinite that would be placed by definition in the nonbanded coals in the Thiessen-Bureau of Mines system.

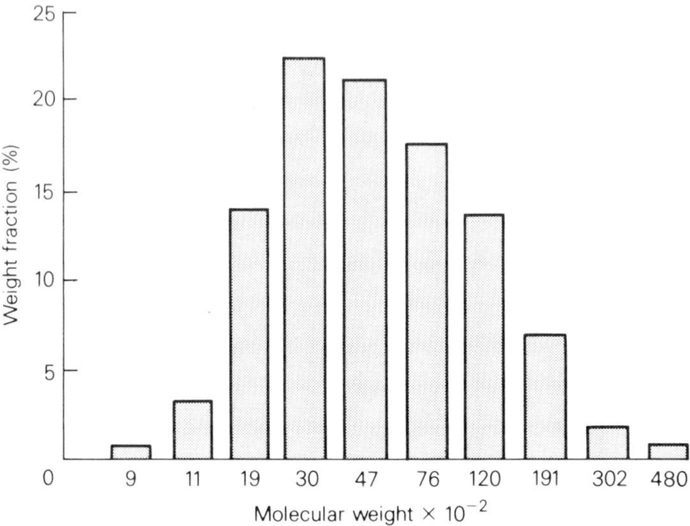

1.6 Molecular weight distribution of ethylated Pocahontas coal

degradation/synthesis processes are used. If it could be determined for coals, the molecular weight would be a measure of rank as well as a fundamental structural parameter. There is no doubt however that a coal is not an aggregation of molecules, all of the same molecular weight, but an assemblage of units whose average size increases with rank; they are however cross linked, which makes it very difficult to separate and evaluate them with consistency and confidence. There are many methods which have been attempted to overcome this difficulty, involving solubilisation through a variety of chemical techniques. These include depolymerisation by means of phenol as solvent and boron trifluoride as catalyst; reductive alkylation in which coal is treated with an alkali metal in tetrahydrofuran in the presence of a small amount of naphthalene followed by treatment with alkyl halides; acylation through Friedel Crafts treatment yielding soluble products; increasing the solubility in pyridine by treatment with sodium hydroxide in ethanol.

As an example of molecular weight distribution, the results obtained by Steinberg from the ethylation (reductive alkylation) of a Pocahontas Coal (90% C), followed by molecular weight determinations by gel permeation chromatography and vapour pressure osmometry are illustrative; the average molecular weight was found to be 3300 (Fig. 1.6) but the distribution was wide showing significant fractions (>5%) outside the main range, which was between 1900 and 19 000.

All these methods give estimates of size of coal fragments rather than true molecules — fragments released by mild breakage of cross links — and whilst the fragments represent the smallest units of coal that can be obtained without vigorous chemical degradation it is understandable that their molecular significance is questioned. However there is broad agreement between the results obtained by these various methods and rough confirmation from other assessments based on X-ray and statistical constitutional analysis based on additive properties such as molar volume.

Coal is not a crystalline substance and X-ray analysis cannot, by itself, show us the molecular structure, although it can provide some useful information. It is necessary, therefore, to take account of all the evidence available from a wide range of physical and chemical methods and attempt a synthesis. X-ray analysis is a great help in establishing a carbon skeleton. Because of their imperfection, coals give diffraction patterns which are diffuse compared to the graphite pattern, but the intensity distributions are not totally unrelated (Fig. 1.7). Broadened peaks are commonly associated with small crystallite size but this is not the whole story here, because a sequence of peaks such as 100, 101, 102, 103 is replaced by one broad, asymmetrical peak. This phenomenon also occurs in other carbonaceous substances such as carbon black, and it can be explained if the layer planes of graphite are stacked without regular order from one to the next. We can then consider two questions:

(1) the size of the layers; and
(2) the number of layers in a parallel stack.

These questions can best be answered by fitting the appropriate part of the intensity curve using a distribution of:

(1) layer sizes; and
(2) stack heights.

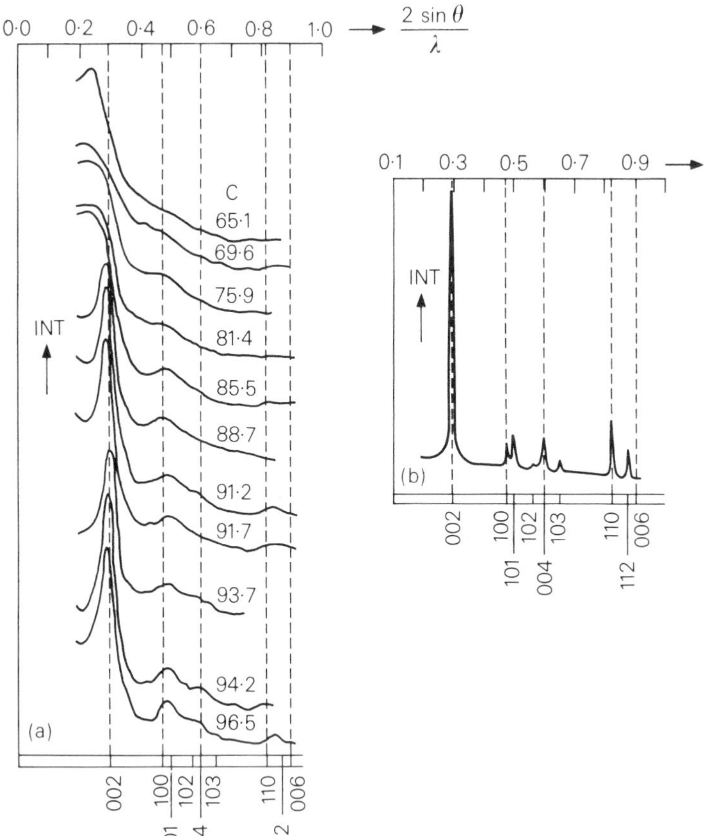

1.7 X-ray diffraction curves of (a) vitrinites, (b) graphite

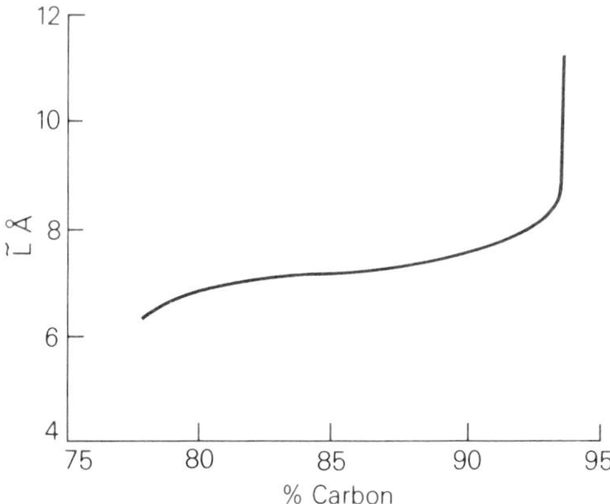

1.8 Variation of average layer diameter

The distribution of layer sizes is based on a number of different atomic groupings in order to average out the characteristics specific to any one molecule. The smallest size occurs with the highest frequency in coals of all ranks from 78 to 94%C and layers larger than 32 atoms are seldom observed except in the coals of highest rank. There is an appreciable content of amorphous material in all coals, the amount decreasing progressively as rank increases. The average layer diameter changes little up to 90% carbon, but thereafter increases rapidly (Fig. 1.8).

The distribution of heights of the parallel stacks of layers can be obtained by the application of a suitable mathematical transformation to the (002) peak in the intensity curve. Analyses of low rank coal and coking coal show that many layers occur singly and that stacks of more than 3 layers are rare; the coking coal layers are better stacked than those in the low rank coal.

The results of X-ray analysis may be summarised, therefore, in terms of three types of structure (Fig. 1.9):

(a) in low rank coals, an open structure of small layers randomly oriented and crosslinked;
(b) in medium rank coals, a "liquid" structure with fewer crosslinks, a moderate degree of orientation and reduced porosity;
(c) in high rank coals, a structure of larger layers with a higher degree of orientation and an oriented pore system.

There have been claims that some of the intensity distribution curves can be fitted just as well by alicyclic structures as by aromatics. To obtain confirmatory evidence of the X-ray structure we must consider a number of methods, generally described as statistical constitution analysis, which make use of the fact that various properties of compounds are additive. The most obvious example of an additive property is molar volume which is equal to the sum of the single-bonded atomic volumes per gram atom minus a term to allow for multiple bonds. Using a number of alkyl aromatic compounds and pitch fractions as reference substances (Fig. 1.10), a calibration can be effected in terms of f_a, the fraction of the carbon in aromatic groups. On this basis, low-rank coals have f_a about 0.7; with increase in rank, f_a increases to approximately 1.0. Other properties, include heat of combustion and sound velocity, can be used in

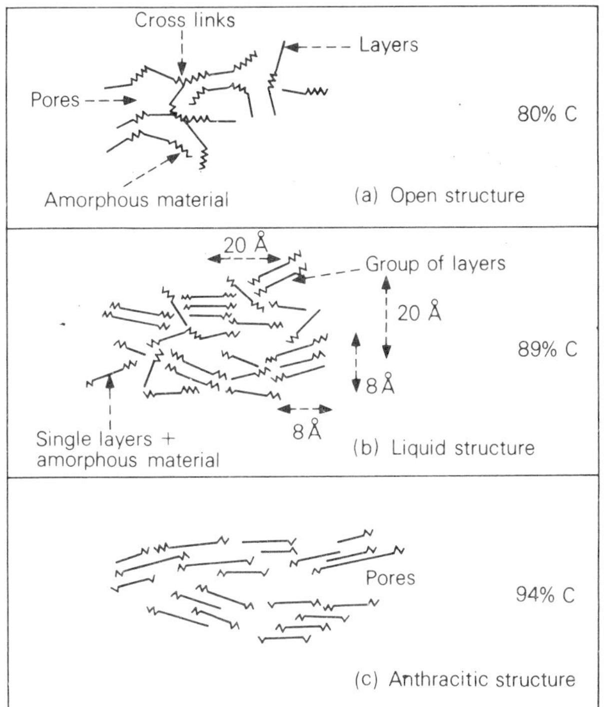

1.9 Structural models of coal

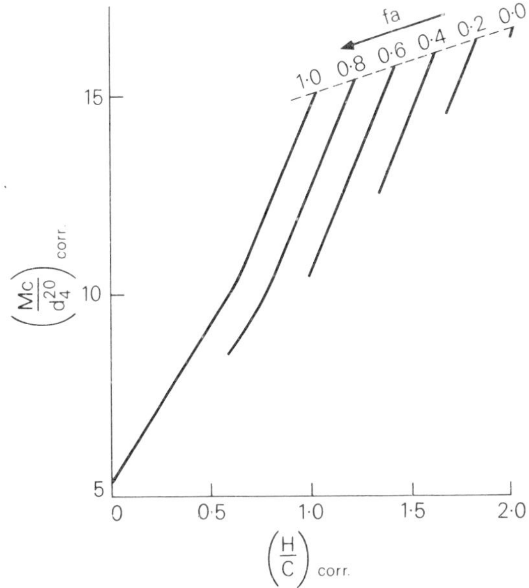

1.10 Molar volume aromaticity correlation

TABLE 1.9 Data on the aromaticity of vitrinites

%C	*Graphical density method*	*Sound velocity*	*Heat of combustion*	*UV spectrum*
70.5	0.70	—	—	—
75.5	0.78	—	—	—
80.0	—	—	~0.82	—
81.5	0.83	0.79	—	—
84.0	—	—	—	~0.78
85.0	0.85	0.82	~0.82	—
87.0	0.86	—	—	—
89.0	0.88	0.85	—	—
90.0	0.90	—	~0.85	—
91.2	0.92	0.90	—	—
92.5	0.96	0.93	—	—
93.4	0.99	0.97	—	—
94.2	1.00	—	—	—
95.0	1.00	—	1.00	—
96.0	1.00	—	—	—

a similar way and give f_a values in reasonable agreement, as shown in Table 1.9. Although these methods confirm the highly aromatic nature of coal, they do not provide direct confirmation of the size of the aromatic units derived from X-ray analysis.

Another additive property has been used in a very ingenious attack on this problem, namely the molar refraction, making measurements using two media with different refractive indices in contact with coal. Estimates of the surface area of the aromatic layers can be obtained, from which can be calculated the number of carbon atoms per aromatic layer. These show reasonable agreement with the values derived from X-ray analysis for coals of up to 87%C (Table 1.10), but not thereafter. A

TABLE 1.10 Size of aromatic clusters in vitrinites. Number of carbon atoms

%C	*Elementary composition and model concept*	*X-Ray diffraction*	*Molar refractive increment*	*Semi-conductivity*	*UV spectrum*
70.5	—	—	12	—	—
75.5	—	—	13	—	—
81.5	—	16	17	—	—
84.0	—	17	—	—	>17
85.0	—	17	21	—	—
87.0	—	17	(23)	—	—
89.0	—	18	—	—	—
90.0	—	18	—	—	—
91.2	—	18	—	—	—
92.5	—	18	—	—	—
93.4	(20)	20	—	(45)	—
94.2	22–40	(30)	—	(50)	—
95.0	43–60	—	—	55	—
96.0	85–100	—	—	>60	—

TABLE 1.11 Data on the hydrogen aromaticity (vitrinites). Fractional aromaticity f_a

%C	*Infrared spectra*	*Nuclear magnetic resonance*	*Elementary composition of residue of primary carbonisation*
76.0	—	—	0.23
79.0	—	0.33	0.33
83.0	0.18	0.45	0.38
88.0	0.28	—	0.54
89.0	0.31	—	0.51
91.0	0.45	—	—
93.0	0.64	0.64	—
94.0	1.00	—	—

possible explanation of this discrepancy lies in the development of charge-transfer characteristics and semiconducting properties in higher rank coals.

Other routes to the same destination (the size of the aromatic layers) are via the elementary composition (very high rank coals contain few hydrogen atoms to saturate the edge valencies of aromatic molecules), semiconduction (the energy barrier is a function of the number of aromatic carbon atoms per molecule in condensed aromatic compounds) and the UV spectrum (which can be synthesized from those of model compounds).

Having thus achieved a fair impression of the carbon skeleton, the arrangement of the hydrogen and oxygen atoms must be considered. To deal with hydrogen first, use is made of infrared spectrophotometry. The spectrum shows a number of distinct bands, the frequencies of which can be assigned to particular groupings such as $-OH$, $-CH_2$, $-CH_3$, $-CO$ and aromatic $-CH$ with some confidence. The absorption bands at 3030 cm^{-1} and 2920 cm^{-1} can be used to compare the amount of hydrogen attached to aromatic carbon and aliphatic carbon respectively; results suggest that four-fifths of the hydrogen in the low rank coals is aliphatic and the proportion falls to zero for an anthracite (Table 1.11).

It will also be noted in Table 1.11 from the few values of f_a determined by nuclear magnetic resonance that there is considerable discrepancy between the infrared and elementary composition derived values but good agreement between NMR and elementary composition values for low rank coals and identity of values from the IR and NMR examination of the anthracite (93.0% Carbon). Recent advances in NMR techniques have made NMR more useful and important than IR. Figure 1.11 gives a derivation of average ring size in a Virginia Coal (90.3%C) and an Iowa (77%C) Coal from data obtained by Gerstein *et al.* using high resolution ^{13}C and ^{1}H solid-state NMR. f_a values for both C and H were obtained from the NMR and used to calculate the H_{ar}/C_{ar} values (Table 1.12). They point out that the average aromatic ring size is a function of H_{ar}/C_{ar} and connectivity. Connectivity is important because the average ring size depends upon the number of side chains connected to the ring and on the degree of ring substitution. The author's preference is for average ring sizes of no greater than three for the Virginia vitrain and a maximum of two in the Iowa vitrain thus providing agreement with van Krevelen that the structural units increase in size with rank.

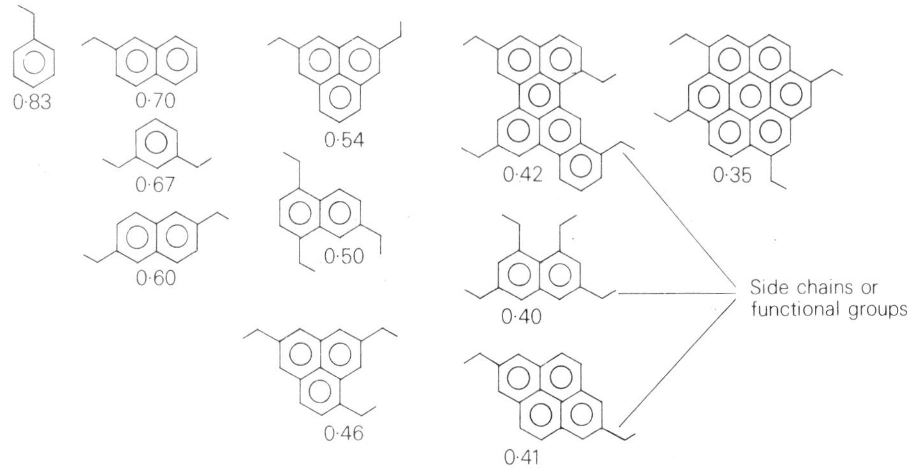

1.11 Average aromatic ring size as a function of H_{ar}/C_{ar} and connectivity

In contrast to the mainly physical methods used in assessing the groupings in which carbon and hydrogen occur, the methods employed to assess oxygen groupings are mainly chemical. The accessibility of groups in solid coal is questionable and some investigators have consequently used solvent extracts of coal, arguing that their infrared spectra are sufficiently similar to those of the coal to justify this procedure. Although, relatively speaking, agreement between chemical and physical evidence is rather poor, it is safe to say that the majority of the oxygen in bituminous coal is in carbonyl and phenolic hydroxyl groups (in roughly equal proportions), the remainder probably being in ether or heterocyclic ring systems.

The state of combination of the nitrogen and sulphur atoms in coal can be deduced. From the presence of heterocyclic compounds in coal tar and retention of both elements in coke, it is likely that the bulk of the nitrogen and sulphur occurs in the condensed aromatic structure, but some may be present also, as –NH, –SH or S–S groups.

There is now fair agreement on the carbon skeleton that exists in coals. There remains the question where the elements that clothe the skeleton fit. It might seem that they could be united into a structure in a vast number of ways. However, many possibilities are eliminated in practice because of the shortage of hydrogen (and especially of aromatic hydrogen) relative to carbon, the smallness of the aromatic nuclei and the steric problems of introducing appreciable cross-linking. One arrangement which fits the facts reasonably well is shown in Fig. 1.12. It has been developed from one proposed by Given in 1959.

TABLE 1.12 Structural parameters for Iowa and Virginia vitrains (Gerstein and others, 1979)

Sample	%C	%H	%C_{fa}	%H_{fa}	H_{ar}/C_{ar}
Virginia "Pocahontas No. 4" vitrain	90.3	4.43	86	77	0.53
Iowa "Star" vitrain	77.0	6.04	71	31	0.40

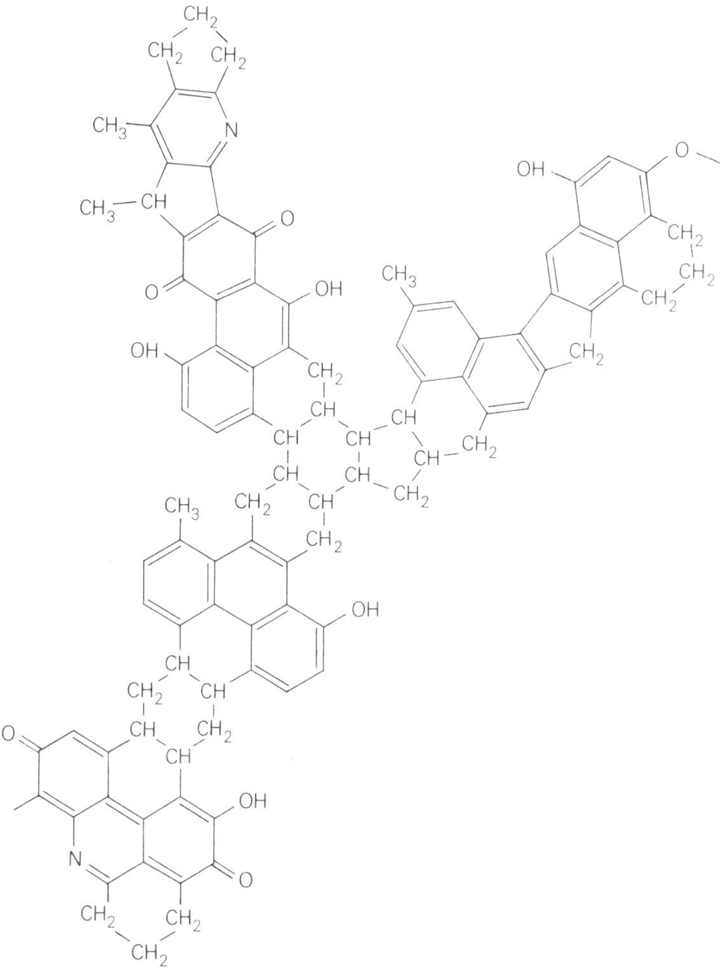

1.12 Proposed molecular model of an 82% carbon vitrinite

An elegant study of the behaviour of polymers undergoing pyrolysis has led to an improved understanding of the structure of coals. The polymers were prepared by condensing various aromatic compounds with ^{14}C-labelled formaldehyde which permitted the subsequent fate of the methylene linkages to be determined. Substituted aromatics containing alkyl and hydroxyl groups were used to prepare the polymers; the pattern of decomposition on pyrolysis bore a marked resemblance to that of coals. Tar, gas and water were evolved between 400 and 500°C and mainly hydrogen gas thereafter. It may therefore be concluded that the type of model structure proposed for coal would behave in at least roughly the right manner during pyrolysis, as far as evolution of decomposition products is concerned.

Another important characteristic of coals is that some, but not all, first become fluid at temperatures between 350 and 550°C, then swell up because of gas evolution and so form a coherent coke. In the same study of the behaviour of aromatic polymers, measurements were made of the dilation curves of these polymers and it was found possible to simulate the dilation behaviour of coking coal with polymers made from phenanthrene and pyrene. However, this was only possible over a limited range of degrees of polymerisation.

THE FINE PHYSICAL STRUCTURE OF COAL

Coal is visibly cracked or fissured, often extensively. These cracks or fissure planes are probably present in the coal seam and influence both the size of coal as mined and the ease with which comminution and further cracking occurs. The relationship between the fissure system and inherent strength, as measured in the Hardgrove Index Test, is difficult to determine. However, in addition, coal always possesses a large number of fine passages and pores. Some may be due to imperfections of fit between compressed particles but others represent atomic scale gaps left during the formation of the structure. Some of these are closed pores but many are open and contribute to the accessible internal surface area, which is surprisingly extensive. Characterisation of the pore structure involves measurement of the adsorption of various gases and the penetration of liquids, from which the surface area and the pore size distribution can be estimated. The available pore surface is of course a vital factor in absorption processes, chemical reactions, solvent extraction and, probably, in thermal decomposition. The surface area of bituminous coals is about 100 to 250 m^2/g, mostly in pores <1 nm in width. The total volume of pores in bituminous coals and anthracites is generally between 5 and 20%, of which pores below 1 nm contribute about one-half by volume. The internal surface and the porosity pass through a minimum at about 88% carbon as a result of changes in the composition and ordering of the fundamental units in the chemical structure during coalification as described above. Coal can be converted into active carbon by partial gasification which develops the existing pore structure in a controlled manner.

CHEMICAL REACTIONS

The principal reactions of industrial interest are discussed in later chapters. Oxidation for the generation of heat is of course the most important; it can however in some circumstances take place at ambient temperature and if the heat is not dissipated, the temperature can increase, with the result that the reaction accelerates, resulting in spontaneous combustion, for example in mines and stockpiles. Slower oxidation of stocks of coking coal can impair coking properties. Oxidation can also promote the formation of humic acids, which can be extracted chemically.

Reactions with hydrogen and steam are of next importance, in the production of gases and liquids, discussed in later chapters. Coal can also react with many other elements or compounds including halogens, sulphur and aqueous solutions of alkali; many compounds are produced but these reactions are not of industrial importance. More interesting is that soluble fractions can be extracted from coal by a wide range of organic solvents. This technique has been widely used to investigate coal structure but is also likely to be one of the key steps in future coal conversion processes and is discussed later, especially in Chapter 19.

Chapter 2

WHY IS COAL IMPORTANT?

The main purpose of this book is to indicate the authors' belief in the supreme importance of coal to the future world supply of energy and feedstocks; this supply in turn is vital to both economic progress and an ordered and peaceful society. It is hoped that the book will add some weight to the case for policies based on these concepts and thus influence those in positions of decision-making and also the general public to adopt sympathetic attitudes.

THE VITAL IMPORTANCE OF ENERGY

Most of the book displays technical and economic information relevant to the importance of coal in providing future energy and feedstock supplies. It may be relevant at this stage to emphasise the importance of energy supplies, especially those provided by fossil fuels, the versatility of which will be an important point of discussion. On a world scale, it is difficult to see how social stability can be achieved without a fairly regular increase of energy consumption per capita, sustained over some decades, possibly with some moderate perturbations as a result of economic cycles. Since there seems little chance of avoiding a continuous increase of world population in the near future, the world energy demand seems certain to grow, by at least a few per cent per year. Failure to meet this demand would have serious social and political repercussions. Most projections of demand growth suggest that energy consumption per capita will continue to increase even in the developed countries and, since there is such a large difference between the absolute levels of consumption in the developed countries (especially in the USA) on the one hand and the Less Developed Countries (LDCs), on the other, any credible rates of increase in energy consumption in the two cases will lead to a larger per capita increase in absolute terms in the developed countries than in the LDCs. However, the needs of the huge populations in the LDCs, even if only increasing by small per capita amounts, will put greatest strain on available supplies. In the probable event of demand exceeding supply, however, economic power will probably determine which countries receive supplies and, unless exceptional international arrangements are made, the LDCs will suffer. This will affect especially the flow of those fuels which are most readily applied, such as oil. At the same time, the LDCs will be least able to obtain their energy needs from sources requiring proportionately large capital investments or a high level of technological competence. In the event the oil producers may decide to influence the distribution of oil in favour of the LDCs.

Various proposals for increased conservation and for changes in life style have been made in order to relieve the demand/supply pressures. The potential of conservation methods and the renewable or "benign" energy sources is considered elsewhere (Chapter 22); however, recent energy demand forecasts have generally taken progress in these directions into account to the maximum prudent extent. Even if the growth of the demand of conventional energy sources could ultimately be abated more rapidly than presently expected, timescales of decades would be involved, especially since the necessary investments in the initial stages of conservation or renewable energy sources might lead to net energy deficits in the short and medium term.

The problems of sustaining some world-wide economic progress with inadequate energy resources can perhaps be emphasised by reviewing the various roles played by energy, especially fossil energy. Electricity (still mostly provided from fossil sources) and transport are perhaps the two main sectors of energy consumption most readily associated with the economic and social status of a country. There can be little progress, in the current understanding of this term, without adequate electrification, especially in the LDCs, both for servicing the needs of industry and for supporting a higher standard of domestic life. It may be that electricity will become somewhat less dominant in these respects, and also less demanding in resources, over the next few decades but in the meantime aspirations in many countries are equated with electricity supply. Transport is equally important, both for the movement of goods (which reflects economic activity) and for the mobility of individuals for commercial and personal purposes. Again, adequate rates of development depend heavily on the known types of machines and on convenience fuels.

Energy is of course essential to industrial development in many other ways. Process heat plays a large part in the manufacture and shaping of many products, especially those, such as constructional materials and the provision of transport facilities, vital to development. In some cases, such as steel production, not only is heat required for carrying out the process but fossil fuels also provide the chemical reducing elements for recovering metal from ores. In the case of plastic materials, which are continuing to increase in importance, the fossil fuel also provides the very substance of the material.

Agriculture is also a heavy user of energy. Apart from the provision of power, the potential demand for fertilisers, made from fossil fuels, seems bound to grow more rapidly than population. Fossil fuel feedstocks find their way also into many other agricultural products, including weedkillers and insecticides. To some extent also, products made direct from fossil fuels relieve the pressure on agricultural resources; man-made fibres are an example. The production of protein foodstuffs from fossil fuels has been technically established on a large scale but awaits economic competitiveness. It hardly needs emphasising that chronic shortage of food is one of the ultimate disasters which adequate energy supplies can at least delay until more complete solutions are found. Nevertheless, we should strive to develop sensible policies to avoid the situation of converting fuel to food on the one hand while using agricultural produce as fuel on the other without valid reasons.

Finally, in this review, the role that increasing availability of energy can play in economic progress generally creates an environmental impact; in order to lessen this impact, it is usually necessary to increase the amount of primary energy which must be provided ultimately to meet a particular end use. Also, the minimisation of other forms of environmental damage, including visual impact (such as the removal of industrial dereliction), requires significant energy input. The input of "capital" energy in the harnessing of benign energy sources has already been mentioned. In the case of the most primitive people, however, shortage of energy leads to overcropping of rough vegetation for fuel and fodder and to the process sometimes called "desertification", whereby man destroys his own environment. This has been described as the "pollution of poverty". It is being increasingly realised that such regional degradation can have severe international effects.

THE ROLE OF FOSSIL FUELS

Having established, at least qualitatively, the importance of energy to the world, the key role of fossil fuels — and, within these energy sources, the crucial position of coal — needs to be understood. At present, fossil fuels — mainly coal, oil and natural gas — supply more than 90% of the world energy demand. This proportion may decline slowly, provided non-fossil energy sources can achieve the rapid growth which has been suggested for them. The total demand for fossil fuels, in tonnage terms, however, is likely to continue to expand inexorably for a long time.

For example, the World Energy Conference (WEC) in September 1977, accepted a central estimate of a four-fold increase in energy demand in the period to 2025. Whether such a demand could possibly be met is debatable, as are questions concerning the effects on demand of inevitably higher prices, conservation and improved efficiencies in energy conversion and consumption. Perhaps an even more worrying doubt is whether economic growth and capital formation will allow the demand and the potential supply to be balanced. Failure to achieve some balance at a level of minimum adequacy will lead to chaos and catastrophe.

The lower limit of necessary energy supply, to avoid disaster, must surely be doubled between 1977 and 2025. Unless the absolute tonnage supply from fossil energy sources increases substantially during this period, the demand on the newer energy sources, which would in that case be required to supply more or less as much energy as the total current supply, may not be achievable.

During the same half-century, however, the output of both oil (normal crude oil) and natural gas will certainly decline, from peaks reached probably in the first half of this period. Ideally, the rates of increase of oil and gas production should be as low as possible, and immediately effective, so that the inevitable plateaux and subsequent declines may be prolonged. Coal is the key factor both for prolonging the reasonable availability of oil and gas and in providing alternatives after the decline has become established; oil shales, tar sands and other "non-conventional oil" are seen to have an increasingly important part to play.

In the European Community, coal at present supplies about 22% of the energy requirement although indigenous coal reserves (with lignite) are more than 90% of the total fossil fuel reserves. These proportions reflect fairly accurately the situation elsewhere in the world. The requirement is for coal production to be increased regularly, over the next few decades at least, at a rate greater than the average increase in fossil fuel demand. The outcome would be that fossil fuels would remain much the largest source of energy for a long time, with coal eventually regaining its position as largest source of supply within the fossil fuel contribution.

Fortunately, both history and future plans support the feasibility of this steady increase in coal output. At the World Energy Conference, 1977, the summation of plans indicated an increase in production of coal and related products from 2.6×10^9

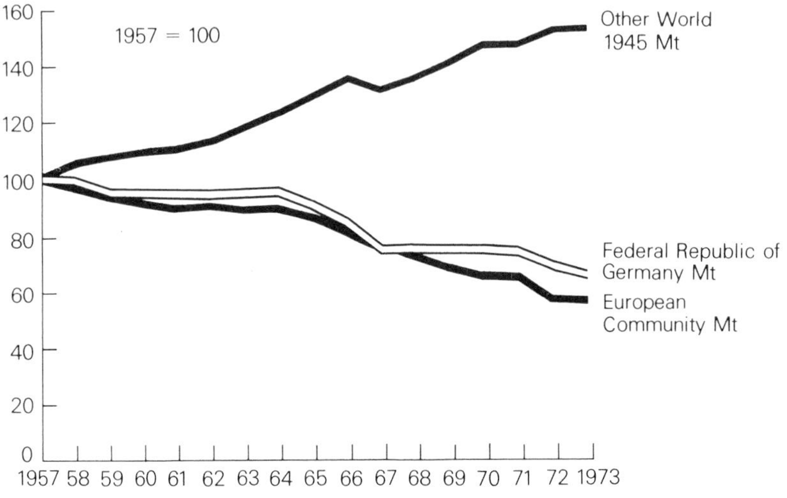

2.1 World coal production trends

tons of coal equivalent (TCE) in 1975 to 8.7×10^9 in 2020. This is an annual rate of increase of about 3%, which is similar to that obtained over the past couple of decades. Figure 2.1 shows this and indicates that a lack of understanding of the steady progress in coal supply and use may partly arise because Western Europe, uniquely, has allowed coal production to decline because of the availability of cheap oil; other countries have taken a more consistent and constructive attitude towards coal. In fact, the steady increase in coal production, discounting perturbation due to wars and depressions, can be traced back many years, to the industrial revolution, and has always managed to grow at about the rate required for the future. Obviously, no parameter can contrive to increase exponentially forever but the reserves, and the ultimate resource base, in the case of coal, are adequate for such continued growth well into the 21st century.

In addition to the strong reserves position and the production expansions planned, the importance of coal is based on:

(a) the flexibility of use, supported by the new technology of advanced coal conversion processes; and
(b) the comparative inflexibility and uncertainties of other energy sources.

It will be seen later that coal can be used in a large number of ways. It can be converted into liquid and gaseous forms, into heat and heat-derivatives (including steam and electricity), into chemicals and materials.

Hydrogen is the key to the conversions of coal to liquids, gases and chemicals. Coal can — and indeed in modern times must — be regarded as a member of the hydrocarbon family, whose properties depend on relatively small changes in hydrogen content. The hydrogen content of coal and other hydrocarbons can be changed, up or down. Where hydrogen needs to be added, the extra hydrogen required can be produced, essentially by reaction between coal and water.

Hydrogen is not only the technical key but is also extremely important economically in coal conversion policies, and in the efficiency of energy supply and distribution generally. The cost of producing hydrogen from any hydrocarbon is significant and energy is also consumed, so that the hydrogen obtained is less in energy potential than the equivalent of the input. Thus, it would be desirable to match the hydrogen content to the end use, ideally using the lowest hydrogen content fuel acceptable for any particular purpose. Coal uses are therefore likely to move up the value scale, initially by releasing oil and gas from simple combustion uses, gradually moving thereafter into the premium fuels and chemical feedstocks fields.

The adjustment in the hydrogen content of coal is only one of the aspects of its flexibility. In addition, coal, in common with other hydrocarbons, is capable of great flexibility with regard to heat release. This flexibility most obviously relates to the scale of the combustion equipment, which may vary from a small fireplace to a large power plant; it is also importantly displayed in controlled changes in the intensity of heat release and in ready adjustments in temperature and total heat output. Storage of coal is also very simple and in this respect compares favourably with other hydrocarbons. Transport is also fairly easy although less so in some respects than oil or gas; new transport methods may erode this difference. An important aspect of efficient energy supply relates to the uneven nature of demand. In order to cater for this situation, it may be necessary to make extra capital facilities available. The flexibility of coal, together with its storage and transport properties, may assist in providing the most economic solutions to these problems.

THE INFLEXIBILITY OF NON-FOSSIL ENERGY SOURCES

The importance of coal as the main fossil fuel of the longer term future, cannot properly be understood in isolation but must be considered in relation to the non-fossil energy sources. These latter include not only nuclear power but also energy sources often referred to as renewable or benign such as solar energy, tidal power and geothermal energy. These energy sources will certainly need to be developed to the maximum extent which is technically and economically feasible; nonetheless they have certain features in common which are in contrast to the flexibility of coal. In general, they are high capital cost systems with, hopefully, relatively low operating costs, (this however remains to be seen). Their product is generally either low grade heat or electricity generated at low efficiency (in terms of primary energy input to electrical output).

Both the timing and location of production can create problems of distribution and load mis-match. Nuclear power stations, for example, are very large and are located some distance from load centres; the incremental addition of one such large unit to a system and the outage risks may create undue technical and economic perturbations. The high capital costs require a high load factor, which (at least in the case of nuclear stations) is probably also desirable technically. Substantial infrastructures connected with, for example, uranium supply, fuel processing and waste management may be even more important than the reactors themselves, both economically and from a real resource point of view. All these factors create uncertainty about the true economics of nuclear power which will become much greater as the proportion of nuclear in the energy system increases.

This is one illustration of the inflexibility and uncertainty of non-fossil sources of energy. In other cases, such as wind, wave and tidal power, large uncontrolled changes in energy output will occur due to extraneous events which have no relationship to cyclical energy demand, thus increasing even further the effective capital cost element in both the production and distribution systems in order to deal with these perturbations. In other cases, such as solar heat, even the simple application of space and hot water heating may require substantial extra capital, not only for the system itself but also for the necessary back-up from fossil fuel systems; more sophisticated solar devices to produce electricity seem doubtful practical propositions of dubious economic importance.

CONCLUSIONS

Energy supplies are vital to human progress. It seems impossible to visualise a world with an adequate energy supply without an expanding supply of coal, used in a variety of ways emphasising its flexibility and employing the more advanced coal conversion technologies now being developed. Failure to achieve timely progress in coal utilisation could lead to social, economic and political unrest, of which the "oil crises" may be modest forerunners.

The economic aspects of coal utilisation will be considered later, as will the policy implications. For the moment, the overwhelming importance of a full and proper use of coal can be postulated on the basis of its reserves and flexibility of use in a world certain to be hungry for energy.

Chapter 3

HOW MUCH COAL?

INTRODUCTION

The question, "How Much Coal?" is not a simple one, nor does it admit of a simple, meaningful answer. On the one hand, it is possible to quote the 1977 World Energy Conference data to indicate that the geological resources exceed $10\,000 \times 10^9$ tons of coal equivalent (TCE), compared with a production in 1975 of 2.6×10^9 TCE, i.e. several thousand years supply at present rates of consumption. On the other hand, in the 1977 Robens Coal Science Lecture, Gibson states:

> "The element carbon plays such an overwhelmingly important role in our lives that its relative scarcity may come as a surprise. In the lithosphere it accounts for no more than 0.04% of the mass and, of this total, only a minute fraction (about one five-thousandth) occurs in a form which can react with oxygen. Most of this reactive carbon occurs in very concentrated form in the fossil fuels, peat, lignite, coal, oil, oilshales, tar sands and natural gas. Of these, coal represents by far the largest proportion."

It would be fair to say that coal use is not likely to be limited strictly by resources. Constraints on transferring resources into reserves and from reserves into production and use are more likely to arise from failures of planning, investment, market development and sheer determination. These failures, if they occur, may stem from prejudice (e.g. "coal is dirty") and ignorance of the potential for coal. This book is intended to help remove these constraints.

That coal should be the most plentiful — and widespread — of the fossil fuels is not surprising, having regard to the vegetable origins of coal, the plant life and decay stages of which took place in such profusion over vast areas and over many years; fortunately, we have witness of the early stages today in the peat bogs. The necessary scale of deposition of plant material can thus be conceived; the coalification processes already described are logical and relatively easy to visualise. Furthermore, apart from gases evolved (which incidentally provide much of the world's natural gas), the deposits, being solid, are likely to remain in place, only moderately affected by geological upheavals such as faulting and volcanic intrusions. Natural gas, which is really a highly volatile by-product, obviously requires a geological trap of some complexity and tightness, surprisingly provided in so many cases by a benign nature. Oil is less volatile, and in fact is more or less solid at room temperature in some cases; nonetheless, it can move considerable distances and it also needs a very favourable geological setting to collect in recoverable quantities in reservoirs capable of being tapped. The "raw material" from which coal was formed may also have been more extensively generated and distributed than in the case of oil formation.

Comparisons of the resources and reserves of coal with those of other fossil forms are difficult enough; the conclusions of such comparisons are nonetheless reasonably valid, so long as the uncertainties of measurement and calculation are appreciated. Comparisons between the fossil fuels and other energy sources are more difficult and much less meaningful. The proportion of the energy contained in various primary energy sources which can actually be made available for use depends on assumptions which cannot be exact: oil can be burned and the heat released can be applied, for example. In the case of uranium, very different considerations apply, so that differences in assumptions about utilisation processes (often requiring uncertain future development) can result in estimates of energy resources differing by an order of magnitude or two. There are difficulties also in comparisons involving secondary energy, energy conversions and the efficiencies of final utilisation methods. Tables of energy resources, or of supply and demand, should therefore always be considered with appropriate reservations, especially when different energy forms are compared.

In this chapter, the question, "How Much Coal?" is addressed in a general and detached way. In later chapters, the economic and industrial implications are considered and reference made to alternative energy sources.

ESTIMATES OF COAL RESOURCES

Procedures and Definitions

There are various ways of establishing the presence (or the probable presence) of a significant occurrence of coal and a number of methods of describing and assessing such deposits. Since coal has been worked for a long time, knowledge of coal remaining in the ground often comes directly from adjacent workings; frequently those workings originally commenced where an outcrop revealed the presence of coal. Active workings, unless their medium-term potential is completely obvious, are normally supported by continuous exploration, including drilling both from the surface and from underground positions, in order to enable a long-term mining strategy to be planned. Where there are a number of workings reasonably close together, a coal-field pattern will probably clearly emerge; this may be more extensive than the existing extraction facilities are capable of exploiting at an optimum rate in present circumstances.

Because of the long history of coal, the geological circumstances associated with coal deposits are fairly well known. In particular, the age and sequence of sedimentary rocks will determine the possibilities of finding coal, as well as the historic presence of a sedimentary basin and weather conditions suitable for peat formation. The percentage distribution of coal over the geological formations is shown in Fig. 3.1, from the WEC Survey of Energy Resources (1980).

The weather conditions for coal genesis, which include humidity and temperatures from tropical to temperate, can generally be related to the geological history of an area. The necessary conditions for the downward movement of an area of the earth's crust at the appropriate rate in order to form a suitable basin for plant growth and sedimentation are also fairly well understood, so that inferences can be made about the possibility of coal-bearing measures. The most important areas are related to the fore-depths of folded zones (e.g. Appalachia); in these cases, coal-bearing layers may extend over large areas and reach great thicknesses (5–10 km):

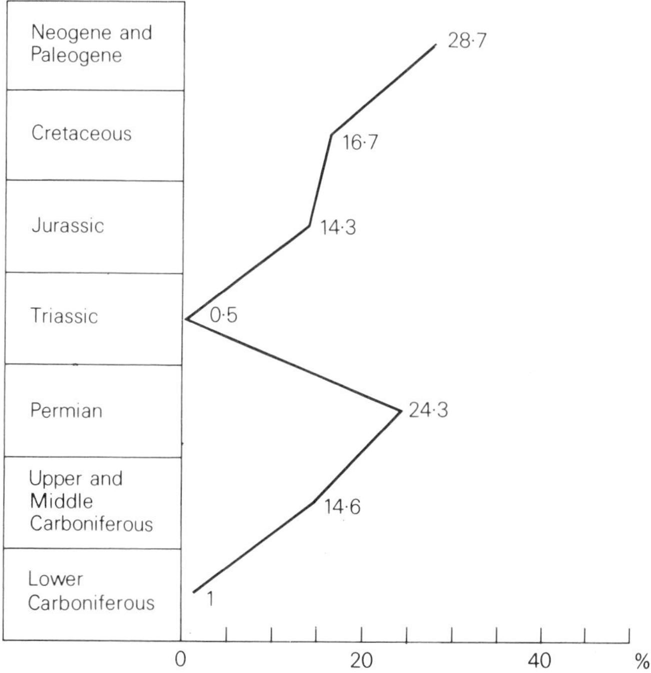

3.1 Percentage of the total coal following the Carboniferous found within each of the geological formations

the coal seams themselves within the coal measures may be numerous and are also likely to be extensive and thick, with the relationship between rank and depth being well displayed. Such fields, however, may be complex at the edges where tectonic movements have been greatest. Other types of movements which may form basins have also been identified. The accumulated knowledge and experience of centuries of mining lead to the following conclusions in the WEC Survey of Energy Resources:

> "It is very unlikely that new, currently unknown coal deposits will be found with the same scale and size as those in the Ruhr area in Germany or the Appalachian Basin. There is however quite a good chance that numerous small deposits possessing several hundred or a thousand million tons of coal will be found, particularly in the case of brown coal."

The recently discovered Elbiston brown coal field in Turkey is given as an example. Other areas mentioned in the WEC document are Western Canada north of the 60th latitude and the Arctic Islands; parts of South America, especially Colombia, Venezuela, southern Brazil and Amazonia; Indonesia; extensions of the Karroo formation in southern Africa, especially in Botswana and Namibia; Antarctica; and southern Europe.

In view of this list (by no means complete) of prospective areas, the quoted statement above may seem surprising and perhaps misleading except to specialist geologists; this is particularly the case with regard to scope for further discoveries. Certainly those most closely engaged in coal exploration and in resource assessment would strongly support the need for a greatly increased exploration effort; perhaps the explanation of the anomaly is that a large part of the exploration need is for more accurate delineation of coal deposits, the presence of which is known or strongly inferred. Figure 3.2, from the WEC Survey indicates the widespread range of coal deposits, many of which still need exploration, and prospective areas or indications.

Despite modern techniques for mineral resource survey, including satellite observation, there is no substitute for a carefully planned series of boreholes to extract cores of the strata including the coal seams. This not only enables the extent of the coal to be determined but also provides information about the coal quality, roof and floor conditions and problems which could affect shaft-sinking; all this information assists in mine planning and investment decisions. Other kinds of information, including magnetic and seismic exploration, assist, especially in determining the optimum positioning of boreholes and, during operation, the detailed mining plan to minimise the impact of faulting or other anomalies.

The borehole information, especially if reasonably closely spaced, is the ideal base for reserve calculation. However, only a small part of the probable coal in place has been assessed in this way and the spectrum of measurement accuracy is accompanied by a wide range of terminology. Care must be taken therefore to ensure that definitions are clear and unambiguous.

Other conventions must also be applied in reserve or resource estimations and need to be stated. In most cases, only coal seams above a given thickness and depth are counted. To calculate tons in the ground, a conversion from volume to weight must be made, based on an assumption about density. A further factor may be applied arbitrarily to estimate "recoverable" coal. Different types of coal may be listed separately but for purposes of comparison or summation, it may be necessary to report in terms of energy content (usually joules, sometimes calories or Btus); lower ranks of coal are often converted into "coal equivalent", based on the energy relationship between the coal assessed and a "standard" coal. Sometimes, attempts are made to assess coal according to its economic merit.

Finally, it must be mentioned that, in addition to metric tonnes, long and short tons (2240 and 2000 lbs respectively) may be used.

WORLD ENERGY CONFERENCE 1977

The World Energy Conference has for some time been regarded as a particularly valuable and authoritative vehicle for the collection and appraisal of energy resources; the great prestige of the WEC commands cooperation from countries of practically all political groupings. In recent years, resource studies have been carried out through its Conservation Commission, involving a broad group of the most eminent experts. Executive Summaries on the main forms of energy resources, including coal, were prepared for the September 1977 Conference and were discussed in a series of "Round Tables". The Commission itself has continued to appraise the findings of the different studies, in the light of the Conference discussion, and to produce updated comprehensive reports. Since then, the WEC has produced further information for its 1980 Conference and various other studies have been carried out and reported upon. However, it is felt that the 1977 WEC estimates established new standards of careful assessment; the results are therefore reported here followed by the most recent WEC findings and other studies.

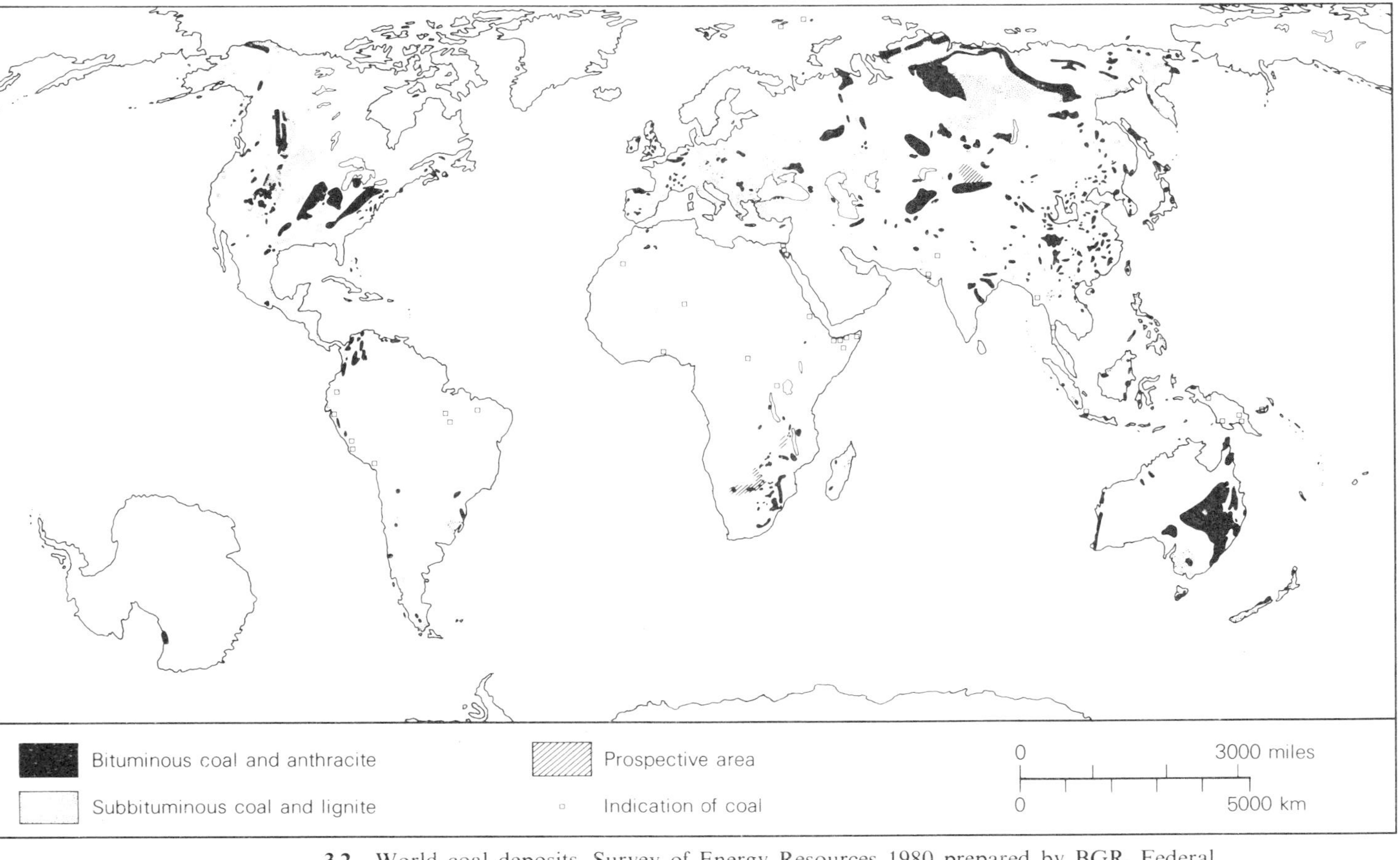

3.2 World coal deposits. Survey of Energy Resources 1980 prepared by BGR, Federal Institute for Geosciences and Natural Resources, Hannover

The World Energy Conference Coal Study of 1977 was carried out under the control of Drs Peters and Schilling, of the Bergbau-Forschung, the Central Research and Development Institute of the Hard Coal Mining Industry of the Federal Republic of Germany. The study was undoubtedly the most comprehensive and careful survey that had been conducted at that time. The study necessarily relied heavily on the responses of the various countries who were all cooperative within the limits of their knowledge. The questionnaire was more explicit and detailed than previous attempts and the approach to uniformity (an ideal extremely difficult to attain) was greatly assisted by personal visits to national authorities and by a certain amount of discretion in the conversion of inconsistencies exercised by the eminent authors of the study.

The authors made the following definitions:

1. Geological resources — resources which may become of economic value in the future.
2. Technically and Economically Recoverable Reserves — reserves actually recoverable under the technical and economic conditions prevailing today.

They further specified that maximum depths for geological resources for hard coal would be 2000m and for brown coal 1500m; these depths would reduce to 1500m and 600m, respectively, for economically recoverable reserves. Generally, they took minimum seam thicknesses of 0.6m and 2.0m for hard coal and for brown coal, respectively.

The main results, by continents, are summarised in Table 3.1.

Within the continental figures, those for the most important countries are given in Table 3.2.

A number of indications emerge from these tables and from the complete collection of data. The first is that coal is not only in very abundant supply but also that substantial amounts of coal are quite widely distributed, on a global scale. These facts have tended to be obscured in the public mind, perhaps for two reasons. One is that, historically, coal has normally been used near to its source and there has been little obvious evidence of large amounts being transported, as in the case of oil. The other is that considerable publicity has been given to the high proportion (c. 85%) of resources within the "big three" — USA, USSR and China. This is true but the other 15% or so includes many countries which are very well or reasonably endowed. Forty-two countries as listed in the WEC 1977 survey as having some coal potential. On a continental basis, even the continents with the least coal have resources for many years, compared with existing production. In Australia, for example, with the smallest continental reserve, a projected six-fold increase in production by 2020 would still leave an economic reserve ratio of 70 years and a resource base of 600 years; these levels of output, incidentally, would be twice the local demand, leaving half for export. In Africa, the Republic of South Africa is known to have great potential for expansion of coal output, both for internal use and for export; even with rapidly accelerating production rates, existing established reserves will last for more than a century. More surprisingly, Botswana is now thought to have resources nearly double those of South Africa. It may be that resources and reserves might be increased in other parts of Africa also, as a result of greater intensity of exploration and improved economic incentives.

Such factors as the above, leading to increases in estimates of resources and reserves, may have been effective in the successive WEC studies; prior to 1977,

TABLE 3.1 The distribution of world coal resources, grouped by continents

Hard Coal (bituminous coal and anthracite)

	Geological resources		*Technically and economically recoverable reserves*	
Continent	*in 10^9 TCE*	*percentage*	*in 10^9 TCE*	*percentage*
Africa	173	2	34	7
America	1308	17	127	26
Asia	5494	71	219	44
Australia	214	3	18	4
Europe	536	7	94	19
Total	7725	100	.492	100

Brown Coal (sub-bituminous coal and lignite)

	Geological resources		*Technically and economically recoverable reserves*	
Continent	*in 10^9 TCE*	*percentage*	*in 10^9 TCE*	*percentage*
Africa	190	—	0.09	—
America	1409	59	71	49
Asia	887	37	30	21
Australia	49	2	9	7
Europe	55	2	34	23
Total	2400	100	144	100

Totals

Hard Coal	7725	76	492	77
Brown Coal	2400	24	144	23
Total	10 125	100	636	100

TABLE 3.2 Coal resources and reserves, major countries ($\times 10^9$ TCE)

	Geological resources		*Technically and economically recoverable reserves*	
Country	*Hard Coal*	*Brown Coal*	*Hard Coal*	*Brown Coal*
Canada	96	19	8.76	0.7
USA	1190	1380	113	64
FRG	230	16.5	24	10.5
Poland	121	3	21	1
UK	164	—	45	—
Botswana	100	—	3.5	—
South Africa	58	—	27	—
Australia	214	48	18	9
China	1425	13	99	NA
India	56	1	33	0.4
USSR	3993	867	83	27

studies were also conducted in 1974 and 1976. In all cases, each study showed an increase over the previous one in all three major measurements:

(a) Quantity of resources;
(b) Quantity of reserves; and
(c) Reserves expressed as a percentage of resources.

The last of these, 6.3% in 1977, is considered by the authors to be a very low estimate and to demonstrate that strict criteria were being applied; since 1977, of course, the economic incentives to translate resources into reserves have become more favourable. The authors of the 1977 report anticipated such trends and expected a considerable enlargement potential for reserves due to economic changes, plus a further possible increase due to technical changes in mining methods. They also thought that the resource base represented a lower limit and wrote of a considerable "potential behind the potential".

This WEC report also considered production plans and export potential. In 1977, annual world production was at the level of 2.6×10^9 TCE; this would represent 250 years production out of the then economic reserves of 637×10^9 TCE. The summation of national plans (supplemented by the authors' estimates where necessary) indicated an increase to 5.6×10^9 TCE by 2000 and to 8.7×10^9 TCE by 2020 — an increase of about 3% per year and a doubling time of 20–25 years. The economically recoverable reserves already established by 1977 were adequate to meet the production level for 2000 and to meet a level of 7×10^9 TCE in 2020. To meet the planned output of 8.7×10^9 TCE in 2020 with a reasonable lifetime, the reserves would have to be doubled. Even if no further resources were located, this would only require the "promotion" of a further 6% of resources into the "economic" class. This seems entirely feasible, requiring only a slight change in prices. In fact, in many places, extra output would in theory only involve a slight increase in the marginal price; in some cases, the "new" coal might enable more efficient and cheaper mining methods to be employed, with capital becoming more important and labour less so. The projected rate of increase in world coal production of about 3% per year is rather similar to that which has obtained for many decades, with only modest perturbations. This continued progress is in spite of, and in contrast to, a perception of the "decline" of coal in Western Europe, where the effects of cheap oil have been particularly marked since World War II.

WORLD ENERGY CONFERENCE 1980

The WEC 1977 estimates have been given in some detail because they represented a high-water mark in method at that time and have not been seriously challenged since. The 1980 figures have not yet had the same degree of scrutiny and discussion. In fact, the latter are in any case fairly similar to those of 1977; unfortunately, the anomalies between methods used by different countries have not been significantly eased. In some cases, subjective factors may have actually increased.

The boundary values used for different ranks of coal are given in Table 3.3. The main results are summarised in Tables 3.4 and 3.5. Table 3.4 gives the tonnages of the different types of coal; in Table 3.5, results are translated to coal equivalent.

These tables show an increase of about 9% compared with 1977; this is stated to be largely due to increases in Australia and Canada from further exploration. The

TABLE 3.3 Boundary values of the most important parameters when classifying coal according to rank

	Natural water content %	*Heat of combustion af, moist, kJ/kg and (kcal/kg)*	*Volatile matter (daf)* %	*Total carbon (daf)* %
Peat				
	70–75	c. 6700 (1600)	60–62	60–64
Soft brown coal, lignite				
	35–40	18 850–19 250 (4 500–4 600)	c. 53	c. 67
Sub-bituminous coal				
	c. 10	23 900–26 800 (5 700–6 400)	45–50	c. 77
Bituminous coal				
	c. 3	35 400[a] c. (8 450[a])	10–14	91–92
Anthracite				

[a] In anthracite stage this value is lower

TABLE 3.4 Total resources of solid fossil fuels *in situ*, and recoverable reserves (in Gt and %), and the calculated lifetime of these reserves based on the production level in 1978

	Bituminous coal and anthracite Gt (%)	*Sub-bituminous coal Gt* (%)	*Lignite Gt* (%)	*Total Gt* (%)
Proved reserves	774.6 (58.7)	221.6 (16.8)	323.5 (24.5)	1319.7 (100.0)
Additional resources	6161.4 (50.7)	3835.2 (31.5)	2159.6 (17.8)	12 156.2 (100.0)
Total	6936.0 (51.5)	4056.8 (30.1)	2483.1 (18.4)	13 475.9 (100.0)
Proved recoverable reserves	487.7 (55.3)	143.0 (16.2)	251.1 (28.5)	881.8 (100.0)
Calculated static lifetime, years	198	780	281	—

TABLE 3.5 Total resources of solid fossil fuels and recoverable reserves in GTCE and %

	Bituminous coal and anthracite GTCE (%)	*Sub-bituminous coal GTCE* (%)	*Lignite GTCE* (%)	*Total GTCE* (%)
1 Proved reserves *in situ*	774.6 (71.6)	172.8 (16.0)	113.2 (10.5)	1081.3 (100.0)
2 Recoverable reserves	487.7 (70.4)	111.6 (16.1)	88.1 (12.7)	693.2 (100.0)
3 Additional resources *in situ*	6161.4 (61.0)	2991.4 (29.6)	848.3 (8.4)	10 102.4 (100.0)
Total of 1 and 3	6936.0 (62.0)	3164.2 (28.3)	961.5 (8.6)	11 183.7 (100.0)

authors point out that only about 10% of the stated resource consists of proven reserves and appear to express stronger reservations than the 1977 authors about the other 90%, which they say consists mainly of "inferred, hypothetical and speculative quantities possessing confidence levels of less than 50% and sometimes considerably less than this".

WAES AND WOCOL

The Workshop on Alternative Energy Strategies (WAES) was an international cooperative study centred on the Massachusetts Institute of Technology, under the direction of Professor Carroll L. Wilson. This project reported in 1977. Subsequently, a similar and more intensive study, The World Coal Study (WOCOL), was undertaken under the same leadership. The WOCOL estimates for reserves and resources are given in Table 3.6.

These WOCOL figures supersede the WAES estimates. They were based on the 1977 WEC figures, "updated" by the cooperating authorities. The results are very similar to the 1980 WEC figures.

INTERNATIONAL ENERGY AGENCY (IEA)

Other studies tend to use the WEC reserves figures as a base. There is wide acceptance of these, at least as a starting point, but there is considerable divergence of view in considering the likely trends of outputs and export availability. The International Energy Agency, for example, in their "Steam Coal" report (1978) give the 1977 WEC figures in a similar form to that presented above but note that many coal experts consider the WEC figures "too low". The importance of the USA is emphasised and it is pointed out that the depth limit (300m) for the US figures was different from the common norm (1500m) and that the recovery ratio is low. In contrast, the IEA report includes notes on a country-by-country basis which

TABLE 3.6 World coal resources and reserves by major coal-producing countries (TCE $\times 10^9$)

	Geological resources	*Technically and economically recoverable reserves*
Australia	600	32.8
Canada	323	4.2
Peoples Republic of China	1438	98.9
Federal Republic of Germany	247	34.4
India	81	12.4
Poland	140	59.6
Republic of South Africa	72	43.0
United Kingdom	190	45.0
United States of America	2570	167.0
Soviet Union	4860	109.9
Other Countries	229	55.7
Total World	10 750	662.9

concentrate mainly on the problems of expanding output. In the case of the UK, the nationally reported figure of 45×10^9 TCE for Technically and Economically Recoverable Reserves is qualified by the IEA: this estimate is considered to include reserves "judged" to be economic to extract only at some future date. This view is firmly rejected by the UK.

FORD FOUNDATION

A report *Energy: the Next Twenty Years* (1979) sponsored by the Ford Foundation and administered by Resources for the Future is an excellent analysis of energy sources, although naturally concentrating on the USA. On world coal reserves, the WEC figures are quoted (although, surprisingly, not including lignite or brown coal). The point is again made that resources and reserves are so large that they constitute no constraint on output.

ESTIMATES FOR THE USA

It is widely accepted that the US reserves and the prospects for exploitation have vital implications for the rest of the world. To some extent this relates to a possibly increased role of the USA as an exporter of coal but, even more importantly, to the degree to which the USA may be able to use coal internally to reduce demand for oil imports. The US Department of Energy Fossil Energy Program refers to coal *reserves* of 3.2×10^{12} tons, which may be compared with the WEC figure of 2.57×10^{12} for *resources* and 178×10^9 for *reserves*. In another US Department of Energy report however the reserves are stated as 438×10^9 short tons (c. 400×10^9 TCE) and this figure is generally quoted. This refers however to coal in place and assumed recovery percentages may variously be in the range 30–90%. R. A. Schmidt in *Coal in America* (subtitled, aptly, *An Encyclopedia of Reserves, Production and Use*) estimates the recoverable reserves as 165×10^9 t, of which 30×10^9 t would be recoverable by underground coal gasification. The total recoverable coal is less than one-tenth of his estimate of identified resources.

The question of the likely US production trends is even more debatable. The maximum feasible output (unconstrained) is considered to be about 1400 MTCE in 1985 and just over 4000 MTCE in 2000, compared with 590 MTCE in 1976–7. In 1977, President Carter announced his National Energy Plan which called for an increase of 363 MTCE by 1985 to 907 MTCE. This was regarded by the National Coal Association as a modest goal if "unnecessary government constraints are avoided". A report to the Federal Energy Administration suggested 1129 MTCE in 1985 but a report to Congress considered 853 MTCE more probable. Predictions from other governmental sources include 841 MTCE in 1985, 983 in 1990 and 1352 in 2000.

WAES considered potential US coal productions in the range 726–944 MTCE in 1985 and between 1104 and 2009 MTCE in 2000; the main differences were related to market demand but policy obstacles were also emphasised. The Ford Foundation report does not make any comprehensive production predictions. However, the report stresses the importance of the Western states, where the output is expected to be between 400–500 Mt by 1985 and as much as 50% of the total US output in 1990.

The main western producing states will be Montana and Wyoming, rising from 74 $\times 10^6$ short tons in 1977 to 160×10^6 in 1990. It seems clear that the Ford Foundation group did not expect total output to grow at anything like the theoretical rate based on reserves or even at the rate required by the National Energy Plan.

CONCLUSIONS

There seems wide agreement that the resources for coal are very large indeed and that the economically established reserves are large enough to sustain an annual increase of about 3% in production rates for a long time, as summarised by the WEC on the basis of national plans. The attainment of these plans seems more open to question.

Further consideration is given to these matters, especially how they may affect availability for export and economics, in Chapter 12. Policy implications are discussed in Chapter 23.

Chapter 4

HOW IS COAL WON?

INTRODUCTION

Coal mining technology is an immense and complex subject with a voluminous literature of its own; it cannot be dealt with properly within the scope of this book. However, some slight familiarity with the general principles of getting coal is a desirable adjunct to a consideration of coal utilisation. It is obvious that the quality of the coal, and therefore its suitability for various uses, can be directly affected by the mining techniques employed; this will apply, for example, to the size distribution of coal and the amount of mineral matter and moisture included. The regularity of these features will be important also and this may be especially relevant when coals from two or more different districts or seams are brought to the surface together or are joint feedstocks. In addition to consistency of quality, regularity of supply and demand are important. There is a high level of capital investment in modern coal mining, preparation and transport systems; this is increasingly being geared to sophisticated and expensive utilisation plant which will probably be continuously operational. All these units have their own pattern of maintenance and reliability. Overall optimisation of the whole system, especially to cover variations either in quality or rate of output, becomes desirable. The fact that coal can readily be stored is an advantage in this respect.

Originally coal was probably dug where the seam outcropped. The seam would be followed as far as possible and, later, simple shafts were sunk nearby and the coal extracted as far as possible around the shaft, creating a bell shaped pit bottom. As mechanical equipment for ventilation, drainage and coal handling improved, roadways were cut into the coal progressively further from the shaft, generally leaving pillars of coal to support the strata.

These primitive systems have evolved, through the development of massive machinery and other equipment, into the main modern methods of getting coal today: (a) opencast (or strip) mining from the surface and (b) underground mining.

UNDERGROUND MINING

Initially underground coal winning was a low productivity and highly manpower-intensive operation. Now it can be a completely mechanised, highly productive operation. Further productivity improvements are being introduced, especially through remote control and automation of the mining systems using computers.

There are two main methods of underground working: "pillar and stall" (or room and pillar), used especially in medium to thick seams at shallow depth, and "longwall", a method adaptable to most mining conditions but which is the sole technique for working at depth. Extraction cannot take place however without access roadways; the provision and maintenance of these is a major cost item of the mining system.

Extraction of coal from the longwall face requires the driving of pairs of roads which link with pairs of access roads into the area being worked. These access roads in turn link with the pair of mine shafts. Access is required for the transport of men and equipment into and out of the mine, for the transport out of coal and rock from the mining operations, and for the maintenance of good healthy environmental working conditions. Shafts and roads are in pairs for the forced circulation of ventilation air. As coal winning is an extractive process, when each longwall panel is worked out — between 6 and 18 months on average — it must be replaced without delay by another. A constant task is therefore the preparation of replacement panels in adjacent or new areas. For this purpose, much underground preparatory work has to be carried out. The determination of workable areas is ascertained from surface boreholes to the seam, by seismic techniques and by the driving of underground exploratory tunnels. These surveys must be carried out in a timely fashion and supported by careful planning in order to circumvent inevitable geological inconformities and permit continuous production.

The pairs of shafts and main roadways provide the main access to the workable areas. From the main roads, two parallel tunnels 100–200 m apart are driven into the coal seam. Between, and at right angles to these, subsidiary tunnels are driven and working faces formed spanning these pairs of roads. Machines cut and load the coal onto the face conveyor which traverses the face. The face conveyor discharges onto a belt conveyor which in turn feeds the main transport system. In "advance" mining, the method most generally practised in the UK, roads are continuously extended at each end of the face as it "advances" away from the main roadway. In "retreat" mining, the roads are first formed to the boundary of the coal panel which is to be extracted; the face is then started at the far end and is worked back to the main road.

The main machinery on a longwall face consists of a cutter-loader, a conveyor and a set of powered supports. The most usual cutter-loader is a shearer, which is an electrically driven drum, equipped with picks and loading vanes which form a scroll. As the cutter-loader moves along the face the drum is rotated and about 0.6 m width of seam is cut and loaded onto a face conveyor. When the coal has been cut, this conveyor is pushed over into the newly formed track ahead so that the machine can then cut another 0.6m slice in the opposite direction. The roof along the coal face directly over the machine and conveyor is supported by self advancing hydraulic chocks. Other forms of power loaders, the "trepanner" and plough, are used in specific conditions. Different cutting systems can produce different qualities of coal — trepanners, for example, produce rather more large coal (suitable for domestic use), but all modern mechanised systems produce a large proportion of fine coal. Because there are usually dirt bands in the seam, the fine coal is produced with fine dirt. Because water is used to damp down dust, the fine coal is also wet.

Powered supports are set at close intervals along the face, near to and back from the face conveyor. They consist basically of telescopic legs which can be extended or contracted hydraulically between a base plate on the floor, and a top plate or canopy. The canopy is designed with forward cantilever to support the roof over the conveyor extending to quite near to where the coal is being cut. The supports are attached to the face conveyor with rams; these are used to move the conveyor to the face as the coal is cut away. Subsequently, the support pulls itself forward to the conveyor. The whole sequence of moving and re-setting proceeds along the face after the cutter has passed; the steps are automatically controlled.

The face has all of the problems of a factory in providing consistency of product and in achieving continuous operation; this has to be achieved safely whilst continuously exposing new ground.

Great improvements in face productivity have been obtained in recent years, mainly through more powerful cutters, by better design and by more robust and reliable construction of machinery. This face improvement has put more emphasis on the efficiency and reliability of transport and coal clearance systems, which have been improved through computer control. Having improved the reliability and capacity of mechanical items, modern sensing devices to control the cutting horizon of the machines are being applied. Nucleonic sensors located near the cutter head can control the position of the cutting drum within the seam and reduce product contamination.

The shaft capacity and transportation layout of existing mines are now often the main bottlenecks to increased production and productivity. Sometimes these constraints can be overcome by major reconstruction of shaft equipment and roadways; this can be an extremely attractive investment when local reserves of coal are adequate to justify it. Even more exciting possibilities exist in new coal fields; these are being discovered even in the UK, where coal has been worked so long. New coal fields lend themselves to better planning *ab initio*, compared with older mines; layouts in virgin seams can be based on the integration of all best modern practices, with the potential for extremely high output and productivity.

Although the longwall system is gaining ground in the world, it is nonetheless the minority system in the US and in many other countries where thick and consistent seams occur at moderate depths. The alternative extraction method adopted is the modern equivalent of the old "pillar and stall" system, operated nowadays by "continuous miners" and modern belt or shuttle car transport. In suitable conditions this system is extremely productive although much coal is left behind in pillars.

SURFACE MINING

In the USA and other parts of the world with thick shallow seams, strip mining can provide up to 50% or more of the total output. These deposits are often of low rank, lignites or brown coal, but can be very thick and near to the surface. They can therefore be very profitably exploited especially when electricity generation is integrated with the mining operation (as is the case with a huge brown coal operation in Germany). In the UK about 10% of coal production is strip-mined and, in contrast to some of the bulk operations producing steam coal in other countries, modest sized individual operations producing special quality coals, such as anthracite or prime coking coals, are to be found. The technology employed in strip mining is very advanced; the size of the deposit, together with the scale and amount of equipment, are generally the limiting factors of production levels.

Strip mining is based mainly on the huge dragline, with buckets scooping up tens of tons at a time. Large mechanical shovels, bulldozers and bucket wheel excavators may also be used. The top soil is first removed and placed on one side, exposing the coal seams, which are then progressively extracted if more than one occurs. Intervening dirt bands may be separately removed and discarded if thick enough. In spite of the massiveness of the mechanical equipment, the separation of individual seams from dirt bands can be conducted in quite a delicate fashion. In some cases, coal measures over a hundred feet thick are available. Depths of overburden of over 200m are quite feasible and total depths of 500m or more can be worked. For the greater depths, however, proportionately larger superficial areas need to be made available and for longer periods; the ratio of overburden to seam thickness is also an important consideration in deciding the optimum area. The most desirable conditions are often more appropriate for remote, undeveloped areas; in these locations, however, new large operations will probably require the provision of infrastructure.

Strip mining also involves restoration of the land after extraction of the coal and stringent requirements tend to be applied now to all sites, even in remote areas. Great progress has been made in the restoration of farm land, by careful preservation and replacement of the top soil, but there is still some controversy about some agricultural aspects of this procedure. In all areas, even when the land is unsuited for farming, the restoration of contours and the diversion or pollution of streams can be extremely contentious issues. Limitation of disturbance during strip mining and requirements of restoration both have direct economic effects; as standards improve, these cost implications are growing in importance. An even more difficult dilemma may arise, in that some of the environmental effects may not have a technical solution at any price; there is then a straight trade-off between amenity and energy which has to be resolved by legal or political means.

COAL PREPARATION

The raw coal that comes out of the mine is a mixture of coal and dirt. At a typical mine the raw coal can contain 35% dirt. In a coal preparation plant, the raw coal is first sized. The coarse coal is separated from dirt either by jigging in water or by dense medium separation; the fine coal below 0.5mm is cleaned by froth flotation. The cleaning process reduces the ash content and to a lesser extent the sulphur content. The washing processes use water and this is largely removed from the products by centrifuges or vacuum filters. For a few specialised markets, or if the clean coal has to be transported great distances, the coal is further dewatered by thermal drying.

Coal for power station fuel may be prepared at about 18% ash content; this is often however within a specification of plus or minus only one percentage point, say, 17–19%, emphasising the quality requirements of modern coal-using equipment. Such fuel is often prepared by cleaning half of the raw coal to about 7–8%, and then blending this with the remaining untreated raw coal. Control equipment must be carefully designed to ensure that the blended produce has a consistent quantity. Coals prepared for coking purposes have a much lower ash content, generally less than 6%, and strict upper limits are placed on sulphur and moisture content. The addition back of fines to coking coals in a consistent fashion is very important as the proportion of fines can affect the coking properties. Coals for power station fuel and coking purposes are generally of small particle size (less than 25 mm), which suits the modern mining methods; it is not unusual for the feed to the coal preparation plant to contain 85% finer than 25mm. The 15% of clean coal larger than 25mm may be prepared especially for domestic and industrial purposes requiring lump coal.

The modern coal preparation plant uses computers and microprocessors to control the processes. This makes it possible to ensure a consistent quality in the product. It also makes it possible for modern coal preparation plant of 1000 t/h capacity to operate with surprisingly few personnel.

In any purification process, in addition to the main product, "tailings" are produced which inevitably contain some thermal values, normally at a level below which further extraction becomes unprofitable. In former times, when processes were less efficient and values different, substantial amounts of coal were contained in the waste materials dumped on the tip, so much so that waste dumps are sometimes now being reworked. Nowadays the carbon content of waste material is quite low but not insignificant, perhaps 1 or 2%. Nonetheless, as values and technology change, these policies will need further re-optimisation. For example, fluidised combustion can consume current quality tailings but usually without generating useful amounts of surplus heat (largely because heat is consumed to evaporate moisture). However, a different "cut" in coal preparation might give a smaller proportion of higher quality coal (perhaps to be used for chemical processing); fluidised combustion for power generation might then be a useful outlet for "middlings", perhaps without producing any tailings for rejection at all, thus getting value for all the carbon mined. Even more important, such a pattern of separation and usage might have economic repercussions on coal cutting procedures; if emphasis on avoiding dirt could be slightly relaxed, production could be increased. This might lead to cheaper extraction and to higher percentage recovery of the coal in the seams, the latter being a factor of increasing importance. Taking the argument further, most gasification and liquefaction processes leave residues which can be burnt (and process heat is nearly always needed). This adds a further dimension to the optimisation of the quality burden designed to be carried by different stages in the extraction and consumption chain. On the other hand, the fluidised combustion process, which could be regarded as the garbage disposal unit, may have limitations on the proportion of fine particles acceptable, possibly affecting optimisation in a different direction.

TRANSPORT AND STORAGE

As noted above, transport systems for coal below ground are very important, both as a cost element and as a potential bottleneck to output. Systems of conveyor belts are now generally used, often with surge bunkers underground to minimise the effects of disruptions at the face. The conveyor belt system can, with advantage, continue to the surface but this is only possible with sloping shafts (or "drifts"). With the usual vertical shafts, coal was once hauled to the surface by means of buckets but continuously operating skips are now greatly preferred.

On the surface, there are some favourable cases where the coal can be moved to the point of consumption by conveyor belts. These cases are generally confined to distances of a few miles; in some instances, conveyor belts may be used where there would be difficulties with other systems. For most inland movements, however, rail transport is preferred. Since World War II, great progress has been made in the unit train concept. In the US trains carrying more than 10 000t are operated but elsewhere the unit size is generally somewhat less. What is more important, unit train systems are now usually equipped with automatic loading and unloading devices, generally not requiring the train to stop. The whole operation, from pit or washery, through stock piles and bunkers, into trains and into highly automated stocking retrieval yards is usually managed with very low manning levels. In some countries, dust blown from coal trains may be an environmental problem and these effects may be reduced by consolidation, by placing a layer of lump coal on top, by tar sprays or chemical wetting agents or by the use of covers.

Where unit trains are uneconomic or cannot be used because of the diversity of delivery points or the lack of rail access, road transport is widely used, again with sophisticated loading and unloading systems, often including pneumatic pipelines. To a great extent, road transport has supplanted barge transport on inland waterways, once very important, but now less so. It is however probably still the cheapest form of transport where suitable and may keep its present market share.

Slurry pipelines have received a great deal of discussion, particularly as a method of improving the competitiveness of rail transport, but only a few have actually been built and only two long distance pipelines are currently in operation. One of these is a 200 km pipeline from the Donetsk coalfield in Russia to a port on the Black Sea. The other is the Black Mesa Pipeline in Arizona; this is 273 miles long and 18 inches in diameter, transporting 5 million tons of coal per year. Another pipeline in Ohio operated successfully between 1958 and 1963 but was mothballed as a result of reductions in rail tariffs coincidental with the introduction of unit trains.

For pipelining, it is generally necessary to reduce the particle size of the coal before mixing it into a slurry with water. Pumping stations are required every 30–60 miles and the flowrate should be about 5 ft/s or more to prevent settling. The biggest problem in slurry pipeline transport however is the dewatering stage at the receiving end, with which may be associated the question of water availability and recycling. The dewatering plants produce an "ink" which may contain up to 25% solids by weight in particles less than 0.04 mm. In the Black Mesa case, no really satisfactory way of recovering this coal has been devised and the loss is substantial (perhaps 7% of the input coal). The environmental impact may be even more serious. At Black Mesa, the ink is dumped into ponds from which the sun evaporates some of the water but this is visually unattractive and the coal in the ink does not appear to be recoverable. It may be possible to recycle ink through a parallel pipeline back to the slurry preparation site, and this solution would appear to have the advantage of saving water. No one has tried this however and the prospects are uncertain. In spite of this, a 1040 mile pipeline is being constructed from Wyoming to Arkansas to move 6 million tons per year.

Coal transport by ship, where appropriate, is relatively cheap; costs, as might be expected, reduce rapidly with size and ships up to 150 000 deadweight tons are now used. In order to take advantage of these low transport costs, of course, port facilities must be available and this generally requires considerable expenditure. On the Great Lakes, self-loading and unloading ships are used.

The transport of coal is an important economic factor, perhaps accounting on average for one quarter to one third of the delivered cost, in some instances being more than half. The optimisation of transport is further considered in Chapter 21.

Whilst the ability to store coal in a pile is a useful attribute, increasing energy supply options, there can be problems, including quality deterioration (especially for coking coals) and spontaneous combustion, often triggered by hot-spots. These problems can generally be overcome provided techniques worked out in the last two or three decades are adhered to. These consist of building stockpiles over a substantial area by adding thin layers evenly, with consolidation of each level. This can also aid homogenisation, particularly important with coking coals. By layering horizontally and then recovering by cutting vertical slices with a bucket wheel, very good mixing can be obtained.

FUTURE PROSPECTS

So far as the development of mining technology on conventional lines is concerned, the parameters within which the optimised integration of mining methods with utilisation will occur are fairly clear qualitatively, if not quantitatively. The possibility of totally new techniques for exploiting coal have been examined, including *in-situ* methods. The latter are discussed in a separate chapter. In processes for mining coal, for utilisation above ground, the most important possibility to be borne in mind is that of hydraulic mining and transport. Very powerful fine jets of water can be used to cut coal and full-scale demonstrations have existed for some time. Coal won in this way would conveniently also be transported hydraulically and the consequences for utilisation would need to be considered. The most obvious point is that the product would probably be very fine and wet; these factors are likely to be disadvantageous unless the coal is to be utilised in a process requiring fine particles and water, such as certain gasification methods. On the other hand, if hydraulic transport becomes widespread, with networks of slurry pipelines, this could have important consequences on the patterns of energy distribution and conversion, as well as on the type of processes favoured.

It has been proposed that human intervention below ground can be eliminated by the development of remote or "telechiric" methods. Such a step could probably not be contemplated until after development of machines, sensing devices and control mechanisms to the ultimate standards of efficiency and reliability; it may then still be difficult to justify, on economic grounds, the complete elimination of the few remaining personnel below ground. In any event, the potential for utilisation of coal may not be particularly affected by such a step.

From the user point of view, it is probably better to think of coal-getting in the following context;

(a) Ample reserves;
(b) Progressive improvements in productivity and consistency;
(c) Potential for constructive interaction between mining and utilisation technologies.

Chapter 5

HOW CAN COAL BE USED?

INTRODUCTION

In subsequent chapters, various ways of using coal are described in some detail and evaluated. A brief review of these methods and a few generalisations may be helpful as a starting point, especially for the lay reader.

The most important factor about the potential utilisation of coal is its flexibility. Although coal is relatively stable as a storehouse of energy and chemicals, it can be treated in a few different but basically simple ways in order to release these values usefully. Coal has been available to man for a long time; it is natural that it should have been used in many different ways and also that the balance of usage has changed from time to time as needs varied and technology developed. The evolutionary process of changing pattern of use is probably now at a critical point; the next decades are likely to see a more rapid adaptation of the role of coal than ever before.

In later chapters, the technology of utilisation processes is described, with emphasis on modern processes and on those which are likely to be developed and commercialised in the next two or three decades. In almost all cases the processes likely to be used are not new, in the sense that a new basic reaction or treatment is involved. Some of the processes that will be important well into the next century are improvements in existing processes; others will be different ways of doing the same duty or will produce products already known as derivatives of coal. The economic merits of these processes will largely determine the directions and patterns of change.

Although the balance of use will probably be modified drastically, it is important to appreciate that the driving force behind such changes will not be any overall shortage of coal or the unsuitability of existing processes, which have been and will continue to be improved, but mainly external factors, especially the changing availabilities and costs of other energy sources and feed stocks, together with general changes in economics and lifestyles. The process of change should always lead to coal moving into higher value uses and into processes of more highly developed engineering character.

COMBUSTION

Originally, coal was used directly as a source of heat but the transfer of heat from coal into another medium for the distribution of that heat has probably been established for a very long time. The industrial revolution gave rise to (or was founded on) two extremely important developments: the raising of steam for motive power and the use of coal for the smelting of metals, especially iron. Both of these latter outlets remain supremely important but nowadays most of the steam is used in central generating stations for the production of electricity. However, although coal usage is dominated, in terms of tonnage, by that used in the central power stations, coal is still used by industry in large numbers of combustion units for the production of steam, both for process heat and for power generation purposes; this industrial market is likely to become the most rapidly growing area in the medium term. In fact, these industrial outlets are likely to be the first example of changing patterns of use as a result of a combination of advanced technology and economic, social and political forces.

In most modern power stations, coal is burnt in the form of pulverised fuel (pf). In this system, powdered coal is injected with air into a hot furnace, where it burns as a flame, heat being recovered in steam-generating water tubes lining the furnace walls. This system has been intensely developed, to a high degree of predictability and efficiency, within rather fundamental limitations. The combustion aspects reached a fairly high level of performance some time ago and in recent years the main emphasis has been on the steam generating side, with the trend to higher temperatures and pressures in order to make improvements (modest but important) in the generating efficiency.

On the other hand, smaller units for industrial purposes have remained based mainly on the grate, one of the earliest devices in the history of coal. A great deal of ingenuity has been employed, via mechanisation, automation, instrumentation and control, to improve the operational characteristics of these furnaces but their inherent disadvantages remain — especially a relatively low specific heat release on a volume basis.

Fluidised combustion is the first real departure for some time from the limitations of grates and pf furnaces. This system — which has a number of versions suitable for different purposes — is considered very promising and will probably become the principal combustion method. Fluidised combustion is so called because the coal particles being burnt are "levitated" by the upward flowing stream of combustion air. These conditions are favourable for good heat transfer and for a high intensity of heat release, although at moderate, controlled temperatures; there are other advantages, including economic and environmental.

The simpler versions of fluidised combustion seem poised to make early and substantial penetrations into the market for small and medium-sized industrial boilers. The large size and high level of technical development of current pf furnaces for electricity production will, however, mean that more complex and larger versions of fluidised combustion will need to be developed in order to compete in this area. These latter developments are currently under way.

A very wide range of fuels, including low-grade coals and residues, can be burnt by means of fluidised combustion. This fuel versatility, allied to potential technical, economic and environmental advantages, is expected to result in fluidised combustion having the greatest impact of any technical innovation over the next twenty years on the scale and type of coal utilisation.

In the UK and many other countries, the domestic space-heating market remains quite important. After a long period of decline, as gas and electricity won ground, it does appear that in the UK there is a relatively small but firmly-rooted residual domestic demand of a few million tons per year, which may continue into the foreseeable future. Again, technology has played a key part. The image of open fires burning bituminous coal in a smoky fashion has all but disappeared. Most coal burnt domestically in the UK is now either naturally smokeless (e.g. anthracite) or de-smoked products such as coke or smokeless briquettes; alternatively, special appliances which provide the cheerful comfort of open fires but which do not exceed permitted smoke emissions are used. Market preference for coal (or, more exactly, solid fuels) in the domestic field may not depend very closely on economic factors but on the advantages in appearance and also on the self-reliance resulting from the ease of stocking. These factors may be effective in regaining smaller markets in other countries also. Where significant domestic markets exist, for example in under-developed countries, it would seem desirable to retain these on energy conservation grounds, rather than adding to the burden of electrification schemes.

CARBONISATION

All the above combustion uses depend on the simple reactions between carbon and oxygen. When coal was substituted for charcoal as the reductant in iron smelting, the process of "carbonisation" became important. In this process, coal is externally heated out of contact with air, in special "slot" coke ovens. The volatile matter is driven off, giving valuable gaseous and liquid products which may be recovered; in most circumstances, however, the coke which remains is the main product. The coke, in honeycomb form, is stable but can react at high temperatures to reduce metal oxides either directly or, more efficiently, by reacting first with oxygen to form carbon monoxide which then acts as the reductant. In practice, some hydrogen may also be produced by reaction between the coke and moisture; this also acts as a reducing agent. One of the great advantages of coke in smelting processes is that it can be made in a strong and stable form which can withstand thermal and mechanical forces and also chemical reactions at low temperatures. Its reactions can then be centralised in suitable hot areas in process equipment such as blast furnaces. Specially selected coals and carefully controlled coking procedures must be used for these higher grades of coke — metallurgical coke. First class metallurgical cokes can be distinguished empirically in use, and to some extent by standard testing procedures, but the basic properties contributing to quality and to the detailed mechanisms of behaviour in blast furnaces are only just beginning to be understood. Hard coke is also used in cupolas for melting iron for castings. Again, technology has already contributed to the way in which metallurgical cokes are produced (especially in widening the range of suitable coals). There is considerable scope for further development on these lines.

In addition to hard, unreactive coke for general metallurgical purposes, more reactive (and often smaller) cokes may be made either in similar slot ovens, or in special retorts, operating at lower temperatures. Coals of lower coking quality may be used in these cases. These cokes can be used in industrial boilers or domestic appliances to minimise smoke production or as a reductant in certain chemical/metallurgical processes.

As an alternative to coke, briquettes (often called "formed coke") may be made from fine coal in a variety of processes. A carbonisation step is normally required, either on the ingredients or on the finished briquette. Such briquettes may be in contention with oven coke for the whole range of applications. It is considered by some experts that briquettes will gain ground at the expense of coke but penetration so far into these markets has been limited.

GASIFICATION

The carbonisation process in coke ovens or retorts has been an important source of fuel gas for many years and so have the off-gases from blast furnaces, the fuel components of which are derived of course from the coke. However, fuel gases have also been deliberately produced as a principal product from coal for a long time; in these cases, the gases produced by the volatilisation process have traditionally been augmented by the addition of steam and air to the reactor vessel at elevated temperatures. Again, the chemical reactions are simple ones, between carbon, oxygen and water; the principal fuel values in the gases so produced come from carbon monoxide, hydrogen and methane.

Fuel gases consisting of such mixtures (often called "town's gas") have been produced and distributed commercially for nearly two hundred years. However, in recent decades, this manufactured gas from coal has been largely supplanted by natural gas, which is essentially methane. Methane has a high calorific value, about twice as high as the former town's gas; the latter often contained some inerts as well as the main fuel components, carbon monoxide and hydrogen, both lower in calorific value than methane. Thus, in modern parlance, the mixtures of carbon monoxide and hydrogen, generally with some methane, are called medium calorific value gas or medium Btu gas (MBG). Gases of this type, with methane eliminated, may also be called "synthesis gas" since various hydrocarbons may be synthesised by reactions between carbon monoxide and hydrogen; the synthesis gases can also be "shifted" to pure hydrogen or "methanated" to pure methane. A particular form of synthesis gas consisting of hydrogen to which nitrogen is added, is used for the manufacture of ammonia.

The versatility of gases produced from coal is therefore obvious. Most major gas distribution systems are now based on methane (natural gas), which on a thermal basis is relatively cheap to pump because of its concentrated fuel value. When gases are manufactured from coal on a large scale again, methane may often be preferred; this will also make best use of existing pipelines and other equipment. However, there will probably also be opportunities for MBG and synthesis gas.

The old style town's gas plant will however be superseded by more modern and efficient plants. Fortunately, a group of new gasifier developments were pioneered in Germany about 50 years ago and although their commercialisation has been held back, except in special circumstances, by the competition from cheap oil and natural gas, these processes form the technical basis of most modern proposals. There are essentially a few different ways of bringing the coal into contact with the gasification reactant; with permutations between these contacting methods, temperatures, pressures and reactants, a range of gases can be made to suit all requirements. It remains to be seen how many gasification processes, and which ones, will become fully established in the future.

LIQUEFACTION

Liquids, as well as gases, can be recovered from the traditional carbonisation processes. These liquids have been used as fuels and they also provided the initial feedstock for the whole of the organic chemicals industry. Subsequently, the growth of the oil and petrochemicals industries dwarfed the supplies of feedstocks arising from coal carbonisation, which was limited by the demand for solid products.

Liquids can however be made from coal without going through the carbonisation route (although combined processes which have an analogous, preliminary step of removing the volatile components first may have a role to play in the future). There are two main classes of coal liquefaction processes, again both pioneered in Germany and used there during World War II; also both — and this is most significant — involve gasification technology.

In one process — synthesis, sometimes called indirect liquefaction — synthesis gas is made and is then catalytically reacted under pressure and moderate temperatures to produce hydrocarbons. This process is generally referred to as the Fischer–Tropsch (F–T) Process, after two of the leading German scientists involved in the work. In the other process, hydrogen (usually made by gasifying coal or residues) is reacted directly with coal, which is usually in the form of a slurry or solution in an organic liquid (often recycled liquid product). The product may be further refined and converted into light hydrocarbons by processes analogous to those used in crude oil refineries.

CHEMICALS

Liquid by-products from carbonisation, including some processes used primarily for gas manufacture, are a rich source of chemicals. The modern organic chemicals and fertiliser businesses started from coal tar and by-products and although similar chemicals are now produced in much larger quantities from petroleum feedstocks, with a consequential (and generally cyclical) impact on the economics of carbonisation, by-product recovery is still normally practised on coke ovens and thus coal still finds its way by this route into many well-known products.

In addition to chemical products based mainly on carbon and hydrogen, carbon itself in various forms is an important industrial material. Coke is an impure form of porous carbon which can be made with a wide range of properties. Active carbons can also be made from coal; these possess fine pores and extremely large surface areas and are used as absorbents. Extracts of coal may be made by dissolution in organic solvents. These extracts can be evaporated to give solid carbon products, which can be quite pure and the extract can even be drawn out into fine fibres. The latter can be "graphitised" to give an ordered structure having great strength and stiffness.

CONCLUSION

It must be clear that the potential uses and outlets for coal, a raw material having a resource base capable of sustaining growth of output for many decades, are amazingly varied and important. Coal and carbon compounds are so important that industrial life could not continue without massive and continuous supplies. In fact, it has even been suggested that carbon could be recovered from carbonate rocks. However, it is impossible to consider such ideas seriously when coal itself is in such abundant and well distributed supply; in any case, recovery from such forms would result in net energy consumption. The particular uses that are employed for coal at any time will change, hopefully in a logical and progressive manner, and these developments will have extremely important impacts on the economic and social well being of the whole world.

SECTION II
Technology of Coal Utilisation

Chapter 6

COMBUSTION

INTRODUCTION

Combustion has for centuries been the principal procedure for utilising coal. Under the influence of the industrial revolution, there has been a long history of the development of devices for burning coal and a very high degree of sophistication has been achieved in these methods. By contrast, fluidised combustion has been under study for a mere two or three decades. Only in the last few years has the work been really concentrated and the first commercial installations established. Although the traditional forms of combustion will be described briefly the main emphasis in this chapter will be on fluidised combustion because the authors firmly believe that this emphasis is appropriate in a book which looks towards future utilisation procedures. This priority is, incidentally, reflected also in the main coal research and development programmes throughout the world, as discussed later.

The potential advantages of fluidised combustion will subsequently be considered in more detail. Meantime, it can be said that the consensus of expert technical opinion suggests certain important advantages (not all applicable in every case), which may be summarised as follows:

(1) Near-ideal configuration for combustion reactions and heat transfer;
(2) High specific heat release at moderate, controlled temperatures;
(3) A wide range of coal types (including residues) may be dealt with;
(4) Sulphur dioxide emissions may be largely suppressed;
(5) Reduced fouling and corrosion of heat transfer tubes;
(6) The formation of nitrogen oxides may be reduced.

It follows that conventional combustion systems, although highly developed within the context of their limitations, may be less than ideal in the above respects. Although fluidised combustion systems (and there will probably be many forms) remain to be fully established technically and commercially, their application is likely to be the most significant combustion trend in the future.

SOME BASIC CONSIDERATIONS

The object in combustion is to convert the elements in coal to their oxides with maximum useful release of heat. Ideally, the carbon and hydrogen should be converted by the combustion air entirely to CO_2 and H_2O; these reactions however may not proceed directly or completely. In addition, nitrogen, sulphur and other contaminants may also be oxidised. The principal reactions are

$$C+O_2=CO_2$$

$$H_2+\tfrac{1}{2}O_2=H_2O$$

$$\tfrac{1}{2}N_2+\tfrac{1}{2}O_2=NO$$

$$S+O_2=SO_2$$

The forward reactions may not proceed to completion in the residence times available but are faster at high temperatures. There may also be some dissociation reactions, providing CO, H_2, OH, and atomic hydrogen, oxygen and nitrogen. Other compounds may also be formed, especially in non-stable conditions; these include NH_3, CH_4, HCN and SO_3. Nonetheless, the proportions of the main constituents of the combustion gases may be calculated theoretically for any combination of temperature, coal composition and excess air. The maximum flame temperature may also be calculated.

The calorific value of coals on a dry, mineral-matter-free (dmmf) basis may vary from as low as 22 MJ/kg for low grade lignites up to about 37 MJ/kg for bituminous coals and anthracites. The following formula has been proposed for calculating the gross calorific value (CV) (dmmf basis) of coals from their composition:

$$\text{CV (MJ/kg)} = 0.336\text{C} + 1.42\text{H} - 0.145\text{O} + 0.094\text{S}.$$

This formula underlines the importance of hydrogen and also, to a lesser extent, the elimination of oxygen during the coalification process as major factors influencing calorific value; as rank increases above the bituminous level, the increase in carbon content and the reduction in oxygen is countered by a loss of hydrogen, so that little change occurs in calorific value at the higher ranks. However, the distribution of heat values between the volatile matter and the char varies with rank and results in different combustion characteristics. Also, the expression of calorific value on a dmmf basis masks the effect of inherent moisture (and to some extent adventitious water) which consumes heat in vaporisation and thus reduces the net heating value. Moisture contents are generally higher in the lower rank coals and this can have an effect on combustion practices.

A high proportion of coal is burnt in pulverised fuel (pf) furnaces, especially for the production of electricity. These furnaces are generally large and their reliability and efficiency have great economic implications. Although a stream of fine particles being burnt in a few seconds can be considered in some ways similar to a gas flame, in practice the behaviour of individual particles is important and the mechanisms involved have been the most highly researched area of coal combustion. In any case, the information obtained by studying the behaviour of particles in pf conditions has relevance in other combustion systems.

In pf furnaces, particles are entrained in preheated air and injected at considerable velocity into a hot furnace; efficient mixing is promoted. Particle temperature increases rapidly but nonetheless, the sequence of generation and release of volatile matter is a key feature of the combustion mechanism. It is possible that some combustion at the particle surface takes place before decomposition starts but the emission of volatile matter soon blankets the particles, preventing access of oxygen. Combustion may therefore be considered as devolatilisation, followed by the separate combustion of the volatiles and the char.

Decomposition in pf conditions is generally accompanied by swelling; there is, however, a range of particle behaviour in any coal because of its heterogeneity. Some particles do not swell but may fuse and become spherical. Others may swell greatly, forming thin walled hollow spherical shells (cenospheres); intermediately, with less extensive swelling, spheres may be formed which are either hollow with thick walls or with high dispersed porosity. Total swelling may be up to 100% or more but does not seem to be directly related to the normal swelling number. Rapid decomposition at high temperatures tends to give a higher yield of volatiles than at lower heating rates. It has been suggested that the increased yield of volatiles may be due to the ejection of molecular species from the coal under rapid heating. With slower rates of heating secondary polymerisation reactions may take place within the coal particles. The change in size of particles through the flame and the fact that the volatile matter has a higher calorific value than the char are important theoretical features affecting furnace design and operation.

The time taken for the evolution of volatiles may be a few milliseconds in typical pf combustion conditions. Subsequently, mixing must occur, followed by chemical reactions, the time limiting step probably being the conversion of carbon monoxide to dioxide; these processes again may be completed in a few milliseconds. However, a complicating feature is the formation of soot which takes place if the temperature reaches about 900°C before contact between the hydrocarbons and oxygen is complete. Soot particles have a high surface area and emissivity and therefore have a large effect on flame emissivity; they also have a slower rate of burn-out than char and so may reduce the efficiency through carbon loss and the coating of heat transfer surfaces.

The combustion of the devolatilised coal char particles is also a complicated process, the rate depending on diffusion and chemical factors. A number of chemical reactions are possible, especially as water vapour will be present and can react with carbon. However, the most important theoretical uncertainty is whether

(a) CO_2 is formed directly, or
(b) CO is first formed at surface and is then subsequently oxidised to CO_2 elsewhere in the gas stream.

Other complications are that the total surface area depends critically on the swelling behaviour and that the temperature at the particle surface (crucial to reaction rates) is both rising rapidly and is typically much greater than that of the gas stream. An expression has been proposed for g, the rate of carbon combustion per unit of geometric external surface area (expressed in g/cm^2 s), as follows:

$$g = P_g/(1/K_{diff} + 1/K_s)$$

where P_g is the partial pressure of oxygen in the gas stream away from the particle surface; K_{diff} is a diffusional factor; and K_s is a surface reaction rate coefficient.

The diffusion term expresses the rate of diffusion of oxygen to the particle surface and dominates when particle sizes are large; this factor is usually determined for the case where CO is the primary product and would need to be divided by two where CO_2 is formed at the surface. For small particles, low temperatures or unusually unreactive carbons, K_s (the chemical factor) is more dominant.

Unfortunately, practically none of the basic factors mentioned are known with any precision so that mathematical modelling of combustion chambers cannot be firmly established so far as the kinetics are concerned. In addition, the theories of aerodynamics are deficient with regard to mixing and heat transfer. The latter depends on radiation and is made more uncertain by lack of knowledge of the emissivity of soot (as we have seen, a large influence) and of ash particles, which are likely to be the principal emitters towards the tail end of the flame. Further large-scale basic combustion studies would be needed for complete theoretical understanding. In the meantime, however, while design and operational improvements must remain largely empirical, the interpretation of practical results has been greatly improved by better understanding of the likely directions of the various influences; as a result, evolutionary progress has been substantial.

PULVERISED FUEL SYSTEMS

Finely ground coal may be entrained in air and combusted in a manner analogous (at first sight) to a gas. Input coal may be transported to the burners in suspension in air and a large proportion of the ash will leave the furnace still entrained. A residence time of 1–2 s is generally adequate for 99.5% combustion of the carbon, provided the coal is ground finely enough. This factor is usually defined by specifying the proportion passing through a sieve of a particular size: e.g. in the UK, a 200 mesh (75 micron aperture) sieve may be used and the proportion required to pass through may be set in the range 60–90% (higher for some industrial applications). The most important application is steam raising for electricity production. Cement kilns are another important application and industrial heating furnaces and shell boilers may also use pf burners.

Large furnaces, such as those used for electricity production, have the combustion chamber lined with steam or water tubes, which are heated by radiation (thus cooling the gases). About half the heat is recovered within the combustion chamber, bringing the temperature down to about 1050°C, sufficiently below the normal softening temperature of the ash. The remaining recoverable heat is removed convectively by passing the gases over tube banks; the degree of this "waste heat" recovery may be limited by capital considerations, stack draft requirements or, more probably, by the dewpoint of sulphuric acid (about 160°C for most British coals). The processes of combustion and transfer of heat to produce superheated steam can be about 90% efficient; it is in the turbo-generator system where losses occur, and for thermodynamical considerations, the gross generating efficiency of modern power stations is limited to rather less than 40%. Power consumption by auxiliary equipment, mainly fans and coal pulverisers, causes a lower nett efficiency. Of the combustion losses (about 10% in total), about 2 percentage points corresponds to unburnt combustibles in the solid and gaseous products; the other eight percentage points results from the sensible heat in the exit gases. This latter is affected by the amount of excess air used — up to about one third in practice. If the stoichiometric quantity of air could be used about another three percentage points could be gained on the system efficiency but this needs to be balanced by the possibility of reducing the combustion efficiency at the same time. The latter can be affected, especially when low excess air rates are employed, by small random variations in coal flow rates or by imperfect contact or mixing. Minor variations in the chemical and physical characteristics of the coal or small changes in the operational condition of the many plant items between maintenance periods can contribute.

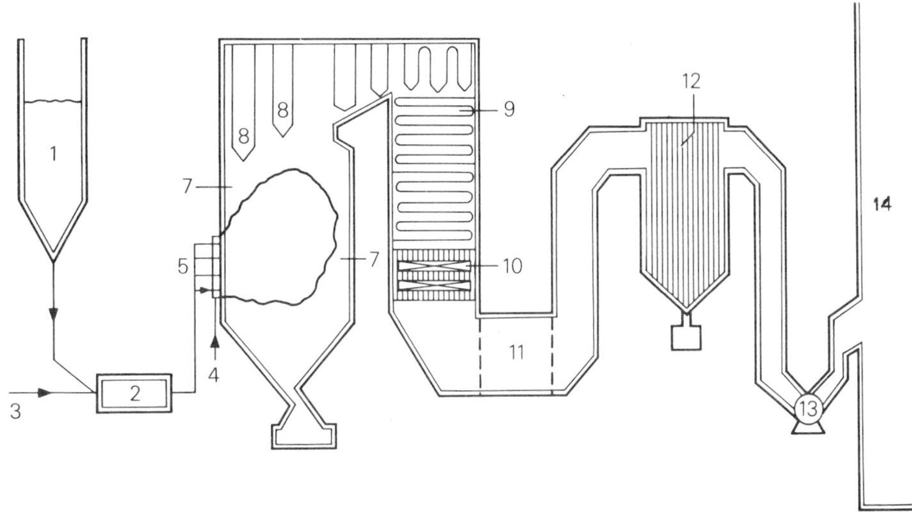

1. Coal bunker
2. Feeder and mill
3. Heated primary-air supply
4. Heated secondary-air supply
5. Burners
6. Ash discharge
7. Furnace wall tubes
8. Radiant superheater panels (platens)
9. Convective superheater tube banks
10. Economizer
11. Air heater
12. Electroprecipitator
13. Induced-draught fan
14. Chimney

6.1 Essential features of a pulverised-coal-fired water-tube boiler plant

A large pf boiler installation incorporates a considerable number of associated plant items, illustrated in outline form in Fig. 6.1. These will be discussed briefly in relation to coal properties. Consistency of grinding is a key feature, both for efficient furnace operation and for overall economics. Both power consumption and maintenance costs of mills affect the economics significantly and both depend strongly on the characteristics of the coal, including the mineral matter. Mills can be of the tube, roll, ball or hammer type. Choice of mills however seems largely pragmatic, based on local experience. Preheated primary air is supplied to avoid adhesion of moist coal particles and mills are sometimes operated in closed circuit in conjunction with classifiers in order to narrow the particle size range. The coal is transported hydraulically from the mill to the burner at about 100°C. The whole of this circuit must be carefully designed since fine coal suspended in heated air is prone to fire and explosion.

In large furnace installations there may be several mills with several burners supplied from each mill; pipelines must be carefully designed to provide an even flow and distribution, both at full output and in situations when some mills or burners are not in use. The mass of the primary air is about twice that of the coal (0.4–0.5 kg/m^3). The mass of secondary air, also pre-heated, is about five times that of the primary air and is supplied in various types of burners which are designed to mix the fuel and air efficiently and at the same time to provide a stable flame and to direct the flame so as to minimise the harmful effects of slagging or flame impingement on the furnace walls. Sometimes all the secondary air is supplied, together with the fuel and primary air, through turbulent jet burners, often placed in rows on the front wall. Sometimes some or all of the secondary air is supplied separately and directed at the fuel jet, in order to promote mixing and turbulence. With high rank coals, the primary air/fuel jets may be directed downwards, so that it may be heated to about 800–900°C before meeting a horizontal flow of secondary air. It will be clear that in most cases the furnace design must suit the coal and ash characteristics to attain the following objectives:

(a) Adequate flame stability
(b) High combustion (burn-out) efficiency.
(c) Minimum slagging or flame impingement on the furnace tube walls.
(d) Adequate cooling of combustion products, to below the slag softening temperature, before exiting to the tube banks.

Once a furnace, with its ancillary equipment, is installed, practically any changes in the coal will result in less than optimum economic results, even if technical performance may be recovered by, for example, changes in the grinding practice. It will also be apparent that there is limited scope for further developments to increase efficiency or reduce capital.

Cement Kilns

Pulverised fuel firing is also generally used in cement kilns, which are rotating cylinders, nearly horizontal, up to about 18 m in diameter and 160 m long. The fuel and primary air enter through a simple axial burner, at the lower end, the secondary air being drawn in after picking up heat from the hot cement clinker. The charge enters at the elevated end and moves downward counter-current to the flow of combustion gases. The ash forms part of the product and its composition must be suitable for this purpose; it should also not induce build-up on the kiln walls. These requirements are not generally onerous.

Cyclone Furnaces

The requirement to quench the ash particles within the combustion chamber, in order to prevent slagging-up of heat exchange surfaces downstream, places a limit on the specific heat release per unit volume in pf water tube furnaces. This limitation also requires a close relationship between coal quality and furnace design characteristics and some inflexibility results, as discussed above. If the ash can be allowed to melt in the furnace and be removed as liquid slag, different design considerations may be allowed and may result in higher combustion intensities. This is accomplished in cyclone furnaces which accentuate the swirling action of jets used in some pf furnaces. This gives a longer flame for a limited volume but, in addition, the centrifugal action is used to fling the ash on to water cooled walls, where it can run down to be tapped into water and granulated. In addition, larger particles of coal are retained temporarily in the slag-covered walls, where their combustion can be completed. Various designs of cyclone furnaces have been developed, with both vertical and horizontal axes, and consist mainly of a relatively small combustion unit, followed by substantial banks of tubes over which the slag-free hot gases pass, rather like some gas or oil fired furnaces.

Cyclone furnaces generally burn coarser particles compared with standard pf furnaces and a much wider range of coal types may be used, including both extremes of rank, from low volatile matter coals to brown coals; particle size may have to be controlled to suit combustion properties. Slagging characteristics of the ash, and especially consistency, are of course extremely important since a slag coating on the furnace wall is essential and the slag must be fluid enough to flow freely into the removal system. Ash levels, as such, are not however very important (though there are economic implications in the sensible heat losses); additions may therefore be made to improve the slagging characteristics. Moisture content may be a sensitive feature of coals in cyclone furnaces since water entering the furnace reduces temperatures and therefore affects slagging.

Substantial progress was made in the application of cyclone furnaces between 1940 and 1970, especially in the USA and West Germany, although application did not proceed to the largest sizes, on economic grounds. Development however came under pressure, as in the case of other coal systems, from oil fired furnaces. Cyclone furnaces, furthermore, are particularly vulnerable in this respect since they can fairly readily be converted to oil firing. Although cyclone furnaces exhibit considerable technical ingenuity, it seems doubtful whether they will ever be adopted very widely, especially since fluidisation may be a better solution to the problems addressed by the cyclone furnace design concept.

FIXED BED COMBUSTION DEVICES

General Considerations

The burning of coal lumps in heaps on simple grates is a traditional procedure, the principles of which remain quite useful for domestic purposes and for certain industrial purposes. Over the years considerable ingenuity has been applied to minimise the inherent disadvantages of fixed beds — low efficiency, smoke production, ash removal and coal feeding — without undue complication or excessive coal preparation requirements.

The main improvements which have taken place over the last century and a half or so have been the provision of controlled and directed air supply to the bed and the development of mechanical means for feeding coal to the bed and for removing ash. These developments not only reduced labour requirements but resulted in much more efficient and controllable combustion devices.

The development of mechanical stokers was perhaps the central feature and a number of basic types emerged. They became established first for water tube boilers in the last century and progressively thereafter became important also in smaller boiler types, especially shell boilers. This overlapped the emergence of pulverised fuel firing in the last thirty or forty years as the preferred system for large water tube boilers; cheap oil and gas in the 1950s and 1960s temporarily removed the incentive for further developments in fixed beds. More recently, renewed interest in coal, together with the availability of modern materials and equipment, has resulted in significant improvements in these systems. Although there are limitations on the rate of heat release per unit area of grate with fixed beds (further reduced by the intrusion of feeding and de-ashing equipment) and stokers generally have mechanical parts requiring maintenance, there are also some important advantages. Fixed beds are simple and easy to control, compared with pf firing; coal can usually be used as delivered, without grinding (a heavy cost item in capital and maintenance); particulate emissions are relatively simple to control; and high turn-down ratios and good recovery rates from the "banked" state can generally be achieved.

There are a number of design types for fixed beds, the main ones being described briefly below. In all cases, in order to achieve high efficiency, it is necessary to be able readily to initiate and maintain stable combustion conditions, allow adequate time for a high degree of carbon burn-out before the residue is discharged and ensure a high degree of combustion of volatiles and CO. The latter requirement is usually met by providing extra air and turbulence at the appropriate point and maintaining an adequate temperature. There is an optimum for excess air, of course, since loss of sensible heat in the flue gases increases with excess air; the optimum is dependent to some extent on the stoker type. In a modern shell boiler fired by a chain grate stoker, with an excess air level of 20%, ideally up to 92% of the fuel may be combusted, with sensible heat losses of about 12%, giving an overall efficiency of 80%. In practice, however, values over 75% may be considered satisfactory. It should be noted that an increase in excess air to 40% in the above case would result in a further three percentage points loss in the sensible heat; if mixing and distribution are not ideal, however, this loss may be justified to avoid undue losses in combustibles.

The designs of different stokers vary mainly in the way the fresh coal is applied to the bed. No stoker can burn all types of coal and the choice of stoker may be determined by the characteristics of the coals intended to be combusted, especially in the way that caking behaviour may be dealt with. Size, moisture and ash properties are other coal factors influencing design choice.

Gravity Feed Stokers

This is a very cheap and simple system, in which coal flows directly into the combustion zone from an overhead hopper. Usually only anthracite and coke are suitable since other coals may swell and stick in the feed system, preventing even downward flow. The higher cost of the fuel is offset to some extent by the low capital cost and these devices are usually therefore confined to small installations (up to about 750 kW), for domestic, commercial and institutional purposes. Ash may be removed manually or automatically and this removal controls the flow of fresh coal onto the top of the bed. Ash characteristics are therefore important to consistent operation; it is preferred for light clinker to be formed and ash fusion temperatures should therefore not exceed about 1300°C. Air is supplied through the grate, plus secondary air above the bed to consume combustible volatiles. Control of burning rate is achieved by regulating the fan supplying the air.

Underfeed Stokers

Underfeed stokers are also simple, cheap devices, in this case not confined to low volatile coals. Fresh coal is forced by means of a screw into the bottom of the combustion chamber (Fig. 6.2). Air is fed through orifices near the top of the bed, where the main combustion takes place; secondary air is also supplied above the bed. Clinker overflows the fuel chamber. Most types of coal can be used (if allowed for in the design) but weakly caking coals (either low or high volatile matter) are generally preferred, in order to avoid a sticky central mass. The coal must be appropriately sized (e.g. 25 to 40 mm) to suit the particular screw drive used. The ash

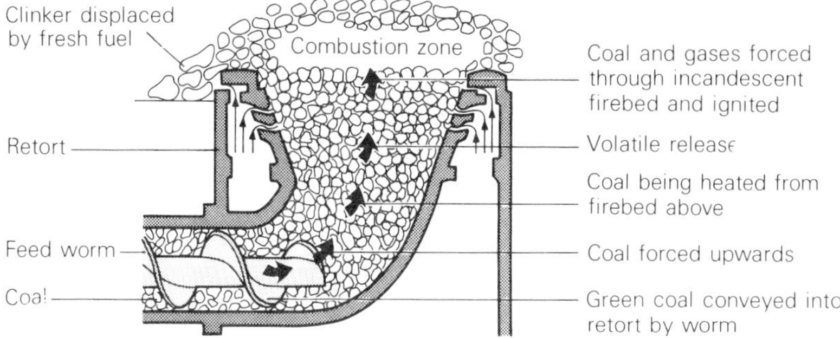

6.2 Underfeed stoker

fusion temperature is important; it should not be so low as to form large clinkers in the body of the chamber or so high as to prevent clinkers being formed for regular overflow and removal. Too high an ash content is also undesirable. The rate of burning may be controlled by linking the rate of air feed to the speed of the screw. It will be apparent that, if the properties of the coal were to vary greatly, regular efficient operation could not be maintained in these simple systems; consistent coal supplies must therefore be arranged. Progress is being made in improving the flexibility and controllability of this stoker type; an automatic de-ashing device has been developed.

Underfeed stokers may go up in capacity to several megawatts but one megawatt has been a more normal maximum; newer developments may make the somewhat larger sizes more attractive.

Overfeed Stokers with Static Beds

In these designs, coal is dropped through a tube and falls freely through the furnace space onto a static bed (Fig. 6.3); distribution may be aided by devices such as cones or by swirling the secondary air. It is claimed that the swelling tendency of the coal is

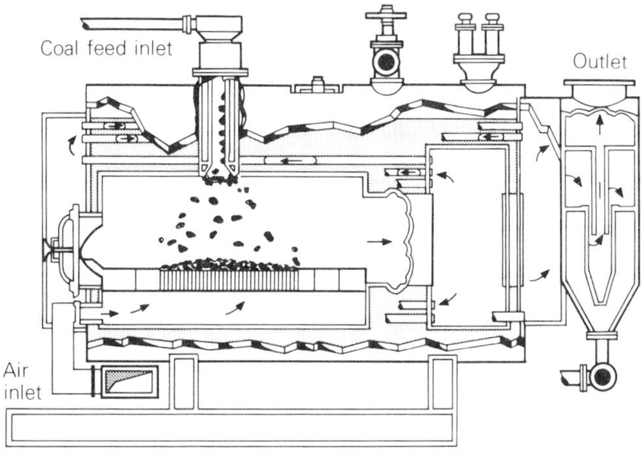

6.3 Vekos-Powermaster boiler with manual ash removal (courtesy Parkinson Cowan GWB Ltd)

reduced by the heating and oxidation the particles receive whilst falling. Some designs employ deep beds, as in gas producers, and in such a unit caking would be especially undesirable. Automatic de-ashing has been developed. In one design some of the flue gas can be recirculated; this is thought to have beneficial effects, including the reduction of alkali salt volatilisation and deposits on boiler surfaces. As with other stokers dropping or spreading particles, entrainment in the exit gases can be a problem. Size grading can alleviate losses but the collection of the larger particles in a cyclone for refiring is usually necessary.

Chain Grate Stokers

This is a popular form of stoker which has been refined over a period of time; a typical design is shown in Fig. 6.4. Coal is fed from a hopper onto a moving grate consisting of bars joined together in the form of an endless chain, driven by a sprocket. The coal enters under a guillotine which controls the feed rate and passes

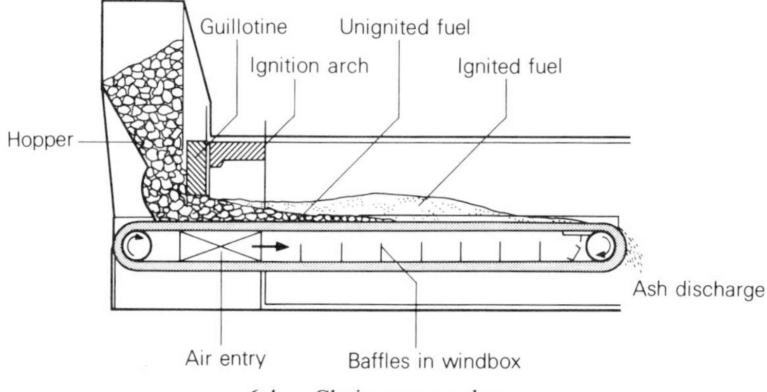

6.4 Chain-grate stoker

first under a refractory ignition arch, in order to commence ignition at the top. As the coal moves away from the feed end, the ignition front moves downwards, leaving burning coke above. Finally, the coke is all consumed leaving a bed of ash which discharges over the far end of the grate. Air is forced upwards through the bed and by providing correctly designed baffles to optimise air distribution over the grate area, burning rates have been increased to about 2 MW/m^2.

Chain grate stokers can operate on most coals, except anthracite. Simple automatic controls can be installed which change the coal and air feed rates, in preset ratios, to meet steam demand. Turn-down ratios of up to 8:1 can be achieved. Response to load changes is rapid; the unit can be banked, with rapid reactivation, which can be automatically provided at pre-set times. Automatic electrical ignition can be fitted to the guillotine.

Chain grate stokers have been built for sizes up to 70 MW water tube boilers but future applications are likely to be mainly in shell boilers.

Vibrating and Reciprocating Stokers

These stokers are somewhat similar in effect to chain grate stokers except that the grate remains nearly stationary and the coal is moved forward not by the continuous movement of the grate but by vibrating or reciprocating motion of small amplitude applied to the grate. Vibrating grates are water-cooled, which increases the heat transfer surface. Recent developments indicate that all the advantages of chain grate stokers may be obtained at much lower capital costs and since there are no rotating parts, maintenance costs are likely to be lower. In addition, the system is tolerant of a wide range of coals, from very low to very high ash. High output rates are being obtained with new developments in reciprocating stokers.

Spreader Stokers

There are a number of designs of stokers depending on distributors which can spread the coal feed from one point across the grate area. One recent design uses a chain grate. The coal is sprinkled by means of a rotating impeller across the grate, which moves towards the feeder. Some of the finer particles burn in suspension; the coarser particles burn out on the bed and the ash is discharged at the same end as the feeder, a distinct advantage in certain circumstances.

Coking Stokers

In this system, the coal is pushed across the grate by a ram but the first part of the travel is across a coking plate where the volatile matter is released and burned above the bed. The coke which is formed then burns as it travels over the grate. The consolidation of the fuel as a result of the coking reaction helps to prevent carbon loss caused by small particles falling through the grate base into the "riddlings". New designs have high unit area release rates, fully automatic operation and ash removal by means of a portable container under the unit.

Recent Developments: Related Equipment

As noted above, automation including ignition, has been applied to a wide range of stoker designs. Automation is also being developed for de-ashing, in some cases involving pneumatic suction and sealed container removal. Coal is also being conveyed pneumatically, with potential cost savings; accurate metering (assisting efficiency) and with the elimination of somewhat troublesome devices such as screw feeders, pushers, mechanical sprinklers and rotary vane feeders.

Domestic Appliances

Gravity-feed and underfeed stokers are often used for domestic appliances, although in both cases domestic requirements (often 5–20 kW) are small compared with industrial sizes. The fact that gravity-fed boilers normally work on coke or anthracite makes these devices suitable for use in smoke control areas. However, there is a strong preference in domestic circumstances for solid fuel appliances where the combustion zone is visible and this feature may be provided either by open grates or "roomheaters", which have a closed transparent front door. Both devices may be fitted with simple water-heating surfaces in the flues to supply partial central heating and hot water. Modern designs are not only simpler to operate and more attractive in appearance than older appliances but are also considerably more efficient. Open fires with boilers have a minimum appliance thermal efficiency of 50% and roomheaters of 65% (in each case 10–15 percentage points are by direct heating to the room and the rest to the boiler); in addition, several percentage points in efficiency may be gained from the heat supplied to the house from the chimney. The ease of overnight "slumbering" in these modern applications is also an economic advantage. Where smoke control is necessary, domestic appliances may burn naturally smokeless fuels, coke or devolatilised manufactured briquettes. Alternatively, designs of appliances are now commercially available which burn bituminous coal smokelessly (as defined in British Standards, for example). This is achieved by using a two-stage combustion system, with down draft through the visible coal bed; uncombusted volatiles leaving the bottom of the fires are burnt in a hot zone in an adjacent compartment supplied with pre-warmed secondary air. This chamber is also refractory lined to facilitate combustion, the hot gases passing subsequently over the water heating surfaces of the boiler. In the first designs, a small fan was incorporated to ensure an adequate supply of combustion air, although more recent designs operate on natural draught alone. Restrictions were originally placed on the coal size (12.5–25 mm) and on quality (non-coking or weakly caking); newer developments have eased these constraints. The combination of visual attractiveness with good efficiency and smoke control is likely to prolong the popularity of solid fuel for domestic heating indefinitely, at least where the tradition still remains. The latter situation of course applies to many underdeveloped or moderately developed countries, where the adoption of improved stoves or grates could make a significant improvement both to energy conservation and to the environment.

FLUIDISED COMBUSTION

General Conditions

Fluidisation is a widely practised chemical engineering technique, the object being to suspend individual solid particles in an upward flowing fluid, which may be a gas or a liquid. When a bed of particles is subjected to such an upward flow, the bed is undisturbed at low velocities and the pressure drop increases directly with increasing velocity. Ultimately, as the velocity increases further, a point is reached where the upward force on the particles is equal to gravity and the particles become levitated. Additional mass flow creates turbulence and "bubbles", giving the appearance of boiling — hence the term "boiling bed" which was once used. The term "fluidisation" more appropritaely describes the phenomenon; the bed finds its own level and has hydrostatic head. Accompanying the bubbling, particles are ejected from the bed and may fall back; fine particles, depending on the velocity, may be elutriated from the bed and carried out of the vessel with the gas stream. Since the large particles require a high minimum fluidising velocity, elutriation will be greater when the range of particle size is increased. There is also a relationship between the size and size range of the particles, the mass flow and the spacing of fluid entry points necessary to achieve complete and ideal fluidisation.

The turbulence of a properly fluidised bed causes rapid mixing of the particles, with greatly enhanced heat and mass transfer rates; the rapid relative movement between particles and fluidising medium is of course ideal for chemical reactions at particle surfaces. In the case of fluidised combustion, air is of course the fluidising medium; the bed can be any conveniently sized refractory particles but, where coal is burnt, the ash may be used to form a natural bed, provided the coal is crushed to the appropriate size. Crushed lime or dolomite may be added to absorb sulphur and then also forms a portion, often a major portion, of the bed. Because of the chemical and thermal properties of the bed, combustion can be sustained with a low carbon content in the bed — often about 1.0% or less when relatively fine coal is used; with lump coal, average carbon levels will be higher.

A range of conditions can be suitable for combusting coal and a choice depends on the objectives, which in turn will influence the basic design of the equipment. Bed depths of 0.5–2.0 m are suitable for most cases and a "freeboard" (height of chamber above the top of the bed) of up to about 3 m is provided to avoid undue loss of particles by splashing. Temperatures are usually between 750–1000°C, the lower level providing a reasonable margin above the minimum required for ready maintenance of combustion and the upper limitation being set in order to avoid sintering of ash particles and/or unacceptable attack on materials. At atmospheric pressures, coal can be burnt at rates of about 40–400 kg/h/m^2 of bed area, giving a heat release of 0.3–3.0 MW/m^2. An air velocity of 0.4–4.0 m/s will be required; a pressure drop of about 1 mm water gauge per mm depth of bed, equivalent to the weight of particles, is necessary to sustain fluidisation. The coal is crushed (not finely ground as with pf combustion) to an upper size of about 2–6 mm, depending mainly on the choice of air velocity. Ash (and other particles) may be removed from the bed by gravity flow; in practice, the carry over of fines may be the main mechanism for ash removal. Some fine coal is also elutriated, usually amounting to 2–10% of the input. Particles are removed from the exit gases by cyclones and may be re-introduced into the main bed or consumed in a separate burn-up bed; a high efficiency of combustion (99% or more) can be achieved (in some industrial boilers, with other design considerations, 97% may be an optimum). The proportion of excess air affects the ease of achieving complete combustion but the greater the excess air, the higher the proportion of heat removed from the furnace by the gases. The heat in the hot exit gases can be retrieved by normal waste heat boiler systems. Cooling tubes may be immersed in the bed and approximately half the heat can be removed and recovered by this means. This feature has the advantage of benefiting from the high heat transfer coefficients (150–500 W/m^2 K) which can be achieved, thus economising on the amount of tube surface. Radiation may account for 20–50% of this heat transfer, the proportion increasing with bed temperature; the major mechanism is a form of convection in which heat is transferred by conduction across a thin gas film between the particles and the cooling surface. Thus, convective heat transfer rates increase as the particle size decreases (more particle surface and thinner gas film) and with higher temperature (higher gas conductivity). Gas velocity is not an important factor in heat transfer in fluidised conditions (provided that the bed is well fluidised as in normal operations).

Specific heat release rates in fluidised combustion are high and are usually limited by the availability of oxygen, which is restricted only by practical gas velocities and operating pressures; coal can readily be supplied in adequate quantities. Provided enough carbon is in the bed, the oxygen in the air is fully utilised; combustion of coal particles can be practically completed in the residence time available (typically, a few seconds for crushed coal), including the successive stages of pyrolysis and combustion of volatiles and char. More air can be supplied and combustion rates increased proportionately by operating at elevated pressure. Pressures of up to 10–20 atm have been used in experimental equipment; the reduction in volume required for a given heat release has been clearly demonstrated. Extra heat transfer surfaces in the bed are needed to remove this heat and this may require a deeper bed. However, under pressure, elutriation of unburnt carbon is reduced for a given combustion rate and mass flow. Another important feature is that the ash does not sinter, and remains soft and friable; also, most of the alkali metal and vanadium salts are retained in the ash. Both these factors should facilitate the passing of the gas, after a particulate removal system, through a gas turbine in order to recover the pressure energy with reduced erosion/corrosion.

Both oil and gas can be combusted in fluidised beds, which may consist of sand or refractory particles. Since lateral mixing of gases is limited by their short residence time, feed points may need to be more closely spaced. Lump coal and other combustible solids, including wastes, may also be combusted in a fluidised bed of refractory particles. The possibility of having a fluidised combustor capable of burning a wide range of fuels is apparent.

General Environmental Considerations

Particulates from fluidised combustors may be easier to remove from exhaust gases, compared with pf furnaces, because the ash is not so fine nor so abrasive. Simple cyclones may well be adequate to collect ash and unburnt carbon; limestone may be more difficult. The criteria may require further detailed examination in the light of developing environmental standards, especially since particulates may be associated with emissions of sulphates; trace metal emissions also may be substantially reduced in fluidised combustion.

However, a most important advantage of fluidised combustion is that, because of the temperature and other conditions in the combustor, sulphur in the coal may be retained in the bed and removed as sulphate with the ash. Although it would in principle be possible to operate beds with a sub-stoichiometric oxygen content and convert the sulphur to sulphide (from which the sulphur may readily be recovered), it is generally preferred for combustion efficiency reasons to operate with excess oxygen, in which case the sulphur is converted to stable sulphates. Some conversion to sulphates takes place normally with coals which have predominantly basic oxides in their ash but lime is generally added continuously as crushed limestone or dolomite in order to control and quantify fixation. Theoretically, 3.15% of the coal weight as limestone or 5.75% as dolomite per 1% of sulphur in the coal would be needed but the efficiency will normally be considerably less than 100%, possibly only 20–60%. It is usually necessary therefore to add two or three times the theoretical minimum when low sulphur emissions are required. This factor may depend on the operating conditions but is also strongly affected by the characteristics of various stones, so that the rate of addition necessary for any particular level of sulphur retention may need to be established empirically. Although the subsequent recovery of sulphur has been considered, so far this does not appear particularly promising economically since calcium sulphate is rather stable. However, the waste material may have some usefulness as a constructional material or, at worst, appears more suitable for dumping than sludges from wet scrubbing systems; further work on the leachability of residues is however necessary to determine satisfactory conditions for disposal. It should also be noted that limestone additions may increase fine particulate emissions; bag filters or electrostatic precipitators would probably be needed.

Because of the low nominal temperature and minimal local variation in the bed, little atmospheric nitrogen is converted to NO_x, in contrast to conventional high temperature forms of combustion, but some conversion of nitrogen in the fuel takes place. The mechanism and factors involved are uncertain although there is evidence that operation at elevated pressure may have a further suppressive effect. Operational variants have been proposed to minimise NO_x further.

The potential for meeting environmental requirements by fluidised combustion is both important and complex. There is an obvious trade-off between costs and environmental standards (the optimisation of which requires much further information) and, for any required level of air purity, the convenience and economics of fluidised combustion can be compared with other processes, as discussed later.

Some Design Considerations

The optimum provision of air is the key feature of design, which must consider velocity, pressure and pressure drop in relation to the objectives. The permutations of these parameters are extremely wide, based on laboratory data. When the current generation of development plants has operated for some years, no doubt evolutionary improvements in design will take place around a few successful models within the whole conceptual range. A low fluidising velocity permits the use of small inert bed material particle sizes, resulting in favourable conditions for heat transfer, combustion efficiency and sulphur fixation. (In practice, the particle size of the absorbent may be more important.) However, an atmospheric pressure bed of 0.1 m depth operating at a gas velocity of 0.3 m/s would require an area of 1430 m^2 (about an acre) to produce 300 MW output (0.21 MW/m^2). Increasing the air velocity to 3 m/s would increase the heat rating to 2.1 MW/m^2 and, in addition, increasing the pressure to 10 bar would produce 21 MW/m^2. In practice, the bed depth may need to be increased with specific heat release to accommodate in-bed heat transfer surfaces, if used, and to ensure adequate combustion efficiency. Slightly more energy would also be consumed in overcoming the pressure drop of a deeper bed but there would be somewhat less elutriation than for a comparable unpressurised combustion capacity in a shallow bed.

It will generally be desirable to remove as much heat as possible from the bed since heat transfer co-efficients here are high, whereas in the exit gas cooling systems, requirements are similar to those in other systems. Although less tube surface is required in the bed than in a pf furnace for the same heat removal, the smaller volume may require some ingenuity in fitting in the desirable amount of tube surface without disturbing fluidisation or distribution. Considerable experimental work is therefore being carried out on the design of tube banks. This may be complicated by the requirements of superheat since transfer to steam is much less efficient than to water; the tube wall metal temperature also rises nearer to the bed temperature when transferring heat to steam whereas water cooled tubes have an average wall temperature nearer to that of the water. The same factors apply in the case of gas-cooled tubes, which have been proposed in some designs.

Both coal and limestone may be introduced to the bed via immersed nozzles fed by pneumatic pipelines, the design of which is well established. An important consideration however is the number of nozzles which may be necessary, especially for the coal feed, in order to ensure efficient mixing; hot spots, losses of combustibles, excessive combustion in the freeboard or slumping of parts of the bed might occur if the nozzles spacing is too great. However, increasing the number of nozzles increases the complexity of design and costs; there is also the problem of dividing the coal into a considerable number of equal flows. The smaller volume of pressurised reactors of course facilitates feeding; deeper beds, associated with pressurised operation, may also be fed more easily from a limited number of nozzles.

Air must be efficiently distributed and the bed supported prior to fluidisation. These functions are generally carried out by a perforated distributor grid, through which the air is introduced. The size and number of holes are important design considerations; "sifting" of bed material into the windbox must be avoided. The separation of holes must not be too great and also a substantial proportion of the pressure drop must take place across the grid in order to avoid channelled flow in the bed. The grid may also be subjected to considerable mechanical and thermal stresses.

Start-up procedures require special design consideration. A flame from an above-bed burner may be used to heat up a gently fluidised bed until the temperature is reached where combustion can be sustained within the bed (about 500°C). Simple methods have been adequate for pilot plants but in large plants it may be desirable to have a special start-up section. In the case of pressurised beds, the compressor is normally driven by the exit gases, which are not of course available until the combustor is operating; a separate compressor drive mechanism for start-up seems inevitable. Control and turn-down are also important design factors. Control at steady operating conditions may be relatively simple since coal feed can be adjusted to correct any small changes in temperature, which are accurately determinable within a fluidised bed; response is rapid and other inputs (air, limestone) can be readily adjusted in sympathy with the necessary coal rates. To reduce output moderately, in designs with fixed cooling surface, bed temperatures may be reduced but this will only reduce output by 20–50% before combustion becomes inefficient. Greater turn-down ratios (up to 1:5) are normally required and these must be sought by allowing the bed depth to be reduced, thus exposing and negating some cooling surfaces, or by arranging the combustion space in a series of separate cells. This latter method of course requires frequent slumping and re-starting of individual cells, with a consequent requirement for suitable and reliable equipment for these operations. Although heat output may be reduced by these means fairly readily, and without elaborate control equipment, more difficult control problems arise in maintaining steam conditions where superheater and boiler tubes are affected differently by changes in the bed volume and conditions. It has been suggested that multi-cell designs, with different tube arrangements in the various cells, will be superior in this respect.

As already noted, the treatment of off-gases may be facilitated by the nature of the ash and by the retention of some potentially aggressive and noxious substances in the bed. Cyclones will play the most important part in clean-up of the gaseous effluent but further clean-up, say by electro-static precipitators or bag-house filters, may be needed, depending on environmental standards and the application of further data from experiment relating different designs, coals and absorbents to emissions. The rate of recirculation of fines from primary cyclones is also an important design feature, which may need specific optimisation for particular types of coal; this factor may also be used to permit some flexibility in increasing air velocity. Off-gas clean-up for pressurised fluidised combustion is much more speculative and crucial since the gases must be suitable for driving a gas turbine. This is an additional and probably a different and more difficult criterion than environmental quality alone. The possible problems with gas turbines are build-up of particles on blades (leading to rapid loss of aerodynamic efficiency), erosion and corrosion (of various kinds, including grain boundary attack). There are various possible ways of dealing with these problems but ideal solutions will require more data, which can probably only be obtained by large-scale trials with equipment of characteristics near to those desired. For example, it may be possible to accept fairly high concentrations of particulate matter in the gas entering the gas turbine, using low performance but rugged turbine designs, regular scheduled maintenance or improved materials; alternatively, it may be better to cope with the dust problem within the combustion system, using operational design to minimise elutriation or by the employment of elaborate clean-up systems. With regard to the latter solution, several systems of cyclones, progressively designed to deal with finer particles, may be employed or more advanced particle removal units, including filters, may be necessary. It will be apparent that the greater the degree of clean-up necessary, the greater will be both the capital investment and operating costs (which may appear as energy losses). Ash removal, which may be mainly or partly from cyclones, in addition to overflow from the main bed, and the subsequent handling of the ash, also require careful design, especially in pressurised systems. The ash removal system will need isolation from the combustors. Hot ash will need to be cooled, probably by quenching and then disposed of suitably; this may be complicated where the product is a slurry.

In addition to turbine blades, other constructional materials, especially for in-bed tubes, may require careful selection, although no insuperable problems are expected. However, it appears that local or general areas lacking in oxygen may suffer deleterious effects; this may need to be dealt with in design. Beds to which a sulphur acceptor is added may pose a particular problem due to sulphide attacks arising from deposits of calcium sulphate.

Some Forms of Application

Before describing some specific designs, some general remarks about the varied forms of application may be appropriate. Since the main requirement for coal combustion for many years has been in connection with electricity generation, a great proportion of the interest in fluidised combustion has been focussed on this application. However, the general conditions for fluidised combustion are so wide that many potential forms and applications can be visualised. Furthermore, some of these applications may permit relatively rapid penetration into new or currently small markets for coal. In the case of major power stations, the time scale for proving fluidised combustion and for replacing a significant proportion of the capital stock is long, although considerable work is being carried out on methods for the conversion of existing boilers. There are substantial difficulties in such conversions, especially with regard to space requirements, turn-down and start-up; where existing boilers are gas or oil fired, coal and ash handling equipment must be provided. Nonetheless, some conversion trials have been successful and this may lead to further efforts, possibly to produce package conversion kits.

For new boilers, the industrial boiler market appears particularly open, both for shell and water tube boilers. In the former, coal has for a long time been at a disadvantage compared with oil and gas; for size reasons standard package units have previously been limited to about 10 MW in the case of coal, compared with about 20 MW for oil. Package units for fluidised combustion may equalise the situation. Shell boilers may be either vertical or horizontal, although the former is a more obvious choise for fluidised combustors and received early development attention. Work is also proceeding to develop horizontal shell designs; this configuration requires some compromise to accommodate the geometry and space needs, especially the lack of height. The height problem was solved by using a shallow bed of refractory material onto which lumps of coal up to 50 mm can be fed and be combusted in suspension. In this arrangement there is little scope for heat transfer from the bed and a higher proportion of excess air is used; in the shallow bed sulphur retention is not very high. However, the difficult requirements of the horizontal shell boiler provided the incentive for the development of lump coal fluidised combustion which is superseding crushed coal combustion in many atmospheric FBC applications where maximum sulphur supression is not a main aim.

For industrial boiler outputs greater than about 10–20 MW (the upper end of the range assuming the most successful development of shell designs), water tube boilers will be required. In moderate sizes, up to about 100 MW, for fairly regular loads, relatively straightforward design concepts seem suitable, with bed depths of about 1 m, freeboards of about 4 m and fairly simple off-gas clean-up facilities. A small number of compartments within a single bed structure can probably be controlled and managed in such a way as to give reasonable turn-down capability and potential for superheat.

For larger water tube installations, such as those required in power stations based on one or more generators of 600 MW or above, there are complexities of arrangement and control, since it is probable that the bed area would need to be subdivided into a substantial number of cells for each generator. The problems, which are by no means insuperable, have economic implications; in any case, for such large investments, the incentives to seek maximum capital savings may lead this market in the direction of the pressurised version.

Large pressurised fluidised combustors, for all their potential advantages, represent much the most difficult development problems, the solution of which will almost certainly not be fully demonstrated and applied until well into the 1990s at the earliest. As noted above, the key engineering development is to show that large gas turbines can be operated on the combustion gases from a fluidised bed, using cleaning systems which themselves may need some innovations and certainly a great deal of careful optimisation based on large-scale practical tests. Moreover, a pressurised system will inevitably operate on a "combined cycle" (electricity generation both by gas and steam turbine drives), a concept which is itself rather advanced, even without the novelty of a pressurised fluidised combustor. Control problems are accentuated by the different characteristics of the boiler and superheater tubes in the combustors and by the necessity to maintain a balance between the steam and gas turbine outputs; some of the energy from the latter will probably be used for driving the compressor, which may complicate control, especially in turn-down and start-up modes. Finally, of course, the efficiencies of combustion and of sulphur retention need to be demonstrated under large-scale operating conditions, as do the other environmental impacts.

Alternative methods of operating a gas turbine from a fluidised combustor involve the heating of air (or other gas) passing through tubes in the bed to drive the gas turbine (Fig. 6.5). This circuit could be quite separate from the combustor gas circuit so the turbine would work in ideal conditions, except perhaps that the temperature would probably be rather low. The problem would be to extract enough heat from the bed; heat transfer characteristics would be relatively poor

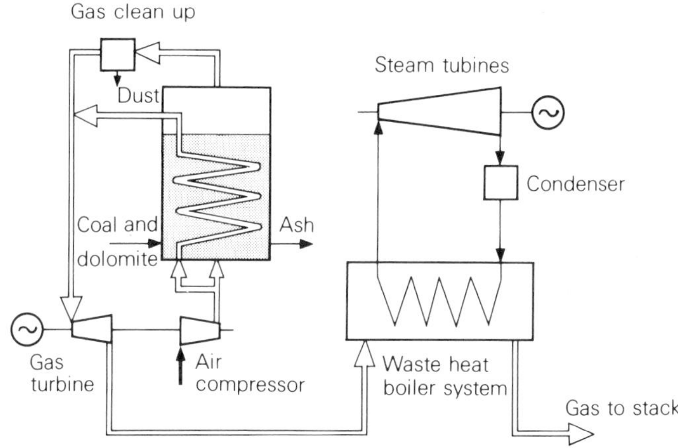

6.5 Air heater combined cycle

because of the low density of the cooling medium. The tube surface in the bed would have to be quite large and the tubes closely spaced. The tube material would have to be capable of being exposed to as high a bed temperature as possible in order to transfer an adequate proportion of the energy. The bed in this case would probably operate at atmospheric pressure; otherwise, a separate gas turbine would be required, causing considerable additional complexity. Heat recovery from the hot combustion gases would be normal, providing steam via a waste heat type of system for a turbine. Provided that an adequate margin of cooling capacity could be installed, some decoupling of the gas and steam turbines might be obtained in such a system, thus allowing an additional means of control. In other proposals (Fig. 6.6),

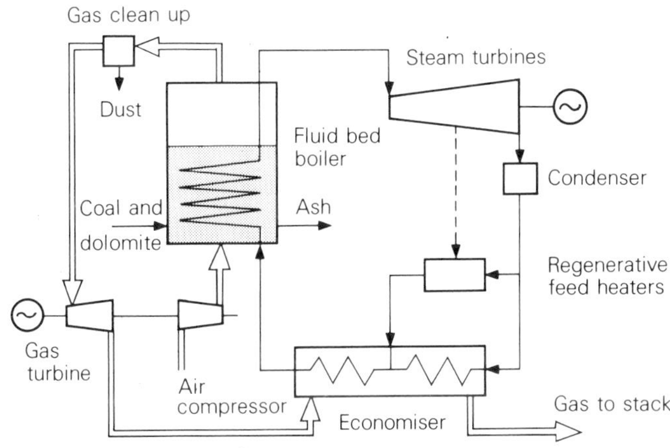

6.6 Supercharged boiler combined cycle

which seem to be more actively pursued, hot combustion gases from the bed are merely supplemented by air from the bed cooling tubes, thus increasing the proportion of energy derived from the gas turbines compared with the steam cycle (in this case, waste heat boilers).

As noted, a considerable amount of the available energy appears in the combustion gases and this can be enhanced, if desired, by blowing extra air through the bed. In practice, this is applicable mainly to atmospheric systems, where the hot gases may be used (in addition to heating boiler tubes) in several ways, especially for drying processes. The drying of grass is an example which has been established commercially (on a scale up to 5 MW); other drying requirements, such as minerals, are being developed.

One of the principal advantages of fluidised combustion is its tolerance to the properties of the fuels. This can be exploited in incineration devices. As noted above, the equilibrium carbon content is generally low because combustion is rapid and efficient. Once suitable temperatures are attained, combustion may be maintained by the addition of fuel, even in a dilute form, provided heat losses through exit gases, steam and solid residues are not so great that a heat balance is not obtained. This has been elegantly demonstrated by work done by the National Coal Board (subsequently by others and also supported by the EEC) on the combustion of colliery washery tailings. This process, using a simple atmospheric pressure bed, was developed as a method of disposal of a rather nasty waste. The tailings are concentrated, in a deep cone for example, and are injected as a thick slurry into or onto a combustor bed; gas or other convenient fuel can be used for start-up or for maintaining the temperatures if the heat content of the residues being burnt is not adequate. In some cases, however, the heat of combustion of tailings is just about sufficient to be used for steam and for electricity generation. The ash product is an easily handleable powder.

Ignifluid System

It may be appropriate at this stage to refer to the Ignifluid furnace, since this system is sometimes mentioned in references to the development of fluidised combustion. In fact, although the combustion took place under fluidised conditions, the concept was not in the main stream of developments which have led to modern FBC devices.

The furnace (Fig. 6.7) was developed in France in the 1950s; it was based on a narrow inclined chain-grate stoker. The air velocity was high through the grate but became progressively lower above the bed because the sloping walls increased the cross-sectional area. The coal particles were fluidised but the ash particles were intended to fall back. Much of the combustion took place above the bed, which was not cooled; all the heat recovery was from the gases. These are essential differences from modern developments. A high rate of fines recycle is also essential.

After some initial interest, this system has not been the subject of serious continuous development, although reviewed sporadically. Any conceivable advantages may be obtained in the more "conventional" fluidised combustors previously described, which in addition have beds with high heat transfer and sorbent properties.

National Fluidised Combustion Projects

Modest experimental work on fluidised combustion, some on its fringes, took place in the 1960s and even earlier. By about the end of the decade, the broad technical

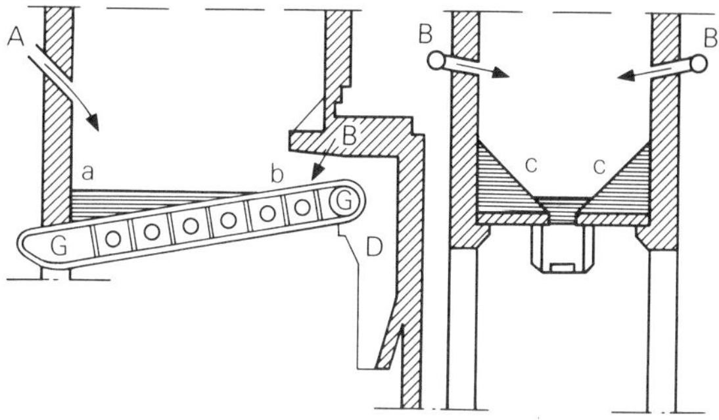

A: coal feed
a–b: bed surface
B: secondary air
c: static fuel layer protecting walls
D: clinker hopper
G–G: narrow chain-grate stoker with compartmented wind box

6.7 "Ignifluid" furnace

feasibility of several types of fluidised combustor systems had been demonstrated; ideas for new versions had been suggested; research on areas likely to prove troublesome or costly had been defined and initiated; reliability, operational characteristics and costs, however, still needed establishment on larger prototype or demonstration plants. This necessary progression in scale and engineering design did not however take place at once; it was not until the middle of the 1970s, in the wake of the oil crisis, that the building blocks for a sensible and serious development/demonstration programme began to emerge. Since then, many projects have been announced but it has sometimes been difficult to distinguish between those which are firm and those which are contingent. Appendix 1 is a list recently compiled by workers at the Coal Research Establishment of the NCB of projects of which firm information was made available to the authors; the list is not intended to be comprehensive, even for the Western countries. A report, *Fluidized Bed Energy Technology* (by Patterson and Griffin, published by INFORM in 1978) also reviews current projects and interests. Some projects considered by the authors to be of key importance are now discussed. These are almost entirely in the UK, USA and West Germany.

UK Projects

Several facilities in the UK were important in establishing credibility in fluidised combustion concepts at a time of some indifference and scepticism. In particular, in the late 1960s, the National Coal Board (after some pioneering work by the Central Electricity Generating Board) decided to establish three experimental units to provide basic design and operating data for the three main outlets then perceived: (a) shell boilers for small and medium industrial purposes (atmospheric pressure) (b) water-tube boilers for larger sizes (also atmospheric) and (c) pressurised combustors for combined cycle electricity generation. The last of these, together with the Grimethorpe project based on it, is described after the notes on the atmospheric pressure developments which are now being commercialised; the pressurised system was always recognised as more complex and longer term.

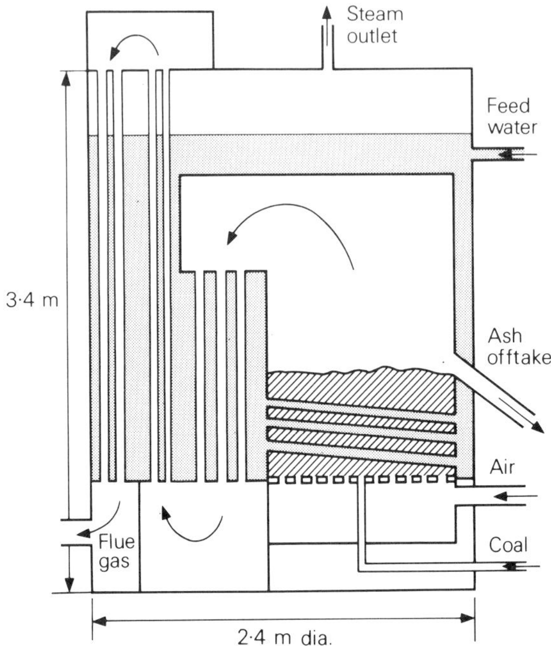

6.8 BCURA vertical shell boiler

The shell boiler experimental unit was a 2.5 MW, 675 mm diameter vertical shell boiler (Fig. 6.8) installed at BCURA (now CURL). This unit had a relatively deep bed and burnt crushed coal; in these respects it has been superseded in UK developments but it operated successfully over a range of conditions and provided information and confidence for later developments. This boiler was converted to burn heavy oil in a fluidised bed, a subject of interest to BP Ltd., who had begun a continuing collaboration with the NCB.

At about the same time, 1968, a 1 MW (0.9 mm square) combustor was commissioned at the Coal Research Establishment (CRE) as a highly instrumented pilot section of a water-tube boiler (atmospheric pressure), containing flexible arrangements of in-bed and above-bed boiler tubes. Typical preparation and feed devices for coal and limestone were provided and the combustion gas exit system was fitted with a cyclone. This unit provided the first information on a reasonable scale on heat-transfer, the effects of boiler tube spacing and configuration, mixing from feed points, temperature control and response, elutriation and sulphur partition. The main combustor was supported by smaller special rigs to study erosion/corrosion effects on a wide range of possible in-bed constructional materials; the results provided confidence and assisted materials choices for later experimental units.

The experimental results from the CRE 1 MW rig were supported by various design studies which indicated the feasibility of various atmospheric pressure schemes and a 200 MW prototype was designed and costed in 1969. However, in view of the availability of cheap oil and the limited interest at that time in the development of coal combustion, this prototype was shelved.

Babcock Renfrew Boiler. In 1974, following a hiatus of a few years, Babcock Power Ltd. (BPL) proposed converting a 13.5 MW cross-fired spreader stoker boiler at Renfrew, Scotland, to fluidised combustion firing, using NCB data under licence from Combustion Systems Ltd., a joint company of NCB and BP; the project went ahead as a joint venture of these three organisations. The converted boiler (Fig. 6.9) was commissioned in 1975.

The Renfrew fluidised combustor has a bed 3.1 m square fitted with immersed hairpin boiler tubes. The boiler has operated satisfactorily for more than 5000 h and has been fired just on coal and then oil. Information has been obtained on a comprehensive list of key design and operating factors, including start-up; load control; uniformity of bed temperature; effects of fuel size; feeding and velocity on carry-over; heat transfer; corrosion/erosion; sulphur retention and NO_x emissions. Boiler efficiency at least as high as with the original stoker could be obtained. With a 3.5% sulphur coal it was possible to retain 90% of the sulphur; NO_x emissions were well below US standards for new sources. By varying the bed temperature and by stopping sections of the bed, output could be varied from 25 to 100%.

Based on the Renfrew demonstration, BPL are offering commercial designs up to 150 MW. Babcock Contractors Inc. (BCI), a US subsidiary of BPL, have contracted to carry out a retrofit of an existing 20 MW boiler at a hospital in Columbus for the State of Ohio Department of Energy; this project was due for start-up towards the end of 1980. The project is part of a programme by Ohio to re-establish the vital economic importance of the vast reserves of high-sulphur coal there in the light of recent environmental legislation. The hospital project incorporates a bag-house filter for additional environmental protection. BCI have carried out other engineering studies for Ohio at selected sites requiring boilers of 30–110 MW. A further study is being carried out for the Tennessee Valley Authority involving a 200 MWe boiler

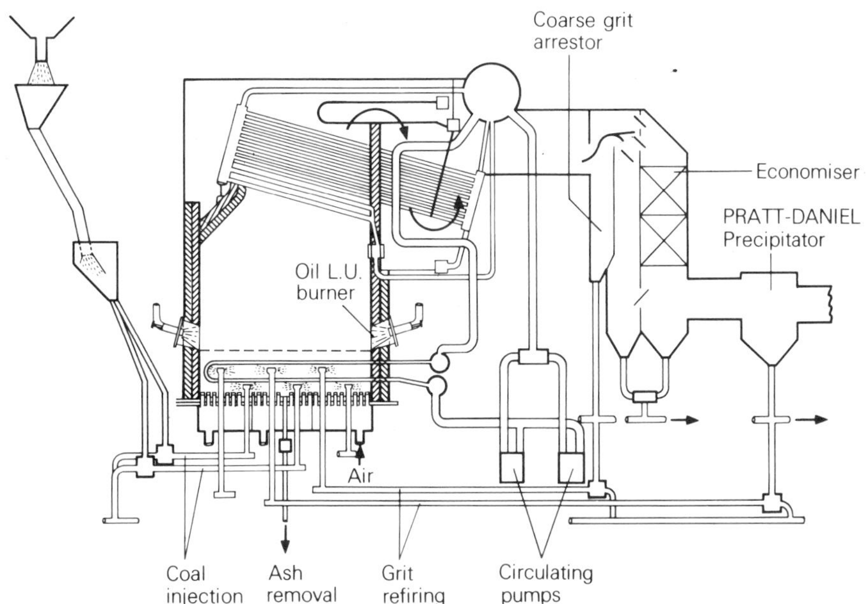

6.9 Babcock boiler converted to fluidised bed firing

to burn West Virginia coal of 4.5% sulphur using local limestone to provide 90% S retention in the bed.

Packaged Boilers. Although the early 2.5 MW vertical shell boiler at CURL was a technical success, the general competitive position of coal at that time did not encourage further exploitation. When the prospects for coal began to look better in 1973 the NCB reviewed fluidised combustion technology and markets. It was decided to increase the emphasis on the industrial boiler market. In this field, coal had been at a disadvantage with oil and gas, chiefly over boiler size, especially height; oil and gas boilers could not only be fitted into smaller boiler houses but could be factory manufactured units of larger output and installed as a package. Even the original designs of fluidised combustors had disadvantages in this respect, employing a tall combustion chamber (about 4 cm) and a deep bed; a substantial blower was needed and the coal had to be crushed and dried. The crushing reduced the size of stone inclusions and the bed consisted of the resultant ash particles.

It was conceived that all these disadvantages (for industrial boilers) could be overcome by burning the standard washed (typically 1.35% S, 3–8% ash) industrial singles (13–25 mm) and smalls (0–13 mm) in a shallow bed consisting of specially added inert particles. Small scale tests showed that this was completely feasible and subsequent developments were centred on the use of a fluidised bed of silica sand, about 150 mm deep, in a combustion chamber of only 900 mm height. Coal, in the form of smalls and singles added by feeding onto the surface, rapidly became dispersed throughout the shallow bed and was burnt at the same efficiency as with deep beds, without recourse to secondary combustion devices (e.g. grit re-firing or burn-up cells). Many commercial designs have now been put forward and developed; only a few examples can be quoted below.

Vertical shell boilers are suitable, using these modern fluidised combustion concepts, for outputs up to 5 MW and two boilers have operated commercially in the UK, since 1977. Based on this experience, a standard 4.5 MW vertical shell boiler has been designed (Fig. 6.10); one has been commissioned and several more are being installed during 1981. The circular bed is about 1.5 m in diameter. A ring of syphon tubes takes away about half the heat liberated; the rest is recovered via convection tubes. Output variations can be achieved by varying the bed height and thus the immersed tube surface; bed temperature is kept constant by automatically varying the coal feed rate. Ignition is by hot combustion gases from an oil burner.

Horizontal shell boilers are in more common use; adapting fluidised combustion to this configuration is difficult since the vertical headroom to accommodate the air distributor, bed and free board is at most 1.5 m and more usually 1.0 m or less. Nevertheless, several manufacturers have successfully designed within these constraints, either by fitting the whole fluidised combustion set-up within the fire-tube or, as in Fig. 6.11, by using an open-bottomed design. Thermal syphon tubes are inserted in the fire-tube and gas ducts run the whole length of the shell. The boiler has legs for easy bed removal and has an overall height of about 4.3 m. A range of boilers up to 20 MW on these lines is being designed.

A traditional method of incorporating a vertical combustor into a horizontal shell boiler is the locomotive configuration. A double-ended 10 MW unit of this type (Fig. 6.12) has operated commercially for two heating seasons. Using UK technology, the Johnston Boiler Co. of Ferrysburg, Michigan, USA has designed and operated since 1977 a prototype 3 MW boiler on these lines (Fig. 6.13). This is now available commercially in sizes up to 15 MW with sulphur reduction capability

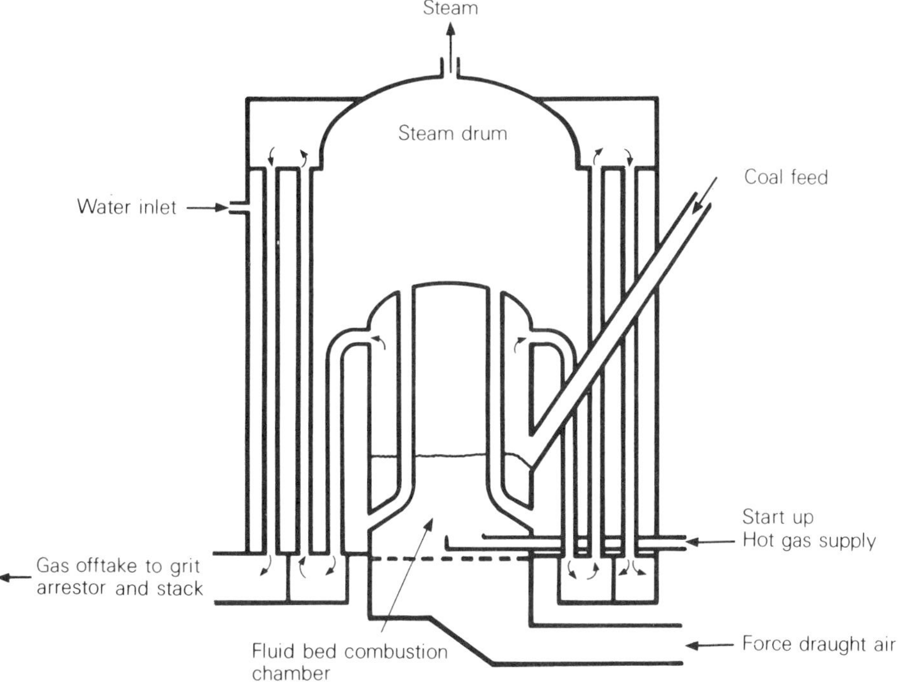

6.10 Diagram of 4.5 MW vertical shell boiler

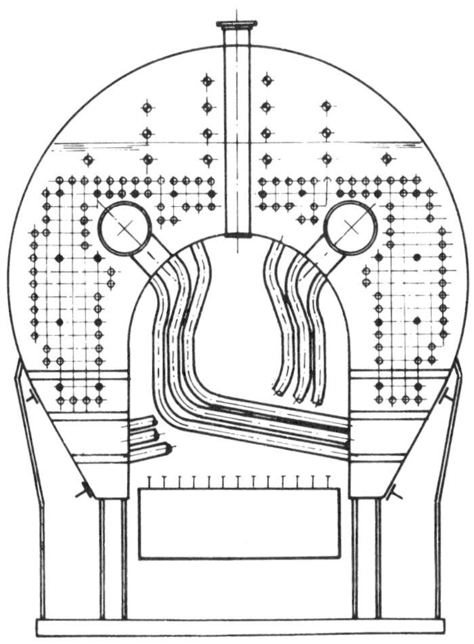

6.11 Cross-section of Vosper Thornycroft (UK) Ltd boiler

Gas offtake
Steam
Fluidised bed combustion chamber
Air

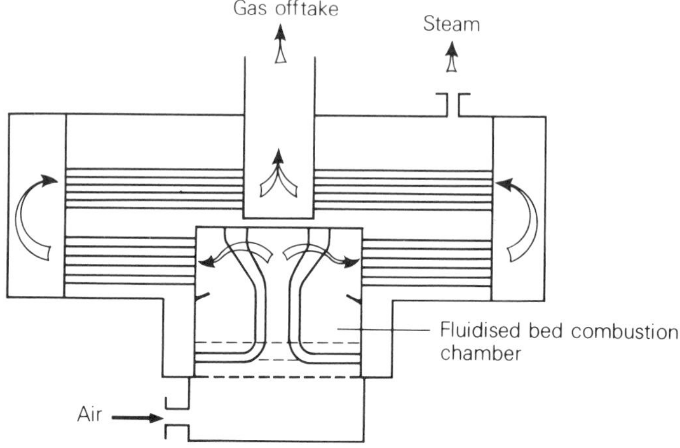

6.12 Diagram of 10 MW double-ended locomotive-type boiler

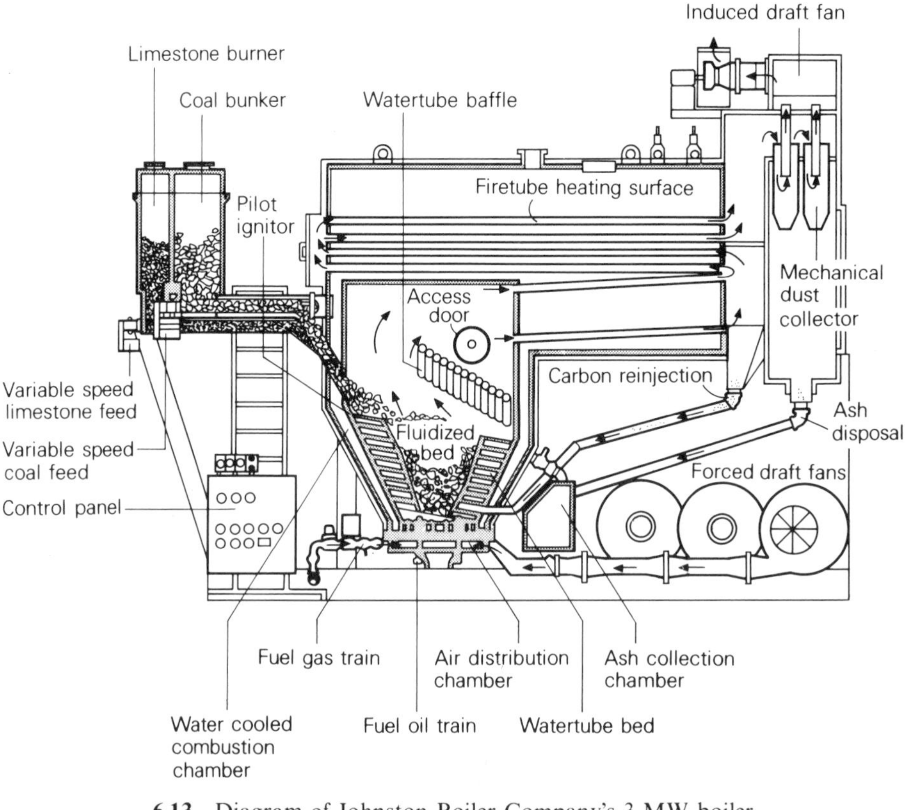

Induced draft fan
Limestone burner
Coal bunker
Watertube baffle
Firetube heating surface
Pilot ignitor
Mechanical dust collector
Access door
Carbon reinjection
Variable speed limestone feed
Fluidized bed
Ash disposal
Variable speed coal feed
Forced draft fans
Control panel
Fuel gas train
Air distribution chamber
Ash collection chamber
Water cooled combustion chamber
Fuel oil train
Watertube bed

6.13 Diagram of Johnston Boiler Company's 3 MW boiler

and about twenty units had been sold by late 1980. In addition to coal, the prototype has been operated on oil, gas, wood, sawdust, shredded rubbers, waste oil and paper sludge.

BPL and the NCB are developing a composite ("Compo") boiler (Fig. 6.14) incorporating a water-tube cooled combustion chamber with a conventional two-pass smoke tube shell boiler, incorporating a thermal storage system. The combustion chamber contains a shallow bed capable of burning top-fired coal up to 30 mm. There is a gas or oil fired overbed burner for start-up. Turn-down (3:1) is achieved by varying the fluidising velocity (0.7–2.0 m/s) and the bed temperature

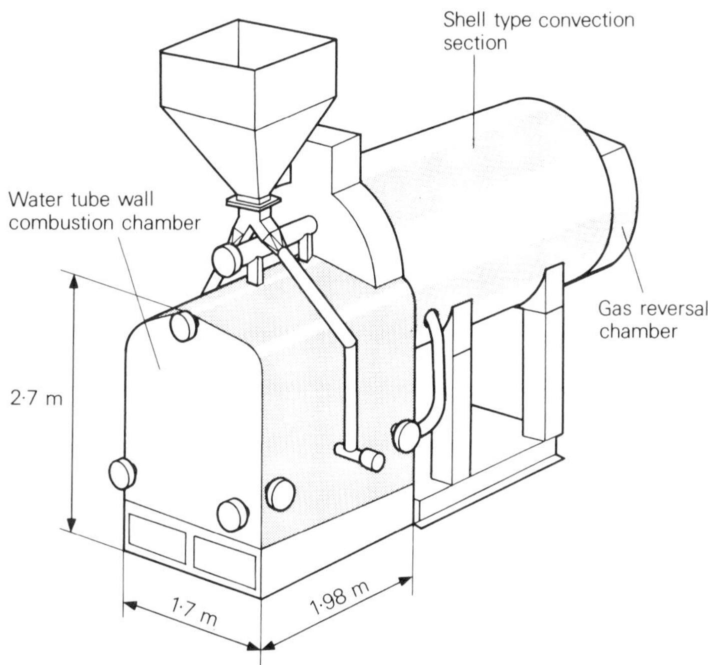

6.14 Babcock Power Ltd's composite water tube/shell boiler

(750–950°C). A 4.5 MW unit is being installed and it is expected that this will lead to a commercial range of boilers from 3 to 30 MW, at lower cost than conventional units.

A 30 MW once-through coil water-tube boiler, incorporating NCB design data, has been built by ME Boilers at the River Don works of the British Steel Corporation. The design is intended both to meet the fluctuating load demand of steam hammers and also provide steam for electricity generation. The 5 m circular bed is divided into six sections, each of which can be expanded to cover more tube surface and burn more coal when required. Commissioning trials are awaited.

Hot Gas Furnaces. An obviously suitable form of exploitation for fluidised combustion is the provision of hot gases for industrial drying purposes; the gases can be taken directly from the combustor, provided a small amount of ash can be

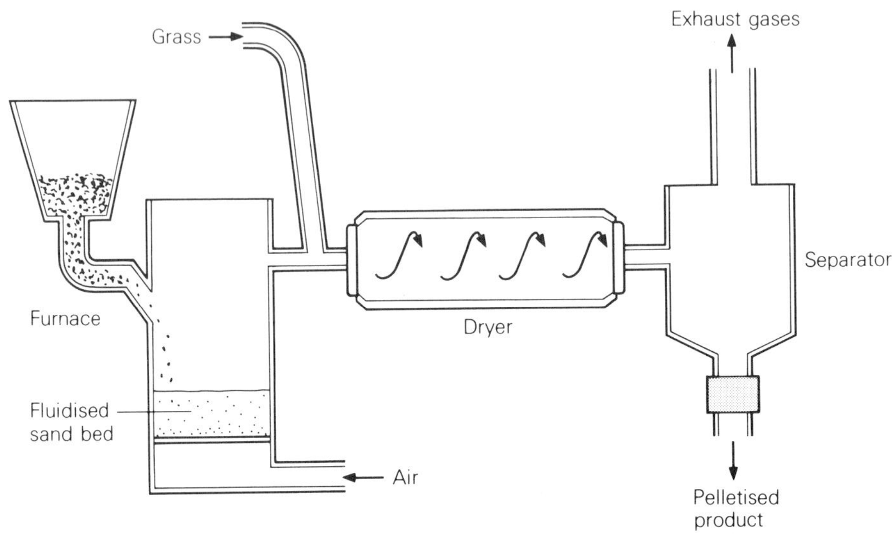

6.15 Diagram of fluidised bed fired grass dryer

tolerated in the product. A system for grass drying is illustrated in Fig. 6.15. This was first demonstrated in 1974. By 1980, five 5 MW furnaces were operating and further units were being installed. A 15 MW furnace is now being used to dry clay for cement manufacture and a further unit has been designed for the drying and heating of roadstones. Other drying applications are being developed.

Incineration. A process has been developed to dispose of colliery washery tailings (concentrated to approximately 50% water, 35% ash, 15% coal), producing a dry friable powder of potential usefulness, possibly as a building material. A 1 t/h unit has operated at CRE for some years and a 30 t/h design study has been completed. This information has also been used to design an incinerator for the disposal of thickened sewage sludge (77–84% by weight moisture). This plant, which has run successfully since 1977, has a combustor with a silica sand bed with an operating depth of 750 mm. The bed runs at a nominal temperature of 750°C with subsequent combustion in the freeboard to raise the temperature to about 800°C to ensure complete destruction of odours.

Pressurised Fluidised Combustion. For the pressurised system there is no doubt that the 2 MW unit at CURL, Leatherhead, has been the leading experimental facility in the world for many years. It was commissioned in 1969 and continues in a pioneering role, having operated throughout the 1970s, to a great extent with funding from US agencies. The unit (Fig. 6.16) consists of a container for the bed (0.6 × 0.9 m) and associated combustion equipment, inside in a pressure vessel, permitting operating at up to 6 bar and 900°C exit gas temperature. A wide range of coals and additives have been used; the general feasibility of combustion under these conditions has been established and confidence has been generated in the type of equipment used. It has been demonstrated that operation under pressurised conditions improves combustion efficiency and sulphur retention while lowering the emission of NO_x. Extensive tests have been carried out on the quality of exit gases,

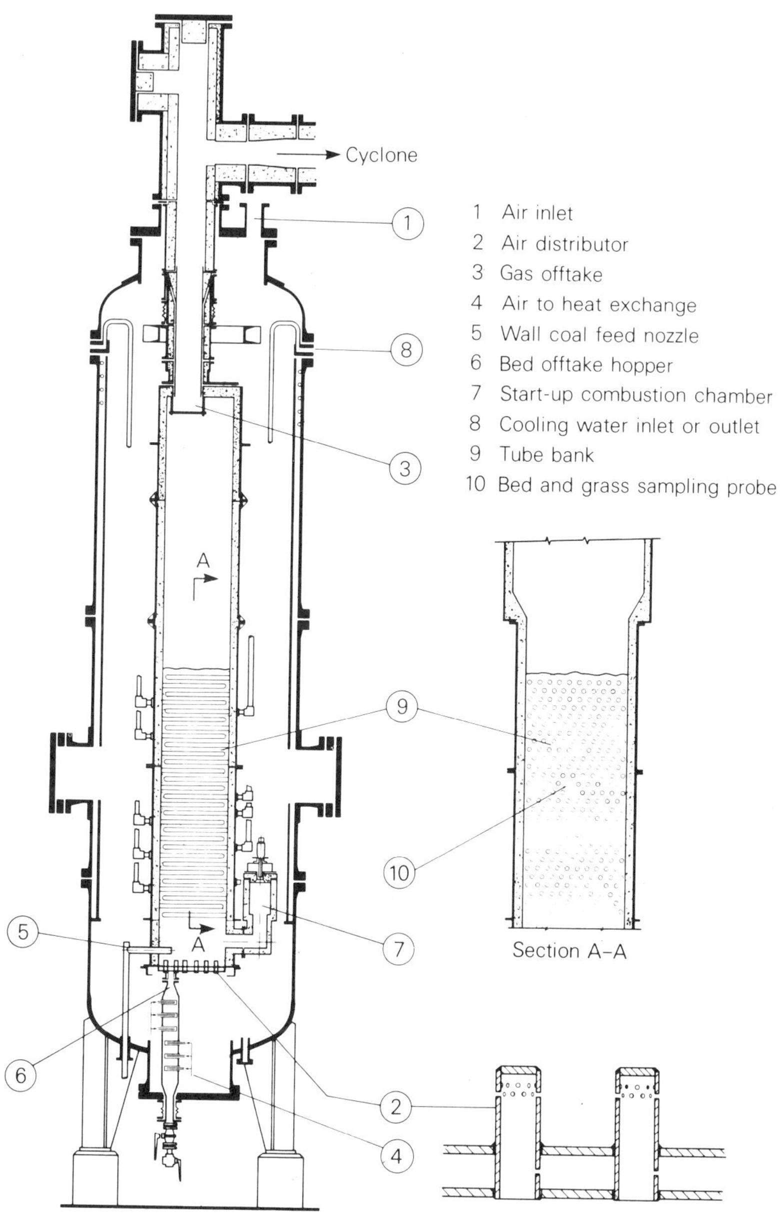

6.16 Arrangement of fluid bed combuster Mk VII

including the exposure of turbine blade specimens and materials; the broad suitability of this system for use with a gas turbine has been indicated. However, much of the work on these aspects has been to crystallise the approach to the necessary larger-scale and more direct experiments to be carried out at Grimethorpe and elsewhere.

The Grimethorpe Experimental Facility. This unit, although internationally funded and controlled, is described in this section since the plant is at Grimethorpe in Yorkshire and also because UK technology, especially from Leatherhead, provided both the design base and the incentive. Following a proposal by the IEA Coal Research Group (located in London and associated with the International Energy Agency), the governments of the USA, UK and the Federal Republic of Germany signed an agreement in November 1975 to build a large pressurised fluidised combustion experimental facility at Grimethorpe. This site was chosen partly because it was already used for a small pit-head generating station; the facilities for making use of steam were a particular advantage. The costs, initially estimated at about $40 million (1975 values), were met entirely by the three countries concerned, in equal shares.

The purpose of the Grimethorpe facility is to provide an essential technical link between basic data from small-scale rigs (up to the Leatherhead unit) and the first fully-integrated pre-commercial designs. It has been suggested for a few years that a medium sized unit (say 60 MW) incorporating an existing gas turbine could be designed and constructed on data currently available. However, it now seems that major steps in this direction will depend on input of data from Grimethorpe, a more sensible progression.

Grimethorpe is therefore not a prototype; a gas turbine was not included in the original schedule of facilities, although provision is made for a possible future addition. Instead, the facility is intended as an experimental unit to extend the range of experience, both in scale and breadth, to make measurements in greater detail and over a wider range of conditions than would normally be feasible in an integrated plant including a gas turbine; it will thus provide a data base for demonstration and commercial plants. The main studies initially required were in the following areas:

(i) combustion performance with various coals, including some with high ash;
(ii) effects of various dolomite additives on sulphur retention;
(iii) study of heat transfer coefficients, geometry and materials;
(iv) gas clean-up for gas turbine use and prior to atmospheric discharge;
(v) off-gas corrosion and erosion effects, with particular reference to gas turbine materials and designs.

The Grimethorpe combustor is rated at 80 MWth and will produce enough steam for a 20 MWe turbine and pressurised hot gases suitable for a 3.5 MWe gas turbine. A cross-section through the combustor pressure vessel is given in Fig. 6.17 showing that a 4 m diameter pressure vessel contains a bed 2 m square and 4 m high (with a further freeboard of 5 m). Maximum operating conditions are 950°C at 12 bar, with an air mass flow of 31 kg/s (68 lb/s) and velocity up to 3.7 m/s. Coal can be fed at up to 17 000 kg/h and dolomite at up to 13 000 kg/h. Steam of 30 bar and 440°C can be produced. The plant is designed to a maximum SO_2 emission of about 0.5 kg/GJ (1.2 lb/10^6 Btu), which will probably imply a retention of about 80% with low sulphur coals (and proportionately more with higher sulphur coals). The coal feed will normally be $\frac{1}{4}$ inch (6.4 mm) top size. The walls of the combustor are liberally supplied with access points to enable a wide range of research measurements to be taken. The combustor walls are water cooled and form part of the steam raising circuit.

The general layout of the Grimethorpe facility is shown in Fig. 6.18. The exit gases are divided into four off-takes each leading to refractory lined primary and

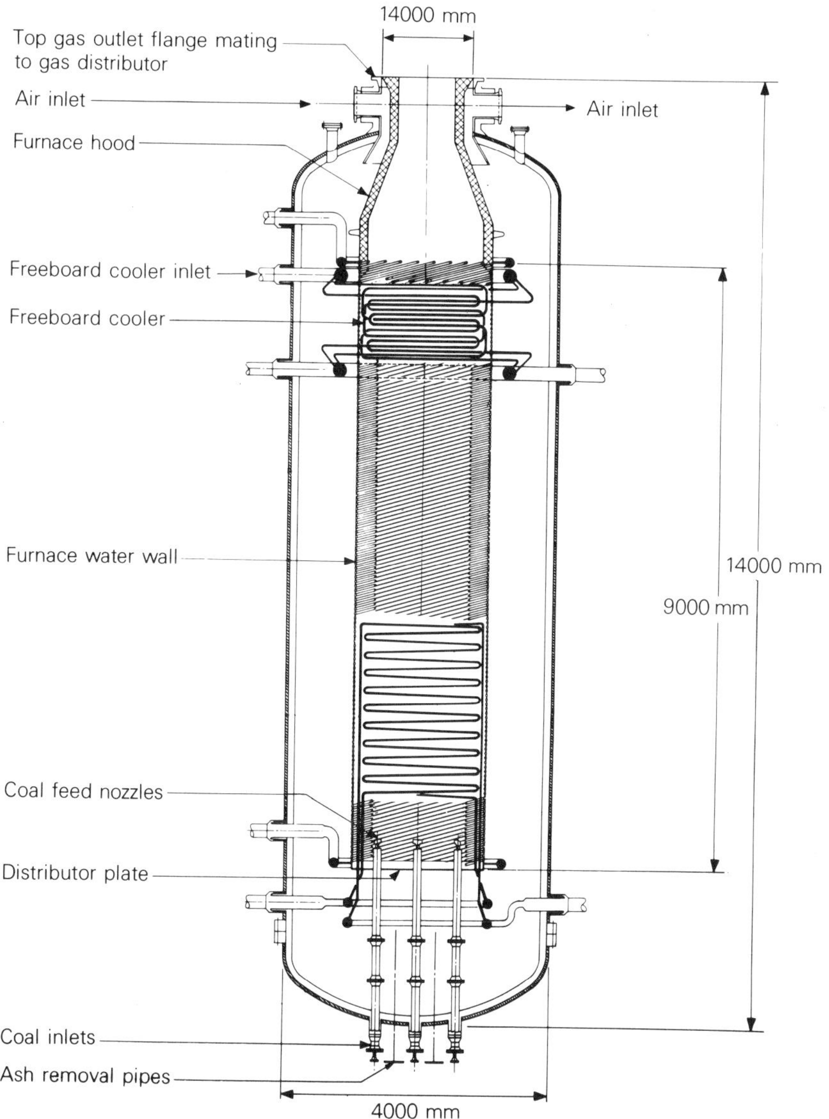

6.17 A section through the Grimethorpe combustor pressure shell

secondary cyclones and thence into a common heat exchanger; dust measuring devices are provided to determine the efficiency of the cyclones and the quality of the cleaned gases. The gases, after various levels of cleaning, can be used for material/component tests, for trials of fine cleaning devices or, later perhaps, to drive a turbine. The fine cleaning devices may be critical items and may include super-cyclones or filters.

The distributor plate forms the floor of the furnace. Coal, dolomite and air are introduced through the distributor plate; hot gas from propane burners can also be

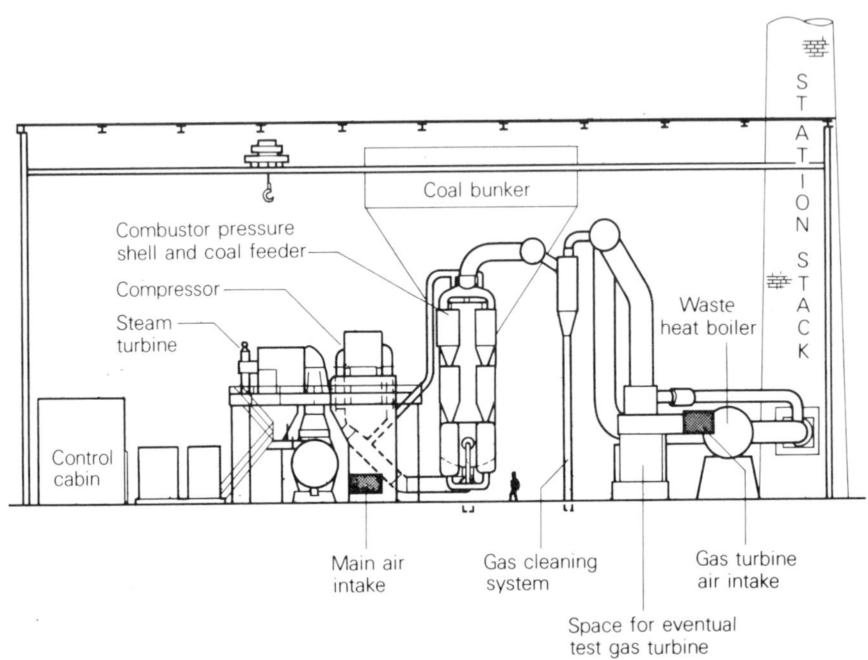

6.18 General layout of experimental fluidised-bed combustion plant at Grimethorpe

introduced for ignition. Ash can be removed through refractory lined caps.

The in-bed tube bundles are designed for easy fitting and experimental replacement; more than one tube bundle design is likely to be studied. Above-bed cooling tubes will control exit gas temperatures. Coal is automatically blended with dolomite and fed from a pressurised hopper by dense phase pneumatic transport through nine feed points to the bed. A compressor powered by a steam turbine (normally driven by the combustor but requiring auxiliary start-up steam) provides the air which passes down between the pressure shell and up into the bed through bubble caps in the distributor plate.

The off-gases are cleaned through two sets of four "Stairmand" high efficiency cyclones, cooled to 300°C in a heat exchanger contained in a similar sized pressure vessel to the combustor and passed to atmosphere through pressure let-down valves, a silencer and a 300 ft high chimney. The main facility for checking gas quality, testing possible "tertiary" cleaners and passing gases over turbine blade simulations is located after the bank of cyclones.

The main construction of the Grimethorpe plant was virtually complete by the end of 1979 and, following re-adjustments and shake-down trials, should be capable of starting the main experimental programme towards the end of 1980. The initial items of work are expected to take about two to three years. Further major additions to the facilities and programme in due course would not be unexpected. An important achievement already is that organisations in all three participating countries have been involved in the design and manufacture of various components; in many cases this has required quite original concepts, thus advancing the design philosophy and generating confidence that problems can be solved as they are met.

This pioneering impetus would presumably be a continuing feature of any extensions to the facilities and test programme.

US Projects

The US Department of Energy's programme on combustion is interesting both for the emphasis on fluidised combustion and for the individual items selected. In the Fossil Energy Program for 1980, out of a total of \$57.4 million, \$25.9 million was for atmospheric fluidised combustion and \$20 million for pressurised. In the former, the principal item was the 30 MWe Rivesville project, described below. This was intended to be supported by a number of other installations, including a Component Test and Integration Unit at Morgantown, West Virginia, due for completion in 1982, consisting of three vertically stacked beds of intermediate size, suitable for scaling up to commercial designs. Three areas of industrial boiler applications are also identified for priority development through industrial contracts to design, construct and operate units, burning 0.75–5 t of coal an hour and producing up to 100 000 lb/h (c. 30 MW) of steam for the following applications:

(a) saturated steam boilers for process and space heating in industrial and institutional facilities;
(b) superheated steam generators, especially for co-generation and industrial process and heating requirements;
(c) indirect-fired process heaters, utilising tubes or other heat exchangers to heat liquids and gases.

In addition, two units are being built to burn waste materials from the mining and preparation of anthracite, at rates between 1–5 t/h. The Atmospheric Fluidised Bed Closed Cycle Turbine system, mentioned earlier, has been developed to the stage of conceptional designs and a small test unit is to be built (300–500 kWe). This will provide suitable data for scale-up to the 5–50 MW range for possible cogeneration applications in the industrial/institutional sector.

In the pressurised system, substantial reliance is placed by the USA on the Grimethorpe project. The main internal US project is an air heater unit being built by Curtiss-Wright at Woodridge, New Jersey, using an available 7 MWe gas turbine/generator system.

The status of the programme items can change. The situation on these and certain other major US projects, as understood in late 1980, is described below. A summary presented by Voelker and Freedman to the International Conference on Coal Research in September 1980, is reproduced as Table 6.1.

Rivesville. In the USA, one of the most persistently active companies in the atmospheric fluidised combustion field has been Pope, Evans & Robins, who have been working on the system since 1965. After operating a 0.5 MWe pilot plant at Alexandria, Virginia, they received support from the US DOE for a 30 MWe scale mutli-cell pilot plant which was completed in 1976 at Rivesville, West Virginia, on a site owned by the Monongahela Power Co. The layout is shown in Fig. 6.19. The unit is designed to produce 300 000 lb/h of steam, equivalent to 100 MW and capable of producing 30 MWe in the host utility's steam turbine equipment. The main combustion area, about 300 ft^2 (30 m^2) is divided into three separate cells which can carry different parts of the evaporation and super-heating load. In addition, there is a burn-up cell fed from the cyclone dust collectors. The main cells operate at 1550°F (845°C) for optimal sulphur control; the burn-up cell is designed to operate at 2000°F (c. 1100°C) and at a lower air velocity than the main cells.

TABLE 6.1 Coal-fired atmospheric fluidised-bed combustors in the United States

Unit	Steam capacity lb/h	Bed size ft^2	Status	Remarks
Alexandria	5 000	9	Operational	R&D Purpose
Rivesville	300 000	480	Shakedown	3 Cells and CBC
Georgetown	100 000	209	Shakedown	One cell 2 beds; Spreader stoker
Great Lakes	50 000	140	Construction; Completion late 1980	In-bed superheater tubes; package boiler
Exxon (Crude oil heater)	10–15 MBtu/h	9.5	Construction	4 inch diameter in-bed tubes
Battelle	5 MBtu/h	2.5	Operational	Multisolids; fast fluidised bed
Johnston	10 000	12–24	Operational	Tapered bed; Multifuel; Firetube/package
Central Soya (Johnston)	40 000	48–96	Operational	Commercial; Firetube/package
IBM (Johnston)	20 000	24–48	Operational	Commercial; Firetube/package
Babcock and Wilcox	20 000	36	Operational	Test facility
Combustion Engineering	2 000	5.6	Operational	Sub-scale for Great Lakes
Fluidyne	2.8 MBtu/h	17.8	Operational	Vertical slice
Shamokin	20 000	100	Construction; Completion late 1980	Anthracite culm
Wilkes-Barre	100 000	213	Construction; Completion early 1981	3 Cell — Anthracite culm
ERCO	20 000	36	Shakedown	Demonstration
Coal Combustor for Cogeneration	6 MBtu/h TBD		Construction; Completion late 1981	AFB for externally fired gas turbine
TVA Pilot Plant	200 000	216	Conceptual Design	Utility Pilot Plant
Babcock Contractors	60 000	120	Shakedown	3 Cells; Retrofit

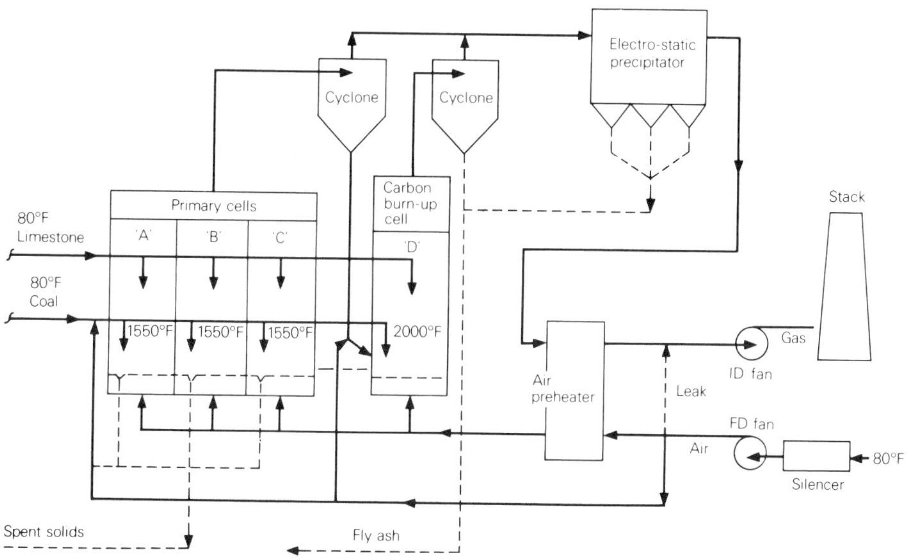

6.19 Fluidised-bed boiler, 30 MWe

Single cell operation on coal was achieved in December 1976, superheated steam was produced in three-cell operation in April 1977 and full nameplate output on four cells was achieved in September 1977. However, there were delays due to the six-months coal strike and repairs necessitated by a fire in the air preheater. There have been problems in certain plant auxiliaries, notably the coal feed system, which have compromised regular operation. The longest continuous period of operation has been about 200 h and there have been several periods of over 100 h. Nevertheless, the operation of the combustor itself, in which start-up, multi-cell operation and flyash re-injection were demonstrated, generated confidence. The design has been supplemented by the Georgetown facility and other units and will probably be closed down in late 1980 or early 1981.

Georgetown University. This is a unit producing 100 000 lb/h (c. 30 MW) of steam. A cross-section is shown in Fig. 6.20. The unit consists of two beds fired by stoker spreaders; only one bed has a start-up burner. The boiler uses conventional stoker coal, $1\frac{1}{4}$ inch (31 mm) with not more than 40% less than $\frac{1}{4}$ inch (6 mm). The unit is a natural circulation boiler with inclined tubes in the bed. The exhaust passes through a cyclone followed by baghouse filters. Fly-ash is reinjected.

The unit has been highly successful. Operation started in the summer of 1979 and in the first six months 1500 h were accumulated; by mid-1980, this had grown to over 3000 h, with several continuous runs of over two weeks. Sulphur emissions are held to 0.8 lb/MBtu while burning 2.5% sulphur coal mixed with limestone (about 85–90% retention). The exhaust stack appears as clear as with natural gas burning (opacity at or below 10%); NO_x emissions are 0.3 lb/MBtu, about half those of a conventionally-fired coal combustor.

Some minor problems (re-injection systems and baghouse) have been corrected. The unit is highly "visible" having been seen by over 1500 visitors from 17 countries.

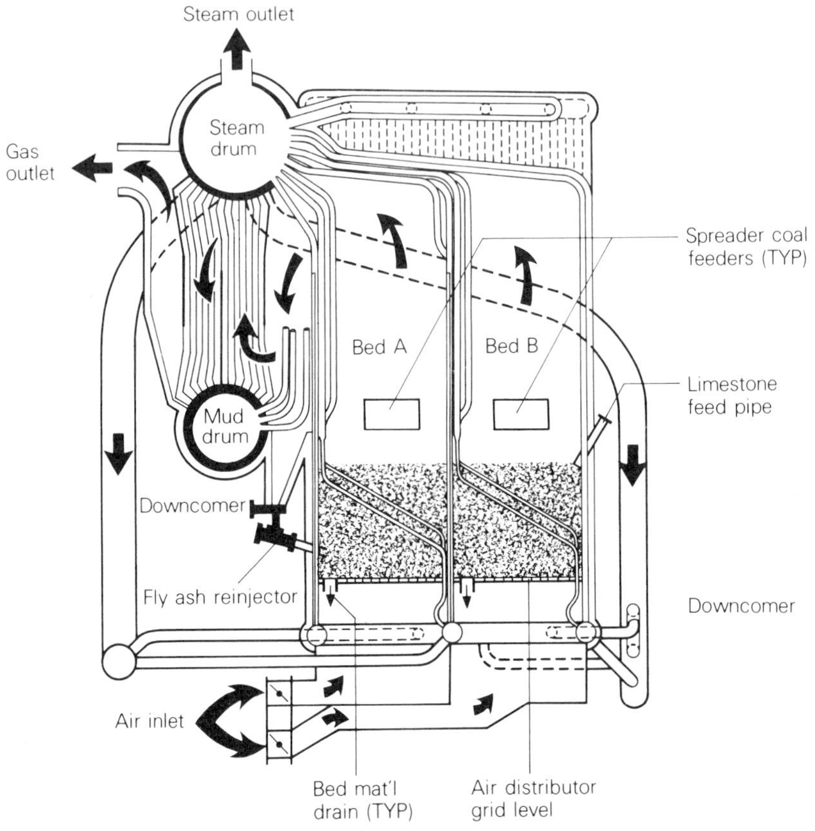

6.20 Cross-section through the Georgetown AFB Boiler

Great Lakes. A 50 000 lb/h (15 MW) atmospheric pressure boiler is being constructed by Combustion Engineering at the Great Lakes Naval Training Station. A feature of this unit is that it has superheater tubes in the bed. Pneumatic fuel injection is used. Start-up is scheduled for December 1980; performance similar to Georgetown is expected. The purpose of the programme is to develop a packaged steam generator which may be built and shipped as a modular unit.

Alliance. The Electric Power Research Institute (EPRI) have sponsored at 3 MW development facility built by Babcock & Wilcox Inc., not connected with the British Babcock Group, at their research centre at Alliance, Ohio. The unit (Fig. 6.21) has operated since 1978, covering the main operational variables. Fly-ash re-injection (together with pre-mixed coal/limestone, fed pneumatically into the bed) was demonstrated. This gave improved carbon and limestone utilisation.

It was concluded that a significant reduction in capital costs was achievable by the elimination of a burn-up cell due to the successful operation of the reinjection system. It was also projected from the results that 95% sulphur removal could be obtained in a commercial-sized unit with a calcium/sulphur ratio of between 1.5 and 2.5; this would probably make the system economically competitive if the stringent

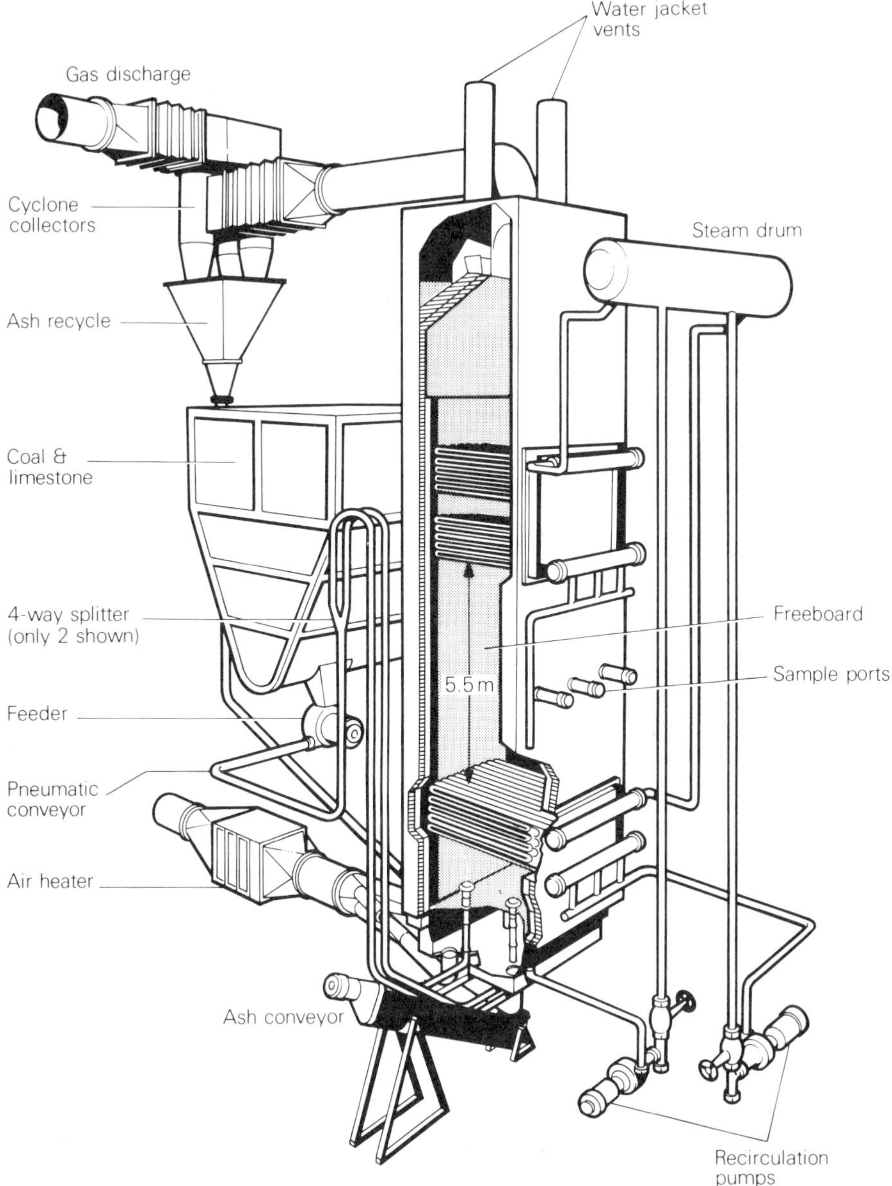

6.21 Babcock and Wilcox Alliance Research Center fluidised bed combustion development facility, 3 MW

environmental controls projected for the mid-1980s materialise. NO_x production will probably be less than 0.4 lb/MBtu and possibly as low as 0.2 lb/MBtu.

As a result of this success, EPRI is collaborating with TVA in a 20 MWe pilot plant at the TVA's Shawnee plant at Paducah, Kentucky. The plant will produce steam at 2400 psi (160 bar)/1050°F (565°C) but no steam turbine is provided. The bed will be 3 m × 4 m, segmented four ways, plus a start-up zone. There will be fines recycle and a high freeboard. The plant is under construction and start-up is expected in 1982.

Information from the 20 MWe pilot plant will be used in the design and construction of a 200 MWe demonstration plant planned by TVA, tentatively scheduled for operation in 1986. It is planned that the demonstration plant should be overlapped by a 800 MWe commercial plant design, possibly to operate in the late 1980s.

Shanokin & Wilkes-Barre. These are two units, located at the places named in Pennsylvania, to burn anthracite wastes (culm); they have capacities respectively of 10 and 50 t/h steam (equivalent to 2 and 10 MWe). Both should start up in 1981.

PFB — Curtiss-Wright. The main USA internal activity is the pilot plant being built at the Curtiss-Wright (C-W) plant at Wood Ridge, New Jersey. The outline of the scheme is shown in Fig. 6.22. The unit is an air-heater cycle, in which air is heated in in-bed tubes and then joins the cleaned combustion gases before entering the gas turbine. Thus, all the energy released from the coal goes into gases which pass through the gas turbine; the combustion gases will contain about one-third of the total energy, with two-thirds in the heated air. Further energy recovery takes place in a steam cycle.

The pilot plant facilities (one-fifth scale of a commercial plant) will include a 12 ft (3.7 m) diameter by 50 ft (15 m) high pressurised combustor, a C-W gas turbine capable of generating 7 MW and a waste heat boiler capable of generating 6 MW of steam at 175 psi (12 bar) and 377°F (190°C). Combustion will take place at 1650°F (900°C) and 8 bar. A three stage cyclone, plus a recirculating cyclone, will be used to clean the combustion gases. The plant will be operational in 1982.

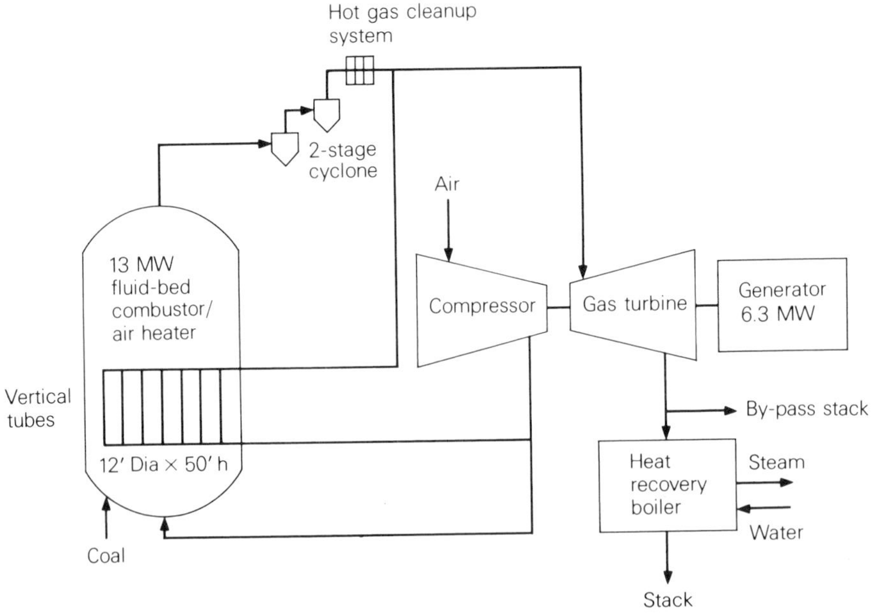

6.22 Pressurised fluidised bed combined-cycle pilot plant, 13 MW

Prior to the building of the pilot plant, a 1 m vertical segment of the combustor was built and was used to supply combustion gas to a 3-stage cyclone and thence to a small gas turbine (175 mm wheel). This rig has operated for 1000 h; data on heat transfer, combustion, SO_2, NO_x and particulate emissions have been obtained (and compared with Leatherhead data). Initial tests on turbine blades after the 1000 h test showed appreciable sulphidation of uncoated turbine blades but barely detectable sulphidation with the best protective coatings. Almost no erosion was found and deposits were thin and generally not tenaciously bonded. These initial observations are favourable; further examinations are being performed.

Proposals have been considered by the US DOE for a large (possibly 200 MWe) demonstration plant on the pressurised system and also for a pilot plant at the Argonne National Laboratory. The latter has apparently been temporarily placed in abeyance, medium term requirements being met by the Grimethorpe facility and also by specific work under contract on the CURL pilot plant at Leatherhead. The construction of a demonstration plant is seen as an essential step but a decision about a government backed plant on the PFB system will probably now await results from the Grimethorpe and Curtiss-Wright projects and an assessment of the relative merits of air cooling versus steam cooling; this may not be possible before about 1983. In the meantime, the American Electric Power Company Inc. has reached agreement with Stal-Laval on a project for a combined cycle pressurised fluidised combustion plant to be built at a site in Ohio. The project would use an existing Stal-Laval design of gas turbine (70 MWe) with a steam turbine of 100 MWe. The feasibility study has been completed successfully and the engineering study is proceeding. Details of the timing and the cost of construction and operation are awaited.

Projects in West Germany

The Federal Republic of Germany has a comprehensive and well-balanced programme on combustion, in which fluidised combustion development is prominent, with close co-operation amongst national and state governments and industry, both in the public and private sectors. In addition to participation in the Grimethorpe project, there are a number of other activities. Using information from Bergbau Forschung, who have worked for some time on small units primarily for heat production, Ruhrkohle AG, with 60% Federal support, have converted an existing power station boiler with a travelling grate stoker to atmospheric fluidised combustion, in a project at Flingern near Dusseldorf. Designed and installed by Deutsche Babcock, it has a thermal output of 35 MW from four cells, individually controlled, horizontally placed in a square formation. Gas cleaning is by cyclones followed by bag filters. The unit is considered as a demonstration for a large fluidised bed boiler with superheat. Start up was planned for 1979. Ruhrkohle have also proposed a 6 MW industrial boiler. A further plant, of 145 t/h of steam, is planned by Elektrizitatswerke, Dusseldorf. This will have six cells and the fluidised bed replaces the radiant section of the boiler.

The West German pressurised programme, like that in the USA, includes pressurised air heater fluidised combustors. One unit, called the AGW Gas Turbine Plant, being built by a consortium including coal producers, a utility (STEAG) and a boiler manufacturer, uses air cooling tubes in the bed. The combustion gases were originally intended to be cleaned (around 869°C) in an electrostatic precipitator being developed but conventional equipment is likely to be used. As in the C-W unit, the two gas streams will be combined before being fed into the gas turbine. The hot electrostatic precipitator is an attempt to provide an alternative to hot-gas clean-up devices being developed elsewhere, especially in the USA, and is continuing as a separate project. A further project provides an even more conservative fallback position, in case gas clean-up proves somewhat intractable. In this concept, the hot combustion gases at 860°C go first to a heat exchanger to provide steam for a steam turbine; the gases leaving the heat exchanger at 430°C may be cleaned conventionally and passed into a gas turbine, where corrosion is not expected. It is noteworthy, nevertheless, that other advantages are considered worthwhile.

Saarbergwerke AG are building an interesting plant to investigate a somewhat complex concept. The main unit is a pf boiler but this is preceded by an atmospheric fluidised boiler. The latter has air heating tubes to preheat the combustion air before burning premium fuel in a gas turbine which in turn drives the compressor and a generator. The turbine exhaust gas, with about 18% O_2, is then used as the fluidising gas in the combustor to burn coal and then passes into the pf combustion chamber to reduce NO_x. The incorporation of the gas turbine improves the overall thermal efficiency by a few percentage points. A second novel feature is that the cooling tower is used as a stack gas scrubber and replaces the stack. The heat recovered here is used in a district heating scheme. The plant, with a total output of 200 MW, is nearing completion.

Projects in Other Countries

There has for a long time been considerable interest in fluidised combustion in Sweden. One source has been the Stal-Laval turbine company, already mentioned in connection with the AEP project, but also interested in a number of other ideas. Another impetus, however, had been the Swedish interest in co-generation and district heating, for which fluidised combustion seems well suited. A 25 MWth fluidised boiler was brought into service in 1978 to provide hot water; it can burn coal and other fuels. Other countries known to have significant projects in hand or under study include Australia, Belgium, Canada, China, Czechoslovakia, Finland, France, India, Ireland, Japan, Norway, Poland, Romania and South Africa.

Fundamental Studies of Fluidised Combustion

In addition to the project work and conceptual studies, there is an important need to fill in the scientific background of fluidised combustion, in its many forms. Developments over the last decade or so have been largely empirical and may have outpaced scientific understanding. This work is being pursued currently, however, in scores of laboratories in many countries of the world. This trend is exemplified by the many scientific and engineering papers presented in recent years, including those at well attended conferences devoted entirely to the subject. Some areas of work, with comments, are mentioned below.

(*i*) *Combustion.* Basic processes of pyrolysis, devolatilisation and combustion discussed earlier are highly relevant. For each of the main steps in the process, variations due to pressure, coal type and size are most important, plus the effects of local oxygen deficiency. Combustion efficiency is a complex subject and no model is yet available to enable elutriation rates and combustion efficiencies to be calculated as a function of coal type and size range, bed geometry and fluidisation characteristics. Such a model would assist engineering decisions to be made about the need for a burn-out cell (now looking less attractive) and about the amount of fines recycle, which affects cyclone design. The possibility of temperature gradients needs to be explored, especially as they may be influenced by the number of coal feed points, in-bed fixtures and deep beds.

(*ii*) *Sulphur retention.* Some data are available relating to particular limestones and dolomites but rationalisations based on factors such as particle size, pore dimensions and density are needed. Further work is also required to relate the necessary addition of limestone, for any specified degree of fixation, to process variables and bed geometry, including factors influencing mixing. Other questions include attrition, regeneration and the effects of various compounds on absorption and regeneration.

(*iii*) *Mixing.* Good mixing is readily achieved in laboratory scale equipment using homogeneous feeds. In practice, both coal and absorbents will have size ranges from the specified maximum down to infinity and the size distribution patterns of the two materials will not coincide, nor will other physical characteristics. Generalisations are necessary to cover minimum velocities, feed point spacing, tube geometry and tube plate design in order to deal with any desired combination of feeds.

(*iv*) *Particulates.* Data are needed on the rates of elutriation of particles in relation to size, especially as affected by attrition. In addition to data for the specification of normal cyclones, new devices (which may include special cyclones) are needed for "polishing" hot gases for turbines. These may include granular bed filters, magnetically stabilised fluid beds, hot electrostatic precipitators and metal or ceramic filters. The effects of changing bed conditions on clean-up efficiency need study. Trace element emissions require much further work. At one time, it was considered that if particulates could be controlled adequately for gas turbines, then environmental standards could readily be met; this may no longer apply and therefore work on fine particulates, difficult to assess accurately, is needed.

(*v*) *Bubbling characteristics.* The formation and characteristics of bubbles are vital to fluidisation phenomena, especially as this is a dynamic phenomenon and there is constant interchange of gases between bubbles and the emulsion phase. The effects of distributor plate design, tube geometry and pressure are important variables.

(*vi*) *Distributor phenomena.* The area near to the distributor is necessarily heterogeneous, the gas and solids entering as jets, the shape and size of which are important. To minimise the possibilities of slugging, hot spots (involving possible clinkering) and dead space, different designs of nozzles and bubble caps have been proposed and local distribution phenomena need investigation.

(*vii*) *Heat transfer.* This is one of the most completely studied areas, at least in simple cases, probably because of the excellent and reproducible characteristics normally obtained. Quantitative data for larger, more complex geometries and a wider range of bed materials, especially in practical dynamic conditions, are needed.

(*viii*) *Corrosion, erosion and deposition.* Generally benign results were indicated in early work on in-bed materials. This general situation has however become somewhat more complex, partly as a result of non-ideal conditions which may occur locally or transiently. There is therefore a need to understand these mechanisms and to determine metal loss or penetration rates over long periods, especially since these rates may not be constant. There is no real doubt about being able to select materials having adequate resistance for evaporative tubes but there are economic implications even here. There are more obvious problems with superheater and air heater tubes, especially with limestone beds.

Even more important is the question of effects on turbine components. Again, early results suggested that erosion would be nominal because of the soft, friable nature of the ash and that deposition and corrosion would be tolerable. Further consideration, especially of longer periods of operation at the higher end of the temperature range and in the presence of alkali metals, sulphate and chlorides and possibly other elements, suggests that a great deal of work will be necessary in these areas although there is complete confidence that the work will clearly identify acceptable materials.

(*ix*) *Gaseous emissions* (*other than sulphur compounds*). The most important gaseous emissions apart from sulphur compounds are NO_x. Oxidation of nitrogen in the combustion air is thought to be avoided because of the moderate temperature but the mechanism by which the nitrogen in the coal is oxidised — or ways in which these reactions may be suppressed — require study, especially as a function of pressure and other operating conditions. It may also be that the reconciliation of results reported in different places suffers from differences in analytical techniques. It has been reported that NO_x emissions are reduced by increase in pressure, by SO_2 or moisture in the combustion gases or by decreases in coal size. CO may also have an effect on NO_x and there may possibly be catalytic effects from solids. All these factors require definition under the potential range of conditions.

CO and hydrocarbons may also be present in off-gases but this may be due to unsatisfactory combustion conditions, the limitations of which require clarification.

(*xi*) *Treatment of Residues.* Proposals have been made for the regeneration and recycling of limestone and dolomite and for the recovery of sulphur although neither process appears economically favourable in current circumstances. The general reactions are well known but the effects of the combustor conditions on the way in which different stones behave is an area for study. If the spent ash and absorbent are to be dumped, leaching tests over a wide range of materials and combustor conditions may be necessary.

(*xii*) *Scale-up and modelling.* Once substantial progress has been made on the main physical R&D items mentioned above, computer models can be developed further and simulations improved. Some hitherto unknown factors which have seriously reduced the value of earlier models include changes in particle size (attrition and agglomeration), hot spots and maldistribution of solids and gases.

Although a complete reconciliation of current experimental work against possible problem areas is difficult, there seems no reason to believe that important gaps will exist as commercial exploitation proceeds; however, the timescales may not be ideal. An assessment of the adequacy of fundamental data will become easier as a result of conferences in the early 1980s.

Environmental Impact of Fluidised Combustion

This factor has been in the forefront of much of the description of fluidised combustion and its importance is also considered elsewhere. However, a recapitulation, especially of the technical considerations of this important area, may be a suitable conclusion to this section.

Suppression of SO_2 has been the primary environmental objective. Retention in the ash of up to 90% or more of the sulphur was readily achieved in early laboratory experiments, by the addition of two or three times the stoichiometric proportion of limestone or dolomite. In the USA, it has been stated that 90% retention can be achieved and that the governmental restriction (in 1977) of 1.2 lb $SO_2/10^6$ Btu (0.51 kg/GJ) could readily be maintained by using Ca/S molar feed ratios of 2:1 for dolomite and 3:1 for limestone. However, more stringent controls than the above may be introduced. The Rivesville plant is reported to have achieved 85–90% retention (presumably in good operational periods) and the Renfrew plant has achieved "80–90%". In practice, on a larger scale, it appears that in some cases improvements beyond about 80% may involve compromises, possibly important, on design and operational factors. It should be appreciated that the fluidised bed is effectively a single stage chemical engineering process and that increases in percentage retention above about 80–85% are likely to be increasingly difficult to achieve. If exceptionally high retention levels are deemed necessary, design compromises which may have unfavourable economic consequences may result; movements in some or all the following directions may be needed:

(a) higher Ca/S ratios (and therefore more solid waste);
(b) lower gas velocities;
(c) smaller limestone particles;
(d) deeper beds;
(e) closer solid feed points and/or devices to accelerate lateral mixing;
(f) constraints on sources of stone and/or pretreatments (requiring development) to increase absorption;
(g) other additives (again requiring development).

The pressurised system is superior in this respect but more detailed evidence, especially relating to mechanisms, is needed.

Emissions of NO_x are definitely much lower in pressurised fluidised combustion than in any other coal burning combustion systems; in atmospheric pressure plants the advantage is still significant but not so much. Although some high CO and hydrocarbon emissions have been reported, these may refer to conditions unrepresentative of commercial design; it should be possible for these problems to be met by adequate excess oxygen and reaction time/space, with negligible penalty in industrial designs.

In theory, fluidised combustors should have an advantage over other systems with regard to particulate emissions because the original particle size of the coal, and thus the ash, is larger. Temperatures are also lower, which may reduce the volatilisation of some elements. However, there are complications, both within the combustion system and downstream. One area of uncertainty is the effect of attrition, especially with heavy recycle rates and additives. Other effects are difficult to assess because future environmental standards are uncertain, in particular for very fine particles, such as sulphates, which may be formed by coalescence after emission of sulphur oxides. There are of course a range of options, but these are on an ascending scale of economic penalty as well as efficiency — cyclones, bags, other filters, precipitators and scrubbers. No doubt a suitable compromise can be found but there will probably be implications on the generation of particles as well as methods of control, so that the whole basis of combustor design could be affected. Uncertainty or unduly repressive standards could seriously affect development. If it is shown, as expected, that particulate standards can be met, trace element emissions may also be adjudged satisfactory but a great deal of further work is required.

Another area of environmental impact requiring further consideration is the disposal of solid waste. Again, at first sight, in this respect fluidised combustion should be considerably better than some coal combustion alternatives, especially taking into account the sludges from flue gas desulphurisation. Work is proceeding, with some considerable promise, to use the ash for constructional or agricultural purposes — in the latter case both chemical and physical benefits might be obtained. Attention is also being given to the leachability of residues and the possible undesirable elements in leachants. The probability that toxic trace elements may not be volatilised and remain largely in the ash could of course be an undesirable factor from this point of view. Consideration is being given to the costs of storage or final disposal in impervious containment. Early estimates suggest that this would not be particularly onerous in itself but the loss of a "bonus" from useful applications would be a disappointment.

FLUE GAS TREATMENT TECHNOLOGY

Flue Gas Desulphurisation (FGD)

The replacement of other forms of burners by fluidised combustion in existing installations, whilst useful in some smaller forms of application, is unlikely to be a significant factor in reducing overall SO_2 levels; emissions come to a great extent from large pf generating stations, which are unlikely to be retrofitted to FBC. Where countries decide that early action must be taken to limit SO_2 emissions, the reduction of input sulphur in the fuel is also a limited option and recourse must be had to methods of removal of SO_2 from the flue gases before release through the stack. The merits and implications of control are considered elsewhere. The USA and Japan have been leaders in the application of FGD by legislation, followed more recently by the Federal Republic of Germany; the greatest experience exists in these countries.

Most FGD plants use wet scrubbers. Some early trials of the injection of dry powdered limestone directly into furnaces were relatively unsuccessful and this method appears to have been abandoned. An alternative dry system is to use a fixed or fluidised bed of an absorber in the flue gas system. Absorbents include char, activated carbon or alumina, which may be impregnated with copper oxide. These systems are regenerable and the sulphur is recovered in elemental form or as sulphuric acid. The theoretical advantages of such a system over wet scrubbing are through increased reliability, decreased maintenance and elimination of a difficult waste disposal problem. It seems however that further development of dry systems may be necessary before they can be widely used and, in the meantime, the wet scrubbers have become increasingly established. Current work in Japan may however bring dry systems back into contention.

Wet scrubbers may be non-regenerable or regenerable. In both cases the flue gases are brought into direct contact, in a spray tower or venturi scrubber, with an aqueous slurry containing an active material such as limestone. In this case, calcium sulphite and calcium sulphate are formed, the former predominating; for disposal, however, it is desirable to oxidise the sulphite in a further stage, since calcium sulphate is easier to dewater and is preferable for landfill (the outlets as gypsum are limited). In the regenerable systems the spent scrubber solutions or slurries are treated with acids and heated to recover SO_2 (which may be converted to sulphuric acid or sulphur) and the regenerated solution is re-used.

There are a number of proprietary variants of the general processes differentiated primarily by the active material: lime, limestone, sodium carbonate, magnesia and mixtures of these, sometimes plus additives such as calcium chloride, are amongst those proposed. The reliability of plants in the USA was poor in the early experience, possibly because installations were made hurriedly under legislative pressures. More recently, the larger number of plants and greater experience have given rise to marked improvements. By late 1979, scrubbers were operating in plants totalling about 9000 MW capacity and a further number representing 12 000 MW were under construction; the total commitment was over 50 000 MW, out of a generating capacity of 200 000 MW. Average operability exceeded 80%, more than doubled from early levels of 40%. New plants were considered likely to achieve 90% after a short shake-down period. Improvements have been achieved largely by the use of better materials to resist corrosion, the opening out of narrow spacings to reduce plugging, higher scrubber liquid to gas ratios and improved designs of mist eliminators and reheaters. Experience of five large scrubber plants in Japan has been even better over a substantial period than in the USA — more than 95% reliability since start-up; these early plants however were operating on oil fired stations, which may be easier to cope with than coal firing.

Overall Flue Gas Treatment (OFGT)

Regulations on emission control are exceptionally severe in Japan. Considerable progress has been made in controlling SO_x and dust emissions. Control of NO_x by flue gas denitration for coal burning is however still in the stage of R & D; no practical full-scale application has appeared, mainly because the presence of large but variable ash contents complicates the denitration process. This led to the concept that all the flue gas treatment processes (including the handling of nitrogen-containing water from treatment plants) should be considered as a whole rather than as a series of separate steps. The Electric Power Development Company (EPDC) are leaders in this work.

Figure 6.23 is an outline of possible processes, presented by Tomozo Kimura *et al.* of EPDC at the ICCR meeting in September 1980; his comments and conclusions are summarised below.

Emissions of NO_x can be reduced by various combustion techniques, including special burners, two-stage combustion and gas recirculation. By these means, NO_x can be reduced to 200 ppm from a former level of 500–600 ppm; new techniques are aimed at achieving less than 100 ppm.

A wet process for simultaneous desulphurisation and denitration was tested in 1972. Although 80% of the target reductions were achieved, the process was considered unsuitable for large volumes of flue gas, chiefly because of its bulkiness. Dry selective reduction processes were then worked upon, because they appeared cheaper and simpler, with no secondary pollution.

Selective reduction methods depend on the use of ammonia to reduce NO_x to N_2 and H_2O. Two processes have been studied: (a) Selective Non-Catalytic Reduction (SNR) and (b) Selective Catalytic Reduction (SCR). In the former, ammonia is injected into the boiler at a temperature range of 900–1000°C. This process was not

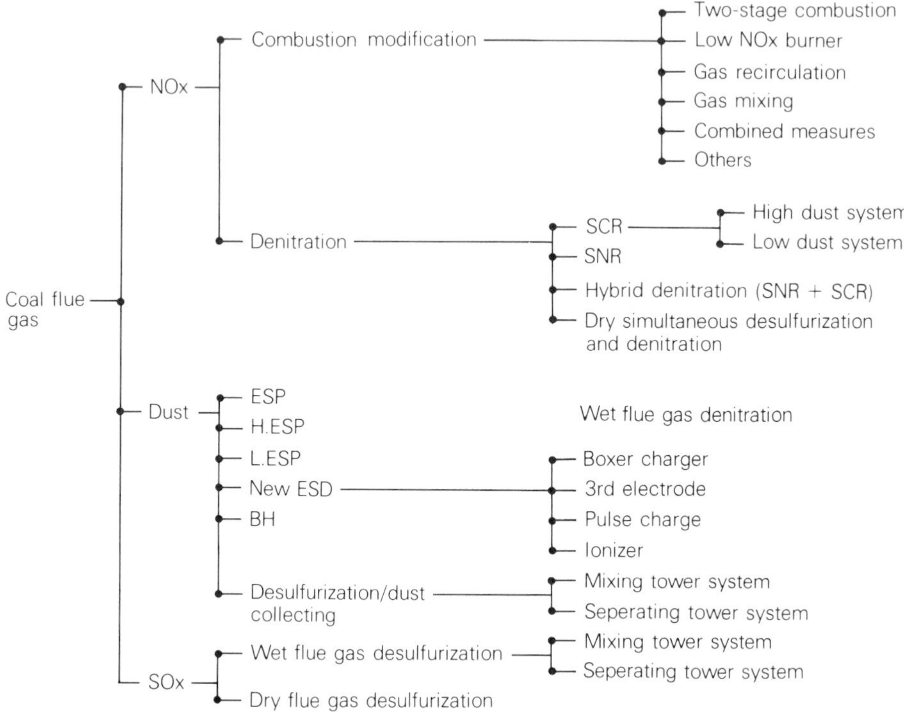

6.23 Flue gas treatment techniques

favoured because denitration efficiencies were low (30–40%) and ammonia leakage relatively high. A hybrid system with a catalyst at the outlet of the economiser gave improved efficiency (50–60%) and would be a possible process. However, SCR is preferred. In this process, the ammonia and catalyst are provided between the economiser and air preheater (AH), in a temperature range of about 350°C. With low dust loadings, there is a hot side electrostatic precipitator (HESP) upstream of the SCR system; with high dust loadings a cold electrostatic precipitator (LESP) follows the SCR unit. In both cases, after denitration and dust removal, FGD follows, plus an after-burner to raise the gas temperature into the stack. The SCR catalyst beds are prone to plugging, which may occur in moving beds; this has been overcome by the choice of shapes of the catalyst, using a fixed bed.

For FGD, both dry and wet systems were tried. Although the former was at first preferred, this system initially proved less economic and in 1972 was completely superseded by the wet system, using limestone powder as absorbent, converted into gypsum. This system has been developed to have 90% desulphurisation efficiency and approximately 100% availability. Technical improvements to achieve this include the use of a Gas–Gas Heat Exchanger (GGH) instead of an after burner to re-heat the flue gas after the scrubber; without reheating, stack flows are inadequate and acids condense causing serious corrosion. In the GGH, the gases from the scrubber are reheated by the incoming flue gas, thus saving energy and avoiding the difficulties of after burners (which include the addition of a further small amount of SO_x and NO_x from the burner to the cleaned gas). Another improvement in FGD is the development of better waste-water treatment; with the wet system, the absorbent liquor contains contaminants originating from the coal. Water must be regularly discharged because the build-up of chlorine promotes corrosion. The water treatment plant is complex, requiring several different processes.

Therefore, although the wet FGD process is technically satisfactory, work on dry FGD has been restarted with the hope of providing economies and improving integration with the denitration and dust removal operations. The dry system uses activated carbon in a moving bed to remove SO_x and NO_x simultaneously from the flue gas; ammonia is added to reduce NO_x. The sulphur is subsequently recovered from the absorbent by heating in an inert gas. The economics of the process are now enhanced, compared with earlier trials, by the availability of improved activated carbon, which not only absorbs more gases but survives recycling better. Other expected advantages, compared with the wet FGD system, apart from simultaneous denitration, are that efficiencies for SO_x are as good or better; elemental sulphur is the product and there is no need for reheating or make-up water.

For dust collection, ESP is the standard method; improvements in the performance of HESP is being sought. New ESP systems, involving alternate a.c. charging, a third electrode and wet ESPs are being studied. Bag filters are also being studied, especially to develop better filter cloth and to examine the possibility of clogging by NH_3/SO_3 compounds if used with SCR.

Several models for anti-pollution flowsheets have been proposed by EPDC (Fig. 6.24). Case 1 would be selected if necessary for an immediate plant. However, it is believed that all technical work will be completed in 1981, including a demonstration test, allowing accurate evaluations to be made. Clearly, Case 3 is preferred on grounds of simplicity, which should result in lower costs.

Already a 700 MW plant is being constructed, for completion in 1982, based on existing technology. This will discharge over 2×10^6 Nm^3/h of gas at 100°C through a stack with a height of 200 m. The SO_x concentration will be 100 ppm, NO_x 60 ppm and dust 0.03 g/Nm^3. The authors refer to this as due to "incomparably high technique for anti-pollution ...".

COAL–OIL MIXTURES (COM)

The concept of burning suspensions of coal in oil, treating the mixture as a heavy fluid, has been studied somewhat sporadically for over a century. It seems that, until recently, the main periods of interest have been associated with the two world wars and the countries concerned have been those most at risk due to their inability to import enough oil. The OPEC oil crisis has provided the incentive to establish studies on a more permanent, and certainly more scientific, basis; again, the countries involved tend to be those having strategic incentives to save imported oil. A great deal will also probably depend on the relative costs of oil and coal, on the one hand, compared on the other hand with the capital costs of necessary equipment to burn the COM, plus any consequential reductions in efficiency or output. The technology is essentially short-to-medium-term, aimed at utilising coal quickly in existing installations, where coal cannot be substituted completely.

The necessary equipment and operational complexity are considerable. Suspensions of coal in fuel oil can readily be obtained with short-term stability if the coal is pulversied to less than 200 mesh (75 μm). For longer term stability, finer

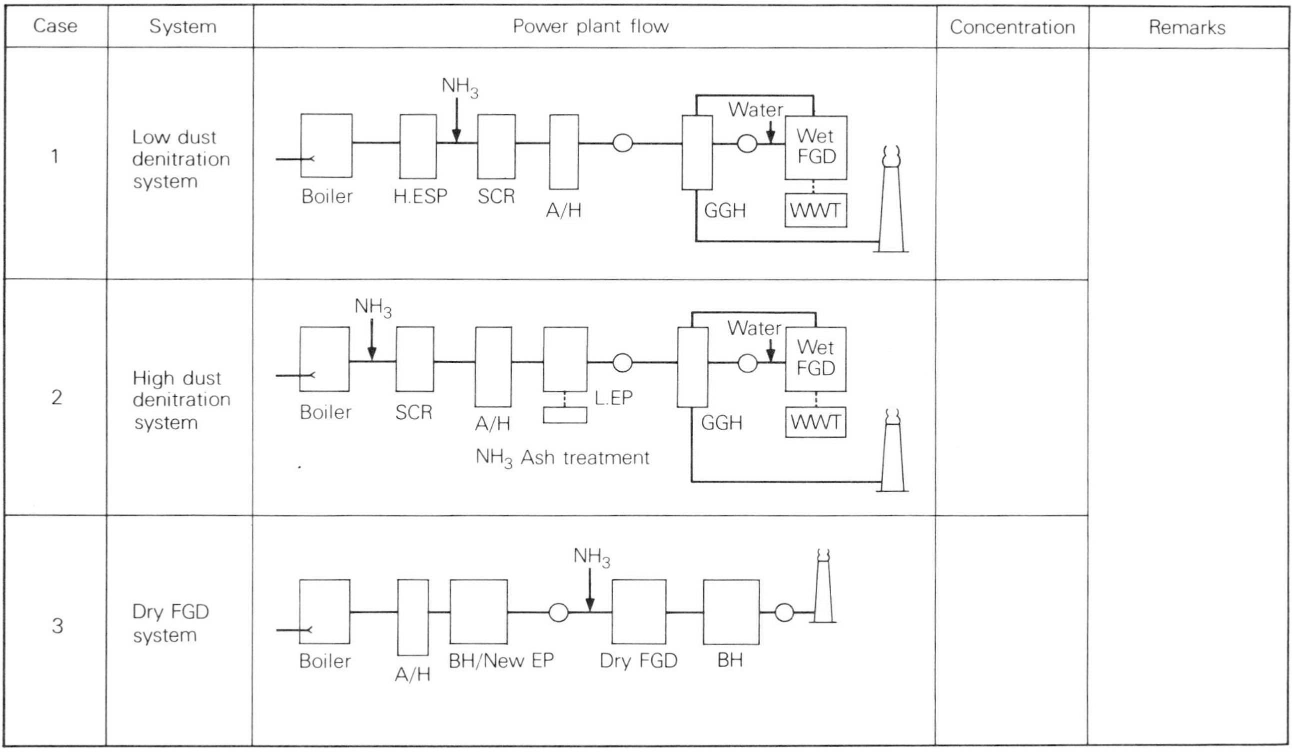

6.24 Model anti-pollution coal-fired thermal power plant

grinding, the addition of surfactants or continuous stirring must be considered. The apparent viscosity of the mixture can be an order of magnitude greater than that of the original oil, with consequent changes necessary in pumps, fuel lines and burners. There will be extra fouling and wear in the fuel supply system. Even more important, if a fuel oil system is converted to COM, the increase in ash, from very little to, say, 5% complicates the whole of the boiler and environmental control systems. A flowsheet for the preparation of COM is shown in Fig. 6.25.

A number of major trials have taken place, or are planned, especially since the mid-1970s and particularly in the USA; the largest appears to be a 383 MW oil fired boiler in Florida, owned by the Florida Power and Light Co. In these trials, which are continuing, coal is micronised to less than 40 μm and up to 45% of coal is used in the COM; there is a mixer in the tank and an additive is also used. The unit has been

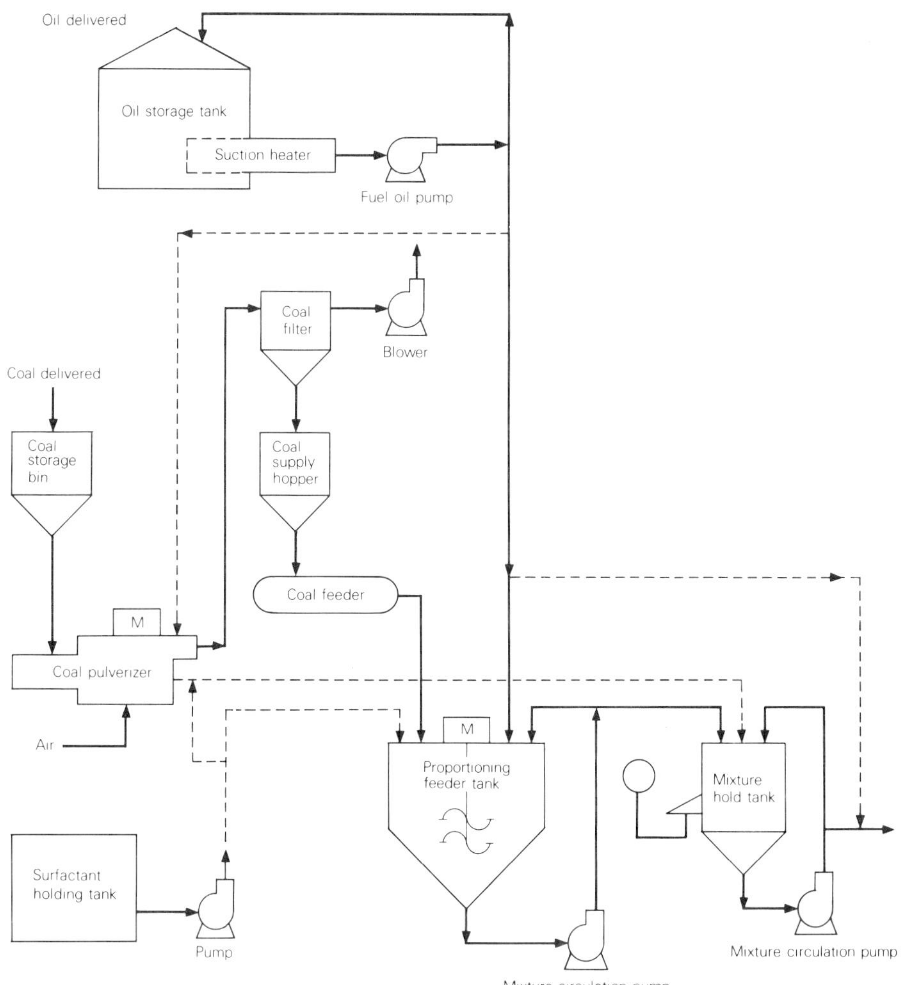

6.25 Coal-oil mixture combustion

run successfully for extended periods without major difficulty; the wall blowers have been adequate to remove the modest deposits. The output was reduced as the coal content increased; this was partly as a result of the lower heat value of the mixture compared with oil and partly because of the different combustion characteristics. There was no apparent effect on thermal efficiency, possibly because heat inputs were cut back to correspond to heat removal potential. Environmental effects were found to be within controllable limits and the size of clean-up equipment could readily be calculated from the ash in the coal and the coal content of the COM.

In another case, an 80 MW unit of the New England Power Service Company at Salem, Mass., was converted to burn COM, containing up to 30% coal. Plugging and erosion were considered the most difficult factors to predict. Coal sizing of 70–90% through 200 mesh was selected as the primary aim, with the use of agitation and additives, but provision was made for decreasing the size to 70–90% through 325 mesh. A satisfactory burn of 8 days has been achieved but coal concentrations in the COM have been limited to 20% due to stack opacity. Improvements in the ESP equipment are being considered. The COM appears stable, having been in storage for over a year. A 2000 h test is planned for late 1980.

COM has also been used successfully for injection into iron-making blast furnaces. It has become fairly common practice to inject oil into blast furnaces; this is said to allow a reduction up to about 15% in coke usage. The economic advantages of this were substantial, especially when oil prices were low compared to coking coal. By substituting COM for oil, using bituminous coal in the COM, there is an additional saving in scarce resources. In one case (Interlake Iron) a 30% loading of coal was used in the COM, which was based on a 5% water-in-oil emusion. Further tests on the same blast furnace were planned in 1980, using 51 t of coal per day in a 50% coal by weight mixture. This has been delayed because of conditions in the steel industry. The preparation plant, which is different from that used in boiler trials, has been tested; wear in some components has been observed.

Trials are also being carried out to evaluate the use of COM in industrial boilers. Initial tests on a boiler at the Saginan plant of General Motors demonstrated that COM could be prepared, stored and burned in an industrial boiler. Unfortunately, the steam demand was not great enough to assess whether derating might be required. The test system was also rapidly assembled from available equipment; as a result, decisive information about wear was not obtained. However, this type of work has been continued on a 24 000 lb/h steam (8 MW) boiler at the Pittsburgh Energy Research Centre. A 500 h run with COM at 40% coal concentration has been carried out successfully; no deposits were found on the tubes.

Canada also has a significant COM demonstration programme. Tests have been carried out since 1977 on a 10 MWe utility boiler in New Brunswick and further tests are planned there on a 100 MWe boiler. An integral (and interesting) aspect of this Canadian work is that it is associated with a coal cleaning and agglomeration process based on mixing a coal/water slurry with oil. The oil wets the coal, which forms small balls which are recovered; the mineral matter is not affected and remains in the water phase which is discarded. The process is especially suitable for cases where the mineral matter is finely dispersed, requiring the coal to be finely ground to release it. This fine grinding is desirable in any case for COMs. Furthermore, the oil used for agglomeration can be part of the COM fuel.

Both the US and Canadian demonstration programmes are well supported by R & D programmes, especially on the rheology of COM and on materials, as affected by wear and erosion by COM.

Analyses of potential areas of application are also important. A review by IEA Coal Research concluded that whilst only boilers designed originally for coal but currently burning an alternate fuel could be retrofitted to burn coal only. COM can be burned in the majority of boilers designed for coal or oil firing with only minor modifications of combustion equipment. The US DOE has also carried out a survey of the potential market for COM in existing boilers, taking account of the characteristics of those boilers, including furnace heat release rate, furnace exit gas temperature and superheater tube spacing. It was concluded that a good fraction of the plants designed for oil firing had furnace volumes and heat transfer surfaces which would facilitate the use of moderate ash fuels such as COM made from properly selected coals. The evaluation of the extent to which units would have to be derated to accommodate COM was considered to be a more complex matter; the main uncertainties were thought to be the deposition of slag on furnace walls or on convective heat surfaces and the changes in flame characteristics and temperature distribution. It was however concluded that the potential market was equivalent to the displacement of 386 000 barrels of oil per day by 35 Mt of coal per year. Of this, about 13% would be in industrial boilers and 3% in blast furnaces; the rest would be in utility boilers.

It seems likely that the demonstration programmes, supported by the scientific and marketing studies, will provide a clear technical opportunity for COM. There do not appear to be insuperable environmental problems. The extent to which this potential is exploited will almost certainly depend on economic factors, discussed in a later chapter.

COMBUSTION OF LOW GRADE COAL

With the growing problems of energy supply, attention is turning increasingly to low grade coals. These fuels have often tended in the past to be disregarded, although in some countries there is a substantial experience in these fields, especially where there are strong strategic energy or supply reasons. It seems likely that trends towards the use of low grade coals will be greatly intensified in the future, not only to increase the total bulk energy supply but also to allow the higher grade coals increasingly to be used in more valuable ways, such as gasification and liquefaction.

In this context, the term "lower grade coals" is taken to mean those coals of low calorific value and high content of inerts (ash or moisture). This includes much of the large reserves of brown coals and lignites, which are fairly widely distributed and also some bituminous or even anthracitic coals of unusually high ash; almost all the coals of India, for example, are of high ash, although almost all are higher in rank than lignite. Some of the same considerations which apply to the low grade coals can also apply to peat or oil shale and possibly also to residues, such as those from colliery washeries or, in due course, those arising from primary conversion steps to separate volatiles from coal. The use of these low grade coals is extremely important in many countries, often where industrialisation and economic standards depend largely on increases in electricity generation; the latter in turn may depend on low grade coals, sometimes almost exclusively. This situation applies to many of the Eastern Europe bloc countries and also to India. In Australia and the USA also these fuels are rapidly increasing in importance.

In some of the older types of boilers, a high ash content could be accommodated fairly readily but the scale on which these coals need to be used makes the use of large pf furnaces desirable. The scale effect, coincidentally, often means that mining methods cannot exclude layers of silt, which is generally the main source of ash. The pf furnace is not basically well suited to high ash coals since most of the ash leaves the furnace in the gas stream, as fly-ash, whereas in many other types of furnace, clinker or slag are removed. The fly-ash, in addition to creating a possible environmental problem, can build up on surfaces, reducing heat transfer efficiencies and possibly creating conditions for corrosion. Both build up and corrosion are strongly affected by ash composition.

Another major problem with low grade coals can be a high moisture content, which reduces the net energy available and also makes it more difficult to consume particles in the time available in the furnace or even in some cases to maintain combustion. The approach to high moisture coals has been through modification to preparation, milling, feeding and firing techniques. Separate driers may be used or drying may take place coincidentally with milling, with hot furnace gases sweeping the mill; some of these gases, carrying off the moisture, may be vented to atmosphere. In some cases, auxiliary firing with another fuel or heat-retaining refractory surrounds to the burners have been used but are not desirable methods. It has been shown that faster ignition can be obtained when the temperature of coal particles is raised as rapidly and to as high a temperature as possible before exposing them to oxygen. This has successfully been achieved by the use of a double concentric swirl burner, where the secondary air enters in a swirling mode through an outer annulus around the fuel jet. This brings a reverse flow of hot furnace gas to the burner mouth. For high ash coals, a spray burner has been developed in India and has a marked thermal efficiency advantage; in this design, the coal and primary air enter through a number of jets grouped in a circle, which is surrounded by an annular band containing secondary air jets which are set so the air issues tangentially.

The main problem with high ash, however, is fouling, which can rapidly reduce furnace performance. The extent and seriousness of fouling seems highly specific and dependent on subtle factors relating to the ash composition. A great deal of work has been done to understand this (sodium is a particular disadvantage) but alleviatory measures, through coal quality control or coal treatment for example, have not so far had much success. Changes in furnace design and operational modes seem more effective at present. Combustion spaces should be larger and tube spacing more generous. The number of soot blowers should be increased and their design and siting carefully selected. Extra large fans may be needed to induce adequate draught when spaces for gas passage become restricted.

Good and regular maintenance is necessary. By these means, coals with inert contents of more than 70% are successfully burned in several parts of the world in pf furnaces, as are some coals of high (c. 40%) ash with unfavourable characteristics. In addition, technically successful demonstrations of burning high ash coals in cyclone furnaces have been carried out but there seems no great momentum behind this system, possibly on economic grounds.

As noted above, the development of fluidised combustion would be particularly advantageous to the use of low grade fuels and it is probable that this will be the main direction of future plans in those areas wishing to exploit such reserves. Much may depend also on environmental standards, which will often have an increasingly important economic effect, having greatest impact on the use of the low grade fuels. This trend will also probably be favourable to fluidised combustion.

ADVANCED COMBUSTION SYSTEMS

The general increase in primary fuel prices will intensify the drive towards high thermal efficiencies in utilisation, especially for electricity generation. One obvious route is through combined cycles, using a combination of gas turbine and steam turbine. For this, the gasification of coal as a preliminary step is technically the most straightforward approach. An alternative is Magneto-hydrodynamics (MHD), for which coal gasification also provides a convenient first stage. The use of waste heat from coal conversion processes for district heating or electricity generation will probably become general practice; so also will the combustion of residues and wastes. Apart from these "economy" measures, a number of more advanced processes and schemes have been proposed — "advanced" being used either in the sense of complex or new technology or because the scheme will not be tested for a considerable time and is speculative.

The concept of using coal directly in an internal combustion engine has been considered for many years, particularly since the combustion properties of fine coal in air have been appreciated; this was the original aim of Diesel in the engine he invented. The ash in coal however makes this impracticable due to abrasion. Various claims have been made for coal treatment processes both physical and chemical, to enable coal to be used for this purpose, but it appears that even if an acceptable fuel could be made, standard coal conversion processes into light liquids would be cheaper. This probably applies also to processes for cleaning ultrafine coal, for which recent claims have been made.

The attraction of using coal in gas turbines is also strong. Again, raw coal appears to be an unsuitable fuel for direct use, although some claims have been made. One long series of tests in Australia in fact established suitable designs and materials but it was necessary to operate at low velocity. This constraint makes the economics generally unattractive; the fact that water is not required in this energy conversion system could make the process interesting in special circumstances. At the other extreme, of course, is the gasification of coal, followed by gas clean-up, for use in a gas turbine, which would probably be part of a combined cycle. Even in this case, however, there is room for compromise in the specification of particulate and impurity limits and for the development of gas turbines of greater tolerance in this respect. If the gas is cooled in order to accomplish fine cleaning, energy is lost. On the other hand, hot gas clean-up requires development and will probably always be sufficiently costly for the cleaning requirements to be as relaxed as possible. In this respect, gasification and pressurised fluidised combustion are similar, but the former may have greater flexibility due to a greater degree of decoupling and the ability to attain high temperatures at the turbine inlet. Possibly some hybrid scheme may be devised by which a gas might be generated separately and used to "superheat" gases from a PFB; the latter will have temperature limits (around 950°C) on exit gases due to constraints in the combustor. Other possibilities for "coal turbines" include the use of solvent refined coal or some form of "minimal" coal conversion (which has been suggested but not apparently defined); these ideas may have either economic or energy conservation disadvantages.

Considerable attention is being given to closed cycle gas turbines, in which the working fluid might be helium or potassium vapour, with an externally fired heat exchanger. The latter could use any convenient fuel but a coal-fired fluidised bed might be most attractive.

Fuel cells are being developed which use hydrogen or in some cases hy-

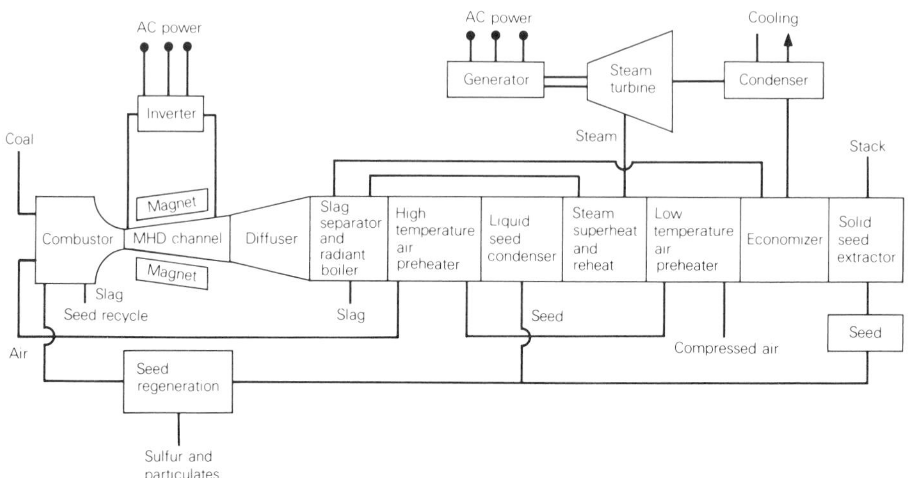

6.26 Open-cycle coal-fired MHD system

drogen/carbon monoxide mixtures (synthesis gas) as fuels. Coal is of course one of the primary fuels which could be used to make these feed gases. The objective with fuel cell development is to achieve the generation of electricity at high efficiency and in an environmentally acceptable manner. The gasification routes would have to be consistent — in fact, a number of different fuel cell sites could be served by a gas distribution system from central gasification plants.

There are two approaches to the use of coal in MHD devices. One indirect system uses clean fuels (which could be derived from coal); the other, and more difficult, route is to burn coal directly in a high temperature furnace (Fig. 6.26 shows an outline of the system). In the combustor the flowing gas passing through the magnetic field acts as the moving element in a generator. In experiments in the UK, in the late 1960s, when different key components were developed separately, a combustor operating at $>2300°C$, using a slag covered water wall container, was operated at the BCURA Laboratory, Leatherhead and demonstrated technical feasibility. Other components, however, not associated with the coal combustor, had serious problems and the project was abandoned. Research on this system has continued elsewhere, especially in the USA, where again key components are being developed separately. Options are being studied for a 250 MWth generator, which might have a 50% thermal efficiency (direct MHD plus steam cycle). It is recognised, however, that severe technical problems remain and a demonstration is not likely until well into the 1980s. With coal-fired MHD, a particular problem could be the separation of the seed from the particulates. However, it has been suggested that potassium carbonate could be used which would also combine with the SO_2 to form potassium sulphate from which sulphur could be recovered and the carbonate regenerated. By control of fuel/air mixtures at the different temperatures it is also believed that low NO_x could be obtained.

Heat pumps are an interesting possibility for "multiplying" energy, as well as being able to work in reverse as coolers. It is often assumed that heat pumps would be electrically driven but this need not be so and there are several possible outlets in these concepts for coal. Coal could be used, for example, in a PFB to drive a compressor, instead of a generator. Gas from coal could be used either in a turbine-driven mechanical drive or in an expansion cycle. Probably these systems would not be so convenient as electricity for small domestic situations but might be suitable for larger premises or district systems. The possibility of pipelining coal to a number of points in an area where it could be burned (possibly without attendance) in PFBs driving compressors for district heating by heat pumps (and possibly also generating electricity) might be investigated.

There are several heat engine systems of high thermal efficiency potential, such as the Sterling engine, where external firing is used. Again, coal in a fluidised bed might be convenient.

Several systems of submerged combustion, in which the oxidation of fuel takes place within a liquid phase have been proposed. One such system (Zimpro) is already used for destroying paper wastes by adding oxygen to a hot (200–350°C) aqueous solution, from which pressurised steam can be recovered. Crushed coal might be used in such a system. Problems of corrosion and obtaining adequate steam conditions may be formidable but gaseous emissions should be favourable.

The Atgas process, which injects powdered coal and oxygen below a slag-covered molten iron bed, was proposed for gas manufacture but could be adapted to the recovery of heat. One initial attraction was the fixing of sulphur in the slag, which can be efficient. However, there are obvious material and chemical engineering problems, especially for large-scale continuous operation (the steel making analogue works in short bursts and still has plenty of problems) and the process seems in abeyance. Another proposal is the "superslagging" combustor, in which coal would be partially oxidised under a lime-containing slag, which would be intended to fix the sulphur and release a weak gas for combustion, with secondary air, externally to the slag bath. Little real work has been done on this idea, which again would have formidable problems, especially materials and control.

One approach to improved combustion, perhaps with high thermal efficiency and environmental control in mind, is to use more than one combustion mode, each functioning within favourable circumstances. Several possibilities have been noted, including partial gasification plus char combustion and "topping up" PFB by gas combustion. In the FRG particular attention has been given to combined schemes; combinations of nuclear power with coal conversion have been noted, as has the Saarbergwerke scheme for AFBC and conventional boilers. The KDV process of STEAG (Fig. 6.27) has also been mentioned. Figure 6.28 shows a simplified version of a scheme by VE Westfalen AG. Partial gasification is used to produce a gas which will contain most of the sulphur and then be desulphurised. This gas is compressed and burnt in a turbine; excess air is used and the hot gases are passed to a combustor where the char, transferred hot from the gasifier, is burnt to produce steam. Attention is paid to the optimum use of heat values. Steps are in hand, through pilot and prototype stages, which could lead to an 800 MW commercial unit by the second half of the 1980s.

CONCLUSIONS

Combustion has been the main process for the utilisation of coal since the beginning of history and will continue to be of great importance indefinitely. Standard processes, such as pf firing, will continue for a considerable period, with further

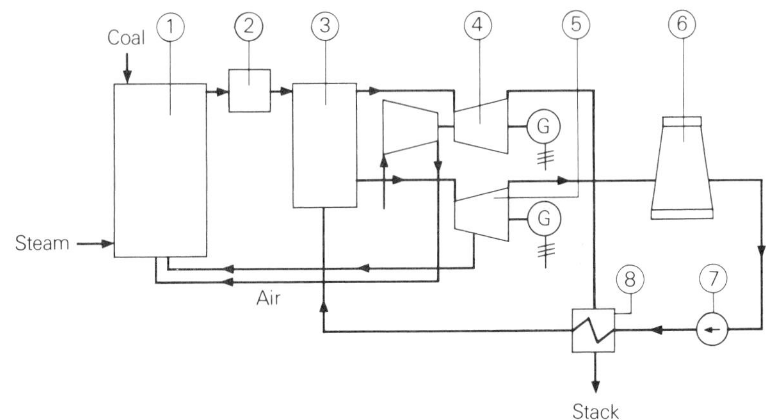

(1) Pressure gasifier
(2) Gas clean up, desulphurisation
(3) Pressurized boiler
(4) Gas turbine
(5) Steam turbine
(6) Cooling tower
(7) Feedwater pump
(8) Waste heat boiler

6.27 KDV-process of Steag AG

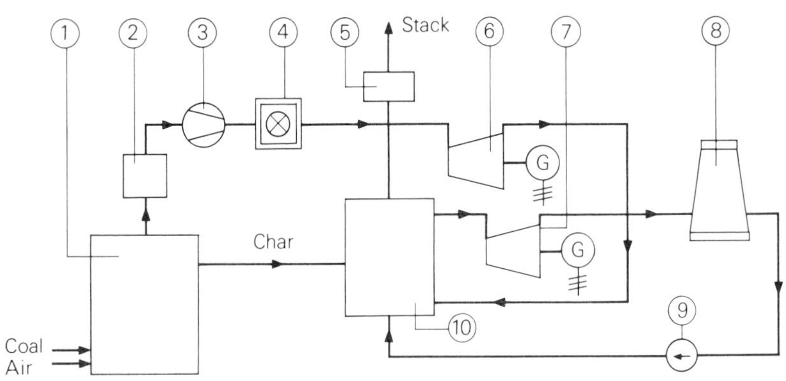

(1) Gasifier
(2) Gas clean up, desulphurisation
(3) Compressor
(4) Combustor
(5) Flue gas clean up
(6) Gas turbine
(7) Steam turbine
(8) Cooling tower
(9) Feed water pump
(10) Atmospheric boiler

6.28 VEW coal conversion process

development; newer processes, however, especially fluidised combustion in a variety of forms, will become more important within the last part of this century and may become the standard processes of the next century. They will be accompanied by greater attention to the use of low grade coals and wastes and to the role of combustion within the whole energy conversion system. This process of change may concentrate attention on smaller, low-load factor equipment, on co-generation and on waste heat utilisation. The impact of environmental issues will however be dominant; this is discussed elsewhere.

APPENDIX 1

Fluid Bed Combustion R&D Projects

Atmospheric pressure boiler

Organisation	*Location*	*Description*	*Size MWth*	*Objective*	*Status*
NCB	Leatherhead, UK	Vertical shell boiler	3.3	Operation and test plant	Operational since 1969
NCB/Combustion Systems Ltd./Johnston Boilers	Ferrysburgh, Michigan, USA	Locomotive type boiler	4	Commercial design	Operational since 1977. A number of commercial orders for this design now being completed
DOE/Exxon Research and Engineering Co.	Linden, USA	Industrial fluid bed boiler	4	Demonstration plant	Operation scheduled for 1979
DOE	Morgantown, USA	Three-cell fluid bed combustor	4	Test unit to provide information for building industrial plants	Operation scheduled to begin in 1979
NCB/Babcock (Shell Boilers)	Cheltenham, UK	Babcock Compo Boiler	4.3	Development boiler to prove new design	Commissioning 1980 for proving trials
NCB/Vosper Thornycroft	Cheltenham, UK	Vosper Thornycroft open hearth horizontal boiler	5	Development boiler to prove new design	Commissioning 1980 for proving trials
NCB/Vosper Thorneycroft	Edmonton, UK	Vosper Thornycroft vertical boiler	5	Commercial prototype field proving trials	Commissioned 1980 6 further boilers ordered
Northern Engineering Industries	Annan, UK	Horizontal shell boiler	5	Works test unit to obtain design data	Operational as a test unit 1978. (Two commercial units at 1.2 MW in manufacture for site installation 1980/81)
Ruhrkohle AG	Konig Ludwig, FRG	Coal-fired power plant	5.8	Demonstration plant for district heating	Construction completed. Plant commissioned
EPRI/Babcock & Wilcox Co.	Alliance, USA	Fluid bed boiler	7	To collect information	Operational
NCB	Newcastle-under-Lyme, UK	Locomotive steam boiler	10	Testing distributor design	Commissioned 1978

Energy Equipment	Birmingham, UK	Retrofit unit for water tube boiler	10	Commercial unit used for testing and demonstration	Operational 1979
Combustion Systems Ltd./Babcock & Wilcox Ltd.	Renfrew, UK	Babcock cross-type steam boiler	16	To obtain operating data	Operational since 1975. Results have confirmed advantages of FBC on commercial scale
DOE	Morgantown, USA	Fluid bed boiler, 0.1 t/h input	approx. 0.3	To burn anthracite waste	Operational
NCB	Cheltenham, UK	Hot water boiler	0.6	To obtain design data	Commissioned 1974, used as test installation
DOE/Battelle	Columbus, USA	Industrial fluid bed boiler	1	Demonstration plant	Operation scheduled for 1980
NCB	Dudley, UK	Conversion of horizontal shell boiler designed for conventional coal firing	1.2	To demonstrate the application of fluid bed combustion to horizontal shell boilers	Commissioned 1975, used as test installation to obtain data for commercial design
NCB	Peterborough, UK	Fluid bed test rig	1.5	Obtaining design data for 30 MWth boiler	Work completed, rig dismantled
Northern Engineering Industries	Derby, UK	Horizontal shell boiler	1	Works test unit to obtain design data	Operation as test unit 1978
Energy Equipment	Olney, UK	Horizontal shell boiler	1.5	Works testing before site installation	Operational as test unit 1979. Move to site with two similar boilers 1980
NCB	Bury, UK	Vertical shell steam boiler	2.5	Field trial demonstration. Used to develop alternative load control and start up systems	Commissioned 1977. Meeting factory heating requirements
NCB	Hereford, UK	Vertical shell hot water boiler	2.5	Field trial demonstration. Used to develop load control systems and test distributor designs	Commissioned 1977. Meeting site heating requirements
NCB	Cheltenham, UK	Horizontal shell boiler	3	Further development for retrofit on gas- or oil-fired boilers	Commissioned 1979. Used as test installation to obtain data for commercial designs

Organisation	Location	Description	Size MWth	Objective	Status
Stone Platt Fluidfire	Keighley, UK	Packaged water tube boiler	3	Demonstration and test installation on site	Operational 1980. Test programme
Whites Combustion	Widnes, UK	Vertical shell boiler	3	Works test unit to obtain design data	Operational 1979. Commercial units under manufacture
DOE/Combustion Engineering Co.	Great Lakes Naval Training Centre, USA	Two industrial fluid bed boilers	16	Demonstration plant	Plant under construction. Operation scheduled to begin 1980
Ohio dept. of administrative services/Babcock & Wilcox Ltd.	Columbus, Ohio, USA	Atmospheric fluid bed boiler, retrofit	18	Demonstration of Babcock Contractors design	Construction nearly complete. Commissioning 1980
AB Enkoepings Vaermeverk	Enkoeping, Sweden	Coal and oil atmospheric pressure boiler	25	High pressure water for district heating scheme	Plant operational since February 1978
NCB/ME Boilers	Sheffield, UK	High pressure steam unit using ME "coil" type boiler	30	Demonstration plant for commercial production of high pressure steam	Construction nearly complete. Commissioning 1980
DOE/Georgetown University	Washington DC, USA	Industrial fluid bed boiler	33	Demonstration plant	Commissioned 1979. Undergoing trials
Ruhrkohle AG	Dusseldorf, FRG Flingern	Coal-fired power plant	35	Demonstration plant for combined production of power and heat	Commissioned 1979. Undergoing trials
DOE/Foster Wheeler/Pope Evans & Robbins	Rivesville, USA	Multiple chamber fluid bed combustor	100	To collect information for the construction of a 200 MWe plant planned for 1980	Operational since 1977
Saarbergwerke	Volkingen/Fenne, FRG	Combined cycle fluid bed and pf fired boiler using air cooled bed coils to a gas turbine with supplementary coke oven gas firing. Flue gas desulphurisation plant in cooling tower. Also supplies heat to d.h. network	667	Large scale demonstration plant	Under construction

Pressurised combustors

Rolls Royce	Ansty, UK	Pressurised fluid bed operating with 0.1 MWe Rover gas turbine	0.3	To collect information on operation of pressurised adiabatic fluid bed combustor	Programme completed
EPA/Exxon	Linden, USA	Pressurised fluid bed unit — "Miniplant"	1	Test rig for combustion, heat transfer and acceptor regeneration	Test programme believed completed
Curtiss-Wright	Wood-Ridge, USA	Pressurised fluid bed air heater, coupled to a Rover gas turbine	2.3	Technology development unit	Operational
NCB	Leatherhead, UK	Pressurised fluid bed boiler	5	Tests on corrosion and erosion of turbine blades	Operational since 1970
DOE/Combustion Power Co.	Menlo Park, USA	Pressurised fluid bed combustion system	4	To test gas clean-up techniques and turbine blade erosion	Plant operated
DOE/Argonne National Laboratory	Argonne, USA	Pressurised fluid bed component test and integration unit	10	To collect information for operation on larger scale. Investigate gas cleaning techniques	Recently suspended
Bergbau-Forschung GmbH/Vereinige Kesselwerke AG (Fluid bed combustion group)	Bottrop, FRG	Pressurised fluid bed air heater with 2 MWe Sulzer gas turbine	22	To collect information for construction of a 100 MWe coal-fired power station using fluid bed combustion	Planning nearing completion
DOE/Curtiss-Wright	Wood Ridge, USA	Open cycle gas turbine plant with pressurised fluid bed combustor	43	Information for construction of demonstration plant	Plant to be operational in 1980
NCB(IEA) Services Ltd. (supported by UK, USA, FRG governments)	Grimethorpe, UK	Pressurised boiler, flue gases cooled in heat exchangers, expanded and removed via stack	85	To collect basic information for the operation and design of pressurised fluid bed combustors	Operation of test plant scheduled for 1980
America Electric Power/Stal Laval/Babcock & Wilcox Ltd.	Pennsylvania, USA	Open cycle gas turbine plant with pressurised fluid bed combustor	200	Demonstration plant for electricity generation	Design study

Organisation	*Location*	*Description*	*Size MWth*	*Objective*	*Status*
British Columbia Hydro/CPC	Vancouver, Canada	Air heater cycle	200	Demonstration plant	Conceptual design stage completed
Atmospheric pressure furnaces and incinerators					
NCB	Cheltenham, UK	Materials testing facility	0.1	For corrosion/erosion tests of metallic components	Operational
Babcock-BSH	Essen, FRG	Tailings combustor, 0.1 t/h input	0.2	To demonstrate combustion of pre-dried tailings	Operational
Fluidyne	Minneapolis, USA	Atmospheric pressure furnace	0.3	Test unit	Operational
DOE/MIUS	Oak Ridge, USA	Atmospheric fluid bed combustor heating air for closed cycle gas turbine	0.3–0.5	Test unit to provide information for the construction of commercial scale plant	Operation scheduled to begin in 1979
NCB	Cheltenham, UK	Fluid bed tailings combustor, 1 t/h input	0.7	To demonstrate combustion and drying of colliery waste material	Commissioned 1974, still in operation
CSIRO	Glenlee, Australia	Tailings combustor, 2 t/h input	1.5	Demonstrate combustion of minestone and, later, colliery tailings to generate power	Operational
DOE/Fluidyne	USA	Industrial fluid bed air heater	6	Demonstration plant	Design and development testing completed
NCB/G. P. Worsley & Co. Ltd.	Selby, UK	Hot gas furnace for crop dryer	5	Commercial prototype demonstration crop dryer of alternative design	Commissioned 1976, in commercial use
NCB/G. P. Worsley & Co. Ltd.	Newark, UK	Hot gas furnace for crop dryer	5	As above	Commissioned 1977, in commercial use

NCB/G. P. Worsley & Co. Ltd.	Okehampton, Thetford (now closed); Lincoln, UK	Three commercial furnaces	5	Third type commercial design	In commercial use
Energy Equipment	Essex, UK	Grass drying furnace	10	Commercial plant	Commissioned 1980
Energy Equipment	Nottinghamshire, UK	Grass drying furnace		Commercial plant	Commissioned 1980
NCB/G. P. Worsley & Co. Ltd.	Lincolnshire, UK	Clay drying furnace	15	Commercial plant	Commissioned 1979
Lurgi-Gesselschaft Fir Chemie und Huttenwesen GmbH	Frankfurt, FRG	"Turbulent layer" fluid bed furnace	33	Combustion of low-grade fuels	Operational
Ruhrkohle AG	Gneisenau Coll., FRG	Fluid bed combustion plant for burning flotation tailings	35	Prototype plant	Completion 1980
Deborah Combustion	Avonmouth, UK	Incinerator for acid tars		EEC demonstration plant	Commissioned 1980
Fluid bed incineration of waste					
Dorr-Oliver Inc.	USA, UK, France, Germany, Italy, Japan and South Africa	Three types of new fluosolids disposal system	0.1–14	73 Commercial plants	Commissioned 1970–1980
US/DOE	MERC (Morgantown), USA	Anthracite waste		Pilot plant	
Several	IGT, Westinghouse City University, others	Char from coal			
	Rocky Flats, California, USA	Chemical and low level radioactive wastes	0.005–0.1	Pilot plant	
US EPA, City of St. Louis, Union Electric Co.	St. Louis, USA	Residential solid waste	300	Full scale testing	Operation scheduled for 1977
Nippon Zein Co. Ltd., Japan Gasoline Co. Ltd., Japanese Government	Tokoyama, Japan	Waste tyres	1		Operational

Organisation	*Location*	*Description*	*Size MWth*	*Objective*	*Status*
Heenan Environmental Systems	Caernarfon, UK	Sewerage sludge			Operational
Dorr-Oliver	Esher, UK	Sewerage sludge (wet)	2.3		Operational
Hydrasposal/Fibreclaim Black Clawson Co.	Franklin, Ohio, USA	Solid refuse	6		Operational
Copeland Systems Inc. Copeland Process Ltd.	Several claimed	Pulping waste, pharmaceutical waste, distillery waste, chemical waste, municipal waste, plastic waste, agricultural waste, wood waste.		Commercial plants	Operational

Chapter 7

THE CARBONISATION OF COAL

INTRODUCTION

Carbonisation is the term used to denote the heating of coal in the absence of air. In these circumstances, volatile matter (VM) is removed. The volatiles are higher in hydrogen content than the basic coal, which therefore increases in carbon — hence, "carbonisation". As noted elsewhere, the off-gases can be useful; fuel gases and chemical feedstocks, especially benzene, can be separated. Carbonisation processes in the past sometimes had the production of these volatiles as a main objective. However, although the volatile by-products are still significant in the economic balance of carbonisation processes, the main incentive in recent times has been the physical, chemical and mechanical properties of the residue (or coke), mainly carbon, remaining after the carbonisation process.

On heating particles of coal, depending on their rank, they may become more or less plastic and swell as gaseous products are generated by the breakdown of the coal substance, resulting in bubbles or cenospheres. When finely ground coal is heated in bulk over about 400°C, it forms a coherent porous plastic mass, which can swell to fill a container larger than its original volume. On further heating and subsequent cooling the mass contracts and some cracking occurs but quite large strong lumps of coke can remain. Although less reactive, and therefore more difficult to ignite, than the input coal, the coke can be a useful domestic or industrial boiler fuel, which burns smokelessly. More importantly, however, the coke can be used in a blast furnace to reduce iron oxide ores to metallic iron; the coke provides both fuel and reductant (C or CO), as well as ensuring an open permeable structure in the active zones. Cokes are also used in other metallurgical processes, normally for reducing oxide ores to metal.

In this chapter, carbonisation will be discussed with particular reference to the properties of coals, and blends of coals, necessary for the manufacture of good coke. "Good coke" is of course coke which does a particular job satisfactorily and this assessment is an essential facet of the subject of carbonisation. Since manufactured briquettes can also be used to some extent as a replacement for coke, this subject is also treated here.

Pyrolysis is essentially similar to carbonisation, in that volatile matter is driven off leaving a residue (in this case a soft coke-like substance, often called "char"). However, since the volatiles, especially liquids, are the principal anticipated product, these processes are treated elsewhere.

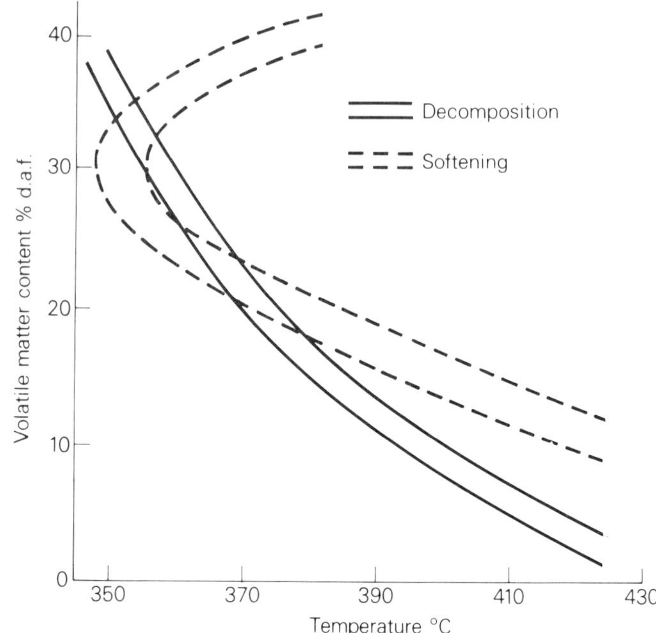

7.1 Volatile matter of coal in relation to softening and decomposition temperatures

SCIENTIFIC BASIS

When all the scientific factors which are relevant to the coking process are considered (not all are fully understood) it seems a minor miracle that it is possible to produce strong coke of a suitable size range on a consistent industrial basis. The most important phenomena are probably the decomposition of the coal structure and the development of plasticity. These factors, and others which play a part, are dependent on the rank of the coal and, since coal is heterogeneous, on the mixture of macerals present. In addition, the swelling (dilation) and contraction behaviour is both complex and critical; rank and maceral content are again involved.

The key relationship between softening temperature (plastic zone) and decomposition is illustrated in Fig. 7.1. Prime quality coking coals have volatile contents between about 20–32% (dmmf). They become plastic before active decomposition, appears to occur. When gases are evolved, the rate of devolatilisation and the viscosity of the plastic mass are such that relatively small, fairly evenly sized, pores are formed, with thick cell walls. At the same time, the individual neighbouring coal particles are brought into contact on all surfaces, much as balloons being blown up in a confined space; they are also able to adhere strongly. Outside the range of prime coking coals, the sequence and quality of events are less favourable. Gases may be evolved without adequate forces for intergranular bonding being available; alternatively, the gases may be evolved rapidly into a plastic mass of low viscosity. In the latter case, strong foaming may occur, resulting in fewer but larger, irregular pores and thin cell walls, which will have little abrasion resistance. Therefore, although there are other factors in the determination of coke quality, the basic

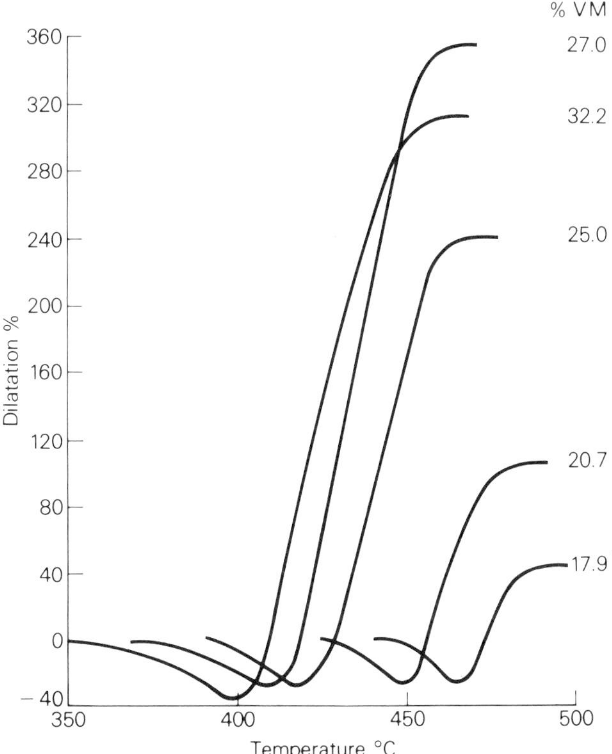

7.2 Dilatometer curves of coals of different rank

strength of coke, whether determined by the crushing strength of lumps or its abrasion resistance, is mainly determined in the plastic state and the early stages of resolidification.

Dilatometry is an important technique for studying coking phenomena. Figure 7.2 shows how, after an initial small contraction, coals in the coking range swell substantially around 400–450°C; both the temperature and extent of the swelling are strongly affected by the VM content. However, after the decomposition expansion, on further heating of the resolidified mass, contraction takes place but in a non-uniform manner, varying from coal to coal, and depending on the rate of devolatilisation and the re-orientation processes in the solid. It is normal to plot temperatures against coefficients of contraction (i.e. the instantaneous rate of contraction at that particular temperature) in order to show the phenomena more clearly. Figure 7.3 shows such plots for a range of coals. Two peaks are observed, the first at about 500°C, soon after resolidification, and the second in the region of 750°C. These are quite sharp changes and it must be remembered that they take place (in coke ovens) in a mass having high temperature gradients and thus substantial differential strains, which can lead to fissures in the cokes. Therefore, this contraction behaviour has been widely investigated (this has been a main research theme at the Coal Research Establishment in the UK). It has been observed that the porosity of the semi-coke remains approximately constant on further heating after

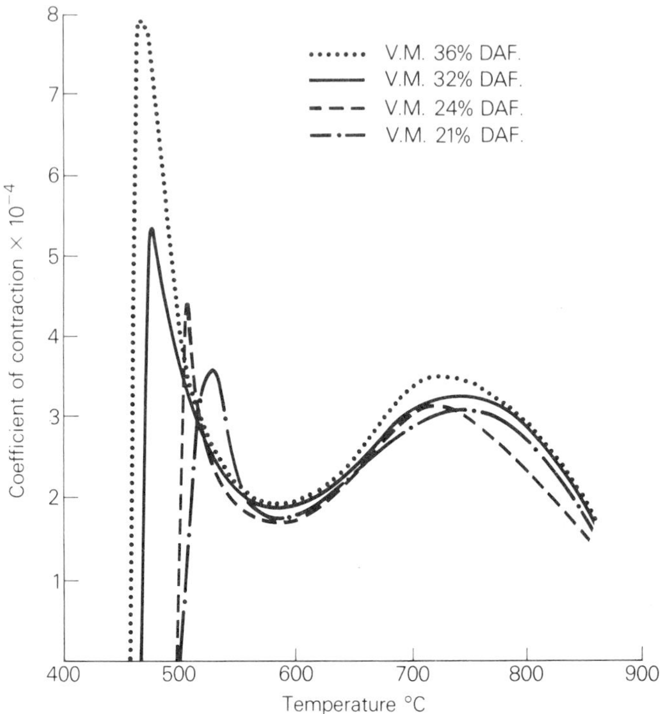

7.3 Coefficient of contraction of coals of differing volatile matter content

resolidification. Contraction must therefore be due to loss of mass and/or increase in density. Both are effective and their efforts are separable; one analysis is shown in Fig. 7.4. It seems that the first contraction peak is mainly due to a continuing high rate of mass loss soon after the plastic zone. This factor however becomes less important as the temperature is increased and the second peak is attributable mainly to the increase in true density during carbonisation, i.e. increased structural ordering. Figure 7.5 confirms that the first (and generally larger) peak is sensitive to VM content but the second is not. It has also been shown (Fig. 7.6) that the increase in true density is associated mainly with the increase in mean atomic weight of the coke, which results from the removal of hydrogen atoms.

In chemical terms, the plastic stage involves the breaking of cross-linkages in the coal structure; these comprise either oxygen or non-aromatic carbon bridges between neighbouring aromatic groups. The more mobile, lower molecular-weight components which are formed can decompose further to yield:

(a) gaseous volatile matter, consisting of methane and other hydrocarbons; and
(b) the highly complex mixtures found in coal tar.

The higher molecular weight fractions remain and form semi-coke on solidification, during which adjacent aromatic clusters join up through a free radical condensation mechanism. The predominant component of the gaseous product is hydrogen; the mechanism of its removal and the linking-up of the aromatic layers is suggested in Fig. 7.7.

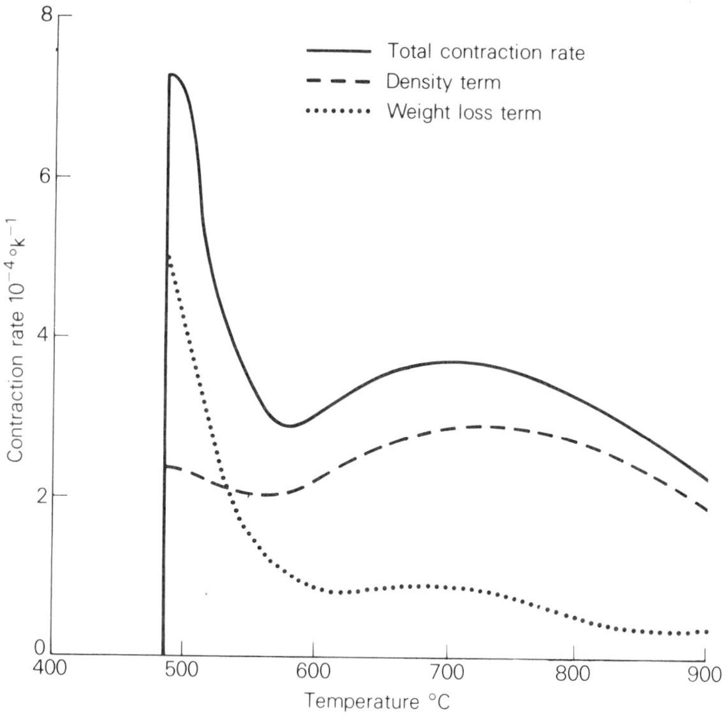

7.4 Analysis of contraction rates

In addition to studies on coals, the structure of carbons and cokes obtained by the pyrolysis of simpler hydrocarbon compounds has been studied to elucidate the mechanisms of the development of strength in coke. It appears that carbons fall into two classes: graphitising and non-graphitising, as indicated in Fig. 7.8. The non-graphitising carbons are deficient in hydrogen and/or rich in oxygen, a classification including coals of high VM content. They form a cross-linked structure not easily re-arranged to an orderly stacking pattern. In the case of carbons which can be graphitised readily (starting at about 1700°C), the crystallites retain mobility in the plastic stage and align themselves in a roughly parallel fashion. Such carbons, which are inherently rich in hydrogen and low in oxygen, include the prime coking coals. X-Ray evidence indicates that the degree of parallel stacking of the aromatic layers increases up to about 500°C, accompanied by a decrease in the proportion of amorphous material. Above 500°C there is growth of layer diameter; up to 700°C, there is also a decrease in stack height but above 700°C the stacks start to increase in height again (Fig. 7.9). It appears that the original (500°C) ordering is only indirectly important and that progress to a higher degree of order is only possible by a partial reorganisation of parallel stacks. However, it seems clear that at all stages the medium rank prime coking coals are more susceptible to the development of ordering than are either anthracites or low rank coals. How important this micro-ordering of the atomic structure is, compared with the macro-effects of cell wall and fissure structure, in the performance of coke seems uncertain but both characteristics may develop coincidentally.

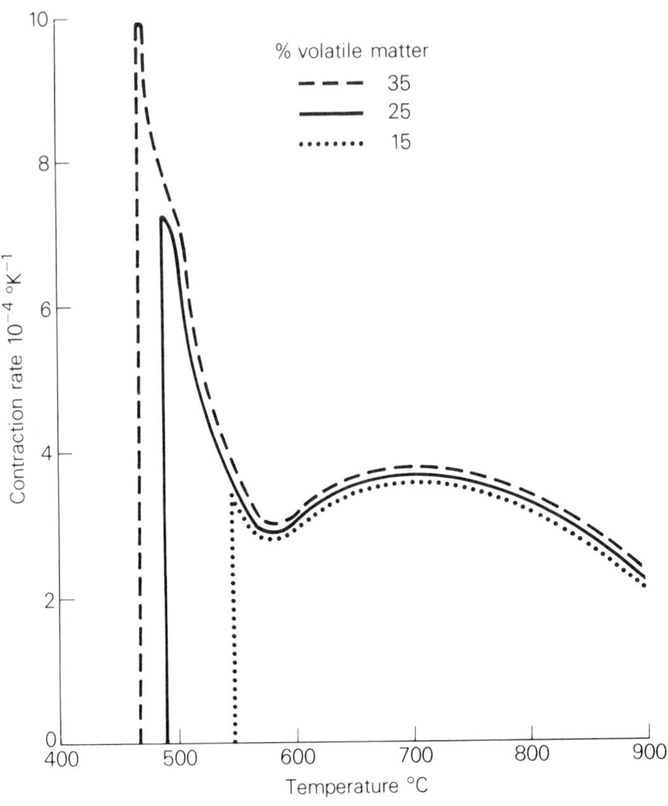

7.5 Variation of contraction rates with temperature

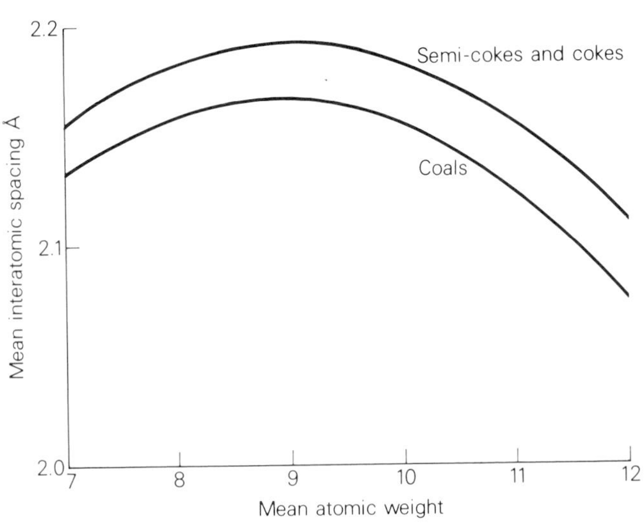

7.6 Variation of atomic spacing with atomic weight

7.7 Hypothetical condensation process leading to layer growth

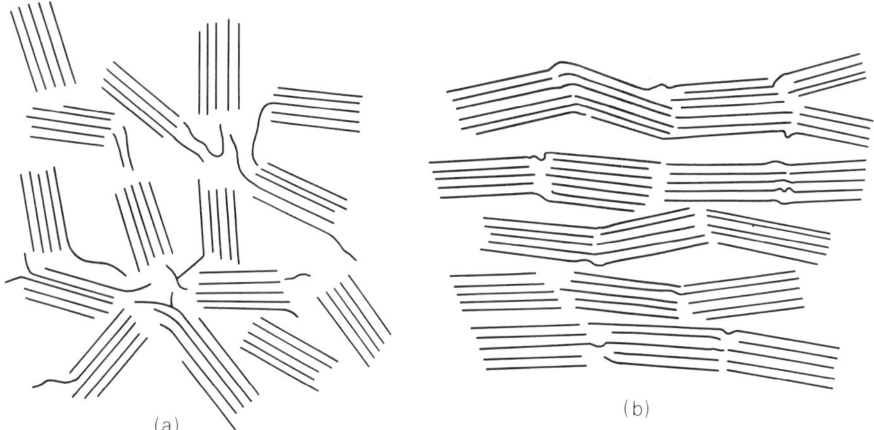

7.8 Arrangement of crystals in (a) non-graphitising carbon: randomly oriented crystallites containing hexagon layer planes, cross-linked, large pores and (b) graphitising carbon: more compact structure, near parallel arrangement of crystallites, little cross-linking, small pores

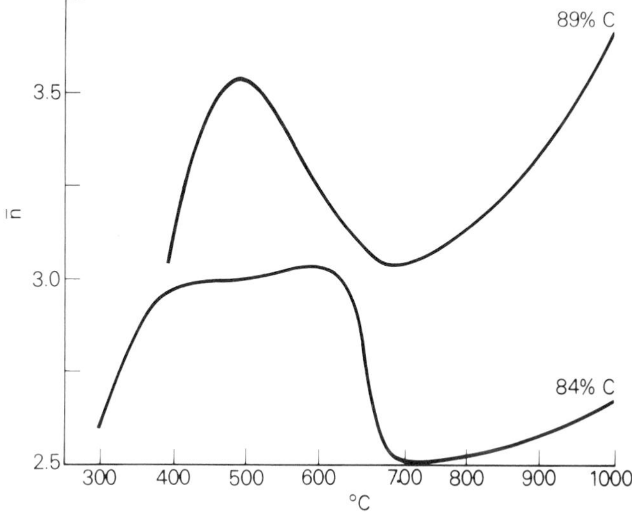

7.9 Variation of stack layer number with carbonisation temperature

SOME FEATURES OF COKE OVEN DESIGN

Most of the world's steel is made from iron produced in blast furnaces fuelled by coke. Steel-making methods which by-pass the blast furnace have been proposed and introduced to some extent; in turn, other fuels have been suggested for the blast furnace. However, in 1979 the world's consumption of coke amounted to about 350 Mt, most of which was used in iron-making in blast furnaces. It seems clear that whatever revolutionary steps may lie over the horizon, large quantities of coke will be made in slot type coke ovens and used in blast furnaces for a long time. Fortunately, in spite of the maturity, both coke ovens and blast furnaces, separately and even more considered in tandem, are still capable of significant technological improvements, through largely evolutionary processes.

Until early in the 18th century, iron smelting employed charcoal; resources of suitable wood were becoming strained at the same time as demand for iron was expanding rapidly. At this time, Abraham Darby demonstrated how coke could be substituted for charcoal and this discovery became a key factor in the industrial revolution — and the manufacture of coke was given great impetus. Initially, coke was produced in a fashion analagous to that used for charcoal — an earth-covered heaped-up mound set on fire and allowed to burn to the desired extent. This system evolved into brick structures, called beehive ovens from their shape, and persisted well into the 20th century. However, during the 19th century a parallel development took place which led to the modern type of "slot ovens". Initially, these were simple rectangular firebrick structures holding a layer of coal about 3 ft deep, heated by allowing some air in to burn the gases evolved. After coking, the end bricks could be removed and the coke raked out by hand. By the middle of the 19th century, in order to increase the rate of coking, the ovens were built as relatively narrow slots with external flues on either side and below to burn the gases and heat the ovens. Doors were added so that the coke could be pushed out by a ram. An important advance was to clean the gases so they could be supplied to each oven in a controlled manner and, finally, to preheat the combustion air in a regenerator system, using the hot exit gases. This development not only recovered waste heat, so that surplus gas could be used elsewhere, but allowed higher temperatures to be achieved in the combustion zone and thus in the oven walls, with consequent increases in coking rates. Ovens with their regenerators, are operated in pairs, with the flow reversed every 20–30 min (Fig. 7.10).

Oven widths have remained at about 16–18 inch, since coal and semi-cokes are not good conductors of heat and it is necessary to get the centre of the oven up to some minimum temperature, often 900°C, to avoid the degradation of quality. To achieve this, oven wall temperatures of up to 1200°C are normally required; all the high temperature areas are constructed of silica bricks, the remainder being fire brick or insulating brick. Recent developments have aimed at improved refractories leading to wall temperatures up to 1500°C. Economies of scale have been sought by increasing the length and more especially recently, the height of ovens. Between the wars, lengths of 40 ft (12 m) and heights of 14 ft (4 m) were common but in recent years lengths have increased somewhat (to the limit beyond which ram design becomes difficult) and heights to more than 20 ft (6 m). Batteries of ovens may contain up to 60 separate ovens. Each oven may be capable of coking several thousand tons of coal per year, or up to 20 000 t or more in the case of the new large ovens, operating on a cycle of about 15 h.

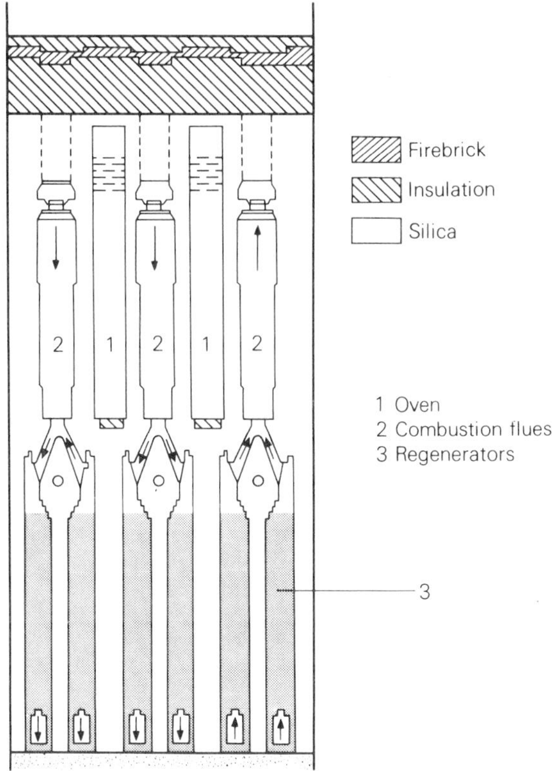

7.10 Flow of gases shown for coke oven gas firing

In addition to the end doors for pushing the coke through, each oven has up to five round openings like manholes along the top for charging crushed coal and there are off-take pipes at the ends for the volatiles. In addition to the ovens, however, a coking plant requires a large amount of other equipment and services. Figure 7.11 shows the layout of the basic coal/coke system; it will be noted that the coke is discharged into cars and then taken to a quench tower for cooling by a water spray before being screened and loaded. The most important ancillary unit is the by-product plant, which takes the off-gases and produces clean fuel gas, ammonia, crude benzole and coal tars. Other, increasingly complex, ancillary equipment is necessary or desirable, especially for environmental protection or improvement of the coking process.

COKE QUALITY

This subject has particular relevance to metallurgical coke — blast furnace and foundry grades. Domestic grades are generally produced at lower temperatures in order to increase their reactivity; other quality aspects include sizing (smaller than metallurgical but with minimum fines), consistency and low proportions of sulphur,

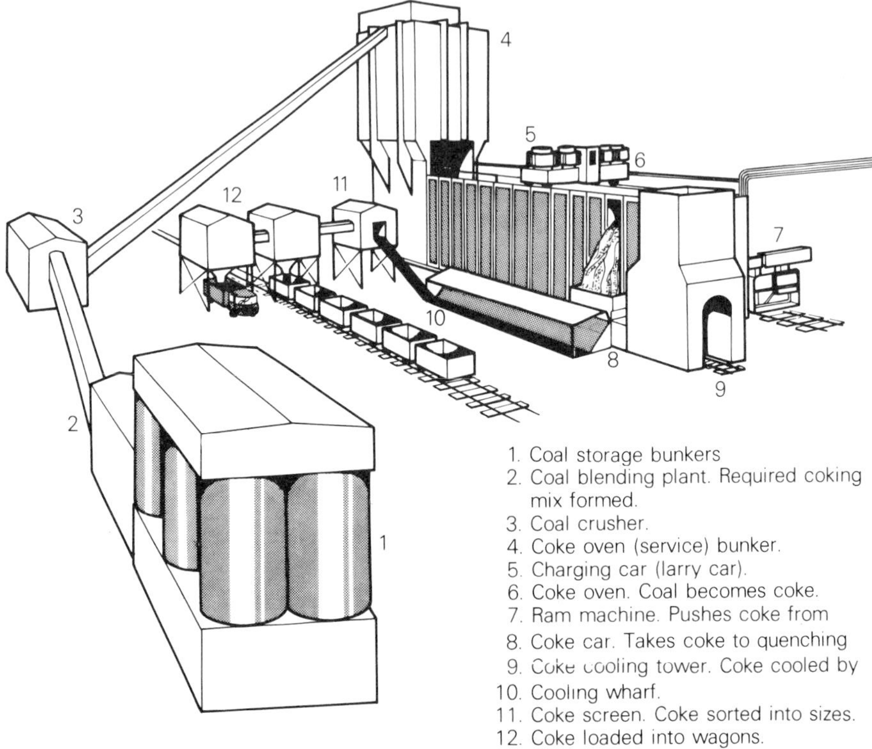

7.11 How coke is produced

moisture and ash. Brand names are often regarded as quality statements. Larger industrial-type boilers may be less sensitive to coke properties, especially reactivity, than small domestic boilers, since the former are likely to have more sophisticated arrangement for adjustment (especially draught control) but consistency and "value for money" (low inerts) are again important. Undersize metallurgical cokes are often used for these boiler fuel purposes.

Quality specifications and control for metallurgical cokes are quite another, and much more complex, matter. Primary attention must be given to blast furnace coke; foundry coke requirements may be explained by differentiation.

Over the last two or three decades, blast furnaces have increased considerably both in size and sophistication and these changes have made a high standard of performance of the coke used, always an important factor, absolutely vital — but still not yet exactly definable. In order to achieve economies of scale and improve thermal efficiency, blast furnaces may now be up to about 14 m in diameter, producing about 10 000 t or more of iron per day. In order to increase production rates and reduce fuel consumption (i.e. "coke rate" — the ratio of coke used per ton or iron produced) blast rates and temperatures have been increased, richer and better prepared burdens are charged and both oxygen and supplementary fuels usually oil may be injected. The latter, of course, whilst reducing the coke rate, substitutes one fuel for another; although the energy saving may not be significant, there may be economic advantages, but probably through improvements in furnace control. It is now not uncommon to use less than 0.5 t of coke per ton of iron, whereas it is not so long ago that double that ratio would have been more usual. Thus, in these huge furnaces with enormous weights of burden in the stack, the proportion of coke is much less than formerly but it is expected to perform an even more stringent task.

The coke is not only a source of heat and a reductant but acts as a refractory column supporting the burden; a high permeability must be maintained so that the reactions can proceed rapidly. At the same time, of course, the burden is descending with obvious abrasive activity, and the coke is also being burnt. There are three main critical stages in the life of coke charged into the blast furnace:

(a) Handling into the furnace. These mechanical systems inevitably impose stresses, both impact and abrasive. In some cases, coke is pretreated by being deliberately subjected to treatments especially designed to cause breakage along easily exploitable fissures and so "stabilise" its size, followed by screening to remove the highly undesirable fines. The difficulty of sampling the coke accurately as dropped into the furnace — a fairly severe operation in itself — adds to the uncertainty of establishing exact parameters of performance.

(b) From the top of the furnace to the bosh or melting zone. Here the coke is subjected to crushing and attrition forces and can also be chemically attacked. If coke reacts in this zone with CO_2, CO is lost to the furnace operation. In addition, chemical attack may make the cell walls excessively thin and fragile. If coke fines are formed, some will be blown out of the furnace, poor horizontal distribution across the stack will also result and affect even blowing.

(c) In the bosh, the coke is heated rapidly to about 1500°C, well above its carbonisation temperature. High thermal stability is necessary since here only the coke is solid in this zone and therefore determines the permeability of the burden. Coke lumps are burnt away completely at the tuyeres but if this final combustion generates excessive fines, these can cause difficulties by segregating at the centre and perimeter and also by increasing the effective viscosity of the slag.

Methods for testing coke were evolved before the recent developments in blast furnace operation and also in times when the availability of prime coking coals facilitated the production of good consistent coke; the relevance of a well-known coke to a particular blast furnace operation was generally well established on an empirical basis. Thus tests often served the purpose merely of showing that all was normal.

Standard tests, in addition to ash, sulphur and moisture (as low as possible) and sizing (close sizing, round about 50 mm), have generally involved the imposition of breaking stresses, followed by analysis by screening. The shatter test is performed by dropping coke under specified conditions but various rotating drum tests have tended to become more established. In these, the pre-screened coke is rotated in a drum with internal vanes so that the coke suffers a series of drops as well as self-abrasion. Two different types of measurements with different significance may be made in drum tests. In the micum test, for example, the M40 index represents the percentage of the original sample remaining *on* a 40 mm screen after a specified procedure; this measure is related to impact strength; the M10 index, on the other hand, represents the percentage passing *through* a 10 mm sieve after this treatment. The M40 index may be said to measure the resistance of the coke to major fractures, possibly due to the propagation of existing fissures. The M10, on the other hand, measures the resistance to abrasion. Both fracture modes are clearly relevant to the conditions experienced in the blast furnace and, when a procedure for satisfactory coke for a particular purpose has been established, these tests may be adequate to monitor quality. Furthermore, the drum tests may be elaborated or extended in experimental procedures to study the factors affecting coke strength and breakage mechanisms.

However, when new sources of coals for coking blends have to be explored, or new more economic methods of operation are being sought for coke ovens or blast furnaces, it now seems apparent that further testing procedures, more relevant to the high temperature portion of blast furnace coke behaviour, are needed. A number of permutations have been suggested, generally involving thermal shock on heating or solution attack in CO_2 at high temperatures, followed by either drum tests or bulk permeability changes.

The effects on coke reactivity of impurities, both metal oxides and alkali metals, is a matter of concern but not yet of complete understanding. In Japan, it is common, as a regular blast furnace control procedure, to react coke with CO_2 at 100°C and, after determining the weight loss, test the remaining product in a standard drum test. In the UK, at the Coal Research Establishment, emphasis has been given to thermal stability, as a result of consideration of the rate of heating up to about 1500°C occurring in the blast furnace bosh area. Tests have been devised for heating bulk samples of coke, at a rate appropriate to blast furnace practice, to 1400°C followed by a study of mechanical properties thereafter. Some interesting differences between cokes, not otherwise detected in ambient temperature tests, have been revealed. This work has also drawn attention to a third contraction peak in cokes, which occurs at temperature above those in the coke oven. Figure 7.12 shows the extent and temperature location of this peak and how it is dependent on the final coking temperature.

It has therefore become clear that it is not possible to predict the thermal behaviour of coke merely from the ambient temperature tests. However, it has also been shown that blend control designed to optimise properties at ambient temperature may also serve to control coke quality as determined in high temperature tests. Unfortunately, although some partial attempts have been made, it has not yet been possible to carry out rigorous scientific tests relating the performance of large modern blast furnaces to specific measured coke properties. Until this can be done, it will not be possible to establish a sound scientific basis either for coke quality or for standard quality control methods for coke. Such large-scale blast furnace trials would of course be extremely expensive and technically difficult but the cost needs to be related to potential economic advantages.

Cupolas are used for melting iron in foundries for making castings. These cupolas are much smaller and simpler than blast furnaces; they are also operated batch-wise, with a cycle of a few hours. However, the coke quality is still a vital factor, perhaps even more so than in the blast furnace. For one thing, since the iron does not have any further refinement before being made into castings, coke impurities affect the product directly and severely; low contents of ash and sulphur (and often phosphorus) are firm requirements. Iron temperatures, and the consequent level of uptake of carbon, are vital and related (somewhat empirically) to coke properties. Since the cupola has little scope for adjustment, regularity of coke quality, including moisture, is also necessary. Finally, since large lumps of metal are charged, the coke must be both large (about 100 mm) and even more resistant to shatter than blast

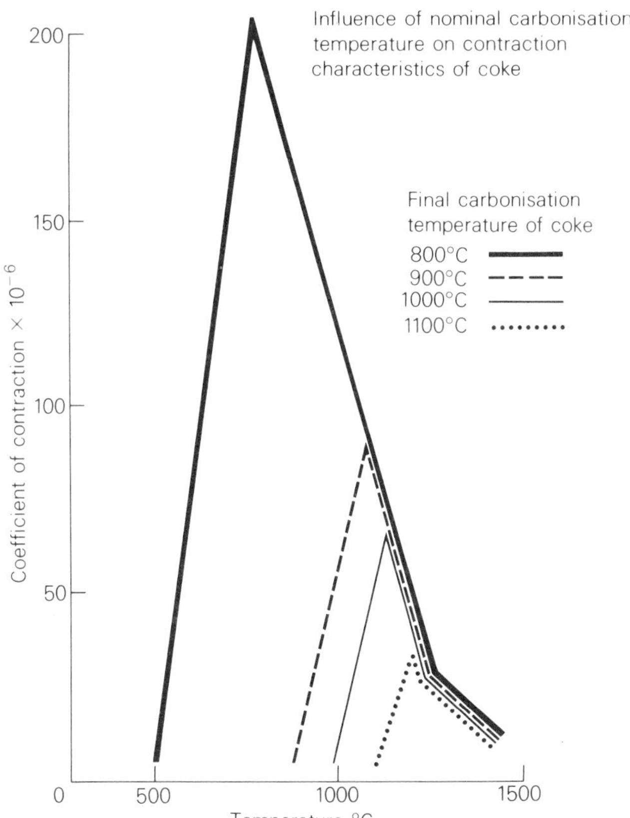

7.12 Influence of nominal carbonisation temperature on contraction characteristics of coke

furnace coke. These properties are obtained by careful selection of coals, with low impurities and very good coking characteristics (which may be assisted by blending) and by operating the ovens at a high temperature and with a long cycle to ensure maximum carbonisation.

For the manufacture of special irons or ferro-alloys in ore furnaces or for zinc/lead blast furnaces, particular grades of coke are specified to suit the purpose. Impurities are generally required to be at a low level and the coke size may be specified to be smaller than foundry or blast furnace coke. Often higher reactivity is desired but consistency and freedom from fines are inevitably key factors.

COKE IMPROVEMENT: BLEND FORMULATION

The shortage of prime coking coals, with the consequent desire to enlarge the range of coals suitable for metallurgical coke, and the trend towards more stringent coke operational requirements have led to considerable research and development. Some of this work has been directed towards coke oven operational factors, e.g. preheating or partial briquetting of the charge. Other work has been aimed at blending coals so that mixtures behave as prime coking coals, i.e. good strong coke can readily be made from them.

Although blending techniques have existed for a long time, they were developed mostly on an empirical basis until the last decade or so, during which reasonably successful attempts have been made to use scientific analysis of the coking process to provide substitutes for nature's bounty — prime coking coals. The first requirement is that the blend should swell enough to give good fusion between particles. It has been established that a minimum dilation of 40–50% is necessary to achieve this in normal oven charging conditions, i.e. crushed to an upper size limit of 3 mm and charged wet. These conditions correspond to a bulk density of 720 kg/m^3, which has a voidage of about 45% (about the amount taken up by swelling). The minimum dilation could be reduced at high charge densities — e.g. 20% at 850 kg/m^3. This higher density can readily be achieved in laboratory conditions by ideal particle size distribution but this is difficult in practice. One method of densification is "stamp charging", in which the coal charge is mechanically compressed into a cake which is transferred as a whole into the oven. This method has been established in a few places for some time but severe environmental and labour problems have restricted its use. Methods more suitable for modern conditions are drying or pre-heating the coal (discussed below); higher bulk densities are achieved, as are higher heat transfer rates, a further factor in minimising the dilation which is necessary to achieve adequate fusion and the production of good coke.

After achieving adequate fusion pressure, coke properties are considered to be related primarily to the post-swelling contraction behaviour. The first, and largest, peak in the contraction co-efficient curve gives rise to differential stresses, bearing in mind the high temperature gradients across the semi-coke mass, and these result in fissuring. This defect structure in turn determines both the size distribution of coke pieces and the bulk strength (which is strongly affected by macro-fissuring). High contraction rates result in small, weak coke; a maximum value of 6×10^{-4} has been suggested, corresponding to a maximum VM value of 32% (under typical British conditions). The original VM has little effect on the second contraction peak but the addition of non-contracting inert particles can mitigate its effects. Also, when finely divided, inert particles can relieve stresses in a manner which does not contribute to crack propagation; this effect may be most evident in creating resistance to abrasion. Naturally occurring mineral particles can have such an influence but small additions of finely ground (<0.5 mm) coke breeze can be added for positive control.

Finally, if the blend were to have an apparently suitable average VM content but achieved by mixing two coals of widely different rank, the swelling zones might not overlap sufficiently and one component might remain solid while the other fuses, or inhibits its fusion. This problem must be overcome by adding a suitable amount of a "bridging" coal, typically a prime coking coal, which is compatible with all the other components and facilitates homogeneous softening throughout the charge.

Blend formulation may be illustrated in Fig. 7.13, which shows the combinations of VM content and total dilation required to obtain cokes having M40 indices at particular levels. Figure 7.14 shows the "target" area again, which can as shown be extended downwards and to the right if the charge density is increased, say by preheating or stamp charging. The principle of "bridging" is also illustrated. Mixtures of coals A and B can readily come within the target area. Coals C and D require the addition of coal E (10%), unless they are previously densified.

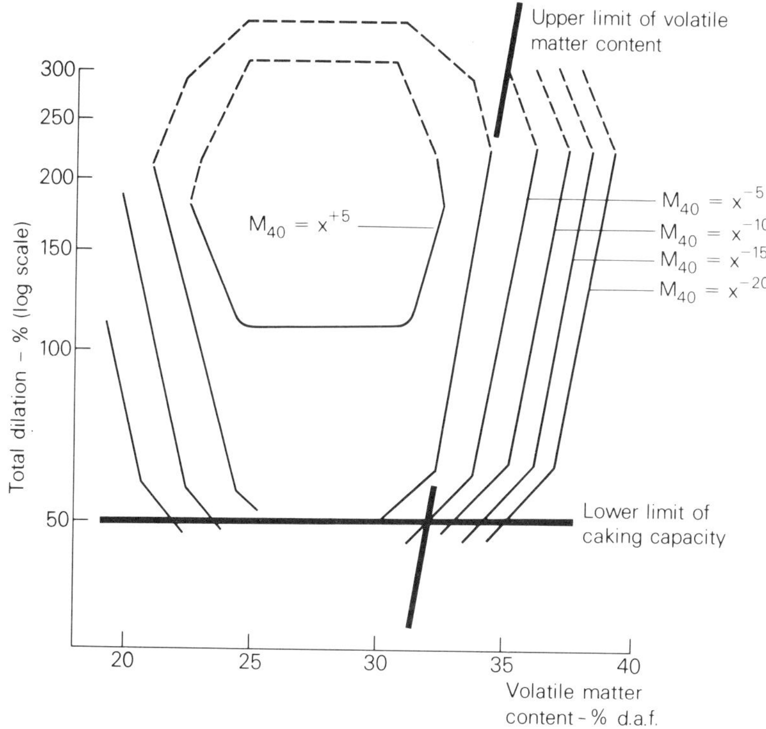

7.13 M40 indices of coke versus total dilation and volatile matter content of charge

OTHER CARBONISATION PROCESSES

As discussed earlier, most of the coal which is now carbonised is processed in slot ovens at 900°C or more to produce primarily metallurgical coke ("high-temperature carbonisation"). Formerly, however, a wide range of carbonisation processes was employed and some still have importance or may have some interest in relation to future developments. Since gas or tar liquids rather than hard coke may be the desired end-product, coals with a significantly higher VM content than prime coking coals are frequently used. "Low temperature carbonisation" may be carried out at 450–700°C, giving a high yield (70–80%) of reactive coke useful as a smokeless domestic fuel and also a high tar yield (7–10%). This is because the VM comes off without the substantial chemical cracking which takes place in the presence of hot coke in the region above 700°C. This cracking is exploited when gas, mainly hydrogen and methane, is required together with reactive coke in "medium temperature carbonisation" (700–900°C). In the latter case the solids yield is only about 55–60% and the tar yield about 5%; however, about 80 therms of gas per ton of coal may be recovered compared with only about 33 in the low temperature processes. Most of the processes and products have proprietary names. In addition to domestic smokeless fuels, special reactive cokes for reduction processes may be produced.

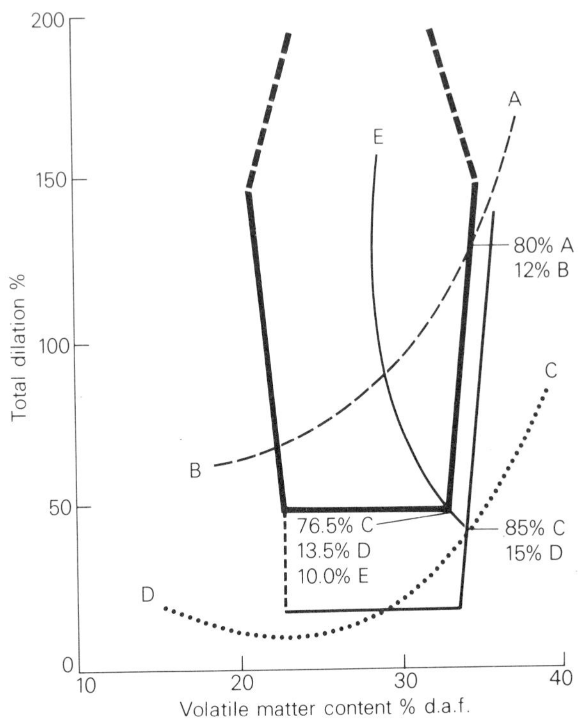

7.14 Blend formulation

A wide range of equipment has been used. In addition to variations of slot ovens, cylindrical vertical retorts are common. Heating in these cases may be arranged by allowing some of the coal to be burnt or by passing hot gases from an external combustor through the charge. Operation may be batch-wise or continuous; in the latter case lock hoppers are required for charging and discharging. Some devices for mechanically transporting the coal through a heating zone, by means of moving circular rotating hearths, conveyors or sloping horizontal kilns have been employed; the equipment is often derived from ore processing, cement or refractories plant. In some cases, sized lumps of coal are charged and retained as individual pieces as far as possible; in other systems, agglomeration is required.

In the UK, in addition to normal slot ovens operated with coals and carbonisation conditions suitable for reactive cokes, other processes are operated commercially, especially for smokeless domestic fuel. "Rexco" is made from sized weakly-caking coal in a cylindrical vertical retort (3 m diameter and 7.6 m high). Heating (to about 700°C) is mainly achieved by introducing hot gases from an external combustor; to complete carbonisation, cool gases are introduced to the hot portions and transfer heat to the remainder. Tars are not produced. A continuous vertical retort has also been developed. "Coalite" is made by agglomerating crushed coal in batches of tapered vertical tubes (11 cm at the top and 15 cm at the bottom) heated to about 650°C by radiation from interspersed combustion chambers. By-products are produced and are quite important to the economics.

There are a number of processes designed for making briquettes from fines, which may be bitminous coal, anthracite, coke breeze or mixtures. Differences lie in the feedstock and binder used, the method of shaping the briquettes and the type and sequence of heat treatment/carbonisation. A later section on formed coke deals with the production of briquettes for blast furnaces. For domestic (usually, but not always, smokeless) briquettes, several processes are used, generally on a small scale to exploit some local feedstock source or demand. Two processes in the UK are operated on a substantial scale and are particularly interesting technically: "Phurnacite" and "Homefire".

Homefire is made by a "char briquetting" process. Char can be produced by partial or complete devolatilisation of crushed coal (up to about $\frac{1}{4}$ inch in size). Devolatilisation may conveniently be carried out in an entrained process or in a fluidised bed. The latter system is used for Homefire, with a mean residence time of about 20 min at about 425°C. Under these conditions, the VM content of the feed coal is reduced to about 20%, a level at which briquettes can be made without addition of binder and which burn smokelessly. If organic binders were used, of course, the smokey products of these would have to be removed. However, it was discovered in the 1950s that the char, produced in the conditions stated, when pressed immediately at the reactor temperature can be formed into handleable briquettes requiring no further heat treatment. In one char briquetting process ("Roomheat", now in abeyance) roll pressing was used. In the Homefire process, hexagonal briquettes are extruded in a reciprocating hydraulic press. The input material is a fine, high-volatile coal, unsuitable for any purpose other than power station combustion; the product is a high grade premium open-grate fuel. Consideration has been given to further heat treatment of the briquettes for metallurgical use but, as produced for domestic fuel, they may contain too many fissures and be too reactive. However, there may be some potential for adaptation of the char briquetting process for a wider range of products.

"Phurnacite" is a product of a medium temperature carbonisation process. Other processes of this type were generally developed mainly for the manufacture of town's gas and have become obsolete. "Phurnacite" uses, as starting material, fines from low VM coals; such fines, produced in relatively high proportion in mining these hard coals, have reduced value unless formed into briquettes. The fines are bound by pitch, roll-pressed into small briquettes (about 30 mm ovoids) and heat-treated up to 750–800°C in special tall slot ovens. These ovens are very narrow (200 mm) because of the poor heat transfer properties of a mass of briquettes, which necessarily have a high voidage; as usual, there are vertical heating flues between ovens. The ovens have a steeply sloping bottom (Figs 7.15 and 7.16) and are discharged through a door at the bottom, with a cycle time of about 4 h. Tar and other by-products are recovered; the gas, after cleaning, can be used to heat the ovens. Phurnacite is a high grade domestic boiler fuel in strong demand. The briquettes have high strength and perhaps with further heat treatment, might be considered for blast furnace purposes. As presently built and located, there are limitations to production on environmental grounds; expansion would also be restricted by supplies of local low VM coals. However, the experience over many years indicates the possibilities of reconstituting low VM starting materials (natural coals, chars, coke breezes, etc.) into strong briquettes. Although the market for solid domestic fuels has declined in many places, this situation might be reversed as other energy sources become more expensive. Furthermore, the role of briquettes in industrial processes may have some potential.

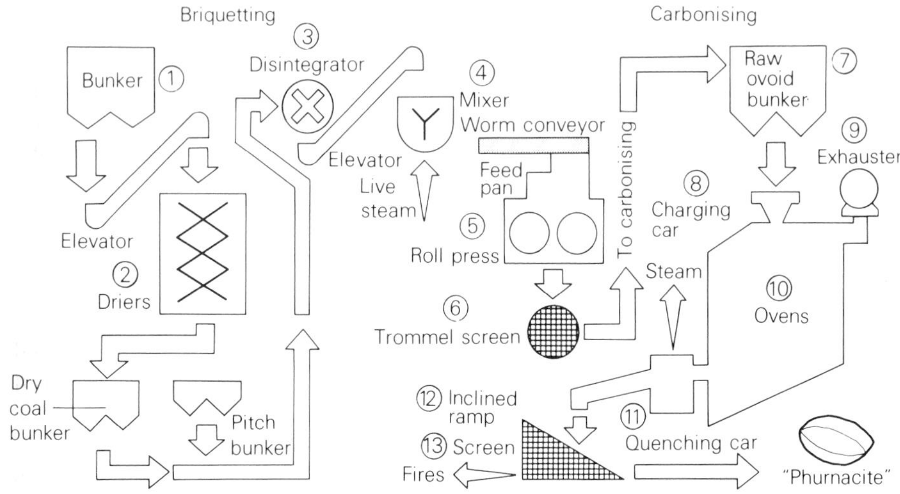

7.15 Flow diagram of the "Phurnacite" production process

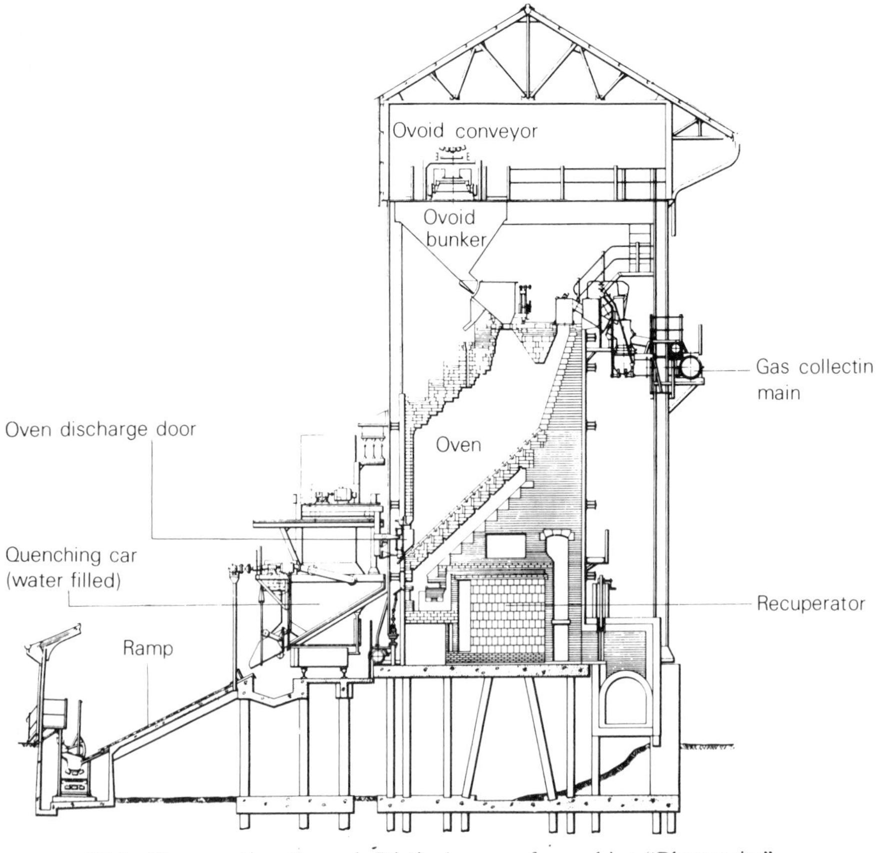

7.16 Cross-section through Disticoke oven for making "Phurnacite"

FORMED COKE

The term formed coke is applied particularly to briquettes intended for blast furnace use. A wide range of processes has been proposed and investigated to some degree. Some are derivatives of processes for domestic fuel; others were designed initially to manufacture blast furnace feedstock. Input coals may be oxidised or carbonised; sometimes more than one coal, differently treated, may be required. In some cases balls may be made by pelleting (rotating the particles under agglomerating conditions on a saucer-like container); mechanical pressing is more common. A distinction may be made between those processes where the briquetting is carried out cold, using binders (often derived from the process), and those carried out "hot" — generally in the softening range of one of the coals. Some of these processes are listed in Tables 7.1 and 7.2, for cold and hot processes, respectively. Carbonisation must be carried out on cold formed briquettes but not necessarily on hot formed briquettes. One proposal is that briquettes of medium strength (generally because they have not been heated to high temperatures) may be incidentally heat-treated and hardened in the top portions of the blast furnace itself; this seems however to rely unduly on providence.

The selection of a process is a complex matter and may not be resolved without making assumptions about future capital and operating cost factors and about the long-term relative values of various types of coals and of by-products. Cold briquettes may be comparatively simple and flexible with regard to feed coal quality but binders are needed and these could become scarcer and more expensive. Furthermore, cold formed briquettes will probably need more costly carbonisation procedures. Hot-formed briquettes may have cheaper carbonisation stages, or none at all; however, they will probably need a proportion of good coking coal and they may be more sensitive and difficult to operate. It has been suggested that pellets may lack strength and size uniformity or have too much moisture; steps have been proposed, apparently feasible, to mitigate these problems.

Direct heating in various forms, is generally preferred; it is simple and flexible and suitable for continuous processes. However, by-products are of low quality and contain dust. Indirect heating gives better by-products and although less flexible than direct heating it is used in a Japanese process.

The large number of processes proposed is partly due to varying local circumstances (feedstock supplies, experience with techniques for other purposes) but may largely be a result of a period of great enthusiasm, which stemmed from the obvious disadvantages of slot ovens — limitations on coal quality, pollution problems, increasing costs and difficult working conditions. Looked at from the opposite point of view, advantages have from time to time been claimed for formed coke, including:

(a) Lower capital and operating costs;
(b) Opportunities for automation and improved working conditions;
(c) Environmental improvements;
(d) Higher coke yield and thermal efficiency;
(e) Flexibility of output to match demand;
(f) Simpler maintenance;
(g) Wider range of suitable coals, some of which may be cheaper;
(h) Improved blast furnace operation due to regularity and predictability of shape, size and other properties (adjustable to suit blast furnace operational requirements).

Most if not all of these potential advantages, still to be proven, are translatable into economic terms, discussed in a later chapter. At this point, it may suffice to point out that substantial improvements in many of these areas have recently also been achieved or may be expected in the operation of slot ovens. Also, those factors which depend on blast furnace performance — and to some extent on the output and reliability of formed coke plants themselves — remain to be determined, since blast furnace trials so far appear to be limited in scale and/or conviction. This is related to the fact that such trials are extremely difficult and expensive to arrange. This applies also, of course, to trials of coke other than formed coke. Some tentative conclusions may nonetheless be drawn:

TABLE 7.1 Cold Briquetting Processes

Process	*Features*	*Raw materials*	*Comments*
INIEX	Briquettes carbonised in 2-stage fluidised sand bed to 550°C, then to 900°C in shaft oven with gas cooling.	90% low volatile coal 10% coking coal. Pitch.	Product tested in small blast furnace. 40 000 t/a pilot plant at Liege 1968–1972 without by-product recovery. No plant now operating (but see Auscoke process).
HBNPC (France)	Single stage carbonisation of briquettes to 1000°C in Fuvo-Erim shaft oven. Gas cooling.	Wide range of coals. Weakly caking non-swelling base coal (or strongly-swelling coals with preliminary treatment). Additive coal with plastic and swelling properties. Pitch.	Maximum productivity with low volatile coals. Product tested in small and large blast furnaces. Pilot plant in N. France treats 150–170 t/d of green briquettes. By-products are tar and low CV gas.
Formed Coke Ltd. (UK)	Cold pelletisation followed by carbonisation in rotary, annular kiln. Gas cooling.	High volatile, weakly- or very weakly-caking coal. Preferred binder is sulphite lye.	No blast furnace tests. 72 t/d pilot plant (without by-product recovery) in existence.
I Ch PJ (Poland)	Lump coal carbonised at 750–800°C. Char cooled, crushed and briquetted with 10% binder. Briquettes hardened by oxidation at 200–800°C and water-cooled.	High volatile weakly- or non-coking coal. Binder is 70% recovered pitch and 30% heavy oil.	Process designed to produce foundry coke. Blast furnace tests unpromising. Commercial plant (800 000 t/a in operation since 1960).
ICEN (Romania)	1/3 of base coal carbonised in fluidised bed. Remainder blended with char, 10–20% binder coal plus 10–18% binder, and briquetted. Briquettes carbonised to 950–1000°C in shaft oven.	Base: Weakly- or non-caking coal, 40% volatile matter. Binder coal; medium-caking gas coal. Binder: pitch.	Product tested in blast furnaces and cupolas 0.5 Mt/a plant planned.

TABLE 7.1 (continued)

Process	*Features*	*Raw materials*	*Comments*
FMC (USA)	Crushed coal oxidised and carbonised to 850°C in multi-state fluidised beds with tar recovery. Cooled char briquetted with 15% tar. Briquettes pretreated at 250°C carbonised at 850°C and gas-cooled.	Wide range of coals. No additional binder needed if volatile matter content is greater than 35%.	Product tested in small and large blast furnaces. 85 000 t/a pilot plant in Wyoming. Only by-product is low CV gas.
Clean coke	Complex process. Char from high-pressure fluidised bed cooled and pelletised with process-derived oil. Pellets oxidised and carbonised to 980°C.	Illinois coal with 3% sulphur and 15% ash.	Laboratory scale only. Aim is to produce blast furnace coke, chemicals, and low-sulphur liquid and gaseous fuels.
SKU (Japan)	Coal briquetted with 10% binder. Carbonisation in indirectly-heated Didier-Hellog ovens. Product water-cooled.	5–10% coking coal. 70–85% low volatile coal. 0–10% coke breeze. Pitch and tar.	Product tested in blast furnace. 45 000 t/a pilot plant in existence. Commercial plant reported to be under construction. By-products similar to those from conventional coke ovens.
Auscoke (Australia)	Modified version of the INIEX process. Final carbonisation to 1100°C.	Weakly caking low- and medium volatile coals.	Blast furnace trials planned 100 t/d pilot plant at Port Kembla.

(a) Formed coke has not led to major problems in blast furnace performance. In some cases, operations have been exceptionally regular due to the smooth descent of the burden.
(b) Productivity in blast furnaces with formed coke has been good, often better than normal, although the tests have either been in small furnaces or of short duration. Due to the better packing and therefore lower permeability, there has sometimes been insufficient blowing power to press output rates to the limit.
(c) Fuel consumption comparisons are not significant. Some dust losses in off-gases can result from abrasion with the softer type of briquettes and with green briquettes a higher yield of gas is obtained (not desirable). Briquettes are generally more reactive and this can result in "solution-loss" (excess carbon monoxide) and in changes in the balance of reactions occurring in the different zones.

It is not simple to generalise about formed coke properties or to relate these to oven coke, since the products are rather different. However, in general, formed coke will probably have relatively high crushing strength but lower abrasion resistance and higher VM content. To some extent these latter two properties result from a desire to simplify and cheapen the process and could probably be mitigated if shown

TABLE 7.2 Hot Briquetting Processes

Process	*Features*	*Raw materials*	*Comments*
BFL (Fed. Rep. of Germany)	Base coal carbonised at 700°C in LR gasifier. Char blended with binder coal and briquetted at 400–500°C. Briquettes may be used as made (green) or carbonised before gas- and water-cooling.	High volatile, non-coking base coal. Coking binder coal. By-product tar may be recycled to reduce binder coal requirements (and to increase coke yields).	Green and carbonised briquettes tested in blast furnaces. Pilot plant facilities at Essen. 100 t/d at Essen until 1972 300 t/d (green briquettes) at Bottrop. 600 t/d (calcined briquettes) constructed at Scunthorpe (development suspended). By-products: low-temperature tar, high CV gas.
Ancit (Fed. Rep. of Germany)	Base coal flash-heated to 600°C and blended with preheated binder coal. Blend briquetted at 450–520°C. Briquettes used green or maybe hardened by heat soaking before air- and water-cooling.	Low volatile base coal and 30% coking coal as binder. Higher volatile base coals can be used after mild heat treatment.	Product tested in blast furnaces. 10 t/h pilot plant at Alsdorf. 30–50 t/h plant proposed.
Sopoznikov (USSR)	Coal flash heated in multi-stage heater and briquetted at 400–500°C. Briquettes carbonised to 800–850°C and gas-cooled.	Weakly-caking high volatile coals, sometimes blended with weakly-caking coals of 20–28% volatile matter.	Product tested in blast furnaces. 240 t/d pilot plant at Kharkov. Commercial plant (0.5 Mt/a by 1973 and 1.5 Mt/a later on) under construction in Siberia.
Consol (USA)	Char from fluidised bed carboniser pelletised at 450°C in revolving drum with added binder. Pellets carbonised at high temperature.	High volatile coals. Coking coal or pitch as binder.	500 t/d plant in Maryland. Large-scale blast furnace trial planned. By-product pitch may be used as binder.

to be technically necessary and economically desirable. The smaller size, together with the usual ovoid shape and regularity, give closer packing and lower permeability in the blast furnace. Both the lump density and bulk density are higher. Formed coke is generally more reactive than oven coke and may have a fine pore structure which can result in a higher moisture content. In summary, it seems that both blast furnace operational procedures and the properties of briquettes may need simultaneous re-optimisation if the best results are to be obtained. This process may take quite a long time.

MODERN DEVELOPMENTS IN SLOT OVEN PRACTICE

A number of developments have been initiated in the design of coke ovens in recent years, mainly to improve reliability and productivity. However, more recently, greater attention has been given to optimising the coke oven operations, with particular reference to the quality of coke, especially from coal blends which minimise the scarce expensive components. Two important lines of development both work mainly through the densification of the charge in the oven: preheating and partial briquetting.

In the case of pre-heating, the initial objective was to dry the coal in order to take some of the heating load off the coke oven and thus reduce carbonisation cycle times. It transpired that this could be done fairly readily but the economic incentives were not great and pre-heating up to about 250°C was then studied. It was soon found that in these conditions the coal flowed readily (not even needing levelling), and had both a higher bulk density and better heat transfer properties. More important, as a consequence, the coke quality was improved or, alternatively, similar coke grades could be made from cheaper coking blends and/or shorter carbonisation cycles.

Fine coal at elevated temperatures is a serious hazard if exposed to air. Pre-heating systems must therefore be carefully designed and tested but there is now good evidence that the operation can be carried out safely, reliably and successfully on the industrial scale, by either of two alternative arrangements. Pre-heating can be carried out in either a flash entrainment type of drier or in a hybrid fluidised bed/entrainment system. Hot coal is recovered from the entrainment gas by cyclones and may be stored in sealed bunkers. The coal may then be transported to the ovens and charged by means of a pipeline, with steam jets as the motivating force. Alternatively, the coal may be transported by a movable charging car or a conveyor system, in both instances being charged into the ovens by gravity.

With pre-heating to about 250°C, both the impact (M40) and abrasion resistance (M10) are improved, the improvement being greatest with poorer coal. The proportion of large coke decreases somewhat but with no increase in breeze and little change in the mean size; these changes are probably advantageous on balance and indicate good stability. Differences in coke quality do not appear to be a significant factor in choosing the pre-heating system and the mechanical equipment; such decisions might be made largely on engineering preferences. Increases in oven throughput of 30–40% can be obtained and since the system is enclosed, the problems of oven top pollution are substantially eliminated (although other methods can be used to deal with this aspect in slot ovens). The main continuing problems with pre-heating are to minimise fine coal losses through incomplete removal of solids in the cyclone system and the carry-over of solids into the oven offtakes (which contaminates the tar). Development work seems to have reduced these factors to acceptable proportions.

The briquetting of charges before coking in slot ovens has been studied sporadically over the years but the concept of a mixed charge of briquettes and crushed coal has been developed recently, especially in Japan, where it is industrially exploited. The main mechanism for improving coke strength by partial briquetting is the increase of bulk density. The latter would reach a maximum at a 50/50 mixture of briquettes and coal but a 30% addition of briquette, the usually selected proportion, increases bulk density by about 17%. When such a density is achieved by means other than briquetting, the coke strength is increased by about 6%; when the same density was achieved by partial briquetting (an easier method of densifying charges), there was a further 1% gain. Thus, although the bulk density factor is paramount, the fact that the coal particles within the briquettes are closer together probably results in a better coking action.

The briquettes in the Japanese process may be made from the same coal blend as used in the rest of the charge, 6% of binder (tar or pitch) is added and the briquettes are formed in a roll press. The green briquettes are carefully handled and must be mixed to give a homogeneous 30% concentration in the coal charged into the ovens. Oven operation is little affected. In practice, instead of seeking further improvements in coke quality (already high in Japanese practice), the process is generally used to reduce the prime coking coal proportion, say from 70% to 50%, allowing also the addition of 10–20% of cheaper non-coking or weakly coking coals. It is worth noting that, with this level of prime coking coal, UK blending practice could produce equally good coke without the expense and the elaborate equipment for briquetting and charging the oven.

In Japan also, work has been carried out on the addition of "coking additives", which are intended to improve fluidity. One additive is a substance rich in aromatics derived from petroleum residues. Others are derived from pitch, asphalt or by-products from coal liquefaction.

The possibility of selective crushing of some of the individual coals in a blend has been shown to be favourable in small-scale tests but this somewhat laborious process may not be widely adopted industrially, especially in the light of favourable results from the other processes.

BY-PRODUCTS

It is well known that the by-products of coal carbonisation played an important role in the establishment of the organic chemicals industry; "family trees" showing coal derivatives have been familiar for a long time. However, these special chemicals represent only a part of the important values from the by-products. For a long time carbonisation products were vital both industrially and economically but in the period of cheap oil came under pressure. Those products which were essentially fuels were in competition with oil refinery products and from natural gas; furthermore, special chemicals could be synthesised more cheaply, and in the larger quantities required in recent decades, from petroleum intermediates. Thus, in some cases, the full recovery of carbonisation by-products has been curtailed in recent years and since their demise, by-products from coal gas retorts have disappeared. The possibility of a complete reversion to former recovery practice depends on economic factors. Here, recovery on traditional lines will be considered.

The main primary product streams, which may be further refined and separated, are crude tar, crude benzole, ammonia (generally ammonium sulphate) and coke oven gas. The total quantities and their distribution and quality vary considerably according to the type of carbonisation process and, to some extent, with the type of process used to treat the off-gases. The latter leave the oven via an "ascension pipe" near the top of each oven end and enter a main where cooling starts by means of sprays of ammoniacal liquor from a later stage of the process. Spray cooling continues in special coolers, which may be augmented by indirect cooling tubes. Tar and ammoniacal liquors are recovered and separated in a tank in which the tar settles to the bottom. Ammonia may then be boiled off the liquors (lime may be added to decompose some of the salts) and collected, usually in sulphuric acid as ammonium sulphate. The gases are further washed with light oils to collect crude benzole, which may subsequently be distilled from the oil. After benzole recovery, the gas may be purified, particularly by removing SO_2 and H_2S; the gas may then be used in the works and/or sold.

A simplified flow diagram (Fig. 7.17) illustrates some of the products which may be recovered; it will be seen that the crude tar is the main source. In the middle 1970s the world production of crude tar was about 16 Mt (about 5% of the carbonisation coal input). A considerable portion of this was probably used, with minimum treatment, as fuel.

The differences between tars made in (a) standard high temperature slot ovens and (b) those from either low temperature carbonisation or in the once-important vertical gas retorts throws an interesting light on the complex chemical processes taking place. The tar yield goes down as the carbonisation temperature (and probably time) increases, from about 10% to 5%. At the same time the proportion of paraffins, naphthenes and phenols decrease and the proportion of aromatics increases from a very low figure to over 90%. The average number of fused rings and the molecular weight of the latter also increase. These changes are believed to be due to secondary reactions occurring after the primary breakdown of the coal and the evolution of VM. Some of these processes are structural simplifications, such as dehydrogenation and the removal of alkyl side chains; however, polymerisation and condensation reactions must also take place leading to the more complex molecules of which coal tar pitch is mainly composed.

In the primary distillation of crude tar, several products are separated at increasing temperatures; the practice varies however and so does the nomenclature. Up to about 200°C a few percent of benzole, naphtha and light oil are recovered. At temperatures between about 200 and 230°C about 15% of the tar comes off as naphthalene oil and above about 250°C, anthracene oils, heavy creosotes and heavy oils are recovered. The pitch, remaining at about 370°C, will normally be about 50% of the input but the proportion may be deliberately increased when the main outlets are fuel oils, road tar, creosotes and pitches. The last may be prepared in a variety of grades for use in briquetting, electrodes, pitch coke, roofing and pipe coatings. The naphthalene oil and anthracene oil fractions from the primary crude tar distillation may be processed in quite complex specialist plants for the separation of a wide range of chemicals. Similarly, the crude benzole may be "defronted" by steam stripping to remove contaminants such as carbon disulphide, hydrotreated to remove other impurities (especially sulphur) and separated mainly into benzene, toluene and xylene.

ACTIVATED CARBONS

Carbons of various kinds, including coal and its derivatives, have for a long time been used for the filtration and purification by adsorption of liquids and gases. With the increasing interest in environmental control, this is a rapidly growing industry and there is considerable activity in developing methods of manufacturing and characterising activated carbons from coal and in studying their performance in industrial schemes. An added interest is that some effluents which may be purified by these means include those produced by coal utilisation processes, especially coke ovens. Furthermore, there is some interest in the use of coal conversion intermediates such as char as the basis of activated carbons.

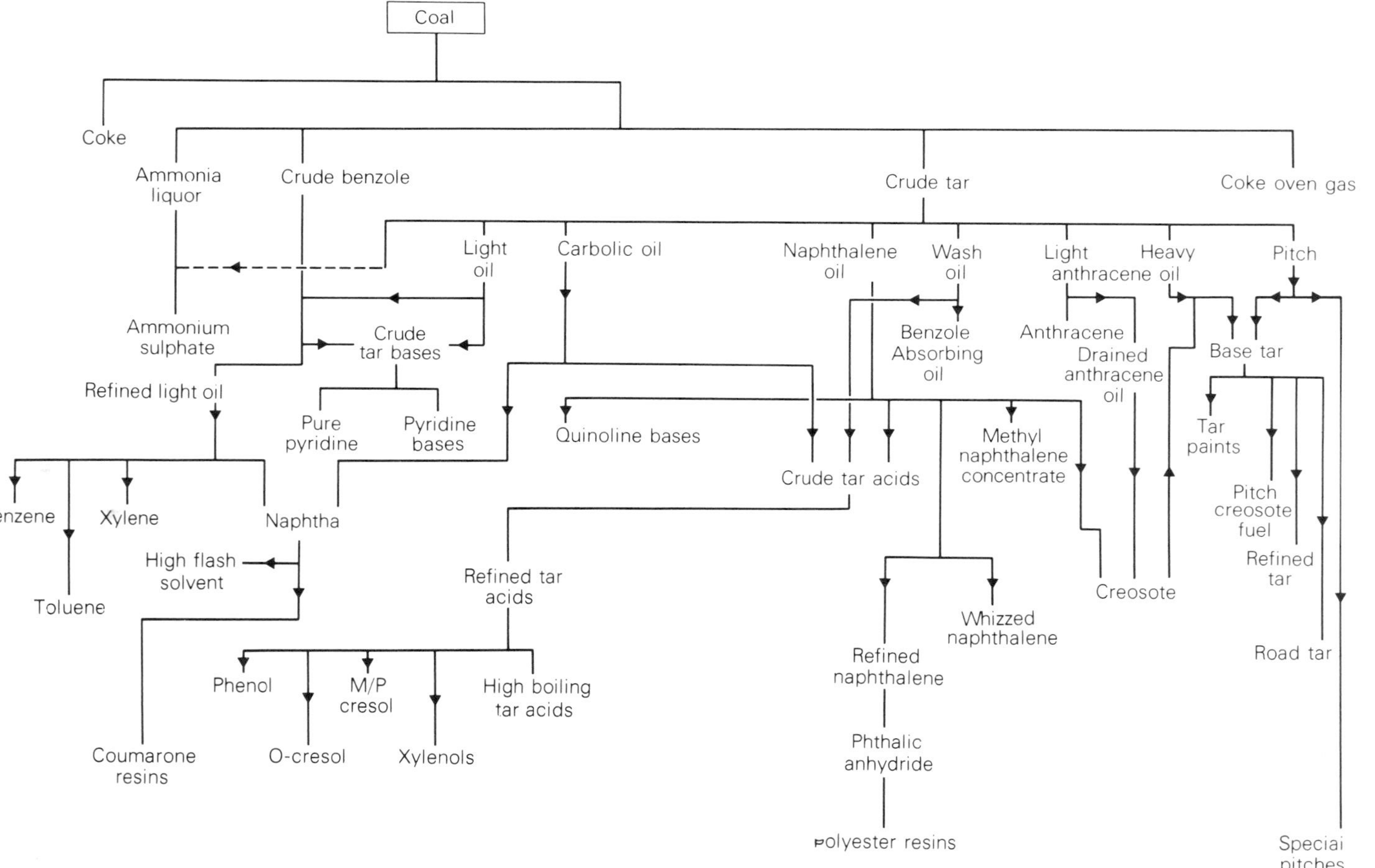

7.17 Compounds taken from the by-products of coal carbonisation

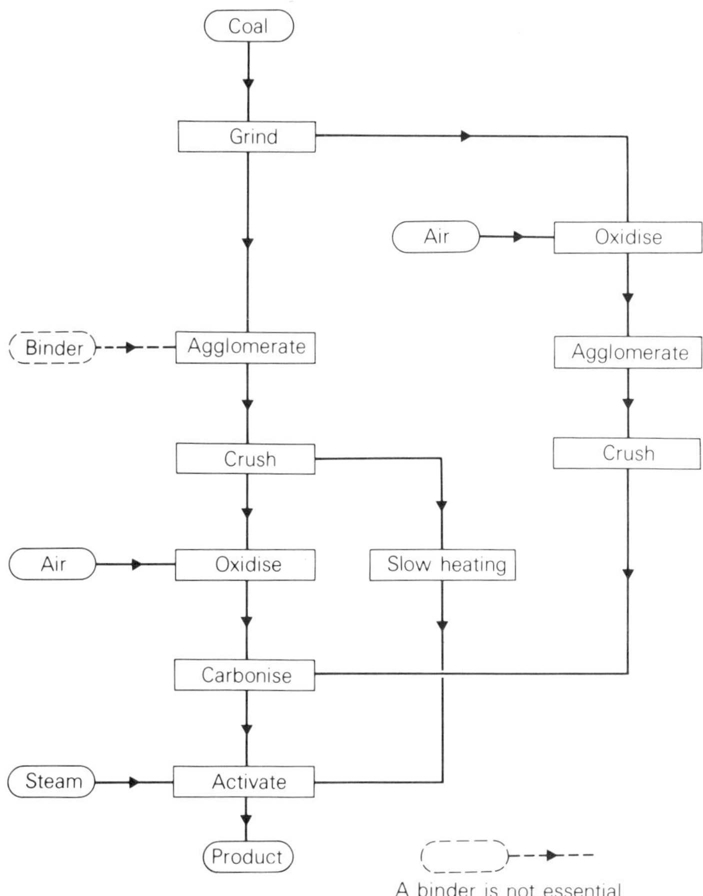

7.18 Manufactured of activated carbon from coking coals: generalised flow diagram

The object of treatments designed to prepare activated carbon is the creation of pores with a high surface area. A number of different starting materials have been used including chars, anthracite, coking coal, brown coal and even peat. A key step usually is "activation" in steam at about 900°C but some degree of oxidation as well as carbonisation may precede or accompany the steam treatment. Crushed and graded coals have been used but it now seems usual to grind the coal at some stage and form pellets. A generalised flow diagram is shown in Fig. 7.18. Heating may be in fluidised beds or multihearth furnaces. The plant used by Thomas Ness Ltd in the UK is illustrated in Fig. 7.19; anthracite was the original feedstock but good results have also been obtained with briquetted char and with the residue from the Supercritical Gas Extraction (SGE) liquefaction process. Figure 7.20 shows the volume and diameter of pores in SGE char, with and without activation. The lignite char was made on the same plant from Australian brown coal. The latter may be made into paste with potassium hydroxide, extended into pellets, carbonised and leached at 900°C; steam activation develops the pore structure further.

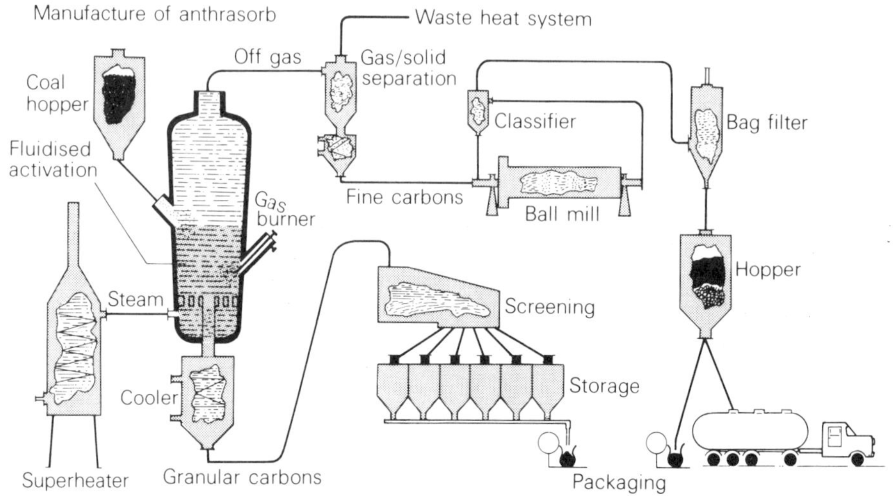

7.19 Flow diagram for manufacture of activated carbon from anthracite

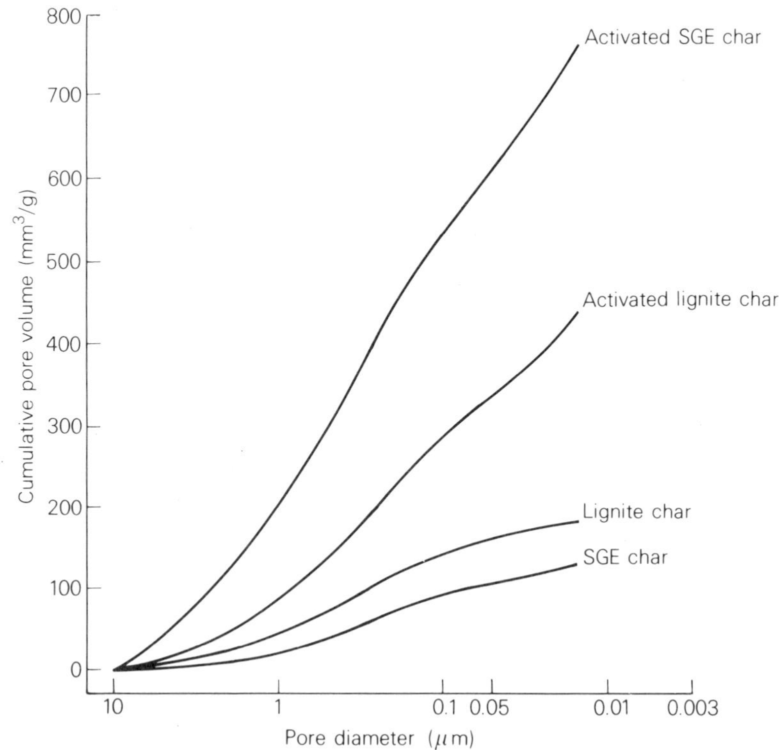

7.20 Pore volume distributions by mercury porosimetry

It will be seen that very high pore volumes may be obtained, with a considerable portion of the volume in fine pores. Details of manufacturing processes tend not to be openly available and products are often marketed under proprietary names; tests are usually empirical and relate one or more of these products to results on specific effluents. It is therefore difficult to obtain results and applications in a generalised form; furthermore, very rapid progress is being made and as a result techniques can be outdated readily. Activated carbon is sometimes used as a stage in some complete systems which may also include biological oxidation treatment and there may be some overlap, deliberate or otherwise, in the effects. The arrangement of these stages, and also filtration, is variable.

A wide range of plant designs is used in liquid effluent treatment systems using activated carbons: fixed beds (either upward or downward flow), downward moving solid beds and fluidised beds are all employed. Powdered carbons may be added directly to the liquids, followed by settling and filtration. Effluents which are treated include those from paper mills, dye-works, coking plants, chemical works and municipal waste-water. Potable water may be treated, especially for the removal of taste, odours and traces of toxic substances. Activated carbons may also be used for the removal of malodorous or noxious substances from gaseous effluents. Industrial processes requiring the recycling of used fluids or separations may also be potential markets.

In principle, activated carbons which have adsorbed contaminants to capacity may be regenerated and such processes are carried out commercially in some cases. However, it appears that *in situ* regeneration may often prove impracticable and in other cases some absorbed species may not be removed very readily. The physical characteristics of the carbon may also be vulnerable to the mechanical, chemical and thermal stresses involved in regeneration.

The future of activated carbons based on coals is clearly going to be very important, not so much in bulk terms but in terms of value and also because of the links formed between this usage and other aspects of coal utilisation. For the moment, a great deal of research and development work is proceeding and is identifying the very wide ranges of properties which can be obtained; this is backed by pilot scale work on specific applications. However, large-scale trials are largely lacking and this inhibits the application of the research effort to overcome problems and improve products in the directions desired. It is clear that the art is complex and that certain markets are in a state of flux; there are however very positive favourable prospects.

ENVIRONMENTAL ASPECTS

The broad aspects of environmental impact from the future utilisation of coal are duscussed in Chapter 19. However, some aspects of environmental control, including health effects, are particularly relevant to carbonisation and are appropriately included here.

Coking plants suffer certain disadvantages common to other large installations handling dusty materials and producing gaseous and liquid effluents. They can never be attractive neighbours. To the work-force, the heavy machinery and general working conditions present an accident hazard. Questions of siting and accident prevention therefore require high standards. However, the particular aspect of special relevance to carbonisation works concerns the potentially unpleasant and harmful features of liquid and airborne pollution.

The liquid effluent arises from the liquor generated in the coke ovens and in the by-product plant from water in contact with coal and coke, especially in the quenching process. This liquid effluent has features common to some other industrial wastes but probably in greater complexity and concentration than nearly any other major industry. The effects, when discharged into rivers or other waterways without adequate treatment, have in the past contributed to the familiar murky, smelly and "dead" conditions, the most important symptom of which is an oxygen deficiency. Some of the chemicals in solution include ammonium salts, phenols, thiocyanates, cyanides and sulphur compounds. The general objectives of treatments are to oxidise these noxious compounds and at the same time to leave adequate levels of oxygen in solution. This process is now greatly aided by biological means; organisms which destroy chemicals by promoting oxidation are employed, usually in an "active sludge". Contact between the liquor and the sludge may be obtained in ponds or by means of columns; oxygen must be provided, for example by stirring or by bubbling air into the liquid. Settling and/or filtration stages may be required and activated carbons probably have an important future role here. It seems that continued improvement in the quality of liquid effluents is technically feasible. The rate of improvement may however need to be related to the economics, especially of old plants, and also to the real requirements; other established sources of water pollution are often also present and may only gradually be alleviated.

Airborne pollution probably presents a more serious problem — and certainly a more obvious one since dust and fumes are readily visible and may also be detectable by smell. Whereas liquid pollution is concentrated into watercourses, airborne pollution can spread radially outwards in all directions. More important still, in addition to the nuisance which may be caused, some components which are generated and may be emitted are damaging to health and include known carcinogens. The many openings in a coke oven battery and the regular use of these, plus the exposure of heated coke on pushing provide plentiful possible sources. The fumes may also be corrosive, abrasive and may be clogging in respect of equipment installed for pollution control.

Nonetheless, in spite of the unfavourable configuration and the nature of the pollutants, substantial progress has been made in the last decade or two in reducing emissions once operators become fully aware of the problems and the public attitude. For example, a recent report compared good practices with averages for emissions published by the Economic Commission For Europe. The latter showed total emissions ranging from 1 to 11 kg/t of coal, with an average of 5.3 kg/t, which includes 1.5 kg/t of dust, 2.6 kg/t of $SO_2$and 1.2 kg/t of other impurities. A German plant without any special emission control equipment (none such incidentally survived into 1978) was shown to have the following dust emissions, by source:

Source	g/t of Coal
Pushing coal	400
Charging ovens	200
Quenching coke	200
Others	20
Total dust	820

The uncontrolled SO_2 emissions were about 2.3 kg/t and the other gaseous emissions about 0.250 kg/t, making a total of about 3.8 kg/t for standard plant. By installing equipment to control dust on charging, pushing, quenching and screening, total dust measurements of only 170 g/t were obtained. Emissions of SO_2 are determined by the sulphur content of gases used for underfiring; by good purification practices this was readily reduced so that SO_2 emissions of 600 g/t were obtained. Other emissions were also reduced by the dust containment devices so that total emissions were reduced to just over 1 kg/t.

Similar progress to that reported from Germany has been made in other countries. It must be said however that there is controversy (strong but constructive) concerning the economic merits, reliability and performance of different designs of equipment and also about the necessary or desirable levels of pollution control. Since gross pollution has been, or is in the process of being, eliminated, discrimination between remaining pollutants becomes more difficult. There is probably a requirement for a period of stability, so that experience with various machines can be obtained and careful statistical measurements made before further standards are set.

Oven charging is a particularly important problem. The rapid evolution of steam and gases can lead to the ejection of both dust and noxious hydrocarbons, in a position where workmen are employed. Two broad methods have been put into operation: washer charging and sequential charging. In the former the charging car contains or is directly attached to a scrubber and a combustor. In the sequential charging system, the amount of coal charged at any time through the various holes in the oven top is measured and arranged in sequence so that there is always a clear channel above the coal through to the gas collecting mains. Both systems seem to satisfy their users. Both require good location of charging hoppers and gas tight connections to the orifices, excellent fitting of doors in the system and good housekeeping in general. Of course, if pre-heating is adopted, this automatically requires and provides pollution-free charging.

For pushing ovens, various devices, including hoods and other forms of containment for the hot coke, have been developed in order to avoid pollution. The hoods collect fumes and pass them through a series of scrubbers. These devices may be incorporated into the movable discharge equipment; alternatively, the hoods may be connected at each oven to a collecting main leading to a stationary scrubber. Quench towers may have baffles or other devices installed which reduce considerably the carry-over of dust in the steam which is generated. However, the most promising system, already well established in some places, is "dry-quenching". In this process coke is cooled by circulating gas or steam through it in a fully enclosed system. There is the further advantage that energy can be recovered through both sensible heat and combustible gases.

Whilst excellent progress has been made in controlling the general environmental impacts of coke oven operations, the possibility of health damage and increased mortality risk was highlighted by statistical studies carried out in the USA from the 1950s onwards. These showed considerable increase in morbidity and mortality, especially from lung cancer, in coke oven workers. Those operating on oven tops appeared to be particularly at risk. Factors of up to seven for the increased risk were indicated in some cases. Similar studies have now been carried out in the UK for about 10 years but with quite different results. For example, whilst US coke workers were two and a half times more likely to develop lung cancer than steel workers, the extra risk in the UK was a small fraction of this. In the USA, those with oven top experience were five times more likely to develop lung cancer than coke operators without oven top experience; in the UK there was no significant difference between these two classes. In the USA, risk appears to depend on length of exposure; lung cancer death rates up to ten times the average have been measured for men with lengthy oven top experience. In the UK, coke workers appear to have a somewhat lower mortality than average but, within this general rate, the rate of lung cancer deaths is about 25% higher than expected. At those levels, however, the relevance of comparisons becomes complex.

The statistics from the UK and the USA, whilst equally reputable, seem to be too different not to be meaningful. Since exposure over some years is involved, it may be that different coke oven practices in the two countries were responsible, especially for cumulative damage, from a considerable period in the past. However, in both countries and elsewhere, current practices are undoubtedly better than the best situations anywhere in the 1960s. This has been established through the careful monitoring of working environments, before and after the drastic improvements which have taken place during the 1970s. There have been mechanical improvements in order to control emissions and greater alertness to housekeeping and approved work schedules. Masks and refuges have been provided and improved medical and hygiene arrangements have been initiated. Based on British experience, it may be that exposures to dangerous substances may not now in good practice constitute a significant hazard; this of course requires proof from the substantial studies now in hand. It follows that if those engaged on oven tops are subject to risks which are certainly at the limit of detection, toxic effects outside the plants, as distinct from nuisance, are extremely unlikely.

GENERAL CONCLUSIONS ON CARBONISATION

For some years after the Second World War, the future of coke ovens seemed doubtful to many. It was suggested that the ironmaking blast furnace, the main user of coke, would rapidly be superseded by other processes (as occurred in steelmaking processes). Further, it was suggested that, even if blast furnaces retained their primacy, oven coke would be displaced to a considerable extent by other fuels, such as formed coke, mainly as a result of the coking coal supply situation and environmental problems.

Both the blast furnace and the coke oven have proved remarkably resilient and capable of substantial development (which can continue further). In the case of coke oven practice the rapid improvements in coking blends and ancillary equipment have been quite remarkable, especially for such a mature industry. It is particularly gratifying that this progress has been based on excellent scientific and engineering work. It now seems clear that the way will be open and the need will be there for continuation of this progress for a long time. The future of the blast furnace and the standard coke oven may be affected by other processes on economic grounds, not yet determinable, but any change in status is likely to be gradual rather than revolutionary.

The subject of carbonisation is closely linked to the basic science of coal. Carbonisation practices, or near relations, may have a role to play in some of the newer coal utilisation technologies.

Chapter 8

GASIFICATION

INTRODUCTION

The gasification of coal is currently the subject of intense development activity. It is widely anticipated that a considerable proportion of the coal consumed in the future, as at certain periods of the past, will first be converted into gases. Motivations for this conversion may include (a) the greater convenience of gaseous fuels compared with solids, (b) the possibility of beneficial changes in environmental impact and (c) possible improvements in the efficiency of use from coal through to final consumption. However, more generally in the recent past, a major objective has been the production on a large scale of Substitute Natural Gas (SNG).

Natural gas, which is mainly methane (CH_4), has in the last few decades become a very important fuel and feedstock practically throughout the world, displacing manufactured gas almost completely. Natural gas is of course now an internationally traded fuel and is transported long distances by pipeline and in liquefied form by tanker. Most developed countries have complex internal distribution systems serving very large numbers of consumers, who use the natural gas in a wide range of appliances. Supplies of natural gas will however ultimately become restricted; it may then be both convenient and economic to provide a direct substitute by manufacturing more or less pure methane from coal in order to take advantage of existing distribution and utilisation facilities, which may not be suitable for other types of gas. The manufacture of SNG on a pilot scale and its admixture with natural gas for distribution to customers has been demonstrated by British Gas at their Westfield development site in Scotland.

Natural gas has a high calorific value (about 1000 Btu/scf; 37 MJ/m^3). Other types of gas have been used in the past and are likely to be important again; they may be grouped into two categories:

(a) Medium Btu gas (MBG) of up to 450–500 Btu/scf (c. 18 MJ/m^3), consisting mainly of CO and H_2.
(b) Low Btu gas (LBG) of about 150 Btu/scf (c. 6 MJ/m^3) the main fuel ingredients (again CO and H_2) being diluted by 50% or more of nitrogen.

Mixtures of CO and H_2 (corresponding to MBG) may also be termed "Synthesis Gas" when intended for chemical synthesis purposes. Such gases may have their CO:H_2 ratio adjusted by "shifting", generally increasing H_2 with the elimination of CO_2; the composition may be shifted more or less completely to H_2, an important gas in many industries (including coal conversion). "Synthesis gas" may be used to describe mixtures of hydrogen and nitrogen, of suitable ratio, used for ammonia manufacture; the hydrogen may be derived from the CO/H_2 mixtures made by gasifying coal and the nitrogen may be provided from the air used in gasifying.

It will be apparent that "gasification" covers a wide range of processes, compositions and end-uses. Most of these have historical precedents.

The first gas to be made from coal and used commercially was produced by the carbonisation process (see Chapter 7). This gas was first used for lighting purposes near the end of the 18th century but became much more widely used in the 19th century. Coke oven gas still remains an important fuel although its output, as a by-product, is limited and its distribution is generally local.

Coke oven gas usually has a calorific value in the MBG category and this became the standard for locally distributed gas mixtures, often called "Town's Gas", for many years. Coke oven gas was a common component of town's gas; it was derived both from slot ovens producing metallurgical coke and from special gas-making retorts. In the latter case the coke was more reactive than metallurgical coke and was used mainly for industrial and domestic purposes. The carbonisation processes, however, were limited by the size of the outlets for cokes, tars and other by-products.

Other possible components of town's gas were producer gas and water gas. Producer gas may simply be made by partially combusting coal with air; water gas is made by substituting steam for the air blown through the hot bed. Since some water is usually present adventitiously in the coal, there is usually some hydrogen in producer gas, which is in the LBG class, containing about 30% CO, 10% H_2 and over 50% nitrogen. Water gas made without any air contains little nitrogen and nearly 50% each of CO and H_2, with a calorific value about twice that of producer gas. In practice, a number of permutations were used in order to produce a consistent gas for sale, depending on practical temperature limits. Blowing with air alone can result in excessive temperatures, leading to clinkering; on the other hand steam cools the bed but this can be overdone, ultimately, to the extent that gasification ceases. Thus, compromises must be reached to suit individual circumstances; in some cases, the gas generating vessel was blown alternately with air and steam. In other cases, hydrocarbon liquids were sometimes sprayed on hot coal or coke to enrich the gas further (carburetted water gas).

The first gas producers were simple brick vessels through which air was drawn upwards through hot coal. Subsequently, water-jacketted steel vessels came into use; these were set upon water-sealed revolving ash grates (Fig. 8.1). Automatic coal feed through a lock hopper was generally provided. Modern versions of this type of design are available and may again become popular since they provide a reliable source of gas, normally for local consumption.

The availability and use of tonnage oxygen provided a breakthrough in coal gas production. The substitution of oxygen for air enabled a gas to be made with the same calorific value as carbonisation gas without the complications of finding markets for the coke. These processes, which were developed in Germany between the wars and subsequently, are described in a later section. They were applied to a wide range of coals, including bituminous and brown coals.

Although fuel and feedstock gases derived from coal have such a long and vital history, they have in recent decades become much less important; indeed, their production has become almost incidental as a by-product of metallurgical coke manufacture. This has been due not only to the discovery and wider marketing of natural gas but also to the availability, cheapness and convenience of liquid petroleum feedstocks for gas manufacture. However, it is expected that coal will ultimately recover this market as natural gas and oil become scarcer and dearer.

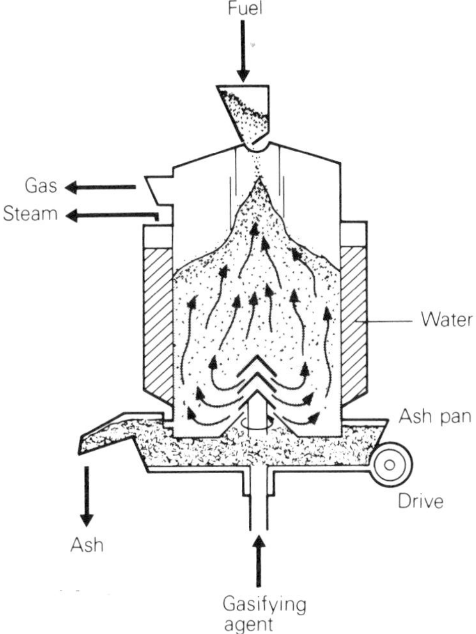

8.1 Rotating grate gasifier

Although the traditional gasification science and technology will undoubtedly have a role to play, certain factors provide strong incentives for newer developments:

(1) The scale of some individual operations in the future may be very much larger than previously.
(2) Distribution systems and consuming plants are now geared to methane rather than other, less rich, gases (although the latter may well also have an important role to play).
(3) Economic constraints may be tighter, with special emphasis on thermal efficiency, as primary fuels become more expensive.
(4) In some cases, large gasification units may be directly associated with new large mining investments in situations where other coal markets are not conveniently placed; such gasifiers will need to accept all the coal produced.
(5) In other cases, gasification plants may be a long distance from coal supplies and should therefore be flexible with regard to coal quality.
(6) Total gasification, without either solid or liquid by-products, may in most circumstances be considered desirable or essential, both because of the strong future demand for gas and for environmental reasons.
(7) Operation at elevated pressures will be more economic and more compatible with associated processes in manufacture, distribution and use.

Some scientific considerations

The basic chemical reactions for gasification processes are simple. However, the reactions are usually reversible and more than one reaction may be taking place simultaneously in the same or adjacent zones. Interactions can be quite complex, the balance being sensitive, for example, to temperatures, pressures and flow rates. The devolatilisation of coal, followed by the cracking of hydrocarbons, are further complications.

The main combustion reaction between carbon and oxygen is:

$$C + O_2 = CO_2 - 394 \text{ MJ (exothermic)}$$

(Heat changes are in terms of megajoules per kilogram-mole; a negative sign shows that heat is released from the chemical system, i.e. the reaction is exothermic.)

However, CO_2 may be reduced again by C:

$$C + CO_2 \rightleftarrows 2CO \pm 171 \text{ MJ (endothermic)}$$

In practice there is an equilibrium between the concentrations of CO and CO_2; higher temperatures (especially above about 700°C) favour CO. There are also two reactions between carbon and water, producing H_2 and either CO or CO_2; both are endothermic in the forward direction:

$$C + H_2O \rightleftarrows CO + H_2 \pm 130 \text{ MJ}$$

$$C + 2H_2O \rightleftarrows CO_2 + 2H_2 \pm 87 \text{ MJ}$$

In the water–gas shift reaction CO reacts with steam to produce further hydrogen exothermically:

$$CO + H_2O \rightleftarrows CO_2 + H_2 \mp 41 \text{ MJ}$$

This hydrogen is, of course, produced at the expense of some carbon loss; the CO_2 can be readily removed.

Carbon can be directly and indirectly converted to methane by hydrogen:

$$C + 2H_2 \rightleftarrows CH_4 \mp 75 \text{ MJ}$$

$$CO + 3H_2 \rightleftarrows CH_4 + H_2O \mp 206 \text{ MJ}$$

$$CO_2 + 4H_2 \rightleftarrows CH_4 + 2H_2O \mp 165 \text{ MJ}$$

All the above methane-producing reactions are exothermic and are enhanced by pressure, since a reduction in volume occurs.

If SNG is the desired final product, the overall reaction may be thought of as:

$$2C + 2H_20 \rightarrow CH_4 + CO_2$$

Since this reaction is on balance endothermic, heat must be supplied, normally by burning some carbon, with the total effect:

$$3C + 2H_2O + O_2 \rightarrow CH_4 + 2CO_2$$

However, since gasification requires high temperatures and the methane reaction is favoured by low temperatures, the two steps are normally carried out separately.

In addition to the heat required theoretically, heat losses from the system are inevitable. The technical and economic consequences are enhanced because the heat input for gasification is at a high temperature while the heat output from the methanation stage is at a low temperature and therefore difficult to recover and use economically. If maximum methane is required, e.g. for SNG, it will probably pay to maximise methane formation in the gasification stage. This is favoured by low temperatures but, since reaction rates are also reduced, this effect cannot be carried too far without bulkier and more complex designs of gasifiers becoming necessary.

Although the basic chemical reactions are simple, both the equilibrium and kinetics are affected by conditions, some of which have already been mentioned, including temperature and pressure. Other factors, including particle size and the relative velocity of gas and solid, affect the mechanisms by which the reactants come into contact. It is thought that the gas must first diffuse through an outer film surrounding the particle; subsequently, the gas diffuses through the pores of the coal. Below about 900°C, the chemical reaction rates determine the overall rate and neither particle size nor velocity have significant effects. In this range, reaction rates increase strongly with temperature and with pressure, although the pressure effect with steam levels out at about 10 bar. Between about 900–1200°C, pore diffusion is the rate determining step and the effects of temperature and pressure are decreased; overall reaction rates are increased moderately with particle fineness. Over about 1200°C boundary film diffusion becomes the rate determining mechanism and both particle fineness and velocity strongly increase reaction rates; temperature and pressure have relatively little effect. Although this idealised picture is generally correct and helpful in understanding the principles of gasifier design and performance, the devolatilisation and swelling behaviour of the coal may also be important complicating factors. This behaviour in turn depends on the rate of heating, the type of coal and other factors; the design of gasifiers therefore requires verification under realistic and near full scale conditions, however well based in scientific theory the conception may be.

GASIFIERS

General features and classification

Some of the gasifiers on which commercial operations or major development work has been carried out in recent years are described in the next section. Before proceeding to individual descriptions, some general observations may be appropriate. First, although development work and discussion tend to focus on the gasifier itself, the total system is a much more complex entity, the elements of which may include coal handling and pretreatment, air separation, steam supply, gas purification and compositional adjustments and environmental control facilities, in addition to the gasifier, which may represent only about a quarter or less of the whole plant investment. Sometimes the utilisation of the gas will be integrated with the gas supply system. In any case, the design of the system will depend on the use for which the gas is required and this may also influence the choice of gasifier, although gasifiers can often be operated in more than one mode. The main process stages and end uses are illustrated in Fig. 8.2; it will be apparent that gasification is a very common unit in coal utilisation schemes. Complete systems incorporating gasification as one unit stage are described in other sections and chapters.

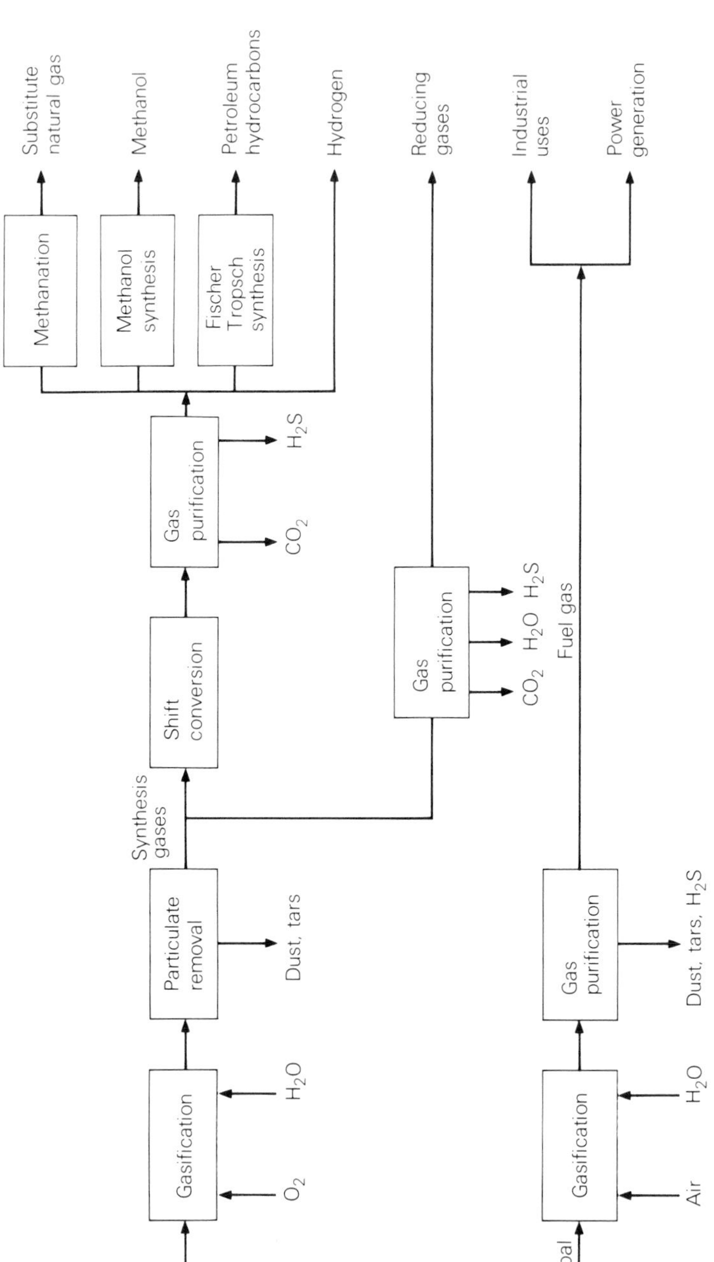

8.2 Process stages and uses of coal gasification

A large number of gasification processes, or variants, have been proposed and several dozen, at least, have been studied practically to some extent. There are good reasons for some diversity of approach; there are also a number of factors which provide some common ground.

The first of the factors leading to diversity is the relationship between inputs and outputs: the type of coal and other resources available on the one hand and, on the other, the products required and how they are to be used. The potential for value enhancement is very important. Where coal is cheap and gaseous products highly valued, a reliable plant for early operation may be more important than the prospects of further refinement. Several processes are already in large-scale operation in commercial circumstances. These processes, "first generation", not only set the targets for second or third generation processes but are themselves capable of improvement, sometimes even to the point of almost revolutionary change. This process of continuing development may incidentally lead to some confusion about the significance of proprietary names.

In most circumstances, new large-scale gasification plants will be justified only as natural gas (or general energy) prices increase, to the level where substitution by gases from coal becomes economic. Thus, a gasification process which promises, say, a 10% reduction over earlier processes will not only be preferable as a choice between processes but may actually advance by several years the date when prospective profitable operation could be expected by any coal gasification process. Probably the most important factor in the long run will be the efficiency of use of the coal input. In assessments to date, emphasis on SNG has often meant that the values of products other than methane may have tended to be ignored but attitudes to this may eventually reverse in time.

Thermal, or carbon, efficiency is usually closely related to the amount of carbon dioxide rejected; this in turn may depend mainly on the thermodynamics of the chemical processes at the operating temperatures selected. However, thermal efficiency will depend also on the heat losses, through vessel walls and in off-gases, for example. Higher temperatures and pressures and the use of oxygen may all be advantageous but will have some limitations or cost disadvantages both in thermal efficiency and investment. The amount and method of removing mineral matter are also important economically.

Coal is in any case at a disadvantage in comparison with feedstocks containing more hydrogen (Fig. 8.3). This factor results in a large requirement of feedstock per unit of gas produced, together with increased amounts of steam decomposed and CO_2 produced, compared with oil for example.

Oxygen demand is an important element both in capital cost and thermal efficiency. As more methane is produced directly (in cases where methane is the desired product) oxygen demand falls (Fig. 8.4); some of the best processes from this point of view use multiple fluidised beds.

Some processes use hydrogen, which at its limit is "hydrogasification". The economic advantages of providing hydrogen in the gasifier, which will require extra cost, may ultimately depend on the efficiency of special in-bed catalysts intended to promote methanation; such catalysts are being developed currently, apparently with considerable prospects of success.

The method of provision of heat for the endothermic reaction is an important design problem. Combustion of part of the carbon is of course the most frequent approach and this may take place either integrally in the gasifier or externally. If the

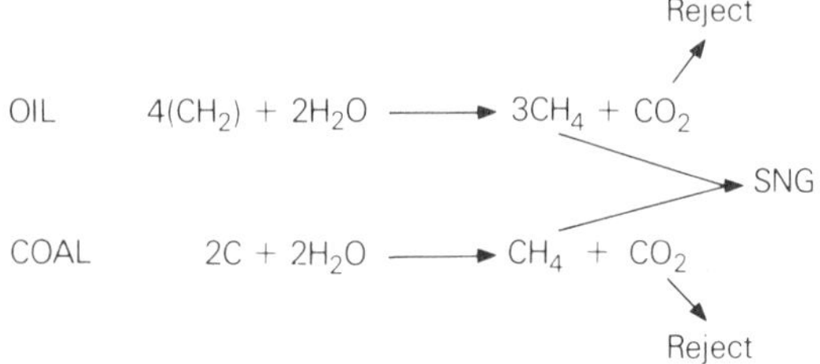

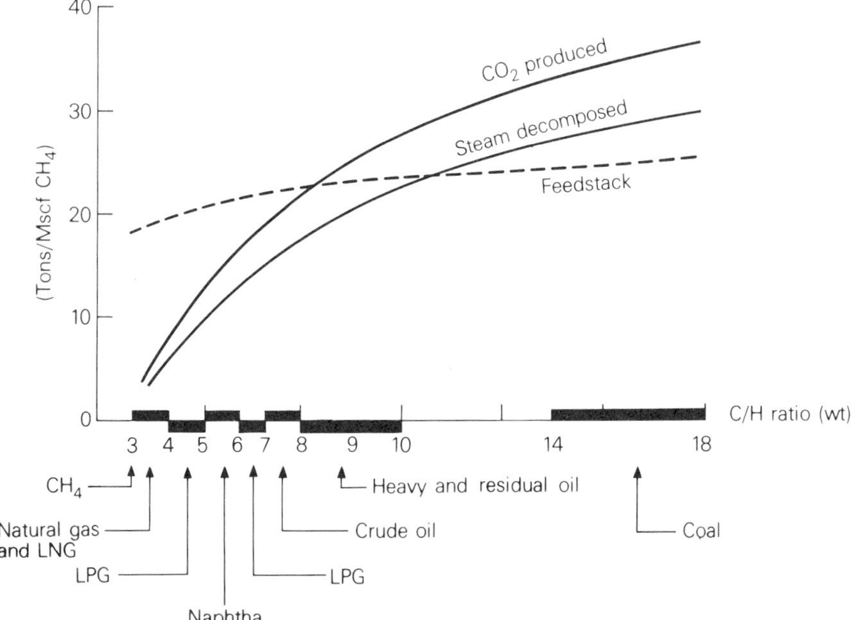

8.3 Steam decomposed and carbion dioxide produced in converting various feedstocks to SNG

former case, the product will be diluted by nitrogen, unless air is excluded and the gasifier is blown with steam and/or oxygen. If the heating is by external combustion, there is a choice of heat-transfer methods. External nuclear heating has also been proposed and, in addition, there is a proposal to use nuclear sources for providing hydrogen, for hydrogasification.

Pyrolysis is dealt with separately but the subject clearly overlaps with gasification. *In situ* gasification is also described in a separate chapter, although the reactions employed are naturally basically similar to standard gasification.

Processes may be classified in a number of ways, one example being in Fig. 8.5 where end products are also illustrated. The breakdown is here based on:

(a) Pressure;
(b) Gasifying agent.

In the USA a particular distinction is made between "low" and "high" Btu

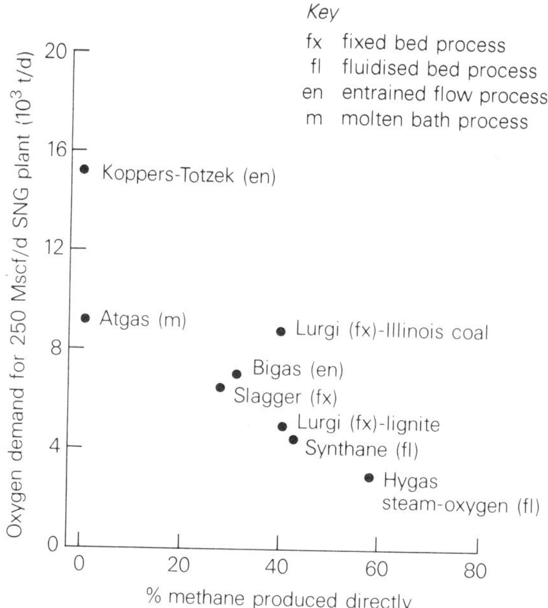

8.4 Variation of oxygen demand with percentage methane produced directly in coal gasification processes

gasification processes and qualitative distinctions are made within the "low" and "high" categories.

Perhaps the basic differences between gasifiers is in the nature of contact between the coal particles and the gas, especially since this choice has consequences on the operating temperature and the rate of heating. The three main types of arrangement are the fixed bed, the fluidised bed and entrainment systems, of which Lurgi, Winkler and Koppers–Totzek, respectively, are the best known examples.

The term "fixed bed" can be confusing, especially as this system is also sometimes called "moving bed" or "gravitating bed". This confusion arises because the coal is fed into the top of a vertical cylindrical reactor and ash is removed at the bottom. The bed level remains static but the individual particles of coal move downwards under gravity, counter-current to the gasifying agents, steam and oxygen (or air). Definite layered reaction zones, each with different temperatures, are established, the highest temperatures rapidly being reached as the oxygen meets the preheated coal.

The general concept of fluidised beds is described in Chapter 6 on combustion. The principle as used in fluidised gasifiers is similar but only partial combustion takes place. The fluidising gases (which are of course also the gasifying agent) rapidly reach the bed temperature, which is relatively uniform throughout because of the good heat and material transfer properties of fluidised beds.

In the entrained system, very fine particles are injected into a vessel co-currently in a stream of the gasifying agents, steam and oxygen (or air). Gasification usually takes place rapidly in a jet flame at very high temperatures and is complete, i.e. no tars are produced and the gaseous produces are all of low molecular weight.

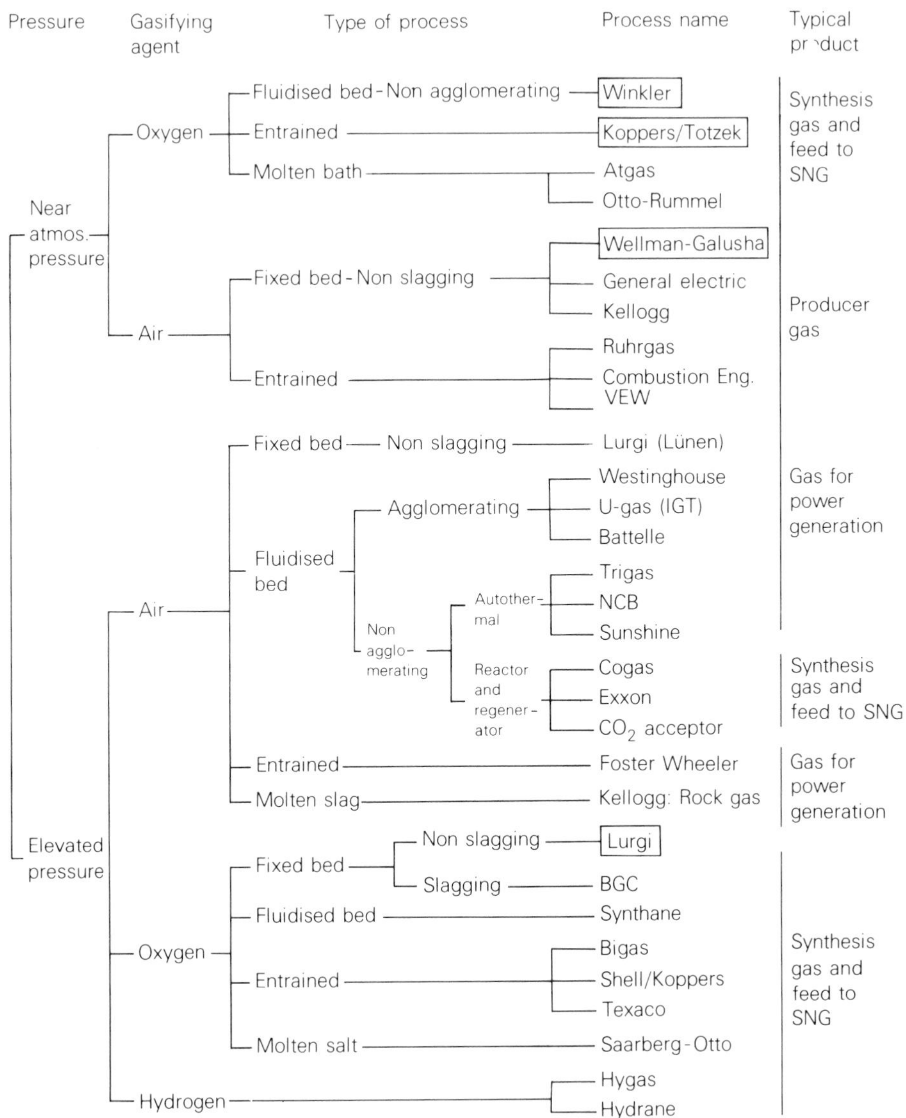

8.5 Classification of gasification processes (names in boxes are commercially available)

Established and under development

Between the two World Wars, considerable development of coal gasification took place in Germany, both as part of the efforts to establish synthetic liquids (requiring synthesis gas or hydrogen) and for fuel gases. It is no coincidence, but rather an indication of thoroughness, that efforts centred on the three main methods of contacting solids by gases described above, giving rise to the Lurgi, Winkler and Koppers–Totzek gasifiers. These processes, which still dominate thinking and development in gasification, are described in general terms below, plus their derivatives and also some newer gasifiers. The status of current development programmes is discussed separately.

Lurgi

The Lurgi Gasifier deserves prominence in any description of gasifiers since it is well established, having been first developed in Germany in the 1930s, and yet is still the subject of further improvements and diversification, both of feedstock and products. As noted above, Lurgi (Fig. 8.6a) is described as a fixed bed gasifier since the bed level remains static. Fresh coal is fed through the top and granular ash is discharged from the bottom; these operations take place through locks since the reactor is operated at pressures up to 32 bar (and versions with substantially higher pressures are being developed). Normally the reactor is blown (from the bottom) with oxygen and steam, for synthesis gas or SNG; air can be used instead of oxygen for fuel gas applications. The coal is spread evenly on the bed by means of a motorised distributor; a stirrer may be incorporated to operate below the surface of the coal to break up incipient caking with susceptible coals. The ash grate is also rotated and is usually operated with a covering of a few inches of ash to protect it from the hot coal.

There are distinct temperature zones, associated with the different reactions. At the bottom, combustion of the coal, already heated in its passage down the reactor by the ascending hot gases, takes place rapidly with the oxygen and temperatures rise to about 1000°C. As the oxygen is used up, the endothermic steam reaction takes over, in a zone at about 800–1000°C. Above this, temperatures continue to fall as the

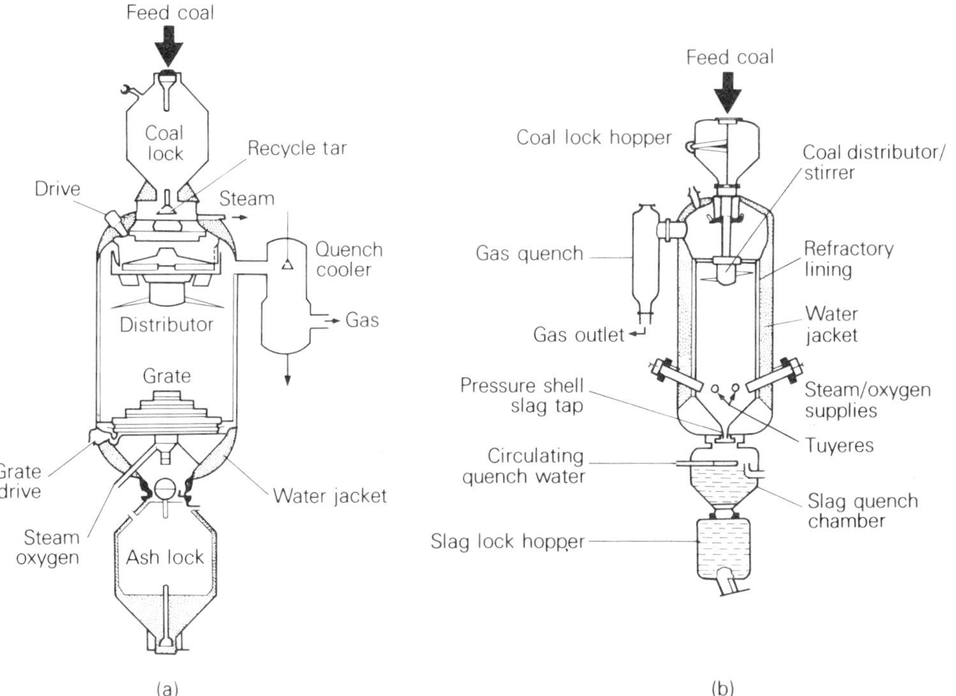

8.6 The Lurgi gasifier (a) dry ash; (b) slagging

input coal removes heat from the gas, which exits at about 500°C. The fresh coal is first dried and then, as it descends, is devolatilised. Tars and phenols are produced and are recovered from the off gases. Provision can be made for recycling these by-products back to the reactor to extinction but it is not clear how much successful experience has been obtained with this procedure.

There are now over 60 Lurgi gasifiers in operation in the world, on about a dozen sites, the best known probably being those at Sasolburg (see Chapter 9). Output per reactor is related to vessel diameter. At Sasolburg the original Lurgi reactors had an internal diameter of about 12 ft, consuming about 50 t/h of coal and producing 39 Mscfd (million standard cubic feet per day). The diameter was increased in newer developments to 13.1 ft (4 m), producing 44 Mscfd and subsequently to 16 ft.

Since the pressure is high and the temperature relatively low, the concentration of CH_4 can be quite high, about 11% in the standard 30 bar reactor, with about 39% H_2 and 19% CO. At 100 bar the CH_4 content may be increased, up to about 18%.

The Lurgi gasifier was initially intended to work on sized (6–40 mm) non-coking coals. These are significant limitations, especially for the wider SNG applications now expected. Progress has recently been made in developing operating techniques to reduce these restrictions, notably in experiments at Westfield in Scotland. British Gas have retained a Lurgi gasifier, formerly out of use, at this site, which has been converted into a development facility. The reactor was fitted with a specially modified distributor and stirrer and, in association with an American consortium led by Conoco, trials have been carried out on a series of American coals. It has been stated that the gasifier operated successfully and for reasonable periods at substantial proportions of full feed rate on US coals, from mildly to strongly coking, and containing up to 25% fines. Some encouraging tests have also been carried out at Sasolburg on widening the range of coals that can be gasified. It is possible that a "universal" basic Lurgi can be developed, with perhaps variations in ancillary parts, to accommodate most coals. Very substantial further full scale experience, however, will be needed in order to predict operating rates accurately with particular coal qualities even if most become technically acceptable.

British Gas Slagging Lurgi

A more fundamental limitation on the so-called "dry ash" Lurgi is that the temperature (and therefore the throughput rate) must be limited to avoid excessive ash agglomeration. This may be overcome by exploiting the fluidity of the slag, by operating at substantially higher temperatures and removing the slag as a liquid. British Gas have modified their Lurgi at Westfield, again with American collaborators, as in Fig. 8.6b. The reactor is now refractory lined and water cooled. Steam and oxygen are supplied through tuyeres and the molten slag is tapped into a quench chamber within the pressure system. Pressures are similar to those in the standard Lurgi. Temperatures within the reactor are 600–1500°C and exit gas temperatures are about 500°C. The tars may be recycled by being sprayed onto the top of the bed (incidentally reducing the dust in the off-gases) or reinjected through the tuyeres with the steam and oxygen.

Continuous runs of up to 22 days have been achieved at Westfield, on British and US coals. High outputs per unit area were obtained, with some increase in thermal efficiency (including lower steam requirement) and less sensitivity to some operational conditions (including coal quality).

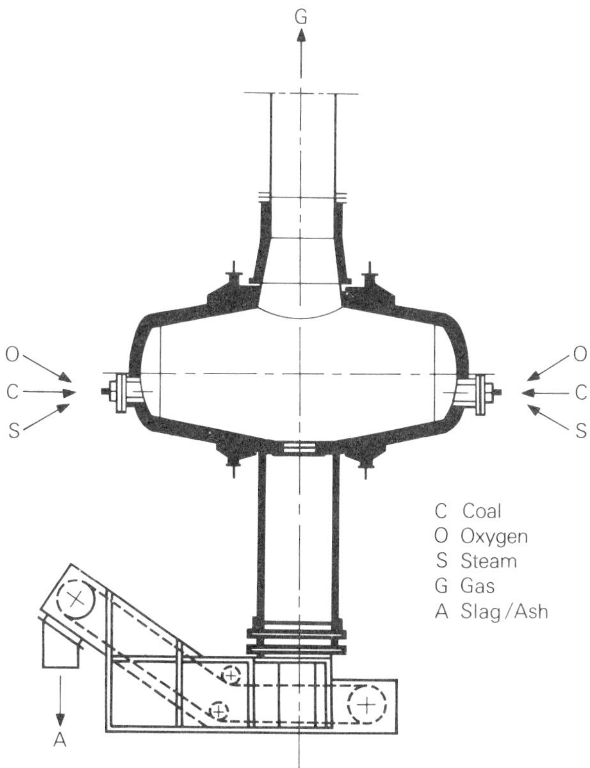

8.7 Entrained dust gasification (Koppers-Totzek)

Koppers–Totzek

There are over 50 gasifiers of this type in commercial operation, in a number of countries, mainly for ammonia production. The Koppers–Totzek gasifier (Fig. 8.7) operates at just above 1 bar, on the entrained system, in which the coal particles are blown, together with steam and oxygen, into a large empty reactor vessel. Reaction takes place in suspension; the temperature at the core of the flame may be more than 2000°C. The temperature of the gas mixture at the centre of the reactor is about 1500–1600°C. The particle size must be fine (less than 0.1 mm) because of the short residence time (about 1 s). The slag forms molten droplets, part of which flows down the sides of the gasifier to be quenched and granulated; the reactor walls become coated with a thin solidified layer which protects the refractory lining. Some of the finer slag particles leave with the hot gas, which is quenched by direct water spraying. This removes most of the dust, to an adequate standard for some applications; further cleaning can be carried out by electrostatic precipitators. In more recent designs, the scrubbers may be replaced by cyclones.

As shown in Fig. 8.7, the reactor has two opposite burners but "four-headed" gasifiers are also in commercial use. Each burner has its own coal feed, which is homogeneously mixed with oxygen and steam (if necessary) in a jet. In the installation at Modderfontein, South Africa, operated by African Explosives and Chemicals, there were initially standard two-headed gasifiers each producing 16

Mscfd of synthesis gas from 10 t of coal per hour (less than half a standard Lurgi unit). Somewhat higher outputs can be achieved with existing two-headed designs and more than twice as much with four-headed gasifiers; it is stated that larger capacities can be designed.

One of the attractive features of the K–T gasifier is that it is not in the least sensitive to coal quality (so long as the coal can be ground finely). Carbon conversion can be very high, up to 99%. No tars or phenols are produced and the raw gas contains only about 10% CO_2 (substantially lower than other types of gasifier). Practically all the rest of the gas is CO (about 55%) and H_2 (29–32%), the composition depending somewhat on the type of coal; there is virtually no CH_4, which is an advantage when synthesis gas is required.

On the other hand, the oxygen consumption is high, up to 2.5 times that of fixed or fluidised bed reactors. The thermal efficiency is not quite so high as with some other gasifiers. It is claimed that 75% of the input energy in the coal appears in the synthesis gas; however, an allowance for the energy input to the oxygen plant would reduce this effectively to about 65% efficiency overall.

Shell–Koppers

This is a pressurised derivative of Koppers–Totzek. In the Koppers–Totzek, heat is transferred to the reactor cooling jacket and the substantial sensible heat of the gas product is also lost (or transferred) on cooling the gas for cleaning. Although every effort is made to recover this heat in the most valuable form, this objective is made very difficult by the large reactor and gas volumes resulting from atmospheric pressure operation. In order to facilitate heat recovery and to improve the thermal efficiency, the Shell–Koppers gasifier is operated at pressure (about 30 bar); the product gases are then incidentally also at a pressure generally more convenient for subsequent uses.

The gasifier arrangement (two burners) is similar to the K–T reactor, except that the coal feed and slag removal systems operate through pressure locks. The pressure vessel shell is cooled by an internal tube wall (integrated with the waste heat boiler system) in which saturated steam of 50 bar is raised. The tube wall is protected by a thin refractory coating. The hot gases leaving the reactor are quenched by recycled synthesis gas (at 100°C) down to 800–900°C in order to solidify the slag particles before the gases enter a waste heat boiler, where superheated steam of 500°C and 50 bar is raised; the gases leave the boiler at 320°C. The main cooling system can therefore be regarded as a high-efficiency steam generator of relatively modest capital cost. In future designs, the reactor tube wall may be replaced by an uncooled insulated refractory lining, thus concentrating the heat recovery operation still further.

From the waste heat boiler the gases pass in turn through economisers, a cyclone (in which 90% of the solids are removed) and a series of scrubbers and separators, which reduce the solids content to 1 mg/Nm^3 and the gas temperature to 40°C. A simplified flowsheet is given in Fig. 8.8.

A pilot plant of 6 t/d has operated since 1976 and a 150 t/d gasifier since November 1978, both very successfully. With hard coal a conversion of 99% is obtained whilst producing a gas of only about 1% CO_2. The H_2 content is about 33% and CO about 63%. With German brown coal, higher concentrations of water (11%) and CO_2 (6%) are obtained; the CO and H_2 contents are proportionately lower. With all coals there is negligible CH_4 content. The calorific value of the gas is about 300 Btu/scf (11.3 MJ/Nm^3).

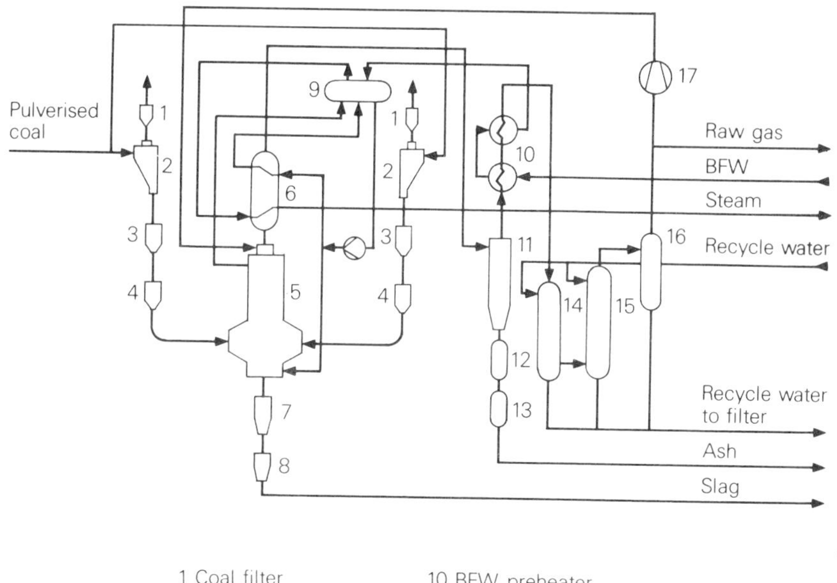

1 Coal filter
2 Cyclone hopper
3 Lock hopper
4 Feed hopper
5 Gasifier
6 Wash heat boiler
7 Slag breaker
8 Slag lock hopper
9 Steam drum
10 BFW preheater
11 Cyclone
12 Ash hopper
13 Ash lock hopper
14 Venturi
15 Scrubber
16 H.P. separator
17 Recycle gas compressor

8.8 Shell-Koppers coal gasification process flowscheme, 150 t/d Hamburg pilot plant

Of the thermal energy in the coal about 82% can be recovered in the chemical energy of the gas and about a further 15% from the steam raised. Allowing for the energy consumption in the oxygen plant and all other equipment, the net thermal efficiency is about 78% for a 12% ash, 7% moisture hard coal. The oxygen demand is about 0.8–1.0 t/tonne of coal (daf basis), the lower figure being for brown coal. About 2000 m^3 of raw gas is produced per tonne of hard coal.

One or two 1000 t/d pilot plants are scheduled for 1983/4 and this would allow scale-up to full-scale (2500 t/d) by 1990.

Winkler

This fluidised bed system was developed in Germany, starting in 1926. The objective was to produce synthesis gas from small friable coals (0–8 mm), not suitable for gasifiers then available. About 40 commercial gasifiers have been built and a number are still in operation, in India, Turkey and East Germany, typically with a bed diameter of 5.5 m, a coal throughput of 35 t/h and producing 50 000 m^3/h of synthesis gas.

The Winkler gasifier is illustrated in Fig. 8.9. Coal is fed via a screw feeder into a fluidised bed blown through a grate with steam and oxygen (air may be used where

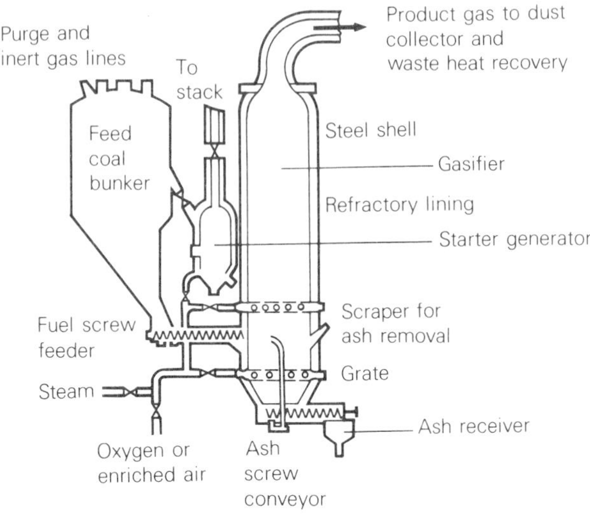

8.9 Winkler gasifier

the product requirements are suitable). There is a high freeboard and a second injection point for steam and oxygen above the bed to improve the gasification of char and tars. There is a scraper which assists the removal of lightly sintered ash.

The gasifier normally operates at atmospheric pressure and the bed temperature may be between 800–1000°C, chosen to avoid undue ash fusion. Because of limitations on temperature, reaction rates are too slow with hard coals and the process works best with younger reactive coals; a relatively high ash fusion temperature is an advantage. Since the fluidised bed composition is uniform, the ash removed from below the bed has the same carbon content as the bed and there may also be significant carbon losses in the fly-ash, in spite of attempts to improve burn-out in the freeboard. The gas leaving the reactor has a high dust loading; it enters first a waste heat boiler and then a rather elaborate system of cyclones and scrubbers. Erosion of the boiler surfaces by the ash can be a problem. The dry dust from the cyclone may be used in a boiler but the wet dust is rejected with the under-bed ash.

When operated on steam and oxygen, the gas contains about 2% CH_4, 16% CO_2, 35% H_2 and the remainder CO.

It is stated that the gasifier can be operated at pressures up to 4 bar although it is not known whether this is actually practised anywhere; it could have considerable advantages, especially on the size of the gas cleaning equipment. However, new versions designed to operate at higher pressures and temperatures are being developed. One such development is in West Germany at the Rheinische Braunkohlenwerke, where the objective is to improve efficiency and the quality of the gas. The temperature is 1100°C and the pressure is 10 bar; recycling of flue char and the feeding of additives (which may help to control ash fusion) are study objectives. Further developments on the pressurised Winkler are in hand by Davy Powergas in Florida. Winkler is also included for residual char gasification in a scheme for direct hydrogasification.

Otto–Rummel and Saarberg–Otto

The development of this type of reactor has been going on for more than twenty-five years and three atmospheric pressure gasifiers were built, with a throughput of 8 t/h, operating at 2400°C. The concept is partly an entrainment system but the jets feeding coal, steam and oxygen are aimed tangentially at a pool of molten slag at the bottom of a vertical, water-cooled, cylindrical vessel. The slag bath, which is caused to rotate by the impingement of the jets, provides a heat transfer medium. Various elaborations of this theme have been proposed, including a rotating vessel and a double shaft version in which the combustion and gasification reaction zones were separate, the rotating slag transferring heat. However, none of the variants became established commercially.

The idea of using molten slag as a heat transfer medium remains attractive and a pressurised (25 bar) version is being developed by Saarbergwerke in West Germany. A pilot plant of 11 t/h of coal was commissioned in December 1979. The nominal output is 18 Mscfd of synthesis gas (30% H_2, 55% CO). The development areas to be studied are stated to be the controlled injection of the coal and the removal of the slag. Other problems may be the recycling of char and the efficient recovery of heat.

Texaco

The Texaco coal gasifier, which is an entrainment system, is based on the well-proven Texaco Synthesis Gas Generation Process for the partial oxidation of heavy fuel oil. In the coal version, finely ground coal is slurried with water and pumped into the gasifier where it is gasified with oxygen (or air) at a high temperature and at pressures which can be up to 175 bar. The jet is directed downward from the top of a cylindrical refractory-lined pressure vessel. At the bottom of the vessel the gas and the molten slag are quenched by a water spray and a slag quench bath. The gas leaving the reactor is cleaned and the carbon-containing fines recovered for recycling to the reactor with the coal slurry. Great care is taken to maintain a constant high solids content. The reactor temperature, which is sensitive to the inputs, is closely controlled in the range 1100–1370°C to suit the ash fusion temperature; too high a temperature results in excessive refractory wear whereas low temperatures give rise to slag lumps which would block the exit to the lock hopper. Both the coal slurry and the oxygen are preheated.

The gas contains about 15% CO_2 and very little CH_4; the rest is mainly H_2 and CO in a ratio of about 3:4. The calorific value is about 300 Btu/scf when blown with oxygen and about half that if air is used. The thermal efficiency is about 70% and the oxygen requirement about 0.6–1.1 lb/lb of coal, depending on the coal type.

A 15 t/d pilot plant has operated at Montebello in California for a number of years and a 150 t/d pilot plant at Oberhausen in Germany since 1978. Continuous test runs of 500 h have been achieved.

Other Gasifiers Under Development in the USA

The status of gasifier development programmes in the USA and elsewhere is discussed in a later section. A wide range of gasifiers have been proposed in the USA; only a few, some of which are based on particularly interesting principles, will be described briefly here. One of the reasons why such a wide range of projects have emerged is that there has been a disinclination to believe that existing processes would be suitable for the establishment of a major industry in the USA. There may have been some valid, but not overwhelming, reasons for such reflections. One factor is that a substantial number of very large plants producing SNG from a wide range of coals is likely to be needed. Existing gasifiers were not originally developed to make SNG; present gasifier sizes are such that multiple units will be required and the capital cost element will be a key factor in economic viability. US coals are often caking or may be low-grade lignites; in either case, a coincidental investment in a new mining operation might depend on the ability of the gasifier to use all the product, including fines. Perhaps the most important element, however, in the proliferation of new ideas was that the three main existing German processes, occupying the strategic theoretical areas had been around for some time; their advantages and disadvantages had been widely discussed. Some ingenious newer concepts based on this analysis arose; it remains to be seen whether any of these will prove in practice to be real improvements.

Cogas This gasification process was developed to use the char from a pyrolysis scheme (COED) for the production of liquids and gas. The gasifier is preceded by several stages of fluidised bed pyrolysis, at progressively higher temperatures, to yield a range of liquids. The char from the last pyrolysis bed is fed to the steam-blown gasifier, also a fluidised bed; some hot gas from the gasifier is fed in the reverse direction to the solids flow to provide the fluidising medium and the heat for the pyrolysis.

The novel feature of the Cogas gasifier, which is now being developed with specific reference to SNG production, is the method of providing heat for the endothermic steam reaction. Elutriated char from the gasifier is partially combusted in a separate air-blown cyclone type combustor; heated particles from the combustor are recycled to the gasifier, transferring their sensible heat. The combustor gases are conducted separately to the stack, so that the product gas is not diluted.

The pyrolysis stages were tested in a 36 t/d pilot plant in New Jersey. A 50 t/d char gasifier pilot plant at Leatherhead, UK, financed by an American consortium, completed a test programme in 1978, validating the technological concepts of the gasifier section.

Hygas The principles of this process, with some variations in detailed procedures, have been studied for a number of years at the US Institute of Gas Technology in Chicago. The gasifier consists of a series of connected fluidised beds vertically arranged within a pressure vessel. The main objective is to maximise hydrogasification at high pressures (up to 1200 psi) in order to increase the proportion of methane in the exit gases from the gasifier, thus reducing the oxygen consumption. Other objectives are to handle all ranks of coal and to avoid the production of tars in the product gas.

The gasifier arrangement is indicated in Fig. 8.10. The coal (which may be pretreated with air at atmospheric pressure and 425°C to reduce caking properties) is crushed, slurried with light oil and fed into the top reactor where it is dried and the vaporised oil is removed by the raw product gas. The coal descends through the beds, countercurrent to the gas flow. The bottom bed is the hottest, about 980°C, and is fluidised with steam and oxygen in order to provide heat and hydrogen for the hydrogasification reactions above, which take place in the two middle beds. The upper of these operates at about 600°C; about 20% of the fresh dry coal reacts here to form methane. The lower hydrogasification bed operates at about 925°C and the descending coal is further gasified here. The ash is discharged as a slurry.

It is obvious that temperature control and heat balances will be very sensitive

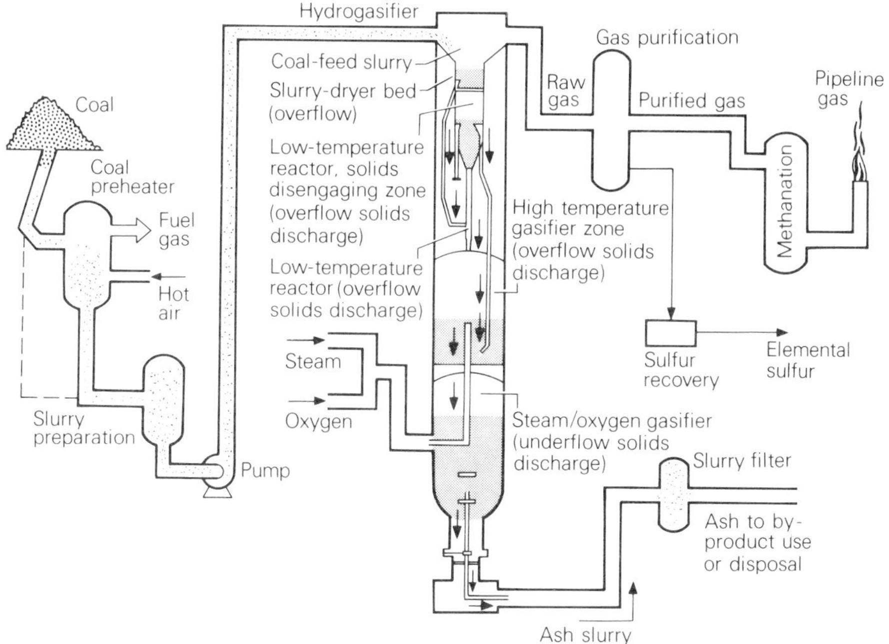

8.10 HYGAS pilot plant configuration

features in a system of this kind and will be affected by the reactivity and other properties of the coal. The balance of the endothermic hydrogen-generating and the heat-producing combustion reactions may be particularly delicate; in some earlier versions the hydrogen production was to be de-coupled; hydrogen generation by an electro-thermal method or by the steam-iron reaction was considered but the integral process seems to be preferred.

Following bench scale tests, a pilot plant of about 3 t/h has operated since 1977. Runs of several days on individual coals are the normal procedure; the longest run (on lignite) is 27 days. One period of 178 h steady state operation on run-of-mine bituminous coal has been reported; char conversion was 75%. Concentrations of methane up to 37.5% have been indicated in experiments, with CO/H_2 ratios suitable for methanation without "shifting". However, on an industrial scale the gas might have about 17% CH_4, 25% CO, 30% H_2 and 25% CO_2, with a calorific value of about 400 Btu/scf (16.0 MJ/Nm^3).

Problem areas might include:

(a) Ash agglomeration leading to clinker formation;
(b) Solids flow problems;
(c) Start-up (external supplies of hydrogen may be needed);
(d) Carbon losses by elutriation and in ash;
(e) Tar formation and cracking of the oil carrier.

Bigas This is a two-stage entrained gasifier, with slagging ash discharge, operated at 83 bar and 1650°C (in the lower hot section) (Fig. 8.11). It also is intended to produce a suitable SNG precursor. Coal in a water-slurry is fed, with steam, into the

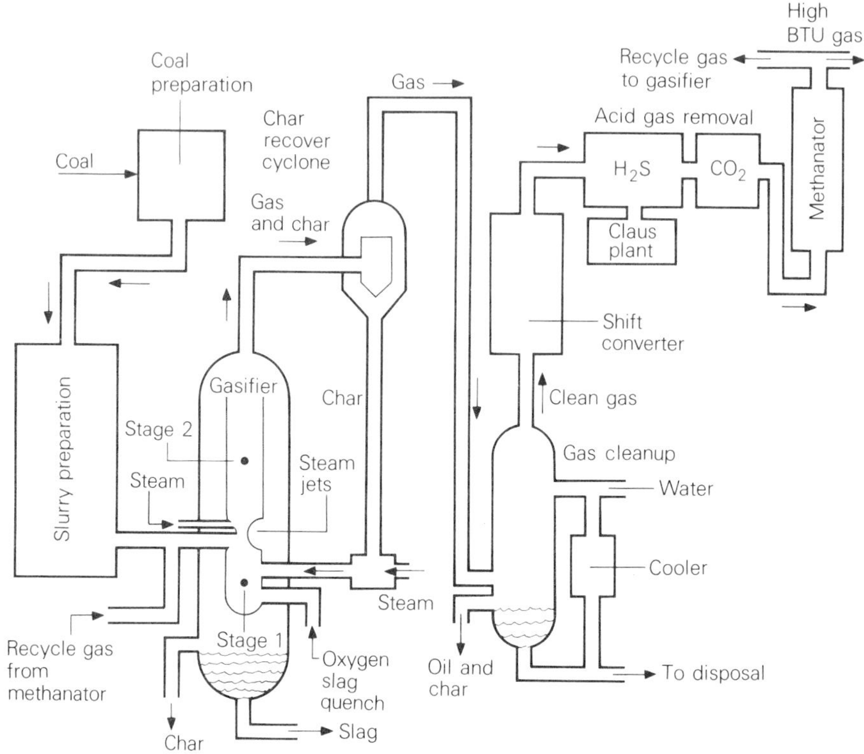

8.11 Bi-Gas process

upper section, which also receives gas from the lower section. The gas is further enriched in the upper section; the entrained char is recovered and fed back to the bottom of the lower section, where oxygen is also injected to provide heat. Molten slag falls down into a receiver below the lower gasifier.

The process has been studied for a number of years on the bench scale by Bituminous Coal Research Inc. and a 5 t/h pilot plant was completed in 1976 at Homer City, Pennsylvania. The pilot plant does not appear to have operated as originally planned and support was at least partially withdrawn in 1979.

Measurement and control of temperature, char separation and recycling are problem areas. Control of particle size is also a difficulty. Fine coal (70% less than 0.07 mm) is necessary because of the short residence time but excessive fines will complicate the char recycle system.

Westinghouse This is a pressurised (18 bar), two-stage, fluidised-bed process (Fig. 8.12) designed to produce low calorific value gas for electricity generation, with integral sulphur removal. After drying, the coal enters a "devolatiliser/desulphuriser", which is fluidised with the raw gases from the air-blown, "combustor/gasifier", fluidised bed. These gases, which contain undecomposed steam as well as hydrogen, carbon monoxide and carbon dioxide, are enriched and a little methane is formed. The gas only reaches 150–160 Btu/ft^3, however, mainly because of nitrogen dilution. This process operates at up to 20 bar.

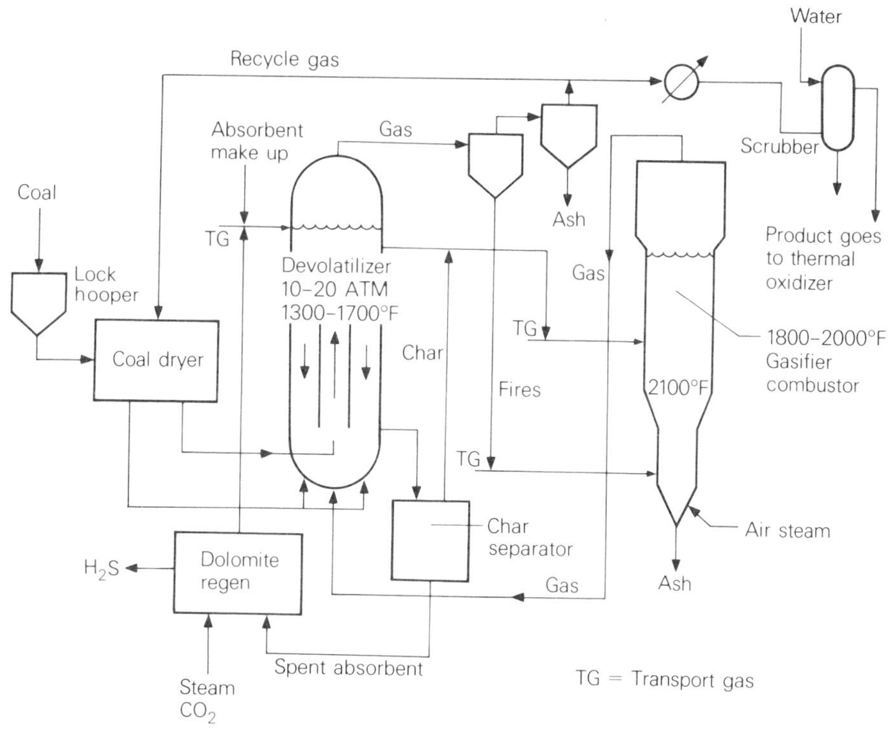

8.12 Westinghouse fluidised bed gasification, two stage pressurised process

The devolatiliser/desulphuriser has some unusual features. The bed is recirculating, with a rapid movement obtained by feeding the coal and fluidising gases into a central draft tube at velocities of 5 m/s or more. The recirculation rate may be a factor of 100 times the coal feed rate. This is intended, firstly, to discourage agglomeration and, secondly, to cool the combustor gases quickly to 700–900°C, so that methane formation is encouraged.

Dolomite is also added to this first vessel and combines with sulphur oxides. Spent absorbent is removed, presumably regularly, from the bed and is separated from char. It is then passed to a regenerator unit (another fluidised bed) where it is treated with steam, which forms H_2S for sulphur recovery. The treated dolomite is then recycled to the bed. The form of the dolomite which allows it to absorb sulphur, be selectively withdrawn from the rapidly circulating bed, separated from char, regenerated and recycled, has not apparently been published in detail. Such information, if based on actual experience on a suitable scale, would clearly be of considerable proprietary importance.

The char from the devolatiliser is fed into the bottom of the gasifier/combustor, together with steam and air, the temperature being maintained at 1170°C in the lower portion and 1000–1110°C in the upper gasifier section. Ash is agglomerated in this vessel and removed from the bottom.

A process development unit of 1200 lb/h capacity has been available since mid-1976 at Waltz Mill, Pennsylvania. This unit has been worked up in stages, with

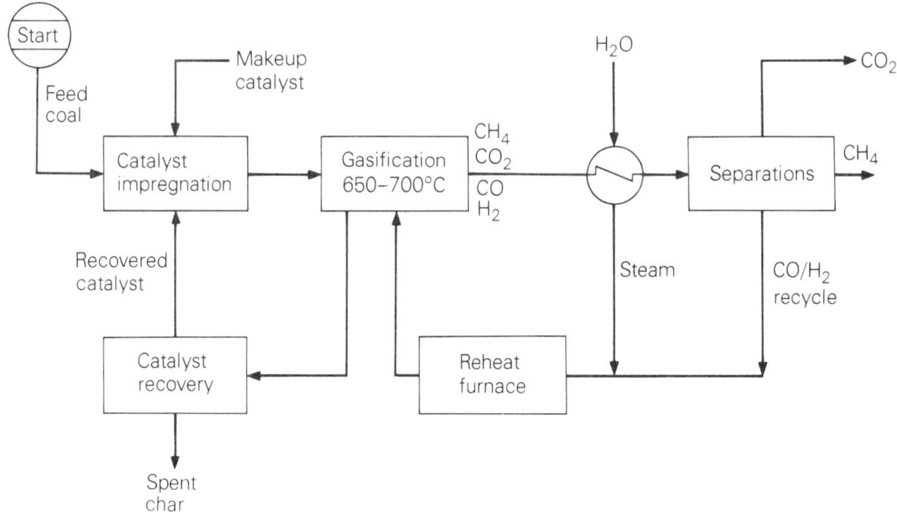

8.13 Exxon catalytic gasification process

integrated working in 1979. In some runs, the gasifier was blown with oxygen, producing gas of up to 270 Btu/scf, suitable for MBG or synthesis gas or for conversion to SNG, in addition to use in power generation. Highly caking eastern coals have been successfully handled and the process produces an agglomerated ash with good disposal characteristics. Problem areas, in addition to the feasibility and reliability of the interlinked systems, may be fines elutriation and recycling, recovery of dolomite and thermal efficiency.

Exxon Catalytic Gasifier In this process the coal is impregnated with catalysts (e.g. the basic and weak acid salts of potassium, such as potassium carbonate) with a view to accelerating the rate of steam gasification and enhancing the methane concentration. This should enable the gasifier to work at a low temperature, possibly 650–700°C, and reduced pressure (500 psi). Other objectives are to eliminate the need for oxygen, shifting and methanation plants. The gases leaving the gasifier contain CH_4, CO and H_2; the CH_4 is removed cryogenically and the CO and H_2 are totally recycled to the gasifier to promote further CH_4 formation. The recycled gases are heated about 80°C above the gasifier temperature in a reheat furnace to make up for system heat losses. The process scheme is shown in Fig. 8.13. The catalyst is partially recovered from the coal by leaching; about 25–50% may be lost, together with some carbon.

Bench scale work has been conducted by Exxon since 1968 and on a process development unit at Baytown, Texas, since 1978. The catalyst impregnation stage is believed also to reduce caking properties but the ability to accept all coals needs to be demonstrated in the pilot plant. Other current development objectives include catalyst recovery at high levels and the avoidance of impurity build-up in the CO/H_2 recycle leg. The potential advantages of the process, however, are so attractive that considerable efforts are being deployed to overcome problems.

Rockwell Hydrogasification Process This process is intended to produce high calorific value gas by the direct reaction of hydrogen with coal, which is

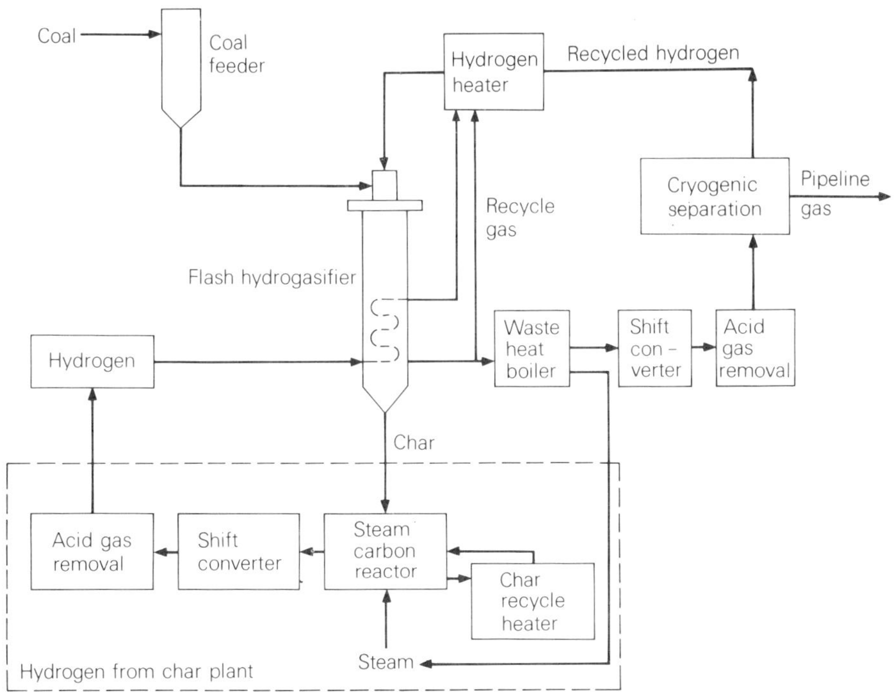

8.14 Hydrogasification process

thermodynamically the most efficient route. A number of hydrogasification processes have been considered; this version uses rocket technology principles to achieve rapid heating and short residence times with a view to achieving high carbon conversion and the acceptability of all coals. A conceptual flowsheet is shown in Fig. 8.14. Useful results have apparently been obtained on a 0.75 t/h scale hydrogasifier but integrated operation with a hydrogen generator has not yet been attempted. The process is of interest also for chemicals production.

Use of Nuclear-derived Energy for Coal Gasification

The general concept of linking nuclear energy to process chemistry has been suggested since the early development of nuclear power but few practical schemes have emerged. One particular problem is that many industrial chemical processes involving heat inputs require high temperatures, above that available from most nuclear reactors. The High Temperature Gas Cooled Reactor (HTGCR) is of particular interest in this connection since exit coolant gas temperatures of up to 950°C are hoped for; this seems a formidable target however for a full-scale reactor and any demonstration plants which may be built this century are likely to have much lower temperatures.

The objective in linking nuclear power with coal gasification is to provide some external energy for the process; this energy would otherwise be provided by consuming some of the coal, inevitably reducing the efficiency of carbon utilisation. The nuclear processes are therefore classified as "allothermic", as distinct from the usual "autothermic" processes. Practical development is more or less confined to

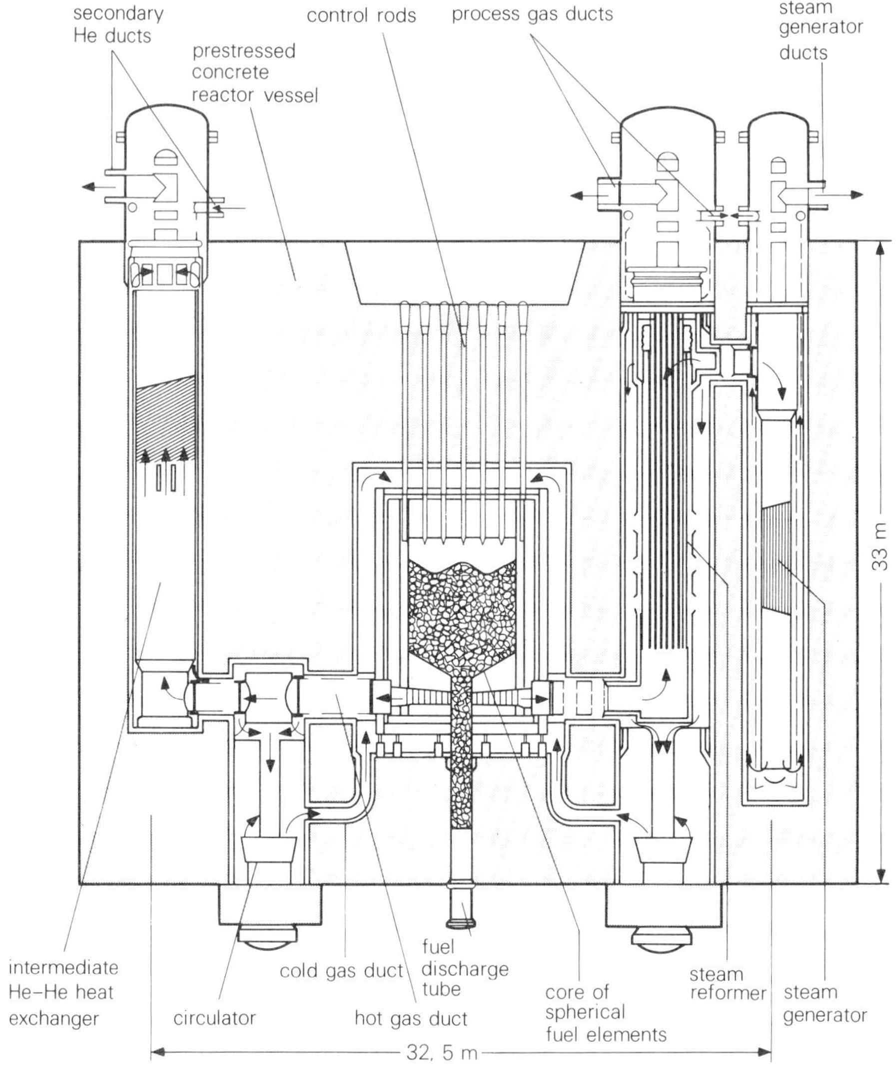

8.15 Cross-section of the nuclear heat supply system

West Germany, where types of HTGCRs have long been of particular interest; brown coal gasification, which can proceed at lower temperatures and is therefore particularly suitable for utilising the marginal temperatures which may be available, has also long been a major German preoccupation. There are two main lines of development being pursued in Germany; both use straightforward gasification chemistry.

Figure 8.15 shows the HTGCR design used for process heat and Fig. 8.16 shows one of the two systems of coupling. The first system merely supplies extra heat to raise steam and to sustain the steam gasification process. The gasifier, shown in Fig.

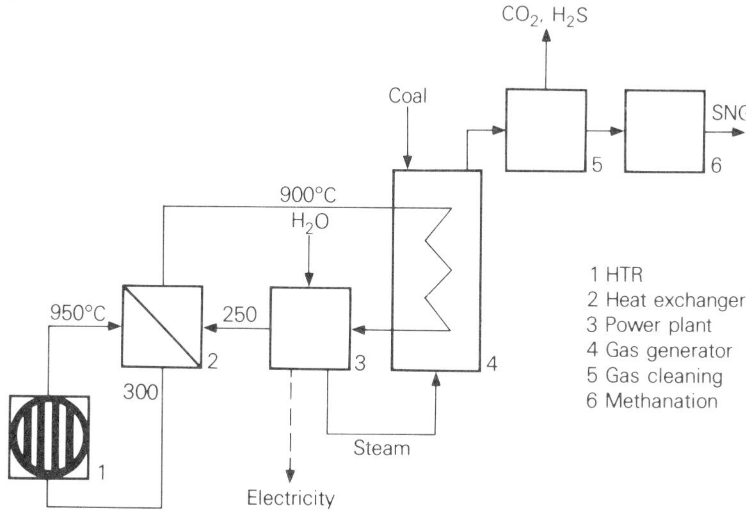

8.16 Simplified process scheme for steam gasification of low volatile bituminous coal using nuclear process heat

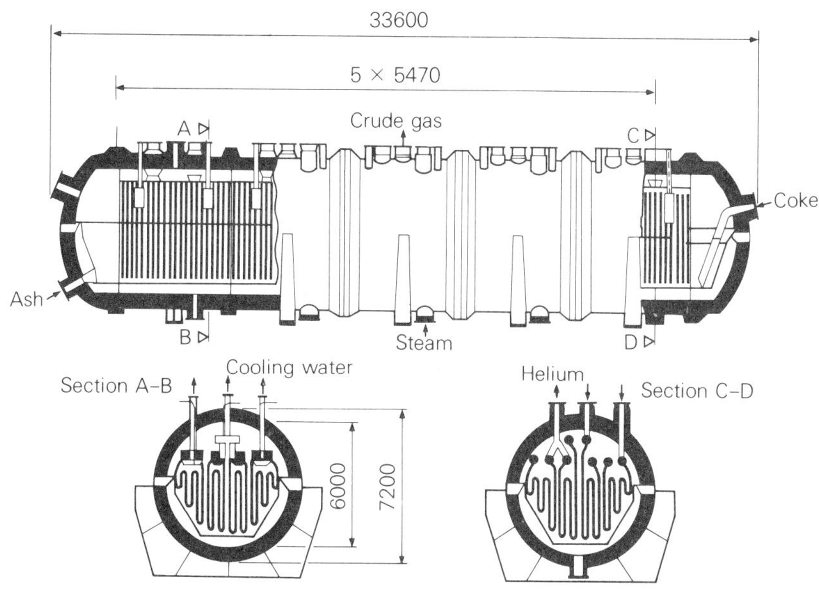

8.17 Preliminary design of a technical scale gas generator suited for steam gasification of coal using nuclear heat

8.17, consists of a helium tube immersed in a fluidised bed; the helium used in this bed however is not heated directly by the reactor but through the helium–helium heat exchanger so that the helium reactor coolant is kept separate from the gasifier circuit. The double heat exchange is at present considered necessary on safety grounds; this results in some drop in available temperature, to an extent unspecified (but perhaps about 50°C). The secondary helium, in heating the gasifier, is allowed to cool to about 800°C for hard coal and about 700°C for brown coal, these being the temperatures below which useful heat supply is not considered to be feasible. It is anticipated that a gasifier as shown could gasify 30–50 t/h; in the laboratory a true section of this gasifier has processed 200 kg/h. The crude output gas consists mainly of hydrogen, carbon monoxide and methane.

In the second process, the nuclear heat is used in a methane reformer, located within the concrete pressure vessel of the reactor; this produces hydrogen which is then fed into a pressurised fluidised hydrogasifier. Some of the methane finally produced is by-passed back to the reactor reformer. Apparently, in the exit gas from the hydrogasifier, there is surplus hydrogen which is separated and recycled. There is also "residual coke" which may have to be burnt or gasified. In a small gasifier (100 kg carbon/h), 70% of the coal could be gasified and the exit gas contained 30% methane.

Pilot plants of 1–2 t/h are proposed for both processes. The next stage would be to couple a 750 MW HTGCR to a 30–50 t/h gasifier. A commercial plant based on a 3000 MW reactor (not available "before 1990") might produce, from 330 t/h of hard coal. 280 000 m^3 of SNG by the steam gasification route and, in the case of the hydrogasification route, 395 000 m^3 of SNG from 2320 t/h of brown coal. In both cases an efficiency of 60%, including gas and electricity, is hoped for. The timetables and expected results appear optimistic.

THE TREATMENT OF GASES PRIOR TO USE

It will have been noted that the gasifier itself relies on a number of associated plant units for its operation, including the handling and treatment of coal and the provision of other inputs such as air, steam, oxygen and hydrogen as required. In addition, the raw gas as produced must be cleaned, purified and possibly have its composition changed. Gaseous and liquid effluents from all these process stages are likely to require treatment before discharge. The whole system is complex and the optimum configuration, including the plant choices for the various stages, will depend greatly on the uses of the gas. Some possible schemes have already been illustrated in Fig. 8.2.

Gas Purification and Shifting: General Considerations

No single system of gas treatment can be applicable to all systems, not only because of the wide diversity of gas specifications for the various uses but also as a result of the different characteristics of the raw gases produced. Exit gas temperatures may be between 250°C and 1500°C, with pressures between 1–100 bar. It has also already been noted that the initial concentrations of the constituents mainly desired (CO, H_2 and CH_4) can vary widely according to the type of gasifier and this applies also to contaminants. CO_2 and H_2O will normally be present, together with nitrogen if air is used in the gasifier; there may be small amounts of higher hydrocarbons. The raw gases will also contain coal fines and residues (char and ash, which may be in the form of slag particles), tars, oils, phenols and various other chemical compounds of which those containing sulphur and nitrogen may be most important. There may also be trace metals present.

First steps in treating exit gases may be to remove some or most of the particulates, the condensable tars and oils and water-soluble materials, including ammonia, phenols, chlorides, cyanides, thiocyanates (and possibly some H_2S) by a combination of cooling, quenching, washing and cyclones. Where the exit gases are hot, heat recovery by separate water or steam cooled surfaces will be more thermally efficient than by, say, water quenching. However, liquid slag particles must be solidified before passing over boiler surfaces and this will require some degree of quenching, by water or possibly by recycling some cooled product gas.

The other main stage of purification is usually the removal of "acid gases" (CO_2 and sulphurous compounds). There will generally also be associated processes for the recovery of sulphur from the products of the acid gas removal system, for the final cleaning of product and tail gases and for the treatment of waters arising. There is scope for great flexibility in the choice and arrangement of process stages. The minimisation of changes in temperature and pressure is often a criterion.

Shifting

It is now quite common for the shift adjustment (where necessary) to be carried out prior to the acid gas removal stage. The process is carried out by adding the requisite amount of steam to the gas and carrying the reaction out over a catalyst at elevated temperatures and pressures. The catalyst is very important. Since the water gas shift reaction is exothermic, the conversion is enhanced at lower temperatures (and by increasing the partial pressure of steam). However, at temperatures where the equilibrium concentration of hydrogen is satisfactory, reaction rates are very slow and must be enhanced by catalysts. The most active catalysts (which enable the reaction to be carried out at lower temperatures) have tended to be sensitive to poisoning by impurities. The process is often operated in two stages: the first stage at about 400°C using a robust iron/chromium catalyst, to carry out the bulk of the conversion, followed by a second stage at about 200–250°C using a more sensitive copper/zinc catalyst to reduce the CO down to low levels (0.3% if necessary). The copper/zinc catalysts are sensitive to poisons and require to be preceded by a good acid-gas removal system (Rectisol is used satisfactorily at Sasol). Sulphur resistant catalysts have recently been developed, including cobalt/molybdenum sulphide types, which enable the shift reaction to be carried out on "sour gas". One proprietary catalyst (SSK) invented by Exxon is stated to be not only resistant to poisons but also to be active over the whole range of temperatures of interest and thus suitable for both stages of a two-stage process.

Acid-gas Removal

There is a wide variety of acid-gas removal systems, reflecting the great range of duties required; many of the processes are fully established. For some uses, very low sulphur concentrations (less than 0.1 ppm for methanation, for example) are required; in other cases (for fuel uses, for example) sulphur concentration limits may be related to local environmental emission standards, which may be based on coal or oil. The acceptability of CO_2 will also vary but not to such stringent low levels. The levels of those contaminants arising in the gasifier product will also vary widely. The sulphur concentration in the gas will depend on the sulphur content of the coal, less that washed out in the primary cleaning stages. The main sulphur component will normally be H_2S, with a small but variable COS concentration and possibly some CS_2 and mercaptans. The CO_2 concentration in the raw gas will be a function of the gasifier process and may range from very low levels to more than 20%. The permutations of acid-gas process selection are complicated by the process chosen for sulphur recovery from the off-gases. The Claus process, which is normally economically preferred, generally requires a minimum sulphur content of 15% in the gas stream; it may be desirable to choose an acid-gas removal process which will give a suitable concentration in the gas after regeneration, even if this may not be the cheapest for that stage. Some absorbers are more selective for H_2S, compared to CO_2, than others and this gives scope for two-stage processes in order to concentrate the sulphur in the first stage.

The acid-gas removal systems may be classified according to the absorber used:

(a) Amine-based systems;
(b) Hot carbonate systems;
(c) Physical solvent systems.

In each of these categories there are a number of processes, generally proprietary. Various amines, such as monoethanolamine and diethanolamine, have been used in oil refinery gas cleanup systems for many years. They are generally used in solution (at about 20%) in water or glycol. After absorbing the acid-gases, the solution is regenerated by heating, driving off the H_2S and CO_2. The amines are not very selective for sulphur and are usually used where the partial pressure of CO_2 is relatively low; some, however, especially the tertiary amines, have a certain degree of selectivity, especially when combined with physical solvents (e.g. Sulfinol). The processes are often known by the acronym representing the amine: MEA, DEA, DGA and DIPA (Adip).

The hot carbonate processes have also been used for many years. Modern versions (Benfield, Catacarb, Giammarco-Vetrocoke) use a 20–30% solution of potassium carbonate plus additives and catalysts, at 100–150°C; the acid-gases are absorbed in a counter-current contactor and are subsequently recovered by flashing and steam-stripping. These processes can have a moderate degree of selectivity for H_2S and can best be used where the partial pressure of CO_2 is fairly high; they are also more stable and have lower energy requirements than amine systems. They need to be operated at moderate pressure, however, and there may be a corrosion problem. Nearly 400 units are built or on order, however; one at Westfield has been operating for 14 years in conjunction with the Lurgi gasifier there.

The physical solvents absorb acid-gases by absorption rather than chemical reaction; the absorbed gases can be removed by heat, pressure reduction or gas stripping. They can be highly selective and thus are suitable where the partial pressure of CO_2 is relatively high. They are characterised by low corrosion and energy consumption and can remove minor components (including CS_2 and mercaptans). The solvents are fairly costly, however, and there are losses so that the processes are less suited when the impure gas has a relatively low impurity level. Rectisol (using methanol) and Selexol (using the dimethyl ether of polyethylene glycol) are well known processes.

Recovery of Sulphur

The acid-gas removal systems, through regeneration of the absorbent, give rise to a gas from which sulphur must be removed, preferably by the Claus process. This process works by combusting one third of the H_2S to give SO_2. This SO_2 can then be made to react with the remaining H_2S:

$$2H_2S + SO_2 \rightarrow 1.5S_2 + 2H_2O$$

The reactions take place in the vapour phase and the sulphur is recovered in the elemental form. The reaction is exothermic and is favoured by low temperatures; it is often therefore carried out in stages with inter-stage cooling, when the sulphur is condensed. A bauxite or alumina catalyst is usually used. When the initial H_2S concentration is relatively low, the proportional combustion is difficult to control. A "split" system may be used, in which the gas is divided; one part is completely combusted and then the two streams are recombined for the sulphur deposition reaction. Recoveries of up to 97% may be achieved in the Claus process but low H_2S concentrations, the presence of CO_2, water or sulphides other than H_2S may all reduce the efficiency of recovery, as may catalyst fouling.

For H_2S concentrations lower than about 15%, liquid phase oxidation processes are used. In these processes, such as Stretford and Sulfax, the H_2S is absorbed in aqueous salt solutions. The solutions contain proprietary oxygen carriers which can oxidise H_2S to elemental sulphur but not to higher oxides. In the Stretford Process, the aqueous solution contains sodium carbonate, sodium vanadate, and anthraquinone disulphonic acid and other additives. The solution is regenerated by contacting with air and heat, the sulphur being recovered by flotation, skimming and filtration.

Tail Gas Cleanup

The tail gases from sulphur recovery processes such as Claus will probably have to be further treated for environmental reasons. In the SCOT process, developed by Shell for this purpose, sulphur compounds are converted wholly to H_2S by hydrogenation and hydrolysis. The H_2S is then recovered using a selective amine solvent and recycled to the Claus plant. The Beaven and Trescor-M processes also start by reducing all sulphur to H_2S but use different absorption processes. Other tail gas treatment processes are based on modifications of the Claus or Stretford processes. The modified Claus process continues the reaction at lower temperatures over a catalyst (Sulfreen, CBA) or in the liquid phase in a polyglycol solvent (IFP-I). There are also thermal oxidation processes where the sulphur is converted to SO_2, which is then recovered as in flue gas desulphurisation (Wellman-Lord).

Assessment of Sulphur Removal Schemes

The possible permutations of choice of unit stages is complex and, since the purification stages may represent up to 30% of the total plant capital cost, this choice is a significant design decision. The type of gasifier will be an important factor, as of course will be the sulphur content of the coal and the upstream processing requirements. Where the latter factors are not fixed, designs will have to cater for the most onerous sulphur removal load envisaged. Sulphur removal costs will increase with coal sulphur levels but not necessarily proportionately, provided correct choices are made. Generally, the acid gas removal system choice may depend on system pressure, with amine systems favoured at ambient pressures, hot-carbonate at moderate pressures and physical solvents at high pressure.

For high sulphur syngas, a selective acid-gas process with a Claus plant is probably preferable to a non-selective process with liquid oxidation of H_2S in the regenerator off-gas. Even a moderately selective process will probably produce gases suitable for a Claus plant.

With low-sulphur syngas, it may be that only a physical solvent process will give a regenerator gas suitable for a Claus plant and this may be the preferable combination. Regenerated gas from the bulk carbon dioxide removal system following a selective physical solvent process may require further sulphur reduction treatment.

Liquid Effluents

Liquid effluents may arise mainly from condensed steam passing through the gasifier, from quench and cooling water or from blowdown water from cooling towers. The quantity and quality of these arisings can vary widely according to the gasifier process but are usually insignificant plant and cost factors. Generally the water phase is first separated from an organic phase containing oils, tars and other hydrocarbons. The organic phase may be recycled to the gasifier, worked up for by-products or used as fuel. The water phase may contain ash, ammonia, phenols and many other organic and inorganic compounds. Clarification and biological oxidation are essential processes, preceded by ammonia removal. It is often necessary to remove phenols before the biological process can be operated satisfactorily.

Dephenolation is usually carried out by solvent extraction using a liquid solvent immiscible with water, such as benzene. The well-known Lurgi Phenosolvan process formerly used isopropyl ether but now uses a proprietary mixture. After contacting the aqueous phase in a mixer settler, the organic phase is distilled to recover the solvent and crude phenols are produced. The aqueous phase then goes for ammonia removal, which may be by steam stripping. The Phosam-W process absorbs the ammonia as phosphate. The gases after ammonia recovery must be treated to remove acid-gases.

After phenol and ammonia removal, the water may be treated in a biological treatment plant using an activated sludge system. This will probably have to be followed by clarification, ozonation and/or activated carbon treatment.

Hot Gas Cleanup

As noted, most systems employ purification methods which cool the gases to near-ambient temperatures. Where the gas is subsequently to be used in a high temperature system, such as a gas turbine, it would theoretically be more efficient if the gas could be cleaned without significant cooling in order to take advantage of the sensible heat in the gas leaving the gasifier; an improvement of up to six percentage points has been indicated. Of course in the case of pressurised fluidised combustion, this hot-gas cleanup is a mandatory requirement, at least to the extent that the impurities in the gas must not seriously affect the operation of the gas turbine. In both cases, effluent gas clean-up for environmental requirements after the gas turbine and other heat recovery units would be possible but such elaboration would significantly reduce the attractiveness of these gas turbine systems.

The main contaminants considered practically for hot gas removal methods are sulphur and particulates; nitrogen compounds and trace elements have been given some thought more recently but not in practical terms. Iron oxides are conventionally used for the removal of sulphur (mainly as H_2S) at moderate temperatures by the formation of iron sulphides; the absorbent can be regenerated by oxidation. This process has been studied for the US DOE for operation up to

815°C. Particulates are first removed in a cyclone and the gases are then passed through an absorber (e.g. iron oxide/fly-ash) in a fixed bed. Two absorber beds in parallel are used, one being regenerated while the other is in use. The sulphur is recovered from the regenerator off-gas. Up to 96% sulphur removal is claimed. Calcium oxide based absorbents have also been studied but appear to be less efficient in absorbing H_2S. Molten carbonates have also been used and has the additional advantage of removing particulates; there are however operating difficulties and probably problems of alkali entrainment.

Progress on particulate removal is somewhat uncertain. This is partly due to lack of data on the details of particulate loading in gasifier exit gases and partly because the requirements of turbines are not well defined. These factors make performance predictions difficult in an area which is in any case relatively unexplored. Cyclones are well established on particle sizes down to about 5 microns and are likely to be used as pre-collectors ahead of sub-micron methods. The latter may include special cyclones employing counter-flow secondary air, ceramic filters (fibres and granular beds) and more novel methods; definitive results are not available.

There are two schools of thought on the potential application of hot gas cleanup. In some quarters, opinion may have moved in the direction that the complications of hot gas cleanup may be more costly than the theoretical gain in efficiency is worth, especially for gasification processes. However, there are some workers and cost analysts who consider that systems can be developed from which a significant part of the theoretical gain can be retained; it may be that these considerations will apply especially with fluidised gasification systems.

The question of hot gas cleanup applies of course with even more force and immediacy in the case of pressurised fluidised combustion. In this system the efficiency of hot gas cleanup has to be related to the tolerance limits of gas turbines and a suitable design compromise established. It may be that work in this area will have a significant bearing on gasification cleanup systems.

STATUS OF GASIFICATION DEVELOPMENT PROGRAMMES

West Germany

The Federal Republic of Germany is conscious of the relatively long and certainly successful coal gasification history of that country. The development programme was sharply accelerated from the mid-1970s and the expenditure from then to 1983 is estimated at 600 million DM, a major part being funded by the Federal Government and the State of North Rhine-Westphalia. The current pilot programme is shown in Table 8.1. All the main types and applications of gasification are represented. The main emphases are on increased throughput (especially through higher pressures) and widening the acceptability of coals. Although emphasis is on German developments, it is notable that the Texaco process is also being studied.

The Dorsten project (Item 4 in Table 8.1) uses the Lurgi gasifier development labelled "Ruhr 100". This can operate at pressures of up to 100 bar at temperatures of 700–1000°C, with coal throughputs up to 12 t/h; the higher pressure should produce a higher methane content (c. 18%) than normal Lurgi gasifiers. Provision is made for a separate discharge of low-temperature carbonisation gas (Fig. 8.18). The Dorsten project is operated by a consortium formed by Ruhrgas AG, Ruhrkohle AG

TABLE 8.1 Pilot plants for coal gasification, in operation or under construction in West Germany

	1	2	3	4	5	6	7	8
Operator Data	Rheinische Braunkohlenwerke AG	Rheinische Braunkohlenwerke AG	Ruhrkohle AG Ruhrchemie AG	Ruhrkohle AG Ruhrgas AG Steag AG	Shell AG	Saarbergwerke AG	VEW AG	PCV (Flick)/ Sophia Jacoba
Coal	Lignite 1 t/h	Lignite 15 t/h	Hard coal 6 t/h	Hard coal 7 t/h	Hard coal 6 t/h	Hard coal 10 t/h	Hard coal 1 t/h	Hard coal 1.5 t/h
Products	Synthesis gas	SNG	Synthesis gas 10 000 Nm^3/h	Synthesis gas Towngas, SNG	Synthesis gas 10 000 Nm^3/h	Synthesis gas SNG	Electricity from coal gas	Synthesis gas 2500 Nm^3/h
Process	High temperature Winkler-process fluidised bed gasification	Hydrogasification	Texaco-Process Coal-dust gasification	Lurgi-Pressure gasification Fixed bed gasification	Shell-Koppers-gasifier coal-dust gasification	Saarberg/Otto-gasification	Partial atmospheric gasification with air	Fixed bed gasification
Total expenses ($\times 10^6$ DM)	37	150	48	150	100	71	25	25
Contributions from public funds	65%	75%	60%	75%	Investments: none Operation: EG	75%	—	80%
Location	Frechen	Wesseling	Oberhausen-Holten	Dorsten	Shell-Refinery Hamburg	Völkingen	Stockum	Hückelhoven
Time schedule	Planning + construction 1974–1978 Test operation 1978–1981	Planning + construction 1979–1983 Test operation 1982–1983	Planning + construction to 1978 Test operation from 1978	Planning + construction 1974–1979 Test operation 1979–1983	Planning + construction 1976–1978 Test operation 1979–1980	Planning + construction 1975–1978 Test operation 1979–1981	Planning + construction to 1976 Test operation 1977–1980	Planning + construction to 1979 Test operation from 1979
Start-Operation	1978	1982	1978	September 1979	1979	December 1979	1977	March 1979

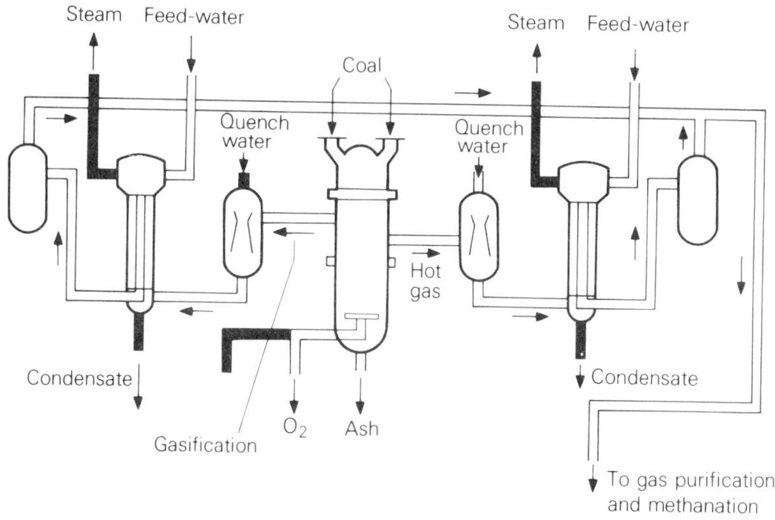

8.18 Flow sheet of the fixed bed pressurised coal gasification plant at Dorsten

and Steag AG, together with Lurgi. The same partners earlier operated an air blown Lurgi system for electricity at Lunen, described later.

The Texaco plant at Oberhausen-Holten (Item 3) has operated successfully since 1978, with continuous runs of up to 500 h; in a total of 4000 h operation, 19 000 t of various coals have been converted into about 350 million Nm^3 of synthesis gas of chemical standard.

The Shell-Koppers gasifier at Hamburg (Item 5) is the property of Shell AG and has not been supported by public funds. The total funding is approaching $50 million. It was commissioned in late 1978 and has made a series of very successful runs, the longest reported being of 108 h.

The Saarberg–Otto project (Item 6) was commissioned in 1978. The test runs have focussed on what are stated to be the "key problems": controlled feedstock injection and slag removal.

In the high-temperature Winkler project at Frechen (Item 1) the pilot plant was commissioned in 1978. Tests are planned eventually to reach 10 bar and 1100°C with air or oxygen/steam mixtures; this is being approached gradually, the latest available reports indicating temperatures of 850–950°C and pressures of 9 bar with air and 5 bar with oxygen.

The plant at Hückelhoven (Item 8) is operated by KGN, a joint company formed by PCV (a chemical company) and Gewerkschaft Sophia Jacoba. It uses a fixed bed process operating at 6 bar to produce synthesis gas from high ash coal residues (middlings), an interesting development.

The VEW plant at Stockum (Item 7) gasifies only part of the coal, with the objective of reducing sulphur emissions more than proportionately; both gas and char are intended for electricity production. In successful tests with different coals, more than 70% of the sulphur could be separated by converting 50–60% of the coal.

Mainly on the basis of these pilot plant operations a number of schemes for industrial scale projects have been drawn up in considerable detail with the active encouragement and financial support of the State and Federal Governments.

TABLE 8.2 Coal gasification projects — W. Germany

	1	2	3	4	5	6	7	8	9	10	11
	Ruhrkohle Ruhrgas	Ruhrkohle Ruhrchemie	Shell	Texaco	PCV (Flick-Group)	Saarberg	Rheinbraun	Rheinbraun	Korf	VEW	Thyssengas
Products ($\times 10^9$ m^3/a)	1.5 SNG	0.7 Synthesis gas; for Ruhrchemie and Thyssengas 50% each (SNG-production)	0.6 Synthesis gas	0.65 Synthesis gas	1.1 Synthesis gas for conversion to SNG	0.8 Synthesis gas for a comp. power plant	1 Synthesis gas	0.7 SNG	Reducing gas for direct reduction of iron-ore	Coke and gas for an 800 MW-Comb. power plant	0.1 SNG
Coal ($\times 10^6$ t/a)	3 German hard coal	0.4 German hard coal	0.3 Hard coal	0.36 Hard coal	0.5 Hard coal	0.4 German hard coal	2.25 Raw lignite	5.0 Raw lignite	0.1 Hard coal	1.8 Hard coal	0.35×10^9 m^3/a Synthesis gas from RAG/RCH-project
Process	Lurgi-Pressure gasification Fixed-bed gasification	Texaco-Process coal dust gasification	Shell-Koppers-Gasifier coal dust gasification	Texaco-Process coal dust gasification	Fixed-bed gasification	Combined process with Saarberg/Otto-gasification	High temperature Winckler-Process Fluidised-bed gasification	Hydro-gasification	Saarberg/Otto-Gasifier	Partial atmosph. gasification with air	Methanation in a fluidised bed
Planning Construction Operation	1980/82 1961/84 from 1984	1980/82 1981/84 from 1984	1980/81 1981/83 from 1983/84	1980/83 1983/85 from 1985	1980/81 1982/84 from 1985	1980/83 1983/84 from 1985	1982/85 1982 from 1984	1984/87 1987/90 from 1990	1980/83 1983/84 from 1985	1980/83 1983/85 from 1985	1980/83 1984/85 from 1986
Location	Ruhrgebiet	Oberhausen-Holten Ruhrchemie site	depending on the coal	Power plant Rheinpreussen Moers-Meerbeck	Hückelhoven Mining area Sophia-Jacoba	Saarland	Berrenrath	Rheinisches Braunkohlen-revier	not yet decided	VEW-Power Plant Gersteinwerk, Lippe and Emsland	Oberhausen-Holten

Fourteen coal conversion projects have been proposed, with a total cost of about 13 billion (10^9) DM. Eleven of the projects are for gasification (Table 8.2). It is anticipated that a substantial number of these gasification projects will definitely go ahead. The processes and the participants have generally become well established and the sizes of the projects represent reasonable scale-up factors. The relationships between the processes and the projected markets seem also to be well conceived. A decision was expected by about the end of 1980. It is thought however that the earliest date for operation may be 1987, rather than the earlier dates given in the table.

Six of the projects are for synthesis gas and are sized appropriately (0.3–0.5 Mt/a of coal) for operation with large chemical plants. The largest proposal (Item 1 in Table 8.2) by Ruhrkohle AG and Ruhrgas AG is for public gas supply and will be built up to 1.5 billion Nm^3/year in stages; the first stage will be based on the established Lurgi gasifier but it is expected that later stages will be able to take advantage of an optimised "Ruhr 100" type. The intended product is SNG but the possibilities of town's gas, synthesis gas and methanol will be examined.

In addition to the autothermal projects, a joint venture company, PNP, between Ruhrkohle AG and Rheinische Braunkohlenwerke AG has been established to do detailed engineering on nuclear assissted processes. A decision on a prototype of this kind is expected in 1984–7.

USA

Massive and widespread support has been given to coal gasification research, development and demonstration in the USA for some years, especially since the energy crisis of 1973. However, the precise manner and timing of full-scale commercial exploitation is uncertain; perhaps the uncertainty had become greater by the end of the 1970s than in the middle of the decade, when it seemed that a large and early unsatisfied demand for gas would push coal gasification into regular industrial use without undue delay or over-sophistication. Subsequently, with the release of more natural gas and other supply/demand developments, the technical and economic choices, compounded by political and environmental considerations, seem to have become more difficult. In particular, the question of how the first major plants, each costing a billion dollars or more, will be funded, and how costs will be recovered in prices, are difficult political problems. The solution to these problems may now emerge with the formation of the Synthetic Fuels Corporation by the US Government with substantial funds to encourage coal conversion developments.

For several years the US Department of Energy annual expenditure on coal gasification (excluding *in situ*) has been of the order of $170 million and there has been substantial non-Federal support. The Federal programme, the organisation and stated objectives of which are most commendable, is divided into the following categories, with average annual funding over 1978–80:

(a) High-Btu gasification for pipeline quality gas (SNG) — $13m.
(b) Low-Btu gasification for industrial and electricity generation — $15m.
(c) "Third generation" processes, covering both (a) and (b) — $16m.
(d) Special projects and support studies — $17m.
(e) Technical support — $20m.
(f) Demonstration gasification plants — $88m.

Under (f), following a competition for a high Btu Demonstration Plant, two projects were selected in 1977 for substantial initial engineering and associated studies, although only one was to proceed to detailed engineering, construction and operation (believed to be on a 250 Mscfd scale). The two processes selected were the British Gas/Lurgi Slagging Gasifier (with Conoco as contractor) and the Illinois Coal Gasification Group's Cogas Gasifier. Conceptual designs of a commercial scale plant have subsequently been completed for both processes. No major problems have emerged and final evaluations were being carried out in 1980 with a view to selecting one process, with construction to start in FY 1981. Other major projects in hand include:

(a) *Memphis Light, Gas and Water Industrial Fuel Gas Project.* This is a project to make gas for distribution in a local network, mostly as 300 Btu/scf fuel gas. Some will be methanated and arrangements will be made to take methane from the natural gas grid for dilution and distribution if the gasifiers are not operational. An oxygen-blown fluidised bed gasifier (U-Gas, developed by IGT) operating at 5 bar and 1024°C will gasify about 3000 t of coal/d to produce 155 Mscfd of gas. The conceptual design and pilot plant work have been successfully completed. The project would cost about $700 million and a decision on whether to proceed should be taken in FY 1980.

(b) *Texaco — Cool Water Gasification Programme.* The 165 t/d Texaco gasification plant in Germany has been mentioned earlier — a plant of similar size is now in the start-up stage for the Tennessee Valley Authority at Muscle Shoals, Ala. A larger plant of 1000 t/d initiated by Texaco and Southern California Edison Company will be built at Edison's Cool Water Generating Station in Mojave Desert, California.

The consortium has been joined by EPRI, Bechtel and General Electric to support a programme for the construction, start up, testing and operation of the first US large scale facility to manufacture synthesis gas from coal and burn the gas in a combined-cycle power unit for the production of electricity. Construction of the gasifier to be used with a turbine in a 100 MW combined cycle system is scheduled to start in 1981 and to be completed in late 1983.

(c) *W. R. Grace Industrial Fuel Gas Demonstration Plant.* This proposal is for a 1900 t/d oxygen-blown Texaco gasifier operating at about 1370°C and 80 bar. The original proposal was directed towards ammonia production and included the following gas treatment features:

(i) Quenching and scrubbing, to remove ash and carbon fines, the latter for recycling.
(ii) Shifting to high hydrogen content.
(iii) Rectisol plus Claus.

Subsequently, the design concept was changed to produce methanol, followed by the Mobil process to gasoline. The redesign will be completed in 1982; if the project went ahead it would cost about $700 million.

(d) *Bigas.* The 120 t/d pilot plant requires further steady-state operation. A commercial design will be available in 1982; further policy will be determined at that time.

(e) *Hygas.* The pilot plant will be operated until end September 1980, in an attempt to establish a firm design data base. Problems include clinker formation, solids flow and breakdown of ancillary equipment. The pilot plant will be converted to operate on peat in late 1980.

(f) *Exxon Catalytic Coal Gasification.* Bench scale research and development will continue through 1981 and work on a Process Development Unit through 1983. A decision will be taken in 1983 whether to go ahead with the construction of a pilot plant, which could be operating in 1986.

(g) *Hydrogasification.* The Rockwell project is being studied on the 18 t/d scale. The integration of hydrogen generation and process economics are major technical issues. It will be 1985 before plans for a demonstration plant could proceed.

UK

Recent UK policy on coal gasification has been dominated by the discovery and continuing development of North Sea natural gas. This resulted in a complete switch from town's gas to methane, involving a new national grid and a several-fold increase in the share of the energy market held by gas, up to about 25%. The supply base has continued to expand, partly through associated gas in the northern North Sea oil fields, and the current supplies may be sustained at the present level or somewhat above for the rest of the century. Understandably, the pressure to establish coal gasification on a commercial scale has not been very high. However, research and planning exercises have continued, with increasing large-scale emphasis, especially since the mid-1970s.

The UK has traditionally been a leader in gas technology, particularly through the British Gas Corporation (BG). In the 1950s and 1960s a range of processes to suit the gasification of all the foreseeable feedstocks, together with appropriate treatment processes (purification, methanation, reforming, etc.) was established on a large laboratory scale, with particular emphasis on catalyst development. Subsequently, some parts of this work, including that on coal, were put temporarily into cold storage. However, the Lurgi plant at Westfield was retained and formed the basis of a Development Site, where the scope of the dry Lurgi gasifier was further explored and gasification integrated with methanation. The British Gas Slagging Lurgi was also developed here. Much of this work was supported by US interests. Similarly, facilities at Leatherhead (a National Coal Board laboratory) were used in the development of the Cogas gasifier for US sponsors.

These developments in co-operation with the USA have been valuable in retaining skills and facilities; at the same time, on the assumption that SNG would not be required in quantity before about the turn of the century, attention has turned to further development before processes are crystallised. British Gas Corporation has made proposals intended to extend the capabilities of the present high pressure slagging gasifier in order (a) to utilise a greater proportion of fine coal, (b) to accept a wider range of coals and (c) to maximise throughput. These objectives would be achieved by the coupling of an entrained flow gasifier to the base of a fixed bed slagging gasifier in an integrated unit called the Composite Gasifier. This could utilise both fine and lump coal, i.e. the whole make. The gas produced in the bottom entrained stage would be fed into the base of the slagger. The composite unit could operate on steam and either air or oxygen, producing either low Btu gas for industry and electricity or synthesis gases suitable for methanation, premium fuel or chemical feedstocks. The proposal when first made in 1977 was for a nominal 100 t/d plant at a cost of £12 million for commissioning in the mid-1980s. Since then, design studies have been successfully carried out and it should be possible for a decision to be taken in 1980 to proceed to construction.

In view of British Gas' considerable background in gasification, the NCB's main interest has been directed towards gasification for electricity generation. Taking account of the UK circumstances and coals it was decided to pursue a system not requiring oxygen, having the characteristics of a high gasification thermal efficiency, the potential for progressive environmental control and for producing gas suitable for a high efficiency combined cycle capable of load changing. The NCB's experience in fluidised combustion was no doubt an influence also.

The process, shown in Fig. 8.19, is a two-stage fluidised bed system comprising partial gasification (at about 1000°C) and char combustion. Dolomite is added to fix sulphur. The system can be operated at pressure and energy is recoverable in gas and steam turbines. A wide range of coals can be used since rapid internal circulation can be promoted and the fresh coal mixes quickly with a large proportion of devolatilised coal; fine grinding is not required and steam requirements are low. Overall carbon utilisation is high and sulphur fixation (at an optimum 2:1 calcium/sulphur rate) highly effective. Tar emissions are low and not a complication for heat recovery.

Following successful laboratory-scale experiments and design studies it is proposed that a pilot-scale plant of 5 t/h should be built by 1985 followed by a prototype of 50 t/h in 1990. The latter would probably be a full-scale unit which could lead to a commercial plant (1995–2000) comprising four such gasifiers. Discussions are proceeding with the UK Government and the Central Electricity Generating Board with regard to implementation of this programme.

Although UK interest has centred on SNG and low Btu gasification/electricity generation for a number of years, the potential of medium Btu/synthesis gas production, closely integrated with industrial and chemicals utilisation, has recently received some attention and a survey of suitable sites and markets is being carried out.

Other Countries

As noted earlier, some of the "first generation" German processes were adopted in other countries and some have survived. These are sometimes relatively small plants or plants whose output is specifically vital to associated enterprises; their economics may not be directly affected by international trends in energy prices. An interesting example is a battery of 24 pressurised Lurgi-type reactors which has been in service in East Germany since the mid-1960s. The reactors are 12 ft in diameter and operate on briquetted lignite gasified with steam and oxygen to produce 500 Mscfd of town's gas at 475 Btu/scf. It does not appear that these "first generation" survivals will have any influence on future developments.

In South Africa, large installations of Lurgi (Sasol) and Koppers–Totzek (AECI) operate successfully and with considerable evolutionary development or adaptation to local requirements.

Japan started a gasification programme in 1975 at the Coal Mining Research Centre, entirely with government funding. The objective is to produce clean fuel gas for combined cycle power generation. The gasifier (Fig. 8.20) is a two-stage fluidised bed type disposed vertically with a narrow throat between. Raw coal is fed into the upper bed where it is devolatilised by the hot gas stream from the lower reactor, where steam and air are introduced, together with char particles recovered from the

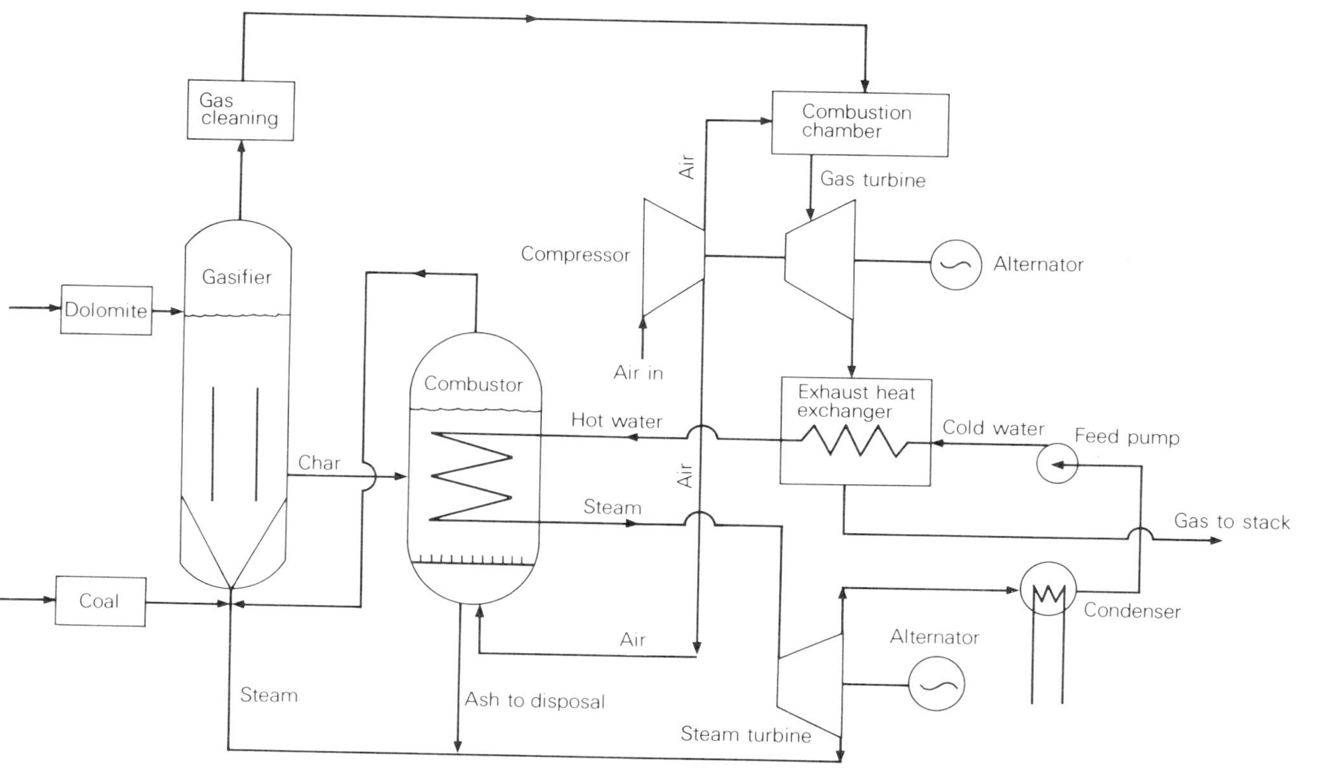

8.19 Fluidised bed coal gasification for combined cycle power generation

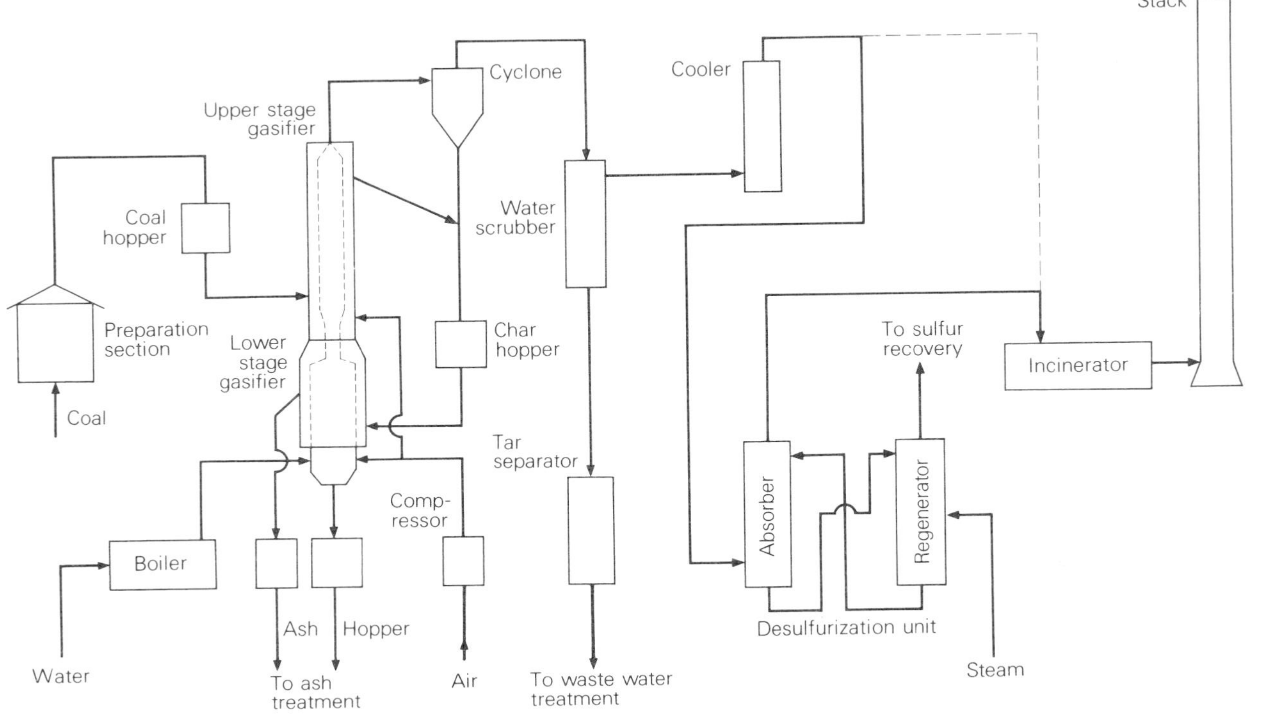

8.20 Schematic diagram of CMRC gasification plant (PDU)

cyclone after the second stage. One problem has been clinkering in the bottom reactor which has been experienced at nominal bed temperatures as low as 800°C even with coals having ash melting points of 1250°C. Clinkering can be largely avoided by increasing gas velocity but this can increase elutriation unacceptably. So far tests have been carried out on a 5 t/d test plant but a pilot plant of 40 t/d is nearing completion and this is expected to be followed by a conceptual design of a 1000 t/d demonstration plant.

GASIFICATION SYSTEMS

Although great emphasis in development is now being given to gasification for SNG, other purposes are also important and may be increasingly so. In all cases, the total system involves substantially more than the gasifier itself, which may be a relatively minor part of the installation. The balance of other plant will vary greatly, mainly according to the purpose of the gas. In addition to ancillary plant directly in the gasification stream there will also be services, some of which may be linked to the main process, e.g. electricity generation from waste heat. Both capital and operating costs may be affected; in the case of the latter, extra complication (for waste heat recovery, say) may theoretically reduce operating costs but may affect availability, especially in an early plant design.

Plant configurations which include a gasification stage are illustrated in Chapters 9, 10 and 20 (Liquefaction, Chemicals and Coalplexes). In several illustrations above of various gasifiers, greatly simplified drawings show the gasifier in relation to associated items where the main product is SNG. Figure 8.21 shows the simplified plant scheme for the slagging Lurgi proposed as a candidate for the US 500 t/h

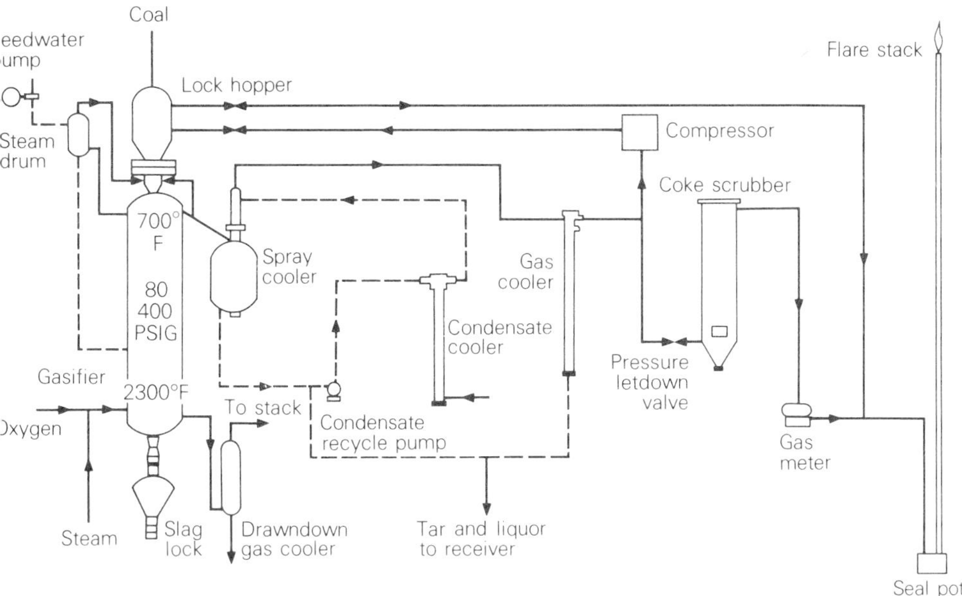

8.21 Fixed bed gasifier, slagging process

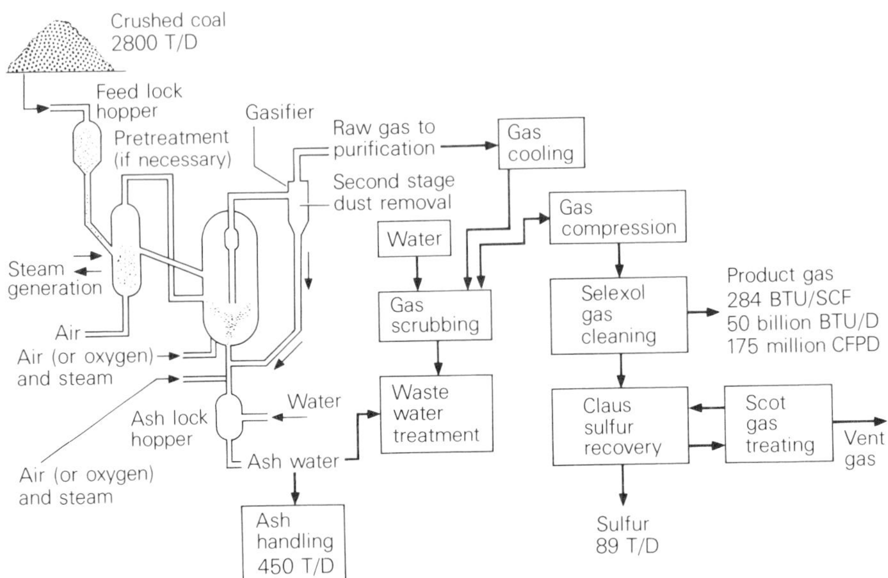

8.22 Low Btu industrial fuel gas process

demonstration plant. In this case, a recovery plant for volatiles is required. In addition to the plant items shown, an air separation plant to produce oxygen is also required, plus a number of service units.

Industrial Fuel Gas

Although fuel gas processes were once well established, their importance diminished greatly due to cheap energy alternatives. Recent revival of interest in this outlet will justify further process refinement, particularly for environmental reasons, but also to demonstrate satisfactory operation in a modern industrial context. Several such demonstration schemes are planned, especially in the USA. Figure 8.22 shows a low Btu fuel gas system (or it could be MBG if oxygen blown) for limited pipeline distribution. Figure 8.23 shows a scheme, with an unspecified two-stage conventional gasifier, intended for an in-house fuel gas supply to fulfil an industrial heat requirement (in this case, pretreatment of iron ore). The proposed size is 500 t/d coal input to produce 7.4×10^9 Btu/d.

Gasification for Electricity Generation

Gas may be produced from coal and burned in a steam raising plant linked to a standard steam turbine. There would not appear to be any advantages in such a route unless there were exceptional environmental requirements; in this case the gas could be cooled and carefully cleaned so that the final effluent would also be very clean. More frequently, some gain in thermal efficiency would also be sought, usually by means of a gas turbine, which might be part of a combined cycle. The theoretical efficiency of gas turbines increases with temperature and new designs under development are intended to increase gas temperatures to 1200°C or even

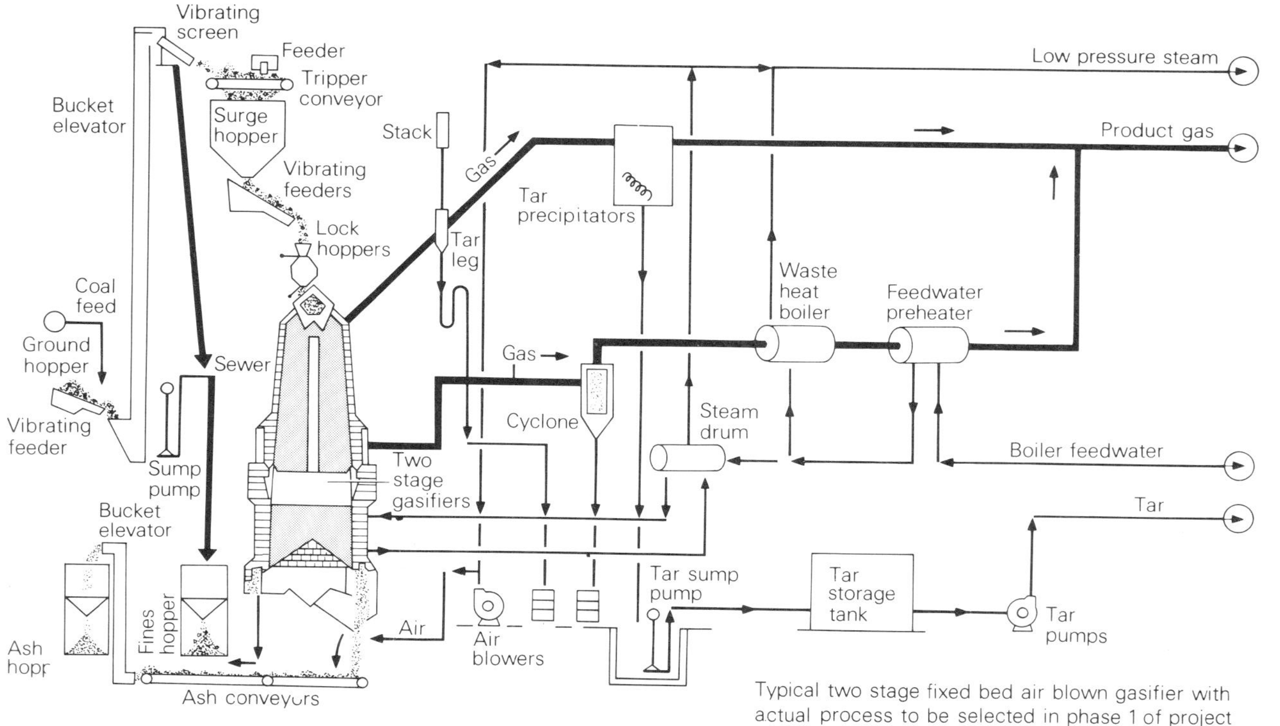

8.23 Small scale industrial fuel gas process

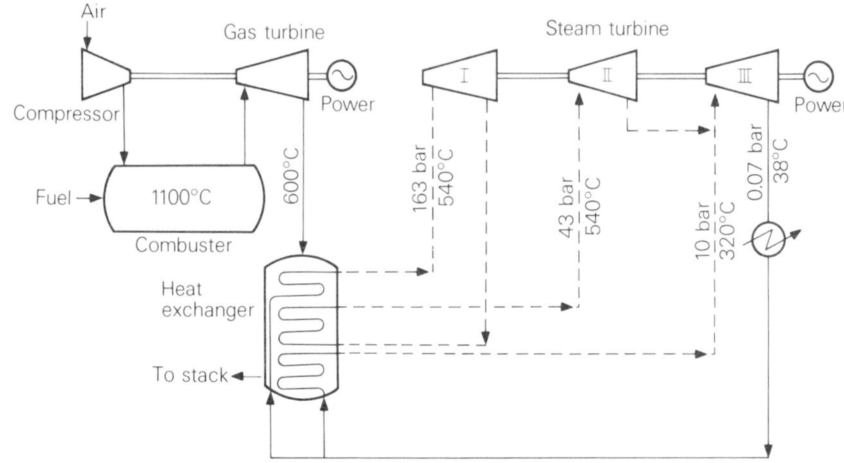

8.24 Waste heat recovery combined cycle

higher. At 1200°C, electrical generating efficiencies of 45–50% are theoretically possible. Gas turbines operating at high temperatures will exploit the gas pressure and reduce the combustion gas temperature somewhat but the residual heat will still need to be used in a steam system. Several basic systems are illustrated in Figs. 8.24–8.26. In Fig. 8.24, the gas turbine exhaust gases are used to raise steam. The gases are too cool however for an efficient steam cycle and, in Fig. 8.25, supplementary fuel is burned in the steam-raising equipment to achieve efficient conditions. In Fig. 8.26, the fuel is burned under pressure in the boiler, the exhaust gases passing to the gas turbine. This system is used at Lunen in West Germany (Fig. 8.27).

The Lunen plant is perhaps the earliest and still the most significant demonstration of coal gasification/electricity generation combined cycle. The plant has five Lurgi gasifiers, each rated at 10–15 t/h. There is a gas turbine connected to a

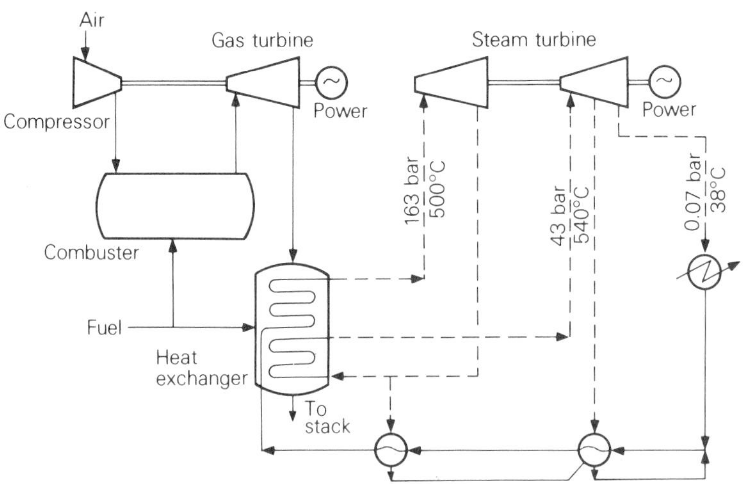

8.25 Exhaust fired combined cycle

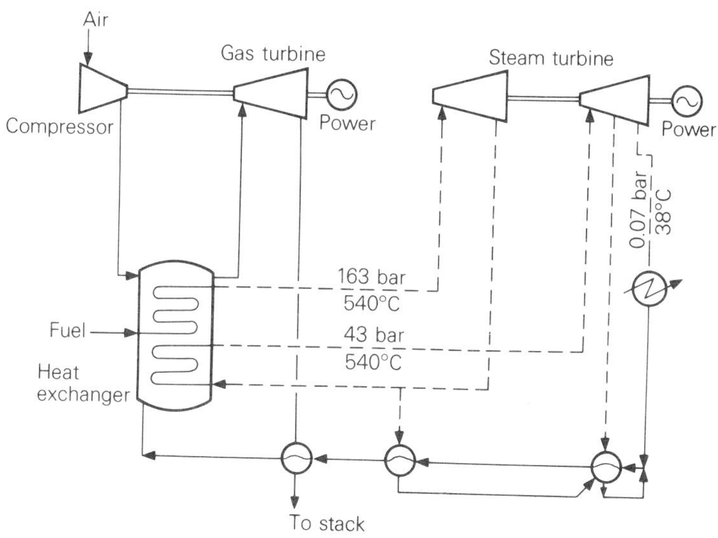

8.26 Supercharged combined cycle

74 MW generator and a steam turbine serving a 96 MW generator. The plant has had operating experience since 1972 but with breaks due to a number of operating problems on the gasification side, especially in gas clean-up and tar recycling. The combined cycle operated satisfactorily. The experience has been incorporated in later plants but the original plant is being refurbished for operation in 1980.

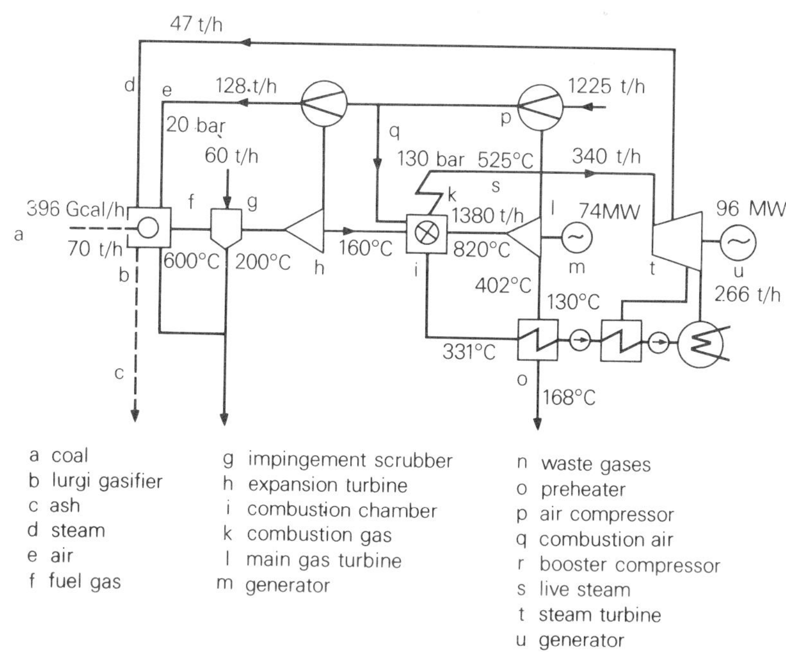

8.27 Gasification for power generation. Plant configuration of Lünen

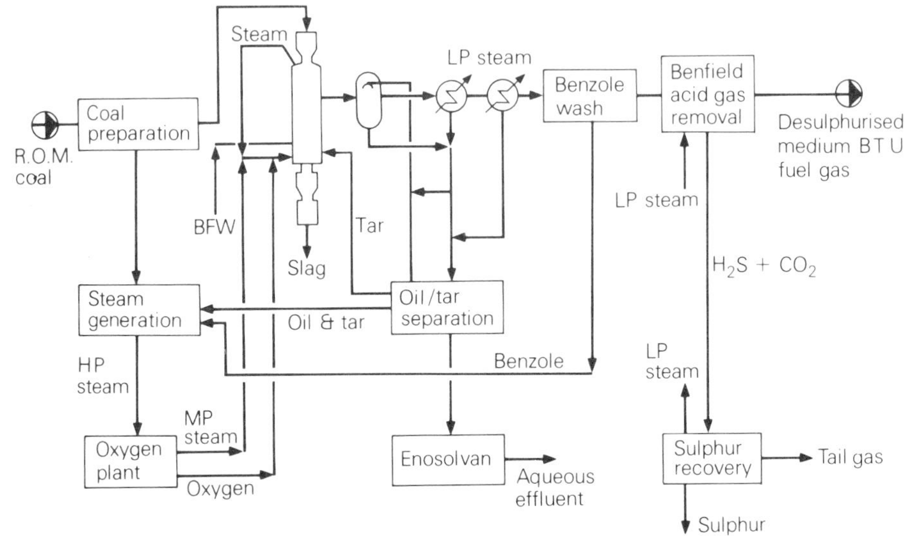

8.28 Medium Btu gas slagging-gasifier route

Medium Btu Gas (MBG)

Figures 8.28 and 8.29 show outline schemes for MBG using the Slagging Lurgi and Shell–Koppers gasifiers, respectively. The main differences are that the former requires organics recovery and acid-gas removal, followed by sulphur recovery. Since the latter has virtually no CO_2 in the raw gas, a single stage sulphur removal is feasible. There are however other and probably more significant economic factors in

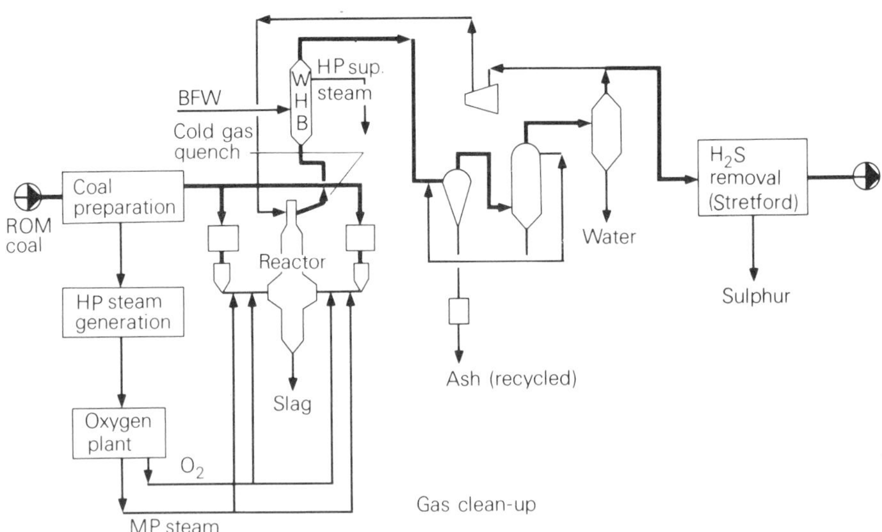

8.29 Medium Btu gas Shell-Koppers route

comparing these two processes; one of these, dependent on the end use, is the desirability or otherwise of methane in the gas.

SOME COMMON DEVELOPMENT AREAS

Over the next decade or two, the rate of progress in certain key areas of development may be critical in determining the acceptability of coal gasification, whichever processes are selected. This may seem surprising in view of the fact that there is a long and substantial industrial experience in this field already. However, the future may be more testing for a number of reasons — environmental, economic and strategic. The first of these may be the most important and persistent. For the future, close attention will have to be paid both to every designed point of access to the local environment and also to the possibility of adventitious leaks; the toxic properties of products, intermediates and wastes will need to be characterised and satisfactory controls instituted. International co-operation will be particularly desirable in this area.

Economic factors will be intensified because many future plants will be on a giant scale with high capitalisation and often with close linkages both with coal input and with product utilisation systems. Reliability will be highly important and this depends critically on materials and on the design of components. An excellent US government regular review publication on these subjects exists; whilst underlining the existing uncertainties, co-operative exchanges are encouraged. The other key economic area is thermal efficiency and this will become even more important as primary energy prices increase. There is need for less ambiguity in the expression of data and for early reporting of actual results on large-scale plants.

Other areas of general interest include the following items.

(a) *Coal handling, sizing and feeding.* Facilities for receiving, stocking and picking-up coal will form a substantial portion of capital costs. Standard equipment exists but the scope of the arrangements and the equipment chosen for homogenisation and for monitoring will be crucial items; their performance will depend, for example, on the sensitivity of the gasifier selected, the surplus capacity built into the purification train and on whether a policy for opportunist purchasing of coal is pursued. Sizing can be a difficult problem. If particle size has to be reduced, comminution costs (including inevitably high maintenance) can be substantial; on the other hand, if fines have to be excluded, logistics could be difficult. Feeding coal into reactor vessels at high temperatures and/or pressures will always be a crucial area; a number of different solutions will be required, depending on whether lump coal, fines or slurries are used. Coal feeding is one of the areas where fresh proving trials will be necessary whenever system conditions are made more onerous. The management of caking potential may also be a key area for development.

(b) *Ash removal.* This also is an intrinsically difficult operation, whether the ash is removed as particles or as liquid slag; rapid cooling and pressure let-down may be involved. Final disposal of ash and the recovery and treatment of water used for quenching are problems. In several processes, ash agglomeration is a process feature, intended to produce easily handleable ash particles, low in carbon and separable from other phases. In the USA a special test rig (6 t/d) has been built at the Institute of Gas Technology, Chicago, to study this process operation. This is essentially a fluidised bed gasifier intended to operate so that agglomeration of ash particles takes place to produce large, heavier particles for selective withdrawal. As might be expected, the acceptable temperature range varies with the type of coal. With a sub-bituminous coal, temperatures were said to be limited to about 920°C (presumably by incipient random clinkering) and controlled agglomeration was not obtained. With other coals, higher temperatures (up to about 1040°C) were possible and agglomeration was obtained but clinker remained a problem.

(c) *Burn-up.* A number of processes do not succeed in completely consuming the coal in one stage and some char (often quite fine) is rejected. This has to be brought back into the energy balance, either by recycling for gasification or by combustion to recover heat values. Combustion may be by entrainment or fluidised bed methods; the recovery and use of the heat may present difficulties, especially in large-scale integrated operations.

(d) *Thermal insulation and heat recovery.* The optimisation of systems for transferring heat from the reactants to water or steam and the subsequent utilisation of that heat is a fruitful field for further study; there are both economic and technical problems. Ideally, heat losses from the system, as distinct from heat transfer recovery, should be minimal. This depends on insulation, where further progress needs to be sought on better intrinsic properties, greater reliability, improved ease of repair or replacement and on methods of treating openings and attachments.

(e) *Instrumentation and control.* Many of the newer processes intending to improve operational efficiency and economics rely on accurate determination and assessment of conditions, linked to sophisticated control devices, often integrating process stages taking place separately. The design and proving of such systems will be a crucial development area.

(f) *Hydrogen production.* Hydrogen will play a key part in coal conversion generally. Standard processes exist for the production of hydrogen from coal via synthesis gas. However, the unique importance of hydrogen has been recognised in the USA, where a special Hydrogen-from-Coal Facility has been proposed, based on the flowsheet in Fig. 8.30, in order to establish engineering and economic data, using lignite on a 1200 t/d scale. Two Koppers–Totzek units will be used and the lignite will be dried and ground. Another, more direct, development proposed for the production of hydrogen is the Steam–Iron Process (Fig. 8.31), which can use char and does not require oxygen. One vessel is simply a gas producer. The other vessel is divided into upper and lower compartments. In the upper, the producer gas (containing CO and H_2) continuously reduces iron oxide. The reduced iron passes to the lower compartment where it splits steam introduced at the bottom, the iron combining with the oxygen; the oxide is lifted back to the top and the released hydrogen is removed at the top of the lower compartment.

(g) *Environmental impact.* The complexities of this subject have been indicated. Some common approach to the solution of problems and the establishment of standards would be valuable.

(h) *Catalyst development.* The recent advances in catalyst technology, especially in the treatment and utilisation of gases, have had a remarkable effect and it is probable that this process of advance is nearer the beginning than the end. A very high level of effort is continuing and there must be a possibility of a breakthrough with dramatic consequences in the economic balance of the different routes. The successful

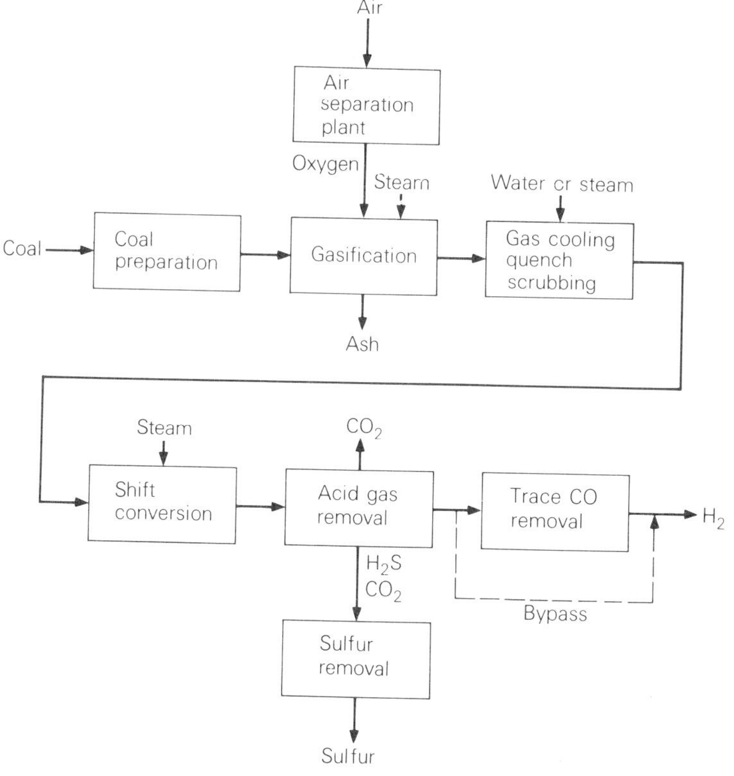

8.30 Hydrogen from coal facility

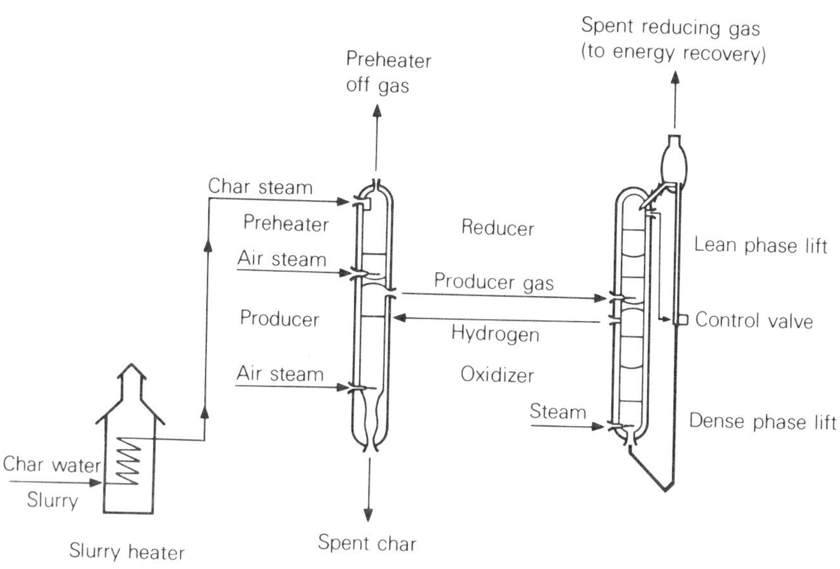

8.31 Steam iron process

development of economic catalysts to enhance methane production under less severe conditions in the initial stages could be extremely important for SNG purposes. In addition to higher specific activity, continuing aims in catalyst development include lower cost, greater stability and reduced poisoning effects. The technology of regenerating catalysts is another important area.

CONCLUSIONS

There can be little doubt that coal gasification will at some time "take-off" on a massive scale with hundreds of large-scale plants in many parts of the world. Gasification is likely to become the largest market for coal (if gasification as part of liquefaction is included) within a few decades and may even become the largest energy conversion process in the world. The circumstances which will trigger off this revolution are difficult to determine exactly but conditions will only need to change marginally to initiate it.

There is a somewhat paradoxical situation in that, on the one hand, large-scale processes are already well established and, on the other, there is a perfectly legitimate requirement for large-scale and wide-ranging development and demonstration of improved processes. Economic issues, based on technical improvements, often marginal but uncertain, become vital; these are discussed later in Chapter 15. Environmental issues are probably the other main factors in determining the direction of progress.

In 1980 it appears that real commercialisation is not likely to commence before the 1990s; this is considerably later than might have been perceived in 1975. Whether the established processes, rapidly being improved, become the major basis of this expansion or whether one or more of the newer processes become established is uncertain. It may be that the varying circumstances and local issues (not always logical) will ensure that the intense and very interesting competition will continue for many years even after the first major wave of commercialisation takes place.

Chapter 9

LIQUEFACTION OF COAL

GENERAL CONSIDERATIONS AND BACKGROUND

Introduction

The liquefaction of coal requires primarily an increase in the proportion of hydrogen to carbon by weight two- or three-fold and the atomic H/C ratio may be increased from 0.7 to up to about 2.5. In addition, the very large complex molecular structure of the coal, containing a few thousand atoms, needs to be broken down into chemical structures containing a few or a few dozen atoms. The liquid products are also generally required in a relatively pure form so this necessitates both the elimination of mineral matter and the removal of atoms other than C and H. It will be obvious that these changes are unlikely to be accomplished without considerable expenditure of energy and other resources.

The different basic methods which can be used to liquefy coal have been classified in various ways and there are subdivisions of the main routes. Names for particular versions of these processes are in common use; these generally relate to the originators or promoters of a particular scheme and are likely to be retained into commercial application. Some explanation of the main principles will therefore be given but significant projects which are described will also be referred to by the name chosen by their sponsors.

One method of classification is to divide the processes into two main types: "synthesis" and "degradation". In the former, the coal is gasified to produce a "synthesis gas" (CO and H_2 in appropriate proportions); from this gas, various hydrocarbons can be synthesised with the aid of catalysts, under pressure and moderate temperatures. In the degradation processes, of which there are several kinds, the coal structure is only partially broken down; additional hydrogen becomes associated with fragments of the original structure. As noted in Chapter 7 on carbonisation, pyrolysis is one method which can be employed for degradation. Some degree of pyrolysis is usually present in other processes, since temperatures in the range of thermal decomposition (around 400°C) are commonly used; this decomposition process can be assisted (or recombination inhibited) by the use of hydrogen and/or solvents.

Nomenclature often reflects the special interests of a particular country. In Germany, with the largest history of major developments, processes are divided into "Synthesis" and "Hydrogenation", reflecting the two main industrial processes used during World War II. Processes emphasising solvent extraction or pyrolysis mechanisms have not been of particular interest until recently. In the USA, the synthesis route, not favoured with much governmental interest, is generally referred to as "Indirect Hydrogenation". The term "Direct Hydrogenation" is used there for the type of hydrogenation initiated in Germany and "Solvent Extraction" is a separate classification, as is "Pyrolysis" (not greatly favoured by the US Department of Energy).

Synthesis

This route starts with a gasification stage, preferably by a process which gives a mixture mainly of CO and H_2, rather than one which maximises methane. However, at the SASOL plant, Lurgi gasifiers, which produce a substantial proportion of methane, are used; much of this methane from SASOL 1, together with other gases, has recently been sold externally as a fuel and adds to the overall thermal efficiency as well as substantially assisting revenue. At SASOL 2 there is no ready market; the methane is converted, substantially reducing the overall efficiency.

For the main process, the synthesis gas is purified from undesirable components, including potential catalyst poisons. The CO/H_2 ratio is then adjusted using a combination of blending with recycled gas and the "shift" reaction, in which steam is reacted with some of the CO in the gas:

$$H_2O + CO = H_2 + CO_2$$

The equilibrium depends on the reactor conditions, which are chosen to provide the desired composition changes; catalysts (usually nickel based) are used to promote the reaction. Carbon dioxide is subsequently removed. The synthesis stage itself, usually referred to as the Fischer–Tropsch reaction, is carried out in reactors in the presence of iron-based catalysts, to produce mainly paraffinic hydrocarbons with the rejection of CO_2 and/or H_2O:

$$nCO + 2nH_2 \rightarrow (-CH_2-)_n + nH_2O$$

$$2nCO + nH_2 \rightarrow (-CH_2O)_n + nCO_2$$

The nature and spread of the products is dependent on a number of factors of a complex nature, including synthesis gas composition, temperature, pressure, residence time and recycling practice and, most of all, the catalyst. There is naturally a considerable degree of commercial confidentiality involved. The original objective in Germany was diesel fuel and in South Africa gasoline. A major problem with the synthesis route, however, can be the generation of a wide spectrum of products, which may include hydrocarbons, from methane to waxes, alcohols and aldehydes. With developments, especially in catalysts, much more selectivity has been made possible. In one version of synthesis, methanol is produced and this in turn can be converted to gasoline in another catalytic process pioneered by Mobil. This, and the SASOL process, are described more fully later.

Degradation

Pyrolysis

Aspects of pyrolysis have been described under carbonisation and in other chapters. In this process, coal is heated in the absence of air and decomposes: fractions are evolved which are richer in hydrogen and smaller in molecular weight than the original coal. In carbonisation processes, where cokes are the main product, the

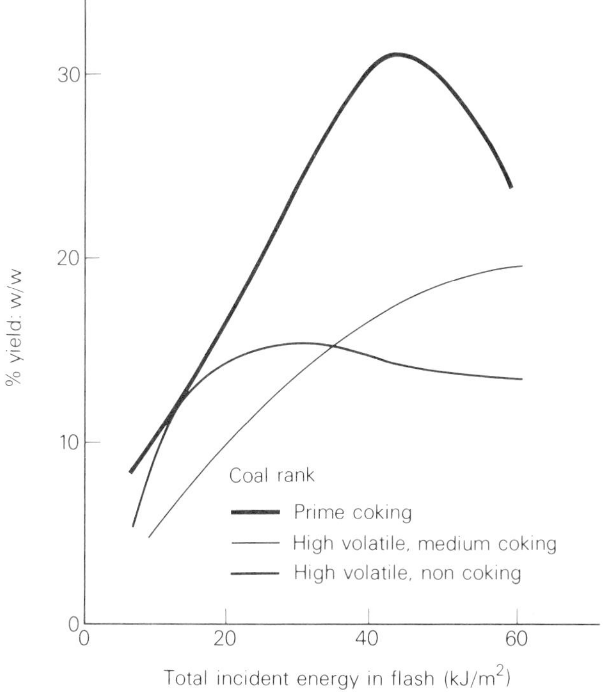

9.1 Tar yields by flash heating

initially generated volatile compounds probably recombine to some extent in the presence of the remaining carbon. More volatiles may be recovered from low-temperature carbonisation processes and these tend to be lighter and often more valuable. However, where commercially valuable liquids are the main objective of the pyrolysis of coal, it becomes necessary both to heat the coal more rapidly than is possible in a solid mass using external heating and also to remove and quench the volatiles as quickly as possible. This is illustrated in Fig. 9.1, where the tar yields of various types of coal are shown as a function of the energy release in a flash tube (and, therefore, rate of heating). In these conditions the coal particles are heated very quickly but, nevertheless, there is still a marked relationship between yield and rate of heating (which is of course related to the energy in the flash). The laboratory conditions and rates of heating would of course be difficult to approach on an industrial scale but various methods of maximising yields and liquid values practically are being studied. One possibility is the use of fluidised beds, in which particles can reach the bed temperature quickly; in one design, a series of beds of increasing temperature is used to encourage the removal of valuable liquids soon after they are released. Other processes use coal particles as a dilute phase in gases, generally in either an entrained or free-fall mode, with rapid heating. In some cases, hydrogen is the gas (flash hydropyrolysis) and temperatures of up to 1000°C or more are being explored. Since substantial quantities of char are inevitably retained, proposals are generally made for the gasification of the char (usually in an integrated plant unit) to produce hydrogen for the process or for its combustion to provide process heat.

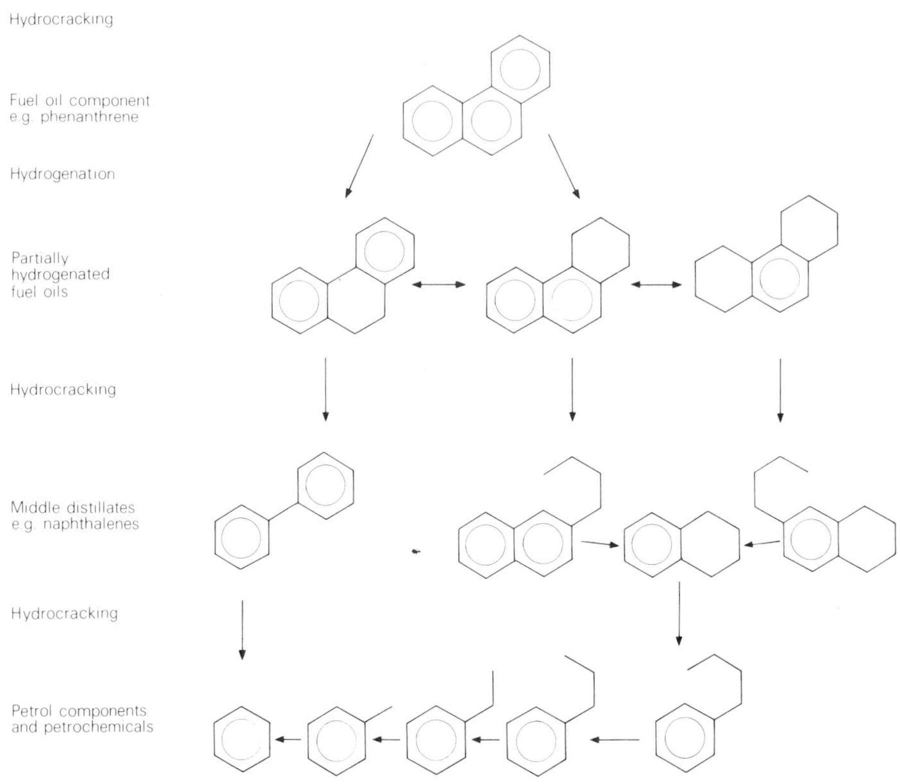

9.2 Complex coal structures and hydrocracking

Solvent Extraction and Hydrogenation

It has long been known that certain organic solvents can be used to extract portions of coal (for research purposes, some extraction can be carried out at room temperature but more typically moderately elevated temperatures are employed). This extraction method has been widely used over many years to investigate chemical structures. It has also been known for a long time that coal could react with hydrogen and that this reaction was facilitated when the coal was mixed into a slurry or paste with organic liquids. Furthermore, once heavy liquids have been derived from coal they can be progressively broken down into still lighter liquids of lower molecular weight by further hydrogenation, in a manner analogous to petroleum processing. It is becoming possible now to see this whole field in a more unified way.

Hydrogen is a key element in these degradation processes: extra hydrogen may either be introduced at the primary stage of degradation or the hydrogen already in the coal may be more favourably distributed, as a result of pyrolysis and dissolution. Hydrogenation contributes to the breaking of links between complex ring structures and prevents primary pyrolysis products from recombining to form larger molecules. Some of these bonds are between carbon atoms and others are through oxygen, nitrogen or sulphur atoms, which may partially be eliminated. The progressive breakdown is illustrated in Fig. 9.2. Here the initial breakdown products are represented by the condensed ring structure of phenanthrene. Hydrogen partially saturates and progressively opens out the ring structure, reducing the number of rings; the end product may be largely composed of single-ring structures which are useful components of gasoline. The extent of these reactions is determined by the temperature, hydrogen pressure, the contact time and conditions but, most importantly, by the catalysts, which may be specifically designed for particular parts of the hydrogenation progression.

It should be noted that most of the products are aromatic, whereas synthesis produces mainly chain compounds. The processes are thus complementary.

Some non-specific solvents will dissolve a few percent of coals, mainly waxes and resins remaining from the original plants, at temperatures below 100°C. More specific solvents can extract up to 40% from some coals at temperatures up to 200°C; again, much of the extracted material appears to have been included as such in the coal and is thought to be closely related to the volatile matter first evolved on heating. These processes are however of little interest in the commercial liquefaction of coal, which is based on degrading and reactive solvents.

Degrading solvents are used, usually at temperatures near 400°C, where pyrolysis is beginning to break down the complex coal structure; the "solvent" (which may or may not actually dissolve the coal fragments) can retard repolymerisation by dispersing the primary products of pyrolysis. In this state, reaction between the coal solute and hydrogen in the presence of catalysts is facilitated, although the degrading solvents do not directly take part in the reactions. Typical degrading solvents are coal tar fractions such as anthracene oil. Despite the fact that the degrading solvents do not directly react with the coal, up to 90% of some coals can be extracted with prolonged or repeated extraction, the maximum extract yield being very dependent on coal type. Figure 9.3 shows the band of British coals on a Seyler chart, different degrees of extractability with degrading solvents being shown by degrees of shading. It will be noted that very high yields correspond to the small portion represented by prime coking coals..

Reactive solvents, which act on the "hydrogen donor" principle, take the breakdown of the coal rather further in one stage. These donor solvents are compounds to which hydrogen can easily be added but which can readily give up hydrogen in combining with coal fragments. The tetralin/naphthalene cycle, shown in Fig. 9.4, is typical. Because this is a reactive process, yields from coals of low carbon content can approach those of prime coking coals.

Although the extracts from degrading and donor solvents are different in molecular size, their chemical composition tends to be similar and extracts from different coals tend towards a remarkably narrow band of composition; this convergence is shown in Fig. 9.5.

The rate and extent of the degradation extraction is affected not only by coal type but also by the maceral content (see Chapter 1) which is to be expected. Exinites are most readily soluble, followed by vitrinites. Fusinite is virtually unaffected and other inertinites are only slightly reactive. In German work, based on the original Bergius process (see the section on Historical Background), it is considered desirable to use coals in which the inertinite does not exceed 5–10%. Geologically younger coals of high volatile content may also be preferred as being more reactive. The ash should preferably be restricted to about 5%, as larger amounts may cause difficulty in the later removal stage. Some of the ash constituents, notably pyrite, have an autocatalytic effect. Because of difficulties of predicting behaviour, it is prudent to test coals for their suitability for hydrogenation in autoclaves before use on a large scale.

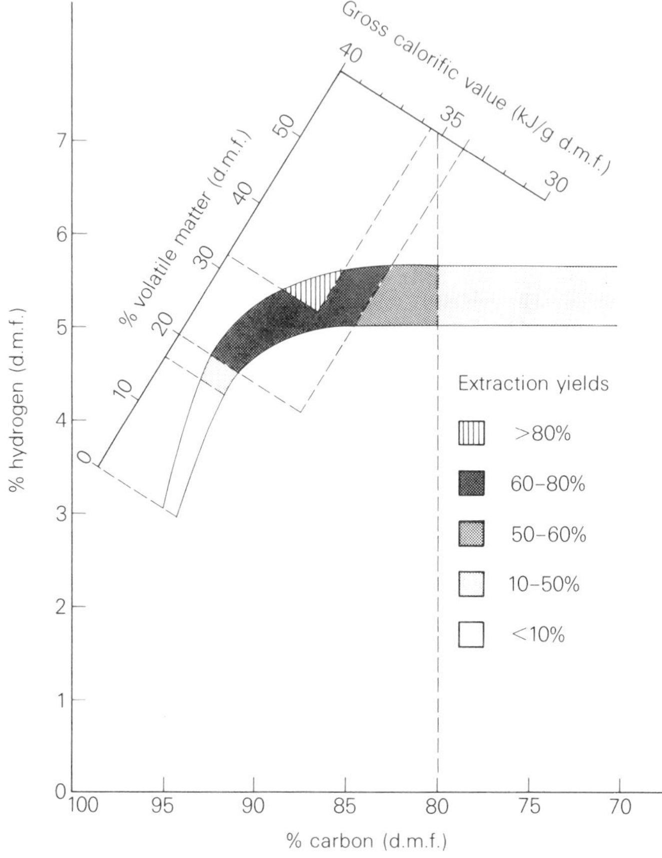

9.3 Extraction yields of British coals

Compressed gases in the supercritical state, where they exhibit the properties of both gases and liquids, can be used as solvents for coal and this elegant process, pioneered with respect to coal by the NCB at the Coal Research Establishment, Stoke Orchard, has some potentially important advantages. The technique of supercritical gas extraction is a general phenomenon and is based on the ability of substances to vaporise more readily in the presence of compressed gas; vapour pressure enhancement effects up to 10 000 times can result. Supercritical gas

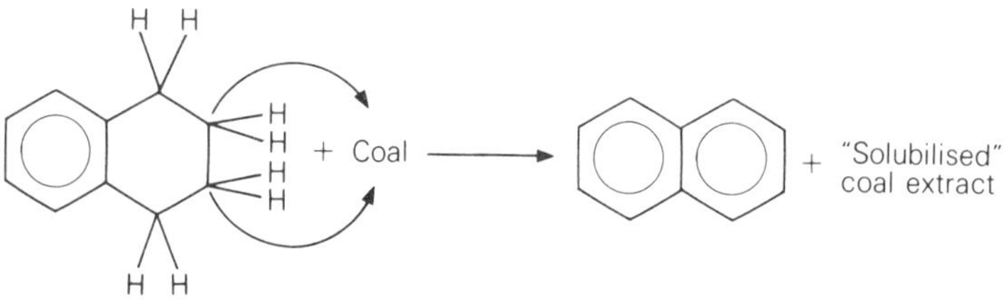

9.4 The donor solvent effect in coal by hydrogenation

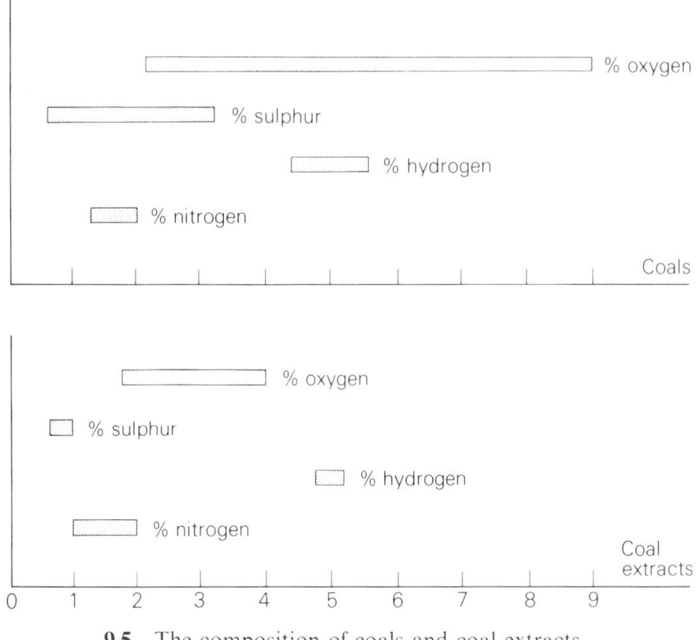

9.5 The composition of coals and coal extracts

extraction uses a gas above its critical temperature (which therefore cannot be liquefied by pressure increase), compressed to densities comparable to light organic solvents. When such gases are in contact with a solid substrate, the concentration of the substrate in the gas phase is far above the saturated vapour pressure of that component at the same temperature but at atmospheric pressure. As the gas density increases, so does the concentration of the solute. Components of the substrate are extracted primarily according to their natural volatility; secondary effects, such as chemical affinity, are minor. The application of this phenomenon to coal is particularly suitable, since the more valuable smaller molecular species (of high hydrogen content) are thought to be fitted into interstices in the more condensed carbonaceous skeleton and have a higher volatility. For a given gas, the greatest density for a given pressure (and therefore the greatest solubilising power) will be obtained at its critical temperature. Thus the extracting gas should be one which has its critical temperature only a little below the selected operating temperature. In the case of coal, it is also desirable that gas extraction takes place at temperatures when decomposition is beginning, but where volatility would normally be low and recombination unlikely. Solvents having critical temperatures in the range 230–480°C are most suitable and some aromatic hydrocarbons such as toluene (present in coal tar) are in this range.

At 380°C supercritical toluene can extract as much as 17% of hydrogen-rich material from coal, although there is little thermal decomposition at this temperature. At higher temperatures, the extract yield can exceed 40%. The diffusion effect is small; the yield does not seem to depend on fine grinding of the coal. Separation of the extract is basically very simple and the solvent can be recovered for recycling by reducing the pressure; the solvent power is lost and the extract is precipitated, as a

TABLE 9.1 Analysis of a typical gas extract, coal feed and residue

	Coal feed	*Extract*	*Residue*
Carbon, % (mmf basis)	82.7	84.0	84.6
Hydrogen, % (mmf basis)	5.0	6.9	4.4
Oxygen, % (mmf basis)	9.0	6.8	7.8
Nitrogen, % (mmf basis)	1.85	1.25	1.9
Sulphur, % as received	1.55	0.95	1.45
H/C atomic ratio	0.72	0.98	0.63
OH, % (mmf basis)	5.2	4.4	4.8
Ash, % dry	4.1	0.05	5.0
Volatile matter, % daf	37.4	—	25.0
Molecular weight	—	49.0	—
Calorific value (MJ/kg daf)	33.7	37.1	33.7

low melting point solid, essentially free of ash and solvent. The extracted residue is a porous solid which does not agglomerate or evolve tar on heating. Although the hydrogen content is reduced, the calorific value of the char is similar to that of the parent coal; it can readily be combusted and is highly reactive in gasification. Some typical properties of an extract, compared with the original coal and the residue are given in Table 9.1.

Some potential advantages of supercritical gas extraction are:

(1) High pressure gas supplies are not required. Energy requirements are low because the solvent is compressed as a liquid and not a gas.
(2) The coal extracts are richer in hydrogen and have lower molecular weights than those obtained using degrading solvents and may be more suited to conversion to light hydrocarbons and chemical feedstocks.
(3) The extract is readily separated from the solvent with virtually complete recovery of the unchanged solvent.
(4) The residue is a non-agglomerating porous solid with an appreciable volatile matter content which appears to be particularly suitable as a gasification feedstock.
(5) Residue separation is readily effected.
(6) Fine grinding is not required.
(7) Integration of the production of liquids with other processes for using the residue is favourable.

A diagram illustrating a simple form of gas extraction is shown in Fig. 9.6. (Further gas extraction applications are illustrated in Chapter 20.)

HISTORICAL BACKGROUND

As noted in Chapter 7 on carbonisation, there is a long history of the use of coal tars and liquids formed by the pyrolytic degradation of coal. This gave an important impetus to the study of coal structure and its reactions. The gas industry also provided an important foundation, including routes to synthesis gas. However, the key experiments leading to the present liquefaction programmes date from German work just before the First World War. In 1913, Bergius demonstrated that brown

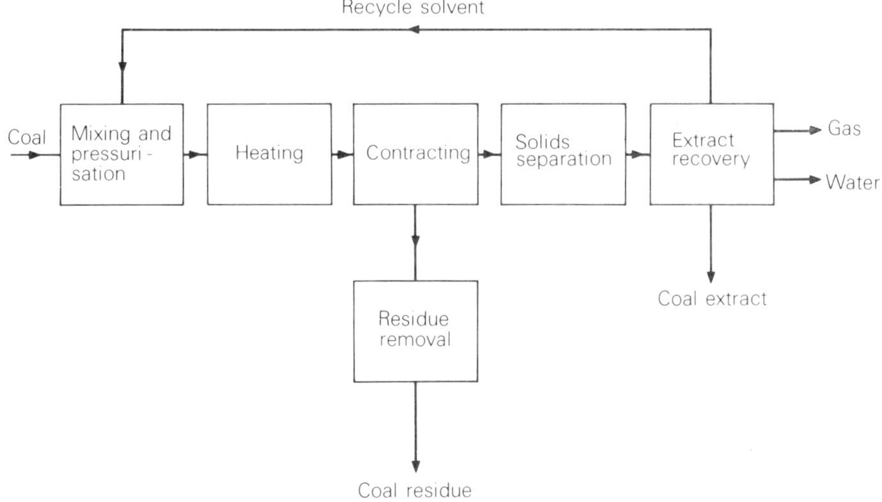

9.6 Unit operations in the SGE process

coal could be converted by hydrogenation into crude oil and in the same year the German company BASF took out a patent covering the catalytic synthesis of organic compounds by reaction between hydrogen and carbon monoxide.

Bergius' first work was carried out by heating coals, solvents and hydrogen in autoclaves. He was aware of and used the hydrogen donor principle but did not use catalysts at first. This came later in small technical plants leading to the setting up by IG Farben of the world's first large-scale plant at Leuna in Saxony. This plant, which was intended for motor fuels, had an annual output capacity of 100 000 t/a of liquids and commenced operations in 1927. At about the same time ICI began work and, after a 10 t/d pilot plant, built a 100 000 t/a plant at Billingham-on-Tees. The process was similar to that at IG Farben and was intended to meet the expected shortage of motor fuel; the plant was expanded to take in addition 50 000 t/a of coal tar and creosote, presumably to improve the economics. However, by 1937, ICI had come to the conclusion that oil from coal was not profitable without government assistance; during the war the plant was used to produce aviation spirit from creosote middle oil.

By contrast, in the years before World War II and subsequently, the German coal liquefaction industry was greatly expanded, presumably because the bare economics were influenced by strategic considerations. By 1938 an output of 1.5 Mt was reached. Twelve plants in all were constructed in Germany and satellite countries, reaching a peak annual capacity of 4 Mt in 1943/4, subsequently greatly reduced by bombing. The first four plants, using bituminous coal, brown coal or tar, operated at hydrogen pressures of 300 bar but subsequent plants were built to operate at 700 bar since it was shown that higher quality motor fuel and aviation spirit could be produced. Interestingly, one of the objectives of recent German work has been to reduce the necessary pressure, without degrading the product spectrum; this is a useful comment on the problems encountered in high pressure operation.

The synthesis process had a pre-war and wartime history in Germany which was parallel and complementary to the hydrogenation process. Hydrogenation was operated on the larger scale and used during the war to produce light motor fuels; the synthesis route was adapted to produce heavier motor fuels, including diesel, and also greatly needed edible fats, soaps and detergents. Fischer, however, working first with Tropsch and later with Pichler, was initially aiming for light hydrocarbons, such as gasoline. In 1923, using an alkalised iron catalyst at high temperatures and pressures (400–450°C and 100–150 bar) he produced synthol (a mixture mainly of oxygenated hydrocarbons such as alcohols, aldehydes and fatty acids with some hydrocarbons). In 1926, by lowering the conditions to 1 bar and 240–300°C, the proportion of hydrocarbons was increased but the catalytic activity was reduced and the degree of conversion of the synthesis gas was low. Thus began the emphasis on catalysts, continuing today.

Commercialisation of the Fischer–Tropsch synthesis route in Germany was associated with Ruhrchemie AG, who built the first plant of 100 t/a at Oberhausen-Holten. By 1936 five plants were operating, producing 150 000 t/a of products; this was doubled by 1939 with a further four plants. The nine plants peaked at 570 000 t in 1943 but were almost completely destroyed by bombing. All operated on the Ruhrchemie process, using mostly cokes for the synthesis gas production; a cobalt thoria–manganese catalyst, which was suitable for wartime product requirement, was used, although iron-based and nickel-based catalysts were also extensively studied.

Several other countries experimented with coal liquefaction before or during World War II, including France, Italy, the USA, the Netherlands, Canada, Belgium and Japan, where four synthesis plants with a total capacity of 260 000 t/a were commissioned between 1939–42. Technical information links were established before the war between German and US companies.

The post-war history of coal liquefaction, until the last few years, has not been encouraging, with the exception of Sasol, described later. In Germany, one of the hydrogenation plants and three synthesis plants were rebuilt after the war and struggled on for a few years against competition from cheap oil but were shut down or converted by the early 1960s. At this time a government commission reviewed prospects in the UK but found the economics so very unattractive that even related research was virtually abandoned. This appeared also to be the attitude in Germany. Several relatively serious efforts took place in the USA, associated particularly with the Bureau of Mines, Pittsburgh Consolidation Coal Company (now part of Conoco), and Union Carbide (with interests in chemicals). Most of these efforts were relatively short-lived and not immediately productive; it can be said however that organisations (and people) involved in that period have played leading parts in the more favourable recent times. Even in Germany, when interest was revived after 1973/4, some of the men working on liquefaction processes during and even before the war, emerged with expert knowledge of the successful plants and also with ideas for improvements which it had not been opportune to try earlier.

Although the 1973/4 crisis was the trigger for renewed activity, the direct involvement of the major oil companies in coal has probably been the main mechanism.

SOME MAJOR PROJECTS AND PROGRAMMES

South Africa

South Africa must take pride of place, not only because the only current commercial liquefaction plant has existed there for some years but also because there are other coal-based enterprises producing chemicals and liquid fuels there. South Africa has the positive incentive of large reserves of low cost coal and also finds access to oil difficult. It may be however that a unified national determination to make a success of coal liquefaction has been the most important element.

The first Sasol plant (Sasol 1) was ordered as long ago as 1951 by the South African Coal Oil and Gas Corporation, a public company but financed by the government through its Industrial Development Corporation. Building started in 1953 and the first oil was actually produced in 1955 (a remarkable time scale); the capital cost was 142 million dollars (1956). There followed, however, five difficult years, ending in very successful integrated operation; success may be measured most appropriately perhaps by the enormous degree of "stretching", debottlenecking and diversification which has taken place since. Sasol 2, a larger derivative, is also based on the successful processes used in Sasol 1; Sasol 3, more recently ordered, is a mirror image of Sasol 2.

Planning for Sasol 1 started from the German experience before and during World War II. Much of the plant design work and manufacture was done in Germany. However, decisions about the main process elements and in particular the extent to which novelty or scale could be pushed were strictly South African. Gasification is based on Lurgi gasifiers which, when designed, were the largest ever made. Originally the design was for 125 000 m^3 of pure gas per hour from nine reactors; by 1975, 270 000 m^3 were being produced from 13 reactors, a productivity increase of 50%; this was being achieved on coal with an average ash content of 34.8%. The nominal coal input is 10 000 t/d. Gas purification methods were selected and commissioned which had never previously worked before on a large scale and the same applied to the Phenosolvan process for purification of gas liquor. Methane in the tailgas from the Fischer–Tropsch reactors was reformed with novel catalysts and under conditions where little practical experience existed.

There was further novelty in the synthesis conditions, where two routes were installed in order to provide a required range of products (Fig. 9.7). The Arge process was based on the German F–T experience with fixed beds but iron catalysts were used at 20 bar in Sasol, compared with cobalt at 1–10 bar in the German plants. About 60% of the synthesis gas is supplied to the Arge reactors; the rest, and the Arge tailgas, is converted in the fluidised bed Synthol reactor, pioneered by Kellogg in the US but first demonstrated commercially at Sasol. This is the plant unit on which the most difficult teething problems occurred, especially because of catalyst bridging and gumming, but these problems were overcome. There is now a relatively high degree of selectivity to meet product slate requirements by virtue of the two different processes and by changes in catalyst and conditions; emphasis remains on motor fuels. However, medium Btu gas, including tail gas from the synthesisers, is distributed via a pipeline and is a very important by-product, in strong industrial demand; it was hoped that by 1976 it would be able to replace 500 000 t of oil per year. Apart from additional revenue, this outlet reduces the complexity of recycling within the plant.

Plans existed during the 1960s for a new plant but these were not approved until

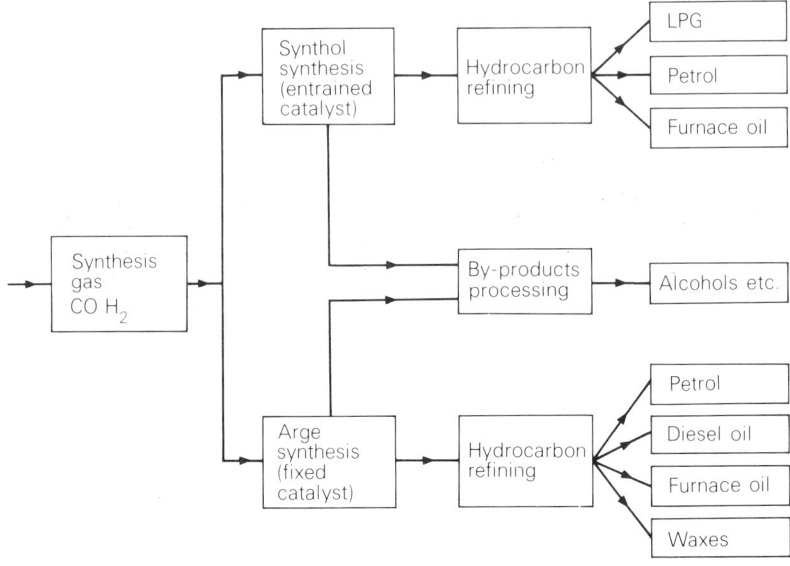

9.7 Fischer-Tropsch synthesis based on current South African processes

the end of 1975. Sasol 2, some 50 miles from Sasol 1, at Secunda in the Transvaal, is partly operational. It will ultimately have a capacity of 38 000 t/d coal input and is due to operate fully in 1981, at a cost of "more than $1 billion (1976)". The process, despite intensive research on alternatives, follows the same lines as Sasol 1. About 36 Lurgi reactors will each have a capacity 50% higher than the stretched Sasol 1 gasifiers. Over 1×10^6 m^3 of pure gas per hour will be produced; all will be converted by the Synthol process, using 8–10 reactors, each $2\frac{1}{2}$–3 times larger than those at Sasol 1. The Synthol process was selected because of easier scaling-up and also because the only desired products are motor fuels; a gasoline yield of 0.2 t (49 gal) per ton of coal is forecast, about three times the rate of Sasol 1. After CO_2 removal, light hydrocarbons are recovered from the synthol product cryogenically and the tailgas is reformed for recycling. The oily condensate is refined and, since the Synthol process gives a high yield of the lower olefins, some of the product is aromatised for adding back to produce premium motor fuels. The aqueous condensate is treated for recovery of the same range of alcohols and ketones as in Sasol 1. A novel feature may be the recovery of organic acids. Substantial quantities of ammonia and sulphur are recovered from the crude gas.

A block diagram showing the feeds and products for Sasol 2 is given in Fig. 9.8. Figure 9.9 shows the gasification and gas purification stages; the Rectisol purification process is shown in further detail in Fig. 9.10. The gas in Sasol 2 is finally air-cooled, whereas in Sasol 1, water cooling is used throughout. Only fluid bed reactors, as illustrated in Fig. 9.11, are used for the Fischer–Tropsch synthesis; the flow diagram is shown in Fig. 9.12. The complexity and scale of the plant are evident from these figures. Not illustrated are several other large sub-systems, for example for steam (4–5×10^6 lb/h), oxygen (12 000–13 000 t/d) and water. Secunda is also a new town, requiring roads and rail connections as well as domestic buildings. The fact that Sasol 2 was partly operational in early 1980 is again commendable.

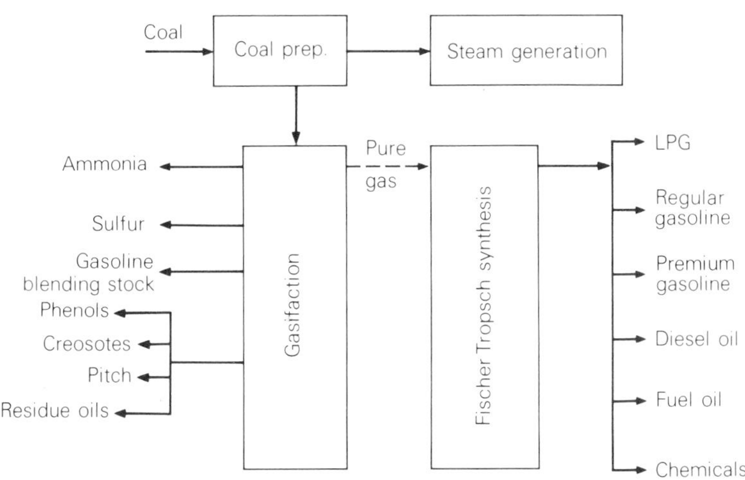

9.8 Sasol 2 feed and products

However, in contrast to the capital cost estimate of "more than $1 billion" which may have been for the plant, a total investment of $2.8 billion was reported at the end of 1978, thought to include all infrastructures.

The thermal efficiency of Sasol 1 is difficult to assess, partly because of the number of products and partly because of confidentiality. The plant has been quoted as having a nominal capacity of 10 000 barrels per day (1 barrel per ton of coal) but

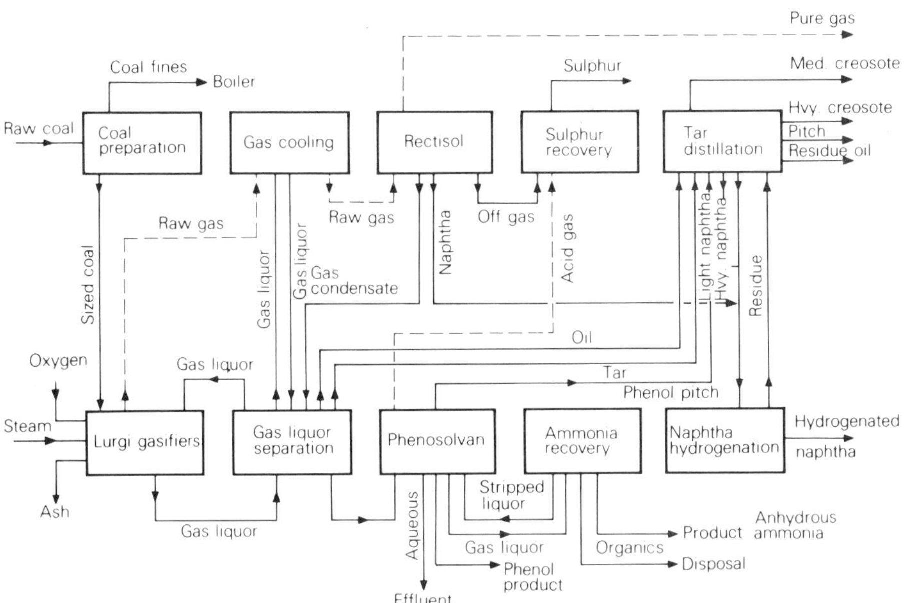

9.9 Block flow diagram of coal gasification and purification

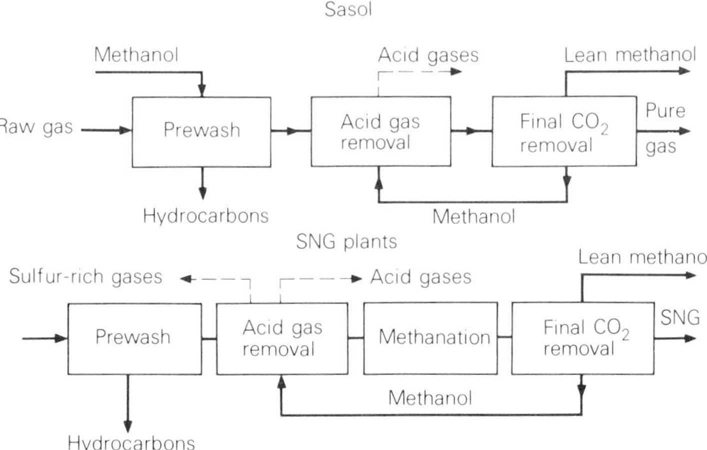

9.10 Rectisol gas purification

unofficial South African sources have referred to an output of 4100 barrels per day; how these terms are defined is uncertain. Sasol 2 is expected to have a somewhat higher ratio of light liquids to coal (about 1.2 barrels per ton of coal). Sasol 2 does not have a market for gas however and this will severely affect thermal efficiency. One estimate is that the Sasol process can achieve up to 60% thermal efficiency with an outlet for fuel gas but only 40% without. There has also apparently been some saturation of the market for petrochemicals. Arrangements have now been made for the Sasol technology to be marketed elsewhere and presumably this will lead to yields and efficiencies being calculated unequivocally for particular circumstances.

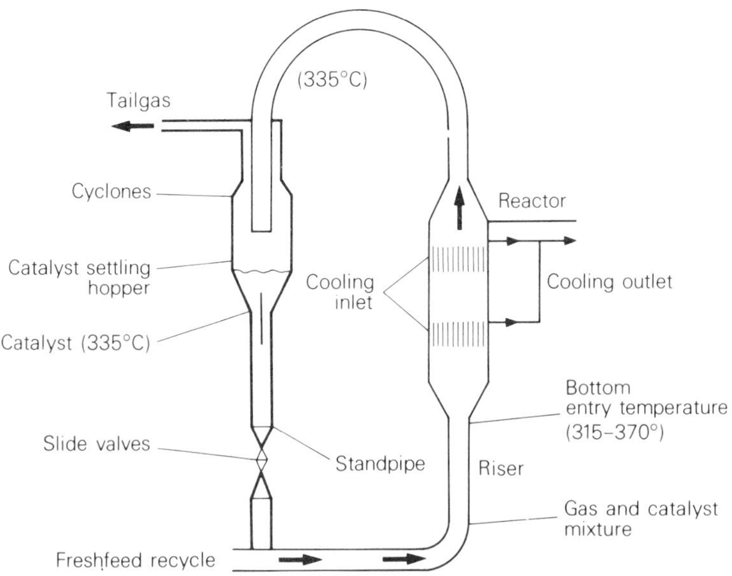

9.11 Fischer-Tropsch fluid bed reactor (Synthol)

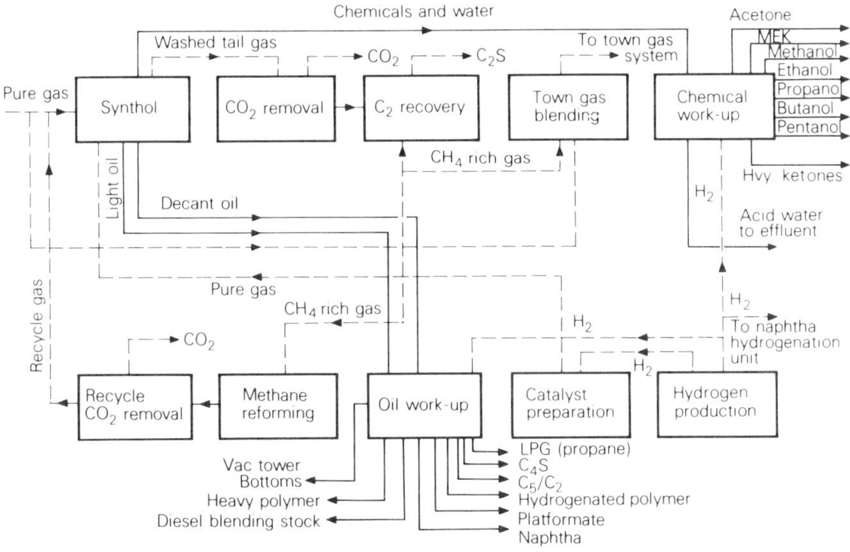

9.12 Block flow diagram of Fischer-Tropsch synthesis

In the South African context, however, the fact that Sasol 1 and 2 are stated to cover 40% of that country's crude oil requirment, and probably more than two-thirds when Sasol 3 is available, may be more important.

The Federal Republic of Germany

The West German programme is of special interest because it reflects the experience of the war time plants. Following the run-down, conversion and abandonment of those plants by 1962, only a little research was carried out until about 1974, when the opportunity for major developments recurred. It is interesting to note that of the substantial R & D programme supported by the Federal government (at an annual rate of DM 140 million in 1979), about 85% is on hydrogenation, with emphasis on reducing the severity of conditions; the remainder of the work, on synthesis, emphasises catalysis. There are several strands to the work, which has been logically planned in close co-operation with industry, and there has been for some time an obvious intention to proceed to demonstration scale plants on one or more processes as soon as possible; for this reason the physical work is strongly supported by design studies and evaluations. Links have been established between the FRG and the USA to share experience on plants of capacity over 5 t/d.

In 1975 and 1976 two technical scale (c. 250 kg/d) testing plants were commissioned for the rapid testing of coals and for refining operational conditions for the hydrogenation process, one at Bergau-Forschung of Essen-Kray and one at Saarbergwerke.

In the Bergau-Forschung testing plant (Fig. 9.13), finely ground coal is mixed with some of the medium oil (from the product) and ferric oxide catalyst to give a 40–50% solids suspension. This is then pressurised (max. 400 bar), hydrogen is added

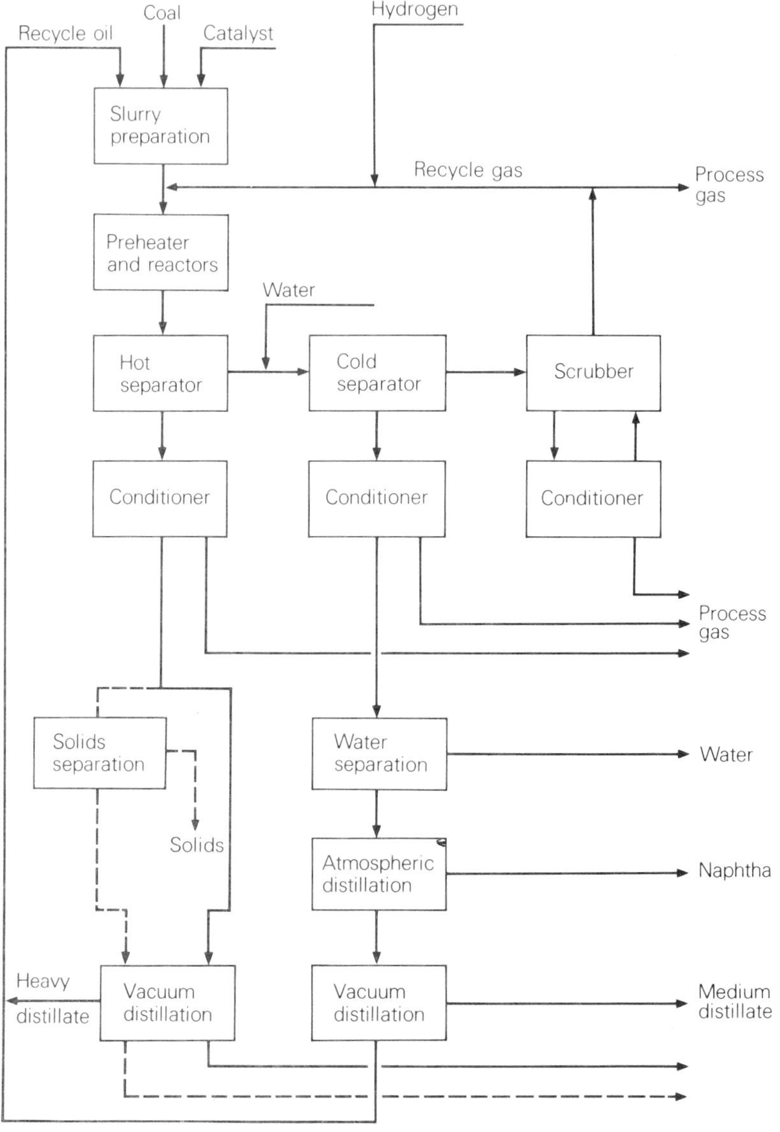

9.13 Further processing of hydrogenation products

and the mixture reacted at temperatures up to 500°C. The products are divided into heavy and medium fractions and various products are distilled. The heavy oils (>325°C) are recycled. A mass flow projection based on this work suggested that, from 100 g coal (daf), about 40 g of medium oil (200–325°C) and 14 g of light oil (<200°C), together with 31 g of residues could be produced, using about 10 g of hydrogen. Substantial excess proportions of hydrogen were however available in the small system and the recycle gas volume was high; the tests may not be a good indication of hydrogen utilisation efficiency in an industrial plant.

The Coal Hydrogenation Testing Plant of Saarbergwerke is rather similar. The residues after vacuum distillation of the bottoms from the hot separator can be coked in a separate plant. The first tests were carried out at 475°C and 285 bar and gave too high a volume of gaseous products (28%) compared to oil (40%); by changing the conditions the gas yield was reduced to 15% and the oils (<390°C) increased to 50%. Based on these results, a 6 t/d pilot plant has been built (Fig. 9.14); the cost was initially stated to be DM 13 million, of which 95% was to be contributed by the Federal government. In an effort to improve thermal and material efficiencies, several special features have been included. The waste heat from products from the top of the hot separator is used to pre-heat the input coal slurry and some oil recovered at this stage in an intermediate separator is also added back to the slurry, relieving the load on the atmospheric distillation unit. It is also hoped that this mixture can be hydrogenated at 200 bar. The bottom product from the hot separator will be treated by flash distillation which gives distillates as free as possible from asphalts; the residue can be coked under hydrogen or gasified.

In 1974 Ruhrkohle AG and Steag AG of Essen co-operated on a design study for a large testing plant, based on former German know-how, modified by the test work then being undertaken at Bergau-Forschung. Figure 9.15 shows a block diagram of the process considered best. It was subsequently agreed to build a 200 t/d plant at Essen on this basis and work commenced in mid 1979. Ruhrkohle AG will manage the plant, in which Veba Oel AG are also participating; the State government will contribute DM 130 million out of the capital cost of DM 145 million. Operation will commence in 1980. This scale was chosen as being both a substantial step from that

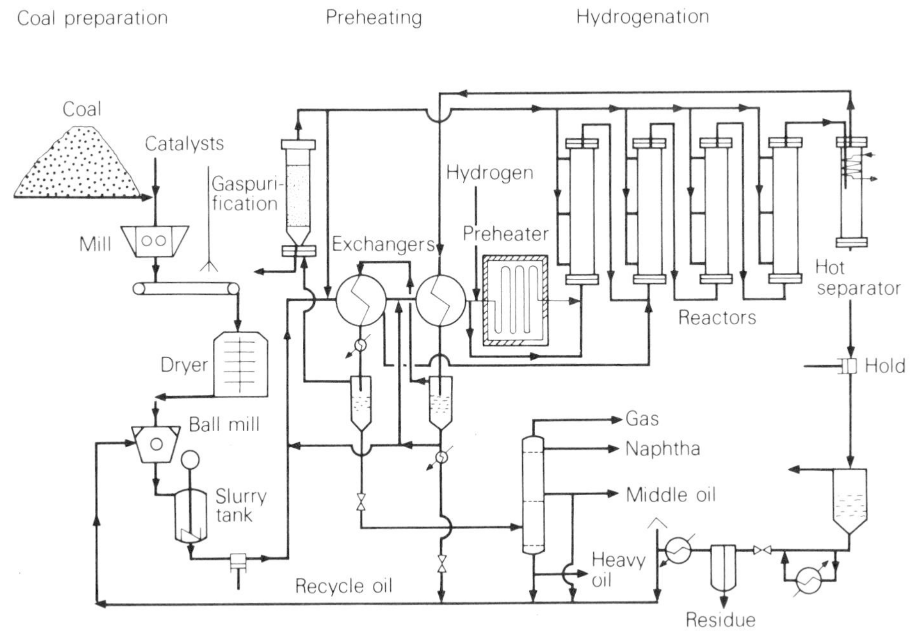

9.14 Pilot plant for coal hydrogenation of Saarbergwerke AG

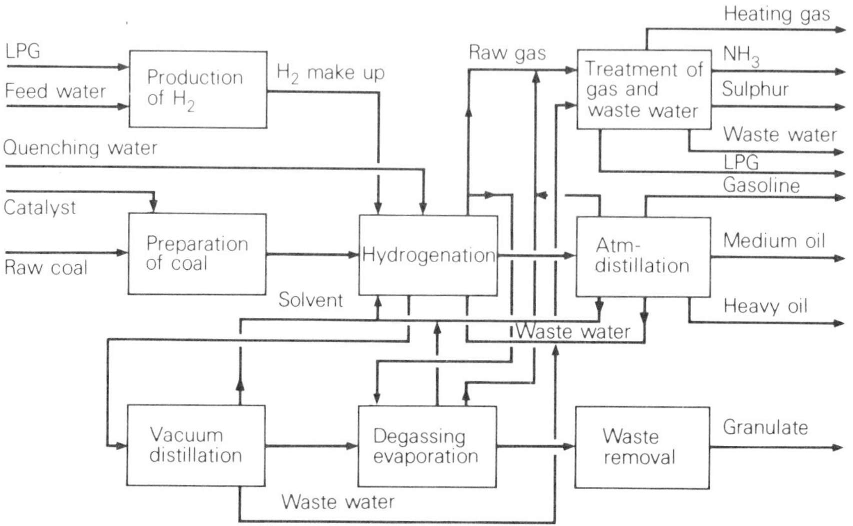

9.15 Block diagram of the "coal oil" large scale testing plant (Ruhrkohle AG, Essen)

on which the basic information was obtained and, at the same time, suitable for scale-up direct to a commercial plant. The process will use a ferric oxide catalyst as in the wartime plants but it is hoped that several other changes will allow the pressure to be reduced from 700 to 300 bar whilst increasing the specific coal throughput by 50% and improving thermal efficiencies and costs. Changes include the use of a mixture of heavy and medium oils for the slurry, separation of the reacted mixture by distillation and the use of the residues for gasification to produce hydrogen. Inputs, in addition to 200 t/d of coal (daf) will include 4 t/d of ferric oxide catalyst and 220 000 m^3/d of hydrogen. Products are expected to include about 13 t/d of LPG, 29 t/d of gasoline, 70 t/d of medium oil and 70 t/d of residues (underlining the importance of using the residues efficiently); ammonia (2.3 t/d) and sulphur (0.7 t/d) will also be recovered. In determining the type of primary products desired and their subsequent working-up, chemical products will also be in mind. The reactor selected was closely compared with the H-Coal process (see later). On German coals the H-Coal process, compared with German technology, gave a lower yield of oils but of a higher quality; it is expected that an H-Coal reactor will also be installed in the testing plant in addition to reactors based on German experience.

Rheinische Braunkohlenwerke AG (Rheinbraun) have also carried out a hydrogenation programme, under Federal Government sponsorship, using brown coal as a raw material. A 5 kg/h pilot plant has been in operation since November, 1978. Tests on brown coal have also been carried out on the laboratory scale by Bergbau-Forschung & Saarbergwerke. It appears that the brown coal is reactive and despite its high oxygen content, hydrogen consumption is not excessive; most of the oxygen appears to be released as CO or CO_2 rather than H_2O.

In addition to the H-Coal study, there are several other German activities in international co-operation. The Federal government has signed an agreement with the US DOE for participation in the SRC-2 project (see later). Ruhrkohle and Steag, who co-operated on the 200 t/a catalytic hydrogenation plant study, also carried out an evaluation, including the question of design transferability, of the 6000 t/d SRC-2 scheme. The German Government has allocated DM 60 million (85% of the total German contribution) for this project. The project is for a "clean" boiler fuel; it may be that the Germans are not thinking of European markets for such a product.

Ruhrkohle AG is also reported to be taking an interest in the 250 t/d Exxon Donor Solvent pilot plant being erected at Baytown. It is also understood that a pilot plant on the Mobil process for coal to gasoline via methanol is to be built at Wesseling with financial co-operation from German government and industrial organisations. These programmes are described more fully in the section on the US programme.

A view of the German programme on hydrogenation was presented to the International Conference on Coal Research, September 1980, in Dusseldorf by Specks (Ruhrkohle). Table 9.2 iiists the relevant pilot plants and Table 9.3 the industrial concepts, including those in the USA in which there may be German interest.

A joint Australian–German liquefaction study programme was announced in April 1979; the programme will involve twenty-six months work at a cost of A$3.6 million, of which 50% will be contributed by the German Federal Government and the rest by the Commonwealth and the state governments of Queensland, Victoria and New South Wales. The work will cover the technical and economic feasibility of a plant nominally producing 2.9 Mt/a of liquids from an input of 9 Mt of either Victoria brown coal or hard coal from the other two states. The liquid products might consist of 45% gasoline, 45% distillate and 10% LPG. The plant would employ a combination of hydrogenation, gasification and Fischer–Tropsch synthesis. The study is in three stages and if the results are favourable, the full size plant might be operating by 1990, possibly in New South Wales.

The West German synthesis programme is on a smaller scale and more research oriented at present, with emphasis on catalysts. However, it is recognised that aliphatic synthesis products would probably be needed as complementary feed-stocks to the aromatic products obtained from hydrogenation when coal starts to replace petroleum products. Shcering AG are one of the organisations involved and have a small pilot plant using slurry-phase Fischer-Tropsch synthesis, with emphasis on short chain olefins, using catalysts supplied by Ruhrchemie, Sasol and the Technical University of Berlin. Schering are proposing to erect a pilot plant, including both slurry-phase and fixed bed reactors, to process 10000 Nm^3/h of synthesis gas. It is hoped to complete design studies by 1982 (indicating that some progress on the current phase is still required) and commence operations in 1984; extensive government support is anticipated.

Ruhrchemie AG are the other major company involved in synthesis work; they are also interested in short chain olefins since a substantial proportion of their output is in polyolefins. Government support is about 75%. A review of the state of the art was carried out first and indicated good potential for the development of selective catalysts. The most promising were promoted iron precipitation catalysts which might have selectivities for C_2–C_4 olefins of around 50%. Some scope for altering the ratio of ethylene to propylene was also demonstrated. However, high selectivities were usually associated with short catalyst lifetimes, probably due to carbon deposition; regeneration of catalysts is being studied.

Several academic organisations are also involved in scientific studies of synthesis. It is probable also that good links exist with Sasol.

TABLE 9.2 Pilot plants for hard coal hydrogenation

Process	*Participating Priv. Comp.*	*Location Tot. expenses Time schedule*	*Process data*	*Products (t/d)*
Modified IG.-Process German technology	Ruhrkohle AG Veba Oel AG	*Bottrop/NRW* 300 mio DM 1975–1983 Operating from 1981	Coal throughput 200 t/d Pressure 300 bar Temperature 475°C	Gas 40.0 Raw Naphtha 30.0 Middle Oil 70.0
Modified IG.-Process German technology	Saarbergwerke AG	*Völklingen/Saar* 30 mio DM 1975–1982 Operating from 1980	Coal throughput 6 t/d Pressure 285–300 bar Temperature 475°C	Gas 0.8 Raw Naphtha 0.75 Middle Oil 1.8
SRC 2 Process	Gulf . Misui . Ruhrkohle AG	*Tacoma/Wash.* 43 mio US-$ 1975–1982 Operating since 1977	Coal throughput 30 t/d Pressure 140 bar Temperature 450–460°C	Gas 4 Raw Naphtha c. 2.5 SRC 2 10
EDS-PROCESS	EXXON, EPRI, Japan CLDC, Phillips Coal Company, ARCO Coal Company, Ruhrkohle AG	*Baytown/Texas* 340 mio US-$ 1974–1982 Operating from 1980	Coal throughput 200 t/d Pressure 100–150 bar Temperature 450°C	Gas 17 Raw Naphtha 26 Middle/ Heavy Oil 48
H-Coal Process	Ashland Synthetic Fuels, Inc., Standard Oil Company, Mobil Oil Corporation, Continental Oil Comp., EPRI, . State of Kentucky . Ruhrkohle AG	*Catlettsburg/Kentucky* 300 mio US-$ 1976–1982 Operating from 1980	Coal throughput 160–180 t/d Pressure 190 bar Temperature 450°C	Gas 17/26 Raw Naphtha 27/— Middle/ Heavy Oil 52/— Fuel Oil —/220

TABLE 9.3 Concepts of industrial coal hydrogenation plants

	Federal Republic of Germany			*United States of America*		
Process	Modif. IG-Process German Technology	Modif. IG-Process German Technology	Modif. IG-Process German Technology	SRC 2-Process	EDS-Process	H-Coal-Process
Coal ($\times 10^6$ t/a)	6 Hard coal	6 Hard coal or heavy crude	2 Hard coal	1.8 Hard coal	7.5 Hard coal	6 Hard coal
Products (10^6 t/a)	SNG max. 1.8×10^9 m^3/a LPG max. 0.6 Raw Naphtha 1 Middle oil 2	Liquid products 2	Gasoline 0.8	SNG 0.115 LPG 0.024 Raw Naphtha 0.146 SRC2 0.642	C_2-Gas 0.504 Propane, Butane 0.192 Raw Naphtha 0.835 Middle/Heavy oil 1.506	Gas 0.140/0.399 $\times 10^9$ Raw Naphtha 0.611/1.071 Middle/Heavy oil —/1.020 Fuel oil 1.711/—
Time schedule						
Planning	1980/83	1980/83	1980/82	1979/82		
Construction	1983/94	1984/87	1983/86	1981/84		
Start-up	1986	1987	1987	1985	1988	1986
Location	Ruhrgebiet	Ruhrgebiet or on shore	Saargebiet	Morgantown	not decided	not decided
Investment[a] ($\times 10^6$ DM)	4000	4000	1500–3000	2600	2800	2200
Operating costs[a] ($\times 10^6$ DM)	—	—	—	190	740	450
Participants	Ruhrkohle AG	Veba Oel AG	Saarbergwerke AG	Gulf, Mitsui Ruhrkohle AG	Exxon	Ashland Oil

[a] 2 DM = 1 US $

The USA

The US Department of Energy programme for liquefaction was funded at $206 million in Fiscal Year 1979 and the estimate for FY 1980 was $122 million. There is also considerable activity funded by private corporations and other bodies, including some of the State Governments, especially Kentucky and Illinois. At one time the government funding favoured diversification but in recent years there has been an attempt to concentrate the programme into a small number of prime candidates for demonstration plants. This has meant that a number of ideas promoted by private interests no longer enjoy federal support although they may be continuing. However willing private sources may be to continue experimental work, even the largest companies would find it difficult or imprudent to go fully commercial without some form of recognition, through direct support or incentives. Thus although the DOE programme is currently relatively narrow, processes outside the sponsored group may still emerge but probably not in a discordant manner.

One of the key elements in US liquefaction strategy has been the implicit belief that only one process is available and proven (obviously Sasol type synthesis) and further, that this process is very high in cost with the products not compatible with US market requirements. There has therefore been relatively little funding for synthesis (none in the latest programme) but the position is being reviewed, possibly because of recent interest in the Mobil process (see below). The market objectives expressed in the programmes for FY1980 state that liquefaction is intended primarily to produce clean-burning synthetic oil so that natural gas and oil may be released; the possibility that this synthetic oil might subsequently be upgraded to gasoline, etc. has been an additional objective. However, a growing requirement for light synthetic products might change R & D priorities.

In FY 1980, the DOE programme contains only one direct hydrogenation project (in US nomenclature), the H-Coal process (Fig. 9.16) although two other "third generation" processes in this category are also being studied. The H-Coal Process, developed by Hydrocarbon Research Inc., is a catalytic hydroliquefaction process using an ebullated bed reactor. Crushed coal is mixed with recycle oil, compressed to 200 bar and added, with hydrogen, first to a pre-heater and then to the reactor. The latter (Fig. 9.17) is of a special stirred design intended to agitate the mixture and keep the catalyst free of contamination. The catalyst is sized so that it should remain in the reactor system, with some being cycled through a parallel leg for regeneration and replacement; there may be some worries about carry-over on the commercial scale. The products and residues leave the reactor and are separated into gases, liquids and solid residues. The last would be used for hydrogen production in a commercial plant. The type of product may be varied by changing the throughput and hydrogen input; high hydrogen use and low throughput is required for refinery type Syncrude. The solids removal stage is not completely resolved; centrifuges and cyclones are not entirely satisfactory and the use of "anti-solvents" is being evaluated. Following a long period of laboratory work and the successful operation of 1–$2\frac{1}{2}$ t/d pilot plant, a 250–600 t/d plant (depending on operating mode) is being constructed at Catlettsburg, Kentucky, for initial operation during FY 1980; the cost, including two years of operation, is $250 million.

Table 9.4 shows the product yield and hydrogen consumption expected from the H-Coal pilot plant when operated in the two modes, Syncrude and fuel oil.

The naphtha product (C_4–205°C) has a naphthene content of about 55% and an

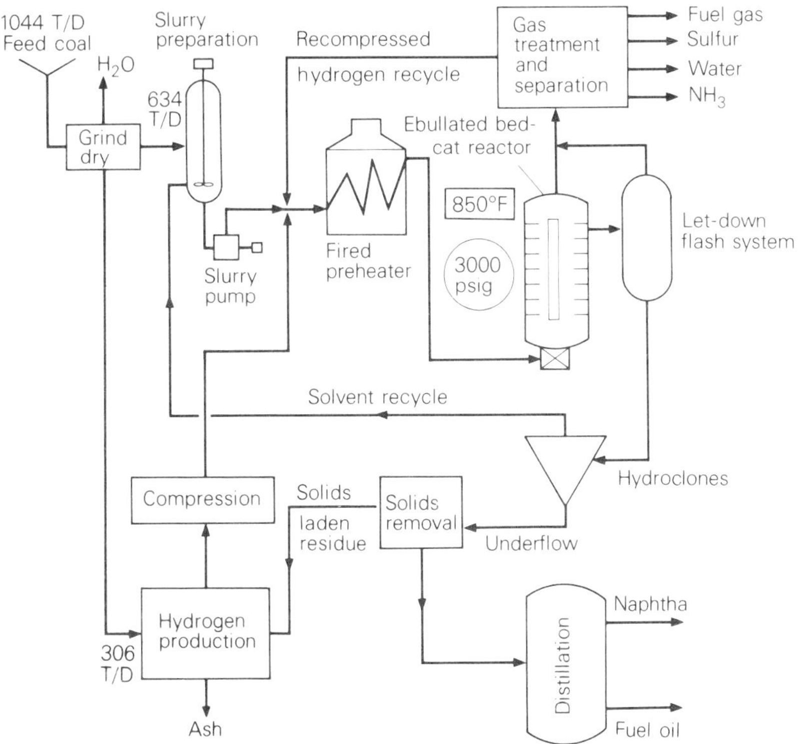

9.16 H-coal process

aromatics content of about 20%; it has been demonstrated that acceptable gasoline can be produced from it but more severe conditions than normal may be necessary. The 205°C+distillate has also successfully been upgraded, again with some departure from standard. Conversion to aviation fuels has also been shown to be feasible and combustion tests in turbines and domestic heating furnaces have been conducted successfully. The residues have also been successfully tested as feedstocks for gasification (by the Texaco process) and gave a carbon conversion of nearly 97%, with lower steam and oxygen demands than for the gasification of heavy oil.

Advantages claimed for H-Coal relate mainly to the reactor design, which is based on commercial experience with oil. Flexibility of both product slate and coal input can be achieved; the catalyst recirculation system facilitates this and also provides a control method. However, a relatively expensive catalyst (generally cobalt/molybdenum/alumina) is in direct contact with coal and mineral matter and can suffer both loss of activity and physical degradation or loss. There are also potential problems with solids removal and in the scale-up of the reactor.

An alternative direct hydrogenation process is Synthoil, Fig. 9.18, developed at the Pittsburg Energy Technology Centre. This is similar to H-Coal but uses a fixed bed of catalyst pellets (cobalt molybdate). Following years of lab-scale work, a 10 t/d plant was built but later mothballed.

The two "third generation" hydrogenation projects concern catalysts, disposable in one case and recoverable zinc chloride in the other. The disposable catalyst effort

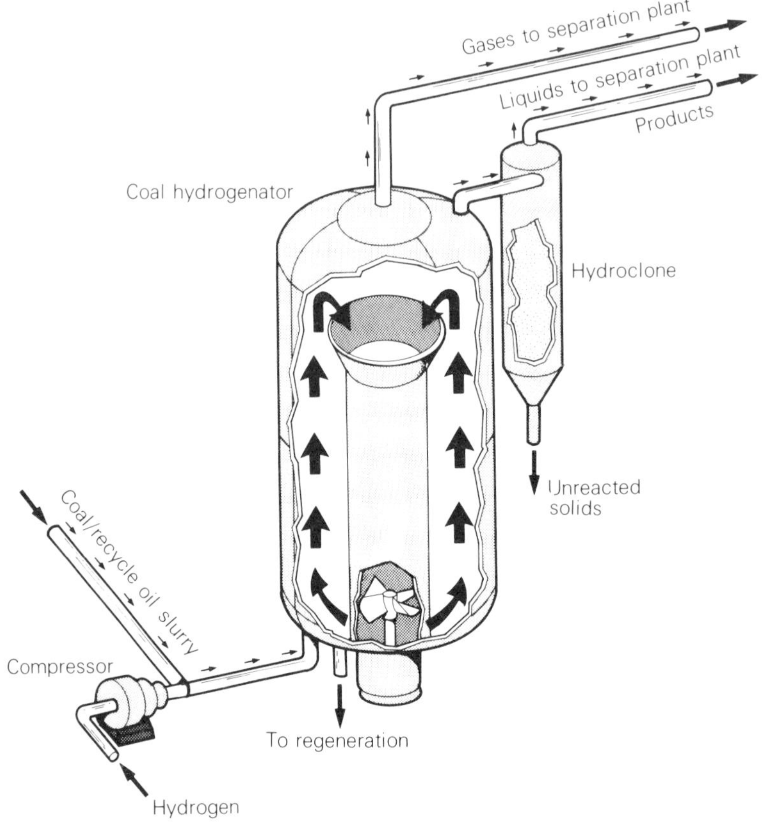

9.17 H-coal process reactor

is centred on the Pittsburg Energy Technology Centre with industrial support. The disposable catalysts may include materials which may be present in coal mineral matter and which might be specifically activated; cheap ores are also being studied. The catalysts would follow the distillation residues, which would finally be coked to recover liquids. Work is proceeding on a 1200 lb/d plant with a view to testing promising catalysts on other pilot plants.

TABLE 9.4 Product yields and hydrogen consumption: H-coal (wt% of dmmf coal feed to reactor)

	Syncrude	*Fuel oil*
C_1–C_3 gases	9.2	8.2
C_4–205°C	20.4	12.6
205–340°C	20.3	4.7
340–525°C	15.3	15.4
525°C +	18.9	30.7
Unconverted coal	4.3	5.9
Heteroatom-containing gases	16.5	16.0
Hydrogen consumed (kg/100 kg coal)	4.9	3.5

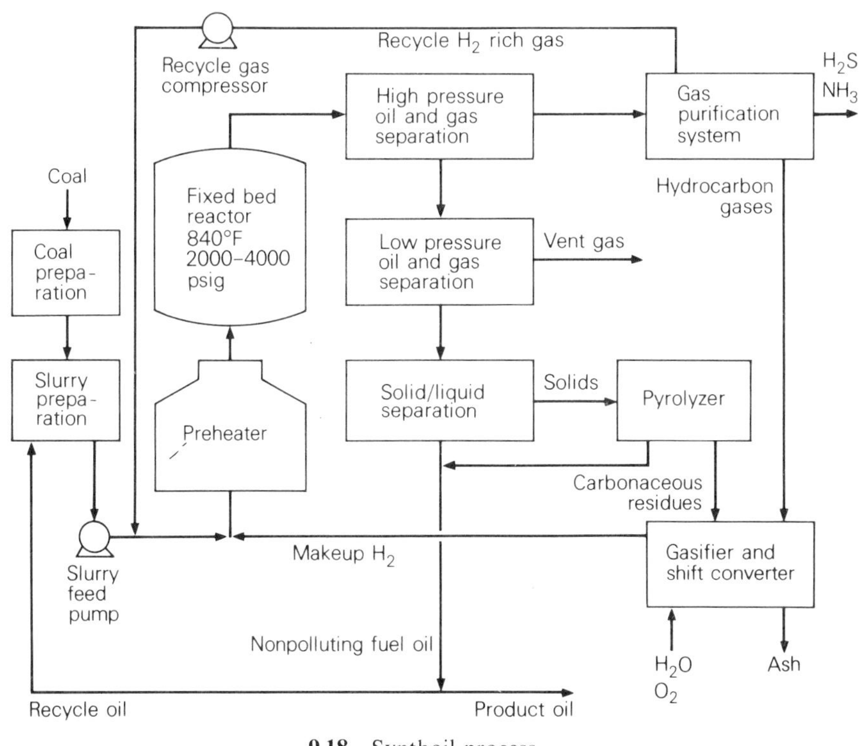

9.18 Synthoil process

The zinc chloride catalyst was originally studied in Japan and later on the Cresap "Project Gasoline" plant operated by Conoco. One configuration of the present proposal is shown in Fig. 9.19. The object is to produce gasoline type products, by severe catalytic hydrocracking, either in one stage as shown or by starting from coal extract. The reactor operates at up to about 450°C and 240 bar. All products are recovered from the top gases; the solid discharge from the reactor is heated in a fluidised bed at 925°C and low pressure, when the $ZnCl_2$ is distilled off and condensed for recycling. A 1 t/d plant has been completed. The original Cresap plant however was not operated very successfully, partly due to corrosion. Coals of high alkali content are unsuitable and this excludes US lignites.

The Solvent Refined Coal (SRC) Process has been studied in the USA since the early 1960s, initially through companies now associated with Gulf Oil. The objective has been to produce a nearly ash-free, low-sulphur product from coals which might be high in ash and/or sulphur. The basic process, now called SRC1, is shown in Fig. 9.20. Pulverised coal is mixed with recycle oil; hydrogen, produced from other sections of the process (off-gases and gasification of residues), is added (about 2% of the coal weight) and after pre-heating digestion proceeds until about 90% of the coal is dissolved. The mixture then passes to a separator where gases are removed and unused hydrogen is recovered for recycle. Part of the sulphur in the coal is converted to hydrogen sulphide which is also recovered from the off-gases. The unreacted solids are separated from the extract; the solids would also be used for hydrogen production in a commercial plant. Solvent for recycle is recovered from the extract

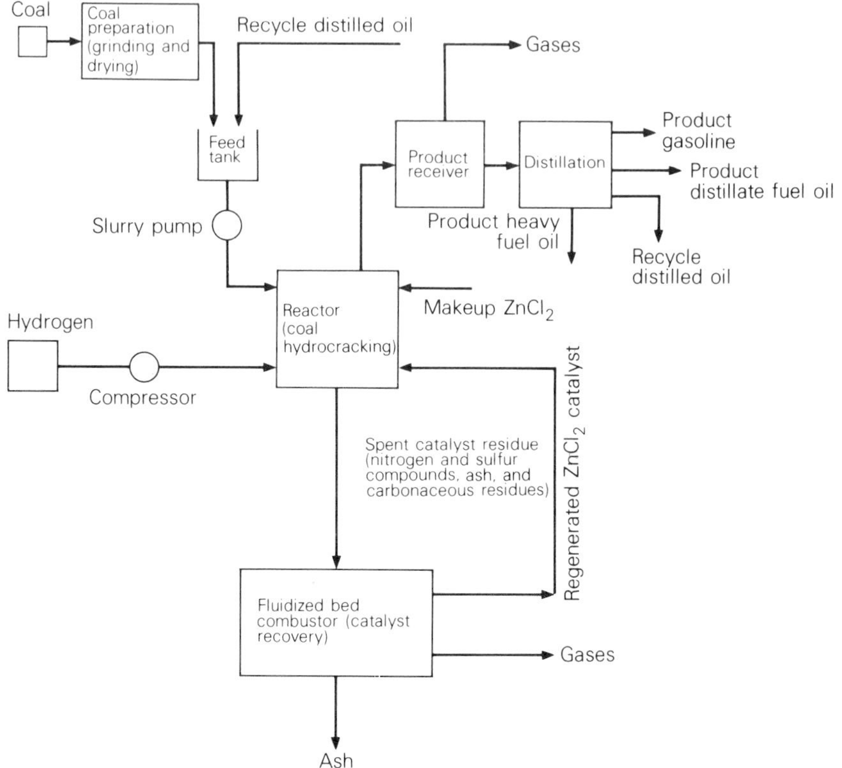

9.19 Zinc chloride catalyst process

by distillation (about 225°C); the remaining product on cooling solidifies at about 200°C to a pitch like solid, with a heating value of about 16 000 Btu/lb. Ash is only about 0.1% but sulphur reductions are less marked (from 3.4% in the coal to 0.8% in the product in one case cited). The fuel has been burned successfully in boilers and in a power station but the sulphur reduction so far obtained may be inadequate to meet newer air pollution standards.

The SRC2 process (Fig. 9.21) is intended to produce a liquid fuel by distillation rather than solid/liquid separation. Instead of the high boiling point distillates, some of the digester slurry is recycled as solvent, increasing the average residence time. The remainder goes for distillation; the residues from this, containing unreacted coal and ash, are used for gasification and it is intended that the quantity will be kept in balance with the hydrogen requirements. This SRC2 version not only eliminates the difficult liquid/solid separation but also, because of the more severe hydrogenation, liquid fuels can be produced and a more complete removal of sulphur obtained.

A 50 t/d SRC pilot plant has operated at Fort Lewis, Tacoma, Washington, since 1977. Regular operation and feed rates above design levels have been obtained. A modification to recycle digest allows the plant to operate on either the SRC1 or SRC2 mode. It seems however that even within each mode the process can be adjusted fairly substantially. A further 6 t/d pilot plant, built at Wilsonville, Alabama, privately, has had government support since 1976 and is used com-

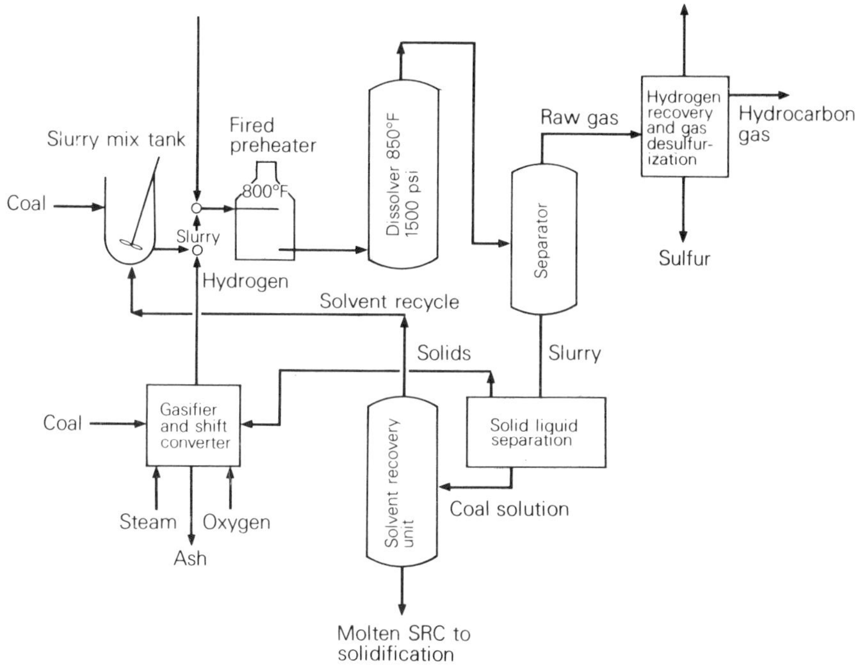

9.20 SRC (solid) process — SRC1

plementarily to the Fort Lewis plant, to screen coals and to test process options. The US DOE programme on SRC will continue into 1980 to seek further improvements, including solid/liquid separation methods (for SRC1). This specific objective is perhaps very significant since nowhere in the programme is any particular method referred to.

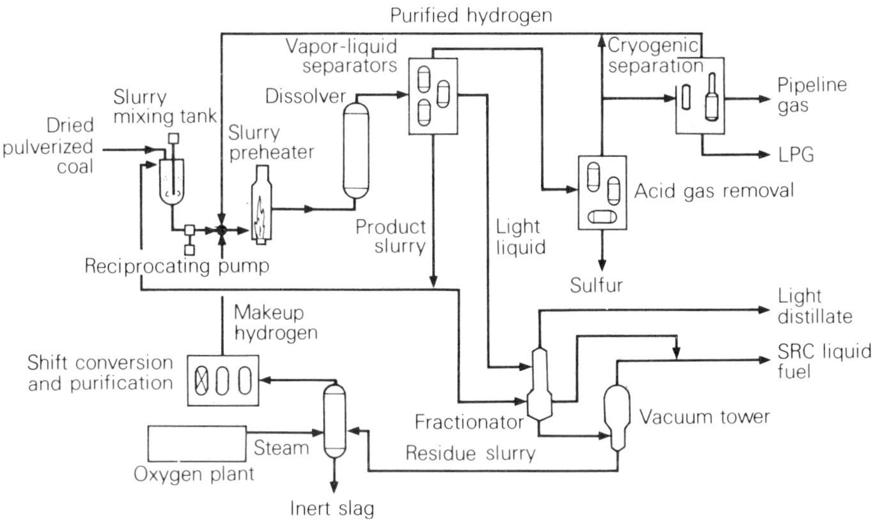

9.21 SRC (liquid) process–SRC2

Technical demonstration plants have been designed and costed for both SRC1 and 2, in order to evaluate large-scale operation. In the case of SRC2, a full-scale plant was envisaged as having a total coal input of 33 500 t of West Virginia coal with the following products:

Methane	50 Mscf/d
Ethane/propane	3 000 t/d
Butane	500 t/d
Naphtha	17 000 b/d
Fuel Oil	56 000 b/d

The cost of this plant was assessed by Gulf in May 1980 at just under $2 billion (in first quarter 1980 dollars). Annual operating costs were assessed at $480 million (including coal at $370 million). This plant would have five parallel streams and it is proposed that the first stream (6000 t/d coal input) should be built as a demonstration plant near Morgantown, West Virginia. The cost of the programme was put initially at $700 million but may since have escalated substantially. Japanese and West German interests have each agreed to take a 25% share. A consortium headed by Gulf will be responsible but most funding is expected to come from governments.

SRC2 is envisaged mainly as a clean fuel oil comparable with that burned in US power stations (No. 6 fuel oil). This appears to have been demonstrated in full-scale trials and no operational or environmental problems were encountered. The nitrogen content is about 1% (compared with 0.23% in No. 6 fuel oil) but NO_x emissions were acceptable. The sulphur content of SRC2 is about 0.25% and presents no problem. Successful tests have also been carried out on the use of SRC2 fuel oil for gas turbines and industrial boilers. The naphtha and middle distillates have shown potential for conversion to high-grade gasoline and diesel fuel. It has also been shown that whilst SRC1 is solid at room temperature it is a satisfactory boiler fuel and can be upgraded by commercially available processes into liquid products. Although at one time it was thought that only one SRC process would go ahead to the demonstration stage, the US DOE is believed to be going ahead with plans for a 6000 t/d SRC1 plant in addition to the SRC2 plant of the same size.

The SRC processes have several advantages, including a relatively simple flowsheet and reactor hydrodynamics. The catalytic action of the mineral matter (mainly iron compounds) enables expensive catalysts or donor solvent regeneration to be eliminated; in SRC2, conversion is sufficiently high to enable solids separation to be achieved by vacuum distillation. On the other hand, the reliance on inherent catalysts may reduce the flexibility for accepting a variety of feed coals and also for adjusting the product slate; gas make may be somewhat high. Typical product yields and operating conditions are shown in Table 9.5.

The Exxon Donor Solvent Process (EDS) appears now to be a front runner in the USA in the solvent extraction field and is supported by other countries. Work was started in 1966 by Exxon and proceeded to a laboratory scale of one ton per day on internal funding. A proposal to build a 250 t/d plant costing $268 million at Baytown, Texas, however, then received DOE backing; industrial partners in other countries, especially Japan and Germany, are also involved.

The process is based on petroleum industry technology and is stated to be simpler than SRC and flexible enough to accept a range of coals and produce a spectrum of products to meet market demands. Figure 9.22 shows the basic system concept. As in other solvent processes, the coal is ground, mixed with recycle solvent oil, preheated

TABLE 9.5 SRC2: Typical yields and operating conditions (yields in wt % dmmf coal input to reactor)

	Illinois No. 6	*Pittsburgh (Slacksville No. 3)*
C_1–C_4 gases	14.4	13.5
C_5–176°C	11.2	11.9
177–287°C	18.6	15.4
287°C + distillate	10.8	6.9
Unconverted coal	5.6	11.9
Solvent refined coal	29.8	36.8
Heteroatom-containing gases	14.2	7.1
Hydrogen consumed (kg/100 kg dmmf coal)	4.6	3.5
Operating conditions		
Temperature (°C)	455	456
Pressure (bar)	130	138
Residence time (h)	0.87	1.0

and added with hydrogen to a plug-flow reactor operating, typically, at up to 465°C and 130 bar; conditions may be adjusted to suit coals and product requirements. The unique feature of EDS, however, is that the recycle solvent is hydrogenated in a separate catalytic reactor before being used for the slurry preparation. The action of the hydrogen, which can readily be donated, is believed to be effective in stabilising coal breakdown products; molecular hydrogen also plays a part. All the products are recovered by distillation, thus avoiding a filtration stage, and the primary products can be hydrogenated further. The distillation residues may be coked to produce further liquid hydrocarbon products and may finally be gasified gas and hydrogen. Extraction may deliberately be limited to less than 70% of the coal in order to provide enough residue to balance the hydrogen requirement.

In addition to the one ton per day Coal Liquefaction Pilot Plant (CLPP), the process has been studied extensively on a laboratory scale, using both a 5 g batch unit for basic studies and a 75 lb/d continuous unit. Results on these plants have emphasised the subtlety of the general process, which might lead to flexibility in choice of design and operating conditions. One area especially studied is that of the preferred level of hydrogenation of the recycled solvent; there are complex

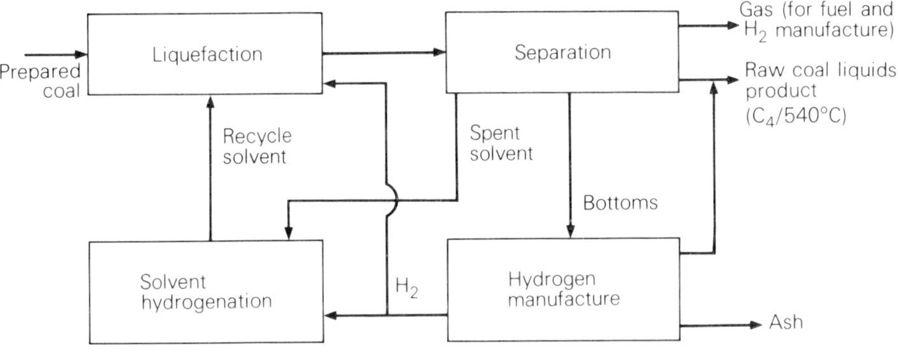

9.22 Simplified block diagram of the EDS process

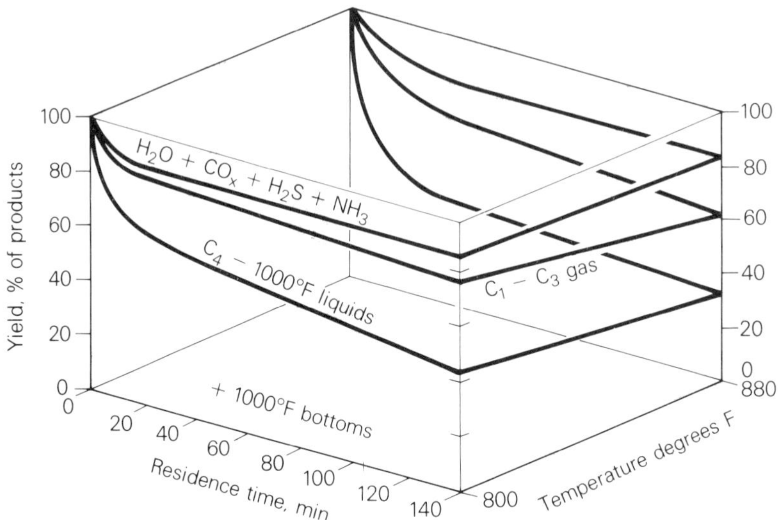

9.23 Conversion of Monterey coal increases with increasing temperature and time

interactions between temperature, residence time and solvent quality, which are different for different coals. Two coals, one Illinois (Monterey) high volatile bituminous coal and a Wyoming (Wyodak) sub-bituminous coal, have been particularly used as typical of the main US candidate feedstock coals. The effects of residence time and temperature on the product slates for these two coals are shown in Figs. 9.23 and 9.24. Total conversion increases with both time and temperature; more oxygenated gases are produced from the Wyodak coal. The lowest practicable operating temperature is about 425°C and, here, longer residence times lead to

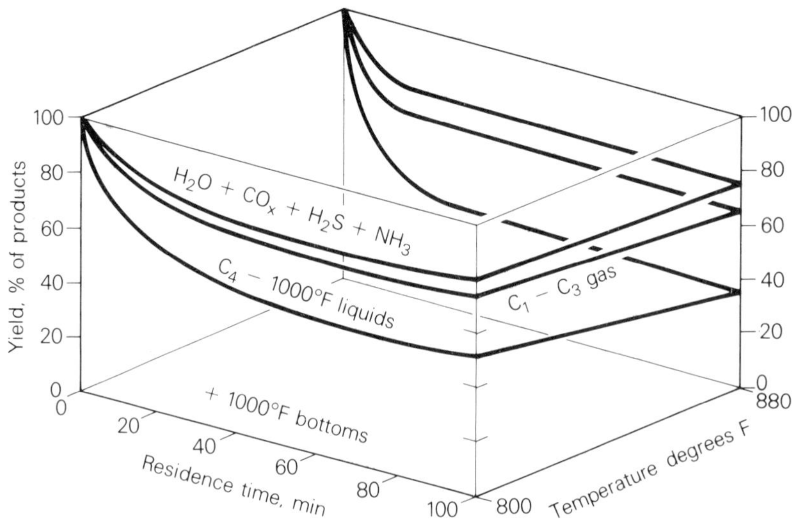

9.24 Conversion of Wyodak coal increases with increasing temperature and time

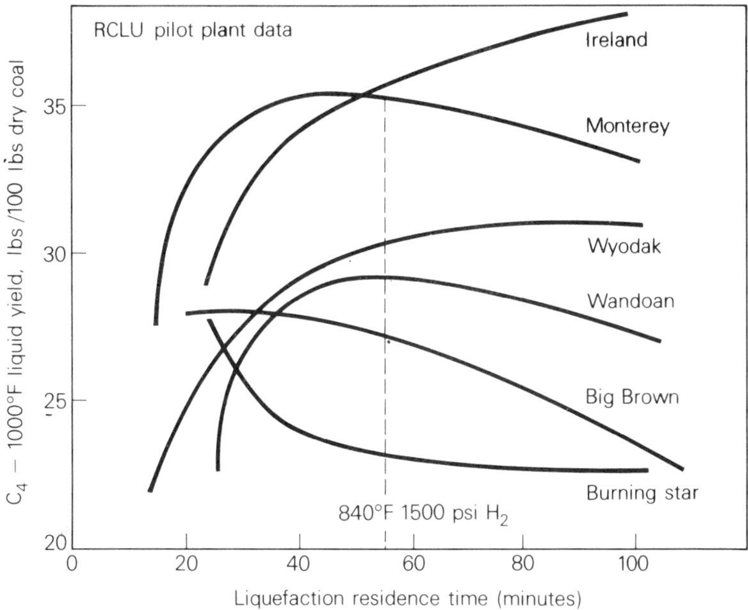

9.25 Liquid yield response differs for each coal

increased production of liquids at the expense of non-distillable bottoms. At higher temperatures, increased conversion of coal is obtained but the additional products are mainly gases. There is clearly scope for substantial and specific optimisation in these areas. In the case of pressure, however, effects are small and 150 psig seems to have been adopted as an adequate level, convenient for the design of equipment.

Solvent quality is important and an arbitrary measure, Solvent Quality Index (SQI) has been established. The optimum SQI for each coal is different, being affected strongly by the amount and form of oxygen in the coal. The development of the optimum SQI for the coal being processed is an important feature and can be controlled in the hydrotreater, mainly by the residence time (actually space-time); thus, there may be more flexibility at the design stage, when dimensions are fixed, than during operations. It should be noted that over-hydrogenation of the solvent can have an adverse effect.

Responses of various individual coals to residence time variations as indicated by yields of liquids, is shown in Fig. 9.25. The optimum balance of liquids from the reactor and the coker also varies for different coals. Considerably more work is necessary to understand the mechanisms by which coal characteristics affect results and in order to be able to predict optimum conditions. Yields of water and carbon dioxide are closely dependent on the organic oxygen content of the coal. Sulphur has an interesting effect, appearing to increase conversion and it is speculated that H_2S may act as a hydrogen transfer agent. Typical product yields for Illinois No. 6 coal are given in Table 9.6 for three different operating conditions:

A — liquefaction;
B — liquefaction with coking of residue;
C — as B but with higher liquefaction severity.

TABLE 9.6 Typical EDS yields for Illinois No. 6 Coal (wt % dry coal — 9.58% ash)

	A	*B*	*C*
C_1–C_3 gases	6	9	12
C_4 and C_5 hydrocarbons	3	4	5
Naphtha	15	16	21
Fuel oil	17	25	19
Liquefaction residue	48	0	0
Coke and ash	0	35	31
Heteroatom containing gases	14	14	15
Hydrogen consumption (kg/100 kg coal)	3	3	4

Exxon have general confidence in the operability of the process. One requirement established is that the bottoms must have a low enough viscosity to be pumped from the fractionator; this is controllable through the reactor conditions. Another problem area is the formation of scale or spheres of calcium carbonate which can adhere to the reactor walls or accumulate in components; these solids may have to be removed before the accumulations can form or some pretreatment of the coal adopted.

A recent innovation is referred to as "bottoms recycle". This has not yet been optimised but significant additional conversion can be achieved; in the case of the Illinois No. 6 coal liquefaction yields could be increased from 56 to 75% and within this the proportion of light liquids also went up. This is believed to be due to increased residence time with fresh donor solvent. However, some bottoms must be bled off and either gasified or coked; the preferred option is to use the established Flexi-coker process for this task in commercial plants but this decision is stated to be subject to revision.

All products have been extensively tested for their suitability for use or upgrading, generally in collaboration with other organisations. The naphtha can be an excellent feedstock for gasoline, yielding 85% of 105 octane material; alternatively a 50% BTX yield can be obtained. Various cuts can be made into excellent jet fuel, turbine fuel or fuel oil. The vacuum gas oil tends to be higher in nitrogen than comparable petroleum products. All liquid products are amenable to upgrading but hydrogen is of course consumed and the economics are not clear.

In summary, the EDS process looks very promising and the organisation of the programme has been excellent. The use of established petroleum technology in several areas is an advantage; also, the process does not rely either on the inherent mineral matter as catalyst nor on an admixture of a special added catalyst requiring recovery. The flexibility of the process has been noted (but may be significantly frozen once a particular plant is designed). On the other hand, there are two separate high pressure areas and there may be a penalty in thermal efficiency as a result of pressure let-down between. The integration of these items with the distillation units and the Flexicoker remain to be demonstrated (now perhaps including bottoms recycle); the quality of the Flexicoker liquids may also not be very high.

However, the 250 t/d plant is due to be completed in mid-1980 and many questions may be answered in the following two years. The sponsors believe that this scale will lead directly to commercial designs. Current sponsors, in addition to the

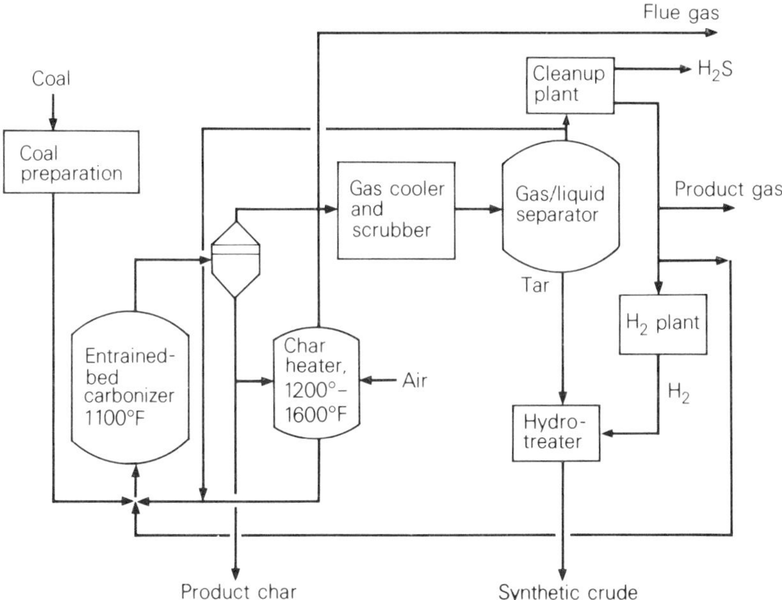

9.26 Entrained pyrolysis process (Garrett/Occidental)

US DOE and Exxon include the Electric Power Research Institute, Phillips, Arco, Ruhrkohle and the Japan Coal Liquefaction Development Company.

Pyrolysis processes do not have a very high priority at present in the US liquefaction programme. The COED multi-stage fluid bed pryolysis process was operated on a 25 t/d scale until 1974. Emphasis has since switched to the gasification of the char (about 60% of coal feed appears as char) in the Cogas process. The latter may proceed to commercial scale. If so, this could revive interest also in upgrading the tar product. COED is considered in Chapter 20 on coalplexes.

Occidental Oil have studied a Flash Pyrolysis Process (originally called the Garrett Process) for some years up to the 3 t/d scale at La Verne, California (Fig. 9.26). Fine coal is heated rapidly (c. 10 s) to up to 590°C at 2–3 bar in an entrained system in the absence of air. The products are quenched and the tar can be hydrotreated. The process has also been studied for the treatment of solid wastes. A more technically advanced process, called Flash Liquefaction, is being studied on a 25 t/d scale by Rocketdyne, Canoga Park, California, under DOE sponsorship. The pulverised coal is injected into an entrained system using hydrogen as the carrier gas. Mixing (using rocket motor technology) and reaction both take place very quickly (reaction time 0.01–0.1 s); temperatures up to 980°C are being used in studies at 35–100 bar. The products are quenched and gases, liquids and char separated. High yields of light liquids (e.g. BTX) are hoped for.

The most interesting US development in the field of coal liquefaction by synthesis is the Mobil Methanol-to-Gasoline (MTG) Process (apparently successor to a process previously called Mobil-M). This is a catalytic process for the quantitative conversion of methanol to hydrocarbons, with a high selectivity for high octane gasoline components. This stage can be preceded by coal gasification and methanol

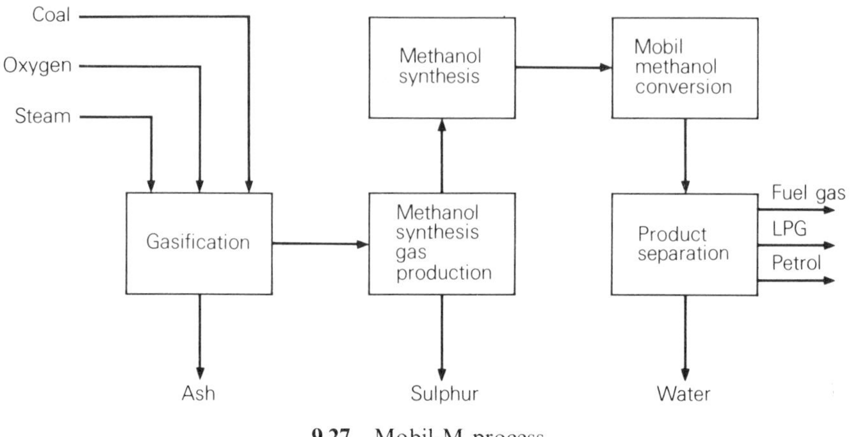

9.27 Mobil-M process

synthesis, thus giving a complete route to gasoline. Although the Mobil research has received some US governmental support, the concept as a major route to liquids from coal has not been given a high priority for large-scale development. Recent work is probably causing a serious re-assessment.

Figure 9.27 illustrates the flowsheet using the MTG process. All the operations upstream of the MTG stage can use established technology, including gasification (Lurgi, Koppers–Totzek or Winkler) and gas purification and adjustment to correct H_2/CO ratio. The synthesis of methanol can be carried out by well-known processes developed by ICI and Lurgi.

The key aspect of the process is the special catalyst. This development can be traced back to the early 1960s when it was shown that oxygenated aliphatic compounds could be converted to aromatic and branched aliphatic hydrocarbons over zeolite catalysts under relatively mild conditions. The molecular geometrics of the catalysts induced high selectivity, especially the limiting of the molecular weight of the products. This led to the development of the special catalyst, ZSM-5, a highly siliceous zeolite compound, containing no precious metals. It is capable of continuous operation for 3–5 weeks and is regenerable, using air for the oxidative removal of residual hydrocarbons. The ultimate life is more than a year, during which each pound of catalyst will have processed 13 000 lb methanol. The reaction takes place in two stages, which can be separated. Methanol is first dehydrated to dimethyl ether, which progresses to a reactive hydrocarbon fragment. These reactions are attributable to the surface acidity of the special zeolites. The hydrocarbon forming reactions occur within the channels of the catalyst which are of such a size that compounds above C_{10} cannot escape. Table 9.7 gives typical conditions and yields.

Table 9.7 refers to a Fixed-Bed system; there is also a fluid bed version. This duality of development arises mainly because the MTG process is strongly exothermic (360–416 cal/g) and an important design feature is the manner in which the heat is removed. The adiabatic temperature rise can be almost 600°C. In the first system, two fixed beds of catalyst are used, the first for dehydration and the second for conversion; this was demonstrated in a 4 barrels per day pilot plant (Fig. 9.28). The reactors are operated adiabatically; about 20% of the overall heat is generated

TABLE 9.7 Fixed bed MTG conditions and yields

Temperature: Inlet	360°C	*Product breakdown by market%*	
Outlet	415°C	Gasoline (including alkylate)	85
Pressure	22 bar	LPG	13.6
Yields (wt% of charge)		Fuel gas	1.4
Methanol and ether	0	*Properties of finished gasoline (wt% after alkylation)*	
Hydrocarbons	43.4	Paraffins	56
Water	56.0	Olefins	7
CO, CO_2	0.4	Naphthanes	4
Coke, other	0.2	Aromatics	33
		Research octane number	96.8 (clear)
Hydrocarbon product composition (wt%)			102.6 (leaded)
Light gas	1.4	Sulphur and nitrogen	nil
Propane	5.5		
Propane	0.2		
i-Butane	8.6		
n-Butane	3.3		
Butanes	1.1		
C_5 + gasoline	79.9		

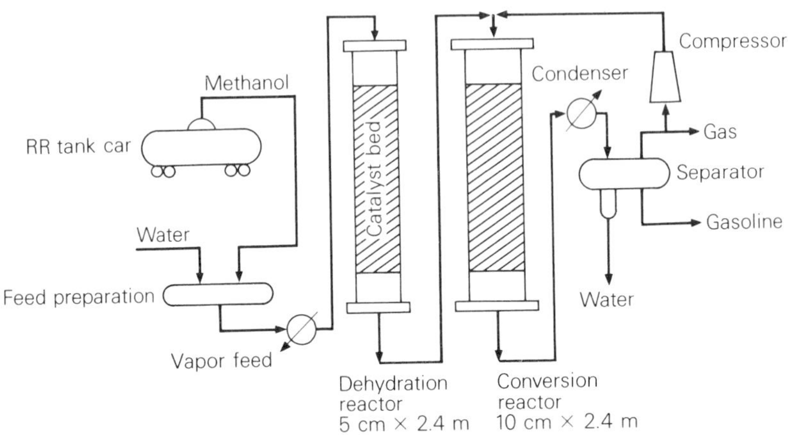

9.28 4 b/d fixed bed pilot plant

in the first reactor. Large quantities of product gas are recycled to the second reactor, where 80% of the heat is removed, in order to keep temperatures down. The catalyst is periodically regenerated by burning off the coke at intervals of about 22 d; after 10 such regenerations over an eight month period, the catalyst was still active. Mobil believe this process can easily be scaled up to commercial sizes.

The second method using a fluidised bed has also now been demonstrated on a 4 barrels per day scale (Fig. 9.29). This system allows cooling to take place via in-bed tubes, which could affect the control of product quality by maintaining optimum temperatures throughout (and could possibly facilitate recovery of higher quality of waste heat). In the pilot plant some cooling is also obtained by using a liquid feed

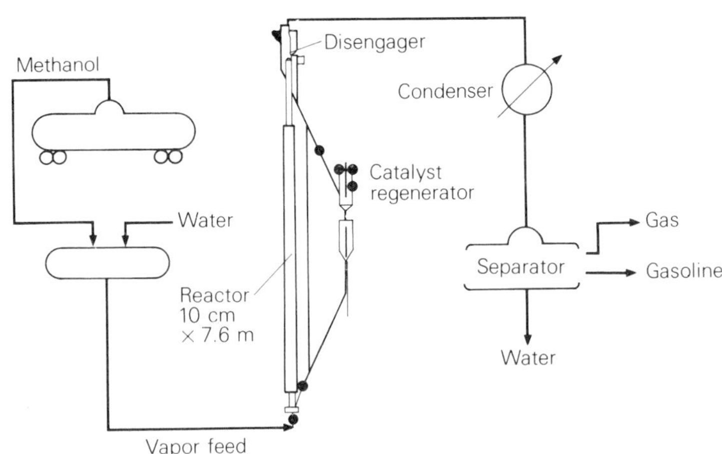

9.29 4 b/d fluid bed pilot plant

which is rapidly vaporised. Catalyst is continuously withdrawn for regeneration in a parallel leg.

There are three major developments in hand. In New Zealand, the MTG process has been selected for a plant to produce 13 000 barrels per day of gasoline from natural gas; the fixed bed version will be used. The US DOE is providing \$12–16 million for initial design of a full-scale (50 000 barrels per day) coal to gasoline plant using the fixed bed version. The fluidised version is to be scaled up in a 100 barrels per day demonstration plant at Wesseling in Germany, in which the US and German governments will participate, together with Mobil, Uhde and Rheinbraun.

There are further possible developments of the MTG process, including the manufacture of chemicals. Figure 9.30 shows the reaction path of feedstock and products; by modifying the catalyst and operating conditions, the manufacture of

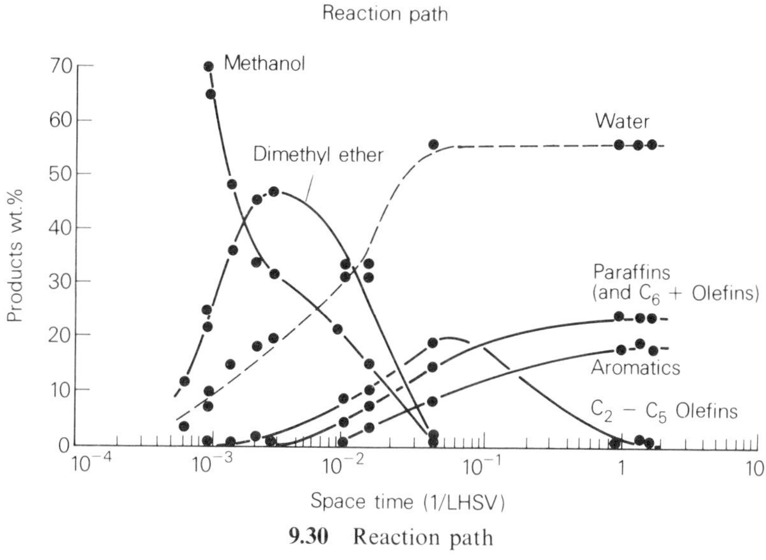

9.30 Reaction path

light olefin (C_2–C_5) or aromatic products can be increased. Light olefin yields as high as 70% of the hydrocarbons have been obtained. Conditions can be chosen so that ethylene is a major constituent of the olefin fraction and xylenes a major portion of the aromatics. Studies are also proceeding on the conversion of the products of a slurry-type Fischer–Tropsch reactor over ZSM-5 catalyst to a high yield of high octane gasoline.

In summary, the MTG process has some attractive aspects. Any coal which can be gasified can be used and most of the technology is mature. Premium products are produced exclusively; these could include chemical precursors. The conditions are relatively mild in the MTG process itself but preceded of course by the gasification train. The catalyst is well proven in other applications. On the other hand, energy conservation may be a problem, although there may be ways of mitigating this. Products are essentially of the gasoline type; it might be desirable to have versions capable of producing diesel and aviation fuels and this does not seem an improbable aim in view of catalyst developments. Provided a market for fuel gas exists, an overall thermal efficiency of about 66% has been suggested; it is claimed that this would be substantially better than the best that Sasol could do and that the costs (for gasoline) would be very much lower.

The US programme on liquefaction is strongly supported by generic studies of associated plant and problem areas. Work is proceeding, for example, on the upgrading of products and on the treatment of residues. The Cresap plant has been re-activated as a component test facility. Environmental and socio-economic studies on the impact of liquefaction plants are being carried out; these are based on measurements being made at pilot plant sites.

UK Programme

In the UK, the conversion of coal to liquids has been regarded as a longer term objective than in some other countries. For a number of years, broadly-based exploratory work on a laboratory scale has been carried out, supported by technical and economic assessments, with a strong presumption that in the UK, (a) high value products rather than bulk syncrude might prove more attractive, and (b) that the utilisation of coal in roles now performed by crude oil might be evolutionary rather than revolutionary. This approach has now culminated in concentration on two processes, work on which has been supported by the ECSC and which are now proposed for stage-wise development up to the full scale. The next stage in each case will be 24 t/d pilot plants; government support has been obtained for design and engineering studies.

One of these processes is liquid extraction using a process-derived solvent which behaves as a hydrogen donor; catalysts and hydrogen are not used in the extraction stage. By careful control of conditions, particularly viscosity and space time, it has been shown that filtration can be carried out in an acceptable manner. The second process uses supercritical gases as the solvent, as described earlier. In both processes, subsequent hydrocracking of the extract has been regarded as an integrated process and has been studied on the 50 kg/d scale.

The liquid solvent extraction is shown diagrammatically in Fig. 9.31. Fine coal

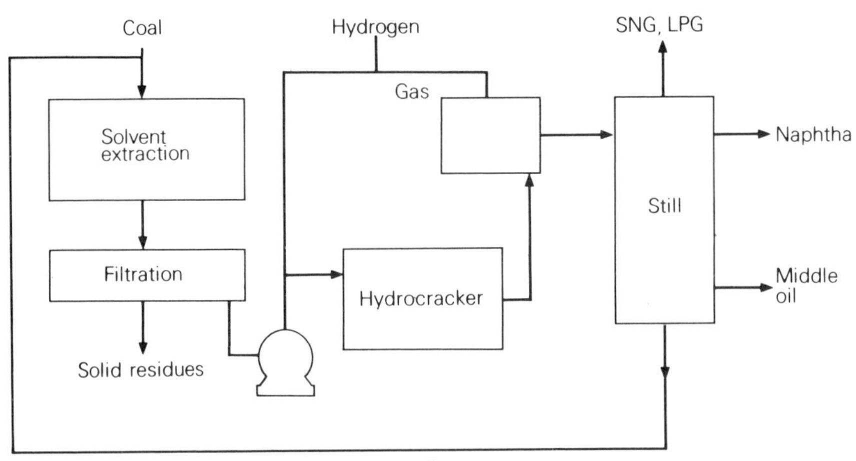

9.31 NCB liquid solvent extraction process

and recycled oil are slurried and pumped to a stirred digester, where extraction (which can be as high as 95%) takes place. After filtration, the solution is concentrated by evaporation to give an extract with about 45% concentration of coal substance in the extract oil, the distillate being recycled. The extract is hydrocracked under pressure and in the presence of a catalyst to give high yields of distillable hydrocarbon oils boiling in the range 60–350°C. The gases are separated; hydrogen, the main component, can be recycled. The liquids are distilled to give light, medium and heavy oils. The light oils are rich in aromatics and have been shown to be suitable for high grade gasoline. They would also be suitable feedstocks for the chemical industry. The medium oils can be further hydrocracked to produce a higher yield of light products. The heavy oils are good coal solvents and are enough to make the process self-sufficient. Excess heavy oils can be coked to produce special cokes for graphite manufacture. The process produces up to 50 gal of gasoline and 80 gal of other distillate oils per ton of coal. The overall yield of petroleum-type liquids can be about 50%, plus other by-products.

The supercritical gas extraction pilot plant is shown in Fig. 9.32. This is based on the successful operation of the 50 kg/d plant. Toluene, produced in the process, is a suitable solvent at 400°C and 100 bar. About 40% of the coal is dissolved and the residue, separated in a cyclone, may be used for hydrogen manufacture, for which it is very suitable. The solvent may readily be separated, cleanly and efficiently, from the extract, which is believed to be very suitable for hydrocracking and distillation, since the extract has a lower molecular weight and higher hydrogen content than normal extracts from liquid solvents.

A site has been chosen for the two 24 t/d pilot plants, each costing £15–20 million, and approval to begin construction is hoped for in 1980. The next stage proposed would be 200–500 t/d demonstration plants, allowing work to start on the first

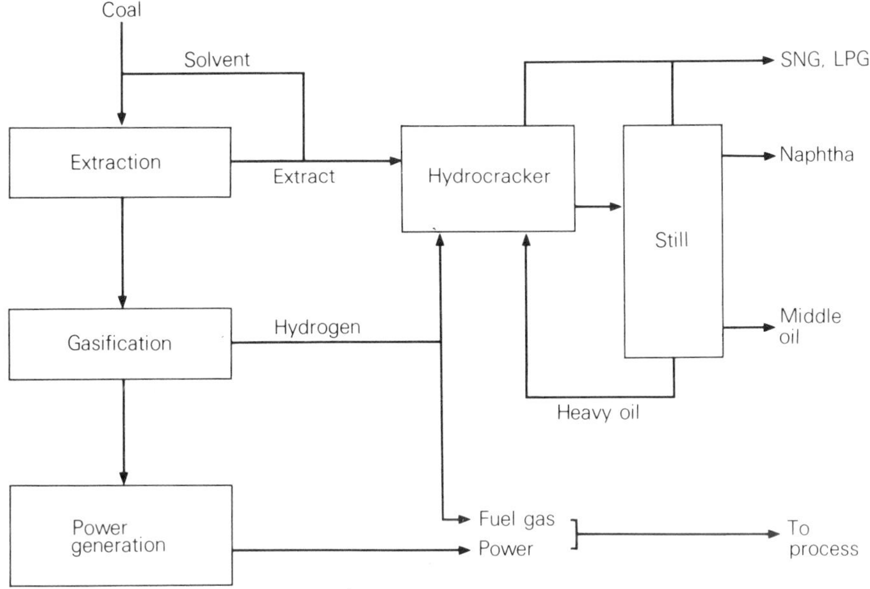

9.32 NCB supercritical gas extraction process

commercial plants in the early 1990s. There is a possibility of co-operating with the USA and West Germany by exchanging information on pilot plants of over 6 t/d.

Japanese Programme

The realisation in the early 1970s that Japanese energy imports would be a crucial factor indefinitely and that imported oil would have to be augmented by imported coal gave rise to a national energy replacement programme, "Operation Sunshine". As part of this policy, coal liquefaction research is being carried out on a 1 t/d scale, with targets for pilot plants on the 40 t/d scale for several types of process by the early or middle 1980s. In practice, it appears that the large trading consortia are forming associations with industrial organisations in other countries. Mitsui have collaborated with Gulf in a 5 t/d SRC pilot plant at Omuta. The use of this process to produce coking additives is one immediate objective; formed coke, binders and carbon products are also potential objectives. The possibility of a 10 000 t/d plant in Australia is being studied. A Japanese consortium is participating with Exxon on the EDS project. Sumitomo have a 2 t/d pilot plant at Akabira and hope to join in sponsorship of a US process. The Osaka Gas Company is studying a process called "Cherry" in which some coal powder is added to crude oil or heavy derivatives before refining and hydrocracking with the objective of increasing light product yields.

Russian Programme

Russian work also seems to be strongly inspired by the objective of improving oil utilisation. Russian heavy oil fractions are often aromatic and good hydrogen-donor solvents for coal. This led to the belief that the combined processing of coal and oil would be advantageous because hydrogen consumption and pressures might be lower than for separate processing and sulphur reduction might be facilitated. Laboratory work is carried out at around 425°C and 100 bar, with an iron/molybdenum catalyst. It has been claimed that a 50% reduction in hydrogen consumption can be achieved whilst converting 90–95% of the coal substance and removing 50–55% of the sulphur in the oil. Technical and economic assessments are being made for a combined plant of 4.2 Mt/a of coal and 8.5 Mt/a of crude oil, with a possible date of 1990 for industrialisation.

Polish Programme

Coal liquefaction would be economically and strategically very desirable in Poland. The current Polish programme began in 1966 and includes a fluidised bed carbonisation process to produce tars and a process for partial hydrogenation, the products of both processes to be further hydrocracked. Laboratory work on solvent extraction has been carried out at 420–450°C at 250 bar with a nickel/molybdenum catalyst. Other work is aimed at direct hydrogenation in ebullated or fixed catalyst beds at 500°C and 300 bar. A 500 t/d plant would be built before proceeding to a fully commercial scale.

Australian Programme

Australia is an attractive location for foreign investment in coal and coal conversion processes, especially liquefaction. A number of liquefaction projects have been announced, including SRC (US and Japanese interests) and Synthesis (German interests). Since access to processes likely to become established earliest in other countries can probably be arranged, internally financed work has tended to look to other routes. Flash pyrolysis is a particular interest, using fluidised and entrained bed reactors on the 0.5 t/d scale.

RESEARCH AND SUPPORTING TECHNOLOGY

All the countries mentioned as having project interest have substantial research programmes both to support processes being developed on a large scale and frequently also to enable a good appraisal of alternatives, including so-called Second or Third Generation Processes, to be made. Thus, as an example, although South Africa has a huge investment in synthesis, which has paid off (for them) and seems likely to continue to do so, there are research studies on methanol, solvent extraction (liquid and supercritical gas) and pyrolysis; close links with steel industry needs have been established, with a view to combined processes. In addition to the countries considered in the project section, many other countries have a liquefaction research interest; India and Canada may be particularly mentioned. The number and diversity of these research programmes are too great for concise summary and categorisation. A few items may however give some indication of trends in thinking and reference to some of the supporting technological work may underline practical difficulties. It may be assumed that projected processes for which the basic research is currently being undertaken or for which serious technical problems involving materials or components have yet to be solved are processes which will not have important commercial impact until well into the 1990s, or later.

In the UK, because of the indigenous energy situation, liquefaction research has centred around two solvent extraction processes, described in the preceding section, which were not expected to be required in a commercial form by an exceptionally early date. Emphasis has therefore been on avoiding potentially difficult steps, on obtaining high value products and on designing processes suitable for integration. In the case of liquid solvent extraction, these principles have led to the following objectives:

(a) High yields of extract from low rank coals;
(b) Successful hot filtration;
(c) High convertibility of extracts in hydrocracking.

The solubility of many coals has been determined, together with the quantities of products obtained by subsequent processing. A typical mass balance for a bituminous coal is shown in Fig. 9.33; the evolution of gases reduces the quantity of coal material actually obtained in solution. Conditions for satisfactory hot filtration have been explored and it has been found that although high yields can be obtained in a few minutes digestion under certain conditions, the optimisation of filtration only occurred after about 30 min and was evidenced by reduction of both viscosity and cake resistivity. The quality of extracts was distinctly higher when prepared using a hydrogenated solvent; the ash and viscosity were lower, the H/C ratio higher and larger proportions of light liquids were obtained. Perhaps even more important, it has been shown that when the extract solution is made with hydrogenated solvent, it can be stripped of solvent without polymerisation or coking. This has enabled the

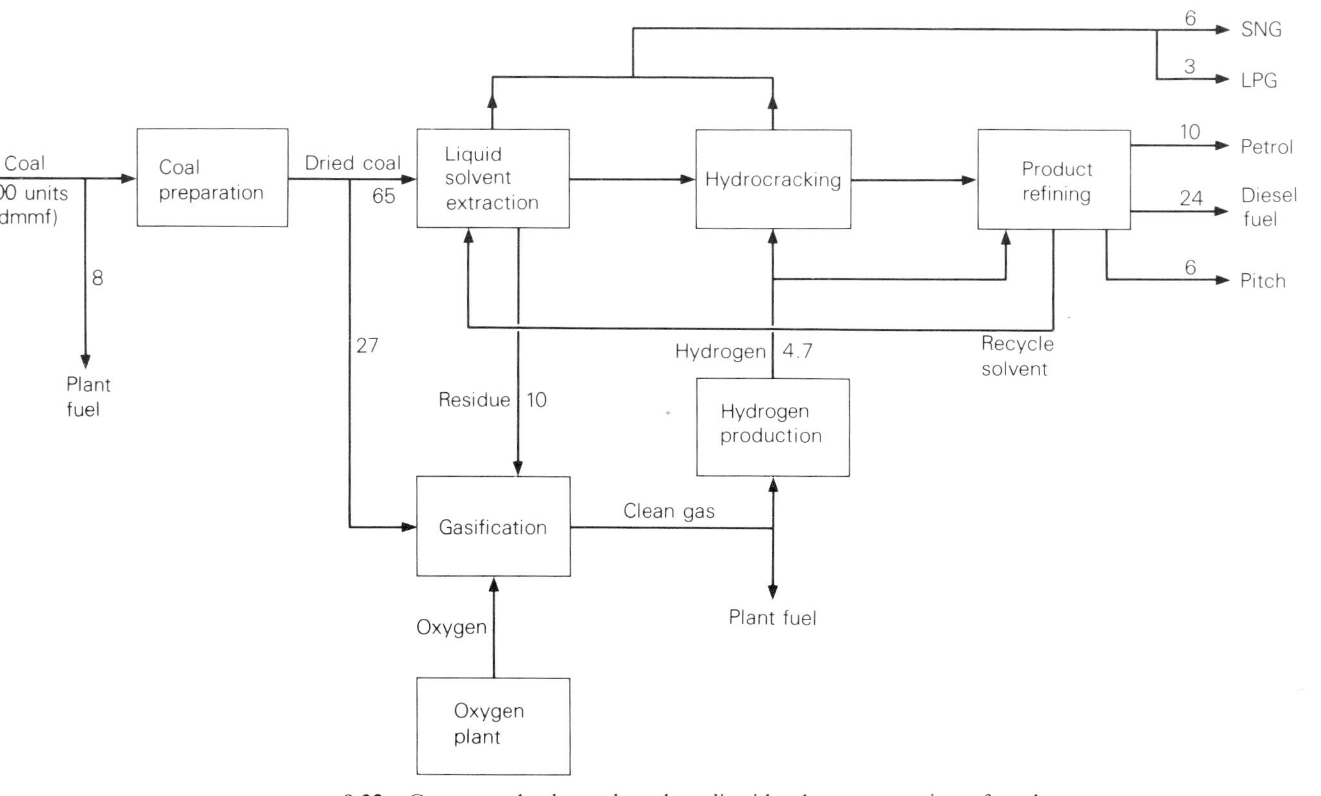

9.33 Conceptual scheme based on liquid solvent extraction of coal

structure of the extract to be studied, confirming the fact that the coal "molecule" has been very considerably broken down and that the hydrogen donated has created points where further breakdown by hydrocracking can be facilitated. Catalysts for this latter stage have been tested and preference shown for a Co/Mo catalyst. On continuous hydrocracking tests, best results were obtained at 440°C and 210 bar when 78% conversion of the extract to distillates boiling below 420°C was achieved. As shown in Fig. 9.33, up to 62% of the coal entering the process can be converted to distillates, mostly in the gasoline/gas oil range since the higher boiling fractions are recycled as solvent. Procedures for establishing this recycling on a continuous basis have been established.

In the Supercritical Gas Process, a basic study of a wide range of variables has identified conditions whereby the extract yield can be increased to between 40 and 50%. Separation efficiencies (ash and char from solution and solvent from extract) have been established at a high efficiency ($<0.1\%$ mineral matter in extract) in operation modes thought suitable for continuous larger scale plants. The extract has been shown to have similar open chain structures to those in liquid extracts but the aromatics are less condensed and the H/C ratio even higher. Extract hydrocracking has led to the projected yields and products shown in Fig. 9.34, a conceptual scheme which includes SNG manufacture.

There is a considerable history of research on pyrolysis in the UK, especially at high rates of heating and in the presence of hydrogen, and this work is continuing.

In the Federal Republic of Germany, research is mainly in direct support of the two routes: synthesis and hydrogenation. In practically all the synthesis work, catalyst research is explicitly or implicitly the main objective; this is linked with such aims as "selectivity" and the preferential production of short chain olefins rather than gases. In one study, various catalysts (including Mn/Fe) were used in both liquid and fixed phase reactors, but the maximum yield was not considered as being up to expectations. In another study, catalysts based on iron plus elements from Groups V–VIII were used and theories about catalyst surface topography and activity or selectivity confirmed. Thermal stress affected performance of catalysts and this was influenced by whether synthesis was carried out in a single pass or by recycling the gases after separating out the short chain olefines. The latter was less destructive but economic means of tapping the desired products needed development. More basic studies of the detailed mechanism of synthesis, using such techniques as lasers, radioactive tracers and ultra-high vacuum, suggest that industrial catalyst choice so far has been largely *ad hoc*, with the implication that the basic approach may in time transform the synthesis processes. The possibility of producing methanol first, to be followed by conversion to paraffins and aromatics, using a new form of zeolite catalyst, is also being studied.

Perhaps because of the large-scale experience during the war, research on hydrogenation in Germany seems less basic, although some work on catalysts (including metal sulphides) and on a comparative study of solvents has been carried out. Emphasis now seems to be on establishing improved operating conditions through the large pilot plants. Some work on the use of coal to enlarge the raw material base for refineries is going on and so also is research on the hydrogenation of lower rank materials, including brown coal and peat, using water based slurries in one case. Following the original UK work, supercritical extraction has been taken up on a small scale; emphasis is on hydrogen additions and catalysts

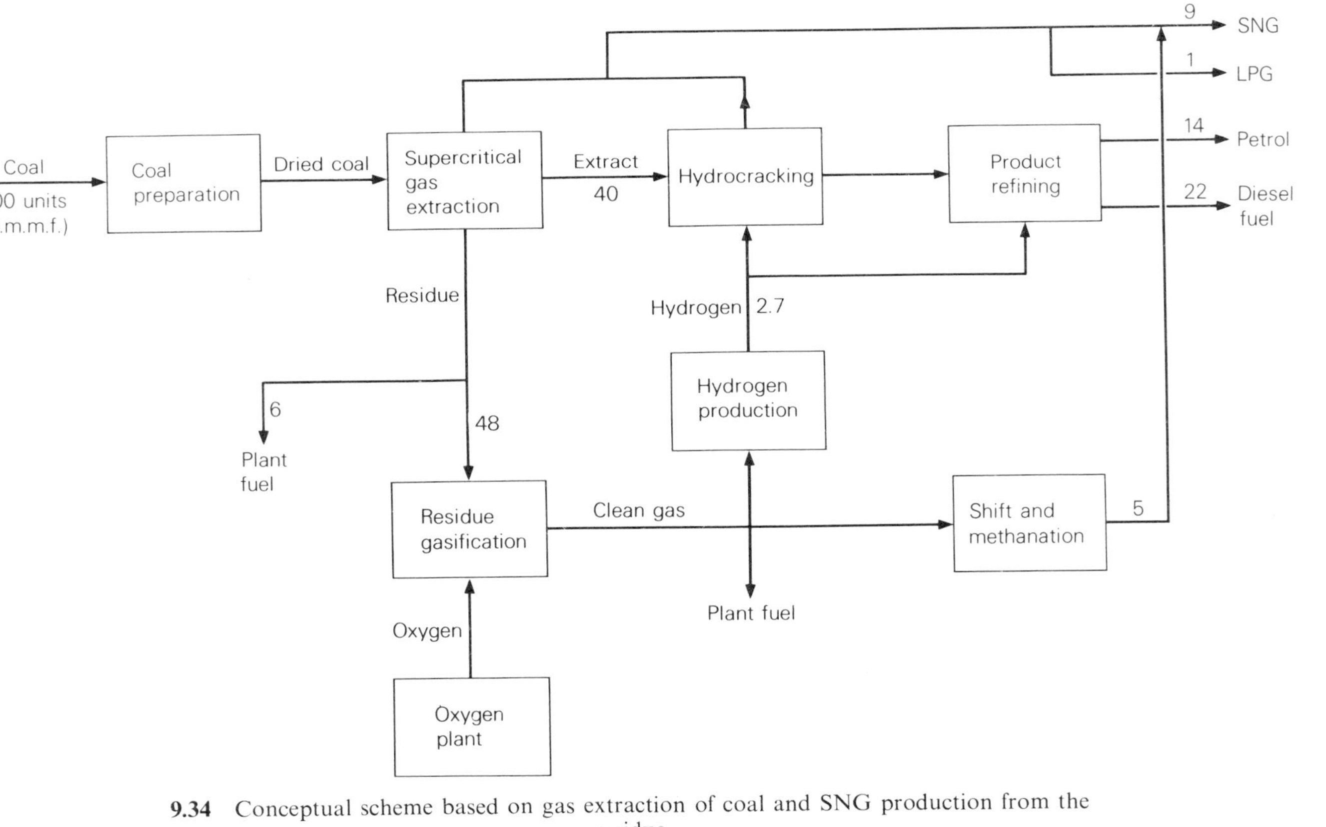

9.34 Conceptual scheme based on gas extraction of coal and SNG production from the residue

to improve extract yield. Some work on pyrolysis, including hydropyrolysis, has been initiated but in a rather low key. Presumably because of wartime experience, there seems to be considerable confidence about the technology of large plants; significantly, perhaps, some work on mechanisms for feeding fine coal and lump coal into pressurised reactors has been undertaken.

An extremely interesting and important optimisation exercise has been carried out by a consortium (85% government funding) involving all the wellknown industrial names. The study is based on the conversion of Australian coals to liquids in Australia and combinations of eleven different processes are included, with the possibility of the variation of product slates. It was concluded that, with bituminous coals, it would be necessary to include a Fischer–Tropsch synthesis plant in addition to a hydrogenation plant in order to produce diesel fuel but the synthesis plant would not be necessary with brown coals. The description of the study indicated that the choice of gasification routes was likely to be a critical factor, both for initial reactions with coal and for the gasification of residues.

In the USA, the number of laboratories engaged in research related to liquefaction must run into three figures. Industrial and governmental organisations involved in the dozen or so principal processes have supporting research efforts and in addition many academic and research institutes are involved, often in rather narrow fields but generally with governmental support, at least initially. The procedure appears to be one of relatively indiscriminate seeding, followed possibly by the selection of attractive specimens for replanting. There appears to be less coherence in the research than in the major process developments. It is interesting that high quality scientists, not previously interested in coal, have become involved; considerable dissatisfaction has been expressed on the existing classification and nomenclature of coals. This was emphasised in one study of the liquefaction behaviour (in tetralin) of over 100 coals, which proved immensely diverse, suggesting a strong influence not only from the normal rank parameters but also from the geological age and history of the coal and the geochemistry of the mineral matter.

Hydroliquefaction is the main process attracting research. One of the main themes of US work concerns the mechanism of the scission of coal "molecules", including factors influencing the release of fragments as gases and the re-polymerisation of low molecular weight radicals; the effects of hydrogen donors, molecular hydrogen, catalysts and changes occurring in recycled solvents are strongly studied. Two stage processes are of interest since hydrogenation can take place relatively quickly and easily up to about 20% conversion, probably with low hydrogen requirements.

Various forms of pyrolysis are attracting research effort, at least in published literature, in spite of the relatively small proportion of major process development in this field. By contrast, basic work on synthesis is not very widely reported and discussed, perhaps mainly because much of the work, which is nonetheless probably substantial, is being carried out in large industrial organisations developing proprietary processes and products. It is known that the Mobil Process (methanol to gasoline) is based on earlier research to improve the selectivity of catalysts through the geometry of their pore structure and this work is being extended.

An area where the US effort is very strong and well-presented is in technological support. It is considered that there are many components (valves, coal feeders, etc.) which are generic to many processes. Large-scale testing facilities are provided and information on components and materials is widely disseminated.

There is a considerable research effort in Japan, which seems to be influenced by the fact that commercial interests have been established in American processes and that foreign coals will largely be used. Thus, much of the work is related to hydroliquefaction, including the effects of coal type (especially Australian), catalysts and reaction conditions. Some interest, as in the USA, is shown in the production of "clean" boiler fuels. The use of asphalt and petroleum residues is being studied. Since most of the coal imported into Japan is presently used for coking, links between liquefaction and solid fuels, e.g. the production of coking additives, are of interest.

Significant research programmes on liquefaction are also being conducted in Canada, Australia and South Africa.

In spite of the considerable amount and range of research projects, however, and a growing inclination to discuss work in international fora, the view has been expressed that a review of basic information is necessary in order to indicate areas where further work is necessary. There is, however, a consensus that the programmes are far from comprehensive and that industrial scale-up is likely to take place without a basic understanding, a situation which usually leads to delays in ironing out difficulties.

ENVIRONMENTAL ASPECTS

Coal liquefaction plants will have much in common with other plants stocking and handling coal and oil refineries for product upgrading. There will also frequently be problems associated with coal combustion and/or gasification plants since these processes may be unit stages integrated into a complete liquefaction plant. Thus there may be a requirement to deal with airborne coal dust and rain water contaminated by contact with coal stack emissions, both particulate and gaseous. In addition, however, within the liquefaction stages proper, there is a need to monitor end products (both the main constituents and potential impurities) and intermediates. This monitoring will have to include both the possibility of contaminants transported outside the plant and the effects on personnel who may be in contact, or come into contact, with intermediates and products of the process. Since the possibility of compounds having a high degree of toxicity (including carcinogens) has to be considered, engineering and monitoring will need to be of a very high standard. Unfortunately, information is currently not at all comprehensive.

CONCLUSIONS

There are a few basic routes to the production of liquids from coal but a large number of variations and "improvements" have been proposed. Apart from SASOL, there is no really large-scale experience and some known difficulties may make the acquisition of this experience a somewhat expensive one.

If strategic requirements were regarded as overriding (as perhaps they should be), successful demonstration of a few alternatives could no doubt be achieved in a few years. The inhibitions are economic and are discussed in Chapter 16.

Although it is not possible to take a firm view because a complete review is not available, it seems likely that basic research is inadequate to support such a massive new and very complex industry, the growth rate of which may eventually be forced. Efforts are being made to anticipate material and component problems.

Chapter 10

CHEMICALS FROM COAL

INTRODUCTION

Much that is relevant to this subject has been discussed in other chapters, notably Chapters 7—9 on carbonisation, gasification and liquefaction. The main feedstocks for the chemicals industry at present are natural gas, liquefied petroleum gases (LPG) and light petroleum distillates, especially naphtha. Direct substitutes for all these can be made from coal, as indicated elsewhere; it would, in fact, be possible merely to follow this line and to produce these same feedstocks from coal. However, since a major switch in primary feedstock towards coal seems inevitable, ultimately, opportunity should clearly be taken to review and re-optimise the whole organisation through the chain of primary chemical feedstock, intermediates and derivatives. Therefore, information discussed elsewhere is recapitulated here from the chemical industry point of view. It may be that there will be greater diversity in future between coal conversion processes for chemicals and light liquid fuels. It should also be borne in mind that hybrid feedstocks will probably be fairly common, especially in the early stages of the transition; a particular possibility might be the use of coal-derived hydrogen in petroleum refining or in petrochemicals manufacture.

Synthesis gas and light hydrocarbons formed by hydrogenation of coal extracts may ultimately become the principal routes to chemical feedstocks from coal, as in the production of liquid fuels. However, pyrolysis (or carbonisation) has historically been an important source of chemicals and may be so again, possibly through more advanced styles of pyrolysis. Acetylene was also once a very important chemical building block; although it has been largely displaced, it still has some attractions and could conceivably have some influence on routes from coal to chemicals. Formerly, acetylene was generally made via the carbide route, using coke to make the carbide; the manufacture of acetylene directly from coal in a plasma arc is a recent innovation which might be attractive if it could be made more efficient. Finally, the Supercritical Gas Extraction route may eventually have an important part to play in chemicals manufacture in view of its ability to extract useful low molecular weight groups from coal without degradation or recombination and at the same time to render the char into a very reactive form.

HISTORICAL REVIEW

Historically, the contribution made to the chemical industry by carbonisation products cannot be overestimated. The rapid and complex development of the

TABLE 10.1 Principal derivatives from benzole and coal tar

Products	*Intermediates*	*Polymer or resin derivatives*
benzene	cyclohexane, phenol	nylons
	styrene	polystyrene, ABS, SBR
	maleic anhydride	alkyds, polyesters
toluene	—	hydrocarbon–formaldehyde resins
	toluene di-isocyanate	polyurethanes
xylenes	—	hydrocarbon–formaldehyde resins
	phthalic anhydride	alkyds, polyesters
	terephthalic acids	polyesters
naphthalene	—	hydrocarbon–formaldehyde resins
	phthalic anhydride	alkyds, polyesters
coumarone/indene	—	coumarone–indene resins
phenol	—	phenol–formaldehyde resins
	cyclohexanol	nylons
	bisphenol A	epoxy, polysulphones, polycarbonates
cresols	—	phenolic resins
xylenols	—	phenolic resins

dyestuffs and pharmaceutical industries merit special mention but in more recent times the bulk markets for organic chemicals have mainly been in the polymer and resin markets. The technical suitability of carbonisation products to meet these requirements is illustrated in Table 10.1.

Unfortunately, the supplies of carbonisation by-products, for a long time small in comparison with petrochemicals, are reducing further, with lower demands for coke from the steel industry and the virtual demise of other sources such as town's gas and low temperature coke. Nonetheless, as noted in Chapter 7 on carbonisation, it is important for the economic health of the coking industry that the highest possible value should be derived from the by-products.

The other historically important bulk chemical from coal is ammonia. This is also produced as a carbonisation by-product but the quantity is not now very significant. The traditional synthetic ammonia industry was based on a combination of coal carbonisation and gasification of the resulting coke. The operations, as then operated, were labour intensive, unpleasant and polluting. The process was replaced by reforming naphtha or natural gas to give synthesis gas. This was more economic because of the low feedstock prices, and was simpler, cleaner and more convenient. However, as mentioned later, the new processes for making synthesis gas from coal can mitigate the disadvantages previously associated with the use of coal.

From the basic raw materials, oil (light fractions) and natural gas (and LPG), the chemicals industry, especially the polymer sector, uses surprisingly few primary intermediates as shown in Table 10.2.

Ethylene is the most important and, of the others, only propylene and benzene have more than a few per cent of the market. From the principal derivatives many individual substances are of course made. The standard relationships are illustrated in Figs. 10.1–10.3.

TABLE 10.2 Primary intermediates and derivatives

Primary intermediate	*Consumption UK 1977[a] (1000 t)*	*Principal derivatives*	*Production UK 1977 (1000 t)*
Synthesis gas	—	phenolics	60
		urea–formaldehydes	178[b]
Ethylene	1195	polyethylene	490
		polystyrene	224
		ABS	39
		SBR	362
		PVC	386
		polyesters (unsaturated)	70
Propylene	800	polypropylene	247
		acrylics	61
		ABS	39
		nitrile rubbers	21[c]
		epoxy resins	24
Butadiene	155	SBR	362[c]
		polybutadiene rubber	45
		polychloroprene	23[c]
		nylon 66	31[d]
Benzene	675	polystyrene	224
		ABS	39
		SBR	362[c]
		phenolics	60
		nylons	31[d]
		polyesters (unsaturated)	70
		alkyds	134
Toluene	na	polyurethanes	79
Xylenes	na	polyesters (unsaturated)	70
		alkyds	134
Phenol	168 (Production: synthetic only)	phenolics	60
		nylons	31[d]
		epoxies	21

[a] CEFIC Statistical Survey: Enquiry 1978;
[b] Covers all aminos, including melamine formaldehydes;
[c] Production capacities quoted;
[d] Figure includes small amount of silicones.

The fact that so many end products can be associated with a few primary intermediates, each of which can then be manufactured on a large scale, creates a stable situation against which the penetration of coal derived feedstocks must be considered. On the other hand, these intermediates have enjoyed a very high growth rate and although this may slacken, the possibility of a demand doubling time of 20–25 years does not seem unlikely. If this growing demand were to continue to be supplied by natural gas and oil, the other outlets for these products would be severely affected. Thus, the gradual penetration of coal into chemical feedstocks seems inevitable. At the same time, the conflicting demands of the transport fuels and chemical markets will apply equally to coal as they do to oil and gas. How these transitions and permutations will work out is complicated by the many alternative possibilities presented by the main coal conversion routes (Table 10.3); where a secondary conversion is necessary this is indicated by brackets.

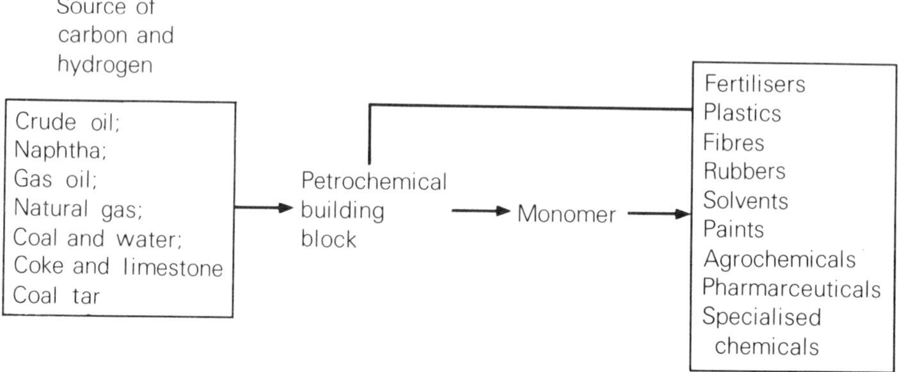

10.1 The petrochemical process

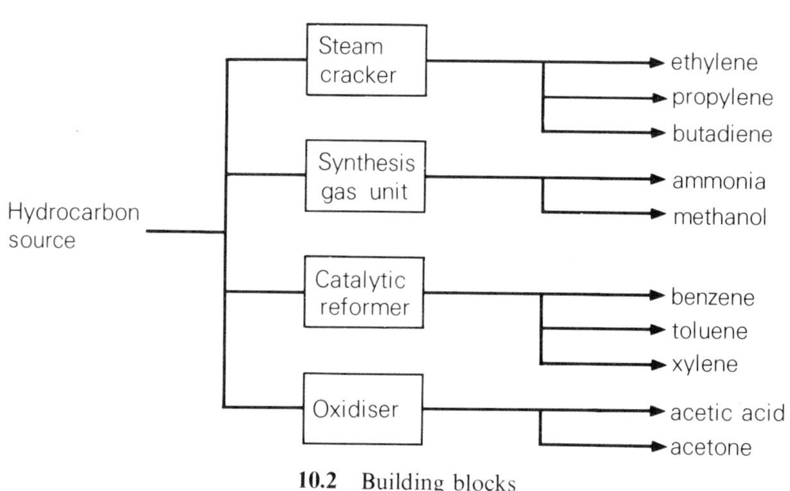

10.2 Building blocks

GASIFICATION/SYNTHESIS

Synthesis is a very versatile process. Some of the classic hydrocarbon forming reactions, for example, are:

Methanation	$3H_2 + CO \rightarrow CH_4 + H_2O$
Paraffins	$(2n+1)H_2 + nCO \rightarrow C_nH_{2n+2} + nH_2O$
Olefins	$2nH_2 + nCO \rightarrow C_nH_{2n} + nH_2O$
Aromatics	$1.5nH_2 + nCO \rightarrow C_nH_n + nH_2O$

All are exothermic.

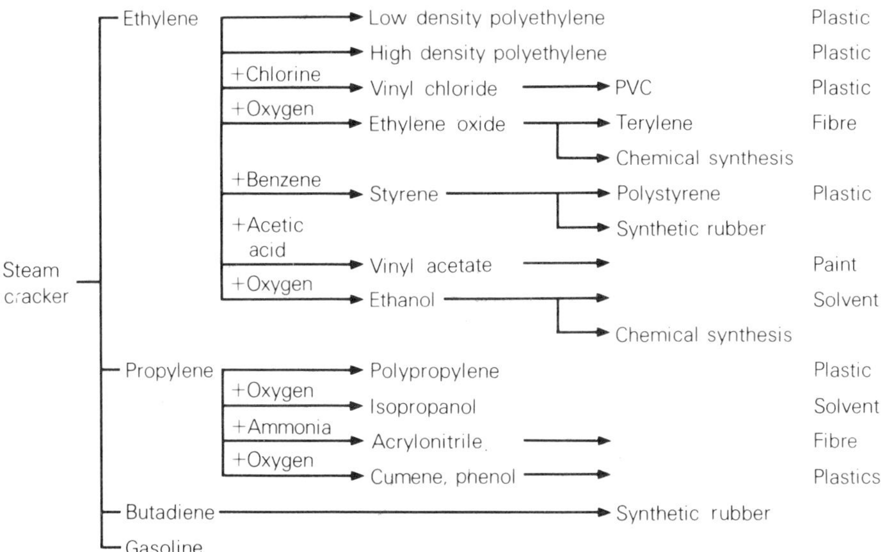

10.3 Steam cracking

In addition, there are possible competing, undesirable reactions, such as the shift reaction and the deposition of carbon or the formation of metal carbides. Methanol and ammonia can also of course be produced by synthesis reactions via coal gasification and these represent the main immediate opportunities.

The wide range of possible products in synthesis and the original lack of selectivity are factors which have been accommodated to some extent in special circumstances. In Germany, in war time, the synthesis process was used in a fashion

TABLE 10.3 Suitability of main coal conversion routes for chemicals production

Feedstock/intermediate	*Gasification/ synthesis*	*Pyrolysis*	*Hydro-pyrolysis*	*Hydrogenation/ reforming*
Ammonia	x	x		
Methanol (ethanol)	x(x)			
Methane (acetylene)	x(x)	x(x)	x(x)	x(x)
Ethylene (ethanol)	x(x)	x(x)	x(x)	x(x)
Propylene	x	x		x
(Butadiene)	(x)	(x)		(x)
Benzene (phenol)	x(x)	x(x)	x(x)	x(x)
Toluene (cresols)	x(x)	x(x)	x(x)	x(x)
Xylenes (xylenols)	x(x)	x(x)	x(x)	x(x)
Phenol	(x)	x	(x)	(x)
Cresols	(x)	x		(x)
Xylenols	(x)	x		(x)

complementary to hydrogenation and provided a wide range of important materials. In Sasol 1, two different synthesis modes are used and these also provide a range of products which have mainly been of value there. However, in future operations to synthesize hydrocarbons, much emphasis will be on large quantities of narrowly defined products, including compounds presently derived from oil for chemical purposes; recent work has demonstrated a great deal of promise in this respect. The control of product is, of course, influenced by the composition of synthesis gas and also by the configuration, temperature and pressure of the synthesis reactor system. It will be noted that all the hydrocarbon forming reactions are exothermic, so the mode of temperature control will be important. Most particularly, however, control of product range is exercised by the catalyst and it is in this area where the most startling advances are being made, both in basic science and in practical application.

Nonetheless, it is in the areas of ammonia and methanol, plus possibly OXO-alcohols to a lesser extent, where synthesis from coal derived gases is likely to make an initial major impact in the chemicals industry. Synthesis routes to make ammonia and methanol from coal via gasification were established commercially in Germany about sixty years ago. For the past thirty years however coal has been a relatively insignificant feedstock for these products. In 1979, capacity for ammonia synthesis based on coal was only about 4500 t/d, as listed in Table 10.4. This is only about 5% of world production; counting ammonia from carbonisation, the contribution from coal only goes up to 12%. The current main feedstock is natural gas (64%), with naphtha next at 13%. The situation regarding methanol is even less favourable: out of a world production of over 14 Mt/a, the only production from coal is 16 000 t/a as a subsidiary product of the AE & CI ammonia plant, as noted in Table 10.4. The implication of these proportions is that processes for ammonia and methanol have now been optimised for feedstocks other than coal; these highly developed plants have reached great size and complexity. It is therefore rather more difficult to substitute coal gasification into existing plants, although as discussed later, systems are available for doing this. The prospects for new construction are

TABLE 10.4 Ammonia plants in operation based on coal gasification (January 1979)

Operator/location	*Plant commissioning date*	*NH_3 production (t/d)*	
		Winkler	*Koppers/Totzek*
Azot Gorazde, Yugoslavia	1953	50	270
Azot Sanayii, Turkey	1959	100	
Nitrogenous Fertilizer, Greece	1959		270
Neyveli, India	1960	280	
Chemical Fertilizer Mae Moh, Thailand	1963		100
Azot Sanayii, Turkey	1966		340
Ind. Dev. Corp., Zambia	1966		100
FCI, India (Ramagundam)	1969		900
FCI, India (Talcher)	1970		900
Nitrogenous Fertilizer, Greece	1970		135
AE & CI, South Africa	1972		1000[a]
Ind. Dev. Corp., Zambia	1974		200

[a] Plant also produces 16,000 te/year methanol.

more straightforward. Fortunately, growth in demand, including methanol for motor fuel, should continue over the medium and long time scales (although not necessarily in the short term, when there may be over-capacity). However, in many parts of the world it would be unwise now to consider basing new chemical capacity on gas or oil; coal would be the obvious alternative for new plants, as well as for the more speculative retro-fitting of existing plants.

The preference for feedstocks other than coal has basically been due to the vital importance of hydrogen in synthesis reactions. Methane has the highest H/C ratio and is therefore ideal for ammonia synthesis. Naphthas high in paraffins are preferred for methanol conversion for reasons discussed under methanol synthesis. In some cases, the nitrogen for ammonia synthesis may be introduced via the air used in the production of synthesis gas. With coal, there will always be a struggle to get a high enough hydrogen content in the synthesis gas; this will require high oxygen inputs, increased carbon dioxide rejection and a large amount of shifting. Other disadvantages of coal as a feedstock are the gas purification (and environmental) requirements and the coal handling and preparation equipment. Many of these disadvantages have energy implications, as shown by the energy requirement for one tonne of ammonia for different feedstocks (the thermodynamic requirement is 4.6 Gcal/t NH_3):

Fuel/feedstock	*Process*	*Energy consumption (Gcal/t NH_3)*
Natural gas	Reforming	8.7
Naphtha	Reforming	9.3
Fuel oil	Partial oxidation	9.6
Coal	Partial oxidation	11.3

A number of commercial coal gasification processes are available as precursors to synthesis plants and others are approaching this category. Typical analyses for raw gas from leading contenders are given in Table 10.5; the top three are fully established processes.

From the point of view of chemicals manufacture by synthesis, about the only important point to note with regard to gasifier preference is that the Lurgi produces a significant amount of methane which would need to be reformed. At the same time, however, the gas from the dry-bottom Lurgi has the highest hydrogen content; this is about adequate for methanol synthesis. The gases produced otherwise would need substantial shifting. Another factor which might be seriously considered in choosing a gasifier for chemicals manufacture is the production of tars, oil, phenols and other organics. They are avoided most readily in the high temperature processes such as Koppers–Totzek and Texaco; these compounds may be destroyed in the Lurgi process by recycling but so far relatively little hard experience has been reported. Finally, gasifiers which operate at pressure fit more readily into downstream stages.

Apart from washing to remove tars and dust, etc. the gas needs purification as well as shifting. These are inevitably substantial and expensive plants, which together may represent 25% or more of the total investment — as much, or nearly as much, as the gasifiers themselves. Several steps are involved and there are alternatives for each stage. The optimum design combination will depend on the coal and the gasifier selected and also on the gas specification for the particular synthesis used. The unit processes available are considered in the gasification

TABLE 10.5 Typical raw gas compositions from coal gasifiers

		Raw gas composition (V/V_1 dry)				
Process	*Fuel*	*CO*	H_2	CO_2	CH_4	N_2
Lurgi (dry bottom)	Lignite	20	39	29	10	<1
	Bit. coal	25	40	25	9	1
Koppers–Totzek	Lignite	56	29	12	0.1	2
	Bit. coal	55	32	11	0.1	2
Winkler	Lignite	46	36	14	2	1
	Bit. coal	35	42	19	3	1
Lurgi (slagging)	Bit. coal	60	28	3	8	1
Shell–Koppers	Bit. coal	64	32	1	—	<1
Texaco	Bit. coal	50	35	17	0.1	1
Winkler (HTW)	Lignite	52	35	9	3	<1

chapter. Normally, it might be expected that purification would precede shifting but there are thermal efficiency advantages in favour of adjusting the CO/H_2 ratio of the hot raw syngas. This recent variant, the "sour gas shift", has been made possible by the development of sulphur-resistant shift catalysts, such as cobalt/molybdenum sulphide. However, for efficient synthesis, a high degree of purification is necessary and, for environmental reasons, the effluent streams will probably also require very effective clean-up.

Ammonia Synthesis

Several standard (catalytic) processes exist for converting stoichiometric mixtures of hydrogen and nitrogen into ammonia at about 500–600°C and 135–300 bar.

The reaction, $N_2 + 3H_2 \rightarrow 2NH_3$, is exothermic so the reactors are designed to abstract heat, by extensive cooling tubes for example. At the same time good contact with the catalyst is required, usually by making the gases flow along a circuitous path. The accommodation of these requirements within a minimum size pressure vessel has led to a few proprietary designs becoming established, including ICI/Kellog and Topsøe. The catalyst is iron oxide promoted by acid and alkaline oxides (e.g. Al_2O_3 and K_2O). Conversion is not complete; the ammonia is recovered by refrigeration and the gases mainly recycled (performing a cooling function).

There are several features of modern ammonia plants which need to be taken into account when considering coal as a feedstock. One is that the minimum size for efficient plants is 1000 t/d ammonia and 3000 t plants are available. The second advantage, linked with this, is that great progress has recently been made in reducing energy requirements; over the last two decades, fuel and feedstocks per tonne of ammonia have gone down from about 50 GJ to the low 30s. This is equivalent to a net requirement of about 1.2 t of coal/t ammonia. Gross energy inputs in the form of coal would necessarily be higher, mainly because of losses in gasification. An oxygen plant would also be necessary for gasification; this would however provide nitrogen, often introduced as air in the partial oxidation stage with other feedstocks. Requirements have been estimated (by Humphreys & Glasgow) for different gasifiers in Table 10.6.

In addition to the successful AECI plant at Modderfontein in South Africa, using six Koppers–Totzek gasifiers to produce 1000 t/d ammonia, other gasifiers are

TABLE 10.6 Daily consumption of coal and oxygen to produce 1000 t/d ammonia

Process	O_2 Plant capacity 10^3 t/d	Process coal 10^3 t/d	Power coal 10^3 t/d	Total coal 10^3 t/d
Koppers–Totzek	1310	1630	800	2430
Winkler	720	1410	410	1820
Slagging gasifier	570	1250	330	1580
Texaco	1400	1550	300	1850

offered with suitable equipment for ammonia manufacture. Lurgi offer an integrated plant (Fig. 10.4). This uses dry-bottom gasifiers (each at 400–800 GJ/h coal input), sour gas shift, Rectisol purification and either liquid nitrogen wash with methane reforming or cryogenic separation. In the latter case, the methane recovered might be sold as SNG. Plants would normally have four (3 on stream) Mk IV Lurgi gasifiers. Inputs, per tonne ammonia, would be:

Coal to gasifiers	35–37.5 GJ
Coal to utilities	20–24 GJ
Oxygen	275–440 Nm^3
Water	33 m^3

(The coal input corresponds to about 2.2 t/t ammonia.)

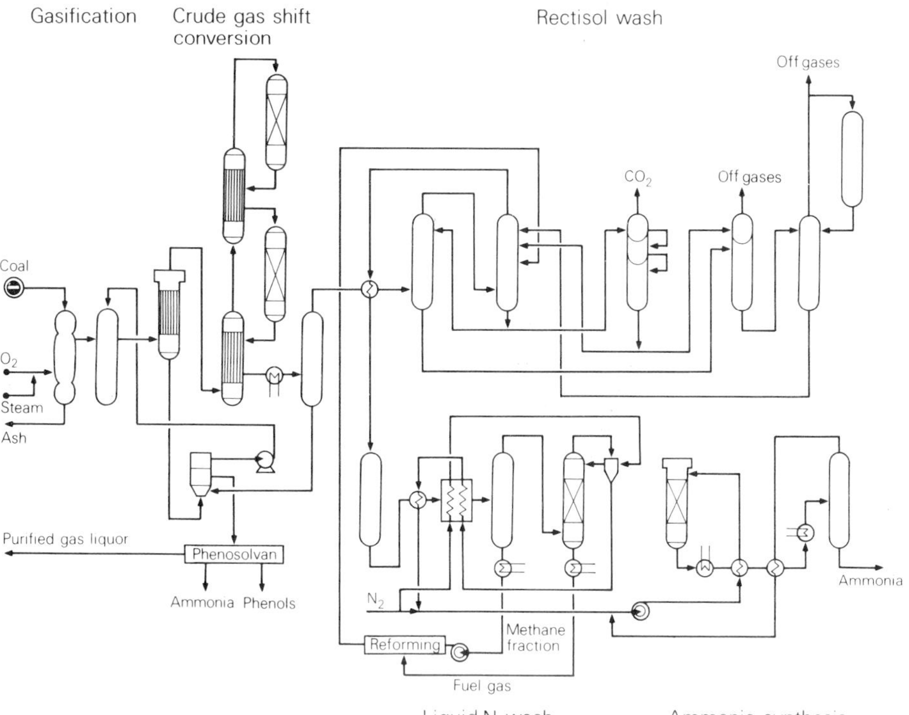

10.4 Lurgi ammonia-from-coal process

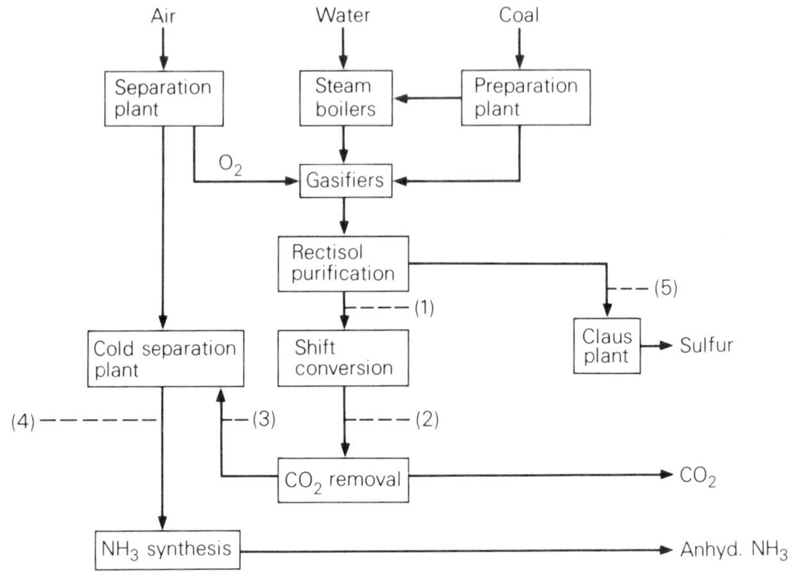

Material requirements
Coal: for gasification 1800 tons/day (at ~8500 btu/lb)
for steam-raising 830 tons/day
Water: 4.8×10^6 gal/day

	Gas stream				
Gas composition, %	1	2	3	4	5
CO_2	12.0	41.3	—	—	75.0
CO	55.0	3.0	—	4.7	—
H_2	32.0	55.0	94.1	75.0	—
N_2	1.0	0.7	1.2	20.3	—
H_2S					22.0
COS					3.0
Gas flow rate, 10^6 scf/day	93.3	97.3	80.2	99.7	21.0

10.5 Flow sheet and related data for a 1000 t/d ammonia plant (with Koppers-Totzek gasifiers)

A scheme developed for Alberta is shown in Fig. 10.5. This is expected to use 2.6 t of coal/t ammonia but the calorific value of the coal is not high. The composition of the gas streams at various points is also shown.

Neither of these schemes indicates the treatment proposed for the unreacted gas from the ammonia synthesis plant. The gas is recycled but some must be purged. This purge gas is often used for fuel but in the case of coal-based plants it may be desirable to recover the hydrogen.

Adaptation of existing plants has been studied, especially where natural gas supplies have become difficult, as in one or two recent winters in the USA. Various gasifiers for this purpose have been compared by Humphreys & Glasgow. One possibility is the standard Lurgi, as in Fig. 10.4. However, this uses very little of the

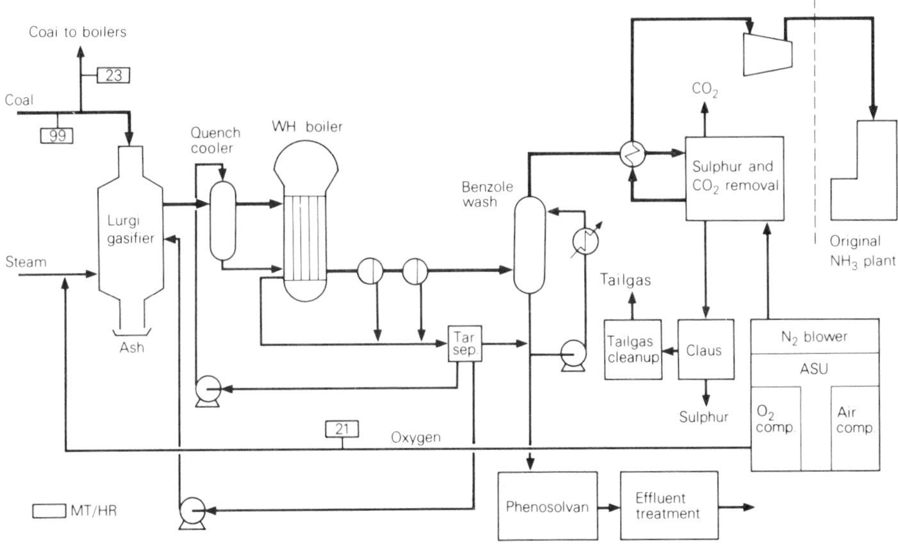

10.6 Modified Lurgi gasification route, 1000 Mt/d

existing plant and unless this is well sited for cheap coal, it might be preferable to build a totally new plant elsewhere. An alternative Lurgi adaptation is shown in Fig. 10.6. In this case the gas is completely cooled and purified, removing all the sulphur and most of the CO_2; this results in gas with a similar composition to that leaving a naphtha steam reformer and can be processed further in the primary and secondary reformers of an existing ammonia plant. In the second stage air is used, as normally, to introduce nitrogen and this uses up some hydrogen. This means that the coal gasifiers and oxygen plants have to be larger than they might otherwise be but there is much less new investment because more of the existing ammonia plant is used. In both cases, some additional compression is probably necessary on the coal gas, which is at a somewhat lower pressure than that used in naphtha reforming plants.

An adaptation based on Koppers–Totzek is shown in Fig. 10.7. This again can use none of the existing plant except for the ammonia synthesis. The gas must be cooled, mostly for dust removal, compressed to 35 bar, heated for shifting and recooled for CO_2 removal. Nitrogen is added and the gases can then be compressed by the ammonia plant compressor. The main disadvantage with K–T is the large oxygen consumption but the minimising of the effluent problem is an advantage.

Similar design studies have been carried out using Winkler, Texaco and the British Gas Slagging Lurgi for adaptations. None of these are able to use any of the existing plant and their relative merits depend on other factors such as pressure and the degree of shifting necessary to eliminate CO. The authors give estimates for a 1000 t/d ammonia plant (Table 10.7).

There are some anomalies between Table 10.6 and Table 10.7, especially for the Winkler coal requirement. This may be due to the differences between a new plant (Table 10.6) and an adaptation (Table 10.7); alternatively, this change may be due to later information or further consideration (Table 10.6 is the later).

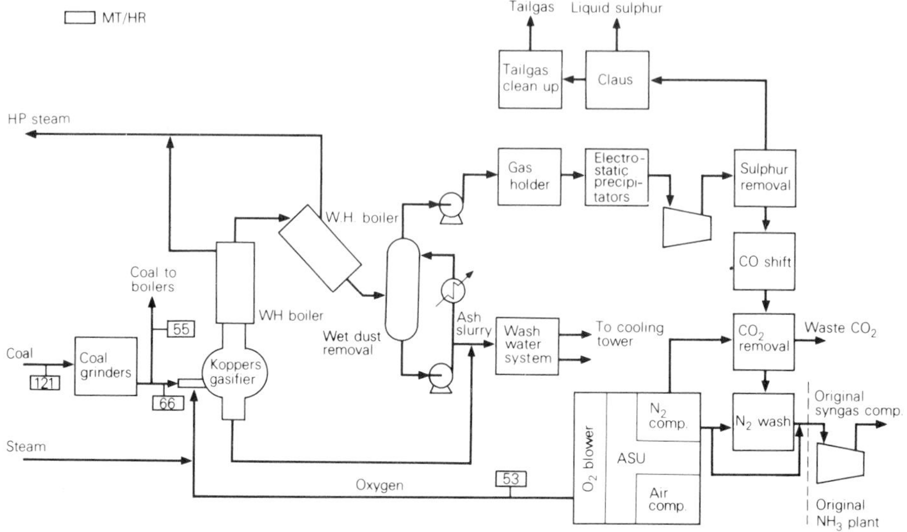

10.7 1000 Mt/d of NH_3; Koppers-Totzek gasification route

TABLE 10.7 Inputs for 1000 t/d ammonia plant

	Standard Lurgi	*Modified Lurgi*	*Koppers-Totzek*	*Winkler*	*Slagging Lurgi*	*Texaco*
Coal, t/d	2350	2380	2900	2950	2270	2400
Electricity, MW	7.4	6.5	13	11	6.5	7.5
Water, t/d	860	885	1100	1090	820	680
Oxygen plant size, t/d	460	500	1265	850	555	1320

Methanol Synthesis

Methanol is synthesized mainly by the reactions:

$$CO + 2H_2 \rightarrow CH_3OH$$

Carbon dioxide can also react to form methanol:

$$CO_2 + 3H_2 \rightarrow CH_3OH + H_2O.$$

When the synthesis gas is made from methane, which is frequently the case, there is more hydrogen than is required by the first reaction. In one variant of the synthesis process this is accepted and hydrogen is recovered in the purge gas, which is then used as fuel. More usually, however, the second reaction is exploited also by adding CO_2 from the reformer flue gas or possibly from an associated ammonia plant. In the case of synthesis gases from coal, of course, an excess of hydrogen is unlikely to arise but some methane could be present and this could take part in the synthesis as a result of the reforming reaction. Again, therefore, the matching of the gasifier to the synthesis plant via gas pre-treatment calls for careful design optimisation.

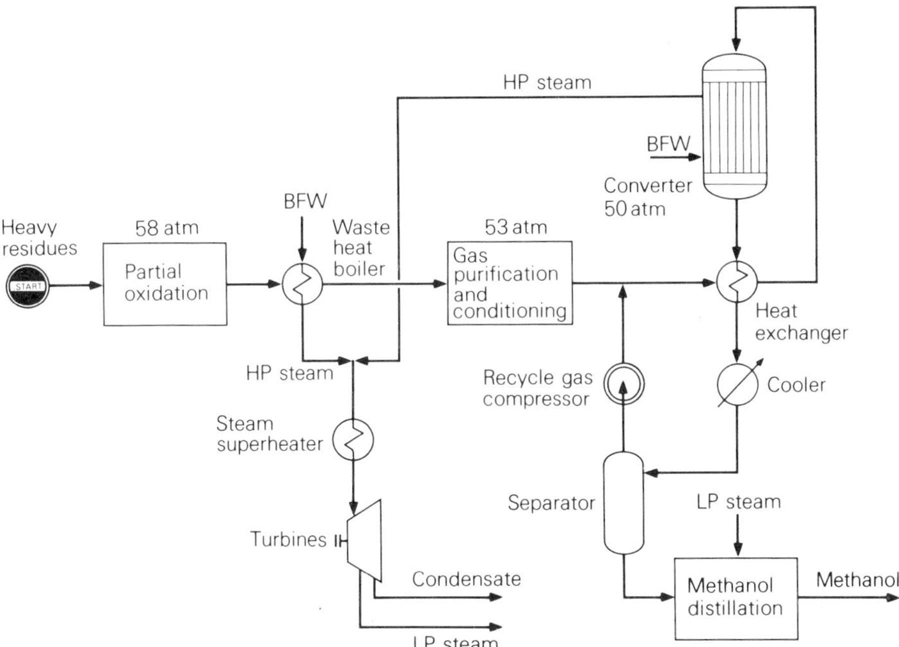

10.8 Lurgi low pressure methanol synthesis process

In early 1979 it was estimated that about 5 Mt of capacity for methanol was newly installed, under construction or planned. Of this, about 60% (including the two largest plants of 2500 t of methanol per day in Russia) were to the ICI low pressure design and 20% to the Lurgi process.

The ICI synthesis converter operates at 200–300°C and 50–100 bar, using a high-yield copper-based catalyst; CO_2 is added when operating on methane. The converter is simple and compact, with injection of cold synthesis gas at several levels and great attention is paid to heat recovery and use at all levels of temperature. The product is distilled in accordance with the particular product specification.

The Lurgi process (Fig. 10.8) features a tubular reactor. Synthesis over copper-based catalyst takes place at 230°C and 50 bar in tubes cooled by being immersed in a boiling water pressure vessel shell. Distillation follows in a steam-saving, three column system. Typical requirements for the ICI and Lurgi processes are given in Table 10.8 (but it would be unwise to compare these too closely).

TABLE 10.8 Feed requirements per tonne of methanol

	ICI			*Lurgi*		
	Naphtha	*Methane*	*Heavy fuel oil*	*Naphtha*	*Methane*	*Heavy fuel oil*
Feed and fuel (GJ)	32.3	31.1	32.8	29.9	29.3	36.8
Power (kWh)	35	35	88	0	50	130
Feedwater (m^3)	1.15	1.15	0.75	0.74	0.69	0.76
Cooling water (m^3)	64	70	88	40	45	75

There would seem to be no particular problem in using coal as a feedstock, except the optimisation of the gasification and gas treatment processes to the synthesis conditions. The small AECI plant is however the only one presently operating on coal; it is understood that a 2500 t/d plant may be built there. There are also plans in the USA for using the Texaco gasifier for producing about 750 t/d of methanol from 1450 t/d of coal. Methanol production will also be associated with the development programme for the Mobil–MTG process for liquefaction, described in Chapter 9. It will be noted also that the MTG type of process can be adapted to make a whole range of products from methanol, so if methanol synthesis based on coal becomes important, either as a fuel in its own right or as a step to gasoline, this would give a great boost to the whole concept of chemicals from coal, which could take advantage, jointly, of facilities established with gasoline in mind.

Synthesis of OXO-alcohols

The demand for OXO-alcohols appears to be growing. In March 1979, new or extended facilities for an additional 1.2 Mt/a were planned, compared with an installed capacity of about 1.5 Mt in 1978. Plant sizes range from very small plants of 1500 t/a to 100 times that size but 50 000 t/a seems to be about standard. The products are used in detergents and plastics or as solvents. The OXO synthesis had its origins in the initial studies of Fischer–Tropsch processes from coal but was subsequently developed in the petrochemicals industry. The synthesis process consists of reacting synthesis gas with an olefin to give alcohols via the appropriate aldehyde; the process usually takes place in two stages, on the following lines:

(a) from a linear olefin

$$RCH{=}CH_2 + CO + H_2 \rightarrow \underset{\text{linear aldehyde}}{RCH_2CH_2CHO}$$

(b) the aldehydes are converted by hydrogenation to the corresponding alcohols:

$$RCH_2CH_2CHO + H_2 \rightarrow RCH_2CH_2CH_2OH.$$

Similar reactions take place with branched olefins. Several commercial processes are available, using cobalt or rhodium catalysts; the latter are claimed to give higher selectivity and operate under less severe conditions: 60–120°C and 20–50 bar compared with 100–160°C and 200–300 bar. A wide range of alcohols can be made; quite high molecular weight alcohols can be produced from the lower olefins through the intermediate formation of dimers, trimers and tetramers. Coal could of course be the feedstock both for the synthesis gas and the olefins.

Integration of Ammonia and Methanol Manufacture

The integration of ammonia and methanol manufacture based on hydrocarbons is established in the Veba-Chemie plant at Gelsenkirchen, W. Germany, (Fig. 10.9). The normal rating is 1150 t/d ammonia and 410 t/d of methanol but these individual figures can be increased to 1270 or 550 respectively, with corresponding reduction of the other. This trend to combined production may be increasing. ICI have announced a "Metham Project" based on natural gas; it is thought that the rating may be 2000 t/d methanol and 1000 t/d of ammonia. The methanol may be used

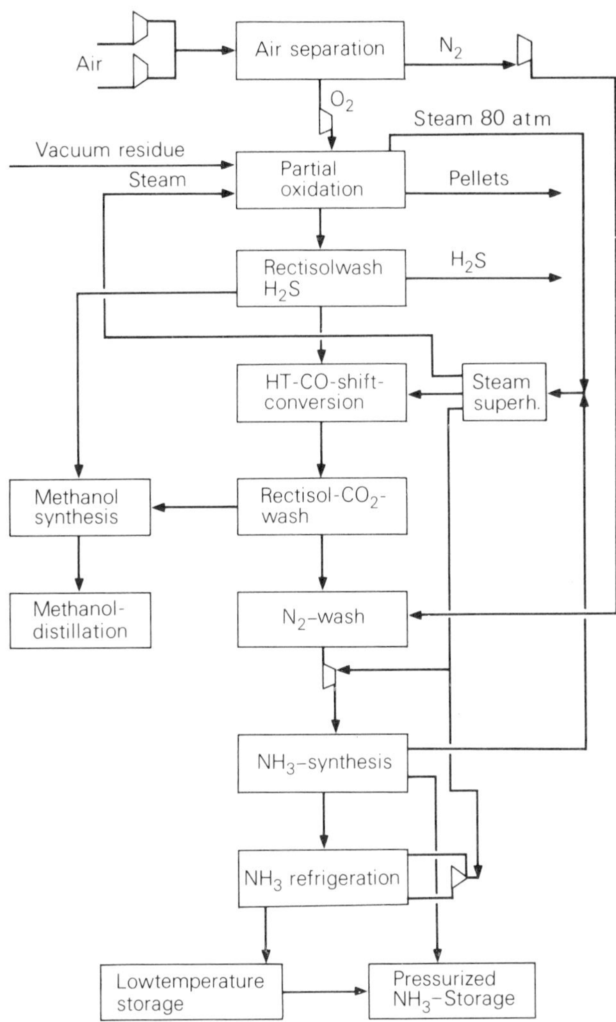

10.9 Veba-Chemie integrated ammonia/methanol process

partially to support a 50 000–70 000 t/d single-cell protein plant. Integration for coal based plants may be even more attractive in view of the costs, especially those associated with handling and using coal.

Synthesis: General Considerations

Further synthesis possibilities are illustrated in Fig. 10.10 (due to S. Andrews of ICI). Efficiencies are estimated on the appropriate lines, with a query where the process is not yet fully developed. A commercial process exists for carbonylating methanol at high efficiency to give acetic acid. The MTG process for converting methanol to gasoline at very high efficiency has been discussed. The type of catalyst employed

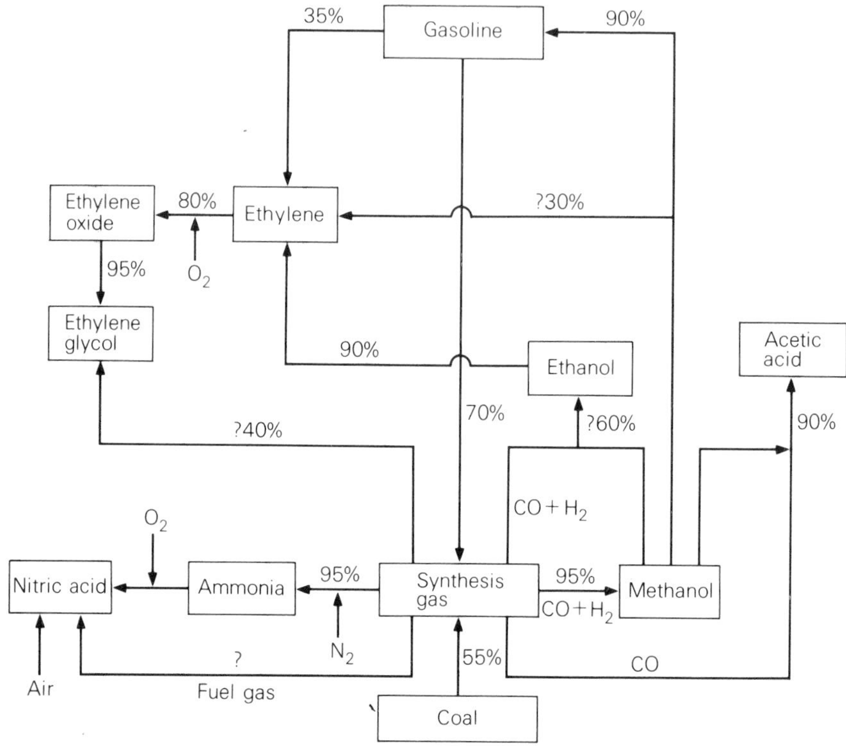

10.10 Petrochemicals from synthesis gas; efficiency of each step shown

might also be used to dehydrate methanol to ethylene, a direct route which might be competitive with that through gasoline. Another route to ethylene might exist if the transformation of methanol to ethanol could be carried out cheaply and effectively; modified cobalt catalysts are being experimented on, with some promise. The dehydration of ethanol to ethylene is well established. A direct route from synthesis gas to ethylene glycol is also being investigated.

An interesting concept is that of a centralised regional coal gasification plant capable of supplying synthesis gas to a number of large customers, for chemical, fuel gas and reducing gas for metals production. Projects of this kind are being examined in the USA by Dupont and Exxon and also in the UK.

HYDROGENATION

Coal hydrogenation processes tend to produce products having a high degree of aromaticity; depending on the particular process, however, the first products may contain only a small proportion of those light products, in the LPG or naphtha range, normally associated with chemical feedstocks. SRC2 products, for example, may only convert 4–6% of the coal input into this range. For these light products (and for gasoline) substantial further processing, especially catalytic hydrocracking,

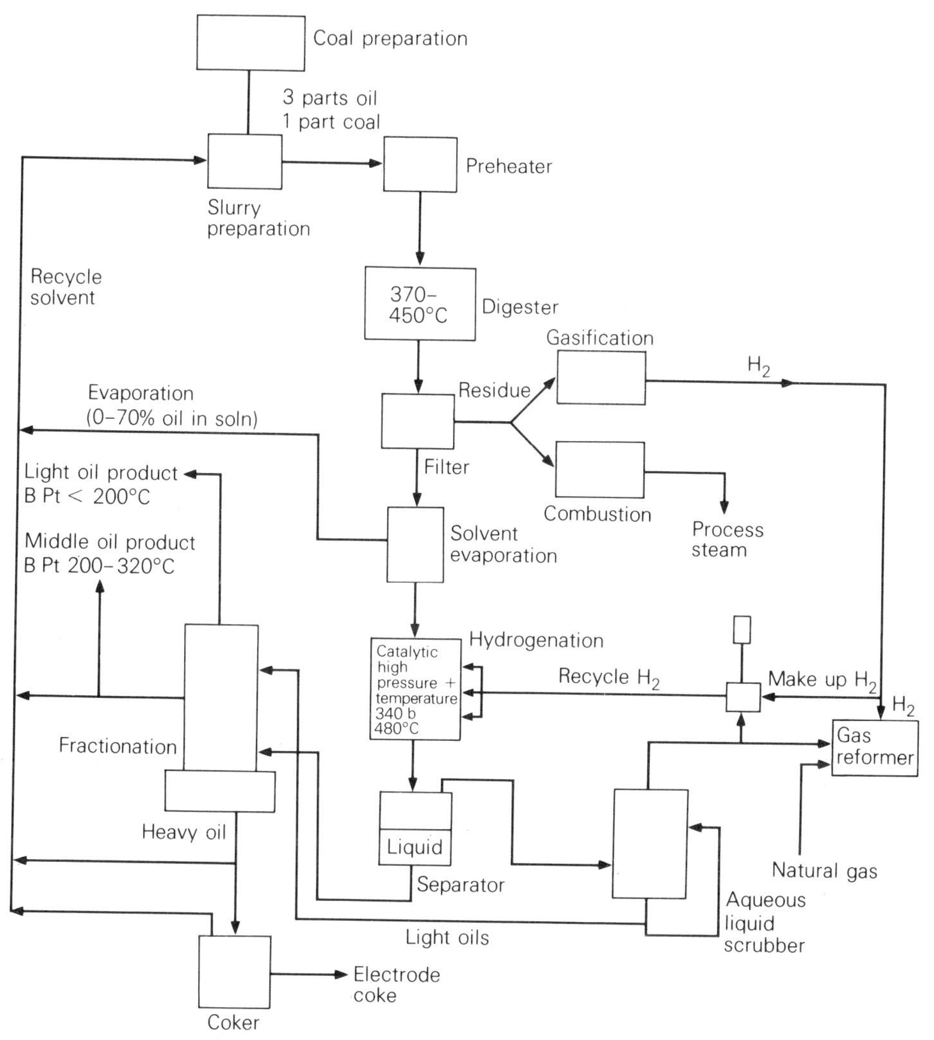

10.11 Proposed scheme for manufacture of liquid hydrocarbons from coal

will be essential. A typical scheme is illustrated in Fig. 10.11. Although such schemes follow petroleum processing lines, it cannot be assumed that results will coincide. Since so much of the processing may take place after the initial liquefaction, it is obviously important to optimise these stages from the point of view of coal liquids rather than merely follow established oil refinery practice.

Substantial work on the processing of coal extracts has been carried out at the NCB's Coal Research Establishment. Basic studies there indicated that kinetic data were similar to those obtained elsewhere on coal derivatives and petroleum extracts; the rate data indicated that it should be possible to convert most of the coal to distillates by extending the hydrocracking time.

The success of the process depends greatly on the activity and life of the hydrocracking catalyst and there appear to be no commercially available catalysts specifically developed for use with coal liquids. Catalysts used in petroleum refining have been tested; none remained active for more than 500 h and selectivity for light products varied. The best one tested in autoclave trials was COMAX 471 (Co/Mo on alumina) and this was used for long trials on a continuous pilot plant. A mass balance based on these runs is given in Fig. 10.12, which indicates that up to 62% of the coal entering the process could be converted to distillates. Characterisation of the light, middle and heavy distillate fractions is given in Table 10.9.

The light fraction was re-distilled to give a coal oil naphtha (BP $< 150°C$). This naphtha, and also some of the middle distillate, were thermally cracked in tests in comparison with petroleum feedstocks. The results are given in Table 10.10.

The olefin yields are, as expected, rather lower than for petroleum feedstocks, which would be preferable for use in ethylene crackers. On the other hand, interesting results (Table 10.11) were obtained by catalytically reforming the coal oil naphtha over a platinum–alumina catalyst.

This product is about twice as rich in aromatics as petroleum and would therefore be an excellent source of aromatics. Because of this, and the associated high Research Octane Number (RON), it would also be a valuable blending component for premium gasolines.

SUPERCRITICAL GAS EXTRACTION

The unique scientific features of this process make it extremely interesting not only as a liquefaction process but also as a potential method of producing chemical feedstocks. Experimental evidence to support this belief is however still slight and a positive assessment should probably await work planned in the next few years.

It is known, however, that the process can be operated consistently and analyses of the extract indicate that it contains small aromatic structures linked by ether or methylene bridges; no highly condensed aromatic structures were present. When the extract was passed with hydrogen over a hydrotreating catalyst, the bridging groups were broken and a distillable oil almost free of heteroatoms was formed. An analysis (Table 10.12) showed a high proportion of low boiling compounds which included alkyl benzenes and cyclohexanes.

PYROLYSIS AND HYDROPYROLYSIS

The conventional carbonisation processes, primarily for solid products, have been discussed elsewhere; valuable chemical feedstocks are produced but their output is limited by coke demand. More modern processes attempt both to optimise liquid production and to use the char in a favourable way, probably by complete gasification followed by shift conversion to hydrogen. In the COED process, using staged fluidised pyrolysis, about 15–20% of the input coal is recovered as tars which are filtered and hydrotreated to reduce heteroatoms and viscosity. The resultant product is a syncrude which is known to be a potential source of aromatic feedstocks. Analyses of COED tars are given in Table 10.13.

The Occidental Flash Pyrolysis process is a low temperature entrained phase process and can produce up to 38% of the coal input as tars; the product gas is

TABLE 10.9 Characterisation of distillates from pilot plant hydrocracking of coal extract solutions

Light distillable (IBP–200°C)		*Middle oil 200–250°C*		*Heavy oil 250°C*	
H/C atomic ratio	1.7	H/C atomic ratio	1.4	H/C atomic ratio	1.2
Oxygen	0.1%	Oxygen	0.2%	Oxygen	0.2%
Nitrogen	0.1%	Nitrogen	0.1%	Nitrogen	0.2%
Sulphur	0.1%	Sulphur	0.1%	Sulphur	0.1%
Paraffins	20%	Paraffins	17%	Mainly polynuclear hydroaromatics	
Naphthenes	40%	Naphthenes	11%		
Aromatics	40%	Aromatics	72%		

TABLE 10.10 Thermal cracking tests on coal oil distillates (Cracking temperature 820°C; residence time 0.5 s)

Feedstock	*Coal oil naphtha*	*Coal oil middle distillate*	*Petroleum naphtha*	*Petroleum kerosene*
CH_4	10.2	7.3	13.9	11.7
C_2H_4	16.9	9.0	23.4	25.0
C_2H_6	2.4	1.6	3.3	2.2
C_3H_6	7.9	3.2	13.2	10.8
C_3H_8	0.3	0.2	0.5	0.3
$C_4H_8+C_4H_{10}$	2.1	0.5	3.5	1.9
C_4H_6	5.6	1.3	3.6	3.8
Benzene	14.4	8.0	8.1	7.6
Toluene	13.7	6.2	5.7	3.9
m and *p* Xylene	2.1	1.7	1.9	0.2
O-Xylene	0.9	0.8	0.8	0.9
Ethyl benzene	2.3	0.6	0.5	0.3

recycled to maintain a reasonable partial pressure of hydrogen.

Coal will react directly with hydrogen under pressure, the products depending on the rate of heating and final temperature; there are optimum conditions for maximum yield of light products. It has become clear in recent work that the process

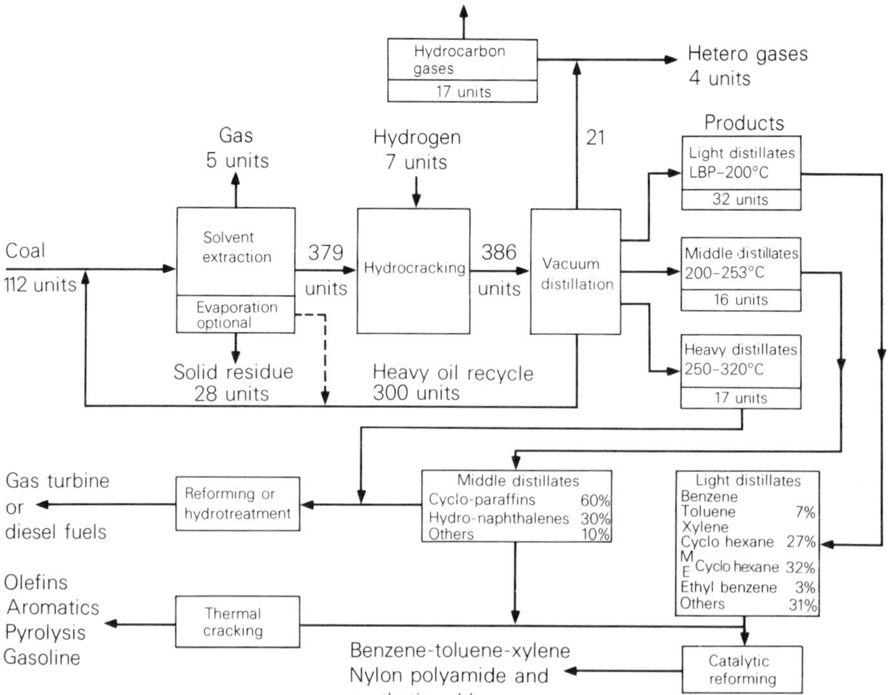

10.12 Liquid solvent extraction material balance and product utilisation

TABLE 10.11 Catalytic reforming of coal oil naphtha

Reformer temperature	480°C	510°C
Yields wt% feed		
Hydrogen	3.0	3.5
Hydrocarbon gases	1.2	1.7
Benzene	22.3	22.9
Toluene	28.2	28.9
Xylene	11.6	12.0
Ethyl benzene	14.7	14.6
C_9 Aromatics	4.4	4.6
Total aromatics	81.2	83.0
Non-aromatics	14.6	11.8
Calc. research octane number of reformate	108	108

TABLE 10.12 Analysis of oil obtained by hydrotreatment of supercritical-gas extract of coal

	Fraction distillation range, °C				
	IBP–170	*170–250*	*250–300*	*300–350*	*350–420*
Chemical type, %wt					
paraffins	17	4	3	6	8
cycloparaffins	36	30	17	12	13
alkyl benzenes	39	20	5	1	0
hydroaromatics	0	43	46	34	6
higher aromatics	0	3	26	46	73
unidentified	8	0	3	1	0
yield, %wt	27	22	20	15	10

TABLE 10.13 Analyses of COED tars

Coal	*Lignite*	*Sub-bit*	*hvCb*	*hvBb*	*hvAb*
Analysis, wt%					
Carbon	82.0	82.7	80.5	83.7	82.6
Hydrogen	8.5	8.0	7.0	8.6	7.6
Nitrogen	1.1	1.0	1.2	1.0	0.8
Sulphur	0.7	0.6	2.0	0.2	2.4
Oxygen	7.4	7.5	9.2	6.5	6.5
API gravity, 60°F	0	−4	−4	−3.5	−5
Pour point, °F	100	120	115	130	94
Viscosity, SUS at 210°F (op)	(−5)	228	1333	300	(74)

involves two stages — pyrolysis and subsequent hydrocracking — and studies have been made using two linked reactor vessels. In the USA, Rockwell International are using rocket motor technology to achieve very high rates of heating followed by rapid quenching of the products, operating at 1 t/h of coal injected into hot hydrogen (up to 1100°C) and pressures up to 330 bar with residence times measured in milliseconds. Results are not yet available for assessment.

ACETYLENE

Acetylene is chemically a useful building block for synthesis of a wide range of petrochemicals and was, until the 1940s, the principal starting point. Oil derivatives more readily obtained from that source, especially ethylene and propylene, subsequently became more important and paved the way to an order of magnitude increase in the size of the petrochemicals industry. It is doubtful whether this growth could have been achieved if the industry had remained based on acetylene, since the classical method of production of acetylene — via the carbide — is unsuitable for massive and rapid scale-up. Acetylene is also a difficult material to handle. However, when oil and natural gas are not widely available and coal resumes the main feedstock role, the acetylene route may deserve to be reconsidered, probably using a process other than carbide for converting coal to acetylene.

The carbide route usually involves the heating of lime and coke (made from low ash coal), generally electrothermally in a resistance-arc furnace, to temperatures above 1000°C. Some production has taken place in blast furnaces, consuming some of the coke for heating. There has also been some experimentation with sodium carbide. Calcium carbide preparation requires a large input of energy. However, there is the advantage that this input can take place near a cheap source of energy and the carbide can then be transported to the chemical plant for reaction with water to release acetylene. The carbide route however is now obsolete, except in South Africa, where the acetylene is used for PVC production.

An updated carbide process is used at Sasolburg, S. Africa. It consumes 105 000 t of anthracite a year to make calcium carbide which is used to produce 300 t of PVC daily. It is interesting that this is an extension to a plant producing 100 t of PVC a day from naphtha made at the SASOL oil from coal plant and can provide useful data comparing two routes to a synthetic product from coal.

When acetylene is required in plants based on oil or natural gas, it is made by partial oxidation or pyrolysis. It would in principle be possible to use coal liquids, SNG or synthesis gas from coal to make acetylene in this way but this route would be unlikely to affect the dominance of ethylene as a precursor of bulk chemicals.

The direct reaction of coal with hydrogen in an arc or a plasma jet seems a more promising procedure. The coal must be heated very rapidly (residence time of 10^{-2}–10^{-4} s) to high temperatures (c. 2000°C) and the product must then be rapidly quenched to less than 400°C; acetylene is the most stable hydrocarbon at high temperatures but in the medium temperature range readily dissociates with the deposition of soot. This process can fairly readily be applied to methane and light hydrocarbon liquids but the direct use of powdered coal is more complicated because it is more difficult to distribute evenly and the pyrolysis reactions are complex. AVCO in the US have been leaders in this field. They strike an arc between a central cathode and the walls of a cylindrical anode surrounding the cathode. The arc is stabilised and made to rotate rapidly by applying a magnetic field at right angles to the arc. Recycled gases, primarily hydrogen, are passed through the arc; powdered coal is introduced into the steam and carried through the arc zone. The methods of quenching and of recovery of the acetylene from the main volume of gases are technical problems (inter-connected) which probably have great economic impact. It has been reported that over 30% of the coal can be converted to acetylene but the energy inputs to the arc and to the circulation and treatment of the gaseous stream may have major impacts on the overall energy balance and economics. The soot can be a satisfactory carbon black but the quantities produced on the scale necessary for a major coal conversion industry would be too great for any conceivable market.

CARBON PRODUCTS

The production of active carbons from coal has been discussed in Chapter 7. Carbon black can also be prepared in a number of ways from coal, deliberately or as a by-product. Carbon can be considered an interesting material for various purposes and it has been demonstrated that coal extracts prepared by solvent extraction can be a satisfactory starting point for such products. In particular, low-ash coal extracts can be converted to cokes for electrode manufacture by a delayed coking process; filaments spun from the liquid extract can be converted to carbon or graphite fibres by carbonisation and graphitisation.

The rank of the parent coal and the extraction procedures have significant effects on the properties of carbons and cokes. Growth and coalescence of a so-called mesophase — a transitional phase leading towards atomic ordering — affect the thermal expansion of graphite rods produced from cokes made by delayed coking of extracts. The low-temperature longitudinal coefficient of thermal expansion (CTE) is used as an index of quality of the "needle" coke, normally made from petroleum, and used in graphite electrode manufacture. High rank coals have a lower CTE (which is desirable). Digestion time and temperature have significant effects on CTE also; increases in either lead to lower CTEs. A suggested mechanism is that the depolymerisation of coal, which continues during prolonged digestion, causes increased elimination of atoms and groups which could hinder the alignment into graphite precursors.

For electrode manufacture, after delayed coking, the green coke is calcined, mixed with pitch (which might itself ultimately be made from extract), extruded into cylinders and subjected to prolonged graphitisation and heat treatment at high temperatures. The electrodes for arc furnace steel manufacture may be several inches in diameter and their behaviour, especially endurance, has critical economic effects. The complex procedures for making good electrodes have been optimised empirically for petroleum cokes and it is clear that coal extract cokes behave somewhat differently, although there is no reason to believe that re-optimisation would not allow the highest quality electrodes to be made from coal. One difference is that heteroatoms are lost rather more readily from coal products but this is accompanied by a higher level of simultaneous loss of carbon atoms; such differences may be clues to the mechanisms involved in developing the desired graphite structure. The market for electrode coke both for the steel and aluminium industries is a large, valuable and highly critical one, where some diversity of feedstocks would be attractive to the users. This may be an early area for the application of coal liquefaction technology. The high purity required seems to be achievable.

A much smaller market, but a high value one, is carbon fibres. Extract can be spun into fibres which are sufficiently handleable to be carbonised. These carbon fibres can be used as such or graphitised to produce fibres with a high strength/weight ratio and very great stiffness; these are commonly called "carbon" fibres although graphite would be a more appropriate name. The final graphitisation takes place at high temperatures while the fibre is being stretched; this improves the ordering of the atomic structure and therefore the properties. In the latter respect the purity is now known to be a vital factor: under high resolution microscopy it can be seen that fractures often take place where tiny inclusions of iron, aluminium and silicon have introduced discontinuities. It is considered possible to eliminate these inclusions by improved filtration and it should then be possible to make fibres at least as good as those currently produced from polyacrylonitrile (PAN). The future of carbon fibres from coal, however, probably depends on the establishment of coal extract production for other reasons; in this event other carbon artefacts made from the same extract might also become attractive.

CONCLUSIONS

The technical possibility of introducing coal into every aspect of petrochemical manufacture, suggested at the beginning of the chapter, has been amply demonstrated; in fact, the proliferation of possible routes may be a temporary embarrassment. However, it may be necessary to rethink the main routes, particularly with regard to thermal efficiencies, as coal becomes again a major feedstock, in the wake of a general steep continuing escalation of feedstock prices. One illustration of this (due to Andrew of ICI) has been given as Fig. 10.10, where many alternative paths are possible; it should be noted that an increase in the thermal efficiency from coal to synthesis gas, above the rather pessimistic figure of 55%, would have favourable implications throughout.

In the event, economics will be dominant; costs may also possibly be reduced by the use of complex plants with more than one coal conversion process. Both subjects are discussed in later chapters.

Chapter 11

IN-SITU PROCESSES: UNDERGROUND COAL GASIFICATION

INTRODUCTION

The idea of recovering the energy values of coal by *in-situ* processes, without the intervention of human effort below ground and without bringing waste to the surface, is one which has exceptional appeal to the public. For this reason alone, a careful analysis of the possibilities is desirable, regularly renewed in the light of developing technology and changing energy economics. Such analysis, however, requires increasingly elaborate experimental and practical data, which in the past has been provided on a somewhat discontinuous and incomplete basis. Work now in progress, especially in the USA, should allow the potential of the process to be defined with a reasonable degree of accuracy by about 1983. Within the last few years, in addition to aesthetic considerations, a new dimension has been added to the possible advantages: the extension of energy resources. In the USA, for example, it is estimated that whilst only 432×10^9 t (out of a total coal in place at less than 6000 ft of 6360×10^9) would be economically recoverable by mining, a further 1800×10^9 t might be recoverable by Underground Coal Gasification (UCG); furthermore, these extra reserves are reasonably widely distributed in the USA. However, in other countries a higher proportion might be considered economic for mining. In Western Europe it is believed that massive coal resources, far beyond the normal assessment, exist either in difficult conditions or deep below land and under the North Sea.As we shall see, the problems of tapping these resources can become very difficult and expensive as the resource base is extended; even so, they may not be more speculative than some other "new" energy sources, such as geothermal heat. Some suggestions have been made for the recovery of the energy values in the coal by other means, for example as liquids (by solvent extraction, for example) but little real basis exists for such ideas, which will be briefly mentioned; and this chapter will concentrate mainly on gasification, at least so far as underground operations are concerned.

HISTORICAL BACKGROUND

The first suggestion for UCG was made by Sir William Siemens in 1868 in a paper to the Chemical Society. Twenty years later Mendeleev in Russia wrote several papers on the subject but the first practical experiments were carried out by Sir William Ramsey in County Durham in 1912. Results were encouraging and patents were filed but World War 1 and Ramsey's death left the project in abeyance. In 1913, Lenin wrote an article in Pravda recommending the process, both on grounds of supposed economics and also in order to relieve labour of an onerous task. This

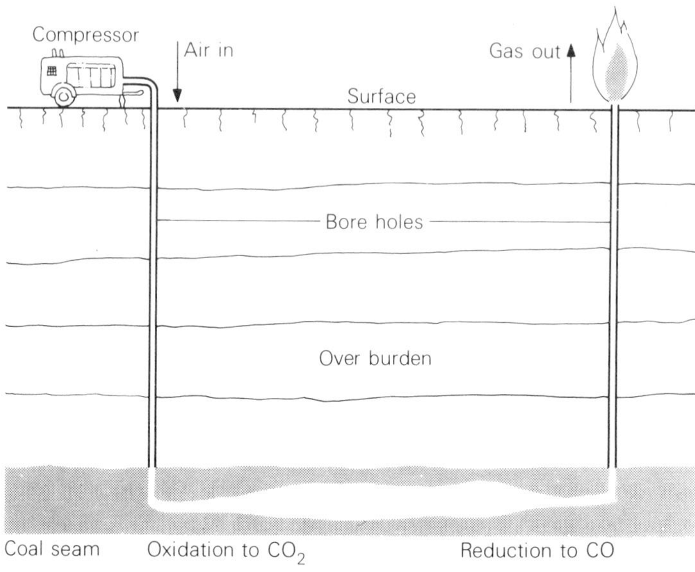

11.1 The basic concept of underground gasification of coal

political backing undoubtedly was an influence in generating the support given to practical work in Russia during the 1930s, including commercial operations, and again after World War 2.

By the 1960s, however, with cheap oil and natural gas, the effort became very low key even there, although after the oil crisis there was some revival of interest, possibly motivated by expectations of licensing revenue.

Substantial work was carried out in the UK between 1949 and 1959 but was then abandoned although the work was fully reported in the open literature and became the main available reference work. Subsequently, British work has been confined to paper studies and assessments.

After World War 2, there was considerable interest in UCG in the USA but the work was somewhat uneven and disjointed until after 1973 when a highly organised co-ordinated effort was mounted, with government projects running recently at ten million dollars per year or more. There are many industrial organisations interested and one group has reached an agreement with the Russians to use their technology in large-scale trials.

Other countries which are known to have carried out work include Belgium, Germany, France (in Morocco), Czechoslovakia, Poland and Italy.

GENERAL PRINCIPLES

The basic concept of UCG is to develop a gasifier underground, using combustion of some of the coal to provide heat (Fig. 11.1). Two boreholes are drilled and a linkage provided within the seam. Air is pumped down one hole and gas withdrawn from the other.

Oxidation first takes place in the flame front, giving CO_2 and H_2O. These are subsequently reduced to CO and H_2 when the hot gases pass over further coal after all the oxygen has been consumed. Adventitious water can also be reduced but too much water can be a disadvantage, consuming heat for its evaporation and reduction and possibly extinguishing the reaction. Some pyrolysis may take place near the flame front but carbonisation is undesirable, since the cracking of volatiles can reduce permeability ahead of the hot zone; also, the coke produced will not be gasified, thus reducing the efficiency of coal use. Because of the dilution of the gas by nitrogen and other factors, in British work a calorific value of 80–100 Btu/ft^3 (3–4 MJ/m^3) was generally obtained under good working and 160 Btu/ft^3 (6.3 MJ/m^3) was regarded as a maximum, difficult to achieve consistently, although some results elsewhere suggest greater success. Special measures, such as blowing with extra oxygen, hot gases or steam, may be used to increase the calorific value, at some cost. Generally, schemes use the gas for electricity generation on site but other applications are considered.

The most critical decision in UCG is probably the layout of the gasifier, together with the method of gaining access. At one time it was considered favourable to use large bores, roadways or drifts in order to gain access but it transpired that such methods were an inefficient way of contacting the coal with air; much of the air by-passed the coal and burnt the gas.

Practically all approaches now are via relatively narrow boreholes from the surface. Rapid and accurate drilling is a great advantage; deviated holes can be drilled from the surface into the seam which they can then follow. The rapid advance of these arts in the oil industry has not been overlooked.

An important method of linking an array of boreholes through the coal seam is the reverse combustion technique, one version of which is illustrated in Fig. 11.2. Several rows of holes are drilled and high pressure air is injected in holes towards which the combustion front must advance, creating a passage. When a suitable path is available, the air, at lower pressure, is switched so that it flows co-currently with the flame, producing gas. The pregasification stage is considered vital and advantage is generally taken of the cleat system (the predominant fracture plane in the coal arising from geological factors); the linkages can be made in a direction exploiting this natural permeability. This pre-existing permeability can be enhanced further in some cases by drying the seam with heated air or combustion gases; many coals will shrink, especially the younger, wetter ones, improving permeability. Explosive and hydraulic fracturing have also been tested. Developing good permeability is extremely important; among other factors, it facilitates operation at a relatively low pressure, which reduces leakage.

Low rank coals appear to be easier to gasify, partly because of the shrinking and partly because they are readily oxidised; this simplifies ignition and the maintenance of steady combustion. High swelling and caking coals are not favoured since they tend to block the gas passages. It was once hoped that UCG would work well in thin seams, too thin for mining, but it has been found that thicker seams work better, 1 m being about the minimum. This is thought to be due to the loss of heat to the strata and the problems of controlling a nearly two dimensional flame system. In-seam drilling, if used, is also more difficult in thin seams. The thickness also determines the volume of coal available to one hole, an important economic factor. With thicker seams it is important to have the flame first advance along the bottom of the seam, so that the coal falls into the void and can readily continue the reaction. If the flame

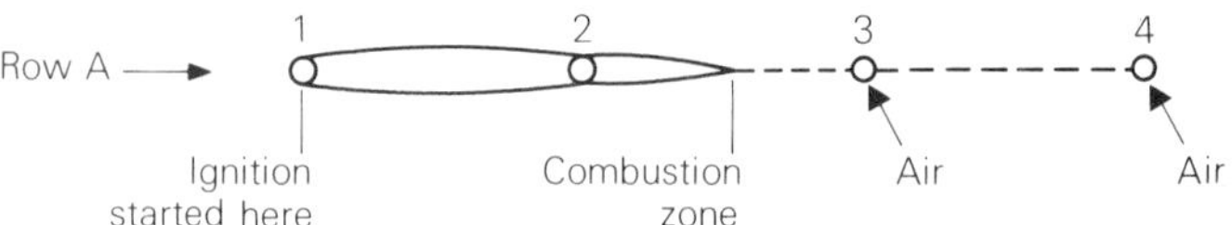

Stage 1. Linking along row A by high pressure air and reverse combustion

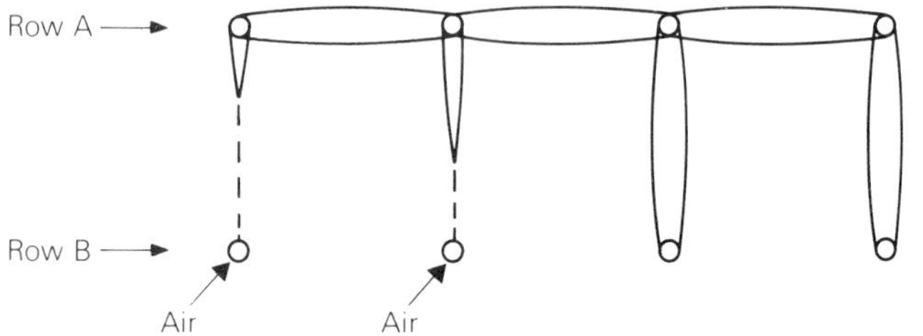

Stage 2. Linking along columns by high pressure air

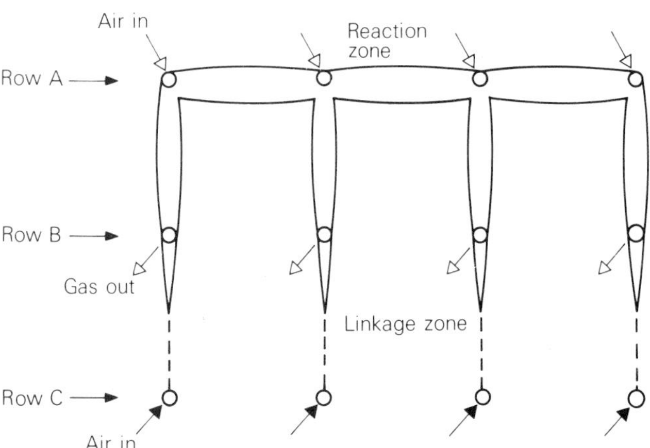

Stage 3. Gasification between row A and row B and linking along columns between row B and row C

11.2 The vertical drilling method showing the three stages of development

advances along the top, the remainder can be sealed off by slag. For this reason, gasification in the updip direction is favoured. It is important to sequence the drilling of boreholes and the creation of linkage between them so that the process can be continuous and the required gas quality maintained. When the calorific value of the gas from one of the extraction holes falls below a certain level, this hole can be capped and that product gas is forced through the seam to the next borehole, so that it becomes more enriched.

The depth, pressure and borehole spacing are interlinked factors. Although high pressures would have economic advantages, a limit is set by the need to restrict leakage through the overlying strata. Losses due to leakage of 10% or more are reported; it is doubtful whether this degree of leakage would be tolerated environmentally in the West for large commercial operations. The allowable pressure varies mainly with the depth but also with the type of overburden. In addition, especially in thick seams, roof falls can disrupt the overlying strata, encouraging leakage. The pressure also needs to be selected to control water ingress; excess of water penetration reduces the thermal efficiency and also the calorific value of the gas. On the other hand, higher pressures permit greater spacing between boreholes. There are some complex optimisations, which will be assisted by better experience and basic understanding. Currently, in vertical well systems, spacings of 20–30 m are common. Typically, about 50% of the energy in the coal may be recovered but taking account of energy expended above ground, the overall thermal efficiency may be about 40%. Favourable conditions may raise these values to 60% and 50% respectively.

NATIONAL PROGRAMMES

USA

Although other countries, notably Russia, did earlier work of historical significance, and some special versions are being studied elsewhere, the current US programme is believed to address the main problems in a comprehensive way which will most readily assist a general appreciation of the prospects for UCG. The objectives, planning, organisation and reporting of the US programme are commendable. It is notable, however, that whilst the objective is to develop commercially viable underground conversion processes for extracting energy from coal, in this stage data will merely be provided to predict the economics of a commercial operation; according to the programme, it is hoped that one will be in being by 1985–7. The strong implication that there is no such process immediately available was certainly not made without consideration by the US authorities of all the Russian data which was available. The US programme also states that some of the significant potential advantages have not yet been proven; the potential environmental disadvantages are also specified. The goals of the programme are very wide, possibly excessively so if taken seriously in relation to the short time horizons; they include low and medium Btu gas, from a wide variety of Western and Eastern coals, for electricity, synthesis gas or conversion to liquid fuels or high Btu gas.

There are several major lines of project work; (a) Low and Medium Btu Projects from Western coals, (b) Eastern (coking) coals and (c) Steeply Dipping Beds. There is substantial back-up support on research and environmental aspects.

The Western Low Btu Project is located at Hanna, Wyoming, on a 9 m thick sub-bituminous seam of about 100 m depth. The system used, called Linked Vertical Wells (LVW) uses reverse combustion for linkage, followed by forward gasification (Fig. 11.3). Hanna 1 was an exploratory test, lasting six months and producing gas of 126 Btu/ft^3 (4.7 MJ/m^3) (with an energy gain of 4 ×). Hanna 2, completed in 1976, demonstrated a high areal sweep efficiency; from a four-well square pattern enclosing 4600 t of coal (within the straight lines); 6700 t were gasified, due, of course, to the arc-shaped edges of the burn pattern. Hanna 3 was an experiment mainly aimed at checking the environmental effects, especially to underground water flows; the work requires extensive continuing analysis. Hanna 4 was an experiment to study the relationship between well spacing and sweep width over long distances in a five-well straight line. The test was interrupted due to a fault and is being restored.

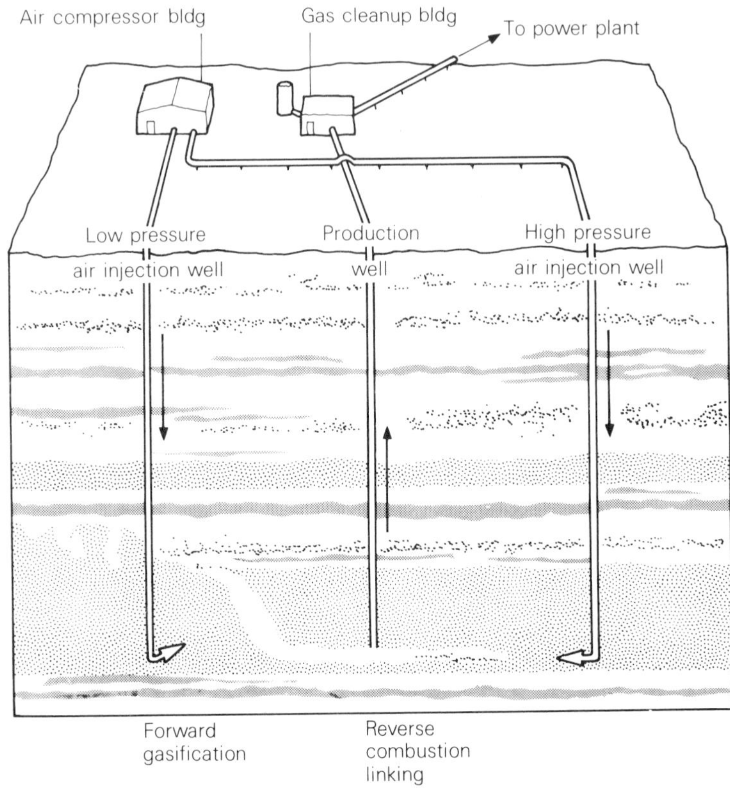

11.3 Linked vertical wells process

The Western Medium Btu Gas Project is being carried out at Hoe Creek, near Gillete, Wyoming, on a 8 m seam, wetter, more reactive and more uniform than at Hanna. The objective is to use oxygen and steam injection to produce medium Btu gas (one test achieved an average of 260 Btu/ft^3 (10.3 MJ/m^3) ranging up to 300 Btu/ft^3 (12 MJ/m^3)).

Explosive fracturing has been tried and, whilst not achieving the permeability sought, allowed reverse combustion linkage to be omitted. Linkage by directional drilling has also been tried successfully (Fig. 11.4). Main interests in this project are to use very deep, thick coal seams to produce gas for upgrading to chemical feedstocks or SNG. Deep Test 1 is scheduled to start in 1982.

The Steeply Dipping Bed Project is designed to exploit seams having a dip greater than 35° relative to the surface; 100×10^9 t of such coal exist in the USA, not mineable by present techniques. Much of these reserves are convenient for supplying the West Coast, otherwise not well provided with energy sources; the main tests will take place in Washington State, although preliminary work is at Rawlins, Wyoming. The general configuration is shown in Fig. 11.5. One hole, for air injection, will be drilled at 40° to enter the seam low down. Another hole at 34°, the product well, will enter the seam much higher and will then follow the coal to within a few feet of the air hole. Following feasibility studies, site characterisation, etc. three field tests were due in

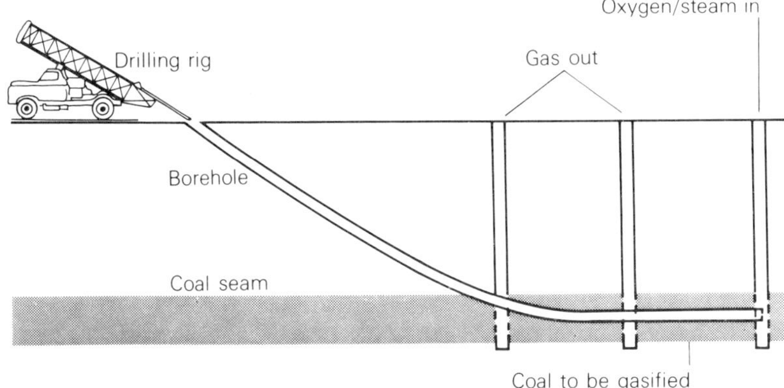

11.4 Hoe Creek Number 3 test using directional drilling

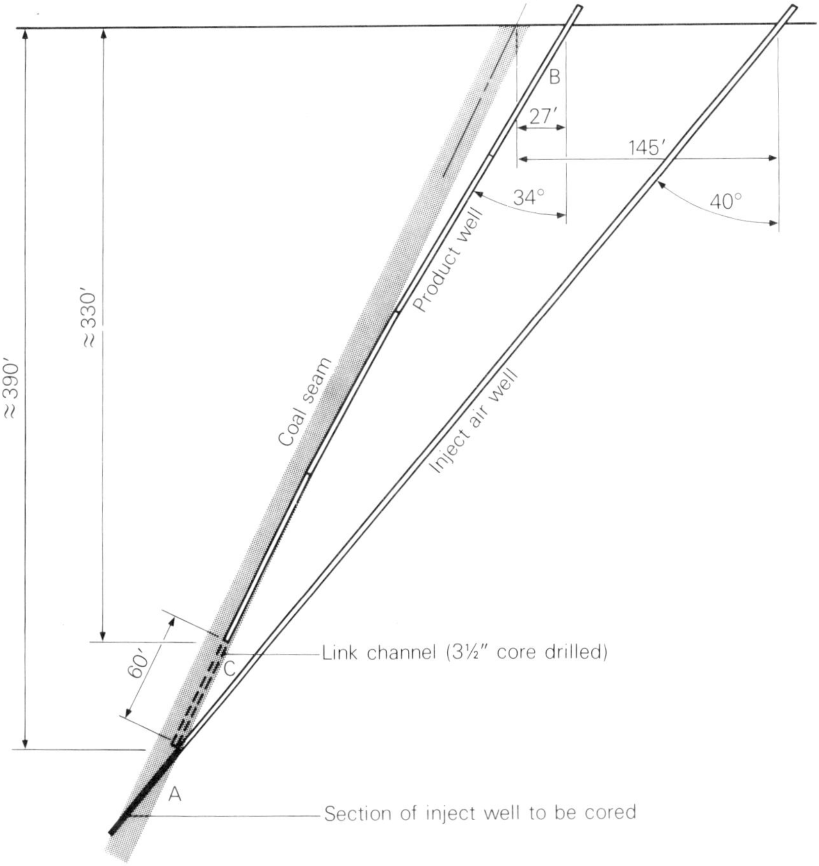

11.5 Steeply dipping beds; Rawlins Number 1 experiment

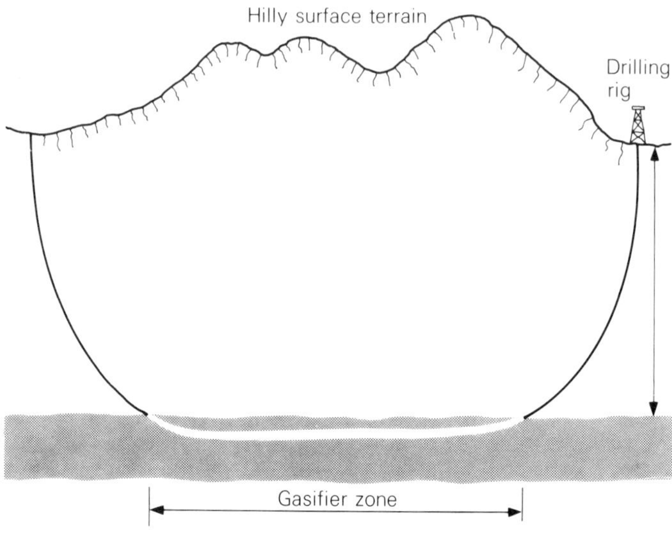

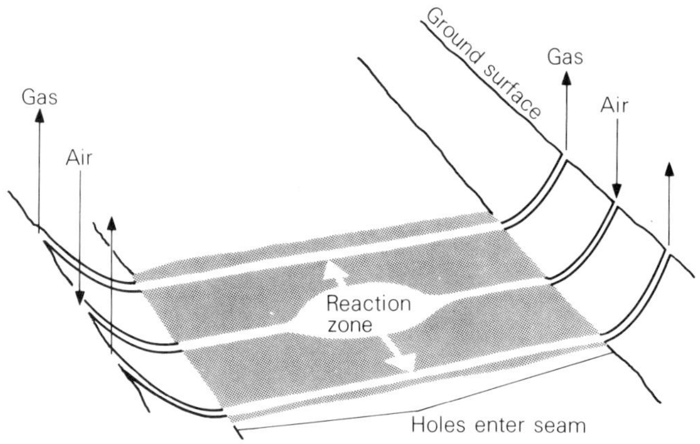

11.6 The original longwall generator concept

the period October 1979–March 1982, ending in the preparation of a design/cost estimation for a Pilot Plant.

The Eastern Coal Technology Project is being carried out by the Morgantown Energy Technology Center (METC). The low-permeability, high swelling coals are difficult to gasify and the terrain and population density are unfavourable to closely spaced vertical wells; it may be desirable to use directional drilling and the Longwall Generator process (Fig. 11.6). Hydraulic fracturing is also being considered (Fig. 11.7). A site has been laid out by METC at Pricetown, West Virginia, to test several of these alternatives in seams at depths of about 900 ft. Following this work, several options will be considered for a further programme to commence in late 1980. The

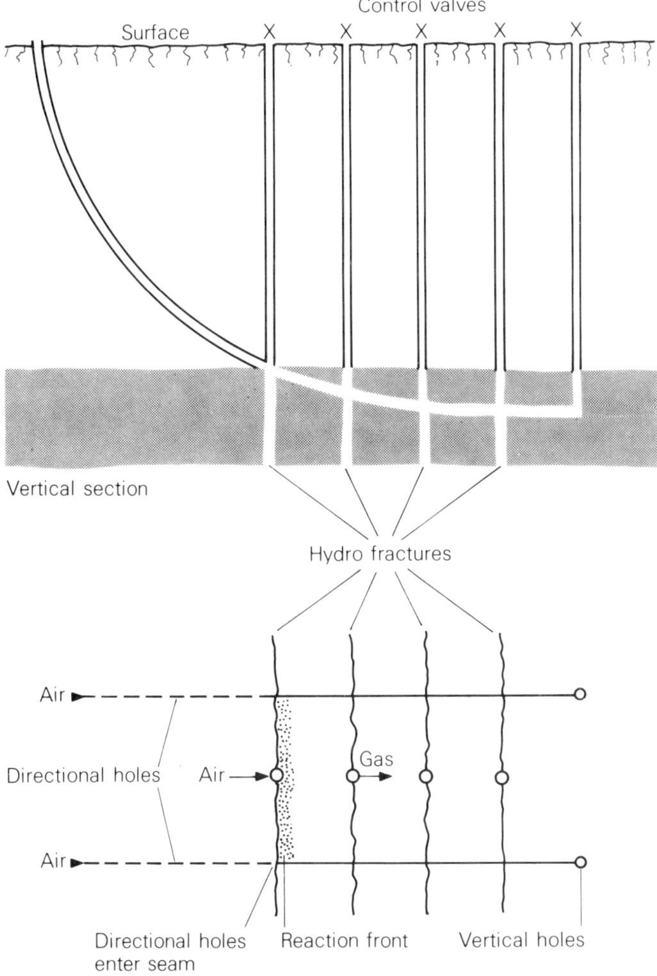

11.7 Modified Morgantown layout using hydraulic fracturing

long term objectives include the application of the system (including directional drilling) to "various thin seams".

General supporting work, in addition to a comprehensive environmental programme on the field tests and in the laboratory, includes work on kinetics, thermal conductivity, permeability (to gases and liquids) on typical coal samples. A technical/economic study is being conducted on a scheme to use CO_2/O_2 for extraction, minimising water requirements. A market study on the uses of low and medium Btu gas is also being carried out.

Several US projects are proceeding mainly within the private sector. The Texas Utilities Co. licensed Russian technology; their second experiment is a nine-well test at Tennessee Colony on a seam 2.5 m thick and 82 m deep.

The Texas A&M University has carried out tests on a 1.5 m lignite seam on their property. ARCO are carrying out tests on a 30 m thick sub-bituminous seam with 180 m of overburden; three process wells have been successfully linked at distances of 23 m and gasification was carried out for a short test period.

Russia

The interest in UCG in Russia was initiated on political/philosophical grounds and given a very high prestige and priority value. Several sites worked commercially for many years. Furthermore, the site operations were supported by large numbers of scientists and engineers working in several institutes. By 1957, there were about 500 scientists, engineers and technologists engaged and considerable publicity was being given to the work. Subsequently, in the light of cheap supplies of natural gas, this emphasis was allowed to wane, although two or three stations have continued to operate routinely but no new sites have been opened in recent years. After the oil crisis of 1973/4, it appears that the special State Bureau for maximising licensing and royalty revenue on Russian technology took a serious interest in UCG and this may have given it a fresh impetus.

The first major sites were operated in the Donbass in the mid-1930s. One, at Lisichansk, was damaged in the War but was reconstructed and continued working for many years. A typical layout is shown in Fig. 11.8.

Initially, the boreholes were drilled from underground workings but the disadvantages of poor air/coal contact were realised and greater concentration was increasingly given to boreholes (Fig. 11.9). The concept of drilling into steep seams was also developed (Fig. 11.10) and improved (Fig. 11.11). It was also realised that all the boreholes could be drilled from the surface, using deviated drilling (Fig. 11.12), which was a very advanced technique indeed at the time. An ambitious layout for a development at Kholmogotsk is shown in Fig. 11.13; this was apparently never actually built but indicates the sophistication of the technology which was reached.

Some of the early coals gasified were very wet which made it difficult to achieve a reasonable calorific value, especially when large cross-section airways were used. With some of the earlier work, using roadways, only 85 Btu/ft^3 (3.3 MJ/m^3) could be obtained even with the use of oxygen but with the borehole system 100–150 Btu/ft^3 (4–6 MJ/m^3) was achieved. The gas was used for electricity generation and industrial heating. In 1940, experiments started in the brown coal at Tula in the Moscow basin

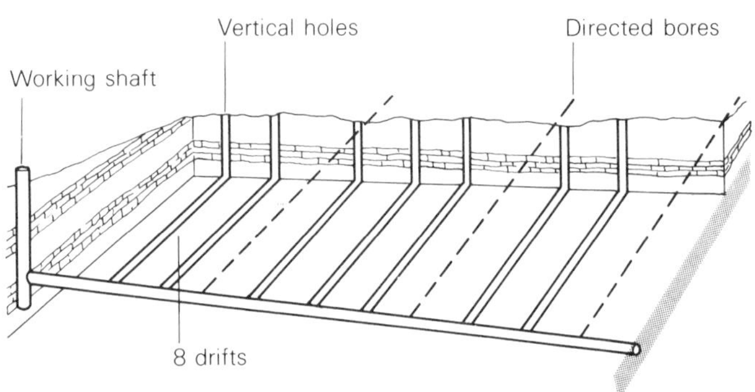

11.8 Typical generator at Lisichansk, USSR

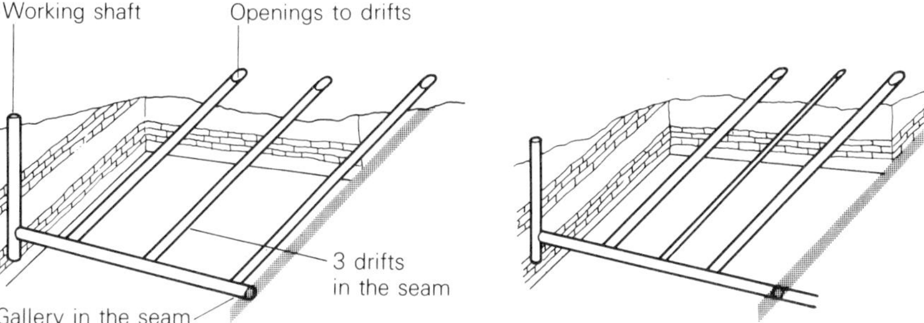

(a) First stage: one gallery, three drifts

(b) Second stage: boreholes, between drifts

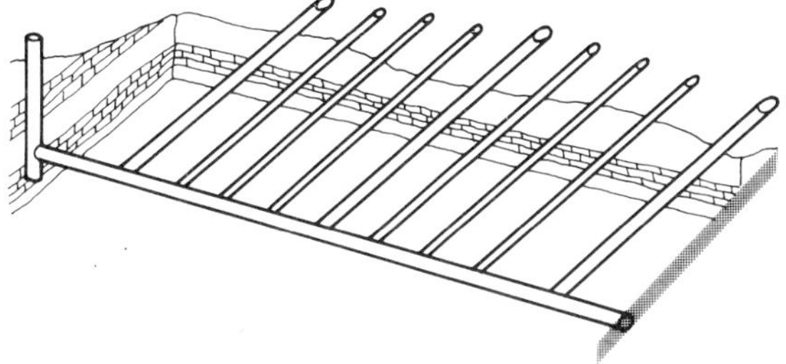

(c) Third stage: longer galleries, fewer drifts, more boreholes

11.9 The development from drifts to boreholes

and the higher permeability facilitated the development of the reverse combustion system, using pneumatic linkages at 6 bar with spacing of about 25 m.

UK

Following the early trials in 1912, work was not restarted until 1949; practical trials continued for about ten years but eventually were abandoned as a result of low oil prices.

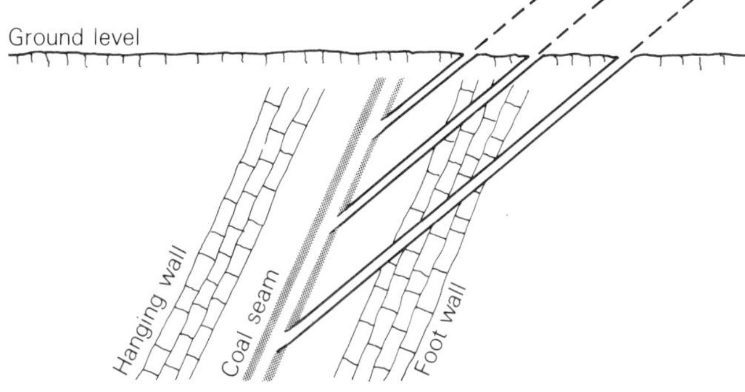

11.10 Drilling into a steeply inclined seam from the surface

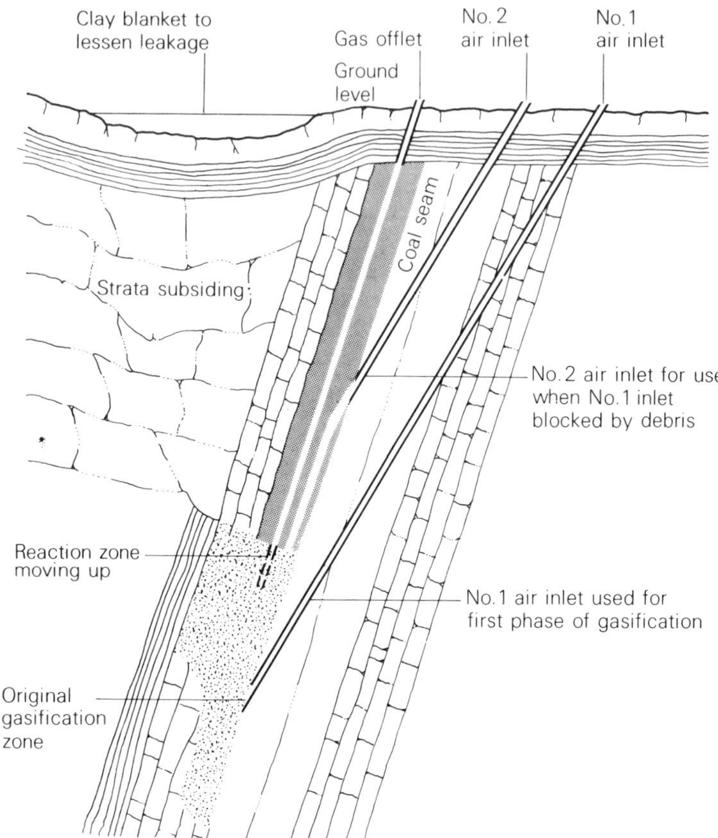

11.11 Use of two air inlet boreholes in a steeply sloping seam

A number of techniques were tried, with emphasis initially on the use of existing roadways or exposed faces as a result of opencast workings. In some cases shafts and roadways were constructed in what was known as the preliminary mining method and boreholes were drilled in the seam, from one gallery to another or from a roadway to a junction with a vertical borehole. One version used blind boreholes into the coal; air was pumped to the far end and ignited, the gas percolating back through the seam (Fig. 11.14). In one trial using this procedure, 1050 t of coal were gasified with an extraction efficiency of 42%, producing gas at 75 Btu/ft^3 (3 MJ/m^3).

The blind borehole method was abandoned and in 1959 the last trial (P5) was carried out at Newham Spinney using the layout shown in Fig. 11.15. A small power station was provided. The trial lasted 118 d and produced gas at the required rate almost continuously but its calorific value was low (average 57 Btu/ft^3 (1.25 MJ/m^3)). This was perhaps due to the thin seam (0.9 m), the implications of which have been better appreciated more recently. Between 1962 and 1966 the workings were exposed by opencast operations and the seam examined. The results confirmed the expected progressive burn-pattern.

In 1976 the National Coal Board carried out a comprehensive review of the subject, including an economic appraisal. As a result, substantial interest has been

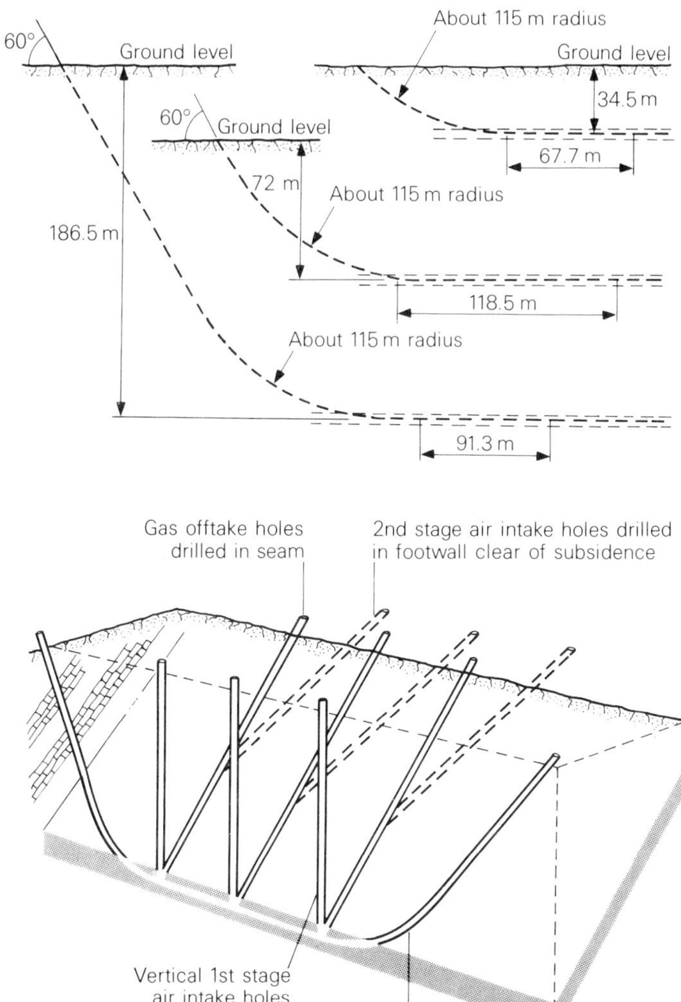

11.12 Early Russian achievements with directional drilling

maintained, as part of an ongoing consideration of advanced mining methods, but so far without further practical programmes.

Federal Republic of Germany

The FRG regard *in-situ* processes as long-term and related mainly to reserves which are not mineable by conventional methods; these are thought to be at least 20×10^9 t and possibly 100×10^9. Gasification is the only active route. The production of medium Btu gas is the preferred objective; upgrading to SNG is envisaged. Hydrogasification is considered an attractive possibility but it is recognised that

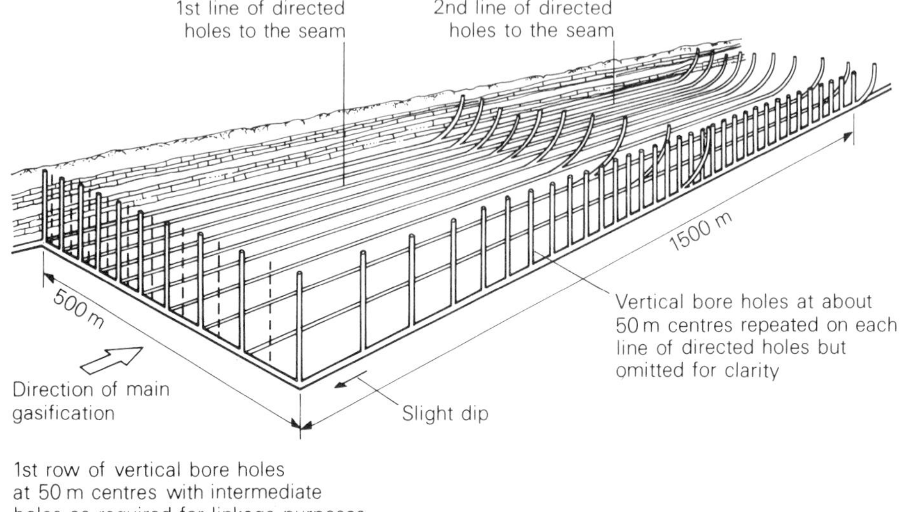

11.13 Proposed method of development of Kholmogorsk, USSR

gasification trials might have to start with air, followed by oxygen before trying hydrogen. The FRG is sympathetic to Belgian ideas on deep seams, described below, which led to the signing of a collaboration agreement between the two countries in 1976.

The German internal programme has been at the level of about 4 million DM/a for several years. Because it is long term and aimed at speculative processes, the work consists of laboratory work on typical German coals, preparation for field studies, system analysis and instrumentation; there is particular interest in the use of high pressure and pressure change to facilitate filtration gasification.

Belgium

In Belgium, it is considered that the gasification of relatively thick, shallow seams, as may be possible in the USSR and the USA, is not applicable. There are no virgin shallow deposits of adequate thickness; experiments on relatively thin shallow seams in Western Europe have resulted in low quality gas, poor use of resources and uneconomic costs.

The possibility of pollution, especially to water, and other environmental effects is also more serious in densely populated countries.

On the other hand, it is considered by Belgian experts that the working of deep seams could increase the reserves in Western Europe from 300×10^9 t to 1000×10^9 if depths were restricted to 2000 m; if it became possible to work to depths of 5000 m, reserves would probably reach $10\,000 \times 10^9$ t. Although the depth of drilling would increase the cost of each hole, this might be compensated by increasing the spacing and by getting more of the energy of the coal seam into the gas. This would be made more feasible by working at high pressure; this in turn is possible because at depth the overlying rocks behave in a plastic fashion, sealing the coal seams. Further advantages of a "tight" pressurised system are that both gas leakage to the surface

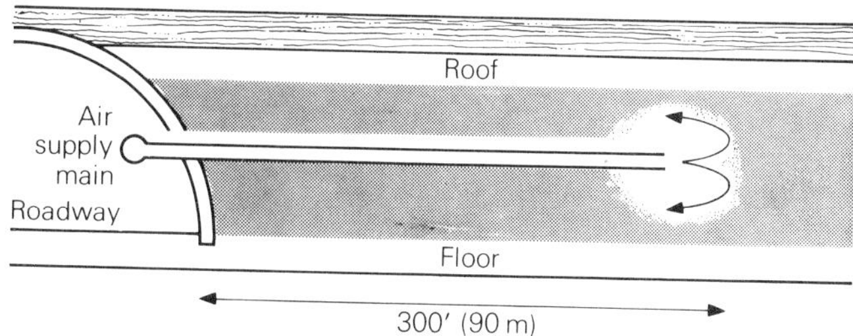

Vertical section through a blind borehole

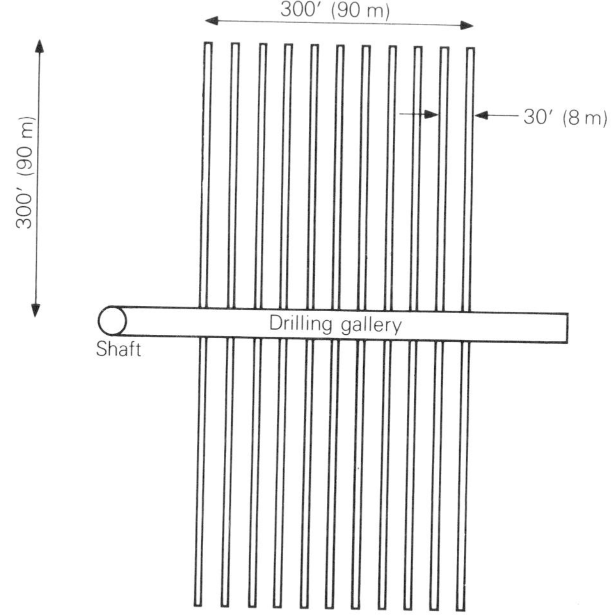

Projected blind borehole array Newman Spinney 1958

11.14 The blind borehole method

and water ingress are prevented. The effects of subsidence are also reduced by working at depth. The gas tightness is also favourable to the use of steam, oxygen or hydrogen. Three stages of development are envisaged:

(1) Lean gas with air and steam for use in a combined cycle (gas turbine plus steam turbine) electricity plant;
(2) Use of steam/oxygen to produce a product similar to that from the Lurgi gasifier;
(3) Gasification with hydrogen to give SNG.

In stage 1, it is envisaged that the air/steam mixture will be injected at 30–50 bar and after percolating through the seam is recovered as lean gas at a pressure of 10–15 bar and a temperature of 200°C; in order to prevent the temperature from rising too far and damaging the tubing, water can be sprayed into the bottom of the

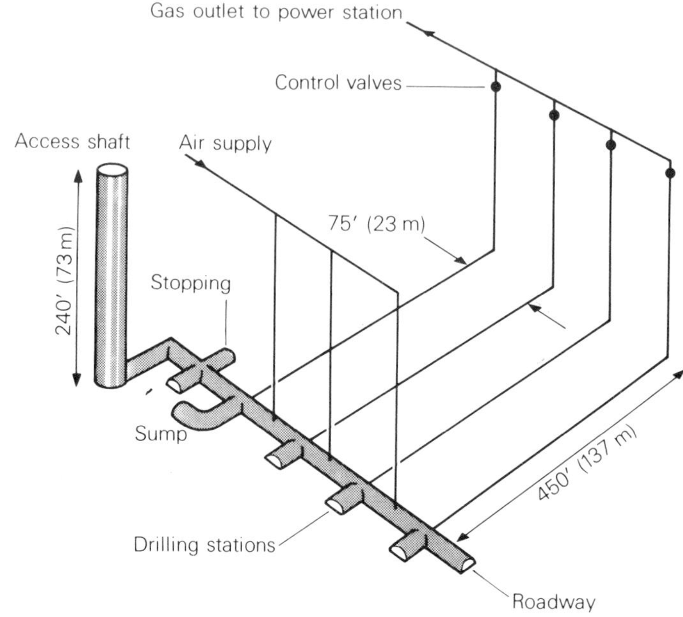

11.15 P5 layout at Newman Spinney

production well. In one mode, a diurnal cycling of pressure is used. In a normal case, a 300 MW generator might have to supply 80 MW to the compressor driving system. If the compressor power were doubled, but only used at night, the station could, with additional generating equipment, export 380 MW during the day and 220 MW at night, thus acting as an energy storage unit. Of course, the initial incentive for pressure cycling is to improve the quality of the gas/coal contact by creating extra fracturing. The energy storage possibility is an interesting one since consideration is being given to the provision of pressurised gas storage for such purposes.

Actual deep gasification trials do not appear to have started although a site has been identified at Thulin over a 1000 m seam. Work to date appears to be theoretical and exploratory but fairly extensive and thorough. Energy balances, presumably based on laboratory work, have been proposed. These suggest that a gross thermal efficiency of gasification of the coal of 70% might be obtained compared with 50% for shallow wells of the Russian type. After allowing for energy inputs for steam, oxygen and pressure (including credit for recovery of pressure in a gas turbine), net thermal efficiencies of 57.5–62% and 40% for the pressurised and unpressurised system, respectively, are indicated.Since a combined cycle is possible, the electricity conversion efficiency is also greater with the pressurised system — giving an overall coal/electricity efficiency of about 24% with only 14% in the low pressure system.

These assumptions are used in theoretical economic assessments, discussed in Chapter 18.

ENVIRONMENTAL IMPACT

Underground coal gasification seems at first sight to create a benign impression. The elimination of underground work is still an attractive feature, with favourable implications on health and safety of labour. Similarly, the elimination of pit-head gear and coal handling, stocking, preparation and transport are attractive features. However, especially in modern circumstances, the environmental impact of UCG would be quite significant and would have to be taken into account. The effects of the most obvious features, land use and visual impact, would depend very much on the location; unfortunately, they would be most important in densely populated areas, where energy needs are most concentrated.

For a 100 MWe site (presumably about as small as might be considered on the basis of providing a significant method of meeting energy demands), with a moderate seam thickness and reasonable coals, about 2–5 hectares would be in use at one time and the active working site would advance at about 0.5 hectare/week. This area would contain about three drilling rigs and many borehole cappings and pipelines.

There would be compressors and gas cleaning and utilisation equipment, including a small power station, and probably a flare stack. There could be considerable noise from drilling and methods of silencing would have to be developed, especially if operations were near to domestic sites. None of these problems would seem too difficult, at least for moderately remote sites. There would be no mining dirt; the small amount of dirt from drilling would not be a large problem. Surface heating effects are negligible and the possibility of the reaction underground getting out of hand is very remote; suitable precautions to isolate the working from any nearby underground operations would have to be carefully worked out and implemented. The effects of subsidence would also have to be dealt with as in underground mining.

The handling of the rather noxious gases above ground would also have to be engineered to a high standard but this is only equivalent to a coke oven or a gas works. (It should be borne in mind however that standards for *new* plants of this kind will be more stringent than for those of an earlier generation.) The most potentially worrying features are gas leakage from below ground and pollution of ground water and aquifers. If 10% or more of the gases produced are leaked, this could be very serious, both to the work force and to nearby residents. Tar and sulphur compounds might create objectionable odours and also health hazards; more important, the gas will be toxic, especially due to the CO content, and the possibility of carcinogens would need careful investigation.

It seems unlikely that controlled losses of 10% or more of such gases would be allowed to percolate to the surface. Much more needs to be known about these effects and, in the meantime, in Western industrial countries at least, severe limitations might be applied, perhaps to the maximum pressure/minimum depth (or permeability) relationships.

The problems of tracing pollution to water supplies may be even more complex. Water in contact with an underground gasifier would certainly be polluted and, although this might not be very heavy pollution and removable from the water concerned, it would not really be acceptable for the extent of this contact and the fate of the water involved to be unknown and neglected.

The US programme has a comprehensive environmental element, which should enable the problems to be assessed and solutions to be determined. It remains to be seen whether these solutions may be serious constraints that might limit the application of UCG, even if other problems were overcome. There have been some relatively re-assuring indications in the US work. In one test gaseous emissions were less or no greater than for surface gasifiers; subsidence did not seem too serious; and the surrounding coal appeared to act as a cleansing medium for water pollution.

UTILISATION OF GAS

The Russian commercial installations appear to have used the gas for combustion and electricity production, with some use for industrial heating.

Most assessments and trials elsewhere are also directed to these ends, with longer term work aimed at higher value gases.

The simplest use of the gas is not however without complications. On the one hand, they can impose on the UCG operation a requirement to produce regular quantities of gas of consistent quality. Apart from the technical problems involved, which appear to be soluble in large installations, some extra investment to provide a margin may be necessary. On the other hand, the electricity or other industries must give a high priority to accepting the gas and this may distort the system economics. Storage of the low calorific value gas or intermittent UCG operations are probably not economic propositions.

Upgrading the gas produced by UCG using air is technically feasible although the end products would at present have to compete with gases made from much more concentrated feedstocks. In the case of UCG with air, the raw gas would have a high level of inerts (about 60% N_2 and 15% CO_2), with only about 10% of CO, 12% H_2 and a small amount of methane. The processing options for upgrading this gas are illustrated in Fig. 11.16. Some of these options are not entirely straightforward but at the present stage an examination of the difficulties is hardly warranted.

Upgrading by adaptation of the UCG process itself has been given greater attention, as noted above. If the enrichment of the gas by oxygen or hydrogen proves technically feasible, a much more attractive starting material would probably result but the economic effects would have to be balanced by the cost of the inputs.

INSTRUMENTATION AND CONTROL

For widespread application of UCG in the Western coungries it is considered essential that there should be a complete chemical and physical understanding of the process. This is necessary for several purposes:

(a) It must be clear that full control can be established and maintained, especially in relation to environmental constraints.
(b) The processes should be fully transferable to different sites and predictable for economic assessment; this requires more than empirical information when political motivation is not the driving force.
(c) The experimental sites in particular need the highest quality of instrumentation.

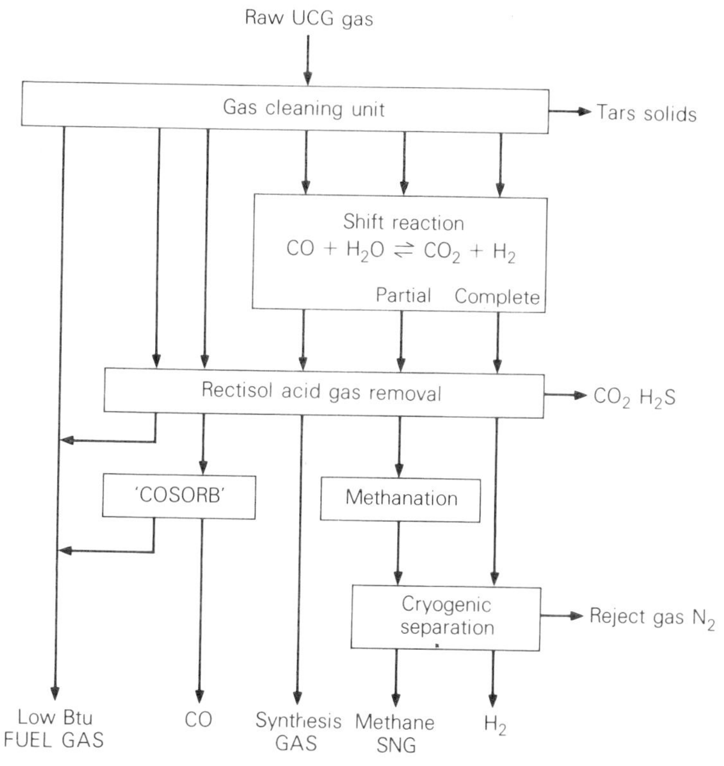

11.16 Upgrading process options

Fortunately, the US programme includes a very high standard of instrumentation development for what is called Reaction Zone Mapping, which must be in three dimensions. Of course, measurements on the input and particularly the output gases give information on the gasification progress, including the amount of coal extracted.

However, such things as the vertical linkage position and the plan position of the gasification position are vital pieces of information in understanding the process and developing a comprehensive model. Models do exist but calibration and checking are necessary.

Three methods are fairly well established but expensive. Specially developed strings of thermocouples may be inserted down boreholes; when arranged in a suitable array, they give three dimensional information about the movement of the reaction front. This may be supplemented by downhole-to-downhole seismic methods and high frequency electromagnetic methods (HFEM). The latter method involves lowering transmitters and receivers down boreholes placed so that the signal traverses the reaction zone. The strength and phase of the signal, taken at different heights, angles and azimuths, enables a three dimensional picture to be drawn. The thermocouple method is the most expensive but the complexity and cost can be mitigated by association with other measurements. Various types of surface measurement, including temperature, seismic, geotechnic, acoustic, electrical resistivity and induction have been studied. Of these the electrical resistivity method appears most promising. This method is based on differences between coal and overburden and on the change of resistivity of a coal seam as a result of gasification. It has been developed to a high degree of sophistication, both on the placing of electrodes and probes on the ground and on the analytical methods.

None of the methods described seem suitable for very deep seams, except at prohibitive cost. However, the establishment of models by shallower work may assist, together with detailed analysis of the gas balance.

OTHER POSSIBLE *IN-SITU* METHODS

The apparent advantages of eliminating underground labour have generated a wide range of more or less unlikely proposals. In some cases, it is suggested that coal as such could be mined and brought to the surface, by means of mechanical moles (terotechnics) and/or hydraulic mining. Chemicals, including ammonia and methanol, have been suggested as a means of fragmenting the coal seam.

At the other end of the spectrum, instead of producing gas with a chemical energy content, it has been suggested that, following *in-situ* combustion, recovery should be entirely through the sensible heat content of the gas, which could include steam from water injection. Complete combustion has been suggested for the recovery of pillars; quenched combustion has been used in the oil industry.

Solvent extraction has also been suggested, with direct solvents, hydrogen donors or supercritical gases. The problems of containment and recovery of these expensive materials seem formidable.

One interesting suggestion is microbiological degradation. The idea would be to develop an organism which could use coal as part of its lifecycle, generating a low molecular weight product, possibly methane. No such organism is known but it might be possible to develop or engineer one. Reaction rates might be very low but it should be noted that methane, spontaneously generated, can often be drained profitably.

All of these suggestions are highly speculative. They are mentioned mainly to indicate that a great deal of thought is being given to the *in-situ* methods and that even quite wild ideas are considered quite seriously. There is currently a greater degree of open-mindedness in the coal industry than there has been traditionally.

CONCLUSIONS

It is clear that UCG can, in suitable circumstances, provide a regular and usable form of energy. This is particularly so for schemes which might use shallow, fairly thick seams, aimed at the production of low Btu gas for industrial heating or electricity generation. The scope may be widened by going deeper and by using more advanced techniques to produce higher quality products but these proposals are more speculative.

Although there are still technical problems to solve, the main barriers to wide application may be environmental and economic. The former needs verification on a much more thorough basis than any so far existing; the latter is dealt with in a later chapter. Perhaps the greatest incentive, however, is the possibility that coal which would otherwise not be exploited may become available at a time when energy needs are very great.

SECTION III
Economics of Coal Utilisation

Chapter 12

GENERAL CONSIDERATIONS OF COAL UTILISATION ECONOMICS

INTRODUCTION

As should already be clear from the chapters on technology, economic considerations are likely to dominate the future of coal conversion practice through such factors as: (a) the end-products most desired, (b) the R & D programmes, (c) the timing of demonstration and commercialisation, (d) fluctuations in the size of the market for coal, and (e) the extent of added value created in a large new industry. Whether the application of economic considerations will be carried out in an ideal manner seems open to doubt; since these imperfections or uncertainties apply also to other energy industries which may be complementary to or competitive with coal, the difficulties of providing an ideal scenario are multiplied. In the first part of this chapter coal conversion is considered generally. In the second part, the cost and availability of coal are discussed.

COAL CONVERSION

Some Generalisations about Coal Conversion

Coal is not used as such but as an intermediate to some other product more directly useful — heat, electricity, gas, chemicals, plastics, etc. In the conversion, not all the coal is converted to the main product and losses occur in various ways. For example, when coal is burned purely for heat in a domestic appliance, much of the heat is lost as sensible and latent heat up the chimney or as uncombusted material in smoke and ash. In more sophisticated conversion processes, in each stage some of the chemical energy in the input coal is lost — generally as waste heat; it must be borne in mind that energy may often have to be provided to a process reactor, as heat, steam or oxygen. Since most liquefaction or gasification processes contain two or more stages, the efficiencies of each stage (each less than 100% of course) must be multiplied. For example, in the case of electricity generation, the combustion of coal, the transfer of heat to the steam and the conversion of steam into electricity are each stages of less than 100% efficiency (much less in some cases); the overall efficiency is the product of these three numbers.

Thus, before any other conversion costs are considered, the cost of a unit of energy in the form of gas or liquid derived from coal will include the cost associated with the input coal, generally equivalent to a much greater amount of energy than that recovered in the product. Capital costs are also bound to be substantial, however, especially since the achievement of high thermal efficiencies will probably require expensive process plant operating in unfavourable conditions at high pressures. Until full-scale plants, probably second or third generation, have operated consistently for some years, capital costs will not be known accurately (especially unit capital costs) since output (actual rather than nominal) will depend both on achieving projected process rates and on plant being available as planned. Quite apart from the specialised process plant, coal conversion requires substantial expenditure on land, coal handling facilities and (increasingly) waste disposal and pollution control equipment. Although modern large-scale chemical engineering plants, such as required for coal conversion, are not labour-intensive, they do require operational surveillance and therefore manning around the clock for almost all the year; the physical separation of the individual process stage units may also set a limit to operational manpower reductions. Furthermore, maintenance facilities of a high order are necessary. Services and utilities are similarly required on an intensive scale, often reinforced for safety or environmental reasons. Since coal is "hydrogen-deficient", hydrogen must be provided and is generally manufactured as an integral part of the process. Taking all such factors into account, it is not difficult to understand why, as a rough rule, it can be assumed that unit energy costs are increased by a substantial factor, perhaps a minimum of $2\frac{1}{2}$ times, in converting coal into more favourable energy forms such as gases or liquids.

Comparative Economics

Obviously if conversion processes that are not yet established on the full scale are being considered which increase energy unit costs by a factor of $2\frac{1}{2}$ or more, there must be some substantial margin of uncertainty. Furthermore, this factor is a multiplier which will apply (with only partial limitation) to future coal prices, which are likely to increase in real terms. Full scale coal conversion processes are not likely to be operated to any great extent within the 1980s; the following two decades are likely to see the major investment. Plants when built will be expected to work for 20 years or more so that in considering economics, coal prices for the next 40–50 years may be involved, explicitly or otherwise.

Of course, these uncertainties about future price levels do not apply only to coal. In fact, if coal output can be expanded regularly and consistently, as expected, coal prices are likely to become the marker for primary energy. The other main energy sources, including oil, gas and uranium, are likely to run into problems of excess demand in the next couple of decades and will probably increase in price not only more sharply but also, as the shortage of supplies bites, more erratically.

Another complicating factor is that often, during conversion, a large number of products result, as with a modern oil refinery. Some liquefaction processes for coal may similarly produce a range of products, which may not be identical with those from crude oil. Since it is difficult to ascribe costs to individual products (and, furthermore, in the case of petroleum, product prices may be linked more closely to market forces than to individual production costs), equating the sums of the values of the two product slates is even more difficult.

A further complication of fundamental importance concerns the financial conventions, including future interest rates, which may be used in assessing processes and determining energy policies. This is particularly the case when comparisons are made between alternatives having different cost structures, especially in the capital charges component. Although interest rates in Western countries reflect market forces, in practice energy policy decisions need to be taken against interest rates a long way into the future, beyond the scope of the market forces and sensitive of course to inflation rates. Therefore, financial conventions are often determined by governments, which may not be influenced solely by a thorough and objective measurement of all relevant factors.

The Value of Economic Assessments

From the above, it could be concluded that economic assessments are open to great uncertainties and that they are therefore not worth the trouble. Nothing could be more incorrect than the second part of this statement. To surrender to the uncertainties would be to abandon the use of all reasoning from future energy development considerations and to accept piece-meal decisions on arbitrary grounds. This would lead to the inevitability that all major decisions will be taken too late — which in the case of energy supplies would have catastrophic consequences. A more constructive attitude is to attempt not only to estimate the important factors as accurately as possible but also to provide a range of values so that different outcomes can be derived on a probability or sensitivity basis. Fortunately, modern calculating techniques facilitate such an approach and can allow regular up-dating of factual information and strategic preferences, the latter of course mostly from governments. Moreover, as suggested in the concluding chapter, governments can, especially if acting in concert, do a considerable amount to narrow uncertainties, to stabilise conditions and to provide the information in a co-ordinated way.

Finally, although uncertainties will remain, economic assessments even at this stage will provide a useful tool for making decisions, not least for eliminating those possibilities which are fairly certain to prove uneconomic.

Types of Economic Assessments

Some economic assessments are based on the type of generalisations discussed earlier — that is, an estimate of the thermal efficiency is made and a capital cost is assigned on general chemical engineering grounds related to the type of equipment and its size. From the capital costs, arbitrary operational and maintenance costs can be assigned. This process can of course be progressively refined as data from, say, laboratory or pilot plant experiments, are obtained. Indeed, the economic assessments, even in a primitive form, may indicate which type of factual information would be of most value in refining cost estimates further. In addition, even in the unrefined stage, it may be possible to make distinctions between processes because of different known efficiencies, which may relate to thermodynamic data, or to different capital costs because a particular piece of equipment, say an oxygen plant, is required in one scheme and not the other. Further, once such differences are established arbitrarily, sensitivity analyses can assess how the differences might be compounded or reduced.

At the other extreme, engineering "hardware" studies have been carried out in considerable detail and cost, especially in the USA. Sometimes these have related to an individual process and sometimes to two or more competitive processes for similar duties. A number of studies of this kind have also been carried out privately in situations where a coal discovery is ripe for exploitation or where an energy need is evident. Vendors of proprietary equipment will of course specify and provide costings (of varying degrees of completeness). In addition to the plant costs, there will be substantial associated investment, including such matters as off-site services, infrastructure and interest during construction.

Although detailed engineering studies are extremely valuable, it should not be forgotten that such estimates are just as vulnerable as less detailed estimates to inaccuracies of technical data (such as those relating to efficiencies) and also to the background uncertainties such as price trends and interest rates. Nonetheless, a detailed engineering estimate relating to a particular set of circumstances — site, coal type, markets, etc. — should be regarded as a necessary precursor for any final investment decision on a full-scale plant, unless the sponsors put a high premium on gaining experience.

In the chapters which follow, on the economics of the different forms of coal conversion, various estimates are sometimes presented and these do not by any means coincide. It is suggested that this diversity, whilst somewhat confusing, may give a realistic impression of the quality of estimates and emphasise the cautions already expressed. First, a general note on the competitiveness of coal conversion is presented.

General Competitiveness of Coal-based Fuels and Feedstocks

The introduction of coal conversion technology will depend on the relative costs of fuels and feedstocks derived from naturally occurring sources compared with coal-based substitutes. The purpose of this section is to indicate a basis for comparing the economics of coal conversion with the other options, and the price relativities necessary for coal conversion to be economic. It should be regarded as an example of a broadly based economic assessment. It relates to UK circumstances and to 1978 prices but is typical of the relationships.

For simplicity, the future fossil-fuel energy market may be considered in terms of the following fuels and feedstocks:

(1) Coal;
(2) Crude oil (including syncrude);
(3) Methane (including SNG);
(4) Heavy fuel oil (including SRC);
(5) Light petroleum fractions (including methanol);
(6) Synthesis gas (including medium calorific value fuel gas).

Various conversions between the naturally occurring and derived fuels are possible, those of most interest being indicated in Table 12.1.

The average delivered prices at March 1978, for UK power station coal, imported crude oil, and industrial natural gas, on an energy context basis are given in Table 12.2.

Estimates of future increases in energy prices differ widely. While energy supplies continue to be dominated by oil, it is widely assumed that the price of oil will remain the energy "marker price". At the lower end of the range of estimates of escalation, there is the OECD forecast of a 50% increase in oil prices (in real terms) by 2000 AD (the "Steam Coal" study). The UK Department of Energy, however, forecast in 1978 that oil prices would increase to between two and three times the 1978 level (in real terms) by 2000 AD.

Real increases in coal prices are also expected, partly as a result of the increased

TABLE 12.1 Fuel conversions

	Product					
Feedstocks	*Coal*	*Crude oil*	*Methane*	*Heavy fuel oil*	*Light fractions*	*Synthesis gas*
Coal		X	X	X	X	X
Crude oil				X	X	
Methane					X	X
Heavy fuel oil			X		X	X
Light fractions			X			X
Synthesis gas			X		X	

TABLE 12.2 Prices of primary fuels, March 1978

Fuel	*Price £/GJ*
Coal	0.9
Crude oil	1.1
Natural gas	1.0

TABLE 12.3 Assumed energy prices in 2010 AD relative to 1978 (for illustrative purposes only)

Fuel	*"Cheap energy"*	*"Expensive energy"*
Coal	+ 25%	+ 50%
Crude oil	+100%	+200%
Natural gas	+ 25%	+200%

investment burden for the expansion of the industry world-wide. In some cases, however, increased productivity, achieved partly through new investment, will act to stabilise coal prices. While coal remains in competition with oil products for the combustion market, coal prices may be related by market forces to those of crude oil. However, this situation will change when oil supplies can no longer meet demand and substitution becomes essential.

Because natural gas prices are dominated by transport costs, the future prices of natural gas will in the short term depend on the location. While the main industrial nations continue to have access to indigenous gas supplies, political pressures are often likely to keep gas prices lower than free market prices. Gas produced with oil away from the main industrial areas may be used as a low cost chemical feedstock to avoid the comparatively high transport costs for LNG. When the indigenous gas supplies of the industrial nations are unable to meet demand, gas prices are likely to rise in line with those of crude oil.

For this study, a time horizon of 2010 AD is assumed. Based on the above analysis and taking a conservative view, two scenarios for fossil fuel energy supply prices can be defined, a "cheap energy" view and an "expensive energy" view, in Table 12.3.

Cost of Energy Conversion

The relative capital costs of the conversion processes identified in Table 12.1 are given in Table 12.4.

The conversion of coal into methane (SNG) using the Lurgi gasification process is taken as the basis for comparison in Table 12.4. A plant producing 250 Mscfd would cost, at March 1978 price levels, approximately \$1500 million, or 140 £/kW thermal input of coal.

The efficiencies of the conversion process are compared in Table 12.5. For simplicity, the data are based on process options producing a single product wherever possible. Combined processes are likely to have higher efficiencies, so that these cost estimates may be conservative, possibly considerably so.

The costs of the conversions can be calculated from the data in Tables 12.4 and 12.5 using the following formula:

$$\text{Product cost (£/GJ)} = [0.57 \times \text{Relative capital cost} + \text{Feedstock price (£/GJ)}]/\text{Fractional efficiency}$$

Implicit in the above formula are the following assumptions:

(1) The capital cost of the coal-based SNG process is 140 £/kW thermal input.
(2) The DCF rate is 5% p.a.
(3) Operating costs are 7% of the capital investment p.a. for continuous operation, and *pro rata* for lower load factors.
(4) The load factor for all processes is 90%.

TABLE 12.4 Relative capital cost of conversion processes (expressed per thermal unit of coal input)

	Coal	*Crude oil*	*Methane*	*Heavy fuel oil*	*Light fractions*	*Synthesis gas*
Coal	0	0.74	1	0.65	0.9	0.85
Crude oil	—	0	—	0.1	0.23	—
Methane	—	—	0	—	0.48	0.2
Heavy fuel oil	—	—	0.53	0	0.33	0.8
Light fractions	—	—	0.25	—	0	0.2
Synthesis	—	—	0.2	—	0.23	0

TABLE 12.5 Efficiencies and yields of conversion processes

	Coal	*Crude oil*	*Methane*	*Heavy fuel oil*	*Light fractions*	*Synthesis gas*
Coal	1	0.7	0.65	0.75	0.6	0.75
Crude oil	—	1	—	0.93[a]	0.85	—
Methane	—	—	1	—	0.55	0.85
Heavy fuel oil	—	—	0.72	1	0.8	0.8
Light fractions	—	—	0.9	—	1	0.85
Synthesis gas	—	—	0.78	—	0.65	1

[a] Refers to a simple refinery configuration producing, typically, 39% heavy fuel oil and 54% light fractions.

TABLE 12.6 Converted energy costs 1978 (£/GJ)

Starting material	Conversion product					
	Coal	*Crude oil*	*Methane*	*Heavy fuel oil*	*Light fractions*	*Synthesis gas*
Coal	0.9	1.9	2.2	1.7	2.3	1.8
Crude oil	—	1.1	—	1.0[a]	1.4	—
Methane	—	—	1.0	—	2.3	1.3
Heavy fuel oil[b]	—	—	1.7	1.0	1.4	1.8
Light fractions[b]	—	—	1.7	—	1.4	1.8
Synthesis gas[b]	—	—	[c]	—	2.3	1.3

[a] Based on a simple oil refinery with light fractions credited at 1.45 £/GJ.
[b] Feedstock at the lowest cost from a primary energy source.
[c] Not applicable.

TABLE 12.7 Energy costs 2010 ("cheap energy" scenario) (£/GJ)

Starting material	Conversion product					
	Coal	*Crude oil*	*Methane*	*Heavy fuel oil*	*Light fractions*	*Synthesis gas*
Coal	1.1	2.2	2.6	2.0	2.7	2.1
Crude oil	—	2.2	—	2.0[a]	2.7	—
Methane	—	—	1.2	—	2.8	1.6
Heavy fuel oil[b]	—	—	3.2	2.0	2.7	3.0
Light fractions[b]	—	—	3.2	—	2.7	3.3
Synthesis gas[b]	—	—	[c]	—	2.8	1.6

[a] Based on a simple oil refinery with light fractions credited at 2.74 £/GJ.
[b] Feedstock at the lowest cost of a primary energy source.
[c] Not applicable.

TABLE 12.8 Energy costs 2010 ("expensive energy" scenario)

Starting material	Conversion product					
	Coal	*Crude oil*	*Methane*	*Heavy fuel oil*	*Light fractions*	*Synthesis gas*
Coal	1.3	2.5	2.9	2.3	3.1	2.4
Crude oil	—	3.3	—	3.0[a]	4.0	—
Methane	—	—	3.0	—	5.9	3.7
Heavy fuel oil[b]	—	—	3.6	2.3	3.1	3.4
Light fractions[b]	—	—	3.6	—	3.1	3.8
Synthesis gas	—	—	2.9	—	4.0	2.4

[a] Based on a simple oil refinery with light fractions credited at 4.03 £/GJ.
[b] Feedstock at the lowest cost from a primary energy source.

The costs of the various forms of energy can be estimated for 1978 market conditions, and also for the "cheap energy" and "expensive energy" scenarios in 2010, defined previously. The results are summarised in Tables 12.6, 12.7 and 12.8.

Discussion and Conclusions on Conversion Economics

Bearing in mind that this study is presented as an example only and uses arbitrary estimates and assumptions, the indications are that the types of fossil energy considered in the above analysis may compete in the various markets in the UK in the following way:

1. Bulk heating. At present coal, oil and natural gas compete for this market, with the latter two possessing an advantage in convenience. In 2010, coal used directly will possess a significant price advantage, although heavy fuel oil or a medium calorific (synthesis) gas would be attractive if severe environmental restrictions were introduced.

2. Premium applications. At present natural gas competes with light fractions from cheap crude oil, the former having a price advantage. In the cheap energy scenario, natural gas (while available) will take over this market, with SNG and light fractions from coal competing with light fractions from crude oil if natural gas cannot meet demand. In the expensive energy scenario, SNG and light fractions from coal emerge as the preferred fuels, with a slight cost advantage to the former.

3. Transport fuels and chemical feedstocks. At present, the preferred source is light fractions from crude oil. The low price of fuel oil derived from crude oil given in Table 12.6 indicates that a simple refinery is still the preferred choice. In 2010, light fractions from coal compete with light fractions from crude oil (advanced refinery) and the synthesis route based on low-price natural gas in the "cheap energy" scenario. In the "expensive energy" scenario, light fractions from coal (by hydrogenation) have a significant price advantage over light fractions from coal by the synthesis route or from crude oil.

4. Synthesis gas for chemical manufacture. The cheapest source of synthesis gas at present is natural gas, and remains so in the "cheap energy" scenario for 2010. If natural gas is not available (as in the "expensive energy" scenario) synthesis gas from coal becomes the preferred choice. Oil based feedstocks (heavy oil or light fractions) have no price advantage in this market at present and are uncompetitive in 2010.

AVAILABILITY AND COST OF COAL

Introduction

The economics of coal conversion are heavily influenced by the cost of coal at the point of use. Even more vital, of course, is that there should be coal available for conversion. An appreciation of the economics of coal conversion, with the potential for producing high value products, may help to generate a greater availability of coal. Many of the areas having the potential for expanding output also have growing energy demands which coal can help to meet. For a large number of countries, however, an expanded role for coal would depend on an increase in the amount of coal available for import; the cost of traded coal would be a major factor in the timing of coal conversion investments based on imported coal.

TABLE 12.9 Projection of major exporters (MTCE)

	1985	*2000*	*2020*	*1978*[a]
Australia	60	100	120	39
Canada	15	40	65	14
China	7	30	50	
India	7	13	32	
Poland	50	50	50	42
South Africa	23	55	60	16
USA	68	90	145	40
USSR	37	50	60	29
Other countries	41	72	74	

[a] Estimated by US DoE, December 1979

TABLE 12.10 World coal trade (MTCE) (Net imports positive; net exports negative)

	1976	*2000*
North America	−50	−143
Europe (OECD area)	55	311
Japan	60	181
Australia	−31	−195
Centrally planned economies	−38	−66
South Africa	−5.6	−90

Availability

The reserves and likely output trends of coal are discussed in Chapter 3, based primarily on World Energy Conference (WEC) estimates. The 1977 WEC report puts the then world coal trade at about 200 MTCE — 7.7% of production. Based on national responses, augmented where necessary by the authors' own estimates, it was considered that the export proportion would be the same in 2020, though marginally higher in intermediate years. This would provide an export availability of 380 MTCE in 1985, 520 in 2000 and 666 in 2020. The major exporters are suggested in Table 12.9.

The estimates given in the last column suggest that, for the countries mentioned, the export targets are within reach for 1985; for other countries without an export tradition, there may be greater doubt. In some cases, notably Poland, exports were then expected to level off, presumably because government policy indicates that internal demand will grow.

The International Energy Agency in their report "Steam Coal" give the general picture shown in Table 12.10.

Figures for intermediate years show progressive trends. A further study by IEA Coal Research suggests the *maximum* export availability of *thermal* coals might be 500 MTCE in 2000, the major exporters being indicated in Table 12.11.

In the WAES study (1977) in all four cases examined, potential world production of coal exceeds demand in 2000 by between 140 and 620 Mt/a. This is regarded as an incentive for greater efforts towards substitution of oil by coal.

The Ford Foundation report (*Energy — The Next Twenty Years* (1979)) suggests that total US coal production in 2000 could be about 1500 MTCE, of which 150

MTCE would be for export (all apparently metallurgical coal); the report remarks that the total production is near the high estimate of various other published models. However, the Ford Foundation report appears to be more concerned about the factors affecting energy supply and use rather than producing specific forecasts.

At the 1980 WEC, coal reserves and future output estimates were marginally higher than in 1977; no update of export capability was given in the WEC survey of Energy Resources. However, Carroll Wilson gave a report based on the WOCOL study and this was the subject of a Round Table Discussion initiated by M. J. Parker (NCB, Great Britain) who presented a paper on "Coal as an International Commodity".

The World Coal Study (WOCOL) was a co-operative study involving 16 major coal-using and coal-producing countries, under the direction of Carroll Wilson of MIT (who was also Project Director of WAES). The report was published in 1980. The main results are mentioned in Chapters 3 and 21; WOCOL views on export availability and prices are discussed here. It is important to appreciate that the WOCOL study was directed towards examining "the role that coal *might* play in meeting the world energy needs during the next 20 years" (italics added) and to what coal "should" do.

Carroll Wilson believes that world steam coal trade must grow 10–15 times by 2000, from 50 Mt to 600–700 Mt. With a 4–5 fold increase in metallurgical coal trade, the total would be one billion tons, requiring a thousand ships of 100 000 DWT costing $40 million each. The US would be the main exporter, increasing 4–8 fold, from 60 Mt to 200–400 Mt.

Parker included the WOCOL results in his presentation. The situation in 1977 is given in Table 12.12. This is compared with the WOCOL "high case" of 980 MTCE in 2000, of which 300 would be coking and 680 steam coal. The proportion of coal traded is shown in Table 12.13.

TABLE 12.11 Major exporters in 2000

	MTCE (net)
USA	140
Canada	20
Latin America	20
Eastern Europe (incl. USSR)	75
China	20
Australia	80
South Africa	130

TABLE 12.12 World coal trade, 1977 (Imports, MTCE)

	Coking coal	*Steam coal*	*Total*
OECD Europe	35	37	72
Japan	60	2	62
Centrally planned economies	18	17	35
Others	17	4	21
Total	130	60	190

TABLE 12.13 Coal trade as a proportion of total coal production (MTCE)

	1977 (actual)	*2000 (WOCOL high case)*	*Increase*
World coal production	2450	7015	4565
World coal trade	190	980	790
Used in country of origin	2260	6035	3775
Coal trade proportion	8%	14%	

TABLE 12.14 Range of possible coal export availabilities (MTCE)

	1977 (actual)	*2000 WEC (1977)*	*2000 WOCOL (High Case)*
USA	50	90	350
Australia	38	180	200
Canada	11	40	67
South Africa	12	55	100
USSR	25	50	50
China	3	30	30
Poland	39	50	50
Others	12	87	83
Total	190	582	930

TABLE 12.15 Summary of steam coal import requirements (MTCE)

	1977 (actual)	*2000*	
		WOCOL (Case A)[a]	*WOCOL (Case B)*[a]
OECD Europe	37	146	333
Japan	2	53	121
Other Asia	—	60	179
Centrally planned economies	17	30	30
Others	4	11	17
Total	60	300	680

[a] Cases A and B are based on OECD economic growth rates of 3% and $3\frac{1}{2}$% p.a. respectively (growth rates in energy, 1.75% and 2.5%)

There is no concentration of reserves in a low consumption area, as in the case of oil which has led to the high proportion of traded oil (55%). The projected output increase (about 4000 MTCE) depends crucially on the USA, as does the export potential. The other large coal producers, USSR and China, have internal demand projections very close to their output projections, thus making "surpluses" more uncertain. The export availability position is summarised in Table 12.14. The import requirements for steam coal are summarised in Table 12.15.

Parker concludes that the extent and geographical distribution of export availability must remain uncertain, even speculative. He stresses the infrastructure and other requirements and also considers demand limitations. Coal demand for conversion processing is not likely to be large by 2000, so demand depends on the building of a large number of power stations, which cannot come into operation until the 1990s. Views were expressed during the discussion indicating that whilst the potential for increases in world output and world trade existed, demand shortfall and delays in provision of infrastructure would delay the build-up, compared with the WOCOL figures for 2000.

Costs of Coal

Predicting the future costs of coal (and prices to consumers) is even more difficult than (and not unrelated to) forecasting the amount of coal which may be available. However, this certainly applies, even more forcibly, to other forms of fossil fuels, where prices will increasingly be determined by market and political factors rather than costs of production. It seems likely that coal will be priced more nearly on a basis related to production costs for the indefinite future. These costs will probably reflect real regular increases in labour wages rates, increased investment and environmental impact costs; productivity increases may not always fully reflect additional capital expenditure.

Traditionally, much of the coal produced has been used fairly near to its source; even so, differences in delivered coal prices, reflecting differences of a few tens of miles by rail perhaps, have been important. In the future, the massive new production centres will often be located a long way, perhaps thousands of miles, from preferred consumption sites. Although sea transport is relatively cheap, long journeys by land will generally also be necessary before shipping. Delivered coal prices in the future will therefore frequently have a very substantial element of transport costs. Some estimates are included in this chapter; the question of the relationship between coal sources and the location of consuming or converting industries is further considered in Chapter 21.

Although coal from other countries, notably South Africa and Australia, can currently be delivered into the principal importing areas (Japan and North West Europe) somewhat more cheaply than from the USA, future American price trends are extremely important internationally. According to US Department of Energy statistics, delivered fuel prices to utilities in the various regions in May 1978, in cents per 10^6 Btu, are given in Table 12.16.

TABLE 12.16 Delivered fuel prices: US Regions, May, 1978 (cents per 10^6 Btu)

	Coal	*Oil*	*Gas*
New England	147	195	184
Mid Atlantic	119	208	163
East North Central	117	262	192
West North Central	87	189	119
South Atlantic	129	194	112
East South Central	116	183	115
West South Central	69	182	136
Mountain	51	226	150
Pacific	78	250	220
National Average	111	210	144

TABLE 12.17 Coal export price competitiveness

Import area	*Source*	*Production cost*	*Price FoB mine*	*Price FoB port*	*Delivered price*	*Price \$/10⁶ Btu (average)*
NW Europe	USA, East, U/ground	15–28	20–35	30–45	39–59	1.85
	USA, West, surface, rail	5–15	8–18	20–35	31–50	2.19
	USA, West, surface, slurry	5–15	8–18	14–30	26–43	1.87
	Australia, U/ground	10–18	15–20	20–25	34–43	1.63
	Australia surface	8–15	12–20	18–25	32–43	1.52
	S. Africa, U/ground	5–8	10–15	15–22	26–35	1.41
	Poland, U/ground			23–31	31–39	1.46
Japan	USA, East, U/ground	15–28	20–35	30–45	44–64	2.05
	USA, West, surface, rail	5–15	8–18	20–35	30–50	2.00
	USA, West, surface, slurry	5–15	8–18	14–30	27–44	1.75
	Australia, U/ground	10–18	15–20	20–25	29–36	1.38
	Australia, surface	8–15	12–20	18–25	27–36	1.33
	S. Africa, U/ground	5–8	10–15	15–22	26–33	1.31
	Poland, U/ground			23–31	36–44	1.67

The considerable differences between the regions emphasises the impact of transport costs, but also the very much lower production costs of Western coal. The national average delivered coal cost is equivalent to about \$30 per ton (1978) dollars); the lowest priced coal, Western, is about \$14 per ton. Coal has a substantial advantage over gas and oil, on a thermal basis, everywhere.

The Ford Foundation report considered the future delivered prices of fuels to utilities. They made very conservative allowances for factors tending to increase real costs and estimated that the national average delivered prices in 2000 would be \$2.60 per 10^6 Btu (1979 dollars). This is equivalent to about \$72 per ton and (allowing for the difference between 1978 and 1979 dollars) represents a real price increase of about 115% between 1978 and 2000. However, it is considered in the Ford report that oil and gas will increase proportionately, or slightly more rapidly; both may cost \$5.20 per 10^6 Btu delivered in 2000, with some substantial up-side uncertainty, especially for gas.

In 1979, the US Department of Energy considered the potential for future exports of American coal. At that time, it was estimated that the USA was at an economic disadvantage in the principal markets; the estimated breakdown of costs and prices (1979 dollars per ton, except last column) is shown in Table 12.17.

On the assumption that the USA would wish to compete strongly, various incentives and measures to improve weaknesses are considered. The desirability of exporting vigorously may not currently be universally agreed in the USA but the assistance to the balance of payments may look increasingly attractive with time and, of course, the support given to the energy supply position and economic health of trading partners and developing countries could be important. The measures which might be considered would include better administrative procedures, improved transport and port facilities and various forms of subsidy. In the longer run, however, it is suggested that the great strength of USA may be her ability, in presumed contrast to other potential exporters, to expand coal production greatly with relatively modest effect on the marginal price: doubling coal production might theoretically only increase prices by about 50%. The political advantages of depending on coal rather than oil as the marginal imported fuel are stressed.

The International Energy Agency, "Steam Coal" report (which strongly favours the growth of world trade in coal) gives the estimated costs in the mid-1980s (1976 dollars per TCE) for various cases in Table 12.18. Other estimates for various cases are given in Table 12.19.

It should be noted that, in general, in the long distance transport cases, delivered costs divide roughly 50–50 as between production costs and transport costs.

In the studies carried out for the WEC survey of Energy Resources (1980), for the first time, broad information was included on mining costs — reserves being stated in categories of less than $15, $15–30, $30–60 and over $60. It was reported than in North America, South Africa, India and Australia, production costs for bituminous coal and anthracite were nearly always under $30/t. In the USA and Canada, substantial reserves would have production costs of between $30 to $60/t. On the other hand, costs in Europe, Japan and South Korea were always more than $30 and more likely to be more than $60/t. Production costs of sub-bituminous coal are nearly always less than $30/t and in the USA and Canada are less than $15. Brown coal and lignite, which are normally mined by open-cast methods, usually cost less than $15/t to mine.

Carroll Wilson confirms the view, stated above, that coal prices are likely to be based on costs and therefore should increase much more slowly than OPEC-controlled oil prices. The price of US steam coal (September 1980) was about $35/t, or more than $100/t cheaper than oil. The cost of imported coal in Japan, at that

TABLE 12.18 Estimated costs of imported steam coals delivered to Western Europe and Japan in the mid-1980s[a] (1976 $US per TCE)

Supply region	Quality (sulphur content)	Estimated CIF cost per TCE	
		Western Europe	*Japan*
United States			
Eastern	LS	$51–$55	$62–$64
	HS	$38–$40	$49–$50
Western	LS	$48–$54	$32–$36
	HS–HMS	$38–$51	$36–$43
Canada			
Eastern	HS	$40–$45	
Western	LS	$46–$50	$42–$46
South Africa	LS–MS	$34–$42	$35–$43
Australia	LS	$44–$47	$34–$46
Poland	LS	$34–$42	
USSR			$37
China			$37–$40
Indonesia			$37–$48
Colombia			$35–$41
Mozambique			$40

[a] First year prices of long-term contracts.
LS: low-sulphur
MS: medium-sulphur
HS-high-sulphur.

TABLE 12.19 Delivered cost of thermal coal in major consuming regions (1977 US dollars per million Btu)

	Recently	*1985*	*2000*
East North Central USA from			
Appalachian–low sulphur	1.09	1.40	1.80
Appalachian–high sulphur	0.89	1.16	1.90
Western USA	1.10	1.18	1.40
Southern Ontario from			
Western Canada	1.78	1.91	2.40
Appalachian-low sulphur	—	—	—
Appalachian–high sulphur	0.93	1.19	1.93
North Western Europe from			
UK domestic (mine mouth)	1.65	1.65–2.16	1.98–2.12
West German domestic (mine mouth)	2.65	2.68–3.72	3.38–4.65
S. Africa	1.37	1.54	1.79
USA-Appalachian–low sulphur	1.56	1.87	2.26
USA-Appalachian–high sulphur	1.37	1.63	1.97
Australia	1.33–1.68	1.48–1.83	1.72–2.07
Poland	1.47	1.71	2.02
Japan from			
Australia	1.14–1.45	1.29–1.60	1.53–1.84
Poland	1.82	2.06	2.37
S. Africa	1.29	1.46	1.72
USA-Appalachian–low sulphur	1.69	2.00	2.40
USA-Appalachian–high sulphur	1.50	1.76	2.10

Note: Costs are for coal unloaded at a power station in E.N. Central USA or at a port in the other regions, except for the UK and West German figures which are minemouth costs.

time, was about $45. Carroll Wilson suggests an allowance of $35/t as the cost of meeting the stringent Japanese environmental standards, making an effective coal cost of $80/t for use in power plants. This is only about half the cost of oil, in spite of the conservative assumptions.

Parker took a more commercial attitude at the WEC, without suggesting any prices. He pointed out the constraints and uncertainties in the supply/demand balance and in future transport costs; he thought this would result in fluctuations in prices and conditions of sale (e.g. long-term contracts). His presentation suggested a future in which periods of surplus demand (which would chase up prices and increase investment in mining) would alternate with periods of surplus coal (which would depress prices and increase investment in consumption capacity).

At the 1980 WEC, no one doubted the inherent ability of the world coal industry to increase output several-fold, at prices which bore some relationship to current prices. There was considerable dismay expressed however about the current lack of progress and the more optimistic forecasts of output in the next twenty years were discounted.

CONCLUSION ON COAL SUPPLY

In spite of the strong position of coal with regard to resources and reserves, the organisation of greatly enhanced production and the delivery of that coal to the preferred consumption sites will require great thought and massive effort. Delivered prices are likely to increase in real terms by about 50–100% over the next couple of decades but this must be subject to considerable uncertainty. Nevertheless, these uncertainties seem less unfavourable than other solutions to the expansion of energy supplies. Perhaps the main lesson is the need to improve efficiencies of supply and use of coal and to improve the deployment of this politically vital energy source. Further consideration to deployment is given in Chapters 21 and 23.

Chapter 13

ECONOMICS OF COMBUSTION AND POWER GENERATION

INTRODUCTION

Combustion has been the main process for the utilisation of coal for a long time and the main application of this combustion for the last few decades has been the generation of electricity. This pattern will not change quickly and to many observers, the question of the future role of coal in the medium term is equated with its competitiveness in the electricity generation field. In the longer term — and perhaps sooner than many experts might agree — the question may be whether it is more valuable to use coal for making electricity or to use it for making gases, liquids and chemicals. This question is dealt with elsewhere but in the meantime the credibility of a world policy for extending the use of coal depends to a great extent on the economics of coal for electricity, however strongly, as the authors believe, the emphasis may shift at a later date. An approach to this assessment is therefore the main point of this chapter.

The main rival for coal, in making decisions about new major electricity generating plant, is nuclear power. Comparisons are therefore made initially between nuclear power and standard coal firing plant for electricity (i.e. pulverised fuel (pf)), with and without Flue Gas Desulphurisation (FGD). Subsidiary comparisons are also relevant between pf and fluidised bed combustion (FBC) stations and also between FBC and gasification/electricity generation cycles. The cost of environmental control measures should also be taken into account, although the uncertainty of future standards makes this a difficult problem. Some estimates are given for the cost of FGD, for example; the question of policy decisions about the way these pollution control processes may be applied, including the issue of value for money, is considered in another chapter.

Only a few years ago a great deal of any assessment of the economics of coal in electricity would have been related to oil as the chief competitor. Although this must still be considered for completeness, oil is scarcely a major option for new investments in the future on political and availability grounds, in addition to the fact that it is unlikely in any case to be attractive on cost grounds. The economics of COM (coal-oil mixtures) may be of more immediate interest, especially for cases where conversions of existing plant are being studied. The possibility of refurbishing existing power stations to burn coal by modern techniques is also an important question, attracting increasing attention; this is however a matter likely to depend on local issues.

Some attention is given to the economics of coal burning in industrial boilers and for domestic heating. The economics of future processes, including MHD, are also considered.

THE ECONOMICS OF COAL BASED ELECTRICITY GENERATION

Coal versus Nuclear Power

Although this question has been closely studied by the authors and some of their colleagues for a long time, the first and most important point to make is that no really satisfactory comparison has yet been made, nor is one likely to be made, until better economic models are developed (and accepted by decision makers) and a great deal more factual data are available (by which time perhaps the whole question may be academic). This is not to say that such comparisons are not worthwhile; those taking the decisions should be aware of the uncertainties, however, and should also be conscious that subjective judgements are involved, implicitly or otherwise. Although some quite detailed *post-hoc* studies of plant now in service have been published, emphasis here will be on cost estimates for future plant, since this is the important area for considering future coal utilisation. The history of the availability of past plant is however relevant.

One of the main uncertainties about nuclear power costs is the obscurity of the meaning ascribed to the term "capital costs"; wide divergencies are also apparent in reported costs. Smaller differences are reported in nuclear fuel and operating costs; these are in any case currently relatively small (especially as a proportion of total generating costs) but this may change as a result of pressures on uranium prices, reappraisals of nuclear fuel cycle plant costs and environmental requirements, including waste disposal and dismantling of old nuclear plant. Some of the uncertainty on nuclear power arises because in neither capital nor fuel costs have free market forces been effective; governmental intervention, especially in the fuel cycle, is one of the main reasons for this distortion. Nonetheless, in any analysis the dominant factors (and those areas which are also especially dependent on judgements), are the cost of capital and the likely trend of coal costs in the future. This situation arises of course because of the different make-up of costs in the two cases: capital appears dominant in nuclear estimates and fuel costs in the case of coal.

The importance of these political or judgement factors is illustrated in two studies carried out for CEPCEO (The European Coal Producers Association) by a committee under the chairmanship of one of the authors (LG), who also presented the results of the first study to an Open Discussion on Nuclear Energy organised by the EEC in November 1977. This committee perused the literature and had private discussions with a number of authorities and organisations; in addition, some of the members of CEPCEO are directly involved with electricity utilities and had up-to-date factual evidence. The various authorities were asked for information relevant to a large new power station (specifically described) to be ordered in 1977, for operation in the mid 1980s. A range of data was produced (for West European conditions); this is summarised in Fig. 13.1, which presents the calculated generating costs for nuclear power and coal (high sulphur with FGD and low sulphur without), as a function of load factor. (Costs are all in mid-1977 dollars.)

Figure 13.1 illustrates some general points. First, there is a break-even or cross over point on the load factor curve where, on the assumptions used, the generating costs for nuclear and coal will be the same; at higher load factors nuclear will have an advantage and at lower load factors coal will be cheaper. The shape of these curves however, together with the small angle they make, means inevitably that differences

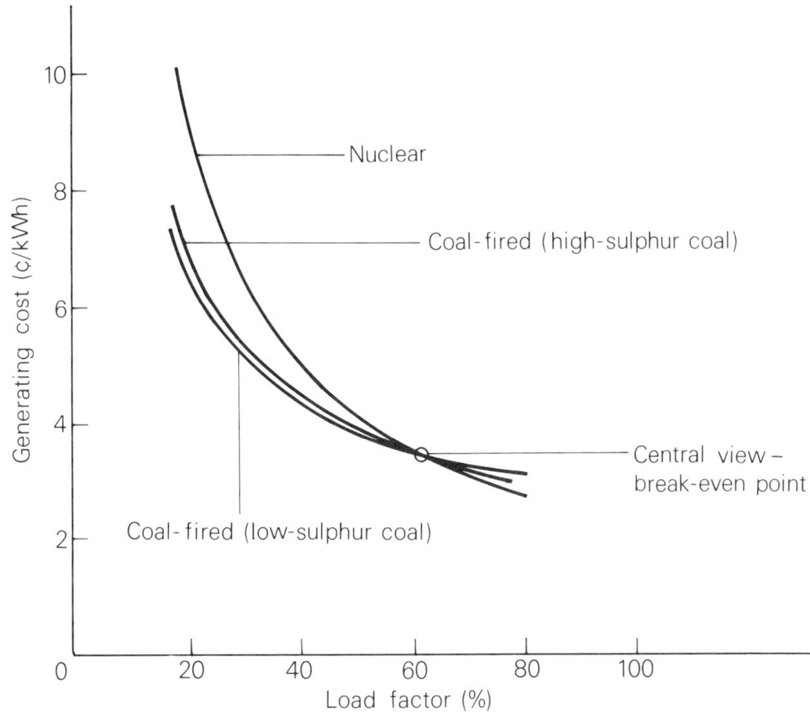

13.1 Electricity generating costs: central view. (*Basic data.* Capital investment: coal \$480/kWe (no sulphur control), \$528/kWe (with sulphur control), nuclear \$860/kWe; Capital charge rate: 15% p.a.; Fuel cost: coal \$2/GJ (low sulphur), \$1.8/GJ (high sulphur), uranium \$40/lb)

in cost are relatively small at load factors near to, and for some distance below, the break-even point; only at fairly low load factors does the high capital cost of nuclear power begin to make a really substantial difference (at load factors, incidentally, which the purchasers of nuclear plant would dearly love to avoid). With the assumptions chosen, the break-even point in Fig. 13.1 is at a load factor of about 63%. A sensitivity analysis was carried out to show the effect on the break-even load factor point of various changes in assumptions and also to relate different sets of assumptions to the maximum coal price in order for coal to be competitive.

Within about a year of the calculation of these numbers, which were at the time based on data supplied by experts and on financial criteria then used by governments, two official reports, one in the UK and one in the FGR, created major changes in attitudes, especially with regard to capital charges and coal prices. Increases in assumed interest rates are of course unfavourable to nuclear power because of their high capital cost. The 15% capital charge rate used in Fig. 13.1 was at that time the central view of the actual or authorised rates in the countries concerned. Subsequently, both in the UK and the FGR, substantially lower rates were taken as more appropriate for investment planning.

In the UK, the later government analysis also considered various annual rates of increase in coal prices (in real terms) between 0.8 and 3.9%. The latter, as an annual rate, is very stringent indeed and would increase the real cost of coal by more than

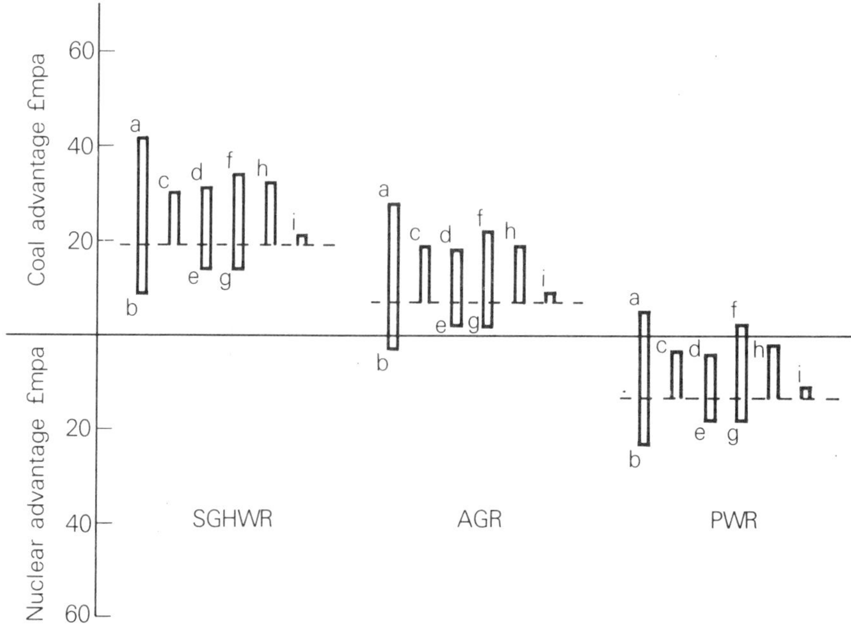

13.2 Economics of coal and nuclear (2000 MW stations) showing the effect of varying the main parameters (discount rate = 7%). (a–b, 20% increase in capital costs (a) nuclear (b) coal; c, nuclear reprocessing costs double; d–e, 10% decrease in station availability (d) nuclear (e) coal; f–g 3 year comissioning delay (f) nuclear (g) coal; h, 20 year life for nuclear station instead of 25 years; i, 35 year life for coal station instead of 30 years; --- central assumptions)

400% by the year 2010. The only justification given is that this matches the government's view of future oil prices and keeps the value of coal just below that of oil, in thermal terms. However, in the case of coal, extra production costs (as opposed to price) must be associated mainly with wages and the implications of a 400% increase in coal prices do not seem credible unless miners have moved into the millionaire class by then. This is even more true if, as expected, coal mining becomes more capital intensive. In the UK government report, one of the more reasonable "profiles" examined takes the case of 0.8% annual increase in coal prices (about 30% total real increase by 2010). The results for a 7% "Required Rate of Return" are shown in Fig. 13.2 (based on a 70% load factor). The results are shown as total financial advantage over the whole life of the stations. The central estimates are shown by the horizontal dotted lines for each of three different reactor systems; only the PWR is shown as having any advantage over coal, the other reactors showing a disadvantage. (The proponents of the other reactors, incidentally, might strongly deny this.) It should be noted, however, that the capital costs are of the order of £1000 million and the total lifetime costs are perhaps three times this; the margins between cases (generally less than 1%) therefore are very small, even if absolutely accurate. For a 10% discount rate the PWR has a disadvantage about equal to the advantage suggested at 7%. In other words, with these assumptions (including 70% load factor and 0.8% annual real price increase in coal) the break-even between coal

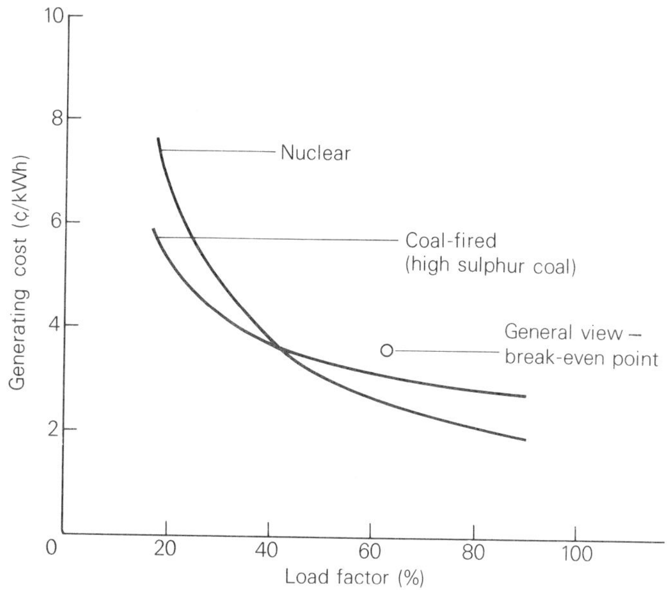

13.3 Electricity generating costs: sensitivity to lower discount rate and higher fuel costs. (*Basic data.* Capital investment: coal-fired \$528/kWe, nuclear \$860/kWe; Discount rate: 7% p.a.; Initial fuel cost: coal \$1.8/GJ, uranium \$40/16; Fuel cost escalation: 1% p.a.; Power station life: coal-fired 25 years, nuclear 20 years)

and nuclear occurs at a discount rate of about $8\frac{1}{2}$%. Substantial real increases in PWR capital costs have been reported subsequently.

In the more recent German official studies, also, interest rates are taken at lower levels than previously (about 6%) and particular attention is given to the differences in prices between German and imported (Polish) coal.

To illustrate the effects of this change in basic assumptions between 1977 and 1978, the CEPCEO committee drew up another curve (Fig. 13.3) on average 1978 assumptions. The original break-even point is shown as the "Central View — Break-even Point" (63%); the new break-even point is just over 40% and at 63% load factor nuclear now has a unit cost advantage of about 16%. Figure 13.4 illustrates, in another way, sensitivity to discount rate and coal price escalation. For three values of the latter (0, 1 and 4% pa "real") the break-even load factor is plotted against discount rate. The area covered by the 1977 CEPCEO study is shown, together with areas covered by the UK and FGR studies in 1978. Figure 13.5 shows the coal prices required for break-even at two load factors (50% and 60%) as a function of discount rate; as shown at the side, the 1978 prices of UK and imported coal cut these lines but German hard coal is well above the whole area.

These results are not presented as having any special degree of accuracy but they do illustrate very clearly how the outcome may be determined very decidedly indeed by assumptions of a non-technical nature, based on judgement, political stance —or prejudice.

Many other generating cost comparisons have been carried out. One of the most widely publicised and responsible studies was that of the International Energy Agency in 1978. The figures in Table 13.1 are average for the OECD area and are

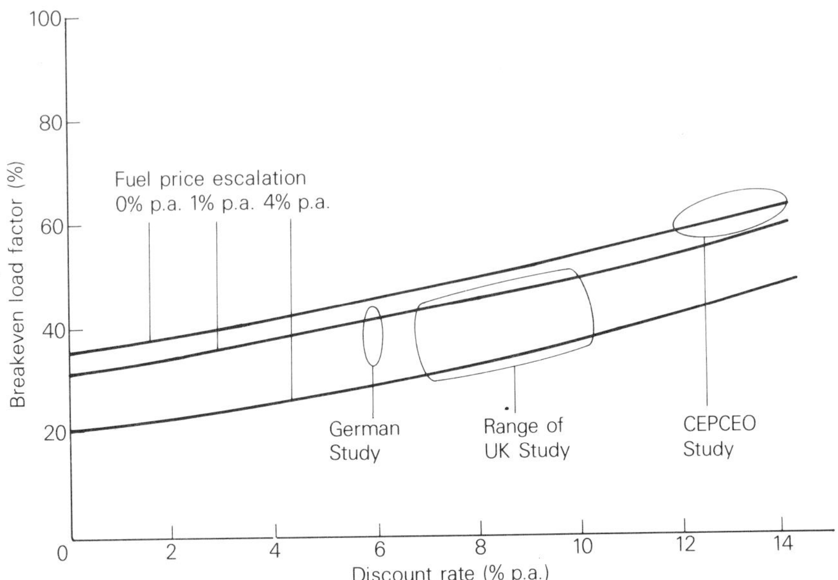

13.4 Breakeven load factor: sensitivity to discount rate and to fuel price escalation. (*Basic data.* Capital investment, initial fuel costs, Power station life as Fig. 13.3)

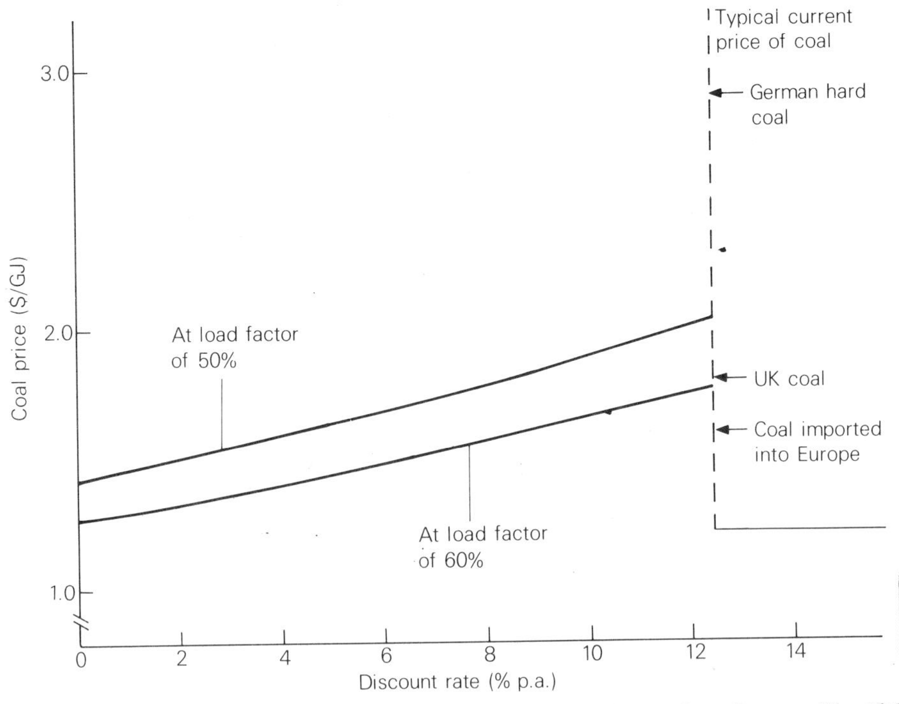

13.5 Coal prices required for coal-firing to be competitive. (*Basic data.* Same as Fig. 13.3 except Fuel cost escalation 0%)

TABLE 13.1 Comparative generating costs: IEA estimates (US mills/kWh)

	Fuel oil		*Bituminous coal*	
Nuclear	*Low sulphur*	*High sulphur + FGD*	*No FGD*	*With FGD*
23.8	40.5	42.8	22.6	28.8

TABLE 13.2 IEA regional estimates (US mills/kWh)

	Japan imported coal	*Western Europe imported coal*		*US domestic coal*
Nuclear	*With FGD*	*With FGD*	*Without FGD*	*With FGD*
23.8	35.8	36.0	30.3	30.3

expressed in mills (USA, 1976) per kWh, for plant commissioning in 1986 and operating for 5500 h/a (63%).

In this analysis, a real discount rate of 10% was taken and coal prices were taken as an average OECD price for coal delivered in 1976 ($32/t) with 1% real escalation per year after 1985. The study however points out that the average coal cost is not representative of any particular region. Regional costs, with the same assumptions otherwise as above, are produced for several areas, based on estimates of the delivered price of coal there in 1986 plus 1% thereafter, are given in Table 13.2.

These figures as they stand, do not present a particularly favourable position for coal and in assessing the report, the following factors are relevant:

(1) The study takes a rather low range for the capital costs of nuclear power ($600–1000 kW) and, within this, uses a less than average point ($700). This is combined (and the effect compounded) with a 10% discount rate.
(2) A uranium price of $35/lb U_3O_8 (1976 dollars) is taken for 1986 deliveries, with no escalation thereafter;
(3) The FGD costs are taken at the high end of a range presented elsewhere in the same report.
(4) The nuclear costs are based on 2 × 1100 MW plants on a site and the other systems on 2 × 600 MW plants; this may cause a distortion of up to 10%.

This report strongly emphasises the advantage of coal over oil, a most important point politically of course to the IEA. To the IEA at the Headquarters (in Paris) coal and nuclear power are regarded as complementary (and equally essential) means of reducing oil imports, which may account to some extent for the choice of assumptions. It can be assumed that nuclear power was considered in need of a public "boost". The IEA Coal Research Group in London (funded entirely by the cooperating countries directly and not supported or controlled by the IEA Headquarters in Paris) also studied the competitive economic position of coal for electricity, more from the point of view of the optimum future use of coal, rather than the overall political advantage of fuel policies within the OECD Group. A wide range of capital costs, fuel costs and coal prices was taken; Fig. 13.6 shows some of the results, for assumptions considered particularly relevant in the present context.

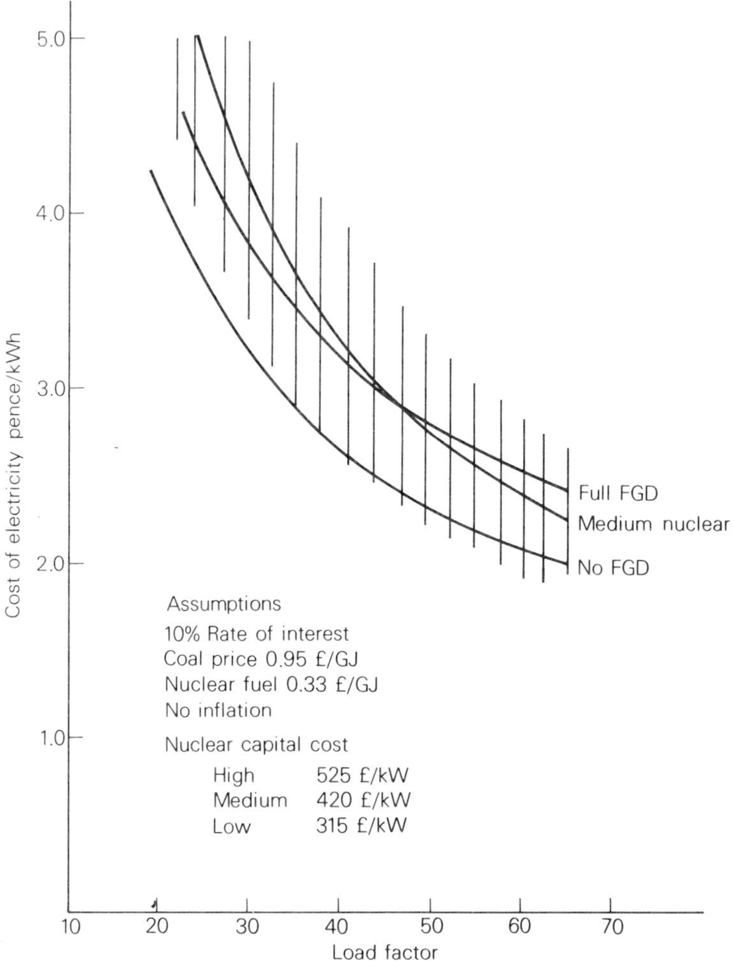

13.6 Costs of coal and nuclear electricity

The hatched area shows the range of nuclear costs (the medium value being shown by a curve); the latter compete with coal, including FGD, at load factors above about 47% but nuclear is never competitive with coal without FGD.

Of many other studies of the economic relationship between coal and nuclear power, the work of the Electric Power Research Institute in the USA is considered particularly useful. The results are broadly in line with the UK results. Higher capital costs for coal stations are assumed by EPRI, to take account of stringent desulphurisation requirements, but this is largely offset by lower coal prices, although results for different regions of the USA are markedly variable, for this reason mainly.

The conclusion of this discussion so far must be that nuclear power is considered to be, by a majority of authorities, up to 20% cheaper than coal firing with 100%

FGD at high load factors and very similar to coal without FGD. The impact of FGD and newer coal combustion methods will be considered next before reassessing these conclusions.

The Effect of Sulphur Control (PF Stations)

Attitudes to sulphur control vary widely; the rationale is considered elsewhere. In the UK, no control is formally exercised but high stacks are used for dispersal. This has a minor effect on costs. In this section, the costs of control to US standards, appropriate to existing power plants, are considered. Only limited relief can be obtained by the choice of coal or by coal washing. In practice, these latter procedures should not be neglected and may be an important element in complex arguments about the time factors over which various levels of sulphur emission may be permitted. However, low sulphur coals will tend to be priced in relation to the costs of using high sulphur coals plus FGD (i.e. scrubbing) and this latter can therefore be considered the datum.

Assuming lime/limestone non-regenerable scrubbing, the capital cost of equipment for a 500 MW station may be about $80/kW, plus $20–30 for off-site disposal facilities. The capital cost is rather sensitive to size of station and the capital for disposal facilities will be greatly influenced by the location of the plant and the amount of waste (i.e. mainly on the sulphur content of the coal). About 5% to 15% of the gross output of the station is used in FGD and is therefore lost from external sales; this would of course add another $25/kW or so to the effective capital cost. Capital charges depend on the financial conventions but might amount to nearly 3 mills/kWh. Operating costs vary, particularly with the sulphur content and local limestone cost, and may be up to 1.2 mills. Labour and other costs may be about 1.5 mills. An overall cost of 5–6.5 mills/kWh is likely; this may mean an addition of 20–30% compared with operation without FGD.

The regenerable systems are more expensive, both in capital (30–50%) and on power consumed (9% of output compared to 5%). The implication is that the regenerable processes (which are less proven) are likely to be at least of the same order of cost and may only score where disposal costs or lime prices are high. Practicability, reliability and convenience may be more important factors in the long run. Costs are not available for the dry (activated carbon) process under development in Japan (Chapter 6).

The IEA Report (Steam Coal) gives sulphur control costs at the high end of the range implied here. A 30% increase in capital cost is indicated, with a generating cost increase of 27.5%. These figures are based on a single coal price and do not take account of any possible price reductions for high sulphur coals. Also, station operating and maintenance costs are shown as increasing by over 130% which may reflect early pessimism with FGD plants. In the 1977 CEPCEO studies, high sulphur coal with FGD and low sulphur coal without were shown to be similar. This was based on a capital increase of only about 10% for FGD plant and a premium of 11% for low sulphur coal.

Economic Potential for FBC

No capital or operating costs based on actual operation are available for utility-scale FBC plants, neither AFB or PFB, and, clearly, actual costs from these large plants will not be available for quite some time. However, some design and cost

TABLE 13.3 Costs of coal electricity (ECAS study)

	Conventional PF	*AFB*	*PFB*
Capital $/kW	835	632	723
Efficiency %	31.8	35.8	39.2
Cost of electricity, mills/kWh (at 65% LF)	39.8	31.7	34.1

TABLE 13.4 Percentage savings in electricity costs by fluidised combustion

	High sulphur		*Low sulphur*	
Coal price	*AFB*	*PFB*	*AFB*	*PFB*
$1/GJ	13–15%	7–8%	<3%	3–4%
$2/GJ	12–14%	9%	<3%	4–5%
$3/GJ	11–13%	10%	<3%	4–6%

studies have been carried out; perhaps the most important (and certainly the most elaborate) was the Energy Conversion Alternative Study carried out in the USA. This study contains many results, perhaps encouraging subjective interpretation to a considerable degree. One set of results is given in Table 13.3.

Results are in 1975 dollars, but include escalation and interest during construction. Full FGD is provided in the case of pf. Similar sulphur suppression is assumed with AFB and PFB. If there were no scrubbers, costs would notionally be reduced with conventional pf to 30.5 millions and efficiency increased to 36.2%; however, all the results are based on 3.9% sulphur coal, which would not be permitted in the USA without suppression. The use of coal with less than 0.65% sulphur would be necessary without FGD and in this case presumably the coal cost would be increased to an extent commensurate with the increased cost of providing scrubbers. The substantial capital saving predicted for AFB and PFB compared with conventional (including FGD) is notable, although costs taken for FGD seem to be rather high. The efficiency figures predicted for FBC may be more important in the long run, especially in Europe, as primary fuel prices increase. In this connection, the standard bed temperature for PFB is taken as 1650°F (917°C); in a subsidiary case 1750°F (972°C) is taken and this is assumed to increase thermal efficiency from 39.2% to 40.0%. These assumptions may be somewhat optimistic.

Some studies have also been carried out by IEA Coal Research to estimate the percentage savings which might be achieved by AFB and PFB compared with pf (with FGD) at various coal price levels (Table 13.4). Again, of course, the question of the actual availability of low sulphur coals may be more important, and, if available, their relative price; this presentation compares the savings at the same coal price.

A CEPCEO study, considering both US and European work, tends to emphasise *potential* and assesses the expectation for FBC as being about 20% reduction in investment cost and 10% reduction in operating cost, on the assumption that the development problems can be solved. (This qualification appears to be aimed at influencing the allocation of development funds rather than any indication that problems may be intractable.)

In forming some conclusions about the economics of FBC, it must be repeated that hard data do not exist. It may be that some plant manufacturers or utilities have better information or estimates than those quoted but this must be doubted; this situation may not be improved greatly before the mid-1980s. The preliminary economic assessments quoted, whilst showing reasonable potential for FBC for power generation, especially in cases where FGD must be practised and high sulphur coal used, the apparent savings on low sulphur coals are negligible. Furthermore, in spite of the potential technical and thermal efficiency advantages of PFB over AFB, this is not significantly reflected in the economic estimates, so far. Analysis of these points obviously requires close collaboration between the economic and technical studies. Apart from better engineering to reduce capital costs, from which the more complex system should in theory benefit most, consideration should be given to ways in which the competitiveness of FBC (and especially PFB) could be improved so that development work can be concentrated on these aspects. In some estimates, for example, PFB is taken as having only three percentage points advantage in thermal efficiency over AFB. If this could be increased, so would the economic advantage and this would be further compounded by future fuel price increases. Thus, great emphasis must be put on ways of increasing temperatures and/or obtaining extra tranches of useful output by combined systems (gas and steam turbines, combined heat and power, etc.).

Apart from technical and engineering factors, however, commercial and non-technical factors need to be pursued in order to exploit the characteristics of FBC. Environmental standards will clearly be extremely important and a great deal will depend on the relative prices of different grades of coal, especially with regard to sulphur content. The ability of FBC to deal with low grade coals or residues may also have particularly important value, depending on commercial and geographical factors.

Other Advanced Systems

The economics of systems which depend on substantial technical progress must clearly be uncertain and based on a presumption that a particular technical advance can be exploited economically in a broadly foreseeable manner.

The most detailed investigation of advanced systems was that carried out under ECAS, referred to above. It is however complex to understand and practically impossible to summarise — this is because any number quoted really needs to be qualified by a wealth of technical data. Two major contractors separately assessed a number of systems, using a range of parameters and the results appeared as ranges of efficiencies and electricity costs for each system. It is perhaps not surprising that the estimates from the two contractors were substantially different, especially since they were free to use their own estimates of the parametric limits (in effect, "guessing" how far R and D would move currently practical limits).

The basic range in the ECAS programme was called "Advanced Steam", embracing pf and both AFB and PFB. Within this group, one contractor estimated the cost of electricity to be 30–38 mills/kWh, with energy efficiencies of 34–40%. (The other contractor arrived at 21–35 mills and 34–43%: these differences, although perhaps surprisingly large, are explicable to some extent and are not especially relevant in the present context.) Compared with these "Advanced Steam" cases, were combined cycles, open and closed gas turbines (with and without "Bottoming"), various types of MHD and fuel cells and systems using supercritical

CO_2 and liquid metals. For the gas turbine and combined cycle cases, "clean" fuels (liquids or gases derived from coal) were used, except in the closed cycle cases, where helium was heated externally, by coal fired AFB.

Of all the systems, only the combined cycle case showed up well on cost (e.g. 23–33 mills compared with 30–38 for Advanced Steam). However, the overall efficiencies, quoted as 21–37%, were low. An important point was that integrated low-Btu gas was best both on cost and efficiency (the latter of course would be expected, provided the gasification stage can be demonstrated to operate at high efficiency). The other gas turbine systems were somewhat less attractive but technical trends which could possibly alter this (to give a small advantage over the combined cycle cases) are identified.

Systems depending on CO_2 or liquid metals as working fluids, heated in many cases by coal-fired FBC, can be neglected, on the basis of results quoted, since they appear to have little economic potential in the foreseeable future; costs are high and efficiencies are not good enough to provide the incentive for future developments. This appears to be the case also with fuel cells, both high temperature and low temperature. Three different types of MHD were studied and, here, although electricity cost estimates were fairly high, at least the potential for increased efficiencies appeared considerable (up to 53–54%), so that comparative costs might be more attractive in the long run. The type of cycle apparently favoured was the open-cycle version, described earlier.

Conclusions on Electricity Generation

On the basis of existing technology — i.e. pf coal fired boilers compared with current designs of nuclear reactors — coal remains broadly competitive in the field of electricity generation. For an individual power station, of large output intended for high load factor use, nuclear power would probably have an economic advantage, albeit relatively narrow, over coal according to a consensus of analysts, especially where local conditions favour low cost of capital and where a significant annual increase is expected in future coal prices. An individual cost study would be necessary in a particular situation — some of the factors to be taken into account have been mentioned above. There are two areas where general speculation may however be worthwhile:

(a) the balance of commitment to the two systems as an indication of future coal markets;
(b) the situations in which coal will continue to remain competitive for electricity production for the longest time.

Both the growth rate of electricity use and the size of the nuclear generating component have been much less than expected (in many quarters) a decade or two ago. Expectations now seem to have been substantially lowered and on a world scale the growth of nuclear power seems likely to be steady rather than spectacular for the rest of the century. There are many reasons for this. In some countries, political opposition has stopped the construction of nuclear stations and even in some cases the operation of completed plant. In other cases, construction has been delayed by inadequacies in design, manufacturing processes or technical data, including those related to health and safety; this has been accompanied by large cost overruns. In the USA, the utilities appear to have difficulty in raising all the capital they would like; this may favour coal burning, especially in refurbished plant.

Experience suggests that there would be economic penalties if a too-rapid build-up of nuclear plant were again attempted. First, if the nuclear plant construction industry were overheated again, costs and delays would probably recur. Second, it now appears probable that uranium prices would respond rapidly to a large increase in programmes; the cost implications of this would be considerable and would be unlikely to be effectively alleviated by a breeder programme for a long time. Third, the front-end and back-end nuclear fuel facilities would have to be expanded greatly with large new capital investment; there are funding problems and technical uncertainties. (A large part of existing fuel cycle plant was funded on military budgets.) Finally, where the economics are in any case relatively closely balanced, it simply does not seem sensible to give overwhelming preference to one alternative. Thus, it appears likely that both coal and nuclear power plants will continue to be built. Against a more modest nuclear programme, the growth in electricity demand is a large uncertainty in predicting the demand for coal for central power stations. However, even with the relatively slow growth rate of electricity established over the last few years, it must be doubtful whether nuclear growth will provide all the new plant margin and it seems probable that there will continue to be a steady growth in coal burnt in large power stations.

After the next ten to twenty years or so, the balance may change as a result of technical progress. Nuclear technology may well have matured so that it becomes possible for standard plants to be built to reliable cost and performance estimates and the whole fuel cycle may be brought under close control. It seems unlikely that these ends will be achieved except at economic levels near or above current forecasts. On the other hand, it appears probable that both FBC and FGD will have become established or improved and be contributing to some modest reduction in coal electricity costs. Other coal technology for electricity appears too remote or speculative to affect broad estimates of coal competitiveness for at least two or three decades, with the exception of gasification/combustion cycles, which may possibly be similar in cost to FBC. Thus, it would certainly be premature to eliminate coal from close contention in this field and this message now seems widely accepted.

Whatever the future role of coal in larger central power stations, coal is likely to be extremely important in stations of smaller size or lower load factor, especially where cogeneration or district heating schemes are concerned; operational developments in these fields, using new technology, would be very desirable. (The following section on industrial boilers is relevant in this context.) The generation of electricity as a by-product in coalplexes may also be advantageous and requires further and sympathetic consideration with regard to system acceptability.

No direct reference has been made to the possible competition between coal and "regenerable" energy resources. Economically, the latter do not appear promising and, in any case, they appear rather to be alternatives to nuclear power than to coal. In the event that substantial amounts of "regenerable" electricity capacity were installed, there would probably be substantial mismatch between the demand and production cycles, the latter depending on extraneous factors; generation via coal might provide a means of adjustment, albeit at some extra unit cost because of an unfavourable load factor.

INDUSTRIAL COMBUSTION

Perhaps the most promising area in which coal may increase its market share in the near future is in the industrial combustion field, where direct heat, steam or hot water are required, possibly in conjunction with electricity production. This is of course a very wide field, with individual units of very different sizes and duties; the economics are currently heavily dependent on local circumstances and fuel prices. Oil and gas are in competition with coal in this field. In the longer run, the availability of coal may be more important than cost differences seen in present circumstances. Another imponderable at the moment is that this market is one where fluidised combustion is likely to penetrate most quickly and thoroughly but it will be a few years before this is established and its economic impact measured.

Since generalised economic studies may not be very applicable in such a diverse field, some studies of the competitiveness of coal fired boilers carried out in the UK over the past few years may be useful examples. Starting in 1976 these were concerned with four basic and representative cases:

(A) Sectional boiler producing hot water, primarily for space heating, with a total rating of 6 million Btu/h, for a school, say.
(B) Shell boiler producing steam at 480°F and 145 psig. Total rating 37 500 lb/h of steam, suitable for heating and process steam in a small factory.
(C) Shell boiler as (B) but rated at 150 000 lb/h.
(D) Water tube boiler producing 300 000 lb/h of steam at 700°F and 600 psig.

At this time in the UK, oil had the major share of these markets. In order to provide a basis, costs calculated for oil boilers for each of the above cases, assuming a ten year life, are given in Table 13.5.

To calculate the competitiveness of coal, the following assumptions were made:

(1) coal fired boilers included stoking and ash removal equipment;
(2) coal fired boilers had the same efficiency as oil;
(3) the costs of flue ducting, electrics and civils (including chimney and storage) were 20% greater than for oil;
(4) labour and maintenance for coal were twice that for oil.

Calculations were made of the relationships between the overall steam costs of coal and oil as a function of (a) the relative cost per therm of oil and coal and (b) the relative capital cost. Figure 13.7 (a–d) shows these relationships for cases of where the overall cost for coal is required to be only 90% of that for oil (i.e. the latter has an amenity value of 10%). In case A, for example, if coal capital costs were 40% more than oil, coal would have to be 23% cheaper per therm than oil, to meet these

TABLE 13.5 Boiler costs

	Capital cost £	*Total cost p/useful therm (including capital charge)*
A	12 900.00	26.6
B	86 600.00	21.7
C	264 000.00	20.8
D	2 600 000.00	19.7

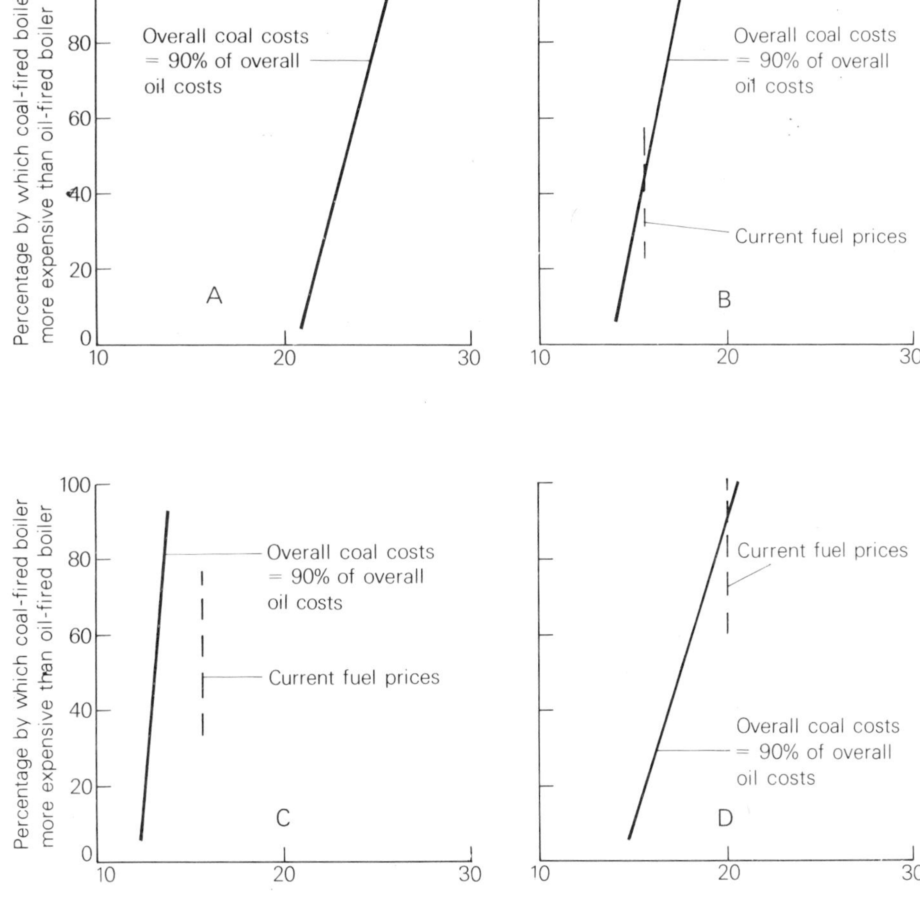

13.7 Percentage by which coal is cheaper than oil. A: Sectional boilers; hot water (Rating: 6 MBtu/h). B: Shell boilers; steam (Rating: 37 500 lb/h). C: Shell boilers; steam (Rating: 150 000 lb/h). D: Water tube boilers; steam (Rating: 300 000 lb/h). (Horizontal axis represents percentage by which coal fuel costs assumed cheaper than oil; vertical axis — capital cost)

conditions. At the time the graphs were drawn, the difference (on that scale) was 12%. For case B, at 40% excess capital, the required margin on coal price was 15.5% and this was the actual case. For case C, where the actual fuel price difference was then 20%, this permitted an excess capital cost of 90%.

The steepness of all the curves shows that relative fuel prices are dominant; the actual differences reflect commercial policies (and probably tax). At the time the original calculations were done, the weakest competitive area for coal was the small sectional boiler; competitiveness decreased slightly again when the water tube sizes were reached. Each of these areas has benefitted from subsequent research, including FBC.

At the same time, the fuel cost advantage to coal in the UK was approximately 40%. In some other countries a wider differential existed. The advantage to coal is increasing and well before the end of the century a 2:1 ratio in favour of coal may be normal; this would completely swamp any credible differences in capital costs. However, the availability of capital to industry may slow down the process of conversion unless positively assisted by governments.

For steam raising, oil is the appropriate marker. Natural gas prices are likely to be linked with gas oil, but commanding a premium; SNG would be grossly uneconomic compared with coal (probably about $3\frac{1}{2}$ times more costly), and electricity even more so. For direct heating, however, the picture is more variable. For cement manufacture, for example, where the ash is compatible with the product, coal is undoubtedly the economic choice. Coal will also be attractive, subject to development, in fluid beds where products can be directly heated or where fluids can be heated in tubes or retorts, indirectly heated by coal. For hot gases for drying, also, coal fluidised beds are already becoming well established (in some cases the ash can even be added to the product with advantage). In some cases (for example, in the heat treatment of bright metal parts) special attention has to be paid to the internal furnace atmosphere and in these cases gas from coal may be preferable, in spite of its higher thermal cost; low or medium Btu gas will be more attractive than SNG in these fields.

DOMESTIC COMBUSTION

Preferences for domestic heating have frequently been expressed in terms of amenity and environment. Coal burning has generally been regarded as hostile to the environment and this was undoubtedly the case at one time; this led to the virtual abandonment of solid fuel domestic heating in some European countries. In the UK, following severe smog damage and the subsequent Clean Air Acts, the environmental problems have been overcome without destroying the market and this may be the case in a number of other countries, especially outside the OECD. In the UK, the use of smokeless fuels (natural and manufactured) or appliances which burn coal smokelessly have been successfully introduced. On the other hand, it is clear that amenity considerations of appearance and convenience often favour solid fuels, both for appearance and for security of supply.

There can be, at the same time, a strong economic argument for coal especially as the efficiency of coal burning appliances has been improved. One calculation of annual costs for different fuels in several cities in the UK in November 1978, is given in Table 13.6.

It will be seen that there are substantial differences according to the location and system used (and also according to who is doing the calculations!). It seems important however for each country where the opportunity still exists to retain coal in the domestic sector to take a fresh look at the economics, taking account of modern coal appliances and also the need for other energy sources in other sectors. In particular, the use of electric resistance heating in the domestic sector would appear most wasteful and a retrograde step in developing countries where industrial development depends on access to limited supplies of oil and electricity. It is appreciated that the convenience of electricity may appear attractive but this may not be so if electricity supplies cannot be guaranteed.

TABLE 13.6 Annual home heating costs (500 useful therms) £'s in UK

	Solid fuel (room heaters)			
	Coal	*Anthracite*	*Gas*	*Electricity*
Appliance efficiency	75%	75%	65%	100%
Leeds	95	127–142	144	234
London	130	137–162	151	250
Cardiff	109	126–131	151	243
Birmingham	113	136–153	142	226
Edinburgh	117	138–157	151	196

COAL–OIL MIXTURES

The elements in the economic analysis of COM are mainly straightforward:

(a) the prices of oil and coal;
(b) the capital and operating costs of the COM preparation plant;
(c) the costs (mainly capital) of modifications to the boiler and associated equipment (or extra plant in the case of new installations) including environmental control requirements.

Any penalties for derating or loss of efficiency as a result of conversion to COM may be more difficult to deal with.

The economics of COM were examined in several different ways by speakers (all from the USA) at an International Conference in Florida in 1978. Perhaps the most categorical and significant statements related to the retrofitting of existing oil fired boilers; for new boilers, it would always be preferable to design for coal firing (although it was not clear whether this would apply to a new boiler in an existing power station).

One analysis concerned industrial boilers (100 000 and 500 000 lb/h; 30 and 150 MW). For the larger boiler, on-site COM preparation was considered, whilst a centralised COM plant was assumed for the smaller boiler; the boiler conversion costs were \$3.4 million and \$0.3 million, respectively. The authors calculated the fuel savings necessary to cover the additional costs of using COM, at various DCF rates, and compared these with projected savings. They claimed an accuracy of only ±20%. In the case of the large boiler, there was a positive gap (i.e. favourable to COM retrofitting) at all but the highest DCF rate (20%). For the smaller boiler the fuel savings were generally not quite sufficient to justify retrofitting but all cases fell within the margin of error. The authors, however, then compared these modest savings for COM with the cost of retrofitting to burn coal. Capital costs in the latter cases would be \$24.5 million and \$5.7 million (compared with \$3.4 million and \$0.3 million) and the overall saving in favour of COM would be \$0.6/10^6 Btu (\$2.7 million/a for the 500 000 lb/h boiler). In spite of this substantial advantage for COM, it was thought that the decision in such a case would depend on the availability of oil and possibly governmental direction, unless substantial differential price escalation for oil and coal in favour of the latter took place.

This differential was a key element in another study; annual rates of escalation (including inflation) of 6.0% and 6.9% were taken for coal and oil respectively. Three boiler sizes were considered: 100, 200 and 400 MW. Conversion to COM from oil was considered likely to be profitable in all cases, the largest plant size being most profitable. It should be noted that investments of $90.9/kW and $141.8/kW were required. The results were presented as the required initial differential between oil and coal prices to "break-even". For the 400 MW plant, with the "base" differential escalation, an initial difference of $0.50/10^6 Btu (oil over coal) was necessary; for the 100 MW plant, on the minimum escalation differential, an initial difference of $0.98/10^6 Btu was needed. The paper gives fuel prices in 1977, from which a differential of between $1.09 and $0.30 can be deduced. Thus, it appears that any investor would have to consider his local coal and oil prices and take a view about differential escalations. Nonetheless, the balance seems definitely in favour of COM according to this analysis and since the number of Btus burned in a power station is large, even a small saving per unit is very profitable.

A further study considered three alternative policies for a 275 MW unit: (a) continuing to burn low-sulphur oil, (b) conversion to COM (low-sulphur oil and coal) and (c) conversion to high-sulphur coal with FGD. At 50% load factor, the respective generating costs were (a) 49.9, (b) 48.5 and (c) 49.9 mills per kW. The utility new capital for COM ($61.6 m) was considerably less than that for coal conversion ($160.7 m). The analysis assumed that coal would escalate annually at the inflation rate (5.7%) and that oil would increase at 10.0% per year.

It will be seen that a strong case can be made for COM conversions, but largely where differential price escalations (oil/coal) are assumed. The effects of environmental controls are somewhat obscure and so are deratings. Each case would therefore require detailed examination. There also appears to be a tendency to use a higher coal concentration in COM in the economic studies than appears to be indicated in the technical reports.

GENERAL CONCLUSIONS

The combustion of coal will remain an economically attractive option for a long time. In particular, coal will retain at least a substantial proportion of its present most important market, electricity, in the face of competition from nuclear power, the most likely alternative in the medium and long term. It is considered possible that nuclear power economics may prove to be a little less favourable than predicted and its growth may be restricted for this and other reasons. Fluidised combustion (and possibly gasification/combustion) on the other hand may improve coal's economic position, though not quickly and spectacularly. The combination of these trends could result in a very strong and growing demand for coal in electricity generation for a long time. Renewable energy sources will have at most a minor influence on coal's position although it may be that these sources will require low load factor coal stations as a back-up. More advanced coal systems are generally of doubtful economic merit; MHD perhaps has the most promise of these outsiders. These comparisons are considered further in Chapter 22.

Prospects in the industrial combustion field are exceptionally promising. In many parts of the world coal should also continue to be an important fuel for domestic purposes; this may require demonstration and education. COM is an interesting possibility requiring further study.

Chapter 14

ECONOMICS OF CARBONISATION

INTRODUCTION

The costs of making coke are dominated by input coal costs. One German source gives, purely as an example, the following cost structure for coke production in existing ovens (Table 14.1).

The author adds that in some circumstances the proportion of costs due to coal could be as much as 88%. This overwhelming importance of coal costs is fairly general although the degree of dominance will vary somewhat. In some places coal is relatively cheap but this may sometimes be countered by transport costs or by the fact that only marginally acceptable coke is produced, thus reducing the value in use. The coincidence of coal of high quality and cheapness is rare and will probably disappear, especially if coals are charged into the coking plant at market prices rather than production costs.

At one time the proceeds from the non-coke products were nearly as great as those from the coke but recently have been only one third as much or even less. In addition substantial costs are involved in recovering and marketing these by-products so that this has become a debatable practice in some circumstances.

The value of different cokes is also difficult to assess and may not be accurately reflected in coke prices determined by market forces. The tendency of course is for blast furnace operators to seek the highest quality possible or at least the sort of coke successfully used in the past. Consistency is highly prized. Value is also influenced by location; coke is bulky and somewhat friable so that transport can be costly both in tonnage terms and in loss of value.

GENERAL ECONOMIC FACTORS

Coking Coal Qualities and Price Factors

The rapid expansion of the steel industry, especially through large new plants having access to deep water harbours, has created pressures on coking coal supplies. This has led to the widening of price differentials between coking and non-coking coals and between different qualities of coking coals. The incentives to widen the range of coals acceptable to coking coal blends (and also, of course, to processes for formed coke) are therefore greater. However, as many of the marginal coals are high in volatile matter, the coke yield is lower per ton of input coal; this is not normally compensated to any great extent by by-product realisations. Also, blends require careful preparation; this applies not only to the final mixing but also to the homogenisation of the individual components. These procedures may require extra

TABLE 14.1 Cost breakdown for coke production

Input coal	70%
Capital cost	9%
Underfiring gas	6%
Maintenance	3%
Wages and overheads	4%
Other energy and materials	8%
	100%

stocking facilities and possibly exceptional grinding equipment, since good mixing of different sized materials is difficult.

Pre-heating may also involve extra capital and operating cost factors, offsetting the advantage of lower cost coals. There may be offsetting savings from pre-heating due to environmental control improvements and also as a result of increased throughput; this latter factor may not be realisable on existing ovens because of constraints elsewhere in the production chain. However, in a new installation having access to indigenous coking coals of moderate quality at a good price differential over imported prime coking coals, pre-heating will be an option which certainly should be evaluated.

Impurities have traditionally been specified at low levels in coke and, since the main effect of carbonisation on most impurities except sulphur is to concentrate them, a high quality coal is usually required. However, this is again an economic balance. Coal cleaning processes become more expensive and less productive of final product as the degree of cleaning increases. The blast furnace can tolerate more impurities than the technical ideal but, again, at extra cost. For certain purposes, however, there are absolute limits, especially for phosphorus. A modern coking installation should ideally have considerable flexibility with regard to coal feedstocks; this implies good facilities for stocking, homogenisation and blending.

Capital Charges

The capital cost of new heavy equipment such as coke ovens has tended to increase rather more rapidly than general inflation and so have interest rates. Capital charges of new plant may therefore be substantially higher as a proportion of total costs than the 9% quoted above; certainly the money equivalent per ton of coke will be high compared with older plant, especially if the latter has been substantially depreciated. Of course, even old plant may require new equipment, especially for dealing with new coal blends or with pollution. Coke ovens are constructed substantially of refractory bricks and are susceptible to damage, especially with irregular operation. Depreciation rates reflect the limited life expectancy, generally about 20 years. However, considerable progress has been made both in restricting wear and damage by means of good maintenance procedures and also by devising techniques for repairing operating ovens or providing extra strengthening. Even where the ovens themselves require complete rebuilding, the supporting plant and the site facilities may make this an option considerably cheaper than construction on a green-field site. By-product plants, however, also tend to have limited lives, at least in their more vulnerable parts.

The location of coke ovens in relation to consuming industries may require assessment on economic grounds. There are of course advantages in integrated operations entirely within the steelworks plant but this will need to be weighed against the advantages and disadvantages of using external coke sources from existing coke ovens, which are usually located on coal fields. Since steel demand is cyclical, the optimum may sometimes be integrated production of the minimum coke demand on the steelworks site, together with imports of requirements for peak demand. There are in fact indications that trade in blast furnace coke is becoming more firmly established. This could perhaps pave the way for a similar trade in briquettes, where production integrated into steelmaking may have no particularly strong advantage; also degradation during transport may be considerably lower.

Other Operating Costs

Wages represent only a small proportion of total costs. This has probably contributed to the relative slowness in the introduction of mechanised or automated methods for carrying out some of the labour intensive tasks. However, more recently, the rate of introduction of such devices has increased, partly because the unpleasant nature of some tasks has made labour recruitment difficult and unreliable and partly because some of the new equipment for environmental control and quality improvement lends itself to automation and in turn demands a high standard of consistency elsewhere in the system. Automatic door cleaning and sealing are typical examples. The additional equipment itself requires operators and maintenance of course; labour costs as a proportion of coke costs are unlikely to change much but total costs and production value will continue to be very dependent on the skill and dedication of a labour force operating twenty-four hours every day of the year; such service is more readily obtained in a good working environment. Energy management will be another key economic factor, both in total energy conservation and in the flexibility with which the type of energy used is adjusted with market changes in order to minimise costs, since coke ovens produce several different forms of fuel or energy, including electricity in some cases.

Formed Coke

An initial incentive towards formed coke was that it might be a cheaper route to blast furnace coke, on a ton for ton basis. It was especially suggested that the use of cheaper coals and more obvious and simpler environmental control would be factors in this competitiveness. These factors have been eroded to some extent by the progress achieved in these same areas with slot ovens but may still retain some credibility, subject of course to full scale demonstration over a period.

Capital costs are sometimes claimed to be lower for formed coke, perhaps by about 20%. However, there seems to be some doubt about this, especially if hard, fully carbonised briquettes are required. It should be remembered of course that the specific production plant is itself only a part of the site capital cost. In Russia, where briquetting has been established for a long time (though with varying reports of its current status), capital charges have been stated to be considerably higher for formed coke plants than for normal coke ovens although the total cost of briquettes is reported as only about 80% of that of oven coke. Such a reduction, if substantiated, must be based largely on differential coal prices which would probably not have universal application. The value of the briquettes in use would also need to be taken into account. Each case would need to be calculated separately, the main difficulty being the acquisition of reliable cost data from full-scale briquetting plants.

Even more uncertain is the value is use, in large modern blast furnaces, of briquettes compared with first-class oven coke, in full-scale blast furnaces.

By-products

For many years carbonisation by-products were the mainstay of certain essential downstream industries; prices for coke oven crudes reflected prices that could be obtained for the refined and derived products, which were often scarce and therefore of high value. In more recent years, and especially from the 1960s, cheap oil provided these intermediates and final products; petroleum processes also came to be operated on a larger and more economical scale. The carbonisation by-products industry made various moves to rationalise production, through centralised tar and benzole distribution plants but this was largely negated through the disappearance of a large part of the feedstocks as a result of the closure of gas-making retorts.

In the best years of the carbonisation by-product industry, crude tar, crude benzole and ammonium sulphate were each many times more valuable on a ton-for-ton basis than coal; by 1971 in the UK all three derived materials were approximately the same price per ton as coal (and all held-down by crude oil prices in the same range).

This price depression for by-products had a number of effects. One was that coke prices, no longer supported by high by-product realisations, had fully to reflect coal prices (themselves strongly increasing); this resulted in the loss of some coke markets and imposed strict economies in others. Another was that by-product recovery was practised much less. Instead of recovering ammonia, new plant units were sometimes installed simply to destroy it. Benzole was often allowed to "slip" into gas streams to be burnt as fuel. Crude tars were also frequently burnt as a substitute for heavy oil.

Immediately following the 1973/4 oil crisis, realisations for by-products improved somewhat, as shown in the UK figures (Table 14.2).

Later in 1974, especially in the case of benzole, prices of by-products rose much higher but subsequently the reduction in real crude oil prices depressed them again. The 1979/1980 oil problems again reversed the trends. It is probable that prices for coking by-products will remain erratic, but generally on an upward trend, so long as potential supplies from crude oil are quantitatively much greater than from carbonisation. However, it seems unlikely that non-recovery of by-products will in

TABLE 14.2 Coal by-product prices per ton (1971 and 1974)

	1971 £	*mid-1974* £	*Ratio 1971/4*
Coal	7.30	16.50	2.2:1
Crude tar	6.10	23.00	3.8:1
Crude benzole	8.70	26.50	3.0:1
Ammonium sulphate	6.50	20.00	4.6:1

future be economically preferable for any substantial period. Indeed, there will probably be increasing scope for more sophisticated methods of recovery, separation and conversion techniques as real prices of crude oil and natural gas increase. Management will have opportunities to adjust energy and by-products policies according to market changes. Although profitability on these by-product operations may be improved, the close linkage between coal and coke prices will remain dominant.

Value in Use

This subject has been mentioned elsewhere but remains a most important and (so far) rather imponderable factor. Modern coking techniques would almost certainly allow the development of cokes with somewhat different, or possibly "better", characteristics if only the specific directions of required improvements could be clearly and quantitatively designated; this may become possible. However, since cost-effective practices have already generally been introduced in coking, any changes in coke specification, including closer quality control, are practically certain to increase costs. It is the extent to which these costs would be recoverable in blast furnace operations that is the more uncertain aspect in cost/benefit analysis. However, as coke prices increase, partly as a result of quality demands, the incentive to reduce coke rates in blast furnaces will increase still further. There is a limit below which the concentration of coke in the blast furnace cannot reasonably go without drastic changes in the permeability and working conditions within the unit; theoretically, there may be scope for a further 20–25% reduction compared with best current practice. Reductions in coke rate however will require the input of additional energy via the blast (heat and/or oxygen enrichment) and by the injection of fuels (e.g. oil) through the tuyeres except to the extent that better recovery of energy from waste products can be achieved. Oil injection was once quite frequently practised but has become less attractive because of increasing costs. Both heating and reducing gases can be produced from coal, concurrently with coke and char. Combined processes, of which the US Steel Clean Coke Process is an example, (see Chapter 20) may well be attractive in the future, especially if products in addition to iron are made simultaneously, using coal derivatives.

A further possibility is that more than one kind of coke may be desirable in order to serve better the various functions currently carried out by coke. In this case, hard unreactive briquettes, of ideal size and shape, would be one likely component specifically for maintaining permeability. Such briquettes would, as pointed out earlier, be an attractive product for international trade.

Foundry coke will remain expensive compared with other cokes, since the best coking coals and lowest impurities are required, and longer carbonisation times are needed. Large sizes are required for foundry use, so losses are incurred through the undersize created either at the oven or in subsequent handling. Electric furnaces are an alternative to cupolas for making cast iron, which is itself in strong competition with other materials. Some slow but steady decline in demand for this type of coke may be expected.

The manufacture of coke to be burnt simply for heat production also seems to have a limited future, although existing specialist installations will probably survive to the end of their natural lifetimes. At the same time, the incentive for making fullest possible use of undersize portions from blast furnace and foundry coke production will be enhanced.

Recent Cost Factors

Coke Oven Capital Costs

Capital cost data (October 1980) for a green-field site coking installation producing about a million tons of coke a year have been provided by a leading contractor.* The data are stated to be "budget" costs which are intended to provide guidance in preliminary assessment of possible investment rather than "hard" tender prices. General data of this kind must of course be treated with caution since costs will be sensitive to site conditions, coal quality, coke specification and other factors. In this case, the following plant performance was assumed (Table 14.3).

The capital cost estimates are given in Table 14.4.

Based on these capital costs, the capital charges per ton of coke from a new plant would be approaching £20. This is a much higher figure than implied in Table 14.1 for an existing plant.

TABLE 14.3 Nominal coking plant performance

Coal throughput, 7% moisture	4200 t/d
Coke output (dry wharf coal, depending on coal VM)	2900–3300 t/d
Gas output	54 000–58 000 Nm^3/h

TABLE 14.4 Breakdown of capital cost estimates for coking plant processing 4200 t/d dry coal (in £ million)

1. Coal handling, including reception, stocking and reclaiming, blending and delivery to oven bins	12.0
2. Two batteries of 53 × 6.2 tall ovens (Underjet, twin collecting mains, rich gas/lean gas firing, coal bin and wet quenching station)	38.0
3. Ovens machines (two sets)	6.5
4. Coke handling and screening	4.0
5. Coke side pollution control	4.5
6. By-product recovery	17.0
7. H_2S removal including treatment of H_2S plant effluent	4.0
8. Other liquid effluent biological treatment	1.0
TOTAL	87.0

Note: Costs are for an erected plant, including civil works. Not included are piling, office/amenity blocks, workshops and steam or electrical generation plant.

Coke Oven Operating Costs

In the UK, in mid-1979, the bare production cost associated with making a ton of blast furnace coke, in a nearly fully written down plant (i.e. negligible capital charges), in a case believed to be typical, was about £90. By-product revenue was equivalent to about £12 for each ton of specification blast furnace coke, making a net cost for the coke of about £78 per ton.

Of the total costs, 65.6% were for coal. The only other important costs were wages-related items (including overheads), 19.4%, and energy, 6.7%.

Of the revenue from products other than prime coke, gas accounted for just over half; tar and benzole provided just over a quarter; the remainder was from fine solid products (coke breeze).

* Babcock Woodall-Duckham Ltd.

Some interesting observations may be made on the basis of these figures, although it has to be remembered that relative cost factors may be different in different places. The first is that capital charges may be quite large, up to 20–25% of the total costs; this emphasises the advantage of keeping existing ovens in service wherever possible. There may, of course, be potential advantages in favour of new integrated ovens, such as coke transport costs and degradation. The balance of these factors may be worth serious assessment rather than an automatic decision for completely new integrated ovens to cover maximum demand *ab initio* for a new steel works.

Another point is that the ratio of cost on an energy basis of the desired main product to that of the input coal is about 2 and is probably significantly lower than for any other coal conversion process. This seems logical since coking is a relatively simple process — and a well-established and developed one. Yet a brief consideration of the simple numbers given above suggests that there is little scope for significant reduction in the cost ratio, product/coke; more detailed examination will support this suggestion. It follows that more complex processes are likely to have higher basic cost ratios; this emphasises the need to obtain the highest possible yield of valuable products.

In the cost considered above, the value of the by-products represents about 16% of the net production cost of the coke. Most of these by-products may in time become more valuable compared with coal and coke, as crude oil and natural gas become more scarce. If the assumption proves correct, it will be increasingly important to maximise the recovery of by-products and to ensure they are worked up in the most value-effective way. By this means, it may be possible to constrain to a modest extent the increases in coke prices which would probably otherwise occur. This cost containment, whilst small, could be a significant factor in both the competition between oven coke and solid reductants made by other routes, on the one hand, and between the blast furnace route and other iron-producing processes on the other.

Operating Costs of Briquette Manufacture

Hard cost data for briquettes suitable for metallurgical use are not available; there are few plants operating regularly on an industrial scale and information tends in any case to be proprietary or projected (generally for promotional purposes).

Some cost estimates for the manufacture of Phurnacite briquettes may provide some perspective. Although these briquettes are used as a domestic fuel they are quite hard; for metallurgical purposes, however, a rather harder product would be necessary. This would require higher carbonisation temperatures, at somewhat increased cost.

For the same period as for the oven coke cost data quoted (mid-1979), the net cost of production was about £70/ton (again neglecting capital cost charges). About £50/ton was due to raw materials, of which about 80% was coal and 20% was for other raw materials (pitch as binder, recycled fines, etc.). Coal costs at the Phurnacite Plant were probably a minimum since the raw material is in the form of anthracite fines, which require no devolatilisation (and therefore no loss); these fines are also cheaper than prime coking coals but are in strictly limited supply. Of costs other than raw materials, wages and salaries accounted for about 40% (8% of total costs) and energy about 20% (4% of total). Other individual cost items were small.

The relevance of these estimates to the costs of metallurgical-type briquettes is not very direct. However, on general grounds, it is not easy to see areas where operating costs would be lower for formed coke than those quoted; the reverse would in fact seem more likely. Coal prices for formed coke in general might be higher and more might be required per ton of product. The other major costs, personnel and energy, seem very low and unlikely to be reduced in a more complex formed coke plant.

For fully written-down plants, based on the above rather tenuous evidence, there may not be a very large difference between the operational costs of oven coke and formed coke. Capital charges may be the key factor. As noted above, these may add perhaps 30% to the operational costs of oven coke. It seems likely that formed coke plants might be more expensive in capital than coke ovens.

It seems obvious therefore that capital cost difference will play a key role in selecting routes and processes; in assessing capital costs, due regard should be had for the reliability of the processes studied since different levels of availability could have significant effects. Other considerations could include:

(a) Cost and availability of feedstocks, especially "cheap" locally available materials.
(b) Value of by-products. In the case of briquettes, this may involve combined processes (coalplexes).
(c) Transportation (including degradation).
(d) Flexibility (especially the ability to follow demand cycles).

CONCLUSIONS

Although carbonisation is a long-established and developed technology, economic factors for the future are complex and rather uncertain. Decision-making would therefore justify a careful and detailed economic analysis for any specific case. It is clear that the main areas for study, in order of importance, may be:

(a) Coal costs (especially opportunities to match coals and processes);
(b) Capital costs (including reliability, flexibility and relationships with other coal conversion processes);
(c) By-product utilisation.

Chapter 15

ECONOMICS OF COAL GASIFICATION

INTRODUCTION

Coal gasification is already well established commercially and of the various possible new coal conversion routes, gasification is likely to be the first large scale application. The reasons for this are partly historical, in that there has been a rapid growth in natural gas consumption over the last few decades; a deficiency of supplies to meet this still expanding demand can be foreseen in general terms but not in chronological detail. Technologically gasification is simpler and also a necessary part of liquefaction whether for synthesis or degradation processes. Already, moreover, there have been serious shortages of natural gas in the USA, although the problem seems to have abated temporarily as a result of price movements. Difficulties seem certain to arise in Western Europe within a decade. Much of the economic analysis has therefore been devoted to the cost of SNG and this is reflected in this chapter. However, the use of MBG (or synthesis gas) or of LBG for fuel or power generation also requires discussion.

Unambiguous quantification of costs or the ranking of different processes should not however be expected. The purpose of the first part of this chapter is to summarise assessments carried out in the mid-1970s in the UK by the Planning Assessment and Development Branch (PADB) of the Coal Research Establishment from published data; although the numbers have since changed and are changing particularly fast at the moment, the PADB analysis may provide a useful general background to relationships. The studies related mainly to processes then commercially available, especially Lurgi and Koppers–Totzek, with reference also to possible improvements or new processes.

Subsequently, some more recent opinions from other sources are reported, in particular estimates made at the World Petroleum Congress in September 1979 by senior officials from the USA and West Germany. These are considered the most definitive, authoritative and objective reports available. Finally, some conclusions are provided. It should be borne in mind that almost all recent economic studies on major coal gasification schemes are for general planning purposes or for the establishment of R & D priorities. Regular investment decisions, which may not be imminent, would require design and costing studies relating to specific circumstances.

PADB STUDIES

Lurgi

In its standard form the Lurgi gasifier produces a gas containing methane and the components of synthesis gas (CO and H_2), together with tars and phenols. The tars can in principle be recycled to extinction and the gas can either be converted totally to methane or its methane content can be reformed to synthesis gas. Alternatively a low CV fuel gas can be produced by air blowing. Capital cost estimates (relative to the standard Lurgi as 100) and efficiencies for the different products are given in Table 15.1.

In 1976 the investment for a standard Lurgi plant (based on a plant size of 250 $\times 10^6$ scfd) was probably about £77 per annual tonne of coal input. Annual operating and maintenance costs were taken as 7% of capital costs. On this basis and using a 10% DCF return and a 25 year plant life, gas costs for the different type of output were calculated as a function of coal price (Fig. 15.1); although price levels have changed, the relationships are probably relatively little affected since coal prices and capital costs have escalated in roughly parallel fashion. Figure 15.1 shows costs at maximum availability; Fig. 15.2 illustrates the effects of lower load factors.

Koppers–Totzek

Data used in this study are given in Table 15.2 and gas costs compared with Lurgi are given in Fig. 15.3. It will be seen (on these data) that the K–T process is more expensive than Lurgi for SNG but is preferable for synthesis gas manufacture, except at high coal prices. These results reflect the methane content in Lurgi gas.

Improved SNG Processes

Considerable attention has been directed towards the development of new gasification processes, particularly in the USA. The economic advantages anticipated are related to:

(a) Increased tolerance to coal characteristics. Many of the processes employ entrained phase systems or fluidised beds, avoiding the limitations of fixed bed systems;
(b) Improved yield of methane in the gasifier;
(c) Improved thermal efficiency;
(d) Lower capital cost.

TABLE 15.1 Lurgi costs and efficiencies

Case	*Product*	*Efficiency*	*Capital cost (per unit coal input) Relative to Lurgi as 100*
(a)	SNG } Tar }	56 } 13 } 69%	100
(b)	SNG only	64%	110
(c)	Syngas	70%	105
(d)	Clean fuel gas	79%	34

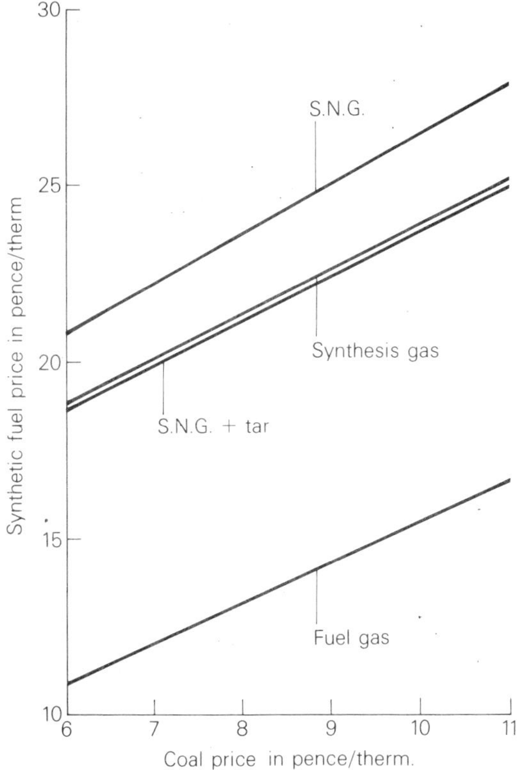

15.1 Synthetic fuel prices for Lurgi gasification process

Savings in the following main areas may be possible:

(1) Elimination of the oxygen plant (e.g. COGAS);

(2) Minimising methanation requirements by encouraging hydrogenation in the gasifier to obtain a high direct yield of methane (e.g. HYGAS);

(3) Improving gasifier throughput (e.g. slagging Lurgi and fluidised systems).

The scope for potential improvements in different areas of the process can be assessed from the breakdown of capital costs for the standard Lurgi SNG process given in Table 15.3.

Relative costs and efficiencies claimed for some of the processes under development ("second generation") are reproduced in Table 15.4.

As can be seen from Table 15.4, little improvement in the overall efficiency (65% for the "Typical" process) appears to be anticipated. However, capital investment savings of 15% to 30% (20% in the "typical" case) are expected, although it must be recognised that these estimates are based on relatively small scale work.

A comparison between the cost of SNG production using the "typical" new process and the Lurgi is illustrated in Fig. 15.4.

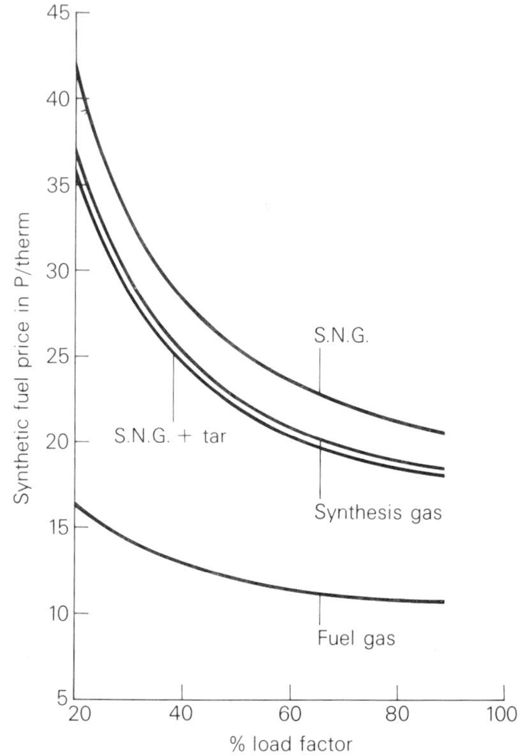

15.2 Synthetic fuel prices for Lurgi gasification process

Power Generation

The production of LBG for power generation enables improved electricity generating efficiencies to be obtained by making combined cycle operation possible. Unless the efficiency with which the gas is generated is high, however, the improved electricity cycle efficiency will be insufficient to compensate for the high efficiency of direct combustion which can be obtained in a conventional system.

An important technical factor affecting the efficiency of a gasification cycle is the loss of latent and sensible heat arising from:

(a) The generation of steam for use in the gasifier;
(b) Sensible heat losses if the gas is cleaned by cooling and scrubbing.

TABLE 15.2 Koppers–Totzek costs and efficiencies

Case	*Product*	*Efficiency*	*Capital cost (per unit coal input) Relative to Lurgi as 100*
(a)	Synthesis gas	56%	70
(b)	SNG	50%	95

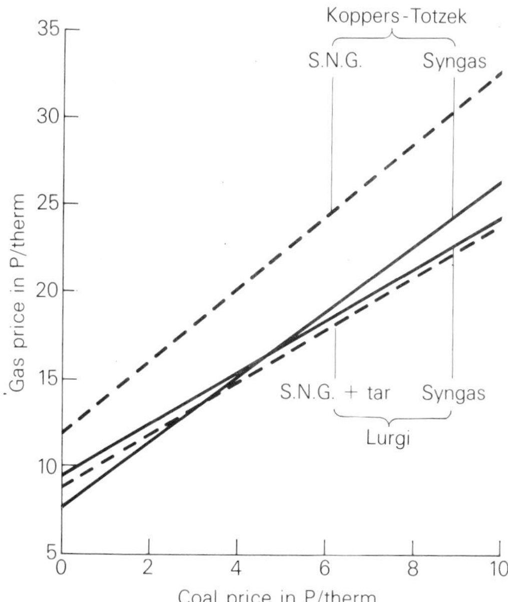

15.3 Cost of gas from Koppers-Totzek process compared with SNG from Lurgi process

TABLE 15.3 Cost breakdown of the Lurgi process, per cent

Coal preparation	4
Gasification	18
Oxygen plant	19
Shift purification } S removal }	25
Methanation and compression	12
Utilities and general facilities	22
	100

TABLE 15.4 Cost and efficiencies of new SNG processes

Process	*Efficiency (%)*	*Capital cost (Lurgi = 100) (per unit of coal input)*
HYGAS	64	70
BIGAS	62	73
CO_2 Acceptor	66	85
Synthane	64	84
"Typical Process"	65	80

Note: The estimates in the table refer to Western USA coals

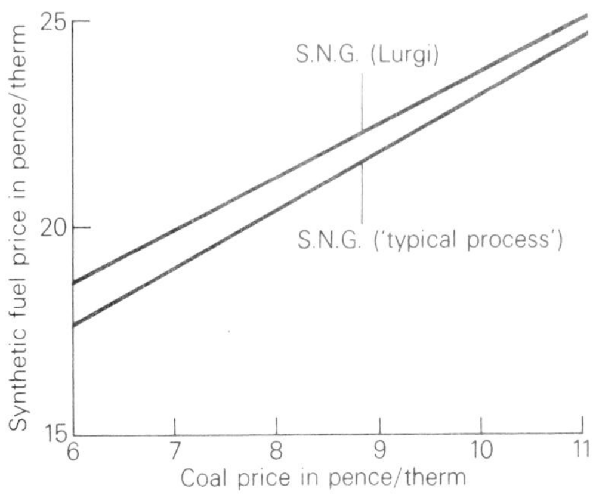

15.4 Comparative costs of SNG

The application of fluidised bed gasification technologies to fuel gas manufacture is expected to lead to lower gasification steam requirements. Several methods of hot gas cleaning are currently under development.

The extent to which the overall efficiency might be affected by combined cycles and hot clean-up is illustrated in Table 15.5.

The efficiencies for combined cycle operation quoted are based on the expectation that gas turbines suitable for continuous operation at a turbine inlet temperature of 1300°C (pressure ratio 18 to 1) will become available.

The greater complexity of gasification/combined cycle systems makes a significant reduction in capital cost per kW of electricity generated unlikely. The relative capital costs have been estimated on the basis of constant investment per kW (electrical output) compared with Lurgi SNG and are shown in Table 15.6.

Most proposals for power generation schemes using gasification have been based on air/steam, rather than oxygen/steam systems. There is little difference in overall efficiency, but the following trade-off for investment costs might provide scope:

(a) Air gasification avoids oxygen preparation;
(b) Oxygen gasification reduces gasifier and gas cleaning plant size and costs.

TABLE 15.5 Generation efficiencies (%)

		Advanced combined cycles		
	PF	*Lurgi*	*Fluidised bed gasifier*	
Temperature of clean up		Low	Low	High
Efficiency	36	41	43.5	47

TABLE 15.6 Comparative capital costs of power generation systems (Lurgi SNG = 100) (per unit of coal input)

	PF	*Lurgi*	*Fluidised bed*	
Temperature of clean up		Low	Low	High
Capital	95	108	115	124

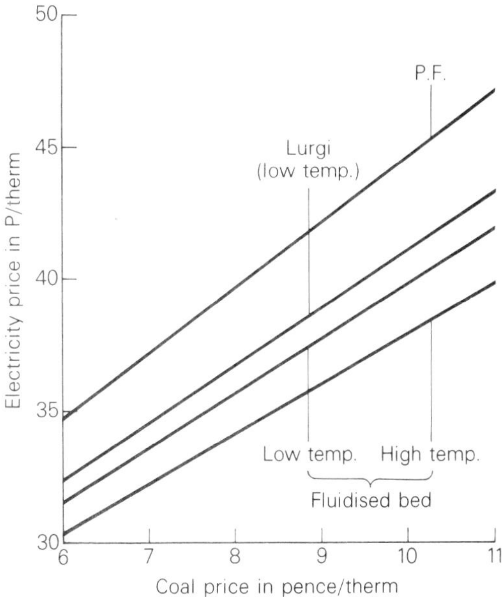

15.5 Electricity prices — Lurgi gasification versus solid fuel combustion

The balance of cost advantage is not clear but other factors, such as flexibility in operation and reliability may favour the less complex air gasification system.

The reductions in power generation costs are illustrated as a function of coal price in Fig. 15.5 and load factor in Fig. 15.6.

RECENT US DEPARTMENT OF ENERGY ECONOMIC ASSESSMENTS

In September 1979, senior officials reported on "Comparative Economics of Synthetic Hydrocarbon Sources". The capital cost of plant for SNG was estimated at about \$5000 per daily GJ output (and the same figure applied also to synthetic fuel oil). A "standard SNG plant" for planning purposes is taken as 250 Mscfd (7×10^6 m^3/d). Capital charges are assumed to be 11.5% of capital investment and input fuel is taken at \$1/GJ (\$27.50/ton of coal). Operating costs are assumed to be \$0.19/GJ of product. The results are shown in Table 15.7. It should be noted that the manufacturing costs are \$3.27/GJ but that taxes and profit (10% return on investment) are assumed to increase this to \$6.05/GJ (≡\$6.40/thousand ft^3).

In the USA, where coal is relatively cheap, emphasis is placed on the reduction of capital costs as the major factor. These authors estimated that the percentage costs of the different plant elements are as follows:

Coal handling and preparation	4%
Gasification System	10%
Oxygen and Steam Plants	29%
Gas and Effluent clean-up	37%
Shift and Methanation	9%
Balance	11%

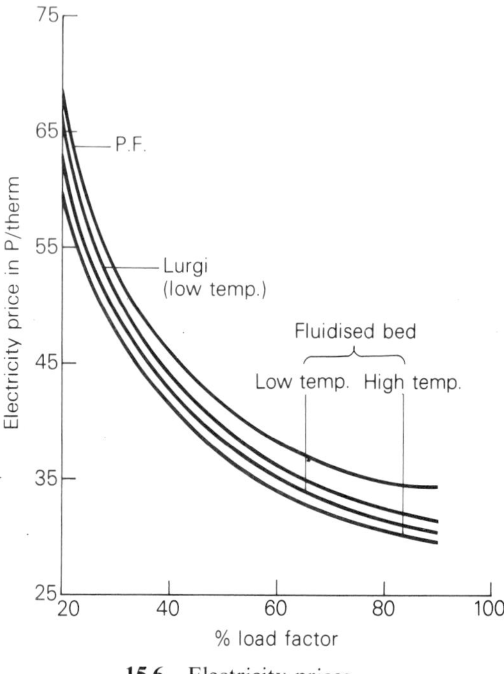

15.6 Electricity prices

These estimates may be contrasted with those in Table 15.3. It is noted that the gasification section in this recent breakdown is even less important than in Table 15.3. Reduction in the complexity of the overall system is desirable (but perhaps not easily attainable). Attention is therefore also given to methods of financing. Figure 15.7 shows comparative cost estimates, with 100% equity financing (at 1 March 1976) of several processes; although some "new" processes are estimated to be cheaper than Lurgi, some other (technically-interesting) processes appear likely to

TABLE 15.7 US basic coal conversion economics

	Million \$ annual	*\$/GJ*	*\$/tonne*
Capital charges 11.5%	144	1.60	77
Depreciation (5.0%)			
Maintenance (4.0%)			
Insurance and taxes (2.5%)			
Coal at \$27.5/tonne	125	1.42	66
Operating cost	16	0.19	9
Manufacturing cost	285	3.21	152
Federal tax	125	1.42	66
Profit at 10% after tax	125	1.42	66
Annual sales	535		
Selling price		6.05	284

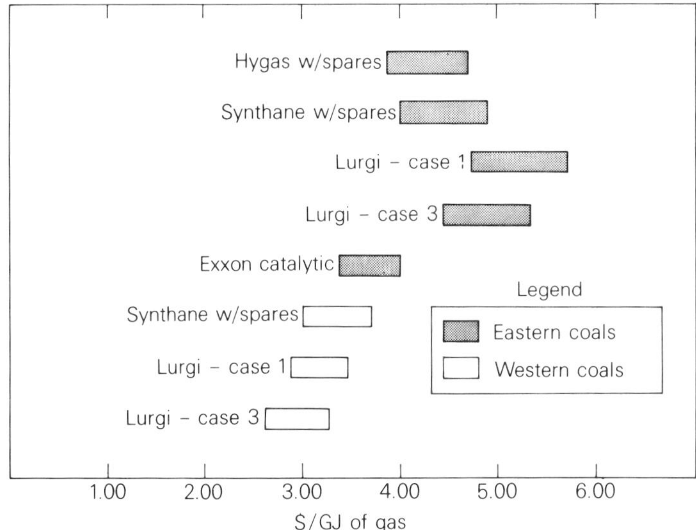

15.7 US comparison of pipeline gas prices (1978 $)

be substantially more expensive. A further representation, this time with the majority of the capital provided by low-interest debt financing, is given in Fig. 15.8. These figures are substantially lower, especially for Western US coals; however it may be that these coals are not yet in actual production, the figures representing estimates for a large new source. If the latter, there will be uncertainties about legal

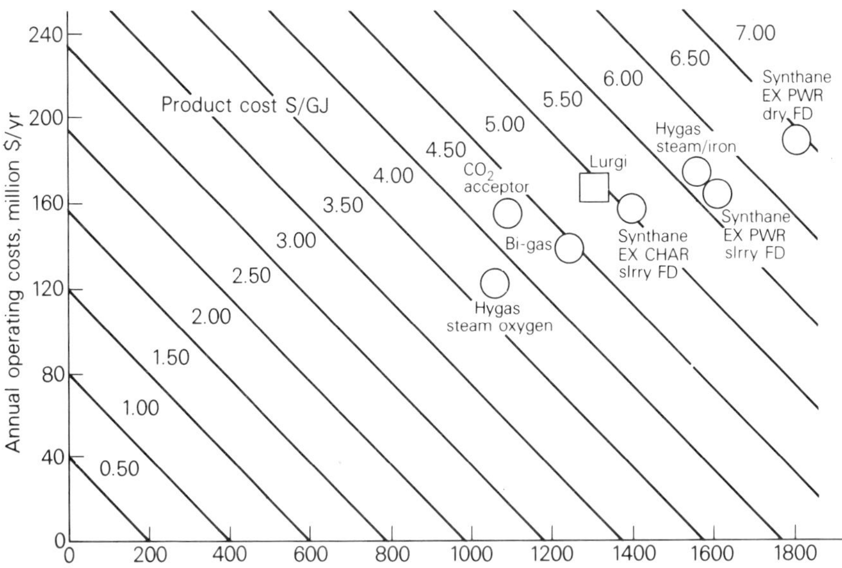

15.8 Total capital requirements, million $ (C.F. Braun estimates 1 March 1976; 100% equity, 12% DCF; western sub-bituminous coal; 71 Mm^3 gas)

TABLE 15.8 US gas cost estimates. Established Lurgi vs. Developmental Effect of Financing Conventions

	Capital *\$/Daily GJ*	*Product \$/GJ*	
		Utility	*DCF*
SNG			
Lurgi	3900	3.55	5.00
Developmental	2900–3900	2.60–3.55	3.80–5.00
LBG/MGB			
Lurgi	3100	3.10	4.50
Developmental	2400–3400	2.35–3.30	3.30–4.75

Basis: 1977 \$, Utility-65% Debt 15% Return on Equity, 10% Interest DCF-15% Rate of Return, 100% Equity

TABLE 15.9 Gasification of Eastern coals

	Capital[a]	*SNG cost \$/GJ*
Lurgi	1478	4.43
Hygas	1170	3.50
Bigas	1110	3.45
Synthane	1270	3.74

[a] Million dollars for "standard" plant (250 Mscfd)

formalities, environmental impact and infrastructure. In the case of Lurgi, case 3 represents a substantial projected up-grading of present capacities — 75% on the gasifiers and 30% on cleanup.

Another study quoted by Mills and Knudsen is given in Table 15.8. This shows again the importance attached to financing method. Also illustrated are the reductions hoped for by further development (up to about 25% reduction) and the lower cost, (up to about 12%) per energy unit, of low or medium Btu gas compared with SNG. Another study, carried out in 1978 but based on 1976 dollars with utility financing, quoted by Mills and Knudsen, is given in Table 15.9.

The official comment is that these investment costs are believed to be accurate to ±20% and that the gas costs of all second generation processes are probably only indicative rather than significant; they "might" be less costly than Lurgi.

In mid-1979, the US Department of Energy were encouraged by reports of an Exxon development, integral gasification/hydrocarbon synthesis, (IG/HS) which uses catalysis to reduce the gasification temperature and produce more methane in one step. Then, by separating the methane cryogenically, the plants for oxygen and the shift and methanation stages can be eliminated. Savings of 15–25% are suggested, with gas costs of \$6.10/GJ for equity financing or \$4.55/GJ on 70/30 debt/equity financing, all at 1978 dollars with an "instant" plant (i.e. free from inflation during construction). It is not obvious how these figures relate to a 15–25% reduction on the DOE estimates quoted earlier. Capital costs were given as \$1.8 billion but mature plants might be 20–30% cheaper. (Again these figures are difficult to reconcile with the earlier plant estimates by the same authors.)

Summing up, the late-1979 US DOE view appeared to be that debottlenecking the Lurgi would give less than 10% cost advantage. Second generation plants appeared to have a potential but unproven advantage over the basic Lurgi. The third generation IG/HS shows a potential cost advantage (15–30%).

These estimates are extremely helpful (though confusing if not accepted in the correct spirit); some further reservations should be introduced, however, especially for non-US observers. One is that location is an important factor in the USA. This is illustrated in Fig. 15.7 (Lurgi case 3 represents target throughput improvements). Although the Western coals appear to be much cheaper, there may be difficulties in exploiting them; furthermore, the product will be a long way from most markets and delivered costs may depend on the availability of existing pipe-lines. Most of all, however, it appears that a great deal of US thinking in the late 1970s was based on the belief that cheap loans will be available indefinitely. The "utility" type of financing (the more expensive) may be more characteristic of real money costs in the 1980s and beyond — unless governments intervene.

RECENT GERMAN ESTIMATES

At the same World Petroleum Congress in 1979 authoritative West German sources (Peters) suggested that synthesis gas for the production of chemicals (e.g. ammonia) would be the first liquid or gas product of coal conversion to be economic. Even then, it was emphasised that the prices of petroleum and natural gas would have to increase more rapidly than those for coal (and, by implication, for some years); SNG appeared to be regarded as being economic only on a long timescale. Table 15.10 gives the results of this estimate; the economic assumptions used are summarised in Table 15.11.

The German figures are equivalent to about $5.0/GJ if the synthesis gas were to be converted to SNG; this is slightly lower than the US standard Lurgi estimate but

TABLE 15.10 Synthesis gas costs ($/1000 m^3; German conditions 1979)

Feed	*Heavy oil*	*Natural gas*	*Bituminous coal*
Specific investment costs	136	90	175
Fuel costs	44	41	66
Capital costs	27	18	43
Other operating costs	4	5	5
By credit	—	—	—
Net product costs	75	64	114

TABLE 15.11 Economic conditions for gas cost estimates

Finance	65% debt (7%); 35% equity (9%)
Depreciation	20 years; straight line
Inflation	4%
Rate of return	7.7%
Coal costs	$94/TCE

similar to that expected for their advanced technology processes. The German report does not relate the costs to particular processes but mentions Lurgi, Koppers–Totzek and Winkler.

If allowance is made for the fact that there are costs associated with the conversion of natural gas to synthesis gas and also with the conversion of synthesis gas from coal to SNG, it can be inferred from Table 15.9 that the relative costs of methane from coal are estimated to be about three times that of natural gas. This equates with a ratio of less than two as synthesis gas illustrating the fact that synthesis gas (or medium CV gas) would be expected to become economic before SNG.

The German report goes on to say that, by the use of nuclear heat, it can be expected that the cost of synthesis gas from coal could be reduced to \$75/1000 m^3. This is an expectation which does not seem reasonable. The cost reduction is about equivalent to that which might be obtained if all the coal were converted to gas at no cost at all for heat, hydrogenation, etc. To be able to save this amount of coal, even at the equivalent cost of that coal in nuclear and ancillary plant, would itself be a remarkable advance.

OTHER SNG COST ASSESSMENTS

The Mills–Knudsen presentation at the 1979 World Petroleum Congress referred to the 1976 C.F. Braun Co. estimates. This was the most comprehensive study carried out on consistent economic bases. The information is neatly summarised in Fig. 15.8, which gives capital and operating costs, plus lines of equal products costs, at 12% DCF rate of return for several candidate processes. (The Synthane Process variants relate to dry or slurry feed and whether char or electrical power is exported.)

Based on the C. F. Braun estimates (for the Lurgi process), Leonard of Chem Systems prepared some further estimates illustrating the importance of financing method and the advantages of an early start-up. An extract from his estimates is given in Table 15.12.

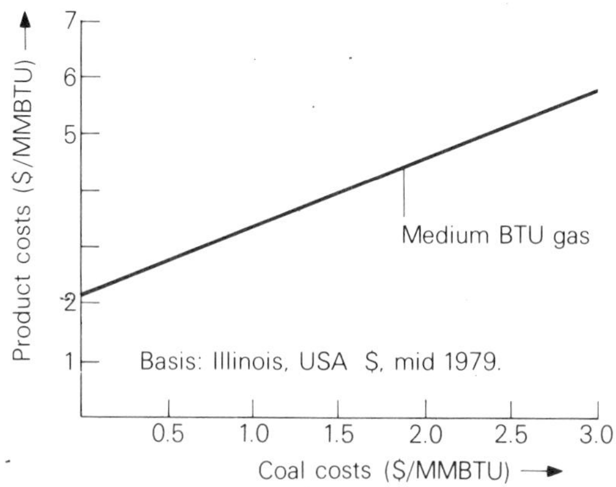

15.9 Shell-Koppers coal gasification costs

TABLE 15.12 Summary of SNG costs ($/million Btu)

	Coal ($/million Btu)	*Total capital $ G*	*Utility financing*[a]	*12% DCF*
1980 plant	0.94	1.48	4.58	6.08
1985 plant	1.20	2.08	6.25	8.36

[a] 75% debt (9%); 25% equity (15%).

TABLE 15.13 Capital and operating costs for SNG (EAS) (coal at $1/GJ)

	Total capital $\$ \times 10^6$	*Gas cost $/GJ*	
		5% DCF	*10% DCF*
Bigas	1100	3.88	4.76
Hygas	1130	4.00	4.91
Texaco	1270	4.85	5.87
Slagging Lurgi	1210	4.51	5.50
Dry-ash Lurgi	1380	4.84	5.96

The Economic Assessment Service of IEA Coal Research (EAS) carried out some tentative assessments of comparative costs for SNG, based on information from all available sources; the results, partly given in Table 15.13, are heavily qualified because of inconsistencies encountered.

Elsewhere, EAS express caution about accepting at face value potential cost savings from new processes compared with established processes; it is pointed out that advantages in critical efficiencies are relatively small. Major differences in costs exist between countries, principally due to financial conventions.

LOW AND MEDIUM BTU GAS AND SYNTHESIS GAS

Fewer cost assessments have been published for LBG and MBG than for SNG, although there is no question that the latter is more expensive to manufacture on a comparable scale. Some cost forecasts are published by the manufacturer of particular gasifiers. Figure 15.9 is an estimate by Shell (1980) of the costs of synthesis gas as a function of coal costs, using the Shell–Koppers Process (under development). Figure 15.10 is a somewhat earlier (1979) estimate by the same source, in this case using German brown coal; an SNG estimate is also included. Earlier still (1977), estimates for the Koppers–Totzek process, the precursor to Shell–Koppers, were published (Fig. 15.11); based on US lignite. In the last case, finance was by 60% debt (9%)/40% equity (12% DCF); in all cases, for cheap coal, the gas/coal thermal cost ratio is about 3.

Hargreaves and Kirk (Humphreys & Glasgow) studied economic factors in MBG manufacture (Institute of Gas Engineers, 1980). They did not predict actual costs for MBG, since they believe this can only be done satisfactorily by a full engineering study taking account of specific circumstances. They do however

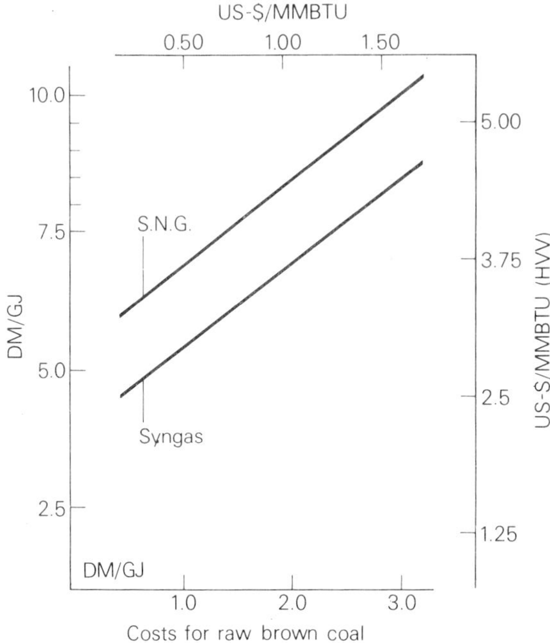

15.10 Cost of SNG + synthesis gas versus raw brown coal costs. Basis: W. Germany 1978

estimate a number of technical factors, including coal and oxygen usage for three different cases:

Case 1 — Standard Lurgi — Western US sub-bituminous coal
Case 2 — Slagging Gasifier — Eastern US, Pittsburgh No. 8 coal
Case 3 — Shell–Koppers — European bituminous coal.

Bases:
15 4-headed gasifier plant
185 billion Btus/day @ 170 psig
Debt/equity: 60/40
DCFRR = 12%
Interest rate = 9%
Jan. 1977 dollars

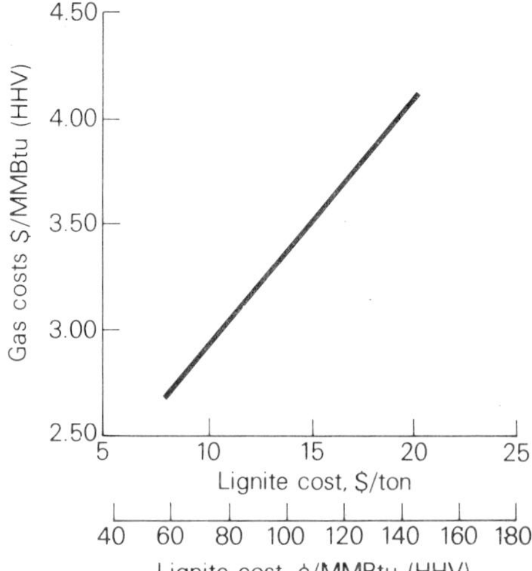

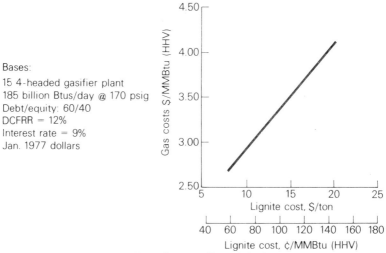

15.11 Koppers-Totzek gasifier — effect of lignite cost on gas cost

TABLE 15.14 Coal and oxygen use for MBG

	Case 1	*Case 2*	*Case 3*
Oxygen requirement, t/d for a 50×10^9 Btu/d plant	840	1020	1900
Coal consumption MJ/MJ of MBG			
Gasification coal	1.23	1.22	1.23
Boiler coal	0.15	0.08	0.16
Total coal	1.38	1.30	1.39

The results are given in Table 15.14.

For comparison, SNG production would require about 12% more coal (MJ/MJ basis) for standard Lurgi and about 18% more in the other two cases (compared with MBG).

Other studies (e.g. EAS) confirm the lower cost factors associated with MBG compared to SNG but emphasise the problems of scale and distribution. On the same ("giant") scale (250 Mscfd (7×10^6 m^3/d)), MBG costs may be only 55–75% of SNG costs but at one-sixth of this scale (perhaps as large an MBG plant as is likely) the MBG costs could be 70–94% of SNG costs, say \$3.3–4.5/GJ from \$1/GJ coal compared with \$4.5–4.8/GJ for giant SNG plants (5% DCF). However, much further work needs to be done on re-optimising cost figures for individual cases, particularly with regard to oxygen supplies.

LBG costs are likely to be in the same range as MBG but perhaps even more susceptible to local requirements since they are likely to be single-user plants, with varying purification requirements depending on the type of process requiring heat and the environmental constraints. Security of supply and possibly load matching potential, may be factors as important as cost per unit of energy. Booz-Allen investigated costs and markets for LBG and MBG in the USA projected to 1985. They identified substantial markets for single-user LBG plants; base costs (in 1978 dollars/million Btu) were \$3.53 for that Raw Gas, \$4.10 for de-tarred de-oiled gas and \$4.56 for clean gas. For MBG, the base costs were \$6.28 for a single-user plant, \$5.08 for a multiple users plant (industry) and \$3.96 for multiple-user (utility). (Presumably scale and coal costs were determined case-by-case.) Ranges of about ±30% on the base cases were indicated.

CONCLUSIONS

From the studies quoted and other sources, it is possible to form some conclusions about the costs of gas from coal, perhaps more in relation to coal prices than in absolute cost terms. Comparisons between different processes can be made, although often these may be between processes at different stages of development. In any case, differences may never be enough to amount to a "breakthrough". The lower manufacturing costs of medium or low CV gas compared with SNG seem more positive though the distribution costs are more uncertain. There are, unfortunately, complications of several kinds about financial conventions. One relates to differences in financial structure, taxes, etc.; this can be by-passed fairly satisfactorily by using a DCF approach, but this approach is too seldom used. The other difficulty is in choosing an appropriate discount rate; costs are rather sensitive to this factor. There may also be considerable sensitivity to the scale of operation, especially where estimates are generalised rather than related to particular circumstances. Finally, when using figures from different countries, although inflation can be dealt with by using the DCF method, changes in exchange rates can be a further complication.

In mid-1978 dollars, the investment for a plant of 250×10^9 Btu/d (250 Mscfd) SNG production will be over one billion dollars, probably in the range \$1.1–1.8 billion, with the standard Lurgi about \$1.4 billion. Using \$1.0/GJ coal and a DCF rate of 5%, a standard Lurgi plant may produce SNG just below \$5/GJ and this estimate is considered reasonably well based. Newer and less well proven processes may be about 20% cheaper, say \$4/GJ; data expected for Bi-Gas and Hygas are in this range. With \$2/GJ coal and 5% DCF, Lurgi may be about \$6.5/GJ and the best aspirants about \$5.5/GJ. An increase in the DCF rate from 5% to 10% increases the gas cost by about 12%. The economic order of the processes, as distinct from the absolute cost level, does not seem to be affected drastically by financial conventions or coal costs, provided the projected capital costs and operating performance are realised.

MBG (or synthesis gas) is significantly cheaper to make and the process is more energy-efficient — probably having an overall efficiency of 70–75%. For \$2/GJ coal, costs as low as about \$3.5/GJ at a 5% DCF rate and \$4.0/GJ at a 10% DCF rate can be predicted for gas from large plants. For \$1/GJ coal, gas costs might be below \$3/GJ. For the same size large plant, MBG might cost only 55–75% as much as SNG. However, it can be questioned whether very large MBG plants will frequently be justified, in view of the extra pipeline distribution costs; even at one-sixth the size, however, MBG may still have at least a 10% cost margin over SNG. The relative economics, including the break-even pipeline distance, need further study and optimisation, probably on the basis of individual circumstances. Some estimates suggest less than 5% increase in delivered cost for 100 miles transport. Second generation gasifiers may have the potential for a somewhat larger saving when making MBG, compared to the standard Lurgi, than in the case of SNG. This is due to lower investment costs. Entrainment-type gasifiers seem to have a particular advantage in this field and produce a gas well suited for use as synthesis gas.

LBG may be produced at about the same price as expected for MBG and with similar process efficiency; this may be the case even at the relatively smaller sizes expected for such gas, which will be mainly for on-site boiler consumption, often with little or no purification. This is largely due to their comparative simplicity of design and operation. Such plants are likely to be economic in a number of places currently or in the near future. Moreover, an important consideration in favour of these installations may be security of supply.

It is difficult to make economic comparisons between natural gas and gas manufactured from coal. The price of the former is generally market-oriented rather than cost-related and is very volatile. In places where coal and natural gas are indigenous, or where an established natural gas supply by pipeline exists, SNG from coal in the early 1980s may be about 50% too costly to be competitive. It is a matter of opinion how soon this gap may close. However, it may become more appropriate to compare SNG with LNG or with light derivates of crude oil. The latter are already at a price level overlapping the lower range of SNG cost predictions. Although some LNG contracts are at low prices, \$2/GJ or less, a prediction of \$4.5/GJ has been made for 1985.

Where a market exists which would enable MBG gas to be made on a reasonable scale and distributed over a limited radius (which may be a maximum of between 20 and 100 miles) costs may within the early part of the 1980s become attractive in many areas in competition with other fuels. Again, it must be stressed that local circumstances will play a vital part in determining economic viability.

Finally, it can be assumed that gas can be made from coal at about half the cost of electricity. This is not always widely appreciated; it may however become an important incentive for early commercialisation of gasification.

Chapter 16

THE ECONOMICS OF LIQUEFACTION

INTRODUCTION

Some general considerations relevant to coal conversion economics discussed in Chapter 12 need to be borne in mind especially in the case of liquefaction. In the case of gas, for example, a few readily identifiable products, which may be natural or manufactured, can be considered. In the case of liquids, the natural crude oils are diverse and the products produced, arbitrarily defined, cover a wide spectrum of properties and prices; the relative prices may be less closely rėlated to the manufacturing costs than to commercial practices. Coal liquids can match all these products, with greater or lesser difficulty, depending on the coal type and conversion process selected. However, this question of diversity of products and the problems of ascribing costs to individual fractions complicates the two main economic considerations:

(a) How do the economics of coal liquids compare with crude oil?
(b) How do the various coal liquefaction processes compare economically?

SASOL

The problems of economic comparisons were well described by Dr P. E. Rousseau, the founder and then Chairman and Managing Director of Sasol, in the 1975 Robens Coal Science Lecture. In discussing Sasol 1, he said that calculations made in the early 1950s indicated that the plant would be marginally economic at the then current prices of \$1.30/barrel for crude oil and \$0.60/t for coal. In the following 20 years crude oil prices were reasonably stable but capital and "other costs" (presumably including coal) escalated. For a long time, these escalations restricted the case for additions to the plant, which presumably would have been unprofitable in a free economy. More recently, substantial investments have been made in a gas pipeline system and this has clearly become an important element in the economics, in addition to the wide range of other products (most strategically important). In 1976, the gas distribution system was expected to replace half a million tons of fuel oil. Such an enterprise could hardly have been made without bringing the whole Profit and Loss Account into credit, however arbitrary the cost distribution may be.

Rousseau confirmed, however, that Sasol 2 (which would concentrate on gasoline) was not attractive until the oil crisis of 1973. Subsequently, with coal at \$4–5/t at the pithead and oil at the then current prices (\$11–12/barrel), Sasol 2 became marginally profitable but would not be attractive for private investment.

Furthermore, the cheap coal could then have been sold for \$10–15/t for export, a price which Sasol 2 would have been unable to meet.

However, it is clear that there was confidence that gasoline from coal would find a ready acceptance in circumstances where alternatives from oil were not readily available. Although coal from oil could never compete with the actual production costs of Middle East oil, the capital investment in Sasol 2 would be only about 25% higher than for a similar amount of liquids via a difficult offshore oil field plus a refinery. Finally, Rousseau pointed out that the final energy cost for liquids from coal were almost exactly the same as for electricity from coal. These statements explicitly confirm the subjective elements in cost assessments.

UK ASSESSMENT

The UK national approach to the economics of liquefaction is based on an energy supply/demand situation which indicates that substitution of oil by coal is likely to be more favourable than conversion of coal to oil for a considerable time, with perhaps the change-over taking place in a gradual way in hybrid plants.

In 1978, estimates were made comparing the costs of products from coal liquefaction (by hydrogenation and by synthesis) with those from a simple oil refinery and a complex oil refinery. The last produces only premium products (82% of the energy input) while a simple refinery produces 51% of the input energy as premium fuels and 42% as fuel oil (with a value related to that of coal but with combustion advantages). Capital and other costs for coal liquefaction were based on published estimates; only Sasol costs, of course, related to a process operating on a commercial scale.

The average price of coal was then £24/t (£0.92/GJ) and the landed price of oil was £50/t (£1.12/GJ). The results were as given in Table 16.1.

A diagram (Fig. 16.1) was also constructed to illustrate the relationships over a range of prices. The triangle to the bottom right indicates the area where coal liquefaction would be economic and it is bordered by a band where complex refineries might be favoured. Figure 16.2 shows the maximum acceptable price of coal for conversion to premium products, as affected by possible oil price rises and by discount rates. Case 1 assumes a discount rate of 3%, Cases 2 and 3 assume a discount rate of 10%; in Case 3 the plant investment costs are increased by 50% over the best estimates. In the most stringent case studied, it will be seen that an oil price of \$40 a barrel would allow a coal price of about £40/t to be accepted.

A study was carried out in 1979 to compare various processes, using published data. The object was to determine for these processes the coal prices and returns on capital which might be achieved, using product slates appropriate to the processes

TABLE 16.1 UK estimates of refinery costs: oil and coal

Cost of products	*Simple oil refinery*	*Complex oil refinery*	*Coal liquefaction by hydrogenation*	*Coal liquefaction by synthesis*
£/t	61.6	74.7	125.5	156.3
£/GJ	1.31	1.59	2.67	3.33
p/gallon	20.4	25.3	42.5	53.0

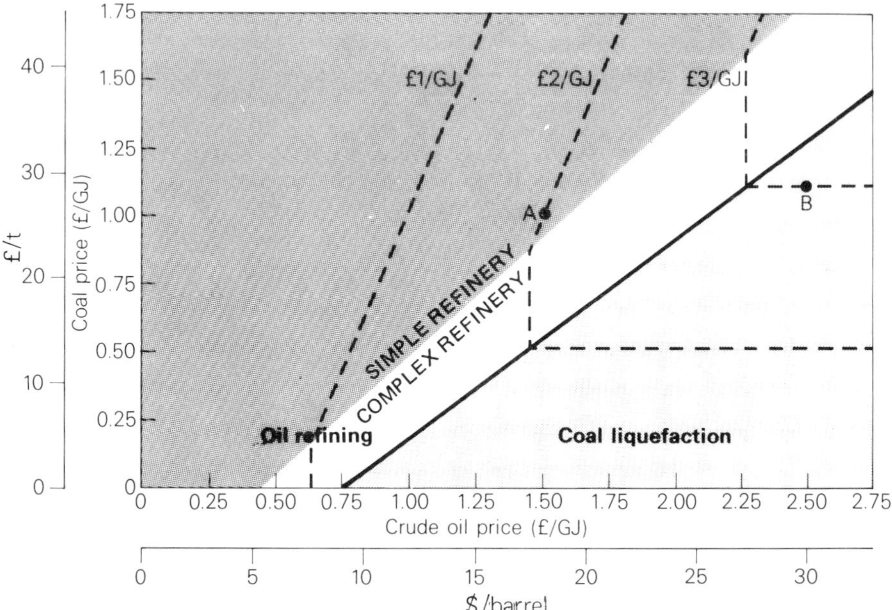

16.1 The economics of producing premium liquid fuels. The dotted lines indicate the prices for premium products in relation to input prices (£1/GJ, etc.). At point A, for example, products at £2/GJ could be made from oil at £1.5/GJ in a simple refinery. At point B products at £3/GJ would be produced from coal at £1/GJ

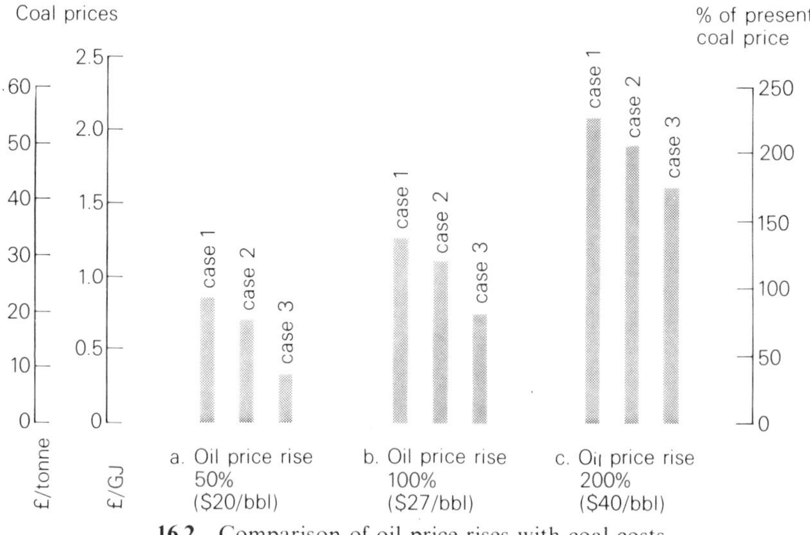

16.2 Comparison of oil price rises with coal costs

and relating the value of these products to prices based mainly on oil refinery practices.

Table 16.2 shows the capital costs and efficiencies selected (Lurgi SNG is included for comparison). Table 16.3 shows the price schedule in the UK in July 1979, on both

TABLE 16.2 Capital costs and efficiencies

Process	*Capital cost (£/kW output)*	*Efficiency (%)*
First Generation		
Lurgi-SNG	244	59
Sasol 2	455	34
Second Generation Synthesis		
FT-Slagger	293	48
Methanol-Slagger	247	57
Mobil M-Slagger	292	52
Second Generation Hydrogenation		
EDS	249	64
HCOAL-FO	161	74
HCOAL-SC	206	69
SRC1	196	70
SRC2	226	70

TABLE 16.3 UK petroleum product prices July 1979

	Value mass basis (£/t)	*Energy content (GJ/t)*	*Value energy basis*	
			(£/GJ)	*Relative*
Crude oil	66.7	45.05	1.48	1.00
LPG and gases	74.9	52.33	1.43	0.97
Naphtha	79.6	47.80	1.67	1.13
Motor spirit	95.2	46.95	2.03	1.37
Kerosine	85.4	46.42	1.84	1.24
Gas/diesel oil	77.5	45.47	1.70	1.15
Fuel oil	54.5	42.84	1.27	0.86
Other	66.7	45.05	1.48	1.00

TABLE 16.4 Product yields (energy content basis) from coal liquefaction

	Sasol 2	*FT slagger*	*Mobil-M slagger*	*EDS*	*HCOAL-FO*	*HCOAL-SC*	*SRC2*
LPG	1	14	5	4			27
Gases	2	14		10	1	2	1
Naphtha				18	19	35	7
Motor spirit	25	17	47				
Kerosine							
Gas/diesel oil	1	1					
Fuel oil	4	2		32	54	32	35
Other	1						
Total	34	48	52	64	74	69	70

TABLE 16.5 Mean product prices for coal liquefaction processess (based on prices for similar products from petroleum)

	Mean product price (£/GJ)	
	At July 1979 oil prices	*Crude oil price doubles*
Sasol 2	1.86	3.72
FT slagger	1.64	3.28
Mobil-M slagger	1.97	3.94
EDS	1.42	2.83
H-Coal-FO	1.37	2.75
H-Coal	1.48	2.95
SRC2	1.37	2.75

a mass and energy basis (in money and relative terms). Table 16.4 shows, on an energy context basis, the product yields assumed and Table 16.5 shows the mean (weighted) product prices which would be applicable in competition with the July 1979, oil price ($20/barrel) and with the oil price doubled (by early 1980 the half-way stage had been reached). Figure 16.3 shows the economics in the UK in July 1979. It will be seen that for nil return on capital, the maximum coal price would be between about £8/t (Sasol 2) and £19/t (H-Coal). At that time average coal prices were about £30/t. Figure 16.4 shows the situation if oil prices doubled; then at the coal price of £30/t all processes except Sasol 2 would produce a positive return. However, even in this case the returns are low and the slopes of the curves steep and therefore sensitive to coal price increases. This method of analysis involves some uncertainties because the coal-derived liquids are "raw" products and have different characteristics from oil-derived fuels. The product values used may not reflect accurately the price that would be realised for coal-derived liquids, now or in the future.

WEST GERMAN ASSESSMENTS

Although the Federal Republic has an extensive liquefaction development programme, there has for a long time been considerable pessimism about the economics of making petroleum substitutes from coal at an early date, at least at German coal prices.

However, towards the end of the 1970s it was becoming recognised that the substantial increases necessary in oil prices in order to close the gap were feasible and that the question became more one of timing. In the meantime, the possible applications of German liquefaction technology in countries having cheap coal has been a great incentive.

In 1978, a view was expressed, based on coal at DM 162/t, that gasoline could be made for DM 0.72/l via hydrogenation and for DM 0.85/l via synthesis, compared with the current price of DM 0.32/l. However, at that time foreign coal might have been available at about half the West German indigenous price and (extrapolating curves beyond those presented) this might have brought the gasoline price down to about DM 0.55/l.

In 1979, Peters *et al.* presented estimates to the World Petroleum Congress; these are given in Figs. 16.5 and 16.6 and in Table 16.6.

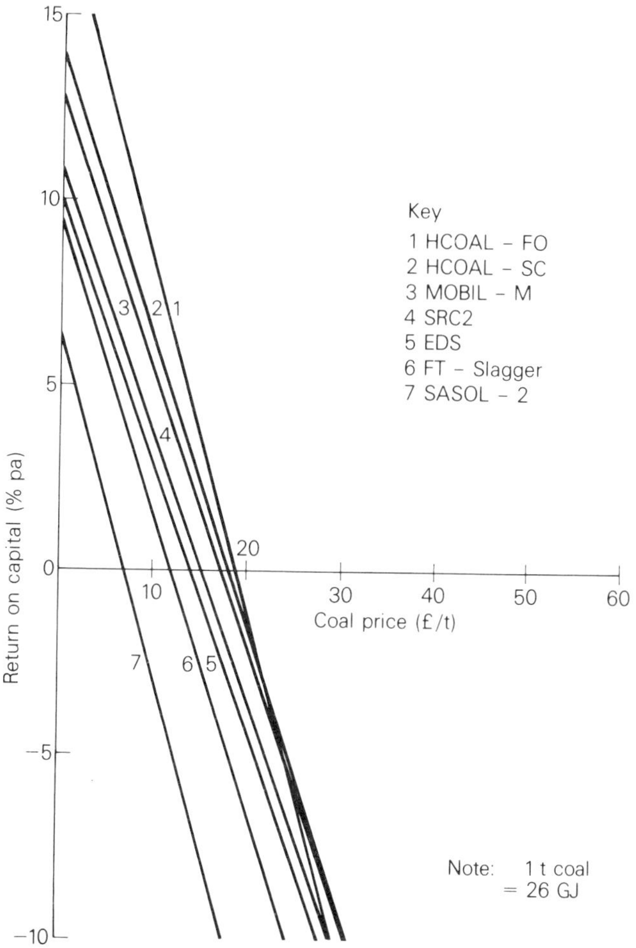

16.3 Economics of coal liquefaction in the UK; July 1979

In these calculations, coal was taken at $94/t. The projects were assumed to be financed on a debt/equity basis. In addition, a rate of return of 7.7% was given and an inflation rate of 4% assumed.

In Fig. 16.5 the horizontal lines on the curves indicate the then-current prices of the appropriate petroleum product; neither the date nor actual prices are given. However, it may be assumed that by the end of 1979, prices for oil products were about 30% higher. If this escalated price were combined with the lowest cost of imported coal (an obviously over-optimistic case), the gap between oil and coal products would be reduced but would still remain at about 20%.

It may be noted that the German assessments consider liquefaction by synthesis to be about 20% more expensive than by hydrogenation. Less than half of this difference is due to capital costs. A large portion is due to coal costs (or in the case of synthesis, coal less by-product credits); this highlights the desirability of increasing

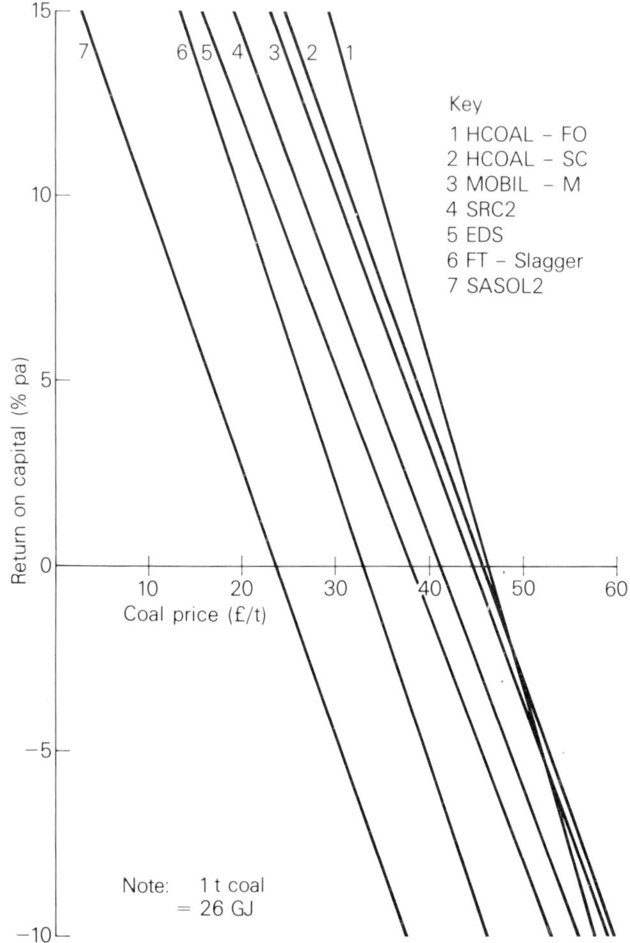

16.4 Economics of coal liquefaction in the UK; oil price doubled

the efficiency of these processes (this could also reduce the specific capital cost). All the plants were assumed to operate at full load for 7500 h/a (about 85%).

US ESTIMATES

At the same World Petroleum Congress in 1979, the US Department of Energy view was presented by Mills and Knudsen. This report summarised various earlier (and often much more detailed) studies critically and is probably the best guide available, especially as some limitations of earlier reports are exposed. For example, the authors regret the absence of a careful study carried out by a single firm using a consistent basis for design features and economic conventions. Most of the estimates to date have been done by the proponents of individual processes using inconsistent

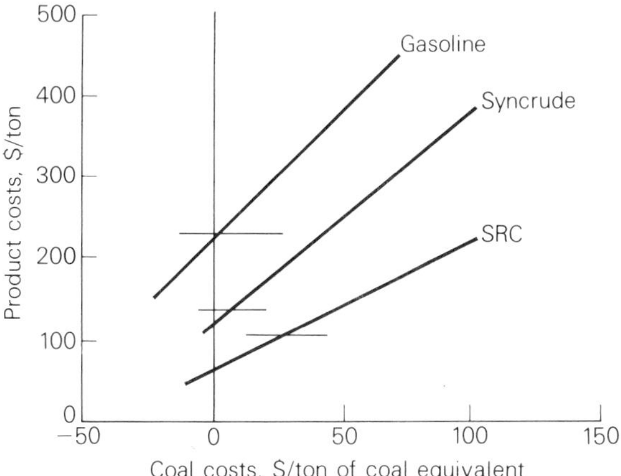

16.5 Effect of coal costs on price of liquids

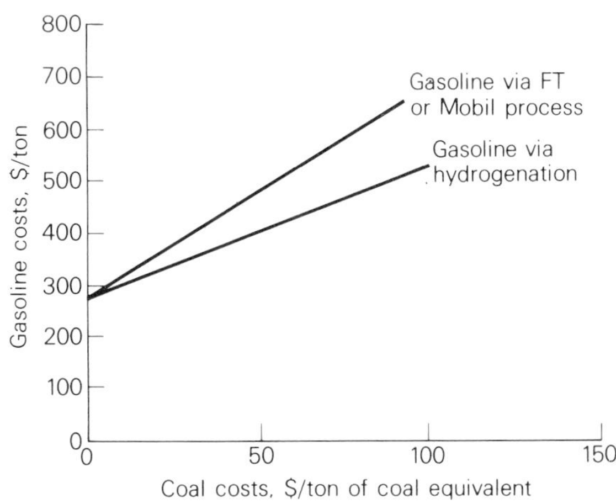

16.6 Gasoline from coal production

TABLE 16.6 Coal liquefaction costs ($/ton product) German conditions, 1978

	SRC heavy liquids	*Syn. crude*	*Gasoline*	
			Hydrogenation	*F–T synthesis*
Specific investment costs	360	540	1050	1350
Coal costs	148	238	262	406
Capital costs	79	115	221	260
Other operating costs	11	15	22	45
By-credit	25	4	2	103
Net product costs	213	364	503	608

procedures. Comparison of processes which produce different end products is particularly difficult. The authors state that estimates for coal-derived liquid fuels range from about \$150–300/t but that the range \$220–300/t is more realistic for first commercial plants (1979 dollars, 10% DCF).

Exxon studied EDS, H-Coal and SRC2, adjusting costs to a common basis. Mills and Knudsen remark that this brought all these process costs to within 10%, an inadequate margin to have real meaning without full-scale plant data. The Exxon study assigned a first year selling price of \$212/t (\$29/barrel) to EDS. The plant used coal at \$21/t and produced half its product as fuel oil, the other half being as lighter products. A thermal efficiency of 63% was assumed, with an initial plant availability of 80%. The capital investment was \$245 000/calendar day ton.

Mobil compared their synthesis route via methanol (Mobil-M) with straight Fischer–Tropsch, for plants of 8 Mt/a coal input (coal at \$7.70/t). Mobil-M was assumed to have a higher thermal efficiency (63% versus 58%) and a lower capital cost (\$293 000/calendar day ton versus \$334 000). The Mobil-M process also gave a higher liquids to SNG ratio in thermal terms (47/53 versus 35/65). This resulted in a selling price of \$6.45/GJ for Mobil-M products and \$7.37/GJ for F–T products.

Mills and Knudsen assessed the hydrogenation route costs as equivalent to \$5.00–6.00/GJ, to which must be added nearly \$1/GJ for further hydrogenation to gasoline, making a total of \$6.00–7.00/GJ which is said to overlap the lower end of the Mobil-M route costs. However, the latter appears to value all synthesis products equally and, as the authors point out, in order to allow for lower prices for SNG and other co-products, the Mobil-M gasoline prices might have to be increased by about 10%. The authors also consider that the Mobil financial assumptions were rather more favourable than in other studies. Nonetheless, they reiterate the general range of \$220–300/t (\$30–40 a barrel) and include both synthesis and hydrogenation. This contrasts with the German estimates where a definite margin for hydrogenation is widely stated; it is also a little difficult to relate to the US DOE policy which has had no commitment to the synthesis route, at least until very recently.

Two further interesting points emerge. Caution is expressed repeatedly concerning plant availability, often taken at 90% in paper studies. This is regarded as optimistic for emerging plants, which can have output restrictions or breakdowns, for which either materials or equipment failures may be responsible. A reduction of 10% in availability is said to increase liquids costs by 18%. A further point is made that the rate of price increase in marginal supplies of petroleum would need to escalate at a rate of 2–3 percentage points more than the costs of synthetic fuels plants to make such plants profitable in the 1980s. Some US experts have however expressed anxiety about the way the projected capital costs appear to be rising.

Various US companies have produced their own estimates of costs, generally for processes in which they might have an interest. The Ralph M. Parsons Co. examined four different routes for the US DOE, including versions of both hydrogenation (SRC type) and Synthesis (F–T). They also considered two "Coalplex" type systems — COED-based Pyrolysis and POGO (Power–Oil–Gas–Other) — which will be discussed further in the appropriate chapter. Economic comparisons were made by bringing all process costs to first quarter 1978 dollars. A captive mine was assumed (about 40 000 t/d) and separate coal prices were not stated. A 12% DCF rate of return was used, in one case with 65% debt financing at 9% interest. In this case the results which emerged are given in Table 16.7.

These results depend on the correct relative evaluation of a number of different products, including electricity and both SNG and LBG. The product slates differ considerably; none give a high yield of gasoline.

Gasoline is the predominant product of a large Mobil-M type plant studied by the Badger Company. The project was based on 1977 dollars, with 63 000 tons of coal input per day at \$25 per ton delivered; the total capital cost is \$7.26 billion. Escalation of 6% on all costs was assumed, with a financial structure of 65% debt (9%) and 35% equity (12% return). A thermal efficiency of 50.7% was assumed; for a high gasoline proportion this is attractive. If the project were initiated in 1980, by January 1990, after one year of operation, gasoline would cost \$1.09/US gal. (\$9.00/MBtu) at the plant gate; it is said that delay in initiation would add to costs, which would reach \$1.46/gal after a five year delay. It is known that this estimate is regarded by the US DOE as optimistic for a first large plant.

TABLE 16.7 Parsons estimates of coal liquids costs

	Hydrogenation	*F–T synthesis*	*COED*	*POGO*
Capital, \$/barrel	1.4	1.8	1.5	2.5
Selling price				
\$/barrel oil equivalent	12.50	16.00	30	13
\$/million Btu	2.10	2.90	5.00	2.17

Badger have estimated the effects of some uncertainties. An increase in capital cost of 20% would result in an 8% increase in gasoline cost. The gasifier section is considered the least proven, at the scale chosen; a 100% increase in the capital cost of the gasifier would add 3% to the gasoline cost. Reducing the scale to one sixth (10 500 t/d) would increase datum costs from \$1.09 to \$1.34/gal; a further halving (5300 t/d, still a large plant) would result in a sharp escalation to \$1.65/gal. Elsewhere, Badger have expressed the view that straight F–T synthesis would be much more expensive than Mobil-M. On the other hand, the Booz-Allen company, also working for the US DOE, evaluated a plant for 50 000 barrels per day of gasoline using a Sasol 2 type of process, and in late 1979 reported a plant gate cost (first year) of \$0.90/gal compared with current refinery gasoline at \$0.60/gal.

Catalytic Inc., a subsidiary of Air Products, carried out a preliminary assessment (and comparison with other processes) of the Supercritical Gas Extraction process, working with the National Coal Board, the pioneers of the process. Using 1974 cost data, the net production costs for liquids were considered to be of the same order as for SRC or COED — about \$10–11/barrel, with coal at \$16.75/t; capital investment costs were lower than SRC but higher than COED. The net liquid costs for both Supercritical Gas Extraction and COED depended substantially on by-product credits — mainly a reactive char in the case of the former. Whilst the value of this residue may be debatable, it might eventually be more valuable than assumed in the study. At the same time, the liquids may have a higher value, especially as chemical feedstocks, compared with those from other processes. However, at this stage the economic assessment of Supercritical Gas Extraction is purely for R & D guidance rather than for purposes of investment in commercial plants.

TABLE 16.8 Generic data of representative routes to coal liquids

	Mobil-M gasoline	*Methanol*	*F–T*	*High hydrogenation*	*Low hydrogenation*
Investment cost (1978 \$/kW output)	500	400	330[a]	380	270
Thermal efficiency (including gas)	0.50	0.63	0.70	0.67	0.77
Products %					
SNG	—	—	45	10	10
Transport fuels	85	100	35	35	15
Heating fuel liquid (including LPG)	15	—	20	55	75

[a] The Fischer-Tropsch figure includes allowances for gas output. An all liquid output would produce a figure of about \$420/kW.

CONCLUSIONS

For a long time economic assessments (and consequent actions) were cushioned by the belief that liquefaction of coal was too expensive to be seriously compared with crude oil. A common statement, even in the late 1970s, was that liquefaction would not be economic until oil reached a price of \$30 a barrel. This (then) apparently remote eventuality has arrived earlier than expected but has not triggered off an immediate construction programme of liquefaction plants, although strategic thinking may have been affected. Factors which have of course tempered the evaluation are the continued rapid increases in coal prices and also presumably in the capital cost projections of the plants, compared with those quoted above.

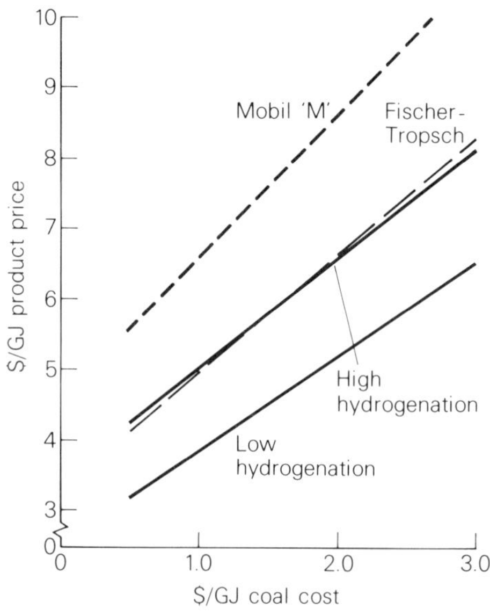

16.7 Product price against coal cost; "base" investment

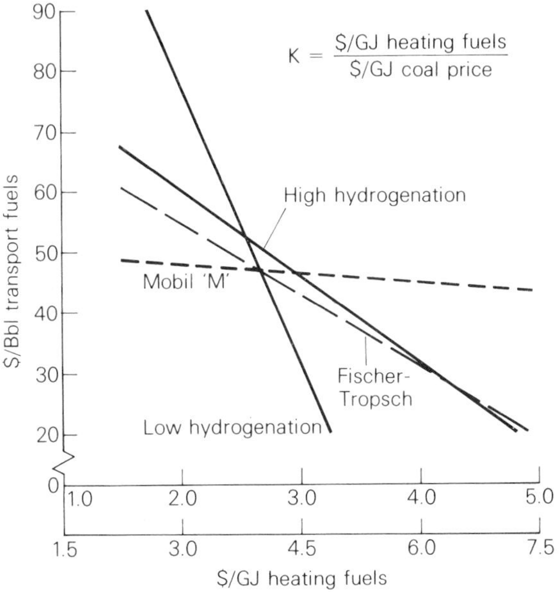

16.8 Fuel costs for $1.5/GJ coal; "base" investment

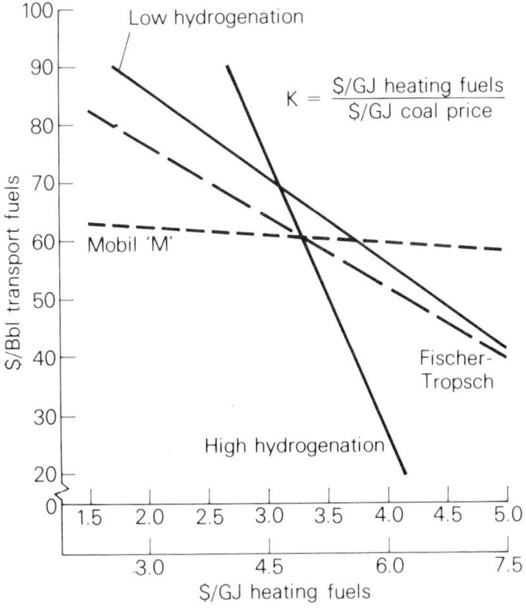

16.9 Fuel costs for $1.5/GJ coal; "base" investment + 50%

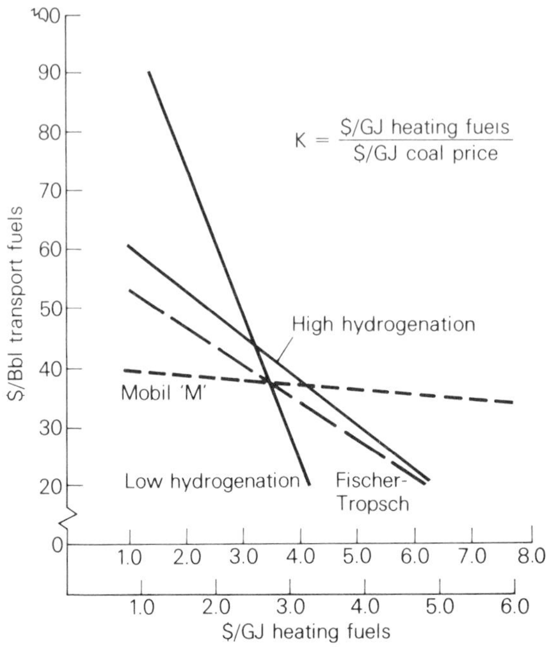

16.10 Fuel costs for \$0.8/GJ coal; "base" investment

In most comparative assessments, hydrogenation routes are considered more favourably than synthesis. However, if a higher proportion of high value products can be made, as in the Mobil-M process, there may be an increase in interest in synthesis; this may be enhanced if a swing to more light products in oil refineries leads to a larger price differential between gasoline and fuel oil.

South Africa is in any case relying on synthesis, presumably on the basis that they are confident of synthesis costs and do not regard the potential savings via hydrogenation as justifying the extra risk of processes unproven on the scale in which they are interested and for the time scales imperative to them. It is certain that the Sasol officials are fully aware of these alternative processes (which might also appear in a more favourable light relatively at higher coal costs).

Most of all, the economics are complicated by the different product slates and the need to ascribe suitable differential values. This complication is also considered in Chapter 20 (Coalplexes) but the methodology needs further development.

An approach to an improved methodology has been made by Baker and others of the Economic Assessment service of IEA Coal Research (operated by the NCB in London). In a paper in early 1979 an attempt was made to evaluate processes on the basis of their different product states in terms of: SNG, transport fuels and heating fuels. Table 16.8 shows the assumed generic data. Figure 16.7 shows the average product prices as a function of coal costs. Figures 16.8–16.10 show the cost of a barrel of transport fuels as a function of the cost of heating fuel and also the ratio (K) between heating fuel and coal prices. Figure 16.8 is for coal at \$1.5/GJ (about \$40/t) and the base investment costs are quoted in Table 16.6; in Fig. 16.9 the base investment costs are increased by 50% and in Fig. 16.10, the coal cost is taken as \$0.8/GJ.

The dependence of the hydrogenation routes on a high ratio of fuel oil to coal price is evident. At lower ratios, synthesis appears better, Mobil-M, of course, being least sensitive to the fuel oil market. At low coal costs and at the base investment costs, transport fuel costs are interesting but this attractiveness is vulnerable to uncertainties in either coal or capital costs.

Finally, it must be emphasised again that the best estimates have large uncertainty factors. This will remain the case until large demonstration plants have operated for a considerable period.

Chapter 17

ECONOMICS OF CHEMICALS FROM COAL

INTRODUCTION

The production of chemicals, and the manufacture of products derived from the basic chemicals, are important components of modern industrial activity. For this reason, there has been considerable speculation about the future of the chemical industries in the event of physical shortages of traditional feedstocks or further large price increases. In these circumstances, the use of coal appears to be the major alternative and the economics of this potential substitution are of considerable importance.

To a first approximation, of course, the economics of the production of chemicals from coal may be assumed to be parallel to those of the liquefaction of coal or to the economics of SNG manufacture. This would certainly be the case if the introduction of coal into the chemicals industry on a major scale were to be confined to providing replicates of the existing feedstocks — methane, ethane, naphtha, etc. Economic viability for coal into chemicals would be reached when the prices of petroleum hydrocarbons and equivalent coal substitutes crossed over; on a thermal basis this has been considered elsewhere. It may, nonetheless, be possible for coal-derived chemical feedstocks to become economic at an earlier date for special reasons. Therefore, the subject deserves separate consideration even though the basic facts are covered in the discussion on gases and liquids.

The economics and policy implications of coal as a chemical feedstock, however, seem to be a more difficult and controversial subject than, say, liquefaction. This may partly be due to the fact that the chemicals industry is only a relatively minor outlet (less than 10%) for petroleum products. Thus, not only will the price of crude oil affect the competitiveness of coal as a chemical feedstock but so will the demand (and price) for other derivatives, especially transport fuels, and changes in refinery procedures. These are expected to move in the direction of a reducing proportion of heavy products and a higher proportion of light products, especially transport fuels. Since the latter will have a high priority and are themselves also suitable feedstocks, petroleum will probably continue to be a major supplier to the chemicals industry long after crude oil output reaches a peak and starts to decline. As it happens, trends towards lighter products and higher efficiencies of conversion of crude into products may be assisted by the utilisation of coal to supply heat and, possibly, hydrogen in refineries. The substitution of coal for oil for heating purposes will also facilitate the conservation of crude oil for premium use and thus help to stabilise the present feedstock supply situation so far as oil is concerned.

Other factors which may help to stabilise the present petrochemicals situation are the existing chemical plant capacities (i.e. overcapacities) for oil-based processes and the fact that oil producers are heavily involved, and often dominant, in the petrochemicals industry. Opinions about the significance of all these factors seem rather sharply divided — and rather volatile with time. Views have been expressed that coal is (or was at the time) on the verge of a major breakthrough into chemicals; on the other hand, there are those who believe that the chemical industry will remain closely linked to oil indefinitely — one official of a major oil company, which is also large in chemicals, forecast that the capital investment for making ethylene from coal would be $2\frac{1}{2}$ times greater than from oil and the energy consumption 3 times greater.

The foregoing remarks apply particularly to the substitution of oil feedstocks by products made by the liquefaction of coal by hydrogenation. Where synthesis gas is used — or can be used — in chemical manufacture, the possibility of substituting coal for methane or oil products as the feedstock for the synthesis gas is probably more favourable and this may be the main potential outlet for coal into chemicals in the near future. The bulk of this chapter is therefore devoted to schemes and assessments based on synthesis gas; short comments on other possible routes follow.

In passing, the relatively modest but significant contribution to chemical supplies made by the coking by-product industry should be noted; in view of the uncertainties, it would appear desirable for the exploitation of tars and benzoles to continue as actively as possible.

Coal products whether derived from carbonisation or direct liquefaction processes are a much richer source of aromatics than the products from synthesis gas. Therefore aromatic feedstocks will eventually be made from coal more effectively and cheaply than from oil and this may provide the stimulus to the chemical industries to develop new products and new routes to materials based on aromatic structures.

SYNTHESIS GAS ROUTES TO CHEMICALS

The technology for converting coal into chemicals via synthesis gas is established; this is one of the main reasons why this is likely to be the main route in the immediate future. Once synthesis gas is produced and purified to the appropriate standard, the synthesis processes themselves are theoretically unaffected directly by the source of the gas. At the same time, the production and purification of synthesis gas is also well established. This is not to say that the combination of coal gasification and synthesis will not present any problems in either a totally new plant or a plant converted to coal in place of natural gas; the coal plant will have more units and more environmental impact. However, the problems can be overcome, as has been demonstrated. In addition, it should be possible to cost plants based on known synthesis processes and on established gasifiers fairly accurately and even to make reasonable estimates of the economic impact of new developments in gasifiers (which are likely to be marginal rather than dramatic). However, the published information does not at present provide as clear-cut a picture as might be expected.

Synthesis gas is an intermediate in the manufacture of SNG from coal; it is more economic, both in financial and energy terms, to make than SNG, probably being about 20% cheaper per unit of energy. At the same time, the manufacture of synthesis gas from natural gas or crude oil involves energy consumption and costs, which increase the cost per unit of energy by about 20%. A general indication of costs is

TABLE 17.1 Estimated costs of energy sources (1978 values)

	DM/GJ
Brown coal	2
Synthesis gas from brown coal	7
SNG from brown coal	8.5
Natural gas	4.3
Heavy fuel oil	4.8

given (in round numbers) in Table 17.1, which is derived from a 1979 study of German conditions.

If 20% were added to the costs of methane and fuel oil for conversion to synthesis gas, the cost would be about DM 5.2–5.8/GJ compared with DM 7.0/GJ for the coal-based synthesis gas. This gap of about 25% may have been typical of areas without cheap coal at that time; the gap may have narrowed substantially in these areas since then. Where cheap coal is available and natural gas or oil are expensive or not readily available, coal may already be economic for synthesis gas.

Synthesis gas may be used for the production of liquid hydrocarbons via the F–T type of synthesis and the outlook for this has already been considered. Synthesis gas may also be used as a premium fuel or it might be "shifted" completely to hydrogen. The hydrogen might be used in petroleum refineries, coal liquefaction plants, in the direct reduction of iron ore or in many applications in the chemicals industry. The largest chemical industry applications for synthesis gas at present, however, are for ammonia and methanol; it is largely against these products that coal has been evaluated. Some of these studies are described below.

Ammonia is of course an important constituent of fertilisers, the demand for which appears likely to grow for a long time without approaching saturation. In addition to fertilisers, ammonia is also used for the manufacture of many other important chemicals, especially nitric acid, urea and amines. Methanol also has important derivatives. Therefore, if synthesis gas from coal became established for ammonia and methanol, this could cover a large and vital sector of the whole chemical industry.

It should be borne in mind that a number of chemical plants around the world are already operating on coal, especially to manufacture ammonia and methanol, as described in Chapter 10. In most, if not all, of these cases however there are special features, of a strategic or policy nature which obscure the economics or make them difficult to translate to Western conditions.

For ammonia, especially, coal is likely to be selected where coal is cheap and access to other feedstocks difficult. This appears to be the case at the Modderfontein plant in South Africa where a 1000 t/d ammonia plant has been operated since 1974. This operation is presumably profitable in the local circumstances but such financial information does not appear to have been made available. Other information has however been reported, including design features and early operational problems. The considerable effort which was deployed to overcome these problems suggests a favourable financial prognosis.

Modderfontein uses six low-pressure Koppers–Totzek gasifiers. Some problems were initially experienced in the gasifiers, especially, it is thought, due to attack on the refractories by the particular properties of the ash of the coal used. The raw gas

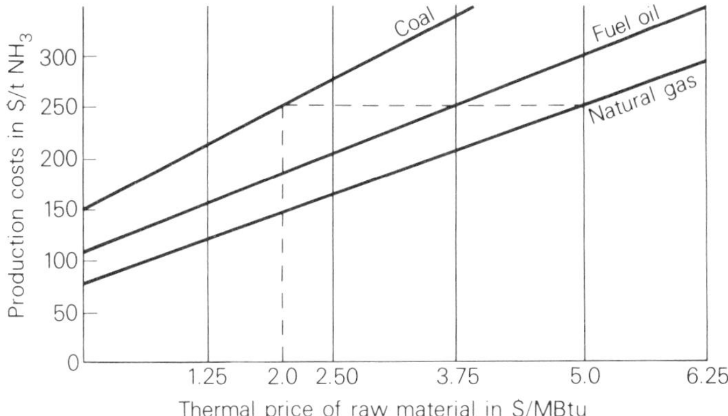

17.1 Production costs of ammonia versus raw material costs (plant capacity 1000 t/d)

contains 58% CO and only 27% H_2 on a volume basis and this called for an unusually large shifting requirement; the shift process was nonetheless successfully adapted. Perhaps more significant is the fact (fully reported) that differences between coal and petroleum feedstocks can persist a long way downstream and create unexpected problems, including corrosion, fouling and catalyst deterioration. These problems appear to have been overcome at Modderfontein. This experience suggests that anyone switching from oil to coal should be prepared for some differences in operation. It seems however that these differences can be dealt with without undue cost penalty. With a plant of this kind, non-availability can be very expensive; the Modderfontein experience may be helpful elsewhere in this respect.

Several companies or organisations have reported or described cost estimates for current designs of plants, either new plants or plants in which coal replaces other feedstocks.

Krupp–Koppers

For a number of years excellent information has been published on the expected performance of Koppers–Totzek gasifiers in a fuels or chemicals context. Beck gave technical and economic estimates to a symposium on "Ammonia from Coal" at Muscle Shoals, Alabama in May 1979. At that time, the investment for a complete, self-sufficient 1000 t/d ammonia plant was estimated at \$200 million for coal, \$135 million for heavy fuel oil and \$90 million for natural gas. Taking 18% for interest and depreciation and 3% for maintenance, production costs for ammonia were given for the three fuels as a function of the thermal price (Fig. 17.1). In Fig. 17.2, the effect of plant capacity is displayed. There is clearly, in this case, a significant advantage in going beyond 1000 t/d; however, this is for a low feedstock price and the scale advantage may be proportionately lower at higher feedstock prices, more typical of Europe.

Beck refers to the current development of the Shell–Koppers pressurised gasifier as an advance but does not suggest the extent of the probable economic gain. Shell forecast technical and operating data for the Shell–Koppers gasifier but do not appear to make a direct economic comparison with the K–T gasifier for chemical plant purposes. It is indicated that methanol could be produced for about \$6/MBtu from coal at \$1/MBtu and for about \$8/MBtu from coal at \$2/MBtu.

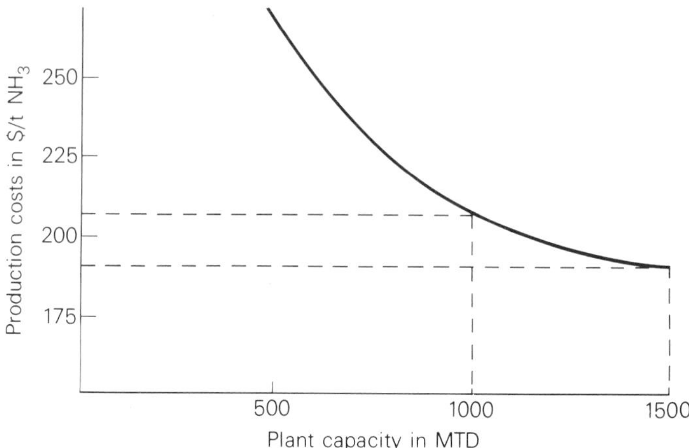

17.2 Production costs of coal-based ammonia versus plant capacity

Chem Systems Inc.

This company has published a number of studies. In 1977, a number of possible coal-chemicals outlets were discussed. It was suggested that methanol could be produced for between about \$130–\$150/t from coal at \$21/t, compared with about \$150/t from natural gas costing \$3.15/MBtu; the conclusions for ammonia were similar. Acetic acid was considered as an attractive derivative, using methanol and CO from coal gasification. By 1990, it was suggested, a direct route from synthesis gas to ethylene glycol could be established and would be competitive with ethylene-based technology. Ethylene itself was considered a long-term aim, requiring further work, none of the best-known processes from synthesis gas appearing very satisfactory for this large-scale and important product. The most promising route to ethylene was considered to be the homologation of methanol to ethanol, which is subsequently dehydrated to ethylene.

In 1978, the company published estimates of costs for various chemicals produced from a large central coal-based synthesis gas plant. An important element was the concept of "leveraged" finance for the gas plant. On a 1985 basis, using coal at \$30/short ton and with full equity financing at a 20% return, synthesis gas costs would be \$3.27/thousand cubic feet; if 70% of the investment could be provided at 9% interest, costs would be reduced to \$2.76. Using these two costs, methanol could be produced for 99.1 and 83.3 cents/gal respectively, compared with 76.6 cents and 90.4 cents/gal from natural gas at \$3.00 and \$4.00/MBtu respectively. This indicates that both a substantial degree of leverage and a projection for natural gas prices at the upper end of the range would be necessary for coal-based methanol to be viable.

A comparison for acetic acid made from coal-based methanol and CO was more favourable; without any leverage, acetic acid costs were 22.3 cents/gal compared with 24.3 cents/gal using natural gas at \$3.00/MBtu. In a combined methanol/acetic acid plant, coal-based costs were similar to those using natural gas if as much as 55% of the investment was debt-financed (9%). The authors concluded that there was at that time no clear cut economic mandate for coal over natural gas; the choice depended on a preference for a capital risk compared with a feedstock supply risk.

TABLE 17.2 Cost data for ammonia production (Existing plant converted to coal from other feedstocks)

	Standard Lurgi	*Modified Lurgi*[a]	*K–T*	*Winkler*	*Slagging gasifier*	*Texaco*
Coal t/h	97.6	99.1	120.9	123	94.5	100
Oxygen plant size t/d	860	885	1110	1090	820	680
Feedstock and utility cost/t, $	16.5	16.6	22.1	21.7	15.9	16.8
Erected cost, million US$	150	125	150	150	115	110
Capital charges/t (at $33\frac{1}{3}$%), $	150	125	150	150	115	110
Total cost/t ammonia, $	166.5	141.6	172.1	171.7	130.9	126.8

[a] The modified Lurgi scheme made more use of the existing ammonia reformer plants. The gasifier, quench system and waste heat boiler (WHB) were as in the standard Lurgi case.

TABLE 17.3 Daily consumption and costs: 1000 t/d ammonia plants

Gasifier	*Oxygen plant capacity t/d*	*Process coal t/d*	*Power coal t/d*	*Total coal t/d*	*Erected cost million US$*
K–T	1210	1630	800	2430	200
Winkler	720	1410	410	1820	180
Slagging gasifier	570	1250	330	1580	165
Texaco	1400	1550	300	1850	145

Humphreys and Glasgow Ltd (H and G)

This firm of consulting engineers has carried out studies of coal gasification schemes, especially for the production of ammonia and methanol, over a number of years; some of this has been for clients but there have been open publications also. Their work is particularly valuable because it has included the comparison of different gasifiers, not only as gasification units but as part of an integrated system with associated plant items selected to suit the particular gasifier. They have also studied the conversion of existing plants to coal from other feedstocks.

In 1977, H and G published a paper comparing six schemes, each using a different gasifier, for converting an existing 1000 t/d ammonia plant to coal. This was in response to an acute shortage of natural gas in the USA. The principal results (using coal at $5.5/t) are given in Table 17.2.

The calculations in Table 17.2 emphasise the importance of capital cost since a short pay-back time (3 years — appropriate to a conversion scheme) is taken, together with a low coal cost. At these capital charge rates, capital cost and coal costs would be about equivalent at a coal price of $30/t. The low temperature processes use more coal and oxygen. The authors point out that the Slagging Gasifier and Texaco (pressurised processes) appear best on economics but the other processes are more established. They also conclude that coal becomes economic when natural gas becomes 3–4 times as expensive, locally, as coal, on a thermal basis.

In 1979, H and G published further estimates for performance and costs for making ammonia and methanol from coal. The data for ammonia are given in Table 17.3. The standard Lurgi was not considered since the slagging version was

TABLE 17.4 Daily consumption and costs: 1000 t/d methanol plants

Gasifier	*Oxygen plant capacity t/d*	*Process coal t/d*	*Power coal t/d*	*Total coal t/d*	*Erected cost million US$*
K–T	1440	1800	540	2340	180
Winkler	800	1560	390	1950	165
Slagging gasifier	1020	1570	290	1860	155
Texaco	1550	1720	170	1890	130

considered superior and would be ready for adoption by the time decisions were necessary.

The authors again point out the advantages of the pressurised processes; also, the lower coal consumption would give the Slagging Gasifier an advantage, even over the lower capital cost of the Texaco system. The K–T system appears least attractive on both capital and consumption grounds but it can handle all coals and its costs are more certain; it is suggested that it may be used for smaller plants or where a wide range of coals has to be handled.

For methanol, H and G estimates are given in Table 17.4.

On economic grounds the Texaco process is considered a clear preference; a very simple flowsheet results from operating at 60 bar. However, it would be prudent for more experience to be obtained before too many projects were committed to this process.

More recent studies by H and G appear to confirm these data. The Texaco gasifier is still preferred (although the Shell-Koppers system would be interesting if projected performance is demonstrated). Capital costs, in the UK in first quarter 1980 money values, might be about $150 million for a 1000 t/d methanol plant; a similarly sized ammonia plant would be about 25% more expensive, reflecting the greater simplicity of the methanol process. Coal consumptions shown in Tables 17.3 and 17.4 remain valid.

NCB: Coal Research Establishment (CRE)

In 1980, CRE carried out a literature survey of synthesis gas process technologies. Some references to the economics of ammonia production were collected (Table 17.5); no attempt was made to rationalise the information, by reducing costs to a common basis of data or financial conventions.

CRE noted the apparent wide range of estimates ($116–283 million in 1978/79 values) for 1000 t/d ammonia plants. It was considered that a coal-based plant would cost nearly 80% more than a plant based on natural gas and nearly 50% more than one on fuel oil. The estimates for coal-based plants to produce methanol were in the range $130–303 million. (The view, by H and G, that methanol plants would be cheaper than ammonia plants does not emerge from this collection of data; it is nonetheless considered valid.)

ICI (Andrew)

S. P. S. Andrew is a highly original and respected Senior Research Associate with ICI. He has expressed views on future feedstock economics which may not

TABLE 17.5 Estimates of investment and production costs: ammonia

Process feedstock	*Scale t/d ammonia*	*Total investment million US$*	*Production cost per tonne ($)*	*Year*	*Source*
Natural gas	1000	55	72	1974	
Naphtha	1000	61	78	1974	
Fuel oil	1000	69	82	1974	
Coal	1000	98	120	1974	
Coal	1090	140	152	1976	Koppers
Coal	1360	185	157	1977	Fluor
Coal	1815	364[a]	260	1980	Chem Systems
Coal	1815	319[b]	223	1980	Chem Systems
Coal	910	140	165	1977	TVA
Coal	1000	127	288	1978	Kellogg
Coal	910	141	198	1979	TVA
Coal	2500	519	315	1985	Chem Systems
Coal	1000	141	160	1979	TVA
Coal	1000	(DM)330		1979	Lurgi
Coal	1000	200	250	1979	Krupp Koppers
Coal	1090	175	196	1979	Davy Powergas
Coal	1045	120	196	1979	Kellogg

[a] K–T gasifier.
[b] Second generation gasifier.

necessarily represent the attitude of his company but which need to be considered. His estimates in 1977 are illustrated in Fig. 17.3, which stress his conviction on the need for large plants and cheap coal. He considers that demand for gasoline and other transport fuels will remain large compared with chemical feedstocks and that the production of synthetic gasoline will be concentrated in large plants near cheap coal. In Fig. 17.3, (all costs in 1976 dollars) the solid line indicates his view of the relationship between coal prices and synthetic gasoline costs: coal at $10/t can be converted to gasoline at $25/barrel (at a scale of 20 000 t/d). One dashed line, using the right hand and upper scales, represents the cost of converting this synthetic gasoline to synthesis gas on a scale of 1000 t/d. The other dashed line (using lower and right hand scales), represents the manufacture of synthesis gas from coal, also on 1000 t/d scale. The intercepts of the three lines on the left hand axis represent the relative unit capital costs, compared with the capital costs for the conversion of gasoline to synthesis gas taken as unity; the relative capital cost for synthesis gas from coal is 1.2 and for coal/synthetic gasoline is 2.2. The dotted lines indicate that if coal at $10/t is used to make gasoline, subsequently converted to synthesis gas, this will compete with synthesis gas from coal at about $37/t. The author concludes that in the UK (and presumably the same conclusions apply to Europe and many other parts of the world) oil will continue to be used for petrochemicals for another 50 years; synthetic gasoline from large plants in cheap coal areas will then take over for another 100 years, followed eventually by UK coal.

There are, however, arguments against this analysis. If coal were available at $10/t in, say, Australia, it is difficult to believe that it could not be brought to Europe for less than $37/t, thus undercutting the imported synthetic gasoline. When imported or relatively expensive indigenous coal is used locally in a chemical plant, it would also be expected that measures to reduce effective conversion costs would be adopted, such as the exploitation of steam and heat and the combination of

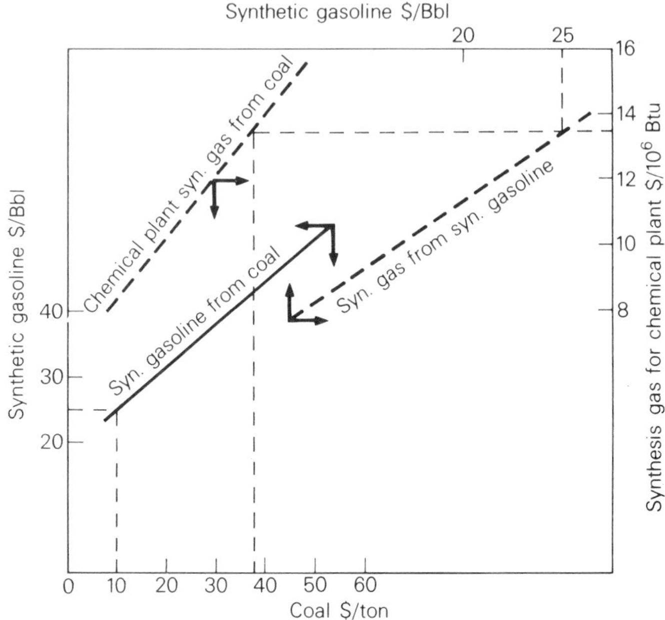

17.3 Comparison of chemical plant synthesis gas costs made directly from coal and indirectly via synthetic gasoline produced in a very large gas + gasoline plant

various fuel and feedstock demands (e.g. coalplexes). In all conversion plants, large or small, outlets for heat or fuels will have an important effect.

In a further paper in 1979 Andrew discussed "The Economics of Fuel and Chemicals from Coal in the Future". This was a broad study, too complex and too dependent on personal judgments to be comprehensively summarised; assumptions about nuclear power seemed excessively (and uncritically) favourable. However, the future price of naphtha in real terms (1978 = 1.0) was taken as the guide to the use of coal for ammonia and chemicals. In UK (and Europe generally), a theoretical breakeven position would be reached when the relative price of naphtha reached 1.4 (presumably with coal remaining constant in real terms; the ratio would have to be adjusted proportionately to take account of real coal price increases). In North America, ammonia and methanol were considered about economic at that time but it is acknowledged that a measure of "overheat" (of unknown degree) would be necessary for investment action to be taken.

The most interesting part (in the present context) of Andrew's 1979 paper is that he considers synthetic gasoline will not reach breakeven until relative naphtha prices (1978 = 1.0) reach between 1.8–2.1 in North America and between 2.8–3.2 in Western Europe. SNG in North America will require a factor of 1.9. These are all very much less favourable than chemicals from coal-based synthesis gas. Andrew ends with an expression of doubt about whether oil prices will increase as rapidly as some forecasters suggest.

Nonetheless, his analysis certainly strengthens the view that ammonia and methanol (and perhaps some associated or derived chemicals) via synthesis gas will be the earliest route from coal into chemicals, whatever the time scale, which will of course be largely dependent on coal/oil price ratios (but also on the penetration of nuclear power).

OTHER ROUTES TO CHEMICALS

Hydrogenation and Liquid Solvent Extraction

Syncrudes made by hydrogenation and similar processes will generally be valued as fuels, at prices equivalent to similar petroleum products. It does not appear that syncrudes would be produced (on the very large scale which is required for cost purposes) specifically for chemical feedstock unless the syncrude product has some technical advantage. The only apparent advantage may be in aromaticity; if the average prices of aromatics increase more rapidly than aliphatics (perhaps because of increasing demand for non-leaded gasoline additives), this could marginally affect the timing of hydrogenation applications for chemicals and possibly also the preferred process design details. The development of uses of coal extract in the production of special cokes and carbon artefacts would also be marginally more favourable.

The Ralph M. Parsons Company has studied the question of feedstocks from coal. In 1978, they presented a complex scheme involving hydrogenation, F–T synthesis and gasification (see Chapter 20). Aromatics were taken from the liquid extraction stream but only represented about 2.5% of the energy input. It may be that if aromatics are required, processes other than standard hydrogenation may be desirable. The overall Parsons scheme was not economically attractive in 1978 by normal financial criteria; the assumption of an annual inflation rate of 6% would however give the project an acceptable return.

Pyrolysis and Gas Extraction

Both of these processes appear promising methods for the preparation of a narrow range of products of interest to the chemical industry, especially if aromatics are required. In the case of pyrolysis, flash pyrolysis or a two-stage hydropyrolysis may be the most appropriate versions for chemical purposes.

Little can be said at this stage about the economics. Preliminary cost estimates have been carried out on gas extraction and on several versions of pyrolysis and have indicated that the processes could be competitive with other forms of liquefaction. However, much more experimental data on a larger scale will be necessary to refine and validate economic projections. The specialised use of these processes for chemicals would also require further work. In any case, the value of co-products such as fuel gas and char will have an important influence, requiring complex schemes.

Acetylene

The economics of acetylene as a route from coal to chemicals, are particularly speculative and convoluted. On the one hand there is the comparison between ethylene and acetylene as precursors for chemicals. Secondly, there is the question of which is the best route from coal to acetylene and, finally, there is the question of when, if ever, coal could compete with other feedstocks for the manufacture of acetylene. At present, it appears that acetylene is unlikely to recover its prominence unless there is some significant change in the relative prices of acetylene and ethylene, sufficient to overcome the technical advantages of ethylene; it is difficult to see this happening or being encouraged by a switch from oil to coal. However, in special circumstances, the coal/carbide/acetylene route is being pursued on a small scale and no doubt any favourable trends will be published and exploited.

However, AVCO and other prominent developers apparently have confidence that the plasma arc process will become economic. In Germany, one study has suggested that the AVCO route is superior to the carbide route and that if coal prices escalate by 6.5% per annum and LNG by 10%, the AVCO coal route will be superior to the normal arc process using LNG by 2000. This forecast makes the assumption that the technical promise of the AVCO process will be demonstrated on the full scale. Some important and possibly controversial assumptions about by-products were also made to reach this conclusion. The competitiveness of acetylene with ethylene is apparently dependent on a shortage of oil supplies but this particular study does not consider whether other routes from coal to chemicals might be superior.

CONCLUSIONS

Most coal conversion processes have some interest for the chemicals industry; it is difficult, however, to identify and separately assess those aspects of greatest importance for chemicals.

In 1977, the Economic Commission for Europe adopted a report, prepared by the UK, on coal in the chemical industry; the conclusions were as follows:

"1. The total quantity of chemical products which could be produced from coal carbonization by-products is limited by the demand for coke in the steel industry.
2. Technology exists or is being developed which could provide coal-based alternatives to the oil-based feedstocks of the petrochemical industry. In general, aromatics would be expected to be provided by purification of liquefaction and pyrolysis products while olefins would probably be provided by gasification and synthesis routes.
3. In order to be competitive with oil as a chemical feedstock, the cost of coal on a calorific value basis would have to be about one-third of that of oil.
4. The upgrading of fuel oil by hydrogenation to lighter fractions provides a firm option for extending the availability of supplies for transport fuels and chemical feedstocks and the process economics are more favourable than coal conversion.
5. In the event of a demand for the upgrading of fuel oil on a large scale, space heating and power generation needs at that time satisfied by fuel oil, could be satisfied by new coal or nuclear capacity.
6. For this reason, large-scale coal conversion to petrochemicals appears unlikely in the near future. When appropriate, the total world market would be about 400 million tonnes per year plus whatever equivalent market growth had occurred in the intervening period. The capital cost of the necessary coal conversion plant would be of the order of $\$50 \times 10^9$ in late 1975 prices.
7. It would seem unlikely that the problem of coal conversion to petrochemicals could be divorced from the much larger question of the conversion of coal to transport fuels."

These conclusions are still broadly valid. However, the last point may need to be modified with regard to the synthesis route. Coal tends to compete in the synthesis route directly on economic terms with oil and gas; the coal hydrogenation route is more complementary with oil because the raw coal products can be integrated with the existing oil refining/petrochemical activities. When hydrogenation is introduced, some coal-derived products will begin to find their way into chemicals. It seems doubtful, until and unless new processes are developed, that large investments will be made in these routes purely for chemical purposes.

It may in fact be perfectly possible to separate the use of coal for chemicals via synthesis gas from transport fuels, especially for ammonia and methanol (although methanol is also a possible route to transport fuels). The economics of coal versus gas/oil in this field are relatively straightforward, depending directly on price relationship. It is generally accepted that relatively small movements in the ratio of coal/gas or oil would be adequate to make coal generally attractive; in some circumstances, it is clearly thought to be so already. Since synthesis gas is also a cheaper fuel than SNG, if these fuel values could be concurrently exploited, the case for the synthesis route to chemicals would be reinforced. The utilisation of heat rejected in the gasification stage would also be an encouraging development.

Chapter 18

ECONOMICS OF UNDERGROUND COAL GASIFICATION

A GENERAL FRAMEWORK

Introduction

As with other coal utilisation processes, economics will be decisive in determining the degree of exploitation of a technically feasible operation. In the case of Underground Coal Gasification (UCG) there are several important economic questions:

(1) How does UCG compare with other methods of exploiting coal resources?
(2) How does UCG compare with other energy sources?
(3) Which are the most economic UCG processes in various circumstances?

These questions are not fully answerable at present but a framework can be developed and some views which have been expressed can be considered. It will be obvious from the technical description in Chapter 11 that the ease and cheapness of recovery by UCG will increase with seam thickness and decrease with the thickness of overburden. Thus, there is a parallel with extraction methods, especially opencast; in each case, qualitative considerations of both coal and overburden are complicating features. It is possible however that ash content may not in UCG be the increasingly negative factor it is in extraction and surface utilisation methods.

In 1976, the National Coal Board carried out a study of UCG, including an economic modelling exercise covering depth, seam thickness and other variables, with some sensitivity analysis. This was based on British experience, which dated back to the 1950s, plus information on more recent work, obtained both from the open literature and privately from many of those then working in the field.

Assumptions

The NCB study considered both the Preliminary Mining Method and the Vertical Drilling Method using reverse combustion with air. Only the latter is considered here since current opinion strongly favours this option.

Gasification to feed a power station was assumed and the scale of the hypothetical project was set at 100 MWe. In passing, it may be said that there appeared to be no factors which would result in a sharp reduction of costs with increasing scale — a 7% reduction in scaling up from the 100 MW basis to 1000 MW was indicated. This appears to be contrary to thinking in the USSR which suggests much greater economies of scale.

Some of the other main assumptions were:

Vertical borehole spacing	30 m (at corners of square)
Proportion of coal gasified	85%
Proportion of heat gained	60%
Overall efficiency	51%
Gas calorific value	80–100 Btu/m^3* (3.2–4 MJ/m^3)
Leakage rate	4%
Operating pressure	2 bar

For the basic case, a seam thickness of 1 m or more was taken, with no excessive faulting or porous strata.

In the project size chosen, about 18 panels would be operated at once and each borehole would gasify 1500 t of coal in a 1.2 m thick seam. It was assumed that drilling could continue on a 24 hour-per-day seven-days-per-week basis. (Departure from this to lower operation factors would be very costly.) Results and costs were all in November 1975, UK money values; a discount rate of 10% and a 20 year life were taken.

Results

The results of the study are shown in Fig. 18.1, in the form of cost contours. The costs, ranging from about 7–24 p/therm (66–168 p/GJ) compare with then-current costs of about 3 p/therm (28 p/GJ) for UK opencast coal and 4–12 p/therm (38–114 p/GJ) for deep mining. At this time, oil prices to commercial users were about 10 p/therm and natural gas about the same. The graph confirms the sensitivity of costs to depth and seam thickness. Up to depths of 500 ft, costs are mainly about 10 p/therm (95 p/GJ) or under and this appears favourable compared with natural gas or oil. However, since the UCG gas is lean, its transport is expensive, so that piping more than about 10 miles would be unlikely; a distance of 50 miles, for example, might add another 6–6.5 p/therm (56–58 p/GJ).

The UCG system is more capital intensive and less labour intensive than underground mining. The cost breakdown, for a 1.2 m seam at 300 m depth, to produce the gas for a 100 MW station is shown in Table 18.1.

Since 1976, capital costs for equipment of the type required has probably increased more rapidly than general inflation, which might tend to make UCG less attractive; this probably applies even more forcibly however to other energy industries. Labour costs in mining may also have increased more rapidly than general inflation in some places and this would somewhat improve the competitiveness of UCG.

Because of the importance of the capital cost element, gas costs are sensitive to gasification efficiency and very sensitive to the distance between boreholes; both of these of course affect the number of holes required to be drilled. Table 18.2 shows the effect of spacing. The effect of overall efficiency (Table 18.3) is less marked. Costs increased at higher pressures (20 bar) (if the pressure energy were not used in a gas turbine) as in Table 18.4. The effect of calorific value (Table 18.5) was smaller.

This analysis may underestimate the extra value of higher quality gas. Similarly, it is suggested that the extra cost of using oxygen would be about 7.0 p/therm (67 p/GJ) (on the datum of 12.4 p/therm (118 p/GJ)) and that this cost would not be recouped.

* On a mole % basis this might consist of $CO=10$, $CH_4=2$, $H_2=12$ and $N_2=60$.

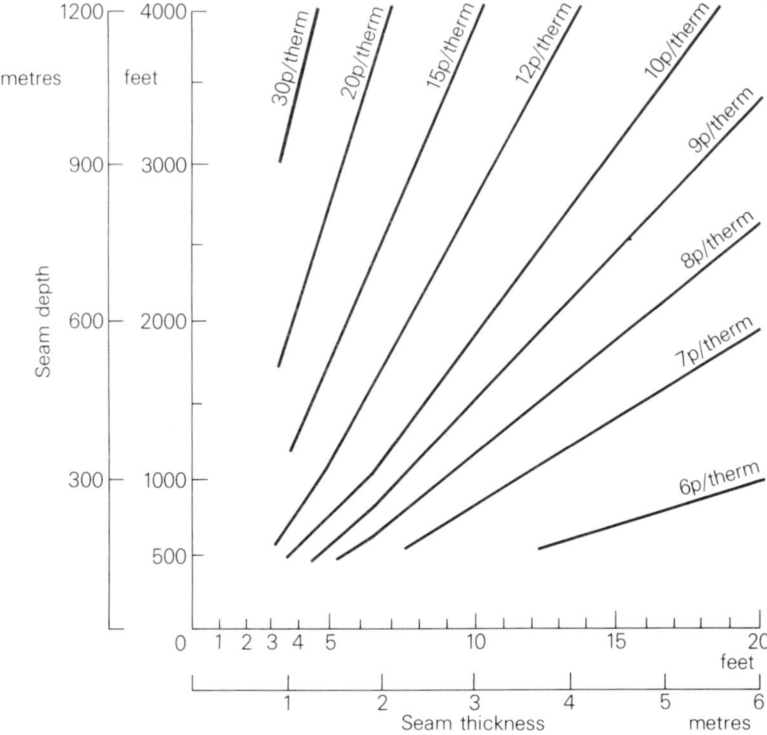

18.1 Cost contours for the vertical drilling method

Table 18.1 Cost breakdown of the UCG system

	Overall plant cost £million	*Cost/unit of clean gas p/therm*
Boreholes	68.8	6.59
Compressors and power for compressors	33.8	3.25
Other capital costs	7.0	0.67
Labour, etc.	20.0	1.91
Land	1.8	0.02
Total	131.4	12.43 p/therm ≡ 118 p/GJ

TABLE 18.2 Effect on cost of distance between boreholes

Distance between boreholes, m	*p/therm*	*p/GJ*
18	27.8	259
70	12.4	118
60	7.3	69
120	5.9	56

TABLE 18.3 Effect on cost of overall efficiency

Overall efficiency, %	*p/therm*	*p/GJ*
40	14.3	136
51	12.4	118
65	11.0	105

TABLE 18.4 Effect on cost of operating pressure

Operating pressure, bar	*p/therm*	*p/GJ*
1.4	12.1	115
2.0	12.4	118
20.0	15.7	149

TABLE 18.5 Effect on cost of calorific value of gas

Calorific value of gas, Btu/cu. ft	*p/therm*	*p/GJ*
70	12.6	120
80	12.4	118
160	11.6	110

However, this is also a matter for continuing review: experience with large oxygen plants might reduce the extra cost and the additional flexibility of use for a medium-Btu gas free of nitrogen may become increasingly more valuable.

One interesting feature of UCG economics is that most of the capital equipment in plant could easily be used elsewhere. Thus it might be possible to close or cut back a site without great investment loss except for holes drilled and not exploited fully. (Holes are not drilled much ahead of requirement.) This applies to the gas production equipment and not necessarily to the electric plant.

The Energy Account

The NCB study also considered the ratio of energy output to energy input. The energy input is of course taken into account in the financial calculations but it is sometimes useful to consider energy "gain". Energy is consumed in drilling, compression, gas purification and site maintenance and servicing. Most of the energy consumption is for compresson and this is divided into the main gasification compressors and the linkage compressors. The former will use the equivalent of about 10% of the output energy. The linkage compressors, whilst smaller and less energetic, are sensitive to depth and seam thickness; in the standard case they may use the equivalent of 5% of the energy produced. Other energy consumptions are about 50% of the total compressor use. Thus, for the standard case, the ratio of energy output to input is about 4:1. This compares with about 20:1 for deep coal mining and 13:1 for opencast working; oil and gas also have similar high ratios. However, other fossil fuels have ratios similar to UCG when they are converted from one form into another; UCG is more representative of a secondary fuel (or a primary plus secondary) energy account.

The energy account ratio for high pressure deep UCG is bound to be lower than 3:1. However, this may not be important in cases where reserves would not otherwise be exploitable, provided there is a net contribution to energy. This seems highly probable, provided the process works at all, but needs verification when experimental data are available.

Comparisons with other Energy Processes

The cost per therm of UCG is likely to be substantially more than that of coal produced by opencast working; the ranges for UCG and deep mined coal overlap but with a general advantage to deep mining. There is a distinct disadvantage on transport costs for UCG (p/therm mile).

UCG	Coal (Rail)	Oil (Pipeline)	Natural Gas (Pipeline)
0.06–0.10	0.02	0.08	0.02

For a UCG site at least 5 Mt and preferably 10 Mt of coal would need to be available. This is less than for a deep mine but substantially more than the minimum that could be economically worked on a reasonable opencast site. Furthermore, the extraction efficiency of opencasting is likely to be substantially higher than UCG, which should however be equal or better than underground mining. The constraints on choosing a seam will be somewhat different from mining but may not be a very critical factor. More important is the strong probability that those seams for which the UCG costs appear reasonable are precisely those cases also favourable for opencasting, which would certainly be cheaper. The land costs and land use restrictions would probably not be significantly more favourable for UCG compared with opencasting. Unless, therefore, there are serious constraints on opencast working, not at present identifiable, it is not easy to see cases where UCG might be preferred. Beyond the maximum depths for opencasting, however, economics are more evenly balanced and UCG could develop an advantage if some of the more advanced procedures are successful. At depths beyond those suitable for underground mining there is a totally new situation which may depend on the prices and availability of natural gas and oil.

A further factor, which may be important in circumstances where the requirement to exploit coal to the maximum becomes urgent, is that of manpower. UCG uses about the same manpower as opencasting but only about one third as many as deep mining; more important, different skills are necessary and perhaps a wider labour market may be available.

Another useful comparison is through electricity costs, which were assessed in 1976 as follows (p/kWh):

UCG	Coal	Oil	Nuclear
1.2–2.5	1.1–1.5	1.3–1.6	1.1

Although the lower part of the UCG range may be open to question, the comparison with nuclear power is interesting, since there may be limitations on expansion in this area; in addition, it seems unlikely that other forms of energy (e.g. renewable sources) will be as cheap as nuclear power.

OTHER FORECASTS AND ESTIMATES

USA

The US DOE clearly considers that an economic assessment is dependent on data from the current experimental work, lasting perhaps into 1982. Nevertheless, a number of estimates have been produced elsewhere in the USA and presumably those licensing Russian technology have satisfied themselves that the economic prospects were attractive.

As in the NCB estimates, gas costs are shown in US estimates to be closely related to the seam thickness and the depth of the overburden. It is considered with Texas lignites that thickness becomes less important over about 2 m, and that a low CV gas can be produced competitively with overburden/seam thickness ratios of up to 150:1 and possibly higher. This compares favourably to strip mining where economic ratios might approach 15:1. However, with a 2 m thickness and a ratio of 150:1, gas costs (in 1976 dollars) are given as approximately $2.5/MBtu. The point of comparison appears to be related to surface coal gasification rather than to regular prices of natural gas or other fuels which could serve the same purposes as UCG gas, at lower prices.

In this study, it was assumed that an overall efficiency (energy in coal in place/energy in gas recovered) of 48% was achieved, with boreholes at 23 m spacing. Capital costs of $200 million were estimated for a UCG facility sized to service a 1000 MW electricity station; of this, only $5.5 million was for compressors and $48 million was for gas clean-up (removal of sulphur and particulates), probably because a gas turbine requiring cleaner gas was included in a combined cycle. A point was made that the combustion of lignites would not be satisfactory from a sulphur point of view for combustion above ground in a power station without FGD. Removal of sulphur from the UCG gas is much simpler than from the flue gas and because of the smaller gas volumes in any case it is believed that half or more of the inherent sulphur is retained in the underground products; the remainder is in the form of H_2S (because of the reducing conditions underground) and is therefore cheaper to remove than the oxides of sulphur.

Belgium

Some indications have been given of the hoped-for economics of the deep, pressurised system. Net thermal efficiencies of about 60% are predicted, compared with 40% assumed for shallow, low pressure systems. It appears, however, that this 40% corresponds to the 48% estimated in the USA. If the latter is achieved, the extra efficiency hoped for from the deep method would of course be reduced but this may not be the crucial point.

The volume of coal which can be gasified from each borehole is taken as the main variable. Two levels are taken — 5000 and 10 000 m^3/borehole. The latter corresponds to working a 2 m thick seam from a network of boreholes 70 m apart. On a thermal basis it is estimated that at a depth of 1000 m and with 5000 m^3/borehole, UCG would about compete with coal from the Ruhr (about $7/Gcal, 1977 prices); with a working value of 10 000 m^3, costs would come down to about $4/Gcal, similar to US and the cheapest British coals. Discussions have indicated that the Belgian estimates are comparable with British estimates given the assumption that higher pressures would lead to greater distances between boreholes.

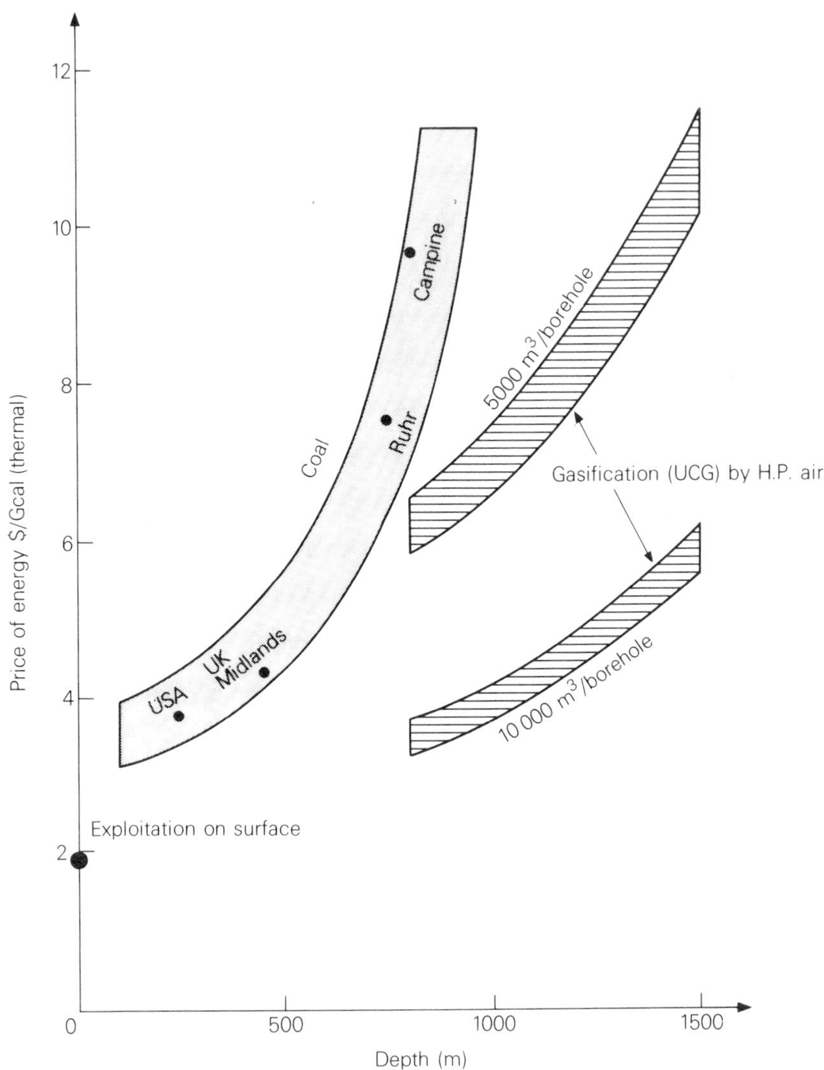

18.2 Comparative costs (thermal) of coal exploitation; surface and UCG

The important point about the potential of the deep system — that it makes available coal which would not otherwise be exploited — is neatly illustrated in economic terms in Fig. 18.2. This shows some thermal costs (1977 basis), intended to be typical, for opencast coal and deep mining in the USA, UK ("cheap" coalfield), the Ruhr and Belgium. The actual numbers are not very important and it would be possible to substitute some better worked figures for the deeper depths. However, the trend is certainly correct: mining costs will escalate very rapidly below a certain level (which may not be far from 1000 m) and there will be an economic barrier. The slope will be less steep, it is suggested, for deep UCG, especially if large volumes can be accessed from each hole.

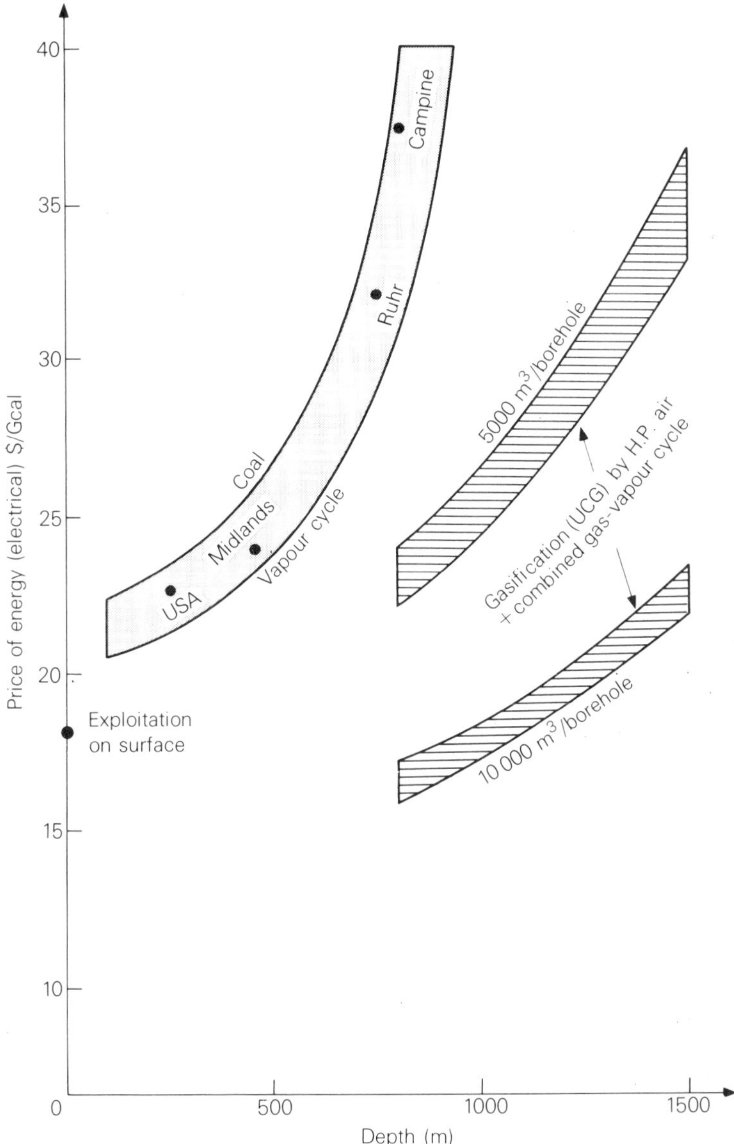

18.3 Comparative costs (electricity) of coal exploitation; surface and UCG

These thermal costs can be translated into costs as electricity, medium Btu gas and SNG, in each case with additional advantages compounding those indicated in the extraction process. The results for electricity, as suggested by Belgian workers, are shown in Fig. 18.3. Several, possibly controversial, factors are assumed. One is that investment and operating costs in the electricity plant would be slightly less using gas rather than coal. Secondly, a combined cycle (39% efficiency) is assumed for the UCG case compared with a steam cycle (35%) for coal. Finally, because of the flexibility of the UCG system, including the possibility of storing compressed air underground at night, a high utilisation rate (7000 h/a) is assumed for UCG compared with only 4500 h for the coal-fired steam system. It could be said that the case for the deep UCG case was being bolstered by comparing it with less than the best alternatives. However, it is commendable that the Belgian protagonists (and their German collaborators) should not restrict themselves merely to the costs of extracting thermal energy but should examine how the utilisation of this different form of extracted energy may be used to particular advantage in conversion processes.

CONCLUSIONS

The general feasibility of UCG has been established (at depths normally exploitable by opencasting or underground mining). It remains to be seen what economic and environmental constraints may limit the widespread adoption of UCG in Western countries. It is not felt that experience in these two areas obtained in other types of economy or by operations set up more than a decade or two ago would be very relevant although some of the technological guidelines undoubtedly are. The necessary information to provide a "base" case for economic assessment and environmental impact statements should be available by about 1982–3.

On present estimates, it appears unlikely that UCG could compete on a thermal cost basis with opencast coal; there may be some overlap between UCG and deep mining but the best of deep mining may be cheaper than UCG. There may however be some factors which could gradually shift the balance in favour of UCG. Also, for reasons not yet discernible, it may prove desirable to exploit a particular packet of coal by UCG even when other methods might in principle be cheaper; such reasons might be environmental, social or labour availability, although such factors may not all be in favour of UCG. At the same time, two disadvantages of UCG should be borne in mind:

(a) Subsidence may be uncontrolled;
(b) Fault-free seams may be even more desirable for UCG than for conventional mining.

There may be technical advances which significantly enhance the prospects of UCG. First and foremost, these might be in the development of cheaper, faster and more highly controlled drilling methods, including linkage procedures; the possibilities here must be rated favourable. This must also apply to the development of sophisticated instrumentation and control equipment, first to improve basic understanding and then, in a modified form, to be applied to practical cases.

The development of better methods of utilising UCG gas would also be important and this has several aspects. Most vital might be a better appreciation of the value of a gaseous form of energy, even such a poor one, at a time when natural gas will need to be supplemented by substitute gases at several times the thermal value of coal. It is difficult to assess the prospects for upgrading the quality of the gas, either after extraction or through the gasification process itself. The economic effects of the use of steam or oxygen in the gasification medium are not well established and the use of hydrogen is very speculative.

The deep, pressurised UCG system, proposed by the Belgians in particular, is also speculative. It would be easy to react to the enthusiastic claims of the protagonists of this idea with scepticism but it should not be overlooked that those enthusiasts are highly competent scientists, whose ability has been demonstrated in other fields.

Finally, it should be reiterated that a special advantage of UCG might be to make available an enormous reservoir of energy, the value of which may be difficult to measure, even in post-1973 terms. This may be the particular appeal of the deep pressurised system. If UCG were to be compared, on a quantitative and economic basis, not with established forms of energy such as coal, oil and gas, but with some of the "newer" forms of energy proposed which are "weak" sources (geothermal, thermal currents), it is clearly a method demanding substantial support.

Chapter 19

ENVIRONMENTAL IMPACT

INTRODUCTION

Issues relating to environmental impact are likely to be amongst the most crucial problems to be overcome in the "second coming" of coal. Public awareness of certain potential problems, in some key countries especially, has been raised to a high level, indeed to a degree which is incompatible with the factual knowledge available. To some extent, this situation has been enhanced by interveners whose interests are not dispassionate or objective; these include "professional" environmental specialists and representatives of other energy lobbies, often stung by criticisms of their own favourite energy source. However, any "success" which the anti-coal environmental lobby has achieved may be largely laid at the door of the coal industry, sections of which have been notoriously ignorant, secretive and indifferent. It is hoped that this chapter will be consistent with a responsible and forward-looking attitude leading to a better public presentation.

It is the firm belief of the authors that there are no environmental impact problems that cannot be solved — or, more properly, for which suitable but progressive compromises cannot be found. Compromises are essential because changes due to environmental impact inevitably have economic consequences, which is why this chapter is placed in the Economics Section, although some technical features of control have been dealt with in individual chapters. The economic effects are not merely those of modest increases in cost; they can seriously affect supply of both primary and secondary energy and thus create a ripple effect throughout the economy. This is one reason why international collaboration on environmental impact is especially desirable.

This chapter will be concentrated on the environmental impact of the utilisation of coal, especially via combustion, which is the most immediately controversial area; various control measures, mentioned in Chapter 6, can be applied, dependent on judgements about necessary standards. Some comments will also be made about the newer coal conversion processes. The utilisation of coal depends on getting it and transporting it; these subjects, including the safety and health of miners, are important but are largely outside the scope of this book.

GENERAL ENVIRONMENTAL EFFECTS OF COAL UTILISATION

The USA and Japan are probably the two countries where environmental impact considerations have so far had the most serious impact on energy systems. Japan temporarily is relatively less dependent on coal than the USA and there are other features of her system which are unique. Attitudes in the USA (reflecting a high degree of public interest and an extremely active but beleaguered bureaucracy) are more advanced (but not necessarily more correct) than in most other parts of the world; this will probably ensure that events in the USA will have a very considerable impact elsewhere. For this reason, some understanding (and assessment) of the US philosophy is very important. The main governmental agencies involved are the Environmental Protection Agency (EPA) and the Fossil Energy (FE) Division of the Department of Energy. The former's role is essentially that of assessment and regulation; the latter's responsibility is that of developing new technology and providing technical assessments. Their relationship is that of two large, unwieldly armies bogged down in trench warfare. The Congress also involves itself in detailed discussion, instruction and amendment; there are also State and local regulations. At one time, EPA directly funded and controlled most government technical development but more recently much of this has been taken over by FE, as shown by the latter's budget on Advanced Environmental Control Technology:

FY 1978	FY 1979	FY 1980
0	\$7 million	\$43.25 million

In addition, much larger sums are devoted to technologies which might not have priority for support except for environmental reasons. FE state the following common environmental impacts with which most coal technologies are associated:

Requirement and techniques for solid waste disposal;
Consumption of water;

Accidental discharge of toxic of hazardous products or by-products;

Carbon dioxide emissions (and their unknown cumulative effects);
Air pollutant effects;

Surface water pollution;
Land and water disturbances from mining;
Occupational and health and safety effects.

The major issues are considered to be:

Application of air and water control technologies;
Potential limitations upon commercialisation posed by environmental protection regulations.

These two issues describe the essential conflict, which may not best be resolved by adversary procedures and institutions. FE state that there are concerns about the environmental and health impacts of new coal conversion processes which have not yet been constructed and operated. In some instances development projects have faced these problems and have had to be handled on a site-specific, case-by-case basis. Some of the specific issues are:

Site suitability (e.g. water and coal availability);
Community effects;
Ability of a plant to meet existing standards (e.g. for trace metals and carcinogens);
Ability of a plant to meet future standards.

Taken literally, these issues require a degree of prescience and indicate little tendency towards compromise. In practice, although the new coal conversion processes are considered, as above, mainly from the point of view of siting development plants, most of the actual experimental work on environmental impact is directed towards combustion.

The EPA have set up a Coal Technology Assessment (CTA) designed to address two basic questions:

(a) Does it matter how much coal the US uses (up to 2030)?
(b) Does it matter which coal technologies are employed?

Three issues were initially identified:

(a) Prevention of significant deterioration (PSD) in air quality;
(b) Possible global build-up of carbon dioxide;
(c) Toxic substances from coal.

Three energy demand scenarios were developed:

Scenario A. Economic growth. Based on technological and management success: the so-called "Hard Path",
Scenario B. Conservation Ethic. Based on Social Change: the "Soft Path",
Scenario C. Business as usual.

The total energy use (in Quads/year; 1 Quad $= 10^{15}$ Btu) in the three Scenarios is different:

	2000	2030
Scenario A	95	125
Scenario B	75	60
Scenario C	80	85

Within these figures, the coal use varies widely and the technology mix for coal use also varies. For each set of circumstances, the gross impact of various types of environmental damage is estimated and control methods considered. These will be referred to under appropriate topics below.

In the UK, there is a long history of progressive compromise, based initially on the Alkali Inspectorate, set up in the pioneering days of the chemical industry and recently incorporated in a Health and Safety Executive. More recently, a Commission of Energy and the Environment has been established and has undertaken a Coal Study. The National Coal Board and Institutions have submitted views and some of the topics considered to be within the field of review of the Commission include:

1. Deep-mined coal production
   (a) subsidence
   (b) dust and grit
   (c) water and effluents
   (d) noise
   (e) spoil disposal and utilisation
   (f) land requirements

2. Opencast coal production
   (a) overburden and coal excavation
   (b) dust and grit
   (c) water and effluents
   (d) noise
   (e) change of surface features
   (f) land requirements and reclamation
3. Coal utilisation
   (a) combustion for power, industrial and domestic purposes
   (b) coke and smokeless fuels
   (c) gasification and liquefaction
   (d) sulphur oxides
   (e) nitrogen oxides
   (f) ash, dust, grit and smoke
   (g) trace elements, including radiation
   (h) carbon dioxide
   (i) hydrocarbons and carbon monoxide
   (j) waste heat
   (k) water and effluents
   (l) land requirements

It will be seen from this summarised list and from the US studies that the environmental impact of coal is a very wide-ranging subject. It is also probably one which has been comparatively neglected until recent years and therefore one in which judgements taken in advance of relevant information, which should surely now be in the process of being collected, might be dangerous and counter-productive.

Coal Production, Preparation and Transport

Deep-mined Coal Production

(*a*) *Subsidence.* When a panel of coal is extracted and the void is left unsupported, a depression is transmitted to the surface; this of course moves with the progress of the workings and the area affected at one time is typically about the same in diameter as the depth of working. The amount of lowering at the centre of the depression may be calculated quite accurately from the depth of the seam, the thickness of the coal extracted and the length of unsupported working; it is never more than 90% of the extraction thickness and may be as low as 20%. Tensile stresses occur at the centre of the depression and compressive stresses at the edge, these passing through a point on the surface progressively. Damage to buildings, road, etc. is less extensive however than might be thought and since newer workings will tend to be deeper, subsidence will become relatively less important. Some preventative work on the surface can be effective but the main approach to control is by leaving unworked panels or reducing the spacing between pillars. This is economically very important, however, and some moderate damage followed by restitution is often more acceptable.

(*b*) *Water and effluents.* Water can penetrate mine workings from the surface or strata and must be pumped to the surface and diverted into water courses; in the UK, for example, three tons of water are pumped to the surface for every ton of coal raised. Much of this water is relatively clean and creates no problem. However, some waters may be contaminated by suspended solids, by chlorides or sulphates and by various metal salts, including iron, calcium, aluminium and magnesium; they may also be acidic and this problem is very prevalent in the USA. Changes in the acidity or state of oxidation may occur after discharge with resultant precipitation, creating for example an unsightly ferruginous deposit in streams. Suspended solids may be removed by settling in lagoons but in severe cases filtration may be necessary. The lagoons may also be used for neutralisation, aeration or chemical treatment. The extent of necessary treatment varies with the site and can be very expensive. In some cases it may be preferable to pipeline the effluent to estuaries or the sea. There is a growing tendency to prevent infiltration by sealing mines or by the promotion of run-off.

(*c*) *Spoil Disposal and Utilisation.* In the UK about one-half ton of waste is brought to the surface for every ton of coal and it is of course cheaper to dump the waste in a simple pile in the most convenient place near the pit. At one time, tips were allowed or even encouraged to burn; erosion and landslides were other problems. Much more acceptable practices are now normal. Tips are constructed and contoured to have minimum visual impact; they are often consolidated and grassed as dumping proceeds; this prevents fires, erosion, slides and severe leaching. Burning tips are things of the past. Where leaching can occur, water treatment of the run-off effluent may be necessary. Attention to all these matters has been greatly intensified since the Aberfan disaster. Underground stowage is seldom practised, since it is not compatible with modern mining methods but some waste material is constructively used as landfill or building material, including lightweight aggregate. It is believed that these uses will grow.

Opencast or Surface Mining

The main problems relating to surface mining are overburden removal and reclamation. Top soil and sub-soil must be carefully and separately removed and stored. Where the ground is sloped, the choice of working contour and compaction can prevent erosion and leaching. Noise, dirt and visual effects are potential problems which can be mitigated by screening and carefully regulated working.

On some sites, previous industrial dereliction can be cleared and the site rendered more valuable as part of the coal-getting operation. For agricultural land, reclamation requires re-contouring to the original or other acceptable shape, the restoration of buildings and other features and the revegetation of the land, after restoring the soils and compacting in stages. Very good results can be achieved within about five years. In the USA, land can not now be worked for coal unless it can be demonstrated that revegetation is possible; in some arid zones this requirement has prevented operations.

Coal Preparation

Coal preparation plants are potentially serious sources of pollution; prevention of these effects is an important feature of modern design and adds substantially to capital and operating costs. Dust must be controlled by cyclones, electrostatic precipitators or bag filters where necessary. However, liquid effluents and the removal of fine solids are generally more potentially serious problems. Some plants now have closed circuits and produce virtually no liquid effluent. In other cases, suspended solids must be removed by flocculation and sedimentation; chemicals used in the process and leachants, such as phenols, must be destroyed. Fine wet material consisting of mineral matter (often clays) and some coal are problems not always satisfactorily dealt with by settling ponds; the use of fluidised combustion has been developed as an alternative, turning the wet waste into a dry friable powder which is disposable, or may have uses.

Transportation

Some problems such as dust and the recovery of "ink" from pipeline wastes are discussed in Chapter 21.

UTILISATION: COMBUSTION

General

The environmental impact of coal is frequently equated with the effects of combustion. This is reasonable since such a high proportion of coal is currently burned and the waste products may be observed; even when combustion is supplemented in the future by coal conversion to gases or liquids, combustion will often be a final stage, applying often to just those portions of the coal involving the most serious polluting effects. Furthermore, the emphasis on combustion may be due to personal experience in the past with black smoke plumes and smogs. In most countries the current visual experience is very different from that of two or three decades ago but the suspicion still exists. Most people associate combustion pollution with sulphur and smoke (or other particulate matter) and these are still prime problems, probably associated in fact as well as memory. However, in most Western Countries great progress has been made since World War II, especially in suppressing particulates; this advance should be borne in mind when considering reports of damage, to health or materials, where the statistical evidence may either be old or the root-cause possibly of long standing. In fact, the greatest problem in assessing environmental problems is the lack of good information about levels of pollution and, more particularly their effects, based on modern conditions. An indication of the changes which have taken place is given in Table 19.1; this is for the UK, but is typical of a universal need to relate conclusions to recent data.

The traditional concerns with coal combustion have been sulphur dioxide and smoke but more recently other problems have been discussed, including oxides of nitrogen, oxides of carbon, hydrocarbons and trace elements.

Sulphur

When coal is burned, most of the sulphur is converted to sulphur dioxide, although some trioxide may be formed; sulphates and sulphites may result, directly or indirectly and there may be associations with particles.

TABLE 19.1 Mean urban sulphur dioxide and smoke concentrations

	Concentration ($\mu g/m^3$)	
Year	*Sulphur dioxide*	*Smoke*
1960/1	158	150
1965/6	116	96
1970/1	100	55
1975/6	71	32
1977/8	60	25

TABLE 19.2 Possible annual sulphur pollution in the USA

	1985		*2000*		*2030*	
	Eastern	*Western*	*Eastern*	*Western*	*Eastern*	*Western*
Ambient SO_2 concentration, microgrammes/m^3	17.0	2.85	30.4	1.64	12.3	2.72
SO_2 deposition, g/m^2	0.79	0.11	1.34	0.07	0.52	0.12
Sulphate concentration, μg/m^3	5.07	0.78	9.89	0.76	3.14	1.34
Sulphate deposition, g/m^2	0.32	0.038	0.55	0.038	0.16	0.08

In most countries coals are available in the range 0.5–1.5% S; in the UK and West Germany average levels are at the top of this range. In Australia and Canada, considerable quantities below 0.5% are produced. These two countries, together with Poland and South Africa may be important exporters and the coal involved in future world trade is likely to contain less than 1.0% S, thus slightly alleviating any problems. A great deal of the recent emphasis on sulphur emissions is due to the fact that Eastern USA coal is generally high in sulphur (more than 2.5% S). Although the majority of the Western coals contain less than 1% S, the bulk of coal mined is still in the Eastern region and the concentration of combustion is much higher there; in addition, the prevailing winds tend to move sulphur pollution from West to East. Emphasis in the USA on coal as the culprit is further increased by the fact that indigenous oil there tends to have moderate sulphur levels. Elsewhere in the world coals may be no worse and possibly better than oil, with regard to sulphur levels. The extent of the total problem in the USA is indicated by the fact that the EPA scenario C predicts emissions of 18.3×10^9 kg of SO_2 in 1985, rising to 22.0×10^9 in 2030. The likely consequences of these emissions are shown in Table 19.2, for the Eastern and Western regions.

It should be appreciated that the 1985 figures imply a prediction of very significant improvements compared with previous years, which would probably be more marked in local measurements; SO_2 levels of several hundred μg/m^3 were not uncommon. This was also the case in the UK, where short-term levels of more than 1000 μg/m^3 were recorded during particularly bad smogs.

The USA has led the way in regulations to control sulphur dioxide emissions. In 1979, the level for new plants was 0.515 kg SO_2/GJ (1.2 lb/10^6 Btu) or 10% of the potential SO_2 emissions, whichever is the lower; for very low sulphur coals, the 10% rule may be waived and emissions need not be reduced below 0.086 kg SO_2/GJ. In West Germany, the limit has been set at 0.27 kg/GJ. In the UK, no controls on emissions are imposed; the Central Electricity Generating Board operates a policy of high stack dispersal which has led to a lowering of ground level concentrations. It has been claimed, however, that SO_2 can be transported in the upper atmosphere from the UK and other industrial areas of Western Europe to Scandinavia, where it may fall as acid rain, reputedly affecting trees and fish in particular. Recent work under the aegis of the OECD suggests that this problem may not be as serious as once thought but work is continuing. If changes in the policies of the UK and other countries were required for this reason alone, the "trade-off" would be a very expensive one, since the economic consequences of pollution transport and remote deposition are very slight compared with the likely suppression costs but this economic argument may not influence attitudes in countries which are net receivers.

TABLE 19.3 Sulphur dioxide standards and guidelines adopted or proposed in various countries (All SO_2 concentrations in the Table are expressed in $\mu g/m^3$ for comparison purposes. Other units may be cited in the original documents)

	Sampling period					
Country	*½ hour*	*1 hour*	*day*	*month*	*year*	
Belgium	—	—	—	—	150	In specified regions
Canada	—	900	300	—	60	Proposed: max. accep.
Canada	—	450	150	—	30	Proposed: max. desir.
Czechoslovakia	500	—	150	—	—	
Finland	720	—	250	—	180	Recommended
France	—	—	1000	—	—	Except Paris
France	—	—	750	—	—	Special zone: Paris
E. Germany	500	—	150	—	—	
W. Germany	400*	—	—	—	140	*95 percentile
Hungary	500	—	150	—	—	Special protec. areas
Hungary	1000	—	500	—	—	Other protected areas
Israel	750	—	260	—	—	
Italy	750	—	380	—	—	
Japan	—	260	100	—	—	
Netherlands	260	—	100	—	—	
Spain	800	—	400	256	150	Proposed
Sweden	625	—	250	125	—	Guidelines
Switzerland	750	—	500	—	—	Guidelines: Summer
Switzerland	1250	—	750	—	—	Guidelines: Winter
Turkey	—	—	150	—	—	Recomm: Residential
Turkey	—	—	300	—	—	Recomm: Industrial
USSR	500	—	50	—	—	
USA	—	—	365	—	80	Primary standard
USA	—	1300*	—	—	—	Secondary standard *3-hour average
Yugoslavia	500	—	150	—	—	

Apart from dispersal by high stacks, emission control by FGD or FBC has already been described and evaluated. Control by selection or prior cleaning of coal may be a limited option, with some application perhaps in the disposition of coals in the USA. It should be noted that the 90% removal requirement in the USA, in addition to the actual control of emission as a function of heat used, may prove to be rather difficult for FBC to meet without some distortion of design from optimal efficiencies, involving economic losses. For difficult cases, coal might have to be converted before combustion to SRC or, much more probably, to clean gas; in the latter case advantages in efficiency in combustion and electricity generation might be a joint objective.

In view of the expense and possibly countervailing environmental impact of control measures, especially as standards become more stringent, the evidence leading to particular standards deserves close attention. Table 19.3 gives the standards and guidelines adopted by various countries; it will be noted that there are large differences between countries not only in levels but in philosophies (as indicated by the different sampling periods adopted). It will also be noted that the UK has not adopted standards. However, a great deal of research has been done on effects in the UK, probably encouraged by the London smog of 1952; one expert opinion is that the lowest levels associated with adverse health effects on severely bronchitic patients is a daily average of 500 $\mu g/m^3$ of SO_2 *in conjunction* with 250 $\mu g/m^3$ of smoke. Since the worst days are generally five to six times the annual average, it was concluded that a level of 80–100 $\mu g/m^3$ may be a desirable standard on an annual average basis. This corresponds to the US standard of 80 $\mu g/m^3$ and proposed standards for the European Economic Community equivalent to 80–120 $\mu g/m^3$, depending on the level of particulates. These levels of maximum desirable ambient SO_2 concentration would appear to be satisfactory also for the avoidance of observable effects on plant life.

The relationship between desirable ambient standards and the emissions from any particular source is controversial. In the UK, the policy for power stations is to employ high stacks for emissions; measurements near to the point of maximum impact at ground level have shown that a particular power station has little effect on ambient air quality. In any case, it is argued, even where 90% of the sulphur originally present in the coal is suppressed, the flue gases will still require dilution by a factor of 100 000 in order to reach ambient standards and this can only be achieved by dispersal into the atmosphere via a stack of suitable design and height.

Emission controls for particular large sources must therefore in logic be related to their general contribution to middle range (more than 20–50 km) effects on health and plant life or to long range effects such as "acid rain". The middle range effects on health are complicated by the possibility of synergistic effects between different pollutants and by the changes which can occur in the atmosphere in the form of the sulphur emissions, such as the formation of sulphates (postulated as being more dangerous than SO_2) or the association of sulphur compounds with particulate matter. An even greater difficulty is the lack of experimental data of acceptable quality. Animal or human clinical studies have been incomplete or generally unsatisfactory at sulphate levels relevant to pollution control. Epidemiological studies suffer both from the lack of monitoring data, especially to differentiate between sulphur in various forms, and from a good statistical population base; nevertheless studies of this type may be the best current guide and possibly the best hope for providing a more logical control basis in the future. In adverse conditions, excess mortality has in previous times been associated with very high SO_2 levels (probably rather more than 500 $\mu g/m^3$) with associated high particulate matter. It seems likely that there is a minimum threshold level below which no health effects at all are experienced; there is in any case a natural SO_2 background level. However, it seems that good data to estimate such a threshold will not be available before the mid-1980s. The best known epidemiological study has been the US CHESS programme (Community Health and Environmental Surveillance System) which attempted to compare health and mortality statistics between areas suffering different levels of sulphate pollution. Most subsequent scientific papers and many national policies have relied on this study but it is now officially considered inaccurate and quite unsuitable as the basis for standards. Subsequent studies of this kind are needed but may face the "problem" that there are few if any areas in the industrialised world where health effects due to sulphur emissions are more than marginal and therefore difficult to disentangle from effects due to other causes.

This difficult discriminatory situation may also be the case with the effects of sulphur emissions on plant life. The effects of sulphates on visibility are also difficult both to determine and to assess economically and socially. Corrosion effects due to pollution seem probable but the exact role of sulphur compounds is uncertain; these effects may however be the strongest argument for maintaining ambient SO_2 standards, possibly at about the US level, so far as medium range effects are concerned.

At distances of the order of 1000 km or more, the impact of individual sulphur sources becomes unimportant. The main consideration is whether sulphur compounds formed in one area are transported to other areas, the SO_2 being converted to acidic sulphates en route; this process might increase the acidity of rainfall in the receiving areas. Certain areas, especially Scandinavia but more recently some parts of North America, have been postulated as places where sulphur pollution from distant industrial areas may be concentrated. The subject has been highly dramatised and some writers have appeared to assume that a causal relationship has been firmly established between adverse ecological effects and distant SO_2 emissions. There are many steps in this argument however and most of them appear open to serious doubt.

The assumption, for example, that rainwater has become more acidic over recent decades in the crucial areas, in line with changes in SO_2 emissions elsewhere is uncertain; some years of further measurements with refined and enhanced instrumentation would be necessary to determine this, especially since it is now known that large changes in acidity, clearly not connected with pollution, can take place at individual sites.

Long distance transport mechanisms are also very complicated. This has been demonstrated by recent studies, including some under the auspices of the OECD; these are remarkable examples, if still rather crude, of the type of international co-operation which will be necessary to make real progress on environmental matters affecting groups of countries. The first results of the OECD work confirmed the belief that the industrialised countries of West Europe are net exporters of sulphur and that other countries including Norway and Sweden, are net importers. However, this does not mean that these receiving areas experienced more sulphur deposition than exporting areas; sulphur depositions are always greatest nearest to the source and diminish rapidly with distance. The transport model so far used clearly requires further development since the results do not by any means yet present a comprehensive picture; local contributions may have been under-estimated.

The reactions which lead to acidity in rain are not understood and there are many factors other than SO_2 from coal combustion which play a part. NO_x may be equally as important as SO_2; in this case automotive exhausts may be a significant contributor.

The ecological effects of acid rainfall are also uncertain. Reduction in the growth rate of forests was once regarded as a probable consequence but recent more intensive research has failed to establish any clear trend. There have been some reports of loss of fish in some areas. Some lakes and rivers may be particularly sensitive, because of geological and other natural causes, to slight intensification of an already relatively high acidity from time to time; this may be enough to affect the survival of fish in conditions which may always be marginal for natural reasons. The link with long-range transport of pollution is however a tenuous one and other factors may be of equal or greater importance.

Although amenity values and other intangible issues from sulphur pollution cannot and should not be underestimated, the economic aspects of sulphur reduction are very serious. It has been estimated that the cost to Europe of a 50–60%

reduction in sulphur emissions would be about \$10 billion per annum; in addition, there might be some physical disadvantages, including the fact that any sulphur removal process inevitably results in another disposal problem. Against this, and still in purely economic terms, there may be some possible local loss of game fish (estimated at 30 t/a) and some unquantified intensification of corrosion; costs related to human health and plant growth seem unlikely to be significant factors. In this situation, it seems essential that objective research should be intensified and applied to modern conditions rather than to retrospective situations before additional legal requirements are established. It should also be borne in mind that there is a tendency to apply restrictions especially to large (and therefore highly visible) coal combustion plants, which are usually quite efficient, whereas small uncontrolled coal combustion units and sources other than coal combustion may in aggregate be more important.

Nitrogen Oxides

During combustion of coal, oxides of nitrogen are formed, both by oxidation of some or all of the nitrogen in the coal and by combination between oxygen and nitrogen in the air at the enhanced temperature.

The environmental impact of coal combustion through emissions of nitrogen oxides is even more complex and less well known than that of sulphur compounds. Nitrogen is of course a major constituent of air and can form a series of oxides; these can have complicated reactions, or display synergistic effects, with many other compounds; these effects may involve both biogenic emissions and fixation, amongst the four major reservoirs: atmosphere, land, oceans and biosphere. The nitrogen oxides are frequently lumped together as NO_x (and therefore reported as ppm), although it may be that separate assessments will be more helpful when better analytical methods and statistical data are available; good information relating to NO_x has however only been available in recent years, which makes data more than about ten years old suspect.

Biogenic emissions of NO_x may in the quite recent past have been about twice as great as anthropogenic emissions but the latter may possibly become the major contributor within the last years of this century. The relative importance of coal combustion and other sources (motor vehicles, fertilisers, etc.) to total emissions is uncertain but the effect of motor exhausts on ground level concentrations is probably dominant. In urban areas, peak concentrations of NO_x in the region of 0.1–0.5 ppm may occur with averages of 0.05 or less. Rural area concentrations may be an order of magnitude lower than urban concentrations and natural concentrations a further order of magnitude lower.

The possible detrimental effects of NO_x, all of which require much further study, include direct health effects, the formation of photochemical smog, effects on plant growth and contribution to acid rainfall.

So far as damage to human health by NO_x is concerned, acute effects probably only occur above 25 ppm. Long term effects from low level exposure are difficult to ascertain since neither epidemiological nor clinical experiments to date are very convincing and voluntary exposures by smokers to much higher levels of NO_x than the ambient atmospheric concentrations complicates the issue. It appears that damage to plant life is only observed at levels of NO_x which are one or two orders of magnitude greater than those normal in urban areas but there may be a synergistic effect with SO_2; this may make urban levels more marginal with regard to reduction in plant growth. The contribution of NO_x to the acidity of rainfall and possible effects on fish may be similar to that of SO_2.

Photochemical smog is formed by very complex reactions, not fully understood, involving NO_x, hydrocarbons (other than methane), oxygen, strong sunlight and stable weather conditions. The main harmful components are probably ozone and organic nitrates, which can damage materials and plants and cause eye irritation in humans. It may be that the availability of hydrocarbons is more crucial than NO_x in the formation of smog. The role of large power stations, unless badly sited, may be minimal.

It is not surprising that air quality and emission standards for NO_x vary widely from country to country; this may reflect the interaction between local weather conditions and NO_x but there are also differences in the way standards are approached (time scales, different oxides specified, etc.); this makes comparisons difficult. The World Health Organisation considered 0.5 ppm of NO_2 to be the lowest level at which NO_2 had an observable effect; their Task Force proposed maximum one hour levels of 0.10–0.15 ppm, which should not be exceeded more than once per month. In the USA, an annual average of 0.05 ppm is in force but additional criteria are being proposed. Japan has an annual standard of 0.01 ppm and a twenty-four-hour standard of 0.02 ppm. Several other countries, including some in the Eastern European zone have strict standards; in the German Democratic Republic, the standard is 0.002 ppm. In the USA there is an early warning and action sequence in the event of high NO_x levels; various steps are implemented, starting from levels of about 0.15 ppm on a one hour sample.

When standards become more firmly based, it may be that current general levels are considered to be too near to acceptable levels, especially to cover surges and locally enhanced concentrations. It must be debatable how far major coal combustion sources might be expected to contribute to suppression. If considerable reductions were necessary, this would create substantial problems. Different fossil fuels (including synthetics) contain different levels of nitrogen and may have different inherent efficiencies of conversion to NO_x but little is known of this. Typical NO_x concentrations in flue gases from power stations without control measures are about 500 ppm. Some further details are given in Table 19.4. The most direct ways in which NO_x output is affected are through flame temperature, residence time and oxygen excess, increases in each being associated with increases in NO_x. These factors may

TABLE 19.4 Levels of NO_x formation

Fuel	*Nitrogen in fuel (weight percent dmmf basis)*	*Type of plant*	*NO_x in flue gas (volume ppm, dry basis)*
Coal	Range 1% to 2%	pf power station boiler	200 to 1400
		Atmospheric pressure fluidised bed combustor	150 to 450
		Elevated pressure fluidised bed combustor	60 to 180
Fuel oil	0.2	Power station boiler	215
	1.0		425
	Range		110 to 800
Natural gas	Negligible	Power station boiler	50 to 1500

be reduced, over a period, by the re-design of furnaces or changes in operating conditions but there are disadvantages, which may include environmental effects. The introduction of FBC, especially PFBC, would probably alleviate the situation but the timescales for this to have a significant effect are even longer than modifications to current combustion equipment. It is in principle possible to employ gas cleaning methods and these are being developed, especially in Japan. Some processes propose to oxidise the NO to NO_2, followed by gas scrubbing; catalytic systems are also being studied. From experience with SO_2 scrubbers, such equipment is likely to be complex and costly; the timescale from conception to full application is likely to be lengthy. In the meantime, it appears that high stack dispersal, properly sited and designed, should be able to provide adequate dilution to meet ambient standards presently proposed. The application of emission standards from existing sources would be a more serious problem.

Particulate Emissions

Much of the coal burnt contains 10–20% of ash and in most cases, especially where the coal is pulverised before burning, more than half this ash enters the flue gases. In pf furnaces only about 10–30% is recovered as semi-fused furnace ash and the rest leaves the furnace as fly ash. The ash is in a chemically inert form, consisting of stable oxides, especially those of silicon, alumina and iron; calcium, sodium, potassium and magnesium are other frequent major constituents. Other elements, generally considered "trace" elements, are dealt with in the next section. The major constituents are in stable form and non-toxic; thus, they can be regarded from a nuisance rather than a hazard point of view.

The main method of fly ash recovery is the electrostatic precipitator, which removes about 99.5% of all the ash, although the proportion of the finest particles collected is lower. Because dust of similar composition arises from natural and other causes, the contribution of an individual large power plant to local deposition is difficult to determine but even a plant sited in open country probably contributes less than 10% to local dust fall. In view of this, standards are usually based on emissions. In the UK, for example, the standard for power stations is 115 mg/m^3; in the USA, the standard is expressed in terms of heat output (0.043 kg/kJ), roughly equivalent to the UK standard. Smaller combustion units are generally not equipped with such sophisticated control plant, usually using cyclones; collectively, they may contribute more particulate matter, the deposition of which will also be more concentrated. Smoke (which consists of fine, probably partly carbonaceous, or uncombusted volatile matter) can be a serious contributor when domestic coal burning appliances are widely used. The application of smoke-control regulations, requiring the use of smokeless fuels, or approved low-smoke appliances, has been a major factor in improving air quality in the UK.

In the future, some further improvements may be necessary in the control of particulates, possibly beyond the range suitable for electrostatic precipitators. The efficiency of the latter devices is assisted by the presence of sulphur oxides. If these oxides are reduced in the flue gas, as seems probable, other methods such as baghouse filters may be necessary for particulate removal. The future standards for particulates may depend not so much on the inert particles themselves but on research concerned with their relationship with trace elements or other pollutants.

Trace Elements

Since coal is derived from plants which live on soil and coal also contains mineral matter, it is not unreasonable that coal may contain or be contaminated with all or most of the elements found in the earth's surface. The combustion of coal can result in a redistribution of those elements, by emissions in the flue gas or by subsequent action on ash. This section deals with metallic elements other than the main ash constituents, which are not thought to be hazardous; gaseous emissions and radioactivity are dealt with subsequently.

Practically all trace elements in coal are present in concentrations similar to their abundance in the earth's crust; only selenium seems to be highly enriched. In spite of the fact that coal is not an obviously rich source of toxic metals, there has recently been a great intensification of work on this subject. So far as is known, however, there appear to be virtually no cases where coal combustion has been positively linked with damage to health. The intensification of interest seems to be part of a general growth of research on the effects of various activities and substances on health and the environment; there are a number of reasons why coal is particularly interesting scientifically in these respects and it is also a relatively virgin field. The main areas of present uncertainty, all requiring a great deal of further work, include:

(a) range of concentration of trace elements in coals;
(b) fate of trace elements in combustion;
(c) distribution after emission;
(d) effects of trace elements on plant and animal (especially human) life.

Great variability of concentration in coals is exhibited in results published for over 50 elements. Tables 19.5 and 19.6 give, for the USA and the UK, values for some elements of particular interest; the variability is apparent, as are the generally very low levels.

In combustion, there is a general concentration into the ash of 5–10 times due to the oxidation of the organic material. The trace elements may further be partitioned according to whether they remain as vapours or are preferentially incorporated into the slag or fly-ash; in the latter case they may be concentrated in the finer particles, which may stand a better chance of escaping the particulate emission control

TABLE 19.5 Distribution in USA coals of trace elements which may be environmentally hazardous

	Concentration (ppm)			
Element	*Powder River Basin*	*Western Interior region*	*Eastern Interior region*	*Appalachian region*
Sb	0.67	3.50	1.30	1.20
As	3.00	16.00	14.00	18.00
Be	7.00	2.00	1.80	2.00
Cd	2.10	20.00	2.30	0.20
Hg	0.10	0.13	1.19	0.16
Pb	7.20	—	34.00	12.00
Se	0.73	5.70	2.50	5.10
Zn	33.00	—	250.00	13.00

TABLE 19.6 Trace element concentrations in British power station coals

Element	*Average conc. (ppm)*	*Range (ppm)*
Antimony	3.1	0.9–9.6
Arsenic	18.0	4–73
Barium	141.6	55–350
Beryllium	1.8	0.4–3.0
Cadmium	0.4	0.3–0.8
Chromium	33.6	12–50
Fluorine	114.0	27–202
Germanium	5.1	<3.0–7.0
Lead	38.0	20–60
Manganese	84.3	35–180
Molybdenum	<2.0	<1.0–3.0
Nickel	27.9	12–40
Selenium	2.8	1.8–4.4
Uranium	1.3	0.5–2.3
Vanadium	76.0	30–154

devices. Figure 19.1 shows one interpretation of behaviour according to the position of the elements in the periodic chart.

The subsequent distribution of trace elements and their possible harmful effects are subjects difficult to assess for a number of reasons; coal combustion may often be a minor contributor to trace element concentration and toxicity, if it exists at all, is at a sub-acute level. Very careful work in the UK, covering 37 trace elements, has shown no anomalous concentrations in the vicinity of a power station; this is not surprising since the contribution of fine particles from a power station at the point of maximum deposition is much less than 1% of the total dust fall. It is believed that airborne concentrations are generally two or three orders of magnitude lower than acceptable concentrations. There are some reports however suggesting dangerous concentrations and accumulations, although these findings may be due to special circumstances or faulty technique. It may also be possible that long-term toxic effects at low levels of exposure may appear more important in the future, leading to reduction in "acceptable" levels, and this may be a reason for additional research. IEA Coal Research, after consideration of the relative importance of coal combustion in distributing those elements and reviewing possible toxic effects, suggested the following elements (in descending order of importance) as worthy of extra research emphasis from the coal point of view: nickel, cadmium, thallium, mercury, arsenic and beryllium. Other elements noted from reports by experts (again in descending order of importance) were: copper, fluorine, vanadium, zinc, cobalt, molybdenum, tungsten, selenium, tellurium and antimony.

If further research indicates that trace elements concentrations may be reaching dangerous levels, possibly locally, control procedures would need to consider all sources and not merely coal combustion. If improvements needed to be made to coal combustion devices, these would probably be aimed at fine particulate emissions and a reasonable first step might be to eliminate the worst sources, probably small, old units and apply the best current techniques to newer, more centralised combustors. Apart from securing the highest possible efficiencies in electrostatic precipitators, the maximum dilution after emission should be sought. If these measures were demonstrated to be inadequate, the next step might be to improve

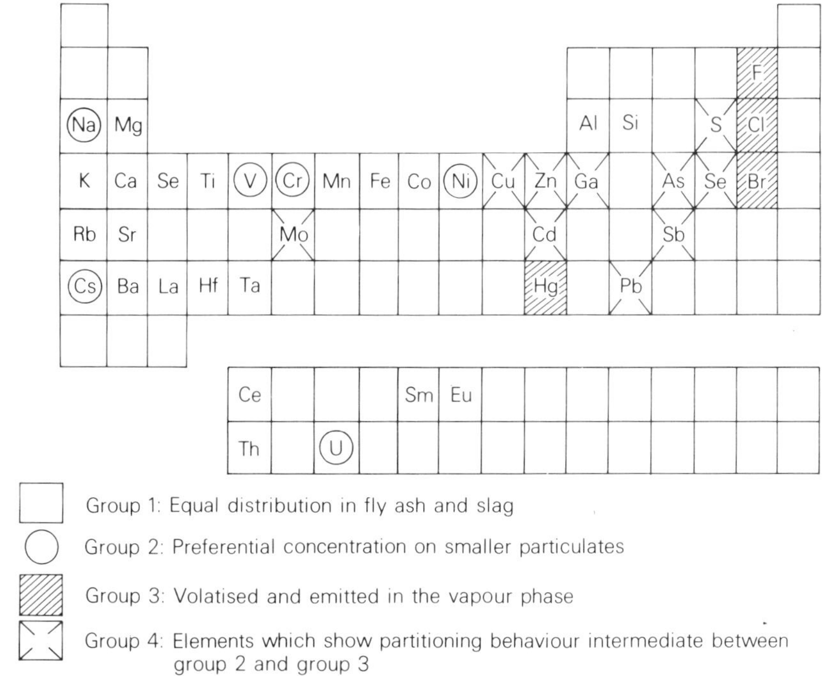

19.1 Trace elements from coal combustion and their position in the periodic table

still further the efficiency of fine particle removal, probably by the use of baghouse filters. A further step might be the accelerated introduction of FBC, thought to produce proportionately fewer fine particles, although this has not been demonstrated on a large scale. Lower stack gas temperatures combined with better removal of fine particles may be a possible means of reducing emissions of metals, like mercury, which condense on fly-ash surfaces.

Radioactive Emissions

The combustion of coal results in the emission of very small amounts of radioactivity. This subject has received a considerable amount of publicity in recent years, the inspiration of which may be more interesting than the facts warrant. The adverse publicity from anti-coal lobbyists frequently takes the form of statements to the effect that "coal power stations give out more radioactivity than nuclear stations of equivalent capacity". Whilst this statement itself is arguable, the relevant facts, based on the best current information, are that the levels of exposure resulting are orders of magnitude lower than either the natural background level or the Maximum Permitted Concentration.

One positive feature of this recent interest is a marked increase in the amount and quality of research to establish the facts, so far as the radioactivity implications of coal combustion are concerned. Coal, as does soil, contains representatives of the main radioactive series; levels of radionuclides in various coals have been studied although with some difficulty because of the very low concentrations. The effects of combustion and dispersal have also been studied. As noted in the previous section, uranium may be slightly concentrated on fine ash particles. Radon can definitely be eliminated as a coal combustion hazard as the total rate of emission from stacks is very much smaller than the natural emanation from the ground.

Recent measurements confirm that the radioactivity of coal ash is concentrated in proportion to the ash content of the coal, say by a factor of 5–10, but is lower by a factor of about twenty than the level defined as a radioactive substance by the International Atomic Energy Agency. On very pessimistic assumptions, the increase in population radioactive dose is about one-third of one per cent of natural background and, for the highest individual, by about 6%. The food chain is considered the most important route and the "highest individual" in this analysis is given extreme dietary assumptions!

An interesting by-product of this work is that the burning of coal, from which carbon-14 has decayed, actually reduces radiation dosage to the global population by a significant amount. The highest occupational doses associated with coal usage are those received by miners from radon but these are smaller than in many other mines and well within the maximum permitted exposure and indeed within the range of exposures in the home. Another finding is that a notional house built with blocks made from fly ash might result in the highest potential individual dose rate from any of the mechanisms studied but within the range of natural materials.

It is fairly clear already that radioactive emissions are a negligible hazard in coal usage. Excellent recent work has defined the outlines of the factual background and the general level of radioactivity to be expected through the various mechanisms. The key areas which have been identified should lead to unchallengeable reassurances in the near future.

Trace Gases and Hydrocarbons

There are a number of possible gaseous constituents or hydrocarbons (which may be condensed onto particles) in combustion gases which need to be considered because they have toxic, carcinogenic or corrosive properties when sufficiently concentrated. For the most part, the concentration of these constituents in flue gases is very low in large modern power stations and dilution by tall stacks should be as efficient for these constituents as for other potential pollutants. However, information is incomplete and, again, there is the complication that many of these substances may arise from sources other than coal combustion, making the attribution of relative contributions very difficult. Furthermore, if coal combustion is a significant contributor requiring alleviative measures, it may be that small and old coal combustors produce proportionately much more and distribute it less efficiently.

Chlorine and fluorine appear to be converted largely into their acid gases (HCl and HF) and are emitted, although some chlorides are also deposited on boiler tubes and are important from a corrosion point of view. Chlorine contents in coal can vary considerably, probably as a result of whether the sea has had access during coal formation or not. Levels of over 1% are known but 0.3% is considered about the maximum that can readily be tolerated for boiler corrosion reasons. At these levels and above it is probable that HCl could contribute significantly to acid rain and corrosion but quantitative relationships do not seem to have been established. Coals may contain about 100 ppm of fluorine, probably as part of the mineral matter; concentrations of 20 ppm of HF have been measured in flue gases. This would not seem to be very important.

Total hydrocarbons have been measured in the UK at a large modern power station at a level of less than one part per million (expressed as hydrocarbons). Somewhat higher figures have been noted in the USA but may not be very representative. Negligible concentrations, of the order of one part per billion, may be expected for nitrous oxide (N_2O), hydrogen cyanide (HCN), carbon disulphide (CS_2) and carbon oxysulphide. Sulphur trioxide concentrations in flue gases may be a few parts per million; the effects of this have been included in earlier discussion. Carbon monoxide (CO) concentrations may be up to 40 ppm or considerably more for short periods. Carbon monoxide is currently highlighted as a health factor and although motor vehicles may be largely responsible for high concentrations in urban areas or near heavy traffic, the possibility of CO emissions must be a factor in determining the acceptability of inefficient coal combustors.

A number of organic substances may be formed during combustion which are thought to be carcinogenic. These are mainly compounds known variously as Polycyclic Aromatic Hydrocarbons (PAH) or Polynuclear Aromatics (PNA), of which benzo(a)pyrene (BaP) is often used as an indicator, although not necessarily the most dangerous substance. PNAs are probably more important as potential occupational hazards in coking (already discussed) and in synthetic fuel manufacture (to follow). Ground level concentrations may vary by two orders of magnitude from rural areas to industrial areas; levels are clearly very much affected by non-coal sources. It has been estimated however that the BaP emission rate for a small industrial boiler may be an order of magnitude higher, on heat input basis, than for a central power station and a domestic boiler may have rates a further two orders of magnitude greater still. The reduction in the number of small combustors with uncontrolled smoke emissions has been a key factor in minimising BaP concentrations.

Carbon Dioxide

Although carbon dioxide (CO_2) is not a "pollutant" (being the fourth most abundant natural constituent of air), concern has been expressed that further increases in the CO_2 content of air, arising from additional fossil fuel use, might have significant effects on climate, due to the so-called "greenhouse" effect. (The mechanism suggested is in fact different from that which warms a greenhouse.) Carbon dioxide is more "transparent" to solar radiation than to the balancing outward radiation from the earth, so that increases in the CO_2 content of the atmosphere might lead to adjustment of the temperature of the earth's surface to a higher level. Other materials, including water vapour and dust, can also affect this radiative balance; the relative importance of different substances is not known.

As a result of fossil fuel combustion about 5 Gt of carbon in the form of CO_2 is injected into the atmosphere each year and the CO_2 concentration is undoubtedly rising slowly. However, this amount should be viewed against the facts that the total amount in the atmosphere is about 700 Gt of carbon and other reservoirs (the biosphere, oceans and sediments) contain far more; the annual reversible exchange rates between the atmosphere and these other reservoirs are of the order of 100 Gt of carbon.

Over the last twenty years, the atmospheric CO_2 concentration has been carefully measured and has increased at about 1 ppm (0.3%) per year from 315 ppm in 1958 to about 335 ppm by the end of the 1970s. This rate of increase probably goes back at least to 1850 but recorded measurements are not very reliable. In addition to fossil fuel combustion for energy, other factors affect the CO_2 levels, including ocean uptake and the growth and decay of plant life; deforestation is believed to have had a large effect recently. Models are very inexact; a common estimate is that about half the CO_2 from fossil fuel combustion stays in the atmosphere but the evidence for any figure is slight. The value of this fraction is however very important in making predictions about future CO_2 levels, which are based on estimates of fossil fuel consumption, combined with this proportion. On such reasoning, it is suggested that a level of 600 ppm of CO_2 will be reached between 2020 and 2040. However, even if no further growth in fossil energy use occurred, the CO_2 concentration will go on rising, provided the assumption about retention of CO_2 remains valid.

Models for predicting the climatic effects of increases in CO_2, whilst very sophisticated, are not very reliable either. A common prediction is that a doubling of CO_2 would result in an average warming at the earth's surface of 2–3°C, but with extreme effects in the polar regions, possibly up to 10°C increase. These predictions are however unable to take proper account of certain possible feed-back mechanisms, such as changes in cloud cover, albedo or precipitation. Unfortunately, the earth's temperature fluctuation "noise" is too great to allow the accuracy of predictive models to be judged in the short term; in any case, there appear to be several substantial natural cycles of temperature with frequencies between 100 and 100 000 years. It is thought that by the end of the century there may be a general warming of about 1°C and about 2°C at the poles. This should be observable and would allow future predictions to be more firmly based.

However, it is suggested that by the time temperatures increase to this extent, it will be very difficult to take remedial action and the upward trend would in any case continue for some time. The most spectacular effects suggested would be polar ice melting and an increase in sea level, perhaps of a few metres for a doubling of CO_2, thus inundating large land areas. The rate of polar ice melting seems uncertain; it might take millenia for the ice to disappear and, again, feed-back mechanisms might come into play. Perhaps a more certain consequence of some warming would be a change of weather patterns and thus of agricultural productivity. Total agricultural output might actually increase but a displacement would be politically very sensitive.

If alleviative measures were necessary, this would of course be a global problem, requiring unprecedented international co-operation. Carbon dioxide can be removed from gases and this is of course practised in some processes, including coal gasification. For flue gas clean-up, carbon dioxide and sulphur dioxide might be removed together by scrubbing but this would be a formidable undertaking if applied over a major part of the combustion activity. Carbon dioxide could be regenerated, in concentrated form, from these wash liquors and liquefied for transport and disposal. One suggestion is that the CO_2 could be injected into downward sea currents leading to great depths. This proposal was accompanied by a very rough estimate that a recovery of 50% of the CO_2 might add an extra cost equivalent to 10% of the fuel cost and a 90% recovery would add 20%. In practice, the feasibility of this scheme would be questionable; the costs could also be very much higher and represented by a real loss of energy.

Greater research efforts are clearly called for. The USA is taking positive steps and there is hope that positive results will emerge within about a decade and that time will be available for corrective measures, if necessary. However, it may be that timescales for increases in CO_2 concentration and temperature changes may be more protracted or causes other than CO_2 will become more prominent; the growth rate of fossil fuel combustion assumed by the climatologists a year or two ago is probably greater than would seem reasonable now. Feed-back mechanisms to slow down any potential climatic changes may come more positively into the future. Nonetheless, the possibility that there could be a serious problem is a further reason for stressing the need for maximum efficiency in the use of the energy available from the combustion of fossil fuels.

Ash Handling and Disposal

The problems of airborne ash particles have been discussed. Most of the ash produced in pf furnaces is retained, mainly in the cyclones and precipitators but a smaller amount is recovered from the furnace bottom and all must be disposed of. A considerable proportion (in the UK, nearly half) is used commercially, for such purposes as lightweight aggregate and building blocks. The handling of ash is a potentially hazardous occupation because of the possibility of inhalation of fine dust. However, extensive tests suggest that this is largely a nuisance problem and that reasonable limitation of exposure does not normally present any special toxic problems. Further work is in hand to confirm the absence of mutagenic and pneumoconiotic effects.

Ash which is not used in other ways commercially may be used for landfill or, if suitable sites are not available, may be discarded in lagoons. This system is now organised largely to avoid dust nuisance and to minimise visual impact, both during lagoon filling and to recover the land in due course. A great deal of work has been carried out, and is continuing, on studies of the effects of leaching of ash on ground water and the possible transmission of trace metals into food chains. Normally, the extent of each of these potential problems seems negligible and a further margin can be obtained by site management techniques which have been developed to limit the concentration of leachants. Some concern apparently still exists in the USA, including the possible extra leachability of FBC ash, which is produced at lower temperatures and is therefore less strongly sintered. If declared a "toxic substance" it would have to be stored without any possible contact with natural waters and costs would be excessive. There may also be problems with FGD sludge storage.

SYNTHETIC FUELS MANUFACTURE

The conversion of coal into synthetic fuels can embrace practically any potential form of pollution and health hazard which can be associated with coal, including combustion products and ash, phenolic liquors and coal liquids which are exceptionally rich in known or suspected carcinogens. This is a field already receiving a great deal of interest, requiring considerable and immediate research effort. However, if it were felt that full knowledge of these possibilities had to be acquired prior to the construction of demonstration or the first commercial plants, considerable delay would result and meantime the doubts about ultimate acceptability could have a destructive effect on planning for investment in coal conversion utilisation.

Perhaps the main area of concern is related to the Polynuclear Aromatics (PNAs) some of which are known or thought to be carcinogenic. PNAs exist naturally and are present in tobacco smoke and automobile exhaust; natural degradation is very slow. The structure of coal however makes it a very rich source of PNAs, which may be released in high concentrations into product and effluent streams. The detailed concentrations of individual compounds, including PNAs, in particular process streams is not well known. Furthermore, although carcinogenicity and toxicity have been demonstrated in laboratory tests with specific compounds, the possibility of synergistic effects is little understood and the likely risk factors to humans is unknown. In the case of coke oven emissions, significant human health damage and excess mortality has been reported, presumably from similar causes. However, more recent statistical evidence suggests that risks can be reduced to negligible proportions with good control procedures and hygiene. Also, some large scale facilities such as the SRC plant at Fort Lewis in the USA have operated for several years and no unfavourable health effects have been reported. So far as PNA contamination arising from major process streams is concerned, it would appear that this could be fairly readily managed by the design of plant items to extremely high standards of leak tightness, as is already the case in some nuclear and chemical plants. Designs would also have to ensure that adventitious leakages can be contained safely. There must be good protection and monitoring for the workforce. There are no likely hazards to the general public from such process operations. However, incidental emissions and the disposal of wastes may be problems requiring attention so far as public health effects due to PNA are concerned.

As noted above, PNAs can be emitted in small quantities with particulates from combustors. It appears that this is controllable by high efficiency combustion and good stack dispersal systems and it may be that a final well-designed incineration system may be an environmentally desirable feature of coal conversion complexes. Otherwise, PNAs could be leached from solid residues and could penetrate soils. The early characterisation of solid wastes and the provision of non-leaching conditions may be an essential design feature. PNAs in aqueous effluents may present a special problem since the biological oxidation treatments normally applied successfully for removing phenolic and some other organic pollutants can apparently remove only marginal quantities. Further processing by ozonation, chlorination, ultraviolet radiation or active carbon processes may be necessary; access of plant intermediates and residues to water should be carefully controlled.

In addition to possible PNA contamination, gasifiers and liquefaction plants will have more traditional pollution control problems. These can generally be dealt with by known processes but it is likely that very high standards may be required in the future for new plants. The economic impact of these engineering standards may be so significant as to affect the order of merit of different gasification and liquefaction processes. For example, entrainment gasifiers may have a substantial advantage in this respect.

CONCLUSIONS

The conclusions must echo the points made in the introduction:

(a) Environmental issues are likely to be controlling factors in the future enhanced utilisation of coal, which is essential to help meet world requirements for energy.

(b) Environmental problems can be solved but economic compromises are involved.

The discussion, which has mostly been concerned with combustion, has indicated how complex these problems may be, without suggesting that there are any intractable areas. A great deal of further factual work and assessment are required. If the right compromises are to be reached, it is essential that those involved in the coal industry play a major role in the research and assessment processes. This is an area for maximum international co-operation and for the most persuasive presentation of the information.

Chapter 20

MULTI-COMPONENT PLANTS: COALPLEXES

INTRODUCTION AND GENERAL CONCEPTS

The world "Coalplex" has recently become fairly familiar as a result of various proposals. Although these are somewhat tentative and speculative, the subject is worth discussing in this chapter. Some general concepts are indicated, examples are given and their potential economics are considered. Finally, some ways in which coalplexes might approach realisation are suggested.

One of the recurring themes of this book concerns the future diversity of use of coal. Not only will more coal be used but it will be used in a wider range of processes than has been the case historically. There will probably be a relative change in emphasis away from direct combustion (although the actual tonnages used in combustion may not decline for a long time) towards the production of hydrocarbon gases and liquids, for use as premium fuels and chemical feedstocks. It is quite possible, within a few decades that 10% or more of the world's primary energy supplies will be processed in coal conversion plants, representing investments totalling hundreds of billions of dollars. It is important therefore that these new systems be operated with maximum efficiency and in a manner compatible with a general improvement in the application of energy from all sources. One possible approach to enhance the flexibility and efficiency of use of coal is to operate more than one process on one site. This complex development of coal processing is analogous to the manner in which oil refineries have developed, including the creation of links between the refineries proper and the petrochemical industry. However, the coal analogue will undoubtedly be more complex and more expensive in terms of capital investment per unit of energy input.

The concepts of the Coalplex, even if not directly or immediately applied, may assist in guiding the evolution of individual coal processes. Moreover, it has been suggested that the competitiveness of coal conversion processes might be advanced by the adoption of the Coalplex concepts, compared with single process schemes.

The Coalplex concept is a loose one, not rigidly defined. One early description (by one of the authors (LG)) is that "the various unit processes are combined together in ways intended to reduce their overall capital costs and to increase overall efficiency while meeting the needs of the energy and raw materials markets, taking account of storage and transport costs". Some general considerations which may be important in devising or assessing Coalplexes include:

(1) Capital savings may be achieved, especially through items such as land, site services and the handling of coal or waste materials, where such items may be a common requirement of more than one of the processes.

(2) By-products, intermediates or energy may be transferred from one process to another in a way which increases overall efficiency.
(3) The output ratio of the different products may be flexible, to some extent at least. This may assist the achievement of a better match with demand, which should have economic advantages since there may be savings on energy storage costs or on the provision of marginal capacity to meet peak-load demands.
(4) Thermal efficiency might be improved. This could be achieved by the recovery or use of energy or materials which might otherwise be wasted. Examples include the further treatment of low-grade residues, the co-generation of electricity or the local distribution of low-grade heat.
(5) The consumption of low-grade residues on the site where they are produced may have environmental advantages.

COMPONENTS OF COALPLEXES AND THEIR RELATIONSHIPS

There are only a few basic components of Coalplexes; all of these (combustion, gasification, etc.) are considered separately in other chapters. There are however a number of variants of processes for each component and the selection of the best process variant may be different considered as a component of a Coalplex compared with a process operating in an isolated situation.

A common situation is that at the tail-end of a process there may be residues containing the original ash, some carbon and possibly unseparated hydrocarbons. These residues may be combusted to produce steam and/or electricity, for export or in-house use. The development of Fluidised Bed Combustion (FBC) makes this a more attractive disposal process, especially if the residues contain the original sulphur, which will now be in a more concentrated form. Steam for gasification may also be produced in this way. A useful combustion outlet could clearly affect the economics of coal selection and preparation and also the design parameters of the processes producing the residues. Alternatively, the residues might be gasified, either completely or partially, followed by final combustion; gasification in these circumstances might preferably be in an entrainment or fluidised type of gasifier. The gas might be sold externally or used in the process, possibly after shifting it first to hydrogen.

In another approach, the initial opening-out of the coal might be designed to remove only a moderate proportion of the coal substance (perhaps up to 20%), rather than seeking a high percentage conversion which might require disproportionate costs. In addition, the aim would be to remove the more valuable material of relatively small molecular weight and containing as much of the hydrogen as possible. The "residues" (80% or more) will probably be more reactive and therefore a suitable feedstock for gasification and/or combustion. Pyrolysis or solvent extraction (especially by supercritical gases) would be suitable methods of carrying out this preliminary devolatilisation. The demand for, and relative values of, the derivatives of the extract on the one hand and the residue on the other could determine the "cut" between them; this might with advantage be adjustable to some extent. Some flexibility might be desirable in any case to adapt to changes in coal quality, especially where imported coal is used. In some cases the overall economics may be assisted by dividing the input coal into two or more portions, based perhaps on size or ash content, for feeding into different initial processes for which the two coal portions have suitable characteristics, thus using the coal with maximum efficiency. Optimisations of this kind could of course involve the mining and preparation stages.

Fuel gases and hydrogen (or synthesis gases) can be regarded as "balancers" especially if an external demand able to accept variable supplies exists. In the case of fuel gas, this acceptability might be assisted by some deliberate manufacture separate from the integrated (and possibly variable) production. Sometimes in coal conversion processes, tail gases are quite rich but recycling to a unit such as a Fischer–Tropsch (F–T) reactor in order to improve efficiencies for the main product alone may be expensive. An external demand for fuel gas may be a more economic solution. Hydrogen is a key raw material not only in coal processing but also in some other large industries, including oil refining, chemicals and metal ore reduction. It might be possible to arrange for surpluses or deficits to be exchanged not only between coal process units in a Coalplex but also externally with these other industries. Synthesis gas is of course, a very flexible intermediate, from which methane, methanol, ammonia and light hydrocarbons may be made.

Oxygen is also a crucial raw material, both in coal processing and in other industries (especially steel). Costs of producing oxygen appear to be sensitive to scale and therefore some sharing between processes or with other industries may be advantageous. Where air separation is carried out for oxygen production, the nitrogen by-product may be used for ammonia manufacture.

Different forms of carbon may also be coal conversion products, suitable for involvement in a Coalplex scheme; coke is an example from an old process. Chars suitable for making into briquettes may be made by pyrolysis processes. Binders for making these briquettes or other carbon artefacts may be made from the liquid products of pyrolysis or from solvent extraction of coal. Purified coal extracts can be coked to produce special cokes, which may be used for making electrodes; the extract can also be spun into carbon fibres, which may be further strengthened by being graphitised.

Examples

Some simplified block diagrams may illustrate a few of the possible combinations based on these components. Figure 20.1 is based on gasification which might be a two stage system; one part designed to produce mainly synthesis gas, with the other using a process which gives a relatively high methane content, suitable for conversion to SNG. The additional coal feed to the combustor might simply be for balancing purposes; alternatively, depending on the steam and power requirements, fresh coal could be the main combustor fuel, the furnace providing a convenient disposal point for the gasifier residues.

Figure 20.2 is based on liquid solvent extraction, with the residues going first to a gasifier and then to a combustor. Ancillary coal supply to the gasifier is provided. The gas in this case is converted entirely to hydrogen, both for the liquefaction stage and for the hydrotreatment of the Solvent Refined Coal (SRC) from the solvent recovery unit.

Figure 20.3 emphasises electricity production, using the clean gas in a combined cycle electricity plant to achieve a high generating efficiency. In addition, the coal is devolatilised before gasification; SNG, SRC or synthetic crude oil may be produced.

Figure 20.4 emphasises SNG production, using a two stage gasifier; electricity

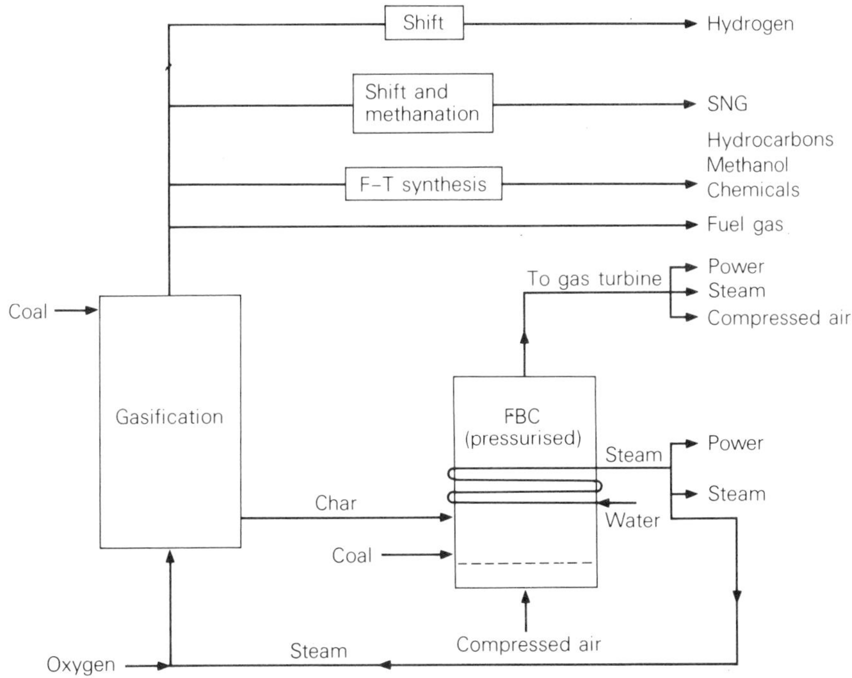

20.1 Coalplex incorporating gasification, synthesis and combustion

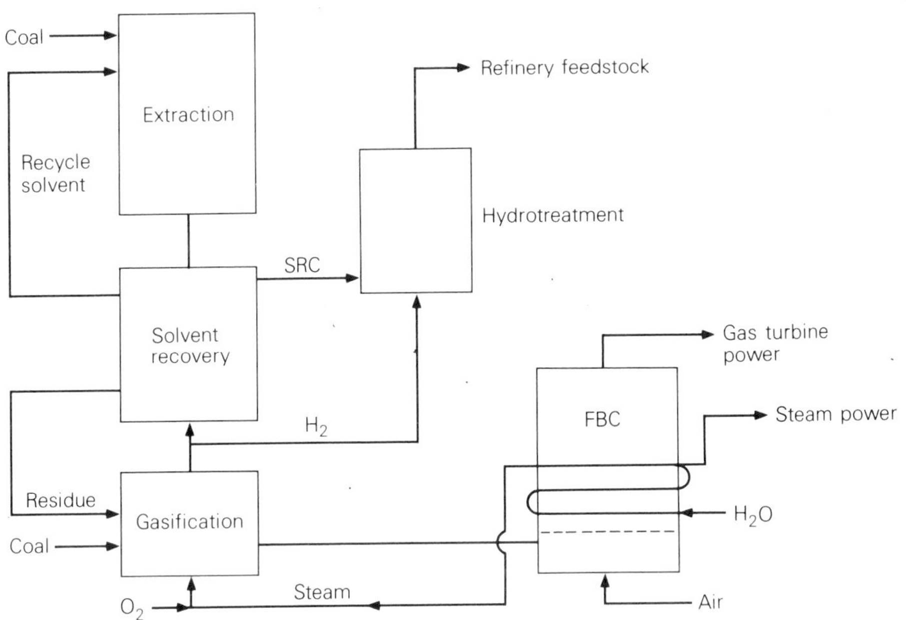

20.2 Coalplex incorporating liquefaction and fluidised combustion

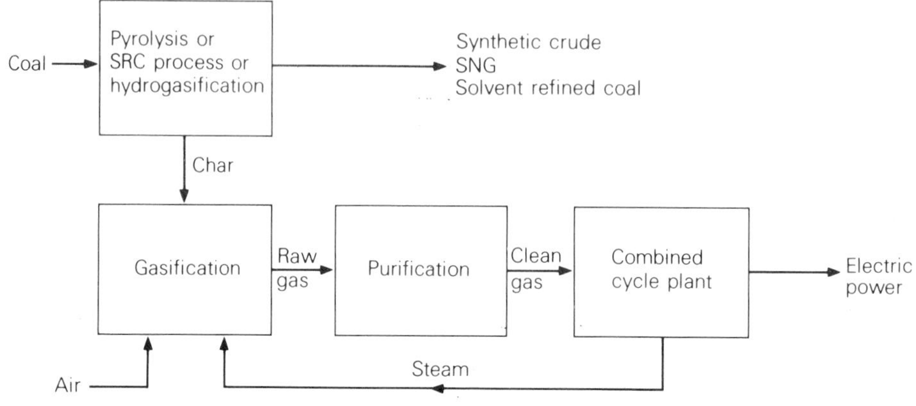

20.3 Integration of power generation in a coal complex

Coal
pressurisation
Pyrolysis/
hydrogenation
Methanator
Ground coal
Raw
gas
Condenser
SNG
SO_2
H_2O
Sulphur
Raw H_2
rich gas
and steam
Char
Fluidised bed
gasifier
Process
heat
O_2 plant
Steam
High carbon
residue(s)
Flue gas
Pressurised fluidised
combustor
Air
Supplementary
coal
Condensers
Surplus
power
Air
Ash

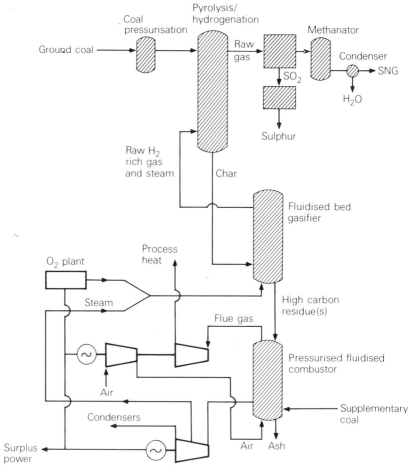

20.4 Integrated plant for SNG and electric power

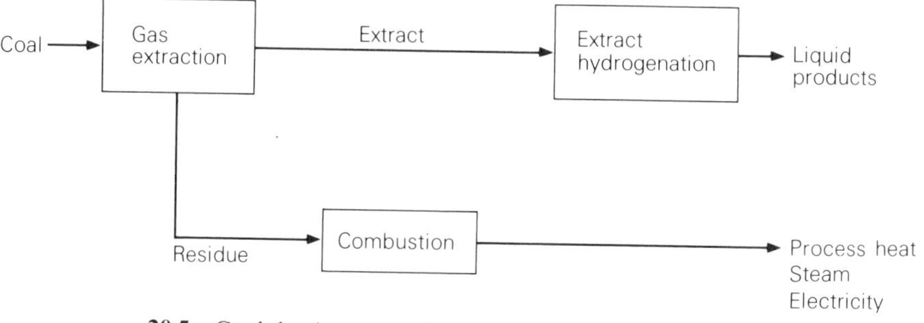

20.5 Coalplex incorporating gas extraction and combustion

generation is based on the gasifier residues. The use of some of the electricity for an integrated oxygen plant is shown.

Figures 20.5–20.7 show schemes based on inital supercritical gas extraction. Figure 20.5 is a simple scheme in which the residue is burnt to provide process heat, steam and electricity. However, this scheme does not take advantage of the possibility that this residue will have enhanced reactivity in gasification processes. In Fig. 20.6, the residue is partly gasified, the gas being converted to hydrogen, some of which is used to hydrogenate the extract. It is likely that the extract will require hydrogen additions up to about 10% of its own weight to produce satisfactory light products; this may only be equivalent to about 3% of the weight of the input coal. This amount of hydrogen can be made from residue representing only a portion of the coal feed, so there is a considerable excess of residue, even taking account of the needs for process steam and power. Substantial further exports of these or of hydrogen as desired are therefore possible. Extra hydrogen may be an attractive proposition, especially if other coal processing plants are adjacent. This underlines the potential importance of the gas extraction process in a Coalplex context.

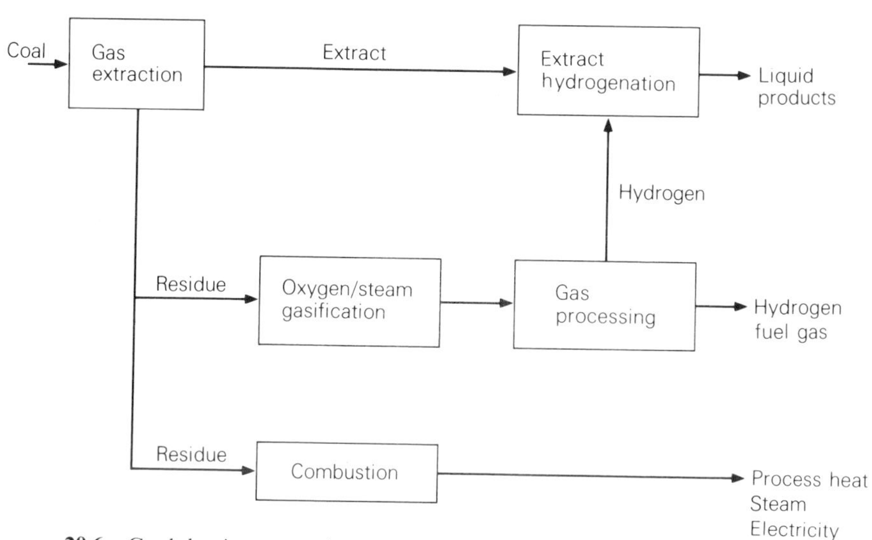

20.6 Coalplex incorporating gas extraction, combustion and gasification

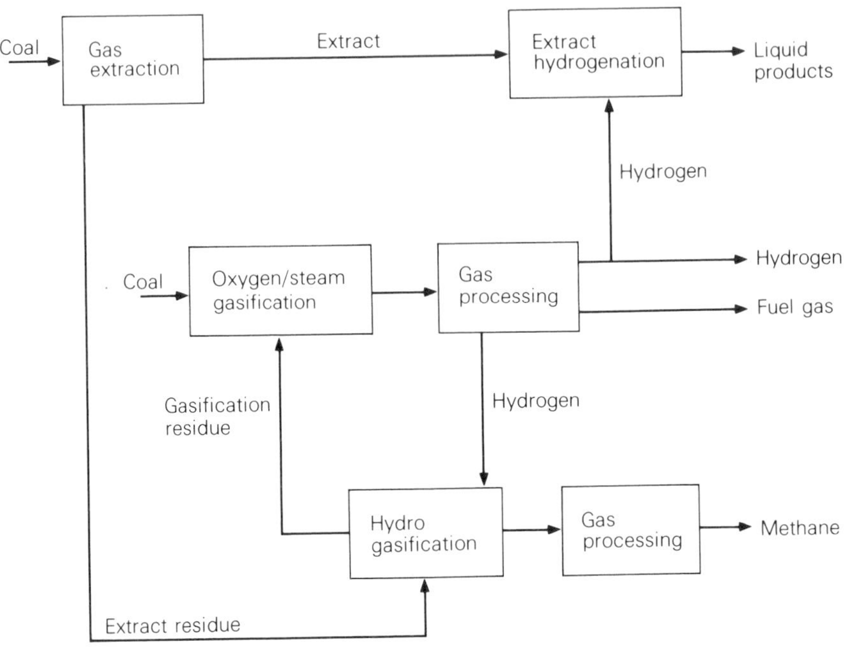

20.7 Coalplex incorporating gas extraction and hydrogasification

A further scheme for exploiting the properties of gas extraction char is illustrated in Fig. 20.7, in which the char is hydrogasified; the char may be exceptionally suitable for these purposes. This scheme might be particularly attractive where methane is especially required; the gas extraction residue is then hydrogasified but complete gasification is not sought. The residue from this stage, together with fresh coal if needed, is completely gasified, probably in a Koppers–Totzek or Shell–Koppers gasifier. Some of this latter gas is converted to hydrogen for the hydrogasification and for the extract hydrogenation. The conversion to hydrogen need not be pushed to the limits if an external requirement for fuel gas is at hand. The hydrogasification of char requires experimental verification.

Figure 20.8 represents a simple form of the Cogas process, which uses a multistage fluidised pyrolysis unit to remove volatiles; the char is then gasified and can be

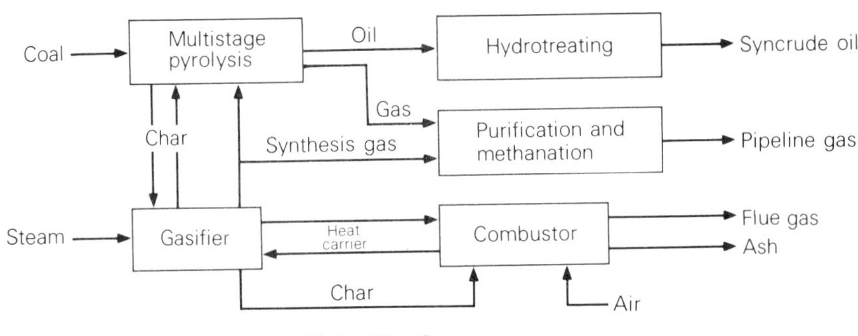

20.8 The Cogas process

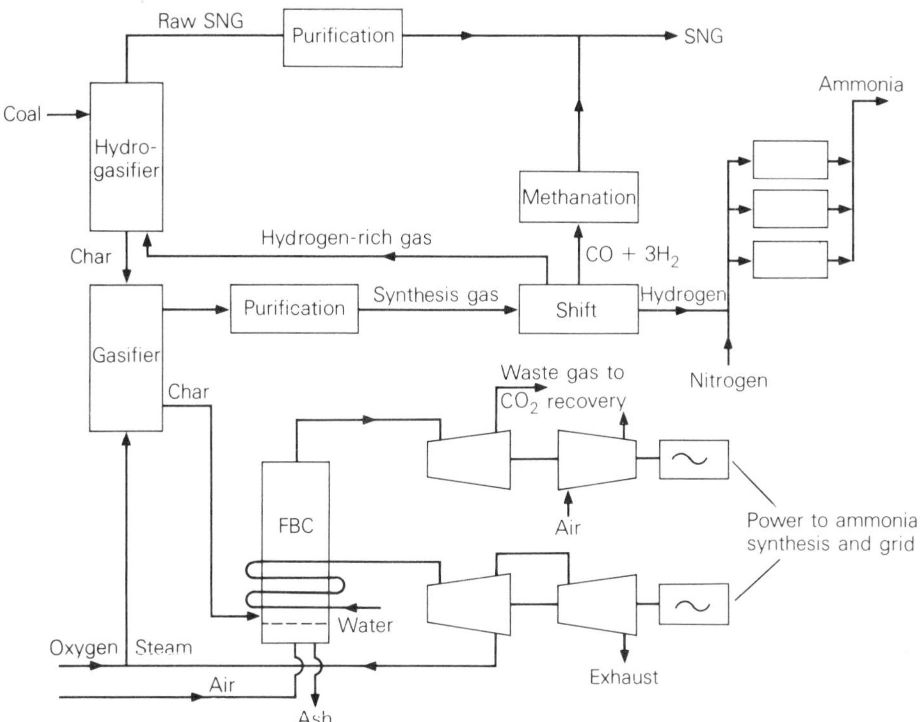

20.9 Gas and chemicals coalplex

converted to methane. Alternatively, of course, some of the gas could be converted to hydrogen for hydro-treating the tars. Figure 20.9 is a scheme which produces ammonia, SNG and electricity. This could be an attractive scheme for fairly early commercialisation, possibly with methanol instead of methane as the alternative product, since this is easier to store.

Design Studies

Although proposals based on Coalplex concepts have been discussed for a long time, interest has not been particularly intense in recent years, attention being focussed rather more on processes intended, at least in their original form, to produce single products. In many cases however, design concepts are broadening and it is being realised that recycling by-products or intermediates in order to maximise the yield of the prime product may become increasingly expensive as this ultimate is approached; the alternative of marketing a second product may therefore be assessed. Similarly, the avoidance of difficult steps such as the filtration of certain coal digests, substituting instead the recovery of values from residues via an ancillary process is an option more frequently considered recently. Thus, there is a definite "grey area". Nonetheless, design studies have continued to be mainly concentrated on single product processes and less on schemes that are declared to be Coalplexes *ab initio*.

An important landmark in the development of Coalplexes however was a design study carried out for the US DOE by the Ralph M. Parsons Company, reporting in November 1978. Conceptual designs were developed for four commercial scale conversion complexes, which were then assessed technically and economically. A brief description of the four processes is as follows:

(1) *COED-based Pyrolysis Plant*. This is similar to that shown in Fig. 20.8. Multiple fluid-bed pyrolysers produce tars, fuel gas and char. The char is gasified according to the Cogas process. Some of the char is cycled through a combustor in parallel with the gasifier; it is heated by being partially consumed, returning to the gasifier to supply the necessary heat. Some of the gas is converted to hydrogen and used to treat the tar.

(2) *Oil/gas*. This is a version of the Solvent Refined Coal Process (SRC) in which the co-production of fuel gases is not discouraged.

(3) *Fischer–Tropsch Synthesis Process*. This again is a particular version suited for multiple product operation. The synthesis reactor is designed so that high pressure steam can be recovered. This steam is used in the process and also for generating electricity for external use. SNG is also a significant export.

(4) *POGO* (*Power–Oil–Gas–Other*) (Fig. 20.10). This is a coal refinery which co-produces electrical power, liquid fuels, gas fuels, chemical by-products and coke. The two main process units, which operate on high grade coal, are SRC-type hydroliquefaction and flash pyrolysis. Lower quality coal (thus maximising the use of the mined output) is gasified and this gas is used for generating electricity in a combined cycle plant. Char from the pyrolysis unit is used to make process gas, some of which is turned into hydrogen for the SRC stage.

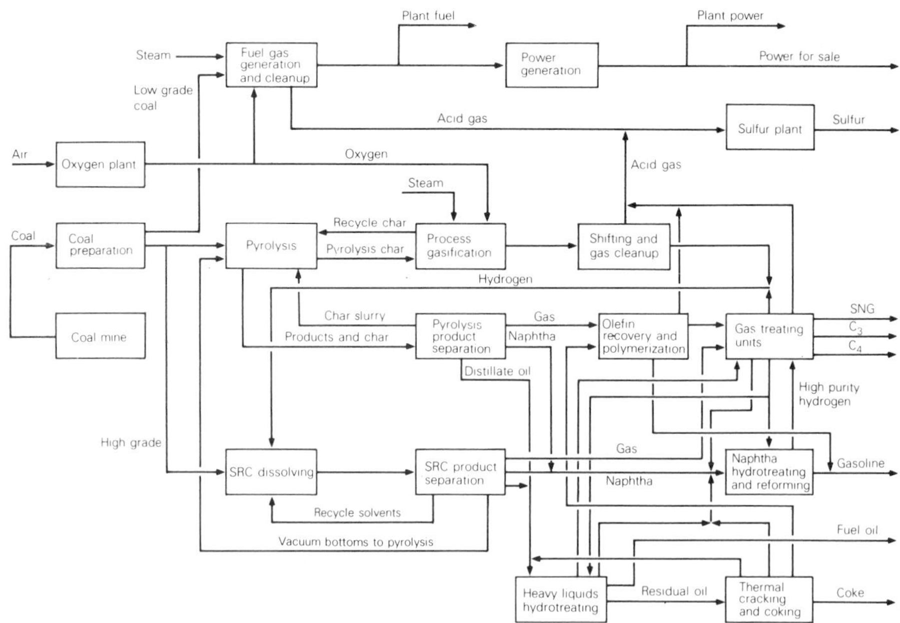

20.10 Simplified block flow diagram of the POGO complex

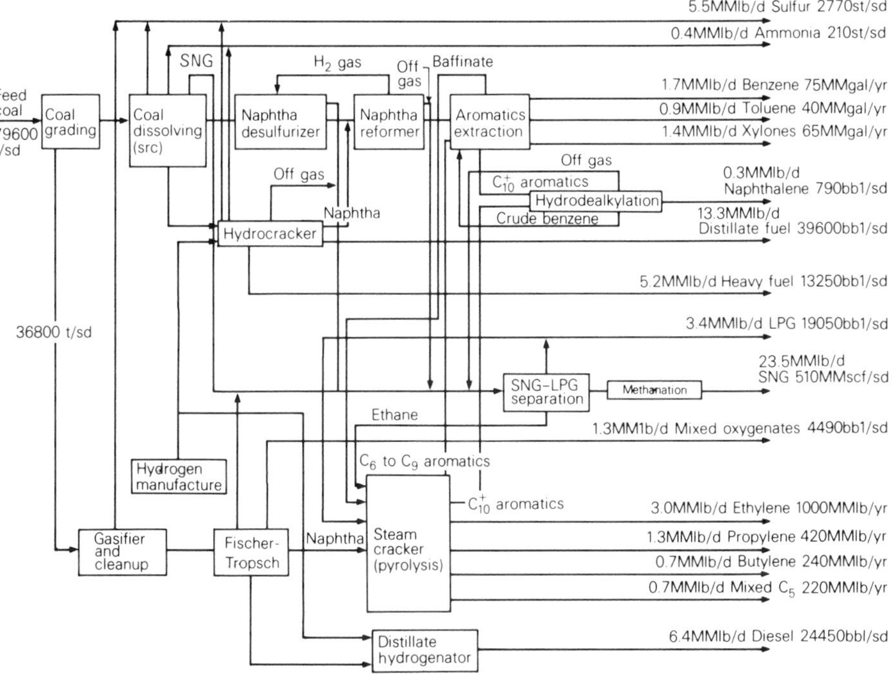

Note: 1. 330 Steam days/year (sd/yr) basis
2. Production rates in lb/sd

20.11 Potential outputs of the chemical-type Coalplex

The general design is capable of considerable variation in the balance of inputs and outputs. In the particular configuration chosen for comparison, about 45% of the coal goes to the SRC plant, 15% to pyrolysis and 40% to gasification.

The same company carried out a further study of a Coalplex intended to emphasise chemical feedstocks (Fig. 20.11). About half the coal is fed to an SRC hydroliquefier and the rest, with residues, to a gasifier, which produces hydrogen and syngas; the latter is converted to liquids in an F–T reactor. The products are hydrogenated and refined. The yield of olefins and BTX (Benzene, Toluene, Xylene) suitable for petrochemical feedstocks is estimated at 5.6%. Chemicals (Ammonia, Sulphur, Naphthalene, Oxygenates, etc.) amount to about 5%. Liquid fuels are about 19% and SNG about 15%. The total weight yield of products is about 45% and the thermal efficiency about 65%.

Udant and Stokes of Catalytic Incorporated, have proposed a simple system (Fig. 20.12), which includes chemicals. Coal is fed into an SRC-type hydroliquefier. From the digest, SRC fuel and distillates are recovered and the residue goes to a gasifier. The gas is converted to hydrogen, some of which returns to the hydroliquefier and the rest is used to make ammonia (the nitrogen coming from the oxygen plant). From coal (10% moisture, 3.2% sulphur, 6.6% ash) fed at a rate of 10 000 t/d 90% is

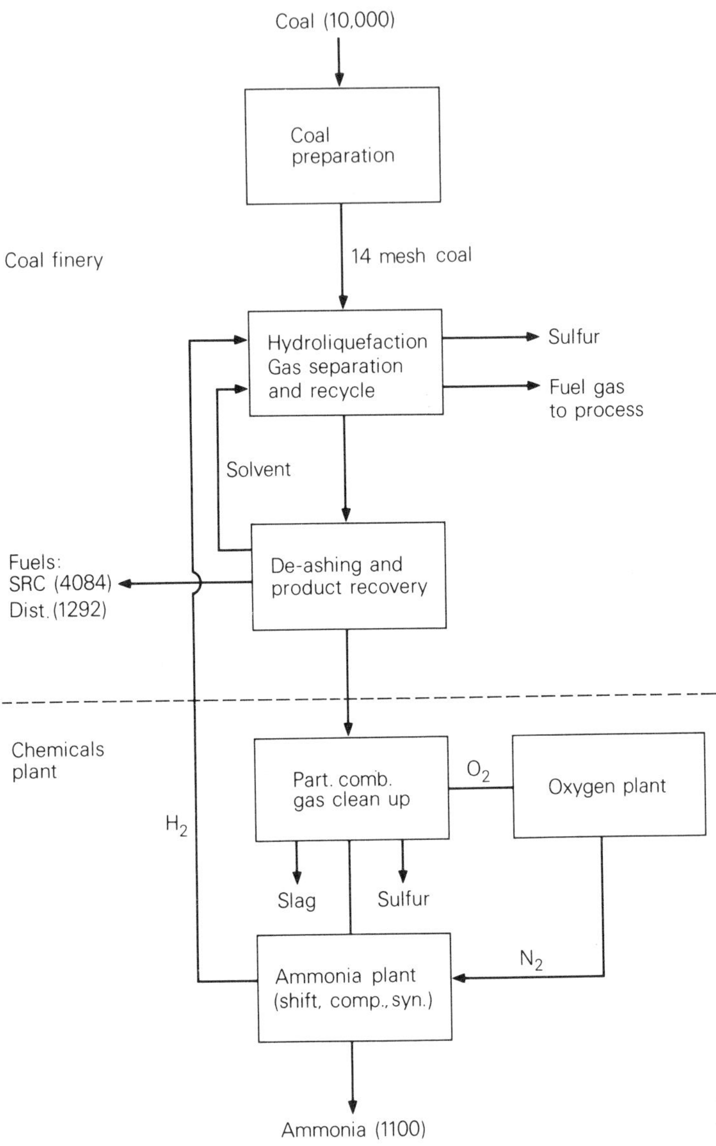

20.12 Coal refinery and chemicals plant

usefully converted and it is claimed that the following products can be made:

	t/d
SRC	4084
Distillates	1292
Ammonia	1100
Surplus Solvent	833
Fuel Gas	333
Sulphur	247

Since 1972, the US Steel Corporation has been studying in detail, with government support, a Coalplex designed to produce high quality metallurgical coke, plus other valuable products, from a low-grade high sulphur coal. This is called the Clean Coke Process and is shown in Fig. 20.13. Part of coal is fed to a fluid bed carboniser; a hydrogen rich gas is recirculated and provides a low-sulphur char which is then pelletised for metallurgical purposes, using heavy liquids from the process as binder. The other part of the coal is fed to a hydroliquefaction unit. The liquids and gases from both processes are worked up to chemical feedstocks and fuel gases.

ECONOMICS

General studies of the economic potential of Coalplexes have been carried out by the Coal Research Establishment of the National Coal Board. Models were established enabling notional combinations of plants and product ratios to be rapidly assessed, using published design data. These studies suggested that the capital costs of plants designed to produce two products could be significantly lower (up to about 20%), per unit of energy output, than for plants producing only one of these products. The average product cost depended on the design of the Coalplex and on the mix and load factors of products required but there is generally a considerable range of product split over which costs are significantly lower than from single product plants (usually SNG or SRC). The comparison takes the form illustrated in Fig. 20.14, where SNG and SRC costs are shown as horizontal and vertical lines. The sloping line for the pyrolysis Coalplex indicates the combinations of costs for the two products which might obtain; in the range indicated by the hatched triangle, at least one product is cheaper and the other no costlier. Figure 20.14 is for a Coalplex

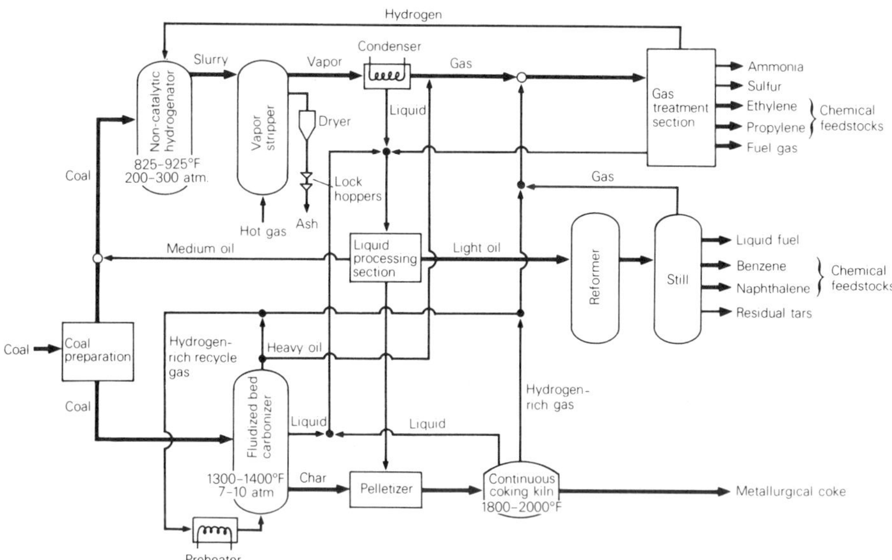

20.13 Schematic diagram of the clean coke process

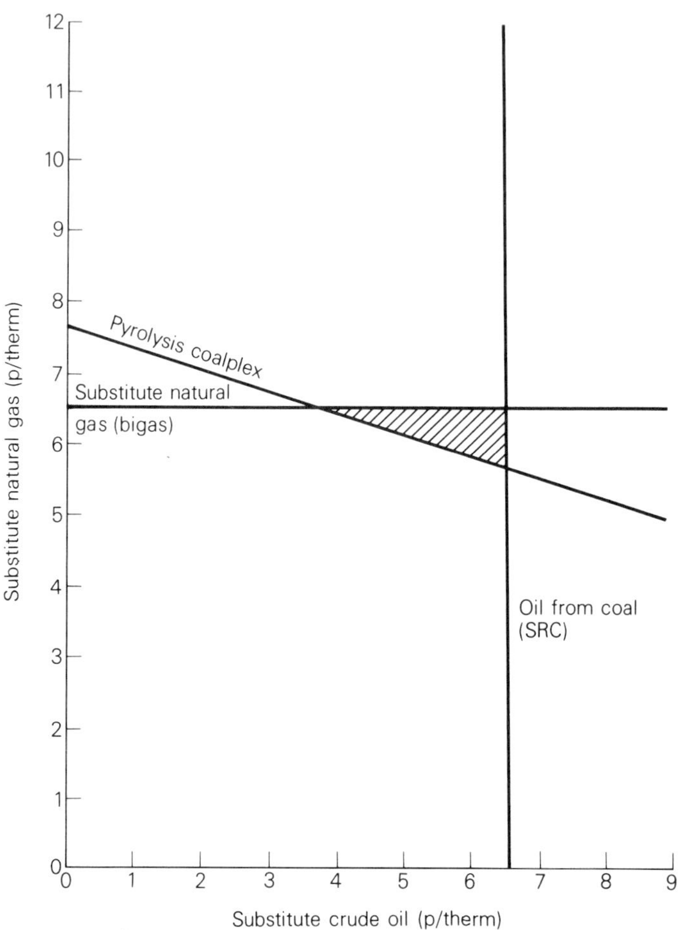

20.14 Cost advantages of a simple coalplex

having a relatively fixed output ratio. Figure 20.15 illustrates a hypothetical case where SRC is required at 80% load factor but SNG is only required for 5% of the time. A standard gasifier would produce very expensive gas in these extreme circumstances and a Coalplex based on pyrolysis would be even more so. However, a flexible Coalplex might be designed to enable the 5% SNG load to be met by discontinuing SRC production temporarily. In this instance SNG might cost only about one-third as much as from a fully dedicated but lowly loaded SNG plant, without affecting SRC costs; if higher SNG costs could be passed on, up to the level for a dedicated plant operating at only 5%, SRC costs would benefit by up to 15%. This is of course an artificial case but, unless cheap storage is available, some production unit in a system may be loaded at 5% or less. Much more work is necessary of course to relate plant design and systems to these peak demands.

The Parsons economic studies were calculated on much more detailed design and operational studies. However they were based on a captive coal mine (without a specific input coal price to the processes being separately shown), which would make

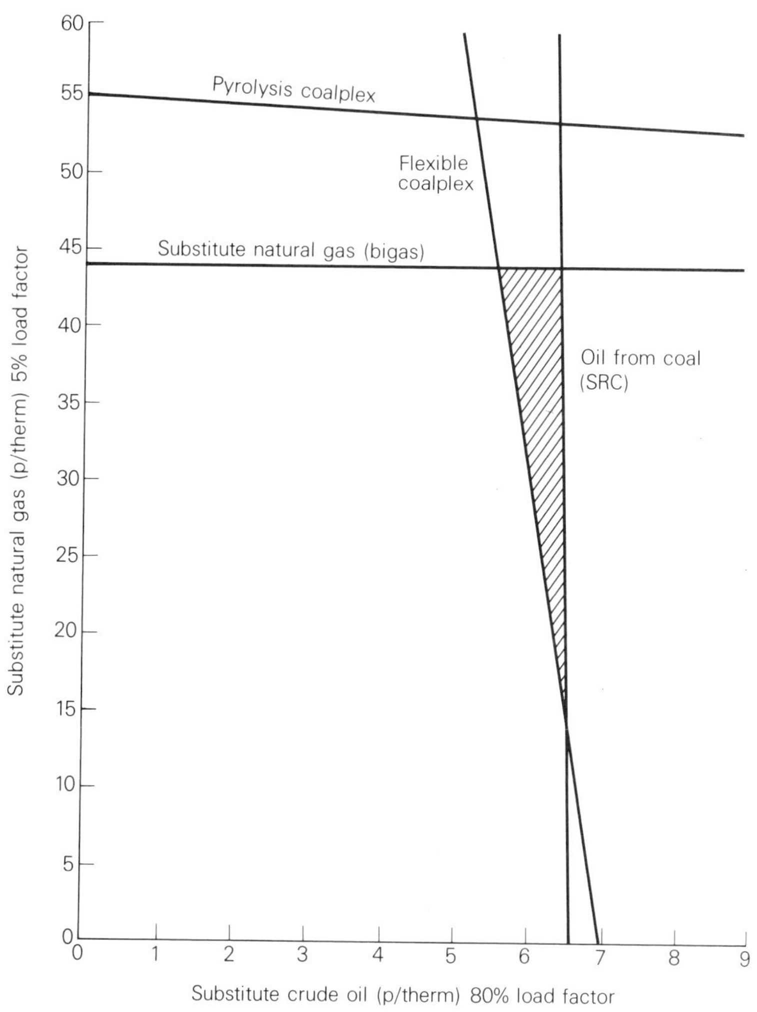

20.15 Cost advantages of a flexible coalplex

translation to other countries somewhat difficult. Nevertheless the comparisons given between the different processes are interesting (Table 20.1).

The authors do not specifically compare these energy costs with those from single product plants. However, the COED costs quoted are probably not very much higher than estimates at that time for SNG or SRC products; the indicated margins by which the other three configurations are lower than COED probably means that all three are projected to be substantially cheaper than SNG or SRC alone. Using "appropriate" prices for all products, the authors claim DCF rates of return between 19–23% for those three cases (using the 65% debt at 9% interest basis); these rates of return are probably considerably higher than practically any other process estimates which have been published.

Of course these calculations necessarily apply to only one set of relative values for

TABLE 20.1 Economic comparison of certain coalplexes (1978 dollars, Financing: 65% debt at 9% and 35% equity at 12% DCF)

	Oil/Gas	*Fischer/Tropsch*	*COED*	*POGO*
Dollars/million Btu	2.10	2.90	5.00	2.17
Dollars/BOE[a]	12.50	16.00	30.00	13.00

[a]BOE = Barrel of oil equivalent

products and to one balance of outputs. However, from the data given, additional calculations or re-optimisations could be carried out to suit different conditions. This would be an excellent starting point for a computerised economic model. In a further description of POGO, a very important claim is made that the projected thermal efficiency from coal to synfuels, by-products and gas (for electricity generation) would be 75%; a combined cycle generator is proposed, using the clean gas, with an efficiency of 44%. The possibility of providing some extra capacity for each product, enabling flexibility of output ratios to be met, was not pursued. However, sensitivities to changes in various cost factors were calculated, of which the most important would be capital. This would be the cost element most affected as a result of providing flexibility. A 10% increase in capital would increase required selling prices by 8%. It would appear that such a trade-off might be worth pursuing if a plant such as POGO were to be built.

The Parsons team also investigated the profitability of the chemical-type Coalplex illustrated in Fig. 20.11, using guidelines from the American Institute of Chemical Engineers for product values and capital charges (15% DCF). Under these stringent conditions, the required annual revenue (1978) including profit and taxes, was \$1890 million; sales values at that time were likely to be only \$1400 million. However, it was stated that modest changes in product prices or financial guidelines (especially the assumption of a 6% inflation rate) would make the 15% DCF rate of return achievable. Both changes look more probable in 1980 than in 1978.

The scheme proposed by Catalytic Incorporated shown in Fig. 20.12 has also been costed. The authors conclude, on the basis of coal at \$30/ton and with 1978 product prices, that the financial targets (50% debt at 10% interest, 12.50% return on equity) could be met with a reasonable safety factor. However, they recognise that there are substantial uncertainties, especially price competition from countries outside the USA and that "floor" guarantees on prices (presumably governmental) would be necessary to induce companies to invest in such a Coalplex.

Price projections are of course at the root of all economic assessments for coal conversion processes; the adoption of a Coalplex concept means that more prices are involved with a greater element of uncertainty. Perhaps more important to the Coalplex concept, is whether the whole slate of products will be readily acceptable into marketing systems and what value can be attached to flexibility of product pattern. It is in these areas that the need for improved economic modelling is most vital. The demand for energy, including that possibly derived from coal, varies widely and in a complex fashion — diurnally, seasonally and as a result of changes in economic activity. These variations are not merely statistics; real resource costs are involved in producers reacting to demand; furthermore, these extra costs are not allocated in any logical way amongst suppliers and customers. The optimum patterns of production, transport and storage (or reserve capacity) will change with time and there will be political as well as purely economic factors. Models to take account of these factors would be complex but are certainly feasible and are essential for the full appreciation of Coalplexes and the way that in-built flexibility of outputs, requiring extra capital, might be recouped. Any departure from the optimum pattern of supply and transport over a large geographical and political area represents a real waste of energy. In considering Coalplexes, the relationship between coal and other energy sources would also be a vital factor but the main effect of such an analysis might be to emphasise more strongly the flexibility of coal and the need to exploit it through variable output complexes.

THE ESTABLISHMENT OF COALPLEXES

Whatever the potential advantages of Coalplexes may be, it seems unlikely that a complete large-scale version of, say, POGO (27 000 t/d of coal input) will be ordered as a single entity in the near future. Finding the right financial and institutional framework for a "straightforward" coal conversion plant of the SRC or SNG type seems difficult enough; the Coalplex will be seen by many as multiplying the uncertainties. The manner in which Coalplexes may become established is therefore worthy of some speculative thought. The process may be a gradual and evolutionary one. Nonetheless it is to be hoped that a clearer visualisation of the long-term optimum would be allowed to guide the development of coal conversion in the short or medium term.

It is possible that there is an inherent tendency for coal processes (and other energy processes) to become more complex and diversified with time. The coking industry may be a case in point; originally, coke was virtually the only product but subsequently the recovery and exploitation of by-products became financially attractive and increasingly complex. Another example may be that although gasoline was the primary objective of Sasol 1, other products are made to utilise fully the by-products and improve the financial returns. It appears that this will continue; more importantly, substantial expenditure has been incurred on a pipeline system to sell fuel gas, obviously for reasons of profitability. It will be interesting to see whether Sasol 2 and 3 develop a more diversified product approach although the initial policy seems against this. It has already been noted that there appears to be a general tendency in the main gasification and liquefaction developments to consider opportunities for profitable exploitation of by-products. The increasing prices of primary energy will provide further incentives for maximum recovery of energy and material values from coal inputs. In addition to this evolutionary progression, however, two scenarios for the establishment of Coalplexes are suggested: (a) co-processing of coal and oil, and (b) growth from synthesis gas as a focal point.

Co-processing of Coal and Oil

Oil refineries and their associated petrochemical complexes are to some extent a model for the development of Coalplexes, although historically perhaps they have been unduly self-sufficient and somewhat prodigal in internal energy consumption. Coalplexes might develop by being "grafted-on" to oil refineries. It might be thought that the simplest situation would be to set up a separate plant for the manufacture of SRC from coal, which would then be used as a supplementary feedstock. This might not be the ideal arrangement however, especially since opportunities for exploiting all the potential values of coal conversion, and not merely the SRC, might be neglected. Furthermore, the coal-derived feedstock might present some problems of compatibility with crude oil. Finally, this procedure would ensure that the economics of coal conversion in this area would depend solely on the timing of very large changes in relative prices for coal and oil.

A more transitional procedure would certainly be better. Initially, to supplement crude oil supplies in a refinery, coal could be used to provide process steam and power. A little later, hydrogen made from coal could be introduced; there is likely to be a growing demand for hydrogen in refineries. At present, low value products such as heavy oils and off-gases are used for these purposes; by replacing these materials with coal, they could be used in more valuable ways, increasing the value of refinery outputs from a given oil input by 5–15%.

Subsequent to the peripheral use of coal in oil refinery practice, various possibilities for inter-mingling of process streams exist. As mentioned above, SRC-type liquids might be added to crude oil streams but since coal derivatives tend to be rich in aromatics, some considerable adjustment to refinery practice would almost certainly be necessary. Coal liquids derived by pyrolysis or gas extraction processes might be added to refinery product or intermediate streams; in these cases the proportion of liquids to be accommodated would be smaller and the large amount of active char could be exploited in gasifiers. Some of this gas might then be turned into electricity or, probably better, into hydrogen for refinery use. Alternatively, the gas might be partially treated in a synthesis unit to produce paraffinic liquid feedstocks or products complementary to the high-aromatic products from coal liquefaction (and from the upgrading of residual oil).

Petroleum products are not normally good solvents for coal, which would otherwise be an obvious possibility for integration. Some success in this direction has however been claimed, notably in Russia; this may be related to crudes which are unusually rich in aromatics. There may also be some refinery streams, especially those associated with delayed coking, which tend to be higher in aromatics than the general refinery flow and these may be useful coal solvents. The addition of coal to these aromatic materials may be a way of extending crude feedstocks. The addition of powdered coal directly to oil intermediates has also been studied in several oil refinery process stages, including hydrogenation, cracking and distillation; extra yields or product value have been claimed, together with a reduction of feedstock costs. There appears to be substance in some of these claims and mechanisms to explain these effects can be suggested. Pyrolysis or solvent extraction of the coal particles may increase the availability of light products; the ash or coal particles may also have catalytic or nucleating properties.

Thus, the prospects seem excellent for the gradual integration of coal into many aspects of the oil refinery process, thereby approaching in this way a Coalplex configuration. This concept would be enhanced if, at the same time, the coal/oil refineries developed closer links with external energy systems, especially through the export of heat, electricity and fuel gases.

Synthesis Gas as a Focal Point

An alternative plan for the evolutionary development of Coalplexes might be through the prior establishment of centres for the production of synthesis (or medium CV) gas on sites which are capable of expansion and which are also convenient for coal supplies and markets for chemical feedstocks, fuel gas, heat and electricity. The manufacture of such gas is already economic or nearly so in many parts of the world, on straightforward competitive terms. Considered more broadly as a method of increasing the value of coal and of conserving supplies of natural gas and light petroleum liquids, the case for such plants seems compelling. Assuming such plants are established, with multiple markets, and in suitable locations, there would be excellent opportunities for further integration at a later stage, both upstream and downstream. Upstream, instead of using raw coal as the gasifier feedstock, plants might be introduced to provide a preliminary separation of liquids by pryolysis or gas extraction; the gasifiers could provide heat and hydrogen. Downstream, chemical synthesis might be developed, with methanol or gasoline as possible products; these synthesis possibilities were well illustrated in Fig. 10.10 (p.272), due to S. Andrew (ICI). Alternatively, if the hydrogenation routes to coal liquids were preferred to synthesis, the gasifiers could provide heat and hydrogen to these processes. The sizes of individual unit stages and the choice of process type could be determined in relation to external energy supply/demand models. For example, one consequence might be that those gasifiers which produce high temperature steam, suitable for electricity generation, might be preferred. The access to demand for low-grade heat might also be a factor in favour of medium sized rather than giant plants and for location near to industry rather than in remote sites.

If cheap heat or hydrogen became available from non-fossil sources such as nuclear power, these could be integrated into a Coalplex based on a "Syngas Core".

CONCLUSIONS

Coalplexes have potential for increasing the efficiency with which coal will make its contribution to energy supplies in the future. The fact that there is a wide range of processes, having different characteristics, under development for each of the major component stages, emphasises this opportunity. The choice of which processes should be further developed may ideally rest partly on their ability to integrate as well as on their characteristics as discrete plants. The analytical tools for energy modelling however probably need considerable improvement if the best choices are to be made and developed.

Major processes under development for single coal conversion products, such as SNG or SRC, may evolve in the direction of greater diversity and overall energy economy. Two other ways, also evolutionary, of approaching efficient Coalplexes have been suggested: (a) "grafting-on" to an oil refinery and (b) the "Syngas Core". In both cases, optimum links with external energy flows should be sought.

An appreciation of the potential of Coalplexes could influence the way in which coal conversion processes become established, with long-term benefits both to the future of coal and the efficient application of energy generally.

SECTION IV
Coal in Energy Policies

Chapter 21

DISTRIBUTION OF COAL IN RELATION TO ENERGY NETWORKS

INTRODUCTION

In this chapter, consideration is given to the way in which coal development may be affected by geographical factors. This subject includes the original source of the coal, its transport and conversion, and the transport of secondary or tertiary products. The general background is first discussed from several points of view. Some relevant economic factors are then considered; the factual basis is not, unfortunately, very satisfactory. Some reflections and conclusions, with policy implications, are then presented.

Although coal is currently important and has been so throughout the industrialised era, it will certainly be even more important in the future. Not only will the amount of coal consumed increase but also its sources may become more diversified; international trade in coal is almost certain to increase. Coal will also be used in new ways which may evolve over a considerable period of time. It is to be hoped that this transition sequence will allow the best possible use to be made of coal at any particular time, in a manner conducive to the optimum exploitation of all sources of energy.

Coal utilisation taking advantage of modern technology should be regarded as a major new industry of strategic world importance; the development of this great industry could be facilitated by analytical assessments in advance. Since coal movements will often be international and the development of new technologies will be of interest to virtually all countries the opportunity for co-operation is exceptional. It is hoped that this co-operation might not only include technical and commercial activities but planning and environmental aspects also.

DISTRIBUTION OF RESERVES

A few countries are particularly well-endowed with coal reserves but it should not be overlooked that quite a few other countries have reserves which are really substantial in absolute terms, even though these may appear small as a proportion of world reserves. Furthermore, the quality of these reserves outside the favoured few countries may at first sight seem disappointing, in terms of both mining conditions and coal characteristics, compared with the most spectacular deposits.

However, this may not be the end of the matter. Some "poor" deposits may become more attractive as a result of successive increases in energy prices and the establishment of new utilisation methods. There will also still be more coal to be found, both in areas of relatively intense exploration and in areas so far little explored. In the former class, the UK is a good example; several major coalfields not previously counted in the reserves have been delineated in the 1970s. Exploration has been very unevenly concentrated over the world, generally being most intense in areas nearest to existing workings. The same situation obtained in the case of petroleum; a wider exploration has in the last twenty years found several major new oil provinces. It is suggested that a similar spread of an increased effort on coal exploration might have fruitful results.

Although there is need for still further expansion of effort on exploration, there has already been a substantial increase in the last few years. Some progress has also been made in the rationalisation of technical terms and the categorisation of reserves on a common basis with regard to coal quality, depth, seam thickness and degree of geological disturbance; this will simplify international cooperation. It should be possible in the next few years to relate the technical factors of resource measurement to economic and commercial features so that an order of merit might be developed to guide plans for exploitation in a progressive manner. Considerable progress has been made in defining terms and developing a logic for such an ordering system; it remains to be seen however whether vital information will be freely provided or withheld, on commercial or national interest grounds.

It seems probable that the decade of the 1980s will be crucial in determining the main patterns of coal production and movement for the first part of the 21st century.

TRANSPORT

Although it is important to establish the main areas for coal production, both large and small, high and poor quality, as soon as possible, the distribution of energy consumption may be less easy to alter than the distribution of exploration and exploitation effort. Therefore, transport will remain a crucial aspect of an integrated coal policy, both on an international and a local scale. Final usage of coal may be in many forms such as local heat, transport fuels, chemicals or fertilisers. Consumption may therefore be highly disaggregated but, on the other hand, transport efficiency generally benefits from increase in scale; the costs of conversion processes may also be decreased by centralisation. There is, therefore, clearly great scope for optimisation in the overall scheme. In addition, however, consumption or loss of energy may occur in transport and, more certainly and significantly, in energy conversion processes, frequently with the rejection of low grade heat. Further complications arise from the fact that the demand rate for most forms of energy is uneven and because there are limitations or high costs associated with storage.

The land transport of coal by rail and road has been highly developed, with near maximum elimination of labour by the intensive use of highly developed handling equipment. There appears to be relatively little scope for further major improvements. However, where new basic facilities have to be developed, such as new railways or roads, much higher transport costs will be necessary. Ocean transport has also been considerably developed in recent years, stimulated by experience with coking coal contracts which have linked large sources of coal with large individual users through specialised harbours and handling equipment. By analogy with oil, further progress will probably be made by increasing ship sizes and by the development of specialised designs, although the results may not be so spectacular as with oil.

ECONOMIC FACTORS

Economic considerations should influence the optimisation of coal transport and usage, in relation to other forms of energy, in a number of ways, including the following:

(1) Relative prices of coal, oil, gas and electricity at various locations.
(2) Relative costs of internal and overseas transport for coal and other energy forms.
(3) Conversion costs from coal into other forms of energy and the effects of scale on these.

Unfortunately, information on these matters is incomplete or inexact. However, it is important to recognise the importance of these factors, not only so that the provision of better information may be encouraged but also so that the form in which assessments and decisions are made can be improved in order to take continuous advantage of the refinement of information. Some information has already been considered in other chapters and some other factors are discussed below.

Long Distance Transport

The price of coal delivered into Northwest Europe or Japan from the USA, Australia or South Africa is typically two or three times that at the mine. The actual ocean shipping costs may be only half or even less of the total transport costs and the prospects for defining and containing shipping costs seem reasonable. The average size of ship seems likely to go up from about 60 000 tons deadweight at present to about 150 000 tons by 2000, a significant change. Fuel costs are probably the greatest uncertainty in the examples of rough estimates in Table 21.1.

Loading and unloading will each add a further dollar or two to these costs. For the much larger tonnages expected in the future, however, very large capital expenditure will be necessary to provide port facilities at either end, if this situation is to be maintained and also if the assumed advantage of larger ships is to be realised.

A greater uncertainty generally relates to the mine to port costs, presently often equal to the ocean transport costs. This element may become even more critical in the future as major new sources of coal may tend to be a long way from existing ports and probably without adequate rail access. In the USA, minimum rail transport

TABLE 21.1 Long distance transport costs of coal (US $/ton)

	1976	*2000*
W. Canada to W. Europe	17	17–23
W. Canada to Japan	7.50	6–8
USA to W. Europe	6.10	6–10
USA to Japan	17	18–26
S. Africa to W. Europe	11.50	9–12
S. Africa to Japan	11.50	9–12
Australia to W. Europe	21	15–21
Australia to Japan	7	6–8
Poland to W. Europe	3–7	3–8
Poland to Japan	21.50	16–23

TABLE 21.2 Costs for pipeline transport of coal (1976 US money values)

	Cents/ton-mile
18 inch diameter pipeline	1.3
36 inch diameter pipeline	0.8
48 inch diameter pipeline	0.7

prices, for hauls of several hundred miles, are about 0.6 cents/ton-mile but 1.0 might be more typical. European prices are somewhat higher. For new railways, however, prices would be at least 30% higher.

Slurry pipelines must be considered as an alternative to rail for transport of coal to ports from distant mines, as well as for inland transport. It is commonly believed that slurry pipeline transport is similar in cost to existing railways, or possibly a little cheaper. This would be consistent with the fact that only two installations are operating but two or three more are under construction and other schemes are being planned; these schemes are generally where rail exists as an alternative. It may be that the major role for slurry pipelines in the next couple of decades will be to connect new coal sources to ports where no rail exists, especially in cases where railway construction would be difficult and expensive. In these circumstances, some suggestions are that slurry pipeline transport might only be about half the cost of rail, but this might be an optimistic view. One estimate (in 1976 US money values) for a 1000 mile pipeline is given in Table 21.2. This indicates the sharp effect of pipeline size. The 48 inch pipe would transport about 33 Mt/a and is the largest anyone is likely to contemplate building, especially as the effects of scale are becoming negligible by this point. At the other extreme, the 18 inch pipeline is about as small as could be considered, since smaller sizes become rapidly uneconomic, at least on the basis of present technology. It should be noted that the above estimates assume that present technology will be acceptable, including the dumping of the "ink" into waste ponds with consequent loss of several per cent of the coal. In addition, the capital charges are about two-thirds of the total costs, so that under-utilisation would be very costly. One potential development of the use of slurry pipelines might be to load and unload ships with slurry rather than a dewatered product. Estimates have been provided for the United Nations (ECE) for the long distance transport of coal over a distance of 4000 km, the first half being overland, Table 21.3.

TABLE 21.3 Long distance transport costs for coal (1978 $US)

	US $/ton	*US $/GJ*	*Efficiency (%)*
Existing rail	22.5	0.83	100
New rail	29	1.07	100
Existing rail and self-unloading ship	17	0.64	100
New rail and self-unloading ship	20.5	0.76	100
Slurry pipeline	13.5	0.49	93
Slurry pipeline and slurry ship	12.5	0.46	93
Slurry pipeline and self-loading ship	13	0.47	93

TABLE 21.4 Transport costs for various forms of energy (1978 $)

	US $/GJ	*Efficiency (%)*
Oil: pipeline	0.18	100
pipeline and ship	0.15	100
ship	0.12	100
Natural Gas: pipeline	0.63	81
LNG ship	1.10	85
pipeline and ship	1.35	76
Electricity (DC)	1.25	79

For comparison, the estimated costs for other forms of energy are given in Table 21.4.

The author also considers the various alternatives available for location of the conversion process when coal is required in a different form at a distance from its source; estimates for a total distance of 4000 km include those given in Table 21.5.

Whilst these estimates are interesting and indicate a reasonable relationship for the different products from coal, the preferences shown between the alternatives of converting before or after transport need to be considered as a guide rather than as definitive. The most important open question is whether lower costs (especially through higher overall thermal efficiencies, higher plant availability or lower capital costs) could be obtained by conversion in the importing country. The answer to this may lie mainly in the ability of the conversion plant operators to make use of "waste" heat now commonly rejected.

Internal Transport

For transport of coal and other energy forms within an individual country or contiguous groups of countries, there are really three cases. One is the smaller countries, such as the UK where the distribution of coal production (even for projected new coalfields) is fairly well settled and the location of industries using coal has historically been related to the coalfields. There are other countries where coal will assume a new and much greater role through imports, which will be centralised (on grounds of port facilities rather than energy demand density); several Western European countries and Japan will be in this category. Finally, there are very large

TABLE 21.5 Transport costs of various forms of energy derived from coal (1978 $)

	US $/GJ	*Efficiency (%)*
SNG from coal		
Transport coal, then convert to SNG	3.3–4.3	56–60
Convert to SNG, then transport	3.7–4.6	46–49
Liquids from coal		
Transport coal as slurry, then liquefy	5.1	60
Liquefy coal, then transport by pipeline and/or ship	4.6	65
Electricity from coal		
Transport coal, then generate electricity	5.5–7.1	35–38
Generate electricity, then by DC line	6.7	30

TABLE 21.6 US estimates of coal transport costs (1978 $)

	Cents/ton-mile
River barge	0.6
Slurry pipeline	0.7–1.7
Railroad	0.6–1.0
Highway truck	3.4
Conveyor belt	3.8

countries, of which the USA is the most important to the West, where coal output is expected to increase greatly, but mainly at very long distances from traditional centres of production or consumption. In the USA there are also important waterways. In this case, the options are similar to those described in the previous section.

In the USA in recent years, rather more than 50% of coal transportation was by rail, with about 25% by water and 10–15% by road. The US Department of Energy gave (Table 21.6) estimates for the "economic" cost (presumably real costs, not distorted by complex tariff structures) in 1978.

The question of whether coals should be prepared before transportation has been studied on economic grounds but without any clear-cut general conclusions. Some coals need some form of preparation to be conveniently transportable; some forms of transportation (e.g. slurry pipeline) impose their own preparation requirements to some extent. On the other hand, preparation may sometimes make certain coals more difficult to handle and transport. It appears that every case needs to be examined on its own merit, taking account of the coal characteristics, transportation options and end use. The availability of water may often be a constraint in considering preparation at the mining site. It has also been suggested that, in the USA, constraints on rapid coal expansion may arise from railroad limitations; coal preparation prior to transportation from the mine might slightly ease this problem.

For shorter distances in internal transport in a country like the UK, it is extremely difficult to generalise, mainly because cost calculations depend very heavily on whether new transport facilities need to be constructed or whether the marginal costs of using existing systems are considered more appropriate. Furthermore, although both overall production costs for gas and electricity and the tariffs to consumers are generally known, the breakdown of the very substantial difference between these two quantities is not very easy to determine. In 1975, relative transport costs in the UK for various forms of energy were estimated as shown in Fig. 21.1.

These costs fall into a similar order to those indicated above for long distance transport and, allowing for changes in money value, are similar quantitatively at the lower end of the scale. However, at the higher end the scale is much extended for the internal transport case, probably because of the costs of "branch lines", i.e. the change in unit costs as the energy leaves main transmission grids into smaller local distribution networks. This factor also affects coal distribution, of course, and especially when individual requirements are too small for unit trains; transport costs for scheduled train loads may be about 10% more expensive than unit trains and individual wagons about 60% more expensive. Road delivery may be 3–5 times as expensive per ton-mile as rail. For typical distances in industrialised areas, the

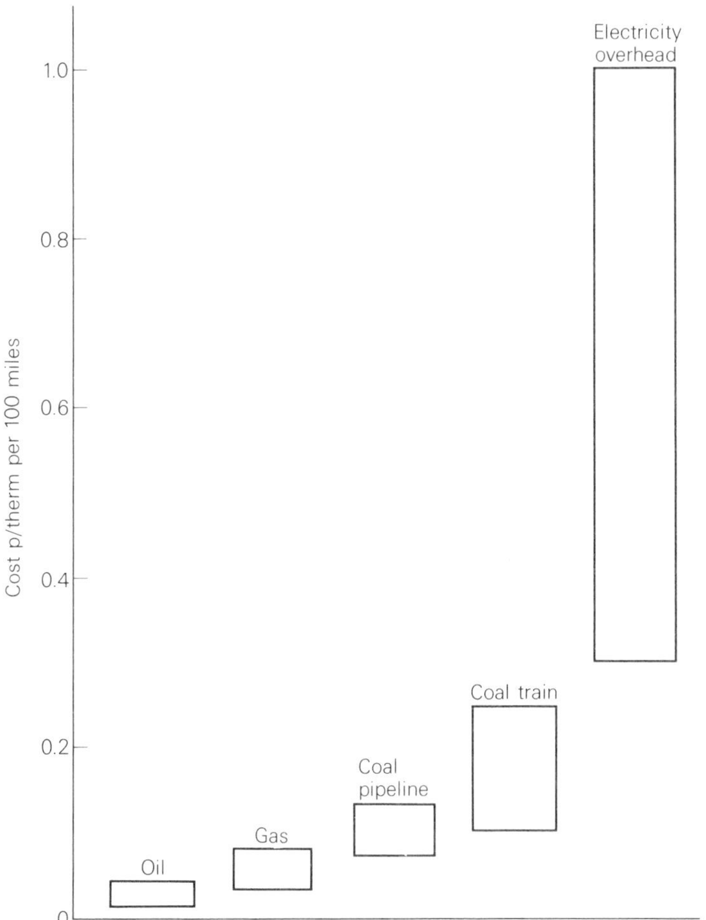

21.1 Cost of energy transmission

delivered costs of coal may be between 10–50% higher than the pit-heat cost for moderate to large scale users.

There seems little doubt that transmission and distribution costs are higher for electricity than for gas. In recent studies by IEA Coal Research, for industrial consumers an average "delivery" cost of \$3.5/GJ was assumed for electricity and \$0.73/GJ for gas from the natural gas grid. From evidence given by the US gas industry in 1978, delivery costs for electricity and SNG respectively of \$5.30 and \$1.59/GJ were indicated. Some earlier estimates from this source are shown in Fig. 21.2.

Delivery costs for gas of lower calorific value on a dedicated pipeline or mini-grid are more complex. For similar quantities, pumping costs for gas are likely to be roughly inversely proportioned to calorific value, say 1:3:6 for SNG: MBG:LBG. It is generally assumed that LBG will be used more or less on the site where it is made and that there is a maximum distance over which MBG can be transported

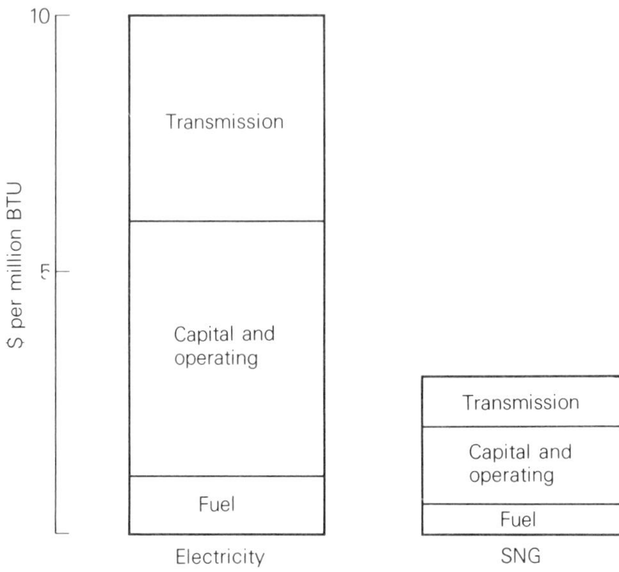

21.2 Cost of electricity and SNG

"economically" — sometimes taken as about 50 miles. However, this is a far from simple matter, depending on the flow rate and complexity of the system, as well as its length. This is illustrated in Figure 21.3. This shows the optional pipeline size and the delivery costs for different flow rates; the standard average national delivery cost is also shown. At above about 15 TJ/d, assuming a single customer, costs are lower than from a national grid, reflecting the diversity of customers for the latter. At the higher flow rates, gas delivery costs are probably similar to road delivery costs for coal but gas delivery costs go up rapidly as scale decreases. In 1977, Koppers suggested a delivery cost of about $0.21/GJ per 100 miles for a flow of about 200 TJ/d, which is roughly in line with the figures indicated above.

Effects of Scale and Load Factor

As shown above, transport costs of various forms of energy may vary from a few percent to perhaps 70% of the delivered energy cost; within this wide range, comparisons are very difficult. Two generalisations may however be possible. The first is that electricity is generally the most expensive form of energy to distribute and the second (which compounds the first) is that costs go up rapidly as the scale of distribution diminishes.

However, scale also effects the costs of production. This will clearly be an important factor together with transport costs in the location of coal conversion plants. For very large coal gasification (or liquefaction) plants, capital charges will be perhaps one third of the total costs, say $2/GJ out of $6/GJ. If the plant size is reduced, the capital cost element will become more important; capital charges might double, say from $2/GJ to $4/GJ (rising from one-third to one-half of total

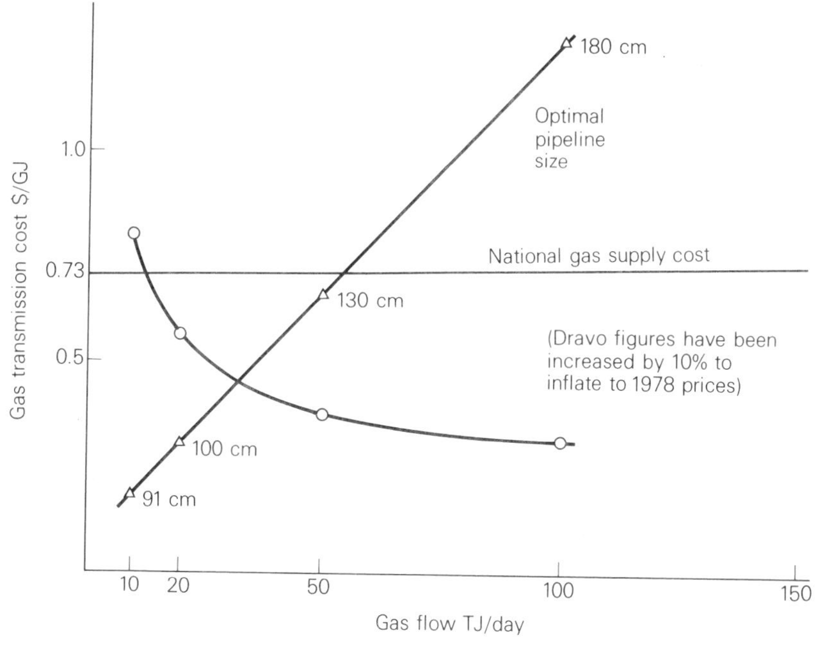

21.3 Industrial fuel gas distribution cost

product cost) by reducing the size to one-sixth of the very large plant. Similar relationships will of course apply to electricity generating plants.

The load factor will have a similar effect to the scale factor; halving the load factor of a very large plant (which might be due either to market mismatch or technical problems) would perhaps have about the same effect on the unit capital charges as a six-fold reduction in size. Capital charges on distribution facilities would also be a factor but perhaps in the direction of favouring relative dispersal of conversion plants.

Summary of Economic Factors

Although economic factors are clearly all important in optimising energy networks, particularly the size and location of plants for converting one form of energy into another, no clear cut guidelines are available. Even in cases where the simplifying assumptions can be made that coal is the preferred energy input and gas is the preferred form of usage, the permutations can be quite complex. In Western Europe, for example, the choice between European and imported coal would determine the starting point or points and this in turn would influence the degree of centralisation of conversion plant; the latter would also be affected by the concentration of demand and its load factor. The type of gas would be a further option. However, although general guidelines are difficult, optimisation in individual specific cases could be a reasonably satisfactory exercise; the broad methodology is available and optimisation should always be attempted wherever possible.

REFLECTIONS AND CONCLUSIONS

Supplies of Coal

Access to supplies of coal in the 21st century may be a crucial factor for economic development, both for the developed and developing countries. As argued in the next chapter, there are no direct replacements for fossil fuels; coal may be the only fossil fuel for which an expansion could be planned on satisfactory and beneficial lines. If these premises are considered plausible, each country should, separately and in co-operation, urgently consider the supply position, both for indigenous supplies and internationally traded coal. A number of countries concluded in the era of the oil glut that indigenous coal had little future. Their view, if it still prevails, should be urgently reconsidered. Each country with the slightest prospects for coal development should re-assess its coal potential by intensified surveys, using international methods of measurement. The abandonment of existing mining facilities, so extensively practised in Western Europe and Japan, should be considered by much more stringent criteria, including national security. Prospects for continuing or re-opening "uneconomic" mines should be reviewed, including the possibilities of underground coal gasification.

However, there will be a need for additional supplies from countries able and prepared to export on a large scale. Assistance should be given to such countries in extending and intensifying coal resource surveys and in preparing outline schemes for exploitation. It is vital that information should be available a long time ahead of the need, and that capability exists for physically developing these resources, so that the sequence of exploitation can be as advantageous as possible to both exporters and importers. Ideally, large importing areas such as Western Europe should know broadly where their supplies will be coming from and in what form for at least 20 years ahead.

International Trade in Coal

The development of new coal areas for export will require the development of transport facilities to ports. A large new fleet of ships will also be necessary. The size and design of these ships (e.g. whether to carry lumpy coal or slurry and whether to burn coal) should be related not only to the port facilities at either end but also the the inland transport to and from ports; at the receiving end this should include considerations of distribution and use.

A related question is whether coal should be converted into liquids before being exported (conversion into gases is probably not relevant because of transport costs in gaseous form). The concept of transporting a refined and concentrated liquid instead of impure coal has attractions. However, there may be some disadvantages also including:

(a) A "standard" coal liquid may not always be the ideal intermediate.

(b) The recovery and use of waste heat and other complex type of possibility may be more favourable in an industrialised importing country than at the exporting end.

(c) Gases may be more useful and efficient products than liquids from coal for some time.

Apart from idealised economics, exporting countries may have a desire to increase employment and add value to their coal assets. Importing countries, on the other hand, may prefer to import and retain a working stock of coal, as such, in order to improve security of supply. This is clearly a matter for discussion and negotiation; solutions might be in a compromise form, especially if assistance in technology and finance for both extraction and conversion facilities were provided to countries willing to see their coal reserves developed, partially at least, for export. The balance between coal exports as such and that of pre-converted coal products may perhaps change with time. The initial exports have been mainly of coking coal and this will probably be followed by greater trade in steam coal before substantial amounts of coal are converted to gases and/or liquids anywhere. If, as expected, gasification developments precede liquefaction on a large scale, ocean transport of coal rather than derivatives may be relatively stabilised for several decades.

Location of Coal Conversion Plants

Assuming that coal in the future becomes an even more important input into the energy supply systems of many countries, the questions of the location, size and type of conversion and consumption centres, together with the transport linkages and provisions for demand variations, are very important and complex. Probably neither factual information nor analytical models are adequate for developing ideal solutions. Even if the latter were available, it is doubtful if the institutional framework would be conducive to their implementation at present. However, this should not be taken as a reason for neglecting suitable studies.

Looking at energy distribution systems from the point of view of coal, the following may be guidelines to the way optimisation should be approached:

(i) Coal should be taken to the point of use wherever possible and used directly for the production of heat, provided this does not result in inefficient use.

(ii) Electricity will not in the long run be the best outlet for coal; in fact, it will probably become eventually the least attractive conversion option. In the meantime, the status quo will prevail; in the near future, the transport of coal to moderate sized generating stations operating on a Combined Heat and Power basis, with other energy forms providing the base load electricity requirement may be a preferred objective.

(iii) Conversion of coal to gas may be economically desirable before liquids. SNG may in some cases be required and best provided by large plants near to existing natural gas pipelines. However, the manufacture of medium calorific value gas in moderate sized plants for local distribution in minigrids should be seriously considered and could be a very early application of coal conversion technology.

(iv) The total costs of energy supply are sensitive to peak load requirements. The ability to store coal, to convert it into gases and to optimise its use during load fluctuations in parallel with natural gas could help to reduce the costs of providing for uneven demands; this will only apply in full if gasification becomes significant before natural gas reserves are depleted.

(v) Emphasis should be given to technologies and siting which increase the prospects of the maximum energy released being available for utilisation purposes.

Importance of the USA

For many reasons, including the size of her coal reserves, the distances involved and the balance of energy supplies, the USA is bound to be in a position of world influence in coal distribution and utilisation. In addition, if the USA adopted a positive policy of substantially increasing exports of coal to her closest trading partners (who are likely to be increasingly oil-deficient), this would provide a great incentive not only for the development of technology but also towards improvements in the analysis and planning of coal utilisation on an international scale.

Final Conclusion

Much further work is urgently required on the optimisation of energy flows with particular reference to coal and its derivatives. Information tends to be concentrated on the production costs of various energy forms and to a lesser extent, on transport costs. The analysis of total system costs does not seem at all well developed; this is not an ideal background for the introduction of new coal-using systems and for much greater world reliance on coal.

Chapter 22

RELATIONSHIP OF COAL TO NUCLEAR POWER AND OTHER ENERGY SOURCES

INTRODUCTION

This chapter starts with a general statement about the importance and role of coal, taking account of other energy sources. Some of these possibilities are then discussed individually and conclusions drawn.

It seems reasonable to assume that fossil fuels will continue to be the dominant source of world energy for a very long time, especially if "energy" is taken to include feedstocks. Nuclear power and other "new" forms of energy will play increasing roles, in absolute quantitative terms, but mainly within rather narrow confines, in particular the provision of electricity and/or local low-grade heat. Within the established fossil fuels, gas and oil have a premium value and, with their limited reserves, will reach declining rates of output before coal. Since the main distinguishing chemical feature between these fossil fuels is the hydrogen content, which may be made from coal and water, liquids and gases will in due course be made from coal by processes involving hydrogenation. The total energy in the fuels produced will of course be less than the input; generally, the higher the hydrogen content of the product, the lower the conversion process efficiency. Although capital and other cost must be considered, it might be desirable to use these fuels in a "sequential" order, based on the use of the lowest hydrogen containing fuel suitable for the purpose of the energy being finally consumed (and not necessarily in the form initially preferred by the customer). The optimum or acceptable level of hydrogen in fuel for a particular purpose may change over time and, ideally, high-hydrogen fuels should be reserved as long as possible. Whether such planning could be achieved in the practical world is extremely doubtful at present. However, the price mechanism may tend, in the long run, to bring this about.

The growing shortage of high hydrogen fuels, which will lead to the upgrading of coal, will also lead to an intensification of interest in other sources of liquid fuels, including heavy oils, tar sands and shales. The exploitation of these sources will probably to some extent proceed in parallel with that of coal. The relative rates of development are difficult to predict but the more general availability of coal will ensure its overall significance.

The proposed use of nuclear heat in coal gasification has been described in Chapter 8. The object of this arrangement is to substitute nuclear heat for heat obtained by the combustion of some of the input coal, thus aiming for a higher utilisation of the carbon in the coal to combine with hydrogen to produce gas. The economics of the substitution are complex but would seem to depend on exceptional progress in high temperature nuclear reactor technology and other factors.

Hydrogen has itself been proposed as a fuel, with suggestions that it could become the standard fuel for distribution and most end uses. Early enthusiasm has subsequently become somewhat tempered but attention now being given to the possible climatic effects of CO_2 may give the concept new life. In view of this, together with the great importance of hydrogen in the chemical industry, and particularly coal (and oil) refining, non-fossil energy methods of producing hydrogen are being studied. Nuclear power is one possible source of energy for splitting water, either by electrolysis or through a complex series of chemical reactions. On the assumption that non-fossil hydrogen could be made in large quantities, the possibility that some of this hydrogen might be used in coal conversion processing has been studied.

The possibility has been suggested of using nuclear heat to carry out endothermic reactions involving coal, with a corresponding exothermic reaction providing useful heat elsewhere in a closed circuit. This has been called "Adam and Eva" in Germany.

Apart from nuclear power, solar energy is perhaps the most important alternative (or adjunct) to fossil energy; solar energy is of course the original source of fossil energy. Many other forms of energy are also derivatives or related to solar sources.

Some other energy factors are now briefly considered, with a view to establishing whether their promise can have any large and early impact on the necessity for a vigorous coal policy. It is not possible or necessary that this review should be comprehensive or even completely balanced. The case for a coal policy must be considered established in the absence of realistic alternatives.

ENERGY DEMAND/SUPPLY BALANCE

As we have seen, predictions of energy supply are very difficult. The demand side of the balance may be even more imponderable. Main uncertainties relate to world population increase, economic growth and the energy coefficients (for different types of national economy) relating energy to economic growth; there are others. It is notable that, at the WEC in 1980, the main rapporteur on demand/supply balances noted that there was little new information on the demand side. Therefore, demand forecasts given at the 1977 WEC are taken as indicative (Table 22.1).

The ranges of demand relate to various scenarios, which vary chiefly in growth assumptions; more extreme scenarios were also considered. The ranges of economic growth rates assumed in Table 22.1, may now seem rather high and so also may the energy coefficients; any overestimates would of course have a compound effect.

At the principal Round Table Discussion at the WEC in 1980, Forster presented figures for the energy supply prospects for the next few decades, based on the Conservation Commission's work between conferences (Table 22.2).

It was stated that Table 22.2 was not a forecast. It was based on what contributors to the study thought *might* be made available from each energy source. The relatively close correspondence between Table 22.2 and Table 22.1 suggests a degree of "hopeful response". Some of these energy sources will be considered below. At this point, however, it should be noted that a very substantial growth in coal is postulated, against supply projections from other sources which may be optimistic; the total supply is nonetheless fairly adequately taken against the 1977 demand estimates.

TABLE 22.1 World energy demand and potential supply (1972–2000)

World energy demand (exajoules $10^{18}J$)				
Average annual percentage growth rates	1972–2000		2000–2020	
Economic growth	3.1–4.2		2.9–3.7	
Primary energy demand growth	2.4–3.4		2.4–3.3	
Energy demand in year	1972	1985	2000	2020
Secondary energy (EJ)				
Fossil fuel	156	199–226	263–351	339–536
Electricity	21.8	34–38	60–77	115–173
Wood, solar, etc.	26	30–33	44–57	78–116
Total final consumption	204	263–297	367–485	527–825
Primary energy (EJ)				
Coal	66	77–88	122–171	230–405
Oil	115	147–169	163–224	145–248
Natural gas	46	50–55	71–92	84–118
Nuclear	2	22–24	85–111	246–380
Hydro	14	23–23	34–34	56–56
Wood and solar	26	30–33	44–57	78–115
Total primary energy demand	269	349–392	519–689	839–1323
Potential total energy (EJ) production	276	350–420	580–680	820–1010

CONVENTIONAL OIL AND GAS

The case for rapid development of the world's coal resource is scarcely affected by the degree of uncertainty associated with estimates of reserves and future production of conventional oil and gas. Estimates by many experts of ultimately recoverable reserves are reasonably close for our present purposes. Because these fuels are relatively easy to recover and use, they will be preferred energy sources. Subject to political action, production will continue to increase, peak and then decline; Table 22.2 indicates the sort of profile which may be expected.

At the 1980 WEC, estimates for the "ultimate recovery" of oil were stated to lie between 240–360 Gt; a figure of 354 Gt was assumed in the official report. Of this, 15% had already been recovered and only 25% were proved recoverable reserves; the other 60% were in the category of "additional resources". The average

TABLE 22.2 Potential world primary energy production (Exajoules)

Resource	*1972*	*1985*	*2000*	*2020*
Coal	66	115	170	259
Oil	115	216	195	106
Gas	46	77	143	125
Nuclear	2	23	88	314
Hydraulic	14	24	34	56
Unconventional oil and gas	0	0	4	40
Renewable (solar, geothermal, biomass)	26	33	56	100
Total	269	488	690	1000

recovery factor, in the estimates, was assumed to increase from 0.25 to 0.4. Natural gas liquids might add $6–12 \times 10^9$ t.

One estimate for oil production rates put forward at the 1980 WEC, after noting the greater recent effort in exploration and the growing desire by producing countries to conserve their reserves, suggested that "technical production capacity" would reach a maximum of around 4 Gt/a toward 1990, before slowly declining towards 3.6 Gt/a around 2020. By the latter date, however, "unconventional oil" (see later) would provide 0.5–1.0 Gt, out of the total given above. However, coal was assumed by the authorities to provide about half of the "unconventional oil".

Natural gas reserves and resources, once of secondary interest, are now seen to be quite large, even by oil standards. Production to date has been about 1000 EJ (25×10^9 TOE (tonnes oil equivalent)) and proved reserves are about 2800 EJ (64×10^9 TOE); remaining undiscovered resources may be about 8250 EJ (187×10^9 TOE). The production profile will be similar to oil but peaking later (round about 2000).

Political action cannot change the resources in the ground but could change the shapes of the profiles, either by accelerating exploitation or by conservation (or embargo). Either way, the requirement for coal remains vital. Ideally, the enhanced production of coal and its conversion to gases and liquids could be used to allow a production plateau for oil and natural gas to be maintained at about 1960 levels for a few decades. However, it appears now that for this option to be most effective, it should have been exercised 10–20 years ago; it might no longer have a major impact.

UNCONVENTIONAL OIL AND GAS

"Unconventional" oil covers oil shale and bituminous (tar) sands (including oil too heavy for recovery by normal oil industry methods). Oil contained in these deposits is thought to be similar in quantity to that in conventional resources, in spite of very inadequate and uneven exploration. Proved reserves may be about 80×10^9 TOE and additional resources may be nearly 400×10^9 TOE (of which about 70% is shale).

However, production currently takes place in only a few places and on a modest scale (perhaps 4 MTOE) in relation to these reserves. This is due to the uncertain technology and economics and also to the certain difficulties. Oil shales may contain from about 10 to over 300 litres of oil per ton. The largest known deposit, in Colorado, has areas of over 100 l/t but the average is lower; values less than 40 l/t are regarded there as uneconomic. Although some deposits are very thick and close to the surface, the "ash" content is usually over 50% (67% in Colorado). Furthermore, the oil cannot be regarded merely as an extension of the normal crude grades. Most important, it has a significantly lower hydrogen/carbon ratio so that, as with coal, hydrogen addition is the major requirement. There are generally also high proportions of asphalts, sulphur, nitrogen, oxygen and heavy metals.

Although recovery by *in-situ* methods (including solvation techniques, as well as underground heating) are being studied, current and near-future methods must be similar to those for coal-mining (opencast normally), followed by processing characterised by the considerable input of energy. Economics are doubtful (and heavily affected by environmental considerations). Forecast costs seem often to be somewhat higher than oil prices and to move further ahead (unfortunately, again like coal) as the oil prices rise above the earlier estimates for "breakeven". It appears that these sources should not be considered an extension of the oil resource but as likely to follow a parallel path to that of coal, requiring time and large capital investments.

Unconventional gas is a speculative area. It includes gas hydrates (methane compounds in solids) and geopressurised gas (e.g. methane dissolved in deep aquifers), for neither of which are there acceptable recovery methods. There are various other hypothetical sources; if they exist, discovery and recovery would both be costly.

LIMITATIONS OF NUCLEAR POWER

The economics of nuclear power as a means of generating electricity have been discussed (Chapter 13). The main point made there is the high capital cost of nuclear power. Even if economic conventions (especially interest rates) are taken which make the cost of nuclear electricity appear attractive, there are still limitations on the rate of installation (even neglecting public attitude). The capital costs refer to real resources, which can only be mobilised at a rate acceptable to a particular country's economic and social situation; high capital costs make anything but base load operation particularly expensive; extra penetration by electricity into the energy market requires further large capital investment in distribution and consumption equipment.

In this section, however, the energy resource base is more particularly considered. This is exemplified by Table 22.3, taken from WEC sources.

The growth of nuclear power at the rate suggested in Table 22.2 would pre-empt all the uranium in Table 22.3, probably before the end of the 20th century. Table 22.3 refers to uranium which could be recovered at less than \$130/kg U (1979 US\$). There are vast amounts of uranium in the earth's surface and oceans which could be recovered at higher costs (much higher in many cases). However, there is a point at which nuclear fuel costs begin to "bite"; furthermore, capital resources, energy inputs and environmental factors become more serious. As an illustration of the industrial problems, a number of estimates have been made of the additional facilities which will be needed. One, by the Conservation Commission of the WEC (1980), suggests that 329 new mines (575 if no nuclear fuel recycling), complete with all grinding and extraction gear would be needed. The facilities for enrichment, fuel element manufacture and, especially, post-irradiation treatment may be even more formidable.

The natural resolution of the fissile resource problem (and to some extent the recycling problem) would be the adoption of the fast breeder reactor (FBR). This

TABLE 22.3 Total world conventional energy resources (Exajoules)

Source	*Proved recoverable reserves*	*Estimated additional resources*
Coal	20 300	296 000
Hydrocarbons	11 200	50 800
Uranium[a]	1 900	2 600
Total	33 400	349 400

[a] Energy contained in U^{235} only.

would allow much greater use to be made of the uranium, since the U^{238} (more than 99%) can be used to generate more fissile material, although the gain would probably not be nearly so high as often claimed. Leaving aside political problems, it is clear that FBR electricity will not be cheap. Capital costs will be substantially higher even than for thermal reactors and expectations of breeding ratios seem to be declining as more detailed engineering studies proceed. Accepting the ultimate desirability of the FBR, there are limits to the rate of exploitation. One is the sheer complexity and capital cost (again, real resources). Another is that above a very limited rate of increase in nuclear electricity based on the introduction of FBR, the contribution to fissile material would be negative for a long time.

Various initiatives have been made to remove nuclear energy from its current utilisation limitations but none have so far made much progress or show any promise. Nuclear power normally provides electricity for general distribution and, therefore, is subject to load factor constraints. Schemes to generate electricity for high-load factor integrated use, such as electrolysis of water to produce hydrogen, seem remote from adoption and may recede further if effective nuclear fuel cycle costs increase. The linking of HTR to coal gasification has been discussed; the HTR seems in any case a long way from large-scale application. Other process heat applications of nuclear power seem even more remote; the use of waste heat for district heating is not favoured by the size and siting of nuclear plant. Applications in transport seem linked to the development of electric vehicles, the path to which appears uncertain and slow.

"Nuclear steelmaking" has been publicised. However, there is no specific, unique link. Nuclear power is merely intended either to provide fuel/reducing gases or electricity. In both cases, the intermediate product is the same as that from other energy sources, which are in competition with nuclear power.

In summary, the growth of nuclear power to conform to Table 22.2 seems to face very great difficulties, both with regard to resources (capital and fuel) and outlets. It has been normal for some years for each successive estimate of nuclear potential, even when made by nuclear protagonists, to be less ambitious than its predecessors and this process may not be over.

However, the potential for a substantial rate of growth of nuclear electricity is already being tested, on a massive scale, by France and this should give a clear indication of the possibilities, within a decade or so. France, more than almost any other country, has a great need for nuclear power and has a highly developed technology and industrial base from which to launch it. France intends to go from 4.5% of her energy needs supplied at present by nuclear energy to 30% by 1990 (this is about the proportion postulated in Table 22.2 for the world by 2020). At the end of 1979, France had 16 operational units representing a net installed capacity of 8300 MWe; 32 units with a net capacity of 31 500 MWe were under construction and expected to be commissioned by 1986. The construction of 14 other units, with a net capacity of 16 700 MWe, was likely to be started in 1981 and 1982. The 1200 MWe FBR, Super-Phenix was expected to be in production by 1983; four further FBRs. 1200–1500 MWe each, may come onto line between 1989 and 1995. Full supporting facilities are planned. The magnitude and depth of France's commitment should answer many questions about the future of nuclear power. Even if successful, however, the date of 2020 for the rest of the world to "catch up" would still seem difficult, for the reasons already given.

Other forms of nuclear energy depend on thorium as a fertile material. Reserves of thorium seem adequate but a comprehensive programme is not apparently being considered and would take several decades to launch and bring to a significant level, in world energy terms.

FUSION ENERGY

Fusion (or thermo-nuclear) energy is often put forward as an alternative to fission (or to practically all other forms of energy), generally on the grounds that it is "clean" and the necessary resource base is virtually limitless. Some qualification of both these factors may be necessary but can be deferred for the present purposes. Timing and economics are more relevant.

It is said that scientific feasibility is "near certain". This appears however to depend on extrapolation of current results, beyond the critical perception of the authors. However, the US government are intending to spend a billion dollars in the 1980s "exploring" engineering devices and an assessment will be carried out at the end of the decade. The problems of extracting useful, economic power seem at present to be formidable. If the engineering assessment in 1990 is favourable, commercialisation could start 10–30 years later.

Fusion power will not be cheap. Even if fuel costs (and this includes all processing) were nil, capital costs would have to be only 130% of light water reactor costs to achieve parity. On the best authorities, such a capital cost is inconceivable. It can be concluded that the promise of fusion energy cannot affect the decisive period for coal development.

SOLAR ENERGY

Solar energy represents by far the most abundant source of renewable energy; it is in fact the "source" of most of the others. It has been described as the "democratic" energy since it falls on everyone (c. 1 kW/m^2) and can be used by individuals and small groups.

Solar radiation can be used in a number of ways, the simplest being to convert it into low temperature heat for space heating and hot water; this is firmly established and expected to grow steadily. Solar radiation can be concentrated by mirrors or lenses and then may be used for process heat or for electricity generation via a thermal cycle. In these cases, because of the capital costs, application is likely to be limited to areas of high and regular insolation. This will probably apply also to photo-voltaic devices for converting solar radiation directly into electricity; these devices are currently expensive and in the foreseeable future are likely to be restricted to installations up to 100 kW. In the case of thermal generating stations, the "Thermal Tower" concept may be aimed at outputs of 10–100 MW; such concepts are still in an early stage of preparation.

Solar radiation may also be exploited by means of biomass — crops which are grown for energy use, mainly or in part. The product may be combusted, pyrolysed to give fuel gases or fermented; the products of these processes may be used directly for heat, motive power or electricity. Organic waste materials may similarly be used.

About 0.2% of the solar energy arriving at the earth is converted into kinetic energy in air fluxes — winds. Various devices have long been used for local motive power requirements and it is likely that these will be supplemented in future by larger installations, including central power stations up to 10 MWe. However, the best sites are generally in coastal areas or islands; these are often some distance from large loads. Environmental impact (including noise) would be considerable. Power would be available in the best locations, 30–60% of the time; these times would not necessarily coincide with demand cycles.

About 0.1% of global wind power is converted into waves. Only a small percentage of these waves are large enough and regular enough to justify attempts to convert their energy for use. An overall conversion efficiency of 10–20% in these cases may be as much as can be assumed and rather complex mechanical, pneumatic or hydraulic devices are required. Work is in a preliminary stage but problems of engineering stability and environmental impact may be foreseen. Again, remote locations and lack of control of output cycles are problems.

Finally, another system depending on solar radiation is the ocean thermal energy system (OTEC). This relies on the fact that, near the equator, there can be differences of about 20°C between the warm surface water and that from the depths (say 750 m down). These two fluids would be pumped through very large heat exchangers, one of which (using the warmer water) would evaporate a working fluid (probably ammonia) to a power turbine; the other heat exchanger using the cold, deep water would condense the fluid. Such an engine could have an efficiency of only 2–3% and about 30% of the gross energy produced would be used in the pumps. Rather remote sites would be desirable. In this case, however, energy is available constantly; the fact that this availability is regarded as a considerable advantage underlines the inconvenience and cost of solar energy devices of erratic output.

The economics of these different forms of solar energy are difficult to assess. Even for simple cases of individual homes, calculations are difficult, although it seems for a new house some degree of heating will be economic in areas of high insolation, especially where back-up supplies of other energy forms are readily available and, possibly, governmental financial inducements are offered. Even in these cases, there may be borderline decisions about the proportion of energy to be provided by solar means and the size of storage systems.

For larger scale applications, especially for electricity, solar energy is probably not economic in typical conditions. In a major study published by the American Association for the Advancement of Science in 1978, it was considered "conventional wisdom" that, compared with current electricity prices, wind generators were within a factor of 2 of being competitive, biomass fuels a factor of 2–4 away, OTEC by a factor of 4–5, power towers by 5–10 and photovoltaic 10–20 times too costly.

TABLE 22.4 Possible solar-derived energy contribution (%) to energy needs in USA (2020)

Hot water and space heating of buildings	2
Process heat	7
Wind energy — electric utilities	3.5
Solar thermal — electric utilities	1.5
Photovoltaics — electric utilities	0.1
OTEC	1.3
Biomass — wood — electricity	0.3
Biomass — wood fuel	2.3

In the same publication, several projections of possible solar contributions to US energy supplies were given; these were prepared for the Department of Energy by various authorities or groups. The studies were generally based on the assumption of vigorous policy backing and tended to the conclusion that solar and solar-related sources *could* individually provide amounts which, if added up (unsound) are equal to 15–20% of US energy by 2020. In one study, the possible individual contributions in 2020 were roughly as given in Table 22.4.

Other uses were looked at but were not considered large enough to be measured. The total is not included since it is unlikely that all could be exploited simultaneously.

GEOTHERMAL ENERGY

The heat stored in the earth's subsurface at temperatures above the annual mean temperature is vastly greater than human requirements. There is a thermal gradient, the intensity of which varies but can commonly be as much as 30°C per km depth. There is a heat flux outward to the surface equal to about five times annual consumption. There are "hot spots", often related to volcanic areas and geopressurised hot aquifers, in addition to obvious signs of heat release, such as geysers and hot springs. Despite this, geothermal energy is little used and future prospects do not indicate a rapid increase in utilisation. There are formidable problems in all but the exceptional cases; exploration and study are nonetheless proceeding at an increasing rate.

The best circumstances occur when hot water or steam can be tapped directly without rapid depletion. For electricity generation, available temperatures must be more than about 150°C and in only a few known places are quantities sufficient to justify a power station; only in California, Italy, Japan and New Zealand are total electrical outputs over 100 MW (and all are less than 1000 MW). For useful domestic purposes, temperatures of more than 65°C are needed and a fairly concentrated demand near to the source is necessary.

The problems of tapping hot dry rocks are greater. The most common approach is to drill two holes, one for injection of cold water and one for removal of hot water or steam. However, hot dry rocks generally have poor permeability and low conductivity, so that paths and surfaces must be created. Some success has been obtained with experiments, notably at Los Alamos, but costs are likely to be excessive except where temperature gradients are anomalously high.

For a very few cases, power generation is cheaper than any alternative. However, in most cases up to 50 wells would be necessary for a 100 MW power station. Because of the low temperatures, turbine costs are high. For the distribution of hot water, costs of $8/GJ (1977 prices) for a thinly populated area of 80 km radius have been suggested.

There are environmental implications also. These are mainly associated with the presence of salts and undesirable gases in water from aquifers and the disposal of water (still warm) from turbines. Underground disturbances are being studied but no immediate serious problems have occurred.

Non-electrical uses in 1979 (space heating, agriculture and industry) were estimated at just over 7000 MW; however, 5000 MW of this was agriculture use in the USSR; the rest was distributed over eight countries. In 1978, electricity capacities of about 1500 MW existed; plans for 1980 are for a total of about 11 000 MW in eleven countries (of which Japan will have more than half).

HYDRAULIC ENERGY

Hydraulic energy, obtained from the fall of water from one height to a lower level is the best known of the "renewables" and possibly easiest to consider, so far as future potential is concerned. Although essentially driven by the sun, hydraulic energy is considered separately because of these features.

Mechanical energy is the oldest form of hydraulic power but electricity is the only form of utilisation important on a global scale. Where applicable, hydro-electricity is usually cheap and may be vital to a country's economy. At the 1980 WEC, figures were given for potential hydraulic energy as in Table 22.5.

TABLE 22.5 Potential hydro-electric output (EJ/a)

	1976	*1985*	*2000*	*2020*
OECD	3.78	4.49	5.37	7.80
Centrally planned countries	0.72	1.20	2.88	8.70
Developing countries	1.17	1.97	4.49	11.80
World total	5.67	7.66	12.74	28.30

The presently operating capacity is estimated to be about 15% of the "technical usable potential" and about a similar amount is under construction or planned. It will be apparent that the estimate for 2020 will virtually exhaust all known potential. In this case, there seems little possibility of any significant breakthrough which could alter the position favourably.

TIDAL ENERGY

The total global energy in tides is about 3 TW but only a small fraction of this occurs in sufficient concentration to be worthy of study. In the majority of these cases, the locations are remote, such as the Bay of Fundy, Alaska and the White Sea.

The only sizable tidal energy plant operating is that at La Rance in France (500 $\times 10^6$ kWh/a). There are one or two small experimental units and several engineering studies. One of the latter concerns the Severn Barrage in the UK, which has an average tidal height of 9.8 m and hydraulic energy of 15×10^9 kWh/a, and is thus one of the most attractive sites in the world. The possibilities have been considered for decades and more intensive engineering studies (costing several million pounds) are in hand. The economic merit of the scheme is however far from obvious although it seems possible that this will eventually be established. Even then, the environmental impact may make a decision difficult.

On the basis of experience to date, tidal energy is not likely to make much impact in the next few decades, if ever.

SUMMARY OF CONTRIBUTION FROM "RENEWABLES"

Some estimates are given above of the possible contributions of solar and solar-related energy sources in the USA to the total energy supply; an upper (and probably unrealisable) range of 15–20% by 2020 is suggested. In Table 22.2, the total world renewables contribution by then is estimated at 10%; this includes some forms not covered by the US definition of "solar". In addition, it is pointed out that in 1972, most of the energy categorised as renewable is non-commercial biomass, mainly wood, plant and animal wastes burnt locally. These forms will probably not increase and should not, since these uses are probably destructive environmentally. The increase projected by 1985 is quite small, leaving a very steep hill to climb in 35 years if the 2020 forecase is to be achieved.

CONSERVATION

It is a commonly expressed belief that energy growth could be contained if "waste" were eliminated or simpler lifestyles adopted. In fact, this is one of the most difficult areas to interpret and the one strong probability is that there are no easy solutions or "soft paths", at least in the short or medium term. Most recent predictions have made allowances for conservation up to and usually somewhat beyond the limits regarded by the authors as prudent. This subject is too detailed and complex for full evaluation here. However, in the WEC 1977 demand forecasts (not significantly changed in 1980), quite drastic assumptions of energy conservation were made. For example, in some cases, a 30% reduction in the rate of domestic and industrial demand is made. In the domestic sector this would involve the full application of known measures for energy conservation (a difficult area); in the industrial sector, technological improvements were assumed, plus a shift in the developed countries away from heavy energy-using industries. In the case of transport, it was assumed that all improvements technically achievable would be implemented, that motor transport would saturate earlier in the developed countries than previously estimated and that growth in the developing countries would be slower. Even with all the above assumptions, a demand of about 1000 EJ in 2020 was arrived at (equal to the projected production in Table 22.2).

Special studies have been carried out to consider more effective ways of reducing energy demand. In the USA some studies suggest that total energy consumption could be kept at about the present level, in spite of population growth; very great changes in lifestyle are however required. In the UK a "Low Energy Strategy" has been proposed by Leach *et al.* in a book published by the International Institute for the Environment and Development. Based on a detailed sectoral analysis, it is concluded that in a "Low Case", compared to 1976, total primary fuel consumption could be down by 7% by 2000 and by 22% by 2025; in a "High Case", consumption might be marginally higher by 2000 but down by 8% by 2025. The methods are plausible but probably no one expects them to be adopted successfully, especially in these time scales. It should also be noted that the "standard" case for energy growth in the UK is lower, and often substantially lower, than in other developed countries.

Thus, in summary, it can be said that conservation is believed to be fully allowed for in most energy demand forecasts. There is clearly scope for much greater progress but this may not be achieved until well into the next century, without extreme social perturbations. In the longer term, control of population growth seems necessary. In the medium term, hopes for conservation should not diminish the need for a positive coal policy.

OTHER ENERGY SUPPLY AND CONVERSION PROPOSALS

In addition to direct conservation, there are a number of possibilities which may favourably influence the energy supply situation. In the view of the authors, the most important development, indicated in various places in this book, would be on improved energy analysis, especially with regard to optimisation of energy conversions and the treatment of uneven demand. These points will not be reiterated here, especially since economic structures, and therefore long times, are involved, going beyond the immediate need for better coal utilisation. This subject also includes such topics as co-generation, waste heat recovery and storage.

An allied and potentially important, development is that of heat pumps. In these devices, based on an inverted thermodynamic cycle, heat is extracted from a low temperature sink by means of an engine driven by motive energy, and supplied to a higher temperature demand. Quite high energy "gains" can be achieved by this means and a number of installations are in use. Further development is however required and this may be relatively lengthy, partly because of the many permutations which may be considered (motive energy, heat source, mode of operation, supplementary energy source and heat distribution method). In particular, low cost, reliable heat pumps need to be developed, especially for small installations without competent supervision; other problem areas include environmental impact, control mechanisms and absorption heat pumps. One forecast suggests that savings of up to 40–50% in primary energy could be made in space heating. There are industrial applications also. One proposal, similar in concept, is for "energy cascades" to convert low temperature heat to higher temperatures.

An interesting proposal for energy transport is that being studied in Germany and commonly called "Adam and Eva", more properly Nuclear Long Distance Energy (NLDE). The method proposed is to convert synthesis gas to methane using nuclear heat (Adam); the methane is subsequently steam reformed (Eva) some distance away from the nuclear reactor to provide heat. The two reactions are connected by a pipe loop. The current programme is funded at DM26 million.

The Adam and Eva concept, as with nuclear heat for coal gasification, depends on the availability of "cheap" nuclear heat, usually at high temperatures. The "hydrogen economy" concept is also closely allied to economic (and surplus) energy from nuclear sources. This prospect now seems much less favourable, with increasing nuclear capital and fuel costs, and with the growth potential of nuclear power probably firmly earmarked for electricity production for established requirements. In most cases, the hydrogen is to be produced by electrolysis, for which the efficiency is low; alternative chemical processes for using nuclear heat have been proposed but these seem excessively convoluted.

If any hydrogen should be available from non-fossil sources at economic prices, instead of attempting to create a new distribution and consumption system (with some difficult technology) it would seem preferable to use the hydrogen chemically on the spot. An obvious outlet would be, as proposed in Germany, to use the hydrogen (or the nuclear heat) in coal gasification processes, producing gases in common use.

The use of non-fossil derived (NFD) hydrogen in several coal processing methods was studied in 1977 by the UK Coal Research Establishment, supported by the EEC. The introduction of NFD hydrogen reduced the number of component stages, increased carbon utilisation and increased thermal efficiency (neglecting the production of hydrogen in the nuclear plant). It was shown that at coal prices then current in Europe ($1.5–2.5/GJ), it would be preferable to use NFD hydrogen rather than conventional processes if the NFD hydrogen could be supplied at $4–7/GJ. It was also shown that conditions can exist where the use of NFD hydrogen in coal processing is preferable to its use as a distributed fuel. These conditions appear likely to be widely applicable.

CONCLUSIONS

An examination of alternative energy supply systems provides no grounds for relaxation of the need for greater and more advanced coal utilisation, certainly over the next four decades. A quadrupling of world coal output over that period is most unlikely to contribute to any overall surplus of energy. The other energy sources appear to be at their limits in demand projections, such as those in Table 22.2.

Quite apart from the quantitative aspects, there are also qualitative matters. For one thing, non-fossil energy sources are almost all unsuited to provide either transport fuels or chemical feedstocks; there are ways of covering these requirements but they seem clumsy.

Almost all the non-fossil energy sources have deficiencies related to load factor or availability. In the case of nuclear power (and other high capital cost systems) high load factors are necessary on economic grounds; in some cases, where energy availability is uncontrolled, this factor compounds the former. In addition, most alternative sources to fossil fuels may only be produced, often in remote areas, in the form of low-grade heat or electricity.

Thus, the fossil fuels, which can be readily stored and moved to the point of consumption or conversion, will continue to be used to counter the mal-distribution in space and time of the non-fossil sources and this requirement will grow. One important by-product of this is that, as certain "renewable" energy sources come to be used more intensively, larger amounts of stand-by energy resources, mostly fossil, will need to be provided. This will effectively detract from the economics of the renewables.

Ideally, a large and continuous build-up of coal utilisation should have commenced at least a decade ago, in order to mitigate the more undesirable effects of the oil and gas production profiles. It is probably too late now for this concept to be very effective (except in emergency conditions imposed by oil suppliers). An alternative attractive proposition is that the increase in coal utilisation should be firmly planned and implemented in order to provide time for the development of alternatives (especially renewables), additional conservation, improved energy modelling and better international understanding.

Chapter 23

COAL UTILISATION IN RELATION TO WORLD ENERGY STRATEGIES

INTRODUCTION

The medium and long term potential prospects for coal emerge rather clearly:

(a) Coal output is capable of increasing, preferably in a regular and steady fashion, with a doubling time of about 20–25 years, continuing over the next 40–50 years at least.
(b) This coal should be used in ways which are complementary to the progressive decline of natural gas and oil and to the growing availability of new energy sources.
(c) These developments will put strains on availability, technology and capital; all these will best be alleviated by international collaboration.

In spite of the clearness of these perspectives, the immediate paths are far from clear. Major research programmes have emerged in a somewhat uneven and haphazard way and strategic objectives have been narrow. Perhaps more worrying, the terms under which first-off, full-scale coal conversion plants might be built and brought into operation have frequently been the subject of intense controversy and uncertainty. There has also been in places some prejudice against coal, on not very rational grounds, which may delay exploitation. In spite of these uncertainties, however, it must be remembered that coal conversion has already been successfully brought out of the laboratory to the development or industrial scale, where strategic considerations have demanded it, notably in Germany during the war and South Africa subsequently. It may be hoped therefore that the urgency resulting from future energy shortages will bring about the adoption of full-scale coal conversion processes elsewhere in the world somewhat sooner than suggested by purely commercial judgements, based on earlier predictions of the ideal rate at which oil and natural gas supplies might be used up if organised for the benefit of the Western world.

EXISTING PREDICTIONS AND POLICIES FOR COAL

WEC

At the World Energy Conference, 1977, encouraging (but not obviously over-optimistic) estimates were given for coal resources and reserves. These indicated that coal output capacities were not likely to be strongly constrained by reserves. Estimates were also given for production and for export availability in 1985, 2000 and 2020 but these are more problematic and hedged with considerable qualifications. These estimates are given in Chapter 3. Estimates made at the 1980 Conference were not greatly different.

Although these WEC estimates were made up from figures which were sought from individual countries, practically all of whom were most co-operative, a number of countries were not able to give figures for 2000 and only three of the main producing countries were persuaded to give data for 2020. There were similar gaps in the information on export potential and in both cases the expert compilers added their own extrapolation based on published information. The estimated production and export tonnages each increase regularly over the years from 1975 to 2020 by a total of about 235%. The compilers in 1977 regarded the production figures as an upper limit that could be achieved with (then) "currently initiated measures". The authors discussed bottlenecks and pointed out that both in the USA and the USSR, extensive expansion of output will require a transfer of coal-mining activities to new areas far removed from existing production centres and markets. This in turn requires a large capital expenditure, on infrastructure even more than on the mines themselves; finance may be difficult to arrange in advance of an obviously established demand. These difficulties are probably magnified in the case of production for export markets even more than for domestic requirements. Another potential bottleneck forseen is a shortage of skilled personnel with experience and understanding of coal.

WAES and WOCOL

In the Report of WAES, the strong reserves position of coal is again emphasised. Estimates for likely production and export availability in 1985 and 2000 are given but these are probably of more interest in the context of national attitudes, rather than for numerical accuracy. In WOCA (World Outside Communist Areas), coal production is estimated to increase from 1308 Mt to between 2708 and 4109 in 2000. (The corresponding WEC figure for 2000 is 3080.) The interesting point is that the wide range results from the use of scenarios varying in rate of economic growth, price of oil, etc. In other words, coal output is by implication considered likely by WAES to be demand constrained. Furthermore, this is said to be the "perceived" demand, which it is thought may well be lower than that eventually realised, especially if oil supplies are inadequate for technical or political reasons. The rate at which the coal demand might change in the event of further oil crises is of course much faster than output capacity could then be increased. To avoid this "whiplash" effect, the logical policy would be to plan for coal output to rise at the technical limit and to ensure that all coal has a place in the market, in preference especially for oil. This latter factor is not merely one of financial incentives, such as taxes or subsidies, but of the provision of actual hardware for coal utilisation.

The WOCOL study, carried out under the direction of Prof. Carroll Wilson of MIT on a similar but expanded co-operative basis to that of WAES has been referred to earlier. He made a statement on this study at the 1980 WEC; this stimulated a considerable discussion, especially about the price differential between oil and coal and about the mid-term rate of expansion in world coal output. Prof. Wilson's comments on the former are mentioned in Chapter 12; on the latter, the WOCOL report suggests that coal output *could* double by 1990 and treble by 2000, compared with 1979.

On the price differential, a representative of the Arab OPEC stated his satisfaction with the current price ratio, oil/coal, and expressed a hope that OPEC would maintain this as coal prices increased (which he clearly expected). There could be no more direct statement of the importance of the challenge facing coal.

By contrast, the consensus at the 1980 WEC was that the rates of increase in coal output suggested by WOCOL were unlikely to be achieved. This consensus probably owed a great deal to the mood generated by the excess coal capacity available at that time, especially in the USA. In fairness to Prof. Wilson, he had made it clear that the WOCOL estimates referred to what *could* happen and outlined the steps which were necessary to make it happen. Since little had been done since the time when the data were being collected (perhaps 12–18 months earlier), it was not surprising that expectations of growth had been lowered.

In concluding the WEC session on WOCOL, one of the authors (LG) made the following points:

(a) The time which had elapsed and the absence of any coordinated drive to provide necessary facilities had put the WOCOL "targets" beyond reach.
(b) Having responded to urgent calls for expansion in the wake of the oil crisis, mining companies are finding themselves with excess capacity and having to make miners redundant.
(c) Coal output will continue to be demand constrained. Growth in the electricity market would not support the growth in coal output capacity proposed in WOCOL.
(d) Demand constraints would only be eased by the establishment on a large scale of new coal conversion processes. The timescale for demonstration and the construction of many new plants was not consistent with the timing of the WOCOL output suggestions; these would probably be required but at a later date.
(e) Exaggerated demand forecasts, unless supported by utilisation developments, could generate scepticism and be counter-productive.

These views appeared to be broadly acceptable.

Energy Policies and Coal: International Energy Agency

The IEA publishes reports on many aspects of energy, especially as these affect IEA members (most of the countries of the OECD), but taking account of the world situation. In the 1978 review of energy policies the possibility of a shortfall of oil in 1985 and 1990 was a matter for concern, even though demand seemed likely to be depressed by low economic activity; shortage of energy and high prices are thought likely to hold back economic growth even further. Within this sombre estimate is built an annual increase of 3.6% in coal production, mostly in North America, up to 1990. There is included however an increase of 80 Mt/year in the European IEA countries, which seems a difficult target. In North America, it is expected that policies on fuel utilisation will reinforce the trend towards coal — in other words, further evidence that output is regarded as demand constrained. Other countries have the potential for increased coal use but only a few, notably Japan, Denmark and Spain, have implemented coal programmes. The Report considers problems associated with increased coal usage but concludes that the absence of a firm commitment is the real reason why many IEA countries have not expanded the use of imported coal "where economically warranted and environmentally acceptable". In a country-by-country survey, in practically every case, recommendations are made for greater efforts for indigenous production and/or more secure coal imports, linked to a utilisation policy.

The IEA has also produced a special report on "Steam Coal — Prospects to 2000". This report stresses the crucial role to be played by a massive substitution of oil by coal, without which severe disruption will occur. The principal impediment is seen to be a failure by authorities to appreciate coal's growing economic attractiveness. Coal will become more cost-effective than alternatives and merits an immensely expanded world trade but this will depend on co-ordinated government action to facilitate coal development and usage. In developed countries, appropriate governmental measures are categorised under the following headings:

(1) Substitution of oil by coal, in electricity, industry and space heating, through financial incentives, more R & D support, prohibitions on gas and oil installations and co-ordination between energy studies and regional planning.
(2) Investment encouragement, through relaxation of environmental constraints, infrastructure development, financial support and removal of "taxes" on coal.
(3) Promotion of international coal trade by removal of restrictions.
(4) Information.

In developing countries, energy policies might reflect a better understanding of the cost-benefit advantages of coal, including spill-over effects from infrastructure development, external financial support and international R & D and information services.

The elementary nature of these policy suggestions indicates how far there is to go before effective pro-coal policies become fully established. This further casts some doubts on the achievement of future expanded targets.

In the IEA Steam Coal report, various scenarios are considered. In the reference case there are two sub-cases: low and high nuclear. OECD coal usage is projected to rise from 986 Mt/year in 1976 to 2081 Mt (low nuclear) or 1854 Mt (high nuclear) in 2000. Net imports within the 2000 figures are 154 or 112. An "enlarged" coal case is also considered which might increase coal use to 2585 Mt/year (low nuclear) or 2291 Mt/year (high nuclear) in 2000. Again, the policy initiatives necessary to achieve the enlarged case are emphasised. In addition to the earlier general policy recommendations, the policies regarded as essential to the enlarged coal case include:

(1) Avoidance of measures which discourage coal imports.
(2) Government planning for large-scale harbour and handling equipment.
(3) Avoidance of measures to restrict coal exports.
(4) Expeditious handling of environmental problems.
(5) Avoidance of delays in granting coal leases.
(6) Lowering of severance taxes and improvement in depletion allowances.

The various rates of increase in coal usage by 2000 predicted by the IEA for OECD countries correspond reasonably well with the WEC forecast for world output increases. The IEA reference case is similar to the WEC prediction and to the low end of the WAES bracket. The "enlarged" IEA case (even with low nuclear) however represents a lower percentage increase in coal usage than the highest WAES coal output increase. If these IEA forecasts could be relied upon, they would suggest that OECD countries, who are most vulnerable to oil and gas shortages, may make less use of increased coal output than the rest of the world.

Perhaps more important, the increase of coal output is clearly going to have its limitations, although a doubling time of about 20–25 years for an industry of such complexity would normally be considered an excellent performance. Furthermore, an increase in annual production of nearly 3000 Mt of coal between 1975 and 2000 (WEC) would be a massive addition to world supplies, the increase representing about 10% of total demand at that time (leaving coal at about 20% of the total demand). An even more interesting and challenging concept would be if the energy made available to the user from each tonne of coal mined could be increased. In view of the low current level of efficiency, a 50% increase (less than 2% improvement a year) would not seem an unduly extravagant target and this would mean that 10% of the world's energy could come simply by using the available coal more efficiently.

RESEARCH AND DEVELOPMENT

General

Research and Development can be regarded as an indication of future prospects. From being a relatively neglected subject for R & D, even in the early 1970s, there has been a large proportionate increase in spending on coal during the second half of the 1970s. Even so, the scale of the work is still not completely satisfactory in two respects:

(a) the level of effort on coal compares unfavourably with that on other subjects, especially nuclear power;
(b) the size and scope of development or demonstration plants are as yet inadequate in relation to commercial scale operations.

The main national programmes, in the Western world apart from S. Africa, judged from the viewpoint of being closest to commercial needs and covering a range of possible processes, are in the USA and West Germany. The UK programme is a well chosen spread at research level but lacks commitment for a drive through pilot and demonstration plants towards commercial scale operation. There are of course R & D programmes in very many countries but in most cases these are either on a modest, scientific scale or lacking breadth. S. African efforts as exemplified by technical achievements by Sasol and AECI exhibit more realism but they are not nationally co-ordinated. Another area where large scale experiments have been carried out over a period of years is in the carbonisation field, especially for briquetting; this, however, does not have broad relevance to progress in coal conversion.

USA Coal Programme

In Fiscal Year (FY) 1980, the USA Government spending on coal R & D was planned at about $670 million, of which $60.3 million was on mining. This level has been approximately maintained for three years and is more than 10 times the expenditure in FY 1973. The US Department of Energy, Fossil Energy Programme is prefaced by an admirable statement of strategic objectives, which are organised around the major applications of coal — direct combustion and conversion to liquids and gases. The industrial sector is emphasised as having scope for substitution of oil or gas by coal. The objectives may be summarised:

(1) *Objective I*: Environmental acceptability of coal for current and future combustion for industry and electricity.
  (a) Coal preparation and physical cleaning to increase the amount of coal suitable for existing installations or with flue gas desulphurisation (FGD) in new or existing facilities.
  (b) Chemical coal cleaning.
  (c) Further developments of FGD to improve performance and reliability.
  (d) Develop atmospheric fluidised bed (AFB), through industrial scale to utility scale.
(2) *Objective II*: Demonstrate capability for production of synthetic liquids and gases in mid-to-late 1980s so that significant capacity could be built in the 1990s if oil prices rise.
  (a) Heavy liquids from coal to replace residual oil.
  (b) Limited investment in alternatives for lighter liquids for transport sector, as a longer term option.
  (c) Detailed design for advanced coal gasification (broader range of coals and lower costs).
  (d) Design for possible future demonstration for medium-Btu gas for industry.
  (e) Accelerated R & D on highly advanced concepts.
  (f) Characterisation of environmental issues of coal-derived liquids and gases.
(3) *Objective III*: Develop advanced systems for 1990s and beyond. This has lesser priority than I and II and includes:
  (a) MHD
  (b) Fuel cells based on coal liquids.
  (c) Turbine development — higher temperatures and heavier, dirtier fuels.
  (d) Advanced cycles and heat recovery.
(4) *Objective IV*: Fundamental improvements in technology base.

The programme also discusses in explicit terms the political and philosophical basis for the work — the need for coal R & D, strategic planning and alternative scenarios for coal, and the rationale for the current programme. The criteria for selection of projects for government support include:

(a) Range of applicability. Projects which could serve a number of functions are preferred, e.g. some gasification processes can produce gases for synthesis as well as hydrogen for liquefaction.
(b) Complementarity and co-ordination. (This presumably means getting a balanced programme in which individual items may contribute to other items and the whole fills the national requirements.)
(c) Potential for environmental, technical and economic success.

The policy is to involve the private sector at all stages, the Federal role being determined by the level of risks and benefits and the pace of development that is required. The world price of oil is a major determinant of the latter. The risk/benefit analysis also seems to depend on this, Federal aid through cost sharing being most appropriate where higher future energy prices are necessary for commercial

TABLE 23.1 Funding of US Government R&D spending on coal (millions of dollars)

Activity	*Actual FY 1978*	*Appropriation FY 1979*	*Estimate FY 1980*
Mining R and D	62.8	76.1	60.3
Coal liquefaction	126.5	206.4	122.3
Surface coal gasification	195.6	159.6	169.3
In situ coal gasification	12.6	15.0	10.0
Advanced research and technology development	37.0	46.4	59.5
Advanced environmental central technology	0	7.0	43.2
Heat expires and heat recovery	42.3	58.0	46.0
Combustion systems	66.1	58.9	57.4
Fuel cells	35.4	41.0	20.0
MHD	71.5	80.0	72.0
Program direction	9.5	10.3	11.0
Unobligated prior year funds	—	credit (78.0)	—
TOTAL	659.3	680.7	671.0

attractiveness. For projects more nearly competitive, tax credits or "a stable Federal policy" (a remarkable statement!) may be adequate incentives. The Federal government needs to determine the appropriate policy instruments for each case and the budget outlay will be set according to the necessary pace of development. The business arrangements are set out and the necessarily complex and decentralised structure of offices and laboratories described. The funding is summarised in Table 23.1.

Some interesting and succinct comments on official attitudes regarding the status of the work occur in the programme document. Demonstration on a near-commercial scale in the early 1980s is the aim for second generation conversion processes. Four coal liquefaction processes (SRC I, SRC II, H-Coal and EDS) are considered to be in advanced stages of development. Gasification processes are commercially available; further development is aimed at costs and efficiencies. Successful sub-pilot field trials have demonstrated *in-situ* processes and the aim is to develop at least one commercial underground conversion process during 1985–7. On fuel cells, the near-term (3–6 years) objective is to establish commercial feasibility of fuel cell power plants for electric utilities, industrial co-generation and total energy systems. MHD commercialisation is expected by the late 1980s or early 1990s. (However, the programme has been thrown into disarray (1981) by the Reagan administration and it will be some time before a clear picture emerges.)

In common with other energy matters, the commercialisation of coal projects is reviewed by a senior wide-ranging committee having explicit criteria and a list of considerations which are used by special task forces to provide a rating matrix of barriers to commercialisation, with recommended strategy. The specific actions which may be taken by the government are not however stated in this document.

The scale and urgency of the US coal programme and the logic of its approach to the R & D stages are admirable, as is the openness with which both the background and results are published. The USA co-operates on several levels internationally, again openly. However, there must be some doubts about whether the timescales for the stated objectives can be achieved. There is also uncertainty about the commercialisation process, with indications that industrial concerns are jockeying for position and legislators are taking a parochial view or responding to superficial analysis which may distort smooth progress. The Synthetic Fuels Corporation, discussed later, may resolve these problems.

Federal Germany

The German Federal Government has recognised for some years the necessity both of having a constructive energy R & D programme and of publicising that programme. These policies took a forward step when a "Program for Energy Research and Technology 1977–1980" was published, followed by an Annual Report instituted in 1977 (although the promise that the annual report would be maintained was confined to the non-nuclear energy field). The Annual Report contains reports on the status of each individual project, including objectives, work programme and interim results, together with administrative and financial information; there are literature references giving, where appropriate, access to more specific information. The Report states that the intention is to spread information and results, with a view to developing contacts between research organisations and industry (the latter is in any case involved both in the work and in funding).

The main objectives are summarised as follows:

"Guaranteeing the continuity of energy supply in medium to long term in the Federal Republic at economically favourable costs considering the requirements necessary for the protection of the environment and population."

An additional aspect of the support policy is stated to be "the development of technologies which are of importance for other countries, specifically for the developing countries". How this is intended to work is not stated but the establishment of an export industry may be expected to be one consequence.

The non-nuclear programme is divided into three subprogrammes, with funding as shown in Table 23.2 for the four years 1977–1980.

The sub-programme on "Coal and other Fossil Sources of Primary Energy" has its objectives:

(1) Improvement in Prospecting and Exploration Methods.
(2) Improved Exploitation of Deposits.
(3) Further Development of Mining Technology.
(4) Substitution of Crude Oil and Natural Gas by Coal based Products.
(5) Development of Technologies for the Ecologically Acceptable Use of Coal.

TABLE 23.2 German Federal energy R&D (1977–1980)

Sub-programme (see notes)	*1*	*2*	*3*	*Total*
Total cost (DM million)	669	1670	200	2539
Government contribution (DM million)	419	981	152	1552
Average support (%)	63	59	76	61

(1) Efficient use of energy in application and in secondary energy.
(2) Coal and other fossil sources of primary energy.
(3) New sources of energy.

TABLE 23.3 West German Energy R&D: coal and other fossil sources of primary energy (1977–1980)

	Total cost millions DM	*Government contribution millions DM*	*Average support rate (%)*
1. Electricity generation from coal	632.6	302.8	48
2. Gas generation from coal	251.9	182.5	71
3. Liquefaction of coal	125.4	101.9	81
4. Coke production and direct combustion	82.8	48.7	59
5. Gallery driving and mining system: logistic system	435.2	248.5	57
6. Coal preparation	54.7	28.3	52
7. Underground gasification	11.7	11.4	98
8. Prospection for oil and natural gas	46.4	32.3	70
9. Extraction of oil and natural gas	19.4	15.8	81
10. Oil shale	10.4	8.8	85

The structure of the programme is however arranged on different lines, according to technologies (Table 23.3).

Under Item 1, there are projects for advanced technology, including fluidised combustion and gasification for gas turbines; there is also a programme for partial gasification followed by fluidised combustion of the coke. A feature is that pilot plant work is generally backed by preliminary design studies for the next stage of scale-up. There is also strong emphasis on pollution control from conventional power stations.

On gasification, the largest projects are the Lurgi gasifier, the Saarberg/Otto gasifier, the pressurised Winkler gasifier and the Texaco gasifier. There is also a substantial programme on the use of nuclear heat for gasification. There are supporting programmes on basic studies and on the development of components.

On liquefaction, work on the synthesis route (Fischer–Tropsch) is directed to basic work, such as catalyst selectivity and slurry reactors. There is also a project for hydrogenation of coal in a slurry with oil, with a pilot plant based on war-time experience augmented with more recent laboratory work on critical aspects.

On Item 4, the major expenditure on coke is on formed coke, although there are items on refractories and preheating for conventional ovens. The direct combustion part is mainly the development of industrial scale atmospheric pressure fluidised bed boilers.

The mining items contain a very large number of individual, mostly small, projects. The underground gasification work consists of laboratory studies, modelling, etc. in preparation for field work. In 1977, Germany and Belgium signed an agreement combining their activities with the intention of carrying out an initial *in-situ* gasification field test in Belgium.

The overall impression given by the Annual Report is that, whilst the then-current coal programme was positive and reasonably comprehensive, it was merely a preliminary to an expansion involving large-scale development and demonstration plants. The Report specifically states that "the current experimental phase will supply criteria so that a few main areas of development may be concentrated on later". The importance of coal to Germany is emphasised on lines similar to those reiterated throughout this book — resources and versatility.

It is pointed out that hard coal and lignite are the only primary energy sources available in large quantities in the Federal Republic of Germany and, consequently, the use of domestic coal should be viewed not only from technical and economic aspects but also "under the aspect of guaranteeing supplies". The importance of international collaboration is emphasised, as is the point that the development of coal conversion methods "offers West Germany industry the possibility to export know-how and the necessary technical facilities". There is no doubt that the German programme is heavily influenced by this (legitimate) aspiration, behind which there is great confidence.

Like many other countries, however, expenditure on nuclear research has over the last couple of decades dominated energy R & D. There is a positive programme in Germany to correct this imbalance. In 1972, the Government's expenditure on non-nuclear R & D was only 1.5% of that on nuclear R & D and by 1975 it had only gone up to 2.3%. In 1977, however, the proportion was 27% and it is intended that the proportion should go up to 50% of that on nuclear R & D (excluding fusion research) after 1979. Whilst respecting the correctness of the German analysis and the firmness of the policy of correction, it should not be overlooked that, even here, the expenditure on nuclear energy will still be twice that on coal R & D in the foreseeable future.

UK

In the UK, R & D on coal has traditionally been carried out by the National Coal Board, without significant external financial assistance. Prior to the energy crisis of 1973/4, work on coal utilisation was on a modest scale, costing about £1 million per year based at the Coal Research Establishment (CRE). A co-operative research organisation, British Coal Utilisation Research Association, experienced dwindling support. A small part survives as an out-station of CRE and is now the Coal Utilisation Research Laboratory (CURL). Nonetheless, it was still possible to establish a scientific programme covering a fairly comprehensive range of items in support of existing coal usage technologies, plus those most likely to be of importance as oil and gas become more expensive.

Some important information on coal gasification had also been obtained and recorded by British Gas (and its precursor), following which in the late 1960s all their work had been discontinued, although key individuals did not lose interest. However, in the early 1970s, in the light of the gathering energy crisis, British Gas created a large-scale coal gasification test site at Westfield in Scotland taking advantage of four workable Lurgi gasifiers, by then rare outside South Africa. In co-operation with US partners, a significant programme was conducted both to enlarge the operational limits of the standard Lurgi and to test the concept of a "slagging" Lurgi. Methanation and purification stages were integrated and the product was successfully fed into the gas grid. The partners and their process are candidates for a demonstration plant in the US. British Gas also have other proposals based on adaptations of earlier work.

In the NCB there was a steady increase in coal R & D from the early 1970s in anticipation of the energy supply difficulties which become obvious in 1973/4; the latter led to further expansion. The mining R & D was particularly emphasised taking account of the reserves and the future requirements for coal. The increase in utilisation R & D was less dramatic since the development of other energy sources available to the UK made an early switch of coal utilisation technologies unlikely. However, plans for scaling up a number of processes were put forward in 1973 on the principle, which was accepted in 1974, that government should provide major assistance. A Report, "The Coal Industry Examination", in 1974 laid down criteria for support and identified the following three areas for early concentration:

(a) fluidised bed combustion;
(b) coal liquefaction by solvent extraction;
(c) pyrolysis.

Proposals in the following fields were favourably considered but were given a lower priority at that time:

(a) liquefaction by supercritical gases;
(b) gasification with oxygen to yield synthesis gas;
(c) gasification with air to yield low Btu fuel gas.

In the event only fluidised bed combustion was given early support, since this coincided with proposals from IEA Coal Research (see below). Continuing definition of the other projects led in 1978 to a further examination of priorities and opportunities (see Fig. 23.1), from which solvent extraction by liquid solvents and by supercritical gases emerged as front runners in the government–NCB programme. Design studies for two 25 t/d (solvent extraction and gas extraction) plants were put in hand with a view to final decision in 1980, with operation scheduled for 1983. Planning envisages the possibility of scaling up to 2500 t/d, i.e. the size of a single stream of a full size plant, in the late 1980s.

A proposal for a plant concept by British Gas may go ahead on a similar scale at about the same time. This is for a "Composite Gasifier" — a fines-entrainment gasifier underneath a British Gas Slagging Lurgi gasifier for the larger coal, thus using all coal sizes. Other proposals are still under active consideration.

The pilot/demonstration plant programme is the vital link between research and commercialisation. During the 1970s, the in-house, internally financed laboratory scale coal utilisation R & D has increased from the £1 million level at the beginning of the decade to more than £10 million in 1979–80. The exact amount depends on the way in which the substantial and complex flows of money involved in co-operative ventures are accounted for.

Whilst it could be argued that the rate of progress is satisfactory in the light of the UK energy resource position, it should be pointed out that even now the expenditure on nuclear power R & D is virtually an order of magnitude greater than that on coal utilisation, a situation which has persisted for many years. This cannot represent the relative potential contribution of the two resources. Furthermore the level of government support is pitifully low compared with US and German funding.

Other Countries

Apart from Germany and the UK, individual countries in Europe do not have large separate coal R & D programmes. France has special interest in chemical activities related to coal and Belgium has an interest in *in-situ* operations. There is an EEC co-operative programme mentioned later.

In the other developed countries, South Africa, Australia and Canada have significant R & D programmes aimed at maximising the benefit to be obtained from

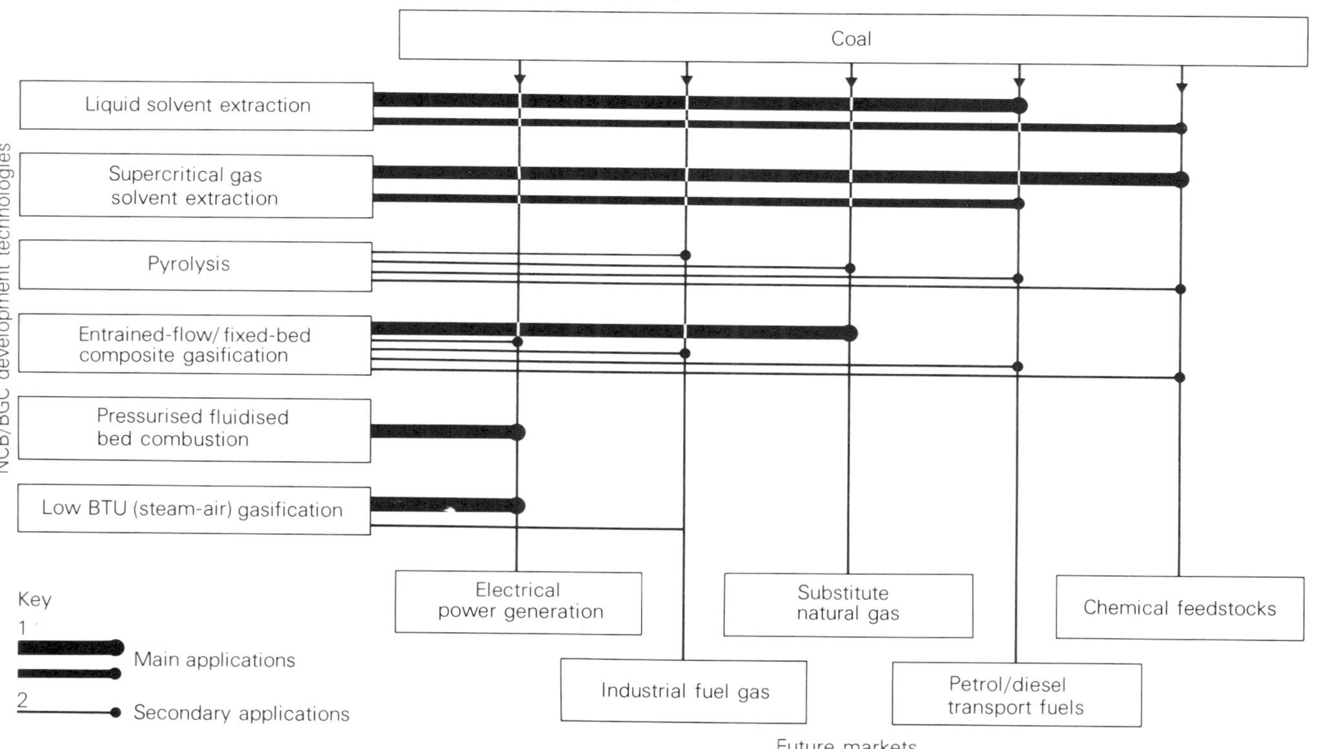

23.1 Development technologies and future markets for coal

their coal reserves. In South Africa, there is of course major R & D work in support of Sasol; this includes both continuing refinement of current and programmed plant and also studies of alternative routes to liquids, including direct hydrogenation. Outside Sasol, however, there are other long-standing projects, notably on behalf of the coking and electricity industries; in both these cases, work is necessary because the properties of the local coals, often somewhat inferior, require modifications in standard procedures used elsewhere. There are also R & D programmes in support of continuing developments in the chemicals industry based on synthesis gas from coal, established by AECI Ltd and Sentrachem Ltd; products include ammonia, methanol, acetylene, synthetic rubber and PVC.

Several other coal R & D groups are active in South Africa including the Fuel Research Institute. The latter are concentrating on hydropyrolysis, supercritical gas extraction, low-temperature fluidised bed pyrolysis and electrode coke. Elsewhere, there are studies on catalytic hydroconversion of coal and coal liquids, acetylene and synthesis.

Australia also has a broadly based research programme on coal conversion; however, transport liquids seem to be emerging as the principal objective. All three of the routes (F–T synthesis, direct hydrogeneration and pyrolysis) are being studied but the best known Australian work is that on pyrolysis carried out by CSIRO. Various types of pyrolysis have been studied; economic analyses suggest that fast pyrolysis is most promising and compares favourably with other liquefaction processes. Work on a 20 kg/h reactor has been completed and suggestions have been made for a 1 t/h pilot plant.

Canada's coal R & D programme is based on a strategic analysis of her future energy situation, which contains some imponderables. Strategic aims include substitution of oil by coal, improved coal-to-electricity processes, industrial coal use including co-generation and district heating, transport fuels from coal and gasification. Actual experimental programmes are carried out on fluidised combustion (both AFB and PFB), COM, gasification (for LBG, MBG and SNG) including work commissioned on the British Gas Slagging Gasifier at Westfield and *in-situ* gasification and liquefaction (both hydrogenation and synthesis).

Of the developing countries, India has the largest and most explicit coal policy. Coal is seen as the key to development and there is also an intention to secure export business. These are considerable R & D programmes, especially (as in South Africa) in support of the use of indigenous low-grade coals in the basic industries (steel, electricity and railways). There is a broad R & D programme embracing most of the more advanced coal processes but specialisation has not yet been discerned.

Japan is a major importer of coal, especially for metallurgical purposes, and is a leading country in research in that field, including formed coke. Japan's indigenous coal output is modest and unlikely to expand. Quality is not high and the coal is used mostly for electricity generation. Recently, policy decisions to increase coal use for power generation based on imported coal have been taken, adding emphasis to the work being carried out on pollution control, which is a particularly serious problem in Japan. (This work has been discussed in Chapter 6.) For some years a national effort, Project Sunshine, intended to diversify energy sources away from oil has been in hand. This has included surveys of improved coal burning technologies, gasification and liquefaction but it appears that alternative "benign" energy sources were initially given highest priority. More recently, a broader approach to coal utilisation seems to have been adopted. Notable progress has been made at the Coal

Mining Research Centre on a two-stage fluidised bed gasifier for combined cycle electricity generation. Work on a 5 t/d unit is virtually complete. Construction of a 40 t/d pilot plant is nearly completed, probably to be followed by a 1000 t/d plant.

Russia has massive coal reserves which have always figured strongly in long-term energy planning. The role of coal did not suffer as a result of cheap oil, early nuclear euphoria or environmental opposition and is therefore deeply rooted in the traditional industries, electricity and steel, and even conventional gas-making. Research and development appears to be strongly linked to improvements in these fields with the more exotic technologies regarded as being applicable in the relatively distant future. However, MHD is an important area, with a collaborative agreement with the USA. *In-situ* gasification is a traditional interest, favoured by Lenin on social grounds and pursued on a larger scale in Russia than elsewhere. During the 1950s and 1960s work on UCG apparently declined and possibly was terminated, to be revived in the 1970s, perhaps at least as much for the export of expertise as for internal use.

Poland has a large and rapidly expanding coal industry and coal export business. Coal used internally meets a strong demand from the traditional industries and this influences the direction of R & D. However, there is a great interest in gasification and liquefaction, which are seen as vital technologies for the future. Polish coal scientists have studied Western developments closely and have formed links with a number of countries. It would not be surprising if Poland hosted some of the earliest full scale plants, taking full advantage of political and financial initiatives.

INTERNATIONAL COLLABORATION

Coal scientists have traditionally been co-operative on an informal basis and although more formal arrangements for technical exchanges and co-operation have recently flourished, the direct discussions in laboratories or at specialist scientific meetings are still very important. Perhaps this tradition became established in the early days of coal science and matured in the period when coal R & D workers felt somewhat neglected. Nowadays, a key laboratory like CRE receives many hundreds of visitors per year from all parts of the world and presents two or three dozen major papers annually at scientific gatherings, as well as many other publications.

More formal international co-operation takes a number of forms. On the one hand there are multi-national agreements, often under the influence of some broader political grouping. On the other hand there are bilateral arrangements between pairs of countries, most of which are for exchange of information, ostensibly to avoid duplication and maximise the effectiveness of resources. In most cases this does not involve either country doing significantly more or less than they would otherwise have done; in the few cases where a substantial new effort is involved, this has usually been under some contract arrangement. Such contracts would usually not of course have arisen without prior contacts engendered by scientific intercourse.

The UK (generally through NCB) has collaborative arrangements on coal R & D with many countries, sometimes with governmental agencies and sometimes with coal operating organisations. The countries most closely involved are the USA, West Germany, Poland, Canada, USSR, Hungary and Australia. The USA in turn formally lists, in its annual report, bilateral arrangements with UK, West Germany, USSR, Poland and Australia. There are certainly other arrangements presumably of a less formal kind or more recent; Japan and China are examples.

European Communities

The European Coal and Steel Community has a collaborative coal research programme covering both mining and utilisation. The funding is about $17 million per year, most projects however having a duration of three years. Although the actual work is done in the laboratories of the member countries under their own local control, certain standards are expected and preference in the selection of projects is given to cases where two or more countries make complementary proposals, which can lead to collaboration. The finance is obtained from a levy on the coal industries, so that in one sense the projects are self-financed. However, benefit arises from the exchange of views during the formation of the proposals, which encourages scientists to consider other people's work, and also in the obligatory publication of results, which normally evokes constructive responses. Thus, the programme can be considered an excellent example of co-operation.

The European Economic Community (EEC) and the associated community, Euratom, also have energy R & D programmes. These are funded from the EEC's budget, which in turn is financed directly by member governments. Proposals are requested under headings laid down by the Commission (the EEC's administration arm) and grants are made. The funding has for years been several times that of the ECSC coal programme and the main share has gone to nuclear power, with the other main recipients being conservation, solar energy, geothermal energy, hydrogen (as an alternative fuel) and the environment. Not only is the volume of support much larger than for coal but the energy systems benefiting do not directly contribute, as in the case of coal.

International Energy Agency

This agency, set up under the aegis of the OECD in 1974, established an R & D Committee, which in turn nominated "lead" countries for each of the main energy sectors. The UK took the lead in coal R & D. It was decided to establish in the first instance five co-operative projects: four of these were office studies, or "services", and one a major "hardware" project. There are no central funds available for IEA R & D projects and the IEA HQ staff in Paris is quite limited. There were of course no precedents or organisational guidelines. It is therefore an indication of a strong desire to co-operate that the dozen or so countries involved agreed to set up and share the costs of the initial projects, all of which were established in the UK. Each project is however solely under the control of the countries participating in that particular project.

For the service projects, the number of participants varies from six to over a dozen and the total annual funding was initially about $2 million. These activities are described below in some detail because their initiation, by agreement of various governments, reflected the perceived needs, not otherwise likely to be met.

(1) *Technical Information Service*: The objectives are to collate and disseminate information both for the benefit of scientific workers in the field and also to increase general awareness of the potential of coal. The main instruments are "Coal Abstracts" and Technical Reviews. The former is issued monthly, each issue containing several hundred abstracts covering the whole field of technology and policy relevant to the coal industry in publications throughout the world. The Technical Reviews are critical appraisals of published

information on specific topics of great interest, such as "Underground Transport", "Carbon Dioxide and the Greenhouse Effect" and "Combustion of Low Grade Coal".

(2) *Economic Assessment Service*: The purposes of this service are to assess coal processes in ways of maximum benefit to its international sponsorship and to relate coal economically to existing and future energy systems. Some early study subjects include:
   (a) Economics of Coal Conversion;
   (b) Cost and Availability of Coal;
   (c) Economic Aspect of Environmental Impact — Gasification Plants;
   (d) Transport of Coal;
   (e) Competitiveness of Coal-based Energy.

(3) *World Coal Reserves and Resources Data Bank Service*: This service was established to provide better and more universal methods of assessing reserves and resources. Information is stored in a computer for use by member countries. Ultimately, it may be possible to develop the system into a dynamic form to take care of factors which change (including economics and energy balances, as well as resource definitions) and which may influence the recovery of coal.

(4) *Mining Technology Clearing House*: The initial purpose of this service was to compile registers of all the mining R & D carried out by member countries. Subsequently, critiques have been prepared on the work within the registers; this has helped to make the value of some work more widely appreciated. Several special Technical Investigations have been carried out. One of these — Hydrotransport of Solids Underground — attracted sufficient interest to lead to the formation of a Working Group and the holding of symposia. The possibility of task sharing programmes was envisaged.

The first "hardware" project, agreed in November 1975, was the Pressurised Fluidised Bed Combustor (described elsewhere) at Grimethorpe, England, at a cost in current terms of about $50 million. The participants were the USA, West Germany and the UK. The objective was to test the combustion, fluidisation, heat transfer, sulphur retention and dynamic response characteristics of the system. The quality of the exhaust gases was considered and tested for its suitability as the input to a gas turbine. The latter was not included in the original agreement but provision was made in the layout for such an addition in a possible later phase. Construction was completed in mid-1980. Following shake-down trials, a minimum exploratory experimental programme extending to about three years should commence early in 1981.

Subsequent to the first five projects other possible projects have progressed to various stages, on such topics as pyrolysis, waste liquor purification, basic coal science and environmental impact. Further technical possibilities will no doubt arise and, in addition to the present system where one country by agreement sets up a management organisation, which then has a contractual relationship to the joint sponsors, other types of arrangements, such as task-sharing are likely to be implemented.

There is no doubt that IEA Coal Research has created a new dimension in co-operation, which will continue to improve within the current limitations. One constraint is that co-operative arrangements require careful planning, organisation, legal study and financial estimation as well as technical optimisation to suit a range of possibly interested parties, before formal documents can be signed. Without initial "seed corn" financial support to carry out these studies, this is a difficult barrier to surmount. The second limitation is that membership of each project is voluntary and restricted to members of the IEA. Each project and the results therefrom are the property of the members supporting that project and have a commercial value. At the same time the legitimate project interest does not end at the member countries frontiers — technical information and world resources estimation are particular examples. Finally, it does not seem likely that IEA, either through Coal Research or its central organisation, could initiate or even take a major active part in a large demonstration type project, although it may by courtesy be kept informed by member countries of progress on such projects.

International Conference on Coal Research (ICCR)

In the early 1970s at the instigation of Sir Derek Ezra, Chairman of the UK NCB, it was decided to set up a forum to discuss coal research programmes and results, initially between the USA and Western Europe. Dr Erwin Anderhagen, chairman of the ECSC coal research committee and Mr Carl E. Bagge, President of the US National Coal Association took the lead in establishing what is now known as the International Conference on Coal Research (ICCR). Representatives from the original countries (USA, W. Germany, UK, France and Belgium) have now been joined by Canada, Australia, South Africa and Japan. The fifth conference was held in September 1980, in Dusseldorf. Authoritative papers are presented on the status of coal research in the participating countries; there is no forum having a similar level of concentration of experts solely dedicated to the subject. Attendance is by invitation and strictly limited. However, communiques are issued and the coming-together of these leading scientists undoubtedly results in improved work and in more and better information becoming available for more general benefit.

SOME CURRENT INITIATIVES

In the final section of this chapter, some comments on current progress in the application of coal utilisation technology are made and suggestions offered. Certain current initiatives may have important implications for the prognosis and are therefore mentioned, although it is too early yet to make an assessment of them.

US Energy Security Act of 1980 and the Synthetic Fuels Corporation (SFC)

Over the past ten years, the coal and other energy programmes have greatly intensified in the USA but there has been great uncertainty about how the main steps to commercialisation would be taken and thus about what real impact the R & D and other activities might have on energy supplies. In fact, a recent statement by the Energy Information Administration of the US Department of Energy indicates that "all government actions taken to date will likely together have only a relatively small impact on energy supply, the average level of energy prices and/or import levels in 1990". Furthermore, it has been calculated that in recent years, with regard to federal cost incentives, coal has received least support of all energy sources to stimulate production and oil has received most. The Energy Security Act, 1980, setting up the Synthetic Fuel Corporation (SFC) is an attempt at a partial rectification of this imbalance. However, the election of President Reagan has disturbed the organisation and the policy vis a vis the SFC is still not apparent although avowedly in favour of sharper and deeper industrial commitment.

The major goals of SFC are to achieve production of 500 000 barrels per day of synthetic crude oil equivalent by 1987 (5–10 plants) and 2 million (20–40 plants) by 1992. The appropriate act was signed at the end of June 1980 and the SFC is charged with making its first solicitation for projects within six months, i.e. end December 1980. However, the November 1980 election result will undoubtedly delay the necessary senior appointments and the expression of policy towards the SCF by the new administration. Nonetheless, there is a legal requirement to pursue the ends of the Act. Means for implementation of the targets may include direct loans, loan guarantees, price guarantees, purchase agreements, joint ventures, acquisition of plants and leasing.

A mechanism is also provided under the Act for a rapid start to supporting synthetic fuel plants. A total of three billion dollars has been appropriated which can immediately be used under the Defense Production Acts to arrange supplies for the Department of Defense by assisting in the establishment of synthetic fuels plants. Any balance from the three billion dollars will revert to the SFC when this is operational. In addition, a similar sum was prepared for interim synfuels support by the Department of Energy. Funds (believed to be about $200 million) are already available under existing Department of Energy budgets for demonstration SNG plants; this activity will also be transferred to the SFC.

Two interesting aspects of the Act are that foreign concerns can apparently make proposals and that provision is made for up to two projects to be constructed outside the United States in the Western Hemisphere. At the same time, however, criteria for judging proposals are laid down fairly precisely and this may mitigate against foreign involvement.

The Synthetic Fuels Corporation and associated initiatives in the USA represent by far the most potentially important activities as yet envisaged in the field of coal conversion. The degree of dedication with which the new administration pursues these aims — some of which have not previously appealed to the Republican Party philosophy — and the success which attends these efforts will have great, possibly even crucial, impact on the US and the world's energy problems.

West German Commercialisation Programme

Industrial organisations have been encouraged to submit projects of a commercial nature for coal gasification and hydrogenation. Fourteen such projects have been put forward — eleven gasification and three liquefaction processes. The total cost of these projects would be about 13 billion DM. All the liquefaction projects are in the multi-billion DM class; the gasification projects cover a much wider range, indicating the potential for relatively small MBG synthesis gas plants. The presumption is that substantial funding will be provided by both Federal and State governments. It was stated in September that a selection of the gasification projects would be made by the end of 1980 and of the liquefaction projects by mid-1981. No information was however forthcoming at that time about the criteria or mechanisms for selection, or about the scale and proportion of public funds. An election has taken place in Germany also and it may be that 1981 will see the resolution of this important programme. The dates for decisions and for plant operation (from 1987) may now seem optimistic but there is every reason to believe that a most important demonstration programme will emerge. In addition, the FGR has committed itself to a 25% stake (now in some doubt) in the US SRC programme and to other internal co-operative exercises for a commercial nature.

IEA Coal Industry Advisory Board (CIAB)

The International Energy Agency, recognising the importance of coal in future energy strategies and aware of the problems of exploiting this resource, have set up an expert committee to advise on necessary measures. It is expected that recommendations, for implementation at government level, could be made in 1981. In this connection, both recent Western "summit" meetings, in Tokyo and Venice, have emphasised the importance of coal but apparently without adopting specific measures. Possibly the CIAB could have a profound influence at this level.

THE STRATEGIC IMPORTANCE OF COAL AND ITS ECONOMIC IMPACT

In this book, the authors have endeavoured to present the most objective picture possible of the economics of coal utilisation and have stressed the importance of economic analysis in the timing and direction of future utilisation patterns. This must remain the case. Taken in its simplest form, this means that the utilisation of coal, both in quantitative and qualitative terms, will depend on future price trends for coal and oil; a growing diversion between the trend lines (with oil increasing in price at a more rapid rate) will be the most favourable circumstance for the widest use of coal. However, the lead time for one complete integrated coal utilisation scheme (say, a new mine and a synthesis gas plant) may be of the order of a decade (especially if infrastructure developments are also needed); a series of such schemes, sufficient to make significant international impact on oil demand, might take a couple of decades.

However, the projection of price trends so far into the future must be essentially a political judgement. More important still must be the consideration that such price trends may be influenced by political action taken on strategic as well as narrow economic grounds. The remainder of this section is based on a statement made by one of the authors (LG) to the Coal and Energy Conference in London in May 1980.

The expansion of coal production and its utilisation in more advanced and efficient ways are objectives which are widely accepted, in general terms, and especially by the OECD countries. Progress however is disappointingly slow. This may be partly because the political urgency of the energy situation (the "moral equivalent of war!") has not yet been fully appreciated, although current international tensions may speed the process. However, slow responses may also be partly due to the fact that economic horizons are too limited. In comparing the costs of coal with oil, especially where substitutions or conversions are concerned, insufficient allowance appears to be made for the potential responsiveness of oil prices to demand levels and, in particular, to the effects of moderated demand for oil. Furthermore, inadequate attention seems to have been given to cumulative international requirements, as distinct from immediate national interests.

The Ford Foundation Report, "Energy: The Next Twenty Years", considered the effect of differences in the levels of US oil imports on the marginal price. The principle is that a small increase in demand may trigger a consequential increase in price, which will then apply to all the oil and not merely to the extra demand. The Report, without taking any quantitative view, discusses the arithmetic, on the basis of two different levels of US demand: 9 and 13 MBD. If the price at the 9 MBD level is $24/barrel and at the 13 MBD is $30/barrel, the incremental cost per barrel is actually $44. Unfortunately, the cost increase will also apply to the rest of the world, so that on an international scale the incremental cost for the extra consumption by the USA of 4 MBD would be nearer $100 per barrel. The same arithmetic, applied at a higher standard price, would produce an even larger incremental cost.

Another illustrative example is as follows. If, in 1970, the USA had decided to invest $10 billion in a crash programme to convert 100 Mt/a of coal into other forms of energy (including, but not solely, gases and liquids), using well established processes, a saving of about 1 MBD of imported oil might have been made. The plants might now have accumulated deficits, on a normal accountancy basis, of about $20 billion (an average of $10/barrel for 5 years). If the total excess cost were therefore $30 billion, this would be paid back in a single year if the effect were to mitigate general price increases in oil by just over $1/barrel. There can be little doubt that the US (and others) would have gone ahead with large coal conversion programmes years ago if they had taken an international view and had envisaged the *marginal* cost of oil in 1980 as possibly being over $100/barrel.

If considered in this way, as a hedge against incremental oil prices, even without counting the political cost, the economic case for coal is already overwhelming. Coal will probably become the "price marker" for traded energy very soon; the OECD countries at least should act as if that day has arrived. The questions are how much coal can we get and which are the best ways to use it? Furthermore, maximising the use and value of coal should take priority over other forms of energy which may be available temporarily and which might be treated as windfalls or conserved.

The US Department of Energy reported that in May 1978, the national average delivered cost of coal to utilities was $1.10 per 10^6 Btu, compared with $2.10 per 10^6 Btu for oil and $1.44 per 10^6 Btu for gas. The Ford Foundation Report suggested that in the year 2000 the costs of coal, oil and gas delivered to utilities (1979 dollars) would be $2.60, $5.20 and $5.20+ per 10^6 Btu, respectively.

In the UK in January 1980, the average prices for the main energy forms, in pence/therm, were:

Coal	13.4
Natural Gas	25.0
Fuel Oil	26.1 (heavy) — 35.3 (light)
Electricity	76.2

Coal is by far the cheapest fuel for direct heat, even taking into account efficiency and utility, and should be used in this way whenever possible. In addition, coal is certainly the cheapest fossil fuel for electricity production and it would seem extremely foolish to use any other fossil fuel for this purpose, even if substantial capital investments were necessary to avoid it. However, turning coal into electricity may not be the best way of using it, especially since nuclear power, another "indigenous" form of energy, may be available in some countries for part of the electricity. In most countries, as in the UK and US, gas from coal must certainly be cheaper than electricity for purposes which can be served by either. Furthermore, the comparison above is on an *average* price basis; if the costs of supplying lower load factors were considered, the advantage of gas from coal over electricity would probably be greater.

The rapid expansion of coal gasification, at least initially to the limits where replacement of electricity by gas is optimised, is probably the most useful step which the Western countries could take. Further penetration by coal gas, in direct replacement of light oil products and natural gas, is often considered to depend on a further widening of the price differentials. However, if the argument about the incremental cost of imported oil is considered plausible, the maximum use of coal gas in these areas should also immediately be promoted.

A good way to start the replacement of natural gas and oil by coal (other than by direct combustion) would be through MBG or synthesis gas, which would in any case be the most favourable route in many situations to take over suitable markets from electricity. Such gas is cheaper and more energy efficient than SNG. It could be a very useful bridging operation since it can be used flexibly and manufactured in smaller plants than SNG. It could be used in parallel with natural gas.

The establishment of coal liquids on a commercial basis may be the most difficult task and it is by no means certain that the hydrogenation route will always be preferred to synthesis. If MBG (or synthesis gas) plants become established, some of them might be sited to promote their further development and integration into processes for producing light liquids. This might be through the conversion of synthesis gas into hydrogen to be used for the hydrogenation of petroleum fractions or syncrude from coal; alternatively, synthesis routes to liquids might be preferred. Both methods may well be used. In any case, the establishment of synthesis gas manufacture on a large scale (not necessarily in very large individual plants), strategically and economically important in its own right, could fit readily into various coal liquefaction schemes.

In all cases, including combustion and electricity generation, coal conversion should be planned for maximum thermal efficiency, which probably implies integration of different processes (coalplexes) and the availability of suitable loads to take low grade heat. This should not be neglected, even for the sake of getting plants going quickly, and should enhance the economic case.

Although the above views are offered with some confidence that they are generally on the right lines, the optimum use of coal, which will change dramatically with time, is extremely complicated. A great deal of detailed analysis is required and this should preferably be done in the closest possible collaboration with those in the coal industry best able to exert pressure at the top level. The objectives should be:

(1) Improve analysis and presentation of the case for the maximum and earliest expansion (quantitative and qualitative) of the role of coal.

(2) Press for greatly increased R & D spending, especially on large scale work and for assessing economic optima, including the costs of environmental control.

(3) Promote measures, nationally and internationally, which would improve the immediate financial prospects of coal utilisation processes.

(4) Improve international co-operation, both technical and commercial, including optimisation of world coal trade.

(5) Consider more positive means of discouraging oil usage by the Western countries, including the use of surcharges; the latter might be applied to provide a fund to finance coal developments.

GENERAL CONCLUSIONS

Although the energy crisis, which became evident to all in 1973/4 and which seems certain to remain chronic indefinitely, gave rise to a flurry of activity, the actions with regard to the exploitation of coal remain inadequate in concept, scale and execution. That seems to be the consensus of those, within or without the coal industry, who have studied the subject. Recognising this has caused the attempt at the new initiative through CIAB. Without a more powerful co-ordinated effort early in the 1980s, coal will not be able to fulfil its potential role and this failure would become painfully evident in the latter part of this century and the beginning of the next. Although some Western countries do have a policy for coal, these are generally more hopeful than determinative in nature; many have none. On the international scale, there are a number of estimates of future output and export availability but these are mainly summations of national aspirations (legitimate but not always fully supported, especially with regard to infrastructure).

The enormous and rapid growth of the oil industry was successful in spite of being fragmented. As demand increased, new sources were fairly easy to find and relatively cheap and quick to develop. Now, of course, the picture has changed but information on maximum production capacities and profiles (at least physical capacities) are remarkably well organised. The total eventual output has been estimated by many authorities and also the year of peak output, with fair unanimity for such a complex subject. The oil companies have shown great resilience in maintaining distribution patterns. Furthermore, within the amounts available to the main Western nations for import, high level negotiations have taken place on the requirements of individual countries, again working to an accuracy of a few percent. Through the IEA, contingency plans exist for meeting emergencies. In the case of uranium also, world reserve data are well documented and collaborative efforts will probably lead to a resource sharing scheme and possibly also to price-support and other procedures to ensure logical and timely investment patterns.

No similar international approach to coal has taken place. Even within the EEC, the closest knit and most bureaucratic association of trading partners ever known, there is no joint initiative to ensure the large supplies of imported coal which will be most essential to this area. Japan has shown the way, first with coking coal and more recently, it appears, with steam coal. No doubt she will also plan in good time supplies of coal for conversion.

Although there is considerable spontaneous activity on a commercial level, often involving national or quasi-national organisations, to develop and market more coal, these efforts lack a high level policy input, particularly with regard to the pace and type of development. Without discouraging such spontaneous local efforts, a massive joint effort on an international scale is needed. The following are suggested as immediate short-term activities:

(a) *Data Bank for Resources and Reserves*: This could be similar in concept to the project presently managed by IEA Coal Research. It should however be on a larger scale, enabling more rapid progress to be made; there should also be unconditional government commitment, over as many countries as possible, to make data available in order to enable economic and logistic priorities to be established as a precursor to direct action.

(b) *World Coal Fund*: Over the remaining years of this century, at least $1000 billion ought to be invested in new coal capacity and several times that on utilisation equipment. This may be difficult to finance on an individual case basis. Coal does not appear to have a high priority in existing world monetary bodies. A World Coal Fund, to be effective, would have to be competent to provide a substantial proportion of the total and to do so on a selective basis, with expert advice. Funds would have to be available not only for mining and consumption but also for associated infrastructure and transport.

(c) *Technical Advice*: An advisory service should have a core of in-house experts and access to the leading mining operational and development centres for taskforce support to countries requiring preliminary advice.

Other desirable activities would subsequently emerge, especially those related to marketing and allocation of coal in emergencies. However, the over-riding priority is to organise the increase in coal getting and coal using capacity at an adequate and steady rate. In the event of over-production, which seems unlikely for any sustained period, a world organisation would be needed to ensure that lasting damage did not occur. In the event of shortages, it would be vital to avoid the different countries bidding against each other in a disorderly way. There have been suggestions that potential coal exporting countries might form an OPEC-like cartel. There would be no great disadvantage in this if co-operation in technical and financial terms were established with the likely importers from an early stage and agreement on pricing principles reached as part of a reciprocity treaty.

Research and Development, whether separately organised or not, should be closely linked to the international policy group. However, many countries will be interested in using indigenous coal, possibly supplemented by some imports, on a relatively modest scale which would not necessitate involvement in a major trading group; a wider base, such as the OECD, might therefore be a more suitable base organisation. A diverse spectrum of expertise and a substantial endowment would be necessary, plus a formula for funding and organising large and small-scale joint projects, preferably by participating in existing programmes. The aim should be involvement in the development of processes of general applicability.

Most essential of all, the R & D organisation should contain an Economic Assessment Group, similar to that in the IEA Coal Research but with a planning function to identify needs and point to the most likely solutions. This group will obviously need to be on a much larger scale and suitable for dealing with diverse industrial conditions. In particular, a requirement of this Group would be to analyse various world energy/supply scenarios so that the great flexibility with which coal can be used is applied over the next few decades in the most effective manner, responsive to the changing availability of other energy sources. This should be done and publicised in a manner which makes all competent authorities and the general public aware that, by using and converting coal in most efficient ways, we could have a "new" energy source commensurate with any others being developed.

BIBLIOGRAPHY

INTRODUCTION TO COAL UTILISATION

(a) Origin and Description of Coal

International Classification of Hard Coals by Type, Economic Commission for Europe, August 1956.
Coal: its formation and composition, 2nd edn, W. Francis. Arnold, 1961.
Coal: typology, chemistry and physics, D. W. van Krevelen. Elsevier, 1961.
The Coal Classification System used by the National Coal Board (Revision of 1964), National Coal Board.
"Origin and development of coal rank", J. B. Caldwell. *Colliery Guardian* **213**, 489 (1966).
"The constitution of coal and its relevance to coal conversion processes." (1977 Robens Coal Science Lecture) J. Gibson. *J. Inst. Fuel* **51**, 67 (1978).
Classification of Coals by Rank. American National Standard ASTM D388-77.
"Chemical and physical structure of coal" *Phil. Trans. Royal Society, London* **A300**, 3 (1981).

(b) Resources and Reserves

World energy resources 1985–2020. World Energy Conference, IPC, 1978.
Symposium on World Fuels, 1970–1990, American Chem. Soc., Division of Fuel Chemistry, Honolulu, Preprints 1979, Vol. 24, (1).
Survey of energy resources. World Energy Conference, Munich, 1980.
Coal in America; an encyclopedia of resources, production and use, R. A. Schmidt. McGraw Hill, 1979.

(c) Mining

Introduction to mining L. J. Thomas. Hicks, Smith and Sons (Sydney), 1973.
Proc. 5th International Conference on Coal Research, Düsseldorf, Vol. 1, 1980.

(d) Coal Utilisation: General Principles

Chemistry of coal utilisation, Supplementary volume H. H. Lowry (ed.). Wiley, 1963.
Carbonization and hydrogenation of coal. UN International Development Organisation, 1973.
Coal conversion technology. I. Howard-Smith and G. J. Werner, Noyes Data Corporation, 1976.
Coal: technology for Britain's future J. Bugler (ed.). Macmillan, 1976.
Coal and energy; the need to exploit the world's most abundant fuel, 2nd edn, D. Ezra. Benn, 1980.
Coal: its role in tomorrow's technology, C. Simeons. Pergamon, 1978.
Coal and modern coal processing, G. J. Pitt and G. R. Millward (eds.). Academic Press, 1979.
"New coal up-grading processes." Vols. 1 and 2 *Proc. Round Table Conference*, Commission of the European Communities. *Colliery Guardian*. 1979.

An introduction to coal technology, N. Berkowitz. Academic Press, 1979.
Chemistry of coal utilisation. Supplementary volume 2, M. A. Elliott (ed). Wiley.

TECHNOLOGY AND ECONOMICS OF COAL UTILISATION

(a) Combustion

The fluidised combustion of coal, D. G. Skinner, Mills and Boon, 1971.
The efficient use of energy, I.G.C. Dryden (ed.). IPC, 1975.
Assessment of advanced technology for direct combustion of coal, National Academy of Sciencies, US Department of Energy Report FE. 1216-1, 1977.
Proc. 5th International Conference on Fluidized Bed Combustion, Washington D.C., Mitre Corporation (for US Department of Energy), 1977.
Fluidisation. Proc. 2nd Engineering Foundation Conference, Cambridge, England, J. F. Davidson and D. L. Keairns (eds.). Cambridge University Press, 1978.
"Combustion of coal in fluidised beds", G. G. Thurlow. *Proc. Instn. Mechanical Engrs* **192**, 145 (1978).
Fludized-bed energy technology: coming to a boil, W. C. Patterson and R. Griffin. Inform, New York, 1978.
"Performance and economics of advanced energy conversion systems for coal and coal-derived fuels", J. C. Corman and G. R. Fox. *J. Engineering for Power* **100**, (2), 252 (1978).
Evaluation of coal-fired fluid bed combined cycle power plant, A. J. Giramonti *et al.*, American Chem. Soc., Division of Fuel Chemistry, California, Preprints, 1978, Vol. 23, (1), p. 173.
Fluidised bed combustion, US Department of Energy, Prepared by Davy McKee, May, 1979.
Power from Coal Conference, Instn. Mechanical Engrs., London, 1979.
"Combustion of low grade fuels in fluidized beds" E. A. Rogers *et al. Proc. Conference on Future Energy Concepts*, Instn. Electrical Engrs., London, 1979.
"Conceptual design and cost estimate 600 MWe coal fired fluidized-bed combined cycle power plant", D. A. Huber and R. M. Costello. *Combustion* **50**, (12), 22 (1979).
"Economic feasibility of pressurised fluidized bed combustor power plant", G. Singh and S. Moskowitz. *Combustion* **50**, (8), 22 (1979).
First International Symposium on Coal Oil Mixture Combustion Proceedings, Mitre Corporation, (for US Department of Energy and General Motors Corporation), St Petersburg Beach, Florida, 1978.
Energy Conversion Alternative Study (ECAS), Summary Reprint, NASA TM 73871.
Combustion of low grade coal, G. F. Morrison, IEA Coal Research, 1978.
Conversion to coal and coal/oil firing, G. F. Morrison. IEA Coal Research, 1979.
Proc. 2nd International Symposium on Coal-Oil Mixture Combustion, U.S. Department of Energy, Danvers, Mass. 1979.
Proc. 6th International Conference on Fluidised Bed Combustion, Atlanta, Ga. 1980.
Proc. Conference on Fluidised Combustion: Systems and Applications, Inst. Energy, London, 1980.

(b) Carbonisation

Coal carbonisation products, D. McNeil. British Coke Research Association, 1966 and 1978.
Carbonisation of coal, J. Gibson and D. H. Gregory. Mills and Boon, 1971.
"The coking principle" (19th Coal Science Lecture), A. M. Wandless. *J. Inst. Fuel* **44**, 531 (1971).
"Coke betterment", D. H. Gregory. *Colliery Guardian Annual Review*, 60 (1974).
"Blast-furnace coke — macro and micro considerations" (6th Carbonisation Science Lecture), L. Grainger. *Coke Oven Managers Association Yearbook*. 1975, p. 282.
Symposium on New Coke-making Processes minimising Pollution, American Chem. Soc.,

Division of Fuel Chemistry, Chicago, Preprints, 1975, Vol. 20, (4).
"Solving the coking coal problem", J. P. Graham. *Coal and Energy Quarterly*, (10), 15 (1976).
"Recent developments in understanding the fundamental aspects of the coking process", J. W. Patrick. *Coke Oven Managers Association Yearbook*, 1976, p. 201
"From rapid devolatilisation to the BFL formed-coke process", W. Peters. (7th Carbonisation Science Lecture), *Coke Oven Managers Association Yearbook*, 1976, p. 303
"The changing scene for coke oven co-products", P. H. Pinchbeck. *Coke Oven Managers Association Yearbook*, 1976, p. 240.
"Formed coke — still waiting in the wings", J. Dartnell. *Iron and Steel International* **51**, 155 (1978).
"The science of coke-making technology and its development in Japan" (9th Carbonisation Science Lecutre), Y. Miura. *Coke Oven Managers Association Yearbook*, 1978, p. 292.
"Philosophy of blending coals and coke-making technology in Japan", N. Nakamura *et al.* *Ironmaking and Steelmaking* **5**, 49 (1978).
"Trends in the profitability of coke oven by-product recovery over the past 25 years", E. Th. Herpers. *Coke Oven Managers Association Yearbook*, 1978, p. 132.
"Blending and proportioning of coal in coking blends; effects on production operations and coke quality", J. Gibson and D. H. Gregory. *Proc. 38th Ironmaking Conference*, AIME, Detroit, 1979.
Commonsense of Carbonisation, J. Gibson, 11th Carbonisation Science Lecture, 1979.
Progress and potential in coking plant technology, K.-G. Beck, 12th Carbonisation Science Lecture, 1980.

(c) Gasification

Symposium on Gasification of Coal to produce low-Btu Gas, American Chem. Soc., Division of Fuel Chemistry, New Orleans, Preprints, 1977, Vol. 22, (1).
"Review of gasification for power generation", B. Robson. *International J. Energy Research* **1**, 157 (1977).
"Economics of six coal-to-SNG processes", R. Detman. *Hydrocarbon Processing* **56**, (3), 115 (1977).
"Economic advantages and areas of application of small gasifiers", R. W. Culbertson and S. Kasper. *Proc. 4th Annual Symposium on Coal Gasification, Liquefaction and Conversion*, University of Pittsburgh, 1977.
"Economic evaluation of synthetic natural gas production by short residence time hydropyrolysis of coal", C. J. LaDelfa and M. I. Greene. *Fuel Processing Technology* **1**, 187 (1978).
Proc. 10th Synthetic Pipeline Gas Symposium Inst. Gas Technology, Chicago, 1978.
Symposium on Advances in Coal Gasification, American Chem. Soc., Division of Fuel Chemistry, Florida, Preprints, 1978, Vol. 23, (3).
"Gas from coal", M. J. Cooke and B. Robson. *Chem. Engr.* (337), 729 (1978).
"From coal to gas", A. Verma. *Chemtech.* **8**, 372, 626 (1978).
"Coal gasification", G. Archer. *Proc. Power from Coal Conference*, Instn. Mechanical Engrs. London, 1979.
"Manufacture of fuel gas for power generation", B. Robson and G. G. Thurlow. *Proc. Power from Coal Conference*, Instn. Mechanical Engrs., London, 1979.
"Gas from coal — full circle", J. A. Gray and H. J. F. Stroud. *Instn. Gas Engrs. Communication* 1088, 1979.
Symposium on gasification and liquefaction of coal UN Economic Commission for Europe, Katowice, 1979.
"Medium-Btu gas again?" H. G. Hargreaves and R. I. Kirk. *Instn. Gas Engs, Communication* 1124, 1980.
Symposium on Nuclear Heat for High Temperature Fossil Fuel Processing. Institute of Energy, April, 1981.
Hot gas cleanup, G. P. Morrison. IEA Coal Research, 1979.

"Small-scale coal gasification plants" A Verman and P. J. Read. *Energy Sources* **4**, 281 (1979).
Symposium on Coal Gasification Chemistry, American Chem. Soc., Division of Fuel Chemistry, Washington, D.C., Preprints, 1979, Vol. 24, (3)
"The future role of gasification processes", C. Timmins. *Instn. Gas Engrs. Communication* 1112, 1979.
"Gas from coal — a British Gas approach", D. Hebden. *Proc. 1st International Gas Research Conference*, Chicago, 1980.
Proc. 7th Annual International Conference on Coal Gasification, Liquefaction and Conversion to Electricity, University of Pittsburgh, 1980.
"Comparison of coal-gasification combined cycle developments in the USA", J. W. Larson. *Modern Power Systems* **1**, 39 (1981).

(d) Liquefaction

The Fischer-Tropsch and related syntheses, H. H. Storch *et al.* Wiley, 1951.
Solvent treatment of coals, W. S. Wise. Mills and Boon, 1971.
"Pyrolysis route to coal conversion", R. T. Eddinger. *Proc. World Coal Conference*, London, 1975.
"The coal renaissance: a South African point of view" (1975 Robens Coal Science Lecture), P. E. Rousseau. *J. Inst. Fuel* **48**, 167 (1975).
"Sasol II: South Africa's oil-from-coal plant", J. G. Kronseder. *Hydrocarbon Processing* **55**, 56F (1976).
"Flash hydrogenation of coal", R. A. Graff *et al. Fuel* **55**, 109 (1976).
"Early coal hydrogenation catalysis", E. E. Donath. *Fuel Processing Technology* **1**, 1 (1977).
"The Fischer-Tropsch synthesis", M. E. Dry. *Energiespectrum* **1**, 298 (1977).
Supercritical solvents and the dissolution of coal and lignite, J. E. Blessing and D. S. Ross. American Chem. Soc., Division of Fuel Chemistry, Chicago, Preprints, 1977, Vol. 22, (5), 118.
Conversion of coal to liquids by Fischer-Tropsch and oil/gas technologies, J. B. O'Hara *et al.* American Chem. Soc., Division of Fuel Chemistry, Chicago, Preprints, 1977, Vol. 22, (7), 20.
Production of liquid fuels from coal, American Chem. Soc., Division of Fuel Chemistry, Chicago, Preprints, 1977, Vol. 22, (6).
Assessment of technology for the liquefaction of coal: Summary, National Academy of Sciences, US Department of Energy Report, FE. 1216-2, 1977.
Fischer-Tropsch conceptual design/economic analysis, oil and SNG production, J. B. O'Hara *et al.* ERDA R & D Report No. 114 Interim No. 3, FE. 1775-7, 1977.
"Engineering development of a short residence time, coal hydropyrolysis process", M. I. Greene. *Fuel Processing Technology* **1**, 169 (1978).
"Reaction paths in donor solvent coal liquefaction", A. M. Squires. *Applied Energy* **4**, 161 (1978).
"Differential economic analysis: gasoline from coal", R. Shinnar. *Chemtech* **8**, 686 (1978).
"Coal refining by solvent extraction and hydrocracking", G. O. Davies. *Chemistry and Industry* (15), 560 (1978).
"Oil from coal", G. G. Thurlow. *Chem. Engr.* (337), 733 (1978).
"The economics of advanced coal liquefaction", E. W. Neben. *Chem. Engineering Progress* **74**, (8), 43 (1978).
Liquid fuels from coal, National Coal Board, 1978.
"A survey of methods of coal hydrogenation for the production of liquids", J. M. Lytle *et al. Fuel Processing Technology* **2**, 235 (1979).
Symposium on Coal Liquefaction Fundamentals, American Chem. Soc., Division of Fuel Chemistry, Honolulu, Preprints, 1979, Vol. 24, (2).
"Solvolytic liquefaction of coals with a series of solvents", I. Mochida *et al. Fuel* **58**, 17 (1979).
"Structural analysis of supercritical gas extracts of coals", K. D. Bartle *et al. Fuel* **58**, 413 (1979).

"Supercritical extraction of coal", R. R. Maddocks *et al. Chem. Engineering Progress* **75**, (6), 49 (1979).
"Flash hydropyrolysis of coal", M. Steinberg and P. Fallon. *Chem. Engineering Progress* **75**, (6), 63 (1979).
"Economics of synthetic hydrocarbon production", Round Table Discussion 2, *10th World Petroleum Congress*, Bucharest, 1979. Heyden and Son, 1981.
Symposium on gasification and liquefaction of coal, UN Economic Comission for Europe, Katowice, 1979.
"Production of oil and gasoline by carbon monoxide hydrogenation", J. C. Hoogendoorn. Position Paper RT1, *World Energy Conference*, Munich, 1980.
"Coal to gasoline via syngas", B. M. Harney and G. A. Mills. *Hydrocarbon Processing* **59**, (2), 67 (1980).
Symposium on Liquid Fuels from Coal, American Chem. Soc., Division of Fuel Chemistry, Houston, Texas, Preprints, 1980, Vol. 25, (1).
"Economics and market potential for SRC II products", B. K. Schmid *et al. International J. Energy Research* **4**, 173 (1980).
Proc. 7th Annual International Conference on Coal Gasification, Liquefaction and Conversion to Electricity, University of Pittsburgh, 1980.
Coal liquefaction fundamentals, D. D. Whitehurst (ed.). American Chem. Soc., 1980 (contains 1979 Symposium papers).
"Motor fuels from coal — technology and economics", K. H. Eisenlohr and H. Gaensslen. *Fuel Processing Technology* **4**, 43 (1981).
"South Africa's Sasol project: how to succeed in synfuels", R. W. Johnson. *Coal Mining and Processing* **18**, (2), 42 (1981).
"Coal liquefaction: direct versus indirect — making a choice", D. R. Simbeck *et al. Oil and Gas J.* **79**, 258 (1981).

(e) Chemicals

"Carbon molecular sieves for the concentration of oxygen from air", S. P. Nandi and P. L. Walker. *Fuel* **54**, 169 (1975).
"Porosity studies on active carbons from anthracite", D. H. T. Spencer and J. Wilson. *Fuel* **55**, 291 (1976).
"Conversion of coal into fuels and chemicals in South Africa", J. C. Hoogendoorn. *Proc. 3rd International Conference on Coal Research*, Sydney, 1976.
Various aspects of the use of coal in the chemical industry, United Nations Economic and Social Council, April, 1977
"Organics for the 1980's — petrochemicals or coal chemicals", S. P. S. Andrews. *Eurochem Conference*, June, 1977, Clapp & Poliak Europe Limited.
"Using coal to replace hydrogen feedstock in existing ammonia plant", Brown. *Eurochem. Conference*, June, 1977, Clapp & Poliak Europe Limited.
"Coal — burn it or convert it?", L. Grainger and P. F. M. Paul. *Eurochem. Conference*, June, 1977, Clapp & Poliak Europe Limited.
"Turning anthracite into an activated carbon", J. S. Batchelor and J. Wilson. *Water Waste Treatment* **20**, 44 (1977).
"Separations using activated carbons", J. Wilson. *Process Biochemistry* (10), 25 (1977).
"Graphitising carbons from coal via solvent extraction: effects of processing conditions on product quality", G. M. Kimber. *Proc. 13th Biennial Carbon Conference*, California, 1977.
"Coal tar pitch technology relevant to carbon and graphite production", C. R. Mason. *Proc. 5th London International Carbon and Graphite Conference*, Society of Chemical Industry, **1**, 344 (1978).
"Economic analysis of coal- and oil-based chemical processes", H. Gaensslen. *Energy Systems and Policy* **2**, 369 (1978).
"Methanol — fuel or chemical?", H. H. King *et al. Hydrocarbon Processing* **57**, (6), 141 (1978).

"Ammonia from coal: a technical/economic review", D. A. Waitzman. *Chem. Engineering* **85**, (3), 69 (1978).
"Catalytic concepts in the conversion of coal to chemicals", G. A. Mills. *Proceedings of Chemistry of Future Energy Resources*, 1978.
Symposium: ammonia from coal, Tennessee Valley Authority, Muscle Shoals, Alabama, 1979.
"Polymer feedstocks from coal: the potential" J. Gibson and P. F. M. Paul. *Proc. Diamond Jubilee Conference of the Rubber and Plastics Research Association*, 1979, p. 27.
"Chemicals from coal", J. Gibson. *Chemistry in Britain*, **16**, 26 (1980).
Coal Chem — 2000, Institution of Chemical Engineers, Sheffield, 1980.
Coal gasification — routes to ammonia and Methanol, F. C. Brown and H. G. Hargreaves. The Fertiliser Society, London, Proceedings No. 184, 1979.
"The use and manufacture of activated carbons", J. Wilson. *Proc. Round Table Conference, New Coal Up-Grading Processes*, EEC, Colliery Guardian, 1979.
"Olefins from coal via methanol", R. G. Anthony and B. B. Singh. *Hydrocarbon Processing* **60**, (3), 85 (1981).

(f) *In-situ* Processes

Liquefaction study of several coals and a concept for underground liquefaction, D. R. Skidmore and C. J. Konya. American Chem. Soc., Division of Fuel Chemistry, Dallas, Preprints, 1973, Vol. 18, (2), 86.
Underground gasification of coal. A National Coal Board reappraisal, P. N. Thompson *et al.* National Coal Board, 1976.
"A laboratory investigation of the chemical comminution process: a method for the *in-situ* fracturing of coal for gasification", P. Datta. *Proc. 3rd Annual Underground Coal Conversion Symposium*, California, 1977.
"The economics of UCG in British conditions and a comparison with other forms of energy extraction", J. R. Mann. *Proc. 3rd Annual Underground Coal Conversion Symposium*, California, 1977.
Proc. 4th Annual Underground Coal Conversion Symposium, Steamboat Springs, Colorado, 1978.
"Underground coal gasification", D. W. Gregg and T. F. Edgar. *AIChE Journal* **24**, 753 (1978).
Underground coal gasification — reaction zone mapping, S. J. Cure. IEA Coal Research, 1979.
Proc. 5th Annual Underground Coal Conversion Symposium, Alexandria, Va., 1979.
Proc. 6th Annual Underground Coal Conversion Symposium, Afton, Oklahoma, 1980.

(g) Economics — general features — distribution and cost

Synthetic fuels processing, comparative economics, A. H. Pelofsky (ed.). Marcel Dekker, 1977.
Symposium on Economics of Coal Conversion, American Chem. Soc., Division of Fuel Chemistry, Chicago, Preprints, 1977, Vol. 22, (7).
Symposium on Economics of Coal Conversion Processing, American Chem. Soc., Division of Fuel Chemistry, Florida, Preprints, 1978, Vol. 23, (3).
"Future role of coal as raw material and energy source", W. Peters. *Gluckauf* **114**, 180 (1978).
"Fuels and chemicals from coal — some energy and resource implications for the UK", A. Stratton. *Chemistry and Industry*, (15), 551 (1978).
"The use of non-fossil derived hydrogen in coal conversion processes", G. Rasmussen *et al.* *Proc. Conference on Future Energy Concepts*, Instn. Electrical Engrs, London, 1981.
"The economics of fuel and chemicals from coal in the future", S. P. S. Andrew. *Chem. Engr.* (345), 414 (1979).
"Economics of synthetic hydrocarbon production from oil sands, oil shale and coal", Paper 2. Coal as an oil or gas replacement *Proc. 10th World Petroleum Congress*, Vol. 3, Bucharest, 1979.

"Long range energy problems and perspectives", T. P. Gerholm. Position Paper RT6, *World Energy Conference*, Munich, 1980.
"The use of non-fossil derived hydrogen; a possible role for coal conversion", J. S. Harrison and others. *Proc. 3rd World Hydrogen Energy Conference*, Tokyo, 1980.

(h) Environmental Impact

Sulfur Oxides: current status of knowledge, Electric Power Research Inst. Report EA-316, 1976.
"Polynuclear aromatic contamination from coal conversion processes", D. H. France. *Jounral Inst. Energy* **52**, 169 (1979).
Carbon dioxide and the greenhouse effect, I. Smith. IEA Coal Research, 1978.
Nitrogen oxides from coal combustion — abatement and control, G. F. Morrison. IEA Coal Research, 1980.
Nitrogen oxides from coal combustion — environmental effects, I. Smith. IEA Coal Research, 1980.
Trace elements from coal combustion — atmospheric emissions, M. Y. Lim. IEA Coal Research, 1979.
Symposium on Environmental Control in Synfuels Processes, American Chem. Soc., Division of Fuel Chemistry, Houston, Texas, Preprints, 1980, Vol. 25, (2).
Environmental and climatic impact on coal utilization, J. J. Singh and A. Deepak (eds.). Academic Press, 1980.
Environmental effects of utilising more coal, F. A. Robinson (ed.). Royal Society of Chemistry, Special Publication No. 37, 1980.

(i) Combined Processes

"Coalcom — a prognosis for coal in an integrated fuel technology", D. W. Horsfall. *Mining Magazine* **136**, 452 (1977).
"Feedstocks from coal: how? when?", J. B. O'Hara *et al. Hydrocarbon* **57**, (11), 117 (1978).
"Project POGO — a coal refinery", J. B. O'Hara *et al. Chem. Engineering Progress* **74** (8), 49 (1978).
"The coal-based chemical complex", J. P. Leonard and M. E. Frank. *Chem. Engineering Progress* **75**, (6), 68 (1979).
"Synthetic gas or chemicals from coal; economic appraisals", J. P. Leonard. *Chem. Engineering* **86**, (7), 183 (1979).

STATUS AND POLICY: FUTURE COAL UTILISATION

Availability of world energy resources, 2nd edn, D. C. Ion. Graham and Trotman, 1980.
"The nuclear issue as seen by a competitor", L. Grainger, *Energy Policy* **4**, 322 (1976).
"Coal and nuclear power", L. Grainger. *Coal and Energy Quarterly* (12), 9 (1977).
Energy in western Europe — vital role of coal, CEPCEO 1977.
World energy outlook: a reassessment of long-term energy developments and related policies, OECD, 1977.
Energy: global prospects 1985–2000, Report of the Workshop on Alternative Energy Strategies, C. L. Wilson. McGraw Hill, 1977.
"The future for coal in western Europe", J. Dunkerley. *Resources Policy* **4**, 151 (1978).
Energy policies and programmes of IEA countries: 1979 Review, International Energy Agency, OECD, 1980.
Coal: 1985 and beyond, UN Economic Commission for Europe. Pergamon, 1978.
Steam coal and energy needs in Europe to 1985, A. W. Gordon. Economist Intelligence Unit, 1978.

Steam coal prospects to 2000, International Energy Agency. OECD, 1978.
International Coal Technology, Summary Document. US Department of Energy, 1978, HCP/P-3885.
Programme for Energy Research and Technologies 1977–1980, Annual Report, 1979.
Project Management for Energy Research (PLE) KFA, Jülich on behalf of Federal Minister of Research and Technology of FDR.
Fossil Energy Programme, Summary Document. US Department of Energy, DOE/FE-0006, 1980.
The potential contribution of coal to world energy futures, L. Grainger. American Chem. Soc., Division of Fuel Chemistry, Honolulu, Preprints, 1979, Vol. 24, (1), 40.
Coal-Bridge to the Future. Report of the World Coal Study, C. L. Wilson. Ballinger Publishing Company, 1980.
"Prospective energy production", C. I. K. Forster. Position Paper RT6, *World Energy Conference*, Munich, 1980.
"The growth of coal as an international commodity", M. J. Parker. Position Paper RT4, *World Energy Conference*, Munich, 1980.
"Coal as an international commodity", C. L. Wilson. Position Paper RT4, *World Energy Conference*, Munich, 1980.
"Coal conversion, coal preparation and coal utilisation", *Proc. 5th International Conference on Coal Research*, Düsseldorf, 1980, Vol. 2.
Report of the IEA Coal Industry Advisory Board, OECD December, 1980.
Proc. Conference on Future Energy Concepts, Instn. Electrical Engineers, London, 1981.
Energy in a finite world, International Inst. for Applied Systems Analysis. Ballinger, 1981.

SUBJECT INDEX

Acetylene 278–279
  economics from coal 385
Advanced combustion systems 117–119
Ammonia
  from coal carbonisation 152,258
  from coal gasification 258,262
    economics of 378 *et seq.*
Analysis and testing 8–13
  Audibert Arnu Dilatometer Test 12
  Crucible Swelling Number 12
  Gray King Test 12,13
  Roga Index 13
Ash handling and disposal
  environmental impact in combustion 417
ASTM Classification 13,14,16
Audibert Arnu Dilatometer Test 12

Babcock Power Ltd. (BPL) 87,91
Babcock Renfrew boiler 87–88
Bergbau-Forschung GmbH 103,150,225
BIGAS Process 179–180,196,360
Briquetting 145–152
  cold briquetting processes 148–149
  hot briquetting processes 150
    economics of 339–340,341,343
British Gas Corporation 161,172,197,475
British Gas Slagging Gasifier
    172,197,206,267,381
B.S. Swelling Number 13

Carbon dioxide
  environmental impact in combustion 415–416
Carbonisation of Coal 58–59
  active carbons from 153–157
    Thomas Ness plant 155
    effluent treatment 157
  by-products 152–153
    economics of: see Economics of coal utilisation
    tar distillation 153
  coke (slot) oven design and practice 136–137,150–152
    historical 136
    modern 150–152
  coke quality 137–141
    behaviour in blast furnace 139

Carbonisation of Coal
  coke quality continued
    behaviour in cupolas 140
    coke from coal blends 140
    coke testing 139–140
    CRE methods 140
    drum tests 139
  economics of: see Economics of coal utilisation
  environmental aspects 157–160
  fluidised bed 145,178,243,370
    role in "Coalplex" 431
  formed coke 147–150
    economics of 339–340
  low temperature carbonisation processes 143–145
    "Rexco" 144
    "Coalite" 144
    "Homefire" 145
  medium temperature carbonisation processes 142,145
    "Phurnacite" 145
  scientific basis 130–133
    mechanism of coke formation 133
Carbon products
  active carbon 59,153–157
  carbon fibres 280
  electrode coke 279
Cement kilns:
  combustion in 70
Chemical reactions of coal 30
Chemicals from coal 60
  acetylene 278–279, 385–386
  carbon products 60,153–157,279–280
  economics of: See Economics of coal utilisation
  historical 257–259
  via gasification/synthesis 260–272, 377–385
    ammonia 264–267,270–271,378 *et seq.*
    methanol 196,243,268–271,379 *et seq.*
    OXO alcohols 270
  via hydrogenation 272–274, 385
  via pyrolysis and hydropyrolysis 274–278,385
  via supercritical gas extraction 274,385
Chemical structure of coal 20–30
  aromatic ring size 27

Chemical structure of coal continued
fraction of carbon in aromatic groups (f_a values) 24,26,27
Heat of combustion 24,26
Infrared Spectrophotometry 27
Molar Refraction 26
Molar Volume 24
Molecular Weight distribution 22
Nuclear Magnetic Resonance 27
pyrolysis of aromatic polymers — analogy with coal pyrolysis 29
solubilisation 22
Sound Velocity 24,26
Statistical Constitution Analysis 22,24
X-ray Analysis 22,24,26
Classification of coal 12 *et seq.*, 21
Classification systems 13–17
ASTM 13,14,16
Australia 14
Belgium 14,15,17
Federal Republic of Germany 14,15
France 14,15
India 14
International system 14,15,17
Italy 14,15
NCB 9,14,15
Netherlands 15
Poland 14,15
Seyler Chart 9,13,14
United Nations (ISO) 13,14,17
USSR 13,14
Coal:
definition of 3
distribution of in relation to energy networks 441–452
environmental impact 398–419
fine physical structure 30
grade 9
importance of 31–36,482–486
energy, importance of 31–35
fossil fuels, role of 33–35
non-fossil fuels, inflexibility of 35–36
nuclear power and other energy sources, relation to 435–465
origin of 5–8
petrology and macerals 14,18–21,26 *et seq.*
rank 3,19
research and development 470–482,488
USA reserves and resources 47–48
utilisation 56–61
historical 4–5
relation to world energy strategies 466–488
winning 49–55
world energy demand and potential supply 455
world production 42
world reserves 37–48
distribution of 441–442
world resources 38–48
estimates of 38–40
Coal distribution in relation to energy networks 441–452
distribution of reserves 441–442
economics 443–449
transport costs 442–448
effects of scale and load factor 448–449
Coal mining 49–52
surface 51–52
underground 49–51
Coal/oil coprocessing 435–436
Coal–oil mixtures COM:
combustion of 111–115
economics of 335–336
Coal pipelines: See Transport and Storage
"Coalplexes" 420–437
components of and relationships 421–431
definition of 420
design studies 427–431
Catalytic Inc. 429
Ralph M. Parsons Co. 428
US Steel Corp. 431
economics of 431–435
establishment of 435–437
coal–oil coprocessing 435–436
evolutionary development 435
centred on synthesis gas 436–437
examples 422–427
Coal preparation 52–53
Coal production, world 92
Coal Research Establishment (CRE) 85,86,88,92,131,140,216,273,345,382,431,475
Coal size 9,11
determination 11
range 9
Coal tar 153
chemicals from 153
distillation 153
Coal Utilisation Research Laboratory (CURL) 86,88,92,103,118,475
Coal winning 49–55
coal mining:
surface 51–52
underground 49–51
coal preparation 52–53
environmental aspects 401
future prospects
transport and storage 53–55
COGAS process 178,197,243,422
Coke: See Carbonisation of Coal
electrode coke 279
Coke-oven gas 152,162
Combustion of coal 56–58, 65–119
advanced systems 117–119
coal-oil mixtures 111–115
economics of: See Economics of coal utilisation
fixed bed 71–76
mechanical stokers 72–75
flue-gase treatment 108–111

Combustion of coal continued
fluidised combustion 76–108,327–329
application of 81–84
"Coalplexes" and 420–431
design considerations 79–81
environmental considerations 78–79,106–111,403–417
fundamental studies 104–106
National programmes 84–104
Federal Republic of Germany 103–104
UK 85–97
US 97–103
Others 104
fundamentals 65–68
low grade coal combustion 115–116
pulverised fuel combustion 68–71
cement kilns 70
cyclone furnaces 70
Clean Coke Process 431
COED Process 178,243,274
COED tars, analysis of 277
economics of 370
in "Coalplex" 428,433
Combustion of low grade coal 115–116
Combustion Systems Ltd. 87
Composite Gasifier 197
Curtiss–Wright Combustor 97,102–103
Cyclone furnaces:
combustion in 70–71

Density 11
Baum jig 12
Domestic appliances 75–76

Economics of coal utilisation
availability and cost of coal 310–317
coal conversion aspects 303–304
economics assessments 305–306
economics of carbonisation 337–344
briquette manufacture 339,343
by-products 340–341
coke ovens 338,339,342–343
formed coke 339–340
economics of chemicals from coal 376–387
acetylene 385
chemicals via synthesis gas 377–385
ammonia 378 *et seq.*
methanol 379 *et seq.*
hydrogenation — liquid solvent extraction 385
pyrolysis and gas extraction 385
economics of combustion and power generation 319–336
coal combustion-based electricity generation 320–331
coal v nuclear power 320–327
fluidised bed combustion 327–329
other advanced systems 329–330
sulphur control (PF stations) 327

Economics of coal utilisation continued
coal–oil mixtures 335–339
domestic combustion 334
industrial combustion 332–334
economics of gasification 345–361
Federal Republic of Germany estimates 355–356
low and medium btu gas and synthesis gas 357–359
other SNG cost estimates 356–357
UK Process Assessment and Development Branch (PADB) studies 345,346–351
improved SNG processes 346–347
Koppers–Totzek 346
Lurgi 346
Power Generation 348–351
US Department of Energy estimates 351–355
Lurgi 354,355
economics of liquefaction 362–375
Federal Republic of Germany assessments 366–368
Sasol 362–363
UK assessment 363–366
US estimates 368–371
Badger Co. 371
Booz–Allen Co. 371
Catalytic Inc. 371
Dept. of Energy 368,370,371
Exxon 370
Mobil 370
Ralph M. Parsons Co. 370
economics of underground coal gasification 388–397
NCB study 388–392
other forecasts and estimates 393–396
Belgium 393–396
USA 393
Energy transport costs 442–448
coal 443–446
electricity 443,447
gas 443,447
Environmental impact of coal utilisation 78–79,106–111,398–419
carbonisation and 157–160
combustion and 78–79,106–111,403–417
ash handling and disposal 417
carbon dioxide 415–417
nitrogen oxides 408–410
particulate emissions 410
radioactive emissions 413–414
sulphur 403–408
trace elements 411–413
trace gases and hydrocarbons 414–415
production, preparation, transport, and 401–403
synthetic fuels manufacture, and 253, 417–418
UK 400–401
USA 399–400
EPRI (Alliance, Ohio) Combustor 100

European Communities:
R&D programmes 480
Exxon Catalytic Gasifier 182,197
Exxon Donor Solvent (EDS) Process 229,238–242
economics of 370

Federal Republic of Germany
coal classification system 14,15
economics of coal based electricity 320–331
economics of liquefaction 366–368
fluidised combustion projects 103–104
gasification projects 177–183,191–201
economics of 354
hot briquetting processes 150
liquefaction projects 225–229,253
nuclear heat 183–186
projects, commercialisation programme 483–484
R&D programme 473–475
underground gasification projects 293–294
Fine-structure of coal 30
pore structure 30
Fischer–Tropsch Process 60,220,221,222,229,247,370,371
in "Coalplex" 428
Fixed bed combustion: See Combustion of Coal
Flue gas treatment 108–111
economic considerations 319,327,328,329
Fluidised combustion: See Combustion of Coal
Ford Foundation 47,311–312,315,484
Fossil fuels, role of 33–36

Gasification of coal 59,161–210,260
basic chemistry of 163–165
economics of: See Economics of Coal Utilisation
gasification systems 201–207
electricity generation 202–205
industrial fuel gas 202
medium btu gas 161,206–207
gasifiers
general features and classification of 165–169
established processes 170–177
Koppers Totzek 173–174
Lurgi 171–172
Winkler 175–176
under development 177–183
BIGAS 179–180,196
British Gas Slagging Lurgi 172,196,197
COGAS 178,196
Exxon Catalytic 182,197
HYGAS 178–179,196–197
Otto-Rummel and Saarberg–Otto 177
Rockwell Hydrogasifier 182–183,197,278
Shell-Koppers 174–175
Texaco 177,196
Westinghouse 180–181
National programmes 191–201
Federal Republic of Germany 191–194

Gasification of coal
National programmes continued
UK 197–198
USA 194–197
Others 198–201
role in "Coalplexes" 422
some common development areas 207–210
treatment of gases prior to use 186–191
purification and Shift Reaction 186–187
acid gas removal 187–188
hot gas clean up 190–191
liquid effluents 190
sulphur recovery and removal 189–190
use of nuclear heat 183–186
Gas treatment before use 186–191
Georgetown University Combustor 99
Geothermal energy:
relation of coal to 461
Gray King Test 12,13
Grimethorpe Combustor 94–97

Hardgrove Index 11
H-COAL Process 228,232–233
economics of 366
Hydrogen production 208
non-fossil derived (NFD) 465
HYGAS Process 178–179,196–197,360

I. G. Farben (modified) Process 219
Ignifluid System 84
Inherent moisture 8
International Conference on Coal Research (ICCR) 482
International Energy Agency (IEA) 46,94,311,316,325,327,328,357,374,412,447,468–470,476,480–482
International System of Coal Classification 14,15

Japan:
gasification programme 198
liquefaction programme 250
R&D programme 478

Koppers–Totzek Gasifier 169,170,173–174,198,208,263,264,267,345,346,357,378,379,382

Liquefaction of coal 60
by gasification and synthesis 211–212
by degradation 212–218
pyrolysis 212–213
solvent extraction and hydrogenation 214–218
supercritical gas extraction 216–218,274
economics of: See Economics of coal utilisation
environmental aspects 256
historical 218–220
projects and test programmes 221–251

Liquefaction of coal
projects and test programmes continued
Australia 250–251
Federal Republic of Germany 225–229
Japan 250
Poland 250
Russia 250
South Africa 221–225
UK 248–250
USA 232–248
research and supporting technology 251–256
role in "Coalplexes" 422
Low btu gas 161,162,168,194,197,357–359
Lurgi Gasifier 169,171–172,191,198,212,221,265–267,346,347,352,355,356,358,359,360,364

Lurgi "Ruhr 100" Gasifier 194

Macerals 3,6,7,14,18–20,21,26,27,28
identification 18
reflectance, Stopes-Heerlen system 18
role in solvent degradation of coal 215
Mechanical properties of coal 11
tensile strength 11
grindability 11
Hardgrove Index 11
Mechanical stokers: See Combustion of coal
Medium btu gas 59,161,182,206–207,357–359
Methanol 196,243,261,263,268–270,370,379,380,381,382
ICI Low Pressure Process 269
Lurgi Process 269
economics from coal 379,380,381,382
MHD (magneto-hydrodynamics) 117
Mobil Process 196,232,243–247,270,370
economics of 370
Modderfontein Plant:
Koppers–Totzek gasifiers 173,264,378

NCB Classification system 9,13,14
Nitrogen oxides:
environmental problems 106,109,110,111,408–410
Non-fossil energy sources:
inflexibility of 35–36
Nuclear heat 183–186,356
Nuclear Magnetic Resonance of coal 27
Nuclear Power and other energy sources:
relation of coal to 453–465
conservation of energy 463
energy demand/supply balance 454
fusion 459
geothermal 461
hydraulic 462
nuclear, limitations of 35,320–327,457–459
oil and gas 455–457
solar 459–461
tidal 462

Nuclear Power and other energy sources:
relation of coal to, continued
other energy supply and conversion proposals 464–465
heat pumps 464
Nuclear Long Distance Energy (NLDE) 464
Non fossil derived (NFD) hydrogen 465

Oil and gas:
reserves and resources, relation of coal to 455–457
Origin of coal 5–8
OXO alcohols 270

Packaged boilers 88–91
Particulate emissions:
environmental impact in combustion 410
Petrology of coal 3,14
and rank 19
Physical properties of coal 8
POGO (Power — Oil and Gas — Other):
"Coalplex" 428–429,434,435
Power Generation
coal v nuclear 320–327,457
economics of coal based 320–331,348–351
economics of coal-oil mixtures in 335–336
integration in "Coalplexes" 421–431
Pressurised fluidised bed combustion 79,82–84,92–93,97,102,103,117
economic potential 327–329
Producer gas 162
Proximate analysis of coal 8
Pulverised fuel combustion: See Combustion of coal
PVC (Flick) Process 193
Pyrolysis 212–213,243,253,274–278,385
flash pyrolysis 243,274

Quality of coal 8

Radioactive emissions:
environmental impact in combustion 413–414
Research and Development 470–482,488
international collaboration 479–482
European Coal and Steel Community 480
International Conference on Coal Research (ICCR) 482
International Energy Agency (IEA) 46,94,311,316,325,327,328,357,374,412,468–470,476,480–482
National programmes 470–479
Federal Republic of Germany 473–475
UK 475–476
USA 470–473
Others 476–479
Rheinische Braunkohlenwerke AG 176,194,228
Rivesville Combustor 97,99
Rockwell Hydrogasification Process 182–183,197,278

Roga Test 12
Ruhrchemie AG 192,220,229
Ruhrgas AG 191,192
Ruhrkohle AG 103,191,192,194,227,229
Russia
coal classification system 13,14
liquefaction projects 250
R&D programme 479
underground gasification projects 290–291,298

Saarberg Otto Gasifier 177,192,193
Saarbergwerke AG 104,177,192,225
Sasol
economics of 362–363,374
Lurgi gasifiers 172,198,212,220,221
Sasol 1 221,222
Sasol 2 221,222
Sasol 3 221
Seyler Chart 9,13,14,215
Shell–Koppers Gasifier 174–175,193,206,264,358,379
SNG 161,165,167,171,172,177,178,179,182,197,201,253,265,306,308,345,361,364,371,422,427,428,429,431,432,433
Solar Energy 459–461
Solvent extraction and hydrogenation 212,214–218,227–229,248
carbon fibres from solvent extracts 280
chemicals manufacture 274,385
economics of liquefaction 362–375
electrode coke from solvent extracts 279
environmental impact of synthetic fuels manufacture 248,256,417–418
role in "Coalplexes" 422
supercritical extraction 216–218,248,249,274,371
SRC Processes 235–239,272
economics of 370,371
role in "Coalplexes" 422,428,429,431,432,433,435,436
Stal-Laval Turbine Co. 104
STEAG AG 104,193,227
Stopes-Heerlen System 18,21
Sulphur in coal 9
environmental problems of SO_2 403–408
Synthesis gas 59,161,171,194,196,211,260–264,357–359,377–385
as basis of "Coalplexes" 436–437
routes to chemicals 197,247,263,377–385
Synthetic Fuels Corporation (SFC) 482–483
SYNTHOIL Process 233

Texaco Gasifier 177,196,263,267,382
Cool Water Gasification Programme 196
gasification of liquefaction residues 233
W. R. Grace Industrial Fuel Gas Demo. Plant 196

Thiessen — Bureau of Mines Nomenclature (USA) 21
Trace elements:
environmental impact in combustion 411–413
Trace gases and hydrocarbons:
environmental impact in combustion 414–415
Transport and storage 53–55
TVA (Tennessee Valley Authority) 101,102

U-GAS Process 196
Memphis Light, Gas and Water Fuel Gas Project 196
UK:
coal classification system 13,14,15
economic potential of fluidised bed combustion 327–329
economics of coal-based electricity generation 320–331
economics of gasification 346–351
economics of liquefaction 363–366
economics of underground gasification 388–397
fluidised combustion projects 85–97
gasification projects 197–198
liquefaction projects 248–250
R&D programme 475–476
underground gasification projects 291–293
Ultimate analysis of coal 8
Underground coal gasification 281–300
economics of: See Economics of coal utilisation
environmental impact 297–298
general principles 282–285
historical 281–282
instrumentation and control 298–300
National programmes 285–296
Belgium 295–296
Federal Republic of Germany 293–294
Russia 290–291
UK 291–293
USA 285–290
other possible *in situ* methods 300
utilisation of gas 298
United Nations System (ISO) 13,14,17
USA:
coal classification system 13,14,15
economic potential of fluidised bed combustion 328
economics of gasification 351–355
economics of liquefaction 368–371
economics of underground coal gasification 393
fluidised combustion projects 97–103
gasification projects 194–197
liquefaction projects 247–248
R&D programme 470–473
underground gasification projects 285–290

VEW AG Process 193

WAES (Workshop on Alternative Energy Strategies) 46,47,467,469
Water gas 162
Westinghouse Process 180–182
Winkler Gasifier 169,175–176,267,382
  High Temperature Winkler 176,193
WOCOL (World Coal Study) 46,312,467–468
WEC (World Energy Conference)
  1977 33,40–44,311,454,463,466–467
  1980 44–48,310,454,455,457,468
World energy strategies, coal utilisation in relation to 466–488

World energy strategies continued
  existing predictions and policies for coal 33,37–48,310–317,454,465, 467–470
  IEA 311,316,468–470
  WAES 467,469
  WEC 33,38–48,311,312,316,454,455,457,463, 466–467,468
  WOCOL 46,312,467–468
  research and development: See Research and Development
  some current initiatives 482–484
  strategic importance of coal and its economic impact 484–486

X-ray analysis of coal 22,24,26

D0915270

DOES SPELLING MATTER?

"[A] lucid and fascinating study."

—Patrick West, Times Literary Supplement

"This book is a sane, comprehensive and authoritative lesson in why we spell the way we do and why, in order to preserve the richness, subtlety and history of our language, it is right that we keep doing so."

—Simon Heffer, New Statesman

"There's plenty of interesting information here…Horobin is a sane, sensible guide."

—Henry Hitchings, The Guardian

DOES SPELLING MATTER?

Simon Horobin

Great Clarendon Street, Oxford, OX2 6DP,
United Kingdom

Oxford University Press is a department of the University of Oxford. It furthers the University's objective of excellence in research, scholarship, and education by publishing worldwide. Oxford is a registered trade mark of Oxford University Press in the UK and in certain other countries

First Edition published in 2013

Impression: 2

British Library Cataloguing in Publication Data
Data available

ISBN 978-0-19-966528-0

Printed in Great Britain by
CPI Group (UK) Ltd, Croydon, CR0 4YY

'Pay attention, girl,' she said at last. 'I want you to tell me how you spell "seasick".'

'With the greatest of pleasure,' said Pippi. 'S-e-e-s-i-k.'

Miss Rosenbloom smiled sarcastically.

'Oh,' she said, 'the spelling book has different ideas.'

'It's jolly lucky, then, that you asked me how *I* spell it,' said Pippi. 'S-e-e-s-i-k, that's the way I've always spelt it and it never did me any harm.'

Astrid Lindgren, *Pippi in the South Seas* (1959)

For Lucy and Rachel, for sharing with me
the joys and frustrations of learning to spell.

Contents

List of figures viii
Phonetic symbols ix

1 Introduction 1
2 Writing systems 16
3 Beginnings 39
4 Invasion and revision 77
5 Renaissance and reform 109
6 Fixing spelling 144
7 American spelling 185
8 Spelling today and tomorrow 212

Further reading 253
Bibliography 256
Word Index 261
Subject Index 269

List of figures

2.1 Bell's Visible Speech applied to English 28
3.1 Scenes from the Franks Casket 44
3.2 The Lindisfarne Gospels 46
3.3 The *Beowulf* Manuscript 71
6.1 The Shavian alphabet 178
6.2 The Initial Teaching Alphabet 182
7.1 Benjamin Franklin's reformed alphabet 189

Phonetic symbols

Consonants

/t/ tip
/d/ dip
/k/ cat
/g/ got
/f/ fly
/v/ very
/θ/ thank
/ð/ then
/s/ sat
/z/ zip
/ʃ/ ship
/ʒ/ beige
/h/ hand
/x/ Scots loch
/tʃ/ chat
/dʒ/ edge
/m/ man
/n/ not
/ŋ/ ring (southern English accents)
/w/ won
/l/ lip
/r/ run
/j/ yes

Vowels

/i/ happ<u>y</u>
/iː/ sh<u>ee</u>p
/ɪ/ k<u>i</u>t
/y/ French t<u>u</u>
/eː/ caf<u>é</u>
/ɛ/ dr<u>e</u>ss
/ɛː/ French f<u>ai</u>re
/æ/ c<u>a</u>t
/ə/ <u>a</u>dmit
/ø/ French p<u>eu</u>ple
/uː/ f<u>oo</u>d
/ʊ/ g<u>oo</u>d
/oː/ Scots/General American c<u>oa</u>t
/ɔː/ th<u>ou</u>ght
/ɒ/ h<u>o</u>t
/eɪ/ d<u>ay</u>
/əʊ/ southern English c<u>oa</u>t

Chapter 1

Introduction

Orthography is boring. It is a subject for elderly folk who love order, vote Conservative, and always keep their dog on a lead.

(Sauer and Glück, 1995, p. 69)

Despite this claim that spelling is of interest only to elderly Conservative dog-owners, it is a topic that has been hotly debated for centuries and continues to provoke passionate arguments today. In the eighteenth century Lord Chesterfield, patron of Dr Johnson's *Dictionary*, argued that a mastery of spelling was absolutely crucial for any self-respecting gentleman:

I come now to another part of your letter, which is the orthography, if I may call bad spelling orthography. You spell induce, ENDUCE; and grandeur, you spell grandURE; two faults of which few of my housemaids would have been guilty. I must tell you that orthography, in the true sense of the word, is so absolutely necessary for a man of letters; or a gentleman, that one false spelling may fix ridicule upon him for the rest of his life; and I know a man of quality, who never recovered the ridicule of having spelled WHOLESOME without the w.

(*Letters to his Son*, 19 November 1750)

This may sound like an extreme and outdated view, but not to Dan Quayle, former Vice-President of the United States of America, who made a spelling mistake that was to haunt his political career. In 1992 he presided over a spelling bee at a school in New Jersey, where he read out words written on flashcards which the children then wrote on the blackboard. One of the words he was given to read out was *potato* and twelve-year old William Figueroa duly wrote the word on the board. 'You've almost got it', replied Quayle, 'but it has an "e" on the end'. The boy corrected his effort, obediently adding the additional <e>. There is, however, no <e> on the end of *potato*; an <e> is added in the plural *potatoes*, but it is not found in the singular.

Dan Quayle's error earned him worldwide derision and his howler was front-page news throughout the USA and the UK. Speaking on the David Letterman show a few days later, William Figueroa (now known as the 'potato kid') said that he knew Quayle was wrong, but thought he should follow the Vice-President's instructions. He even went as far as labelling Quayle an idiot, questioning whether you have to go to college to become Vice-President. The blunder was alluded to in an episode of *The Simpsons*, which opens with Bart Simpson writing 'It's potato, not potatoe' on the blackboard. Dan Quayle evidently considered it a defining moment of his political career; he and George Bush Senior were voted out of office less than five months later. When he launched his unsuccessful bid for the presidency in 2001, his largest obstacle was termed the 'potato factor'. He devoted an entire chapter of his autobiography to the incident, claiming that the word was incorrectly spelled on the flashcard, and complaining at the media's handling of it. But this was not Dan Quayle's first orthographic blunder: the family Christmas card of 1989 reportedly carried the slogan 'May our nation continue to be the beakon of hope to the world', with the misspelling *beakon* for *beacon*.

The potato incident is a nice example of the way that spelling mistakes are treated today. There's a tendency to view correct spelling as an index of intelligence, moral fibre and general trustworthiness. People who can't spell properly are considered to be ignorant and slovenly, and certainly shouldn't be trusted with running the free world. But it's perfectly possible to be very intelligent and a poor speller, just as it's perfectly possible to spell accurately and not be terribly bright. After all, correct spelling is really as much a question of rote learning as intelligence. In the case of Dan Quayle, I suspect that the appeal of this gaffe to the media was as much about the word itself as the misspelling; as Quayle himself said of the incident: 'it seemed like a perfect illustration of what people thought about me anyway'. He could, of course, have tried to defend his misspelling, by pointing out that this used to be an acceptable variant spelling, and that the plural does have an <e>. What's more it's a very common spelling mistake; try typing it into an internet search engine and see how many hits you get.

A similar gaffe that made headline news in 2001 was made by then Prime Minister Tony Blair, when he misspelled *tomorrow* three times in a memo. There was a characteristic attempt by the New Labour spin doctors to cover this up, blaming it on his flamboyant handwriting, which was prone to extra misleading loops, but the PM finally owned up that he 'had a blind spot about the spelling of "tomorrow"':

> There was . . . a very lame attempt by my press office to suggest it was just my writing. I regret I will have to put my hands up fully and say it was indeed my spelling that was at fault.
>
> (http://news.bbc.co.uk/1/hi/uk_politics/1670494.stm)

But what is the significance of a spelling error? Should we be worried that our world leaders cannot spell words like *potato* and *tomorrow*? Is spelling such a reliable indication of intelligence and moral worth that we should be using it to judge the competence of our political leaders?

Even our great literary authors are not safe from censure of their spelling habits. Comments by the editor of an online edition of Jane Austen's unpublished manuscripts about Austen's attitude to style and punctuation led to a series of sensational articles in the press, expressing shock at Austen's inability to spell. On 23 October 2010, articles appeared under the headlines: 'Jane Austen could write—but her spelling was awful' (*The Independent*), and 'How Jane Austen failed at spelling' (*Daily Mail*). Austen's supposed illiteracy was even big news across the Atlantic, with CBS News running with the headline 'Jane Austen couldn't spell'. One of the fascinating aspects of this response is the way that journalists ignored the comments about style, focusing instead upon the relatively trivial matters of spelling and punctuation. Should such details matter? After all, these were unpublished manuscripts, not intended for publication. While we have an obsession with the idea of correct spelling, and judge it an important index of intelligence and education, is it appropriate to apply those same standards to an author writing in the early nineteenth century? Furthermore, standards of spelling have changed over time, so that what we consider to be incorrect today may well have been viewed as acceptable in the past.

But while correct spelling might appear to be an arbitrary and irrelevant social yardstick, there is an economic argument for the importance of correct spelling. Charles Duncombe, an entrepreneur with various online business interests, has suggested that spelling errors on a website can lead directly to a loss of custom, potentially causing online businesses huge losses in revenue (BBC News, 11 July 2011). This is because spelling mistakes are seen by consumers as a warning sign that a website might be fraudulent, leading shoppers to switch to a rival website in preference. Duncombe measured the revenue per visitor to one of his websites, discovering that it doubled once a spelling mistake had been corrected. Responding to these claims, Professor William Dutton, director of the Internet Institute at Oxford University, endorsed these conclusions, noting that, while there is greater tolerance of spelling errors in certain areas of the Internet, such as in email or on Facebook, commercial sites with spelling errors raise concerns over credibility. Online consumers' concerns about spelling mistakes on websites are understandable, given that poor spelling is specifically highlighted in advice on detecting potentially fraudulent email, so-called 'phishing'. Along with technical issues such as fake weblinks and forged email addresses, such lists include the use of generic greetings, such as 'Dear Customer', and poor grammar and spelling. This may be accidental, or it may be a deliberate attempt to bypass spam filters, perhaps by using strange spellings such as *pass.wrd* or *passw0rd*. So the message is clear: good spelling is vital if you want to run a profitable online retail company, or be a successful email spammer.

While bad spelling tends to incur heavy condemnation in modern society, good spelling is seen as being a highly praiseworthy virtue, as witnessed by the success of the spelling bee. The huge popularity of spelling bees in the USA is focused on the Scripps National Spelling Bee, held in the Grand Hyatt Hotel, Washington DC, where 265 children, selected from some ten million contestants who take part in regional spelling bees, compete for $20,000 in total prize money. The contest is broadcast live via ESPN and the winner becomes an overnight celebrity, typically appearing on numerous chatshows and even meeting the president. The first round involves a written test of twenty-five random words designed to reduce the number of competitors quickly and efficiently. Round two begins the oral challenge, in which contestants have to sound out a word letter-by-letter. Here words are taken from a booklet produced by the organisers, known as the *Paideia*, which contains around 3,800 words. The restricted, though still quite extensive, list from which these words are taken, encourages many contestants to memorize the booklet in its entirety. While such impressive feats of memory are also crucial to success in the subsequent rounds, contestants can no longer rely on their exhaustive recall of the *Paideia*. As well as memorizing huge numbers of irregular spellings, the best spellers have an extensive understanding of word roots and etymologies.

The words themselves, however, are unlikely to be ones that they will find themselves writing much in later life. The list of winning words on the Scripps Spelling Bee website includes such obscure terms as *autochthonous* (2004), *appoggiatura* (2005), *Ursprache* (2006), *serrefine* (2007), *guerdon* (2008), *Laodicean* (2009), *stromuhr* (2010). Such arcane terms are considerably more taxing than the word that led to Charlie Brown's elimination from the Scripps Spelling Bee in the film *A Boy named Charlie Brown*. Having won his school spelling bee in a rare moment of success, Charlie Brown crashed out of the national contest when he incorrectly spelled *beagle* 'b-e-a-g-e-l', all the more embarrassing of course since his dog Snoopy is a beagle. The Scripps Spelling Bee began in 1925; comparing recent winning words with those that catapulted the earliest victors to fame is striking. In 1925 Frank Neuhauser won the competition by correctly spelling *gladiolus*; subsequent winning words included such relatively straightforward spellings as *fracas*, *knack*, *torsion*, and *intelligible*. In addition to their relative orthographic simplicity, these words also differ from more recent winning words in being ones that competitors are likely to have heard of and be able to spell without special study. None of the recent winning words could be considered common; all are derived from foreign languages and belong to specialized and technical registers. *Appoggiatura* is a musical term derived from Italian, *serrefine* is a French-derived medical term to describe a surgical clip, while *Ursprache* is a German philological term meaning 'proto-language'. *Stromuhr*, a device used to measure the amount and speed of blood flow through an artery, successfully spelled by the 2010 winner Anamika Veeramani, is so specialized that it does not appear in the *Oxford English Dictionary*.

The BBC's version of this popular American phenomenon was *Hard Spell*, a show broadcast on BBC1, in which 100,000 British school children competed to win the title of *Hard Spell* champion. A spin-off show, *Star Spell*, featured twenty celebrities who competed against each other in an attempt to become the nation's Star Speller. The winner of the 2004 *Hard Spell* competition, thirteen-year old Gayathri Kumar, successfully spelled words like *troglodyte*, *disequilibrium*, *nyctophobia*, and *subpoena*, to defeat fellow finalist Nisha Thomas, who stumbled over the spelling of *dachshund*. Being able to spell such obscure words is indeed an impressive achievement, but one wonders how useful this skill will prove in later life other than as a dinner party trick. A knowledge of the spelling of these words is, of course, quite different from an understanding of their meaning and an ability to use them in correct contexts. Interviewed for the BBC following her success, Kumar explained that she prepared by learning lots of specialized plant, food, and medical terms, highlighting the importance of a good memory to success in spelling.

Because of its tacit support for the English spelling system and its idiosyncrasies and anomalies, the Scripps Spelling Bee is regularly the target of orthographic protests. In 2010 protesters turned up dressed in full-length yellow and black bee costumes, distributing leaflets calling for spelling reform, and badges proclaiming: 'enuf is enuf, but enough is too much', 'I'm thru with through', and 'I laff at laugh'. Are these protestors right? Should we really be maintaining such an unnecessarily complex spelling system, and rewarding correct spelling in the ways we do? The following are some of the most commonly misspelled words in the English language: *accommodate*, *embarrassment*, *occasionally*, *supersede*, *separate*, *desiccate*. When I lecture on this topic to undergraduates at Oxford I read out these words and ask them for the correct spelling. It is striking how many students misspell at least half of these common words.

Some readers may take this as further evidence for the decay in educational standards also witnessed in the annual inflation of A-level grades. But it seems to me that what this shows is that spelling is not a reliable index of intelligence. After all, Oxford undergraduates are amongst the highest achievers for their age group in the country. Many intelligent people struggle with English spelling, while others will find it comparatively easy to master. Learning to spell correctly requires remembering numerous unusual and peculiar spelling forms. Some people are just better at this form of rote learning than others. As Mark Twain provocatively observed in a speech at the opening of a spelling contest in Hartford Connecticut in 1875: 'Some people have an idea that correct spelling can be taught, and taught to anybody. That is a mistake. The spelling faculty is born in man, like poetry, music, and art. It is a gift; it is a talent. People who have this talent in high degree need only to see a word once in print and it is forever photographed upon their memory. They cannot forget it. People who haven't it must spell more or less like thunder, and expect to splinter the dictionary wherever their orthographic lightning happens to strike'.

Anyone who has struggled with the irregularities and eccentricities of English spelling will have some sympathy with Twain's view. Children today learn to read using the method known as 'synthetic phonics' in which the individual letters of the word are sounded out and then run together to formulate the word. This works well in cases where individual letters map onto sounds in a straightforward way, as in words like *cat*, *dog*, *big* and so on. But the relationship between letters and sounds is often more complicated than this. For example, the simple rule that maps the letter <a> onto the sound in *cat*, is fine for words like *apple* and *van*, but what about the <a> in words like: *hate*, *admit*, *car*? The letter <c> has the hard /k/ sound in words like *cat* and *castle*, but it can also represent the sound /s/ in *city* and, when followed by <h> in words like *cheese*, it spells /tʃ/. The letter <g> has the hard /g/ sound in *goat*, but a soft sound in *gentle*; in the word *though* it has no value at all. The principle of a system in which sounds map onto single letters is further disrupted in English by the use of digraphs: pairs of letters that represent single sounds. So the sound /ʃ/ in *ship* is spelled <sh>, the /θ/ sound in *thanks* is spelled <th>, and the /tʃ/ of *chips* is spelled <ch>. Single vowel sounds can also be spelled with two letters, as in the word *meat* where the long /iː/ is spelled <ea>. Elsewhere, the <ea> digraph can represent a combination of two vowels, a 'diphthong', as in the words *hear* and *gear*; in some instances, such as in the words *tear* and *read* it can be pronounced in two different ways, giving two completely different words. A single letter may represent more than one sound, as in <x>, pronounced /ks/, while in southern English accents at least, the two letters <ng> represent a single sound, transcribed phonetically using the symbol /ŋ/. This last example raises a further problem about the relationship between spelling and sound, namely the question of accents. For while southern English accents pronounce the word *sing* with a final /ŋ/, northern accents have two phonemes, /ŋg/. Similarly, the initial <h> in words like *have* and *hope* corresponds to the sound /h/ in some southern accents, but many accents of English tend to omit the initial /h/, what is known as 'h-dropping'. The opposite situation is found in words where <r> appears after vowels, such as *card* and *car*. Here, the letter <r> has no sound correspondence for most English speakers, while Scottish, Irish, and many North American, and other 'rhotic' accents, do pronounce the <r>. To make matters more complicated, English spelling further disrupts the alphabetic principle by employing a number of letters that have no phonetic value at all for any speakers, as in the case of the <k> and the <gh> in *knight*, or the <b> in *lamb*. The final <e> in *wife* has no sound value of its own, but does serve a purpose in indicating the sound of the preceding vowel <i>, as can be seen if we compare the pronunciations of *cut* and *cute*. Frustrated by a system so fraught with pitfalls and anomalies, George Bernard Shaw famously argued that it would be possible to spell the word *fish* as *ghoti*, by analogy with spellings such as enou**gh**, w**o**men, mo**ti**on.

Because of these many irregularities and the inconsistent relationship between spelling and sounds, some educationalists have claimed that English spelling handicaps children learning to read English. In contrast, children who learn to read using languages where the relationship between spelling and sound is more transparent, languages like Finnish and Spanish, are considered to have a distinct advantage. This has led to a number of calls for

English spelling to be reformed, to bring letters and sounds into closer alignment. The Simplified Spelling Society, now the English Spelling Society, founded as early as 1908, remains a vocal campaigner for English spelling to be reformed as a means of improving literacy. An alternative suggestion to changing the way we spell to aid learners has been to ignore spelling mistakes when marking examination papers, so that students are not unfairly disadvantaged by the oddities of English spelling. There have even been suggestions from university lecturers that common spelling mistakes found in undergraduate examination papers should be ignored in the marking process. Such suggestions have sparked considerable passion amongst those for whom linguistic standards are sacrosanct. The newspaper *Scotland on Sunday* responded to the proposal of an amnesty on spelling mistakes in university education in the following way:

> THREE words. World. Hell. Handcart. It's not often I despair of civilisation as we know it.... But when a university lecturer proposes an amnesty on students' 20 most common spelling mistakes, I fear things may have taken a turn for the apocalyptic.
>
> (*Scotland on Sunday*, 17 August 2008)

The worldwide success of Lynne Truss's zero-tolerance guide towards the incorrect use of punctuation, *Eats, Shoots and Leaves*, further demonstrates the extent of prescriptive attitudes towards linguistic correctness. The simple misplacement of an apostrophe in the confusion of the neuter possessive pronoun *its* and the abbreviation *it's* is considered a fault worthy of particularly vindictive punishment:

> *This is extremely easy to grasp*. Getting your itses mixed up is the greatest solecism in the world of punctuation. No matter that you have a PhD and have

read all of Henry James twice. If you still persist in writing, 'Good food at it's best', you deserve to be struck by lightning, hacked up on the spot and buried in an unmarked grave.

(Truss, 2003, pp. 43–4)

But is such condemnation of commonly made spelling errors really warranted? The reason that *its* and *it's* are frequently confused is easy to understand, as they bring the two major uses of the apostrophe in English spelling into conflict. In the case of *it's*, the apostrophe is functioning as a marker of an abbreviation as it does in *can't*, *won't*, *he'll*, and so on. But the apostrophe is also employed as a marker of possession, as in *the boy's book*, or *Rachel's dolly*, and so it is natural to assume that the neuter possessive *its* should also have an apostrophe. Insecurity about where to place the apostrophe increasingly leads to its omission, so that we find examples of *its mine*, as well as instances of its insertion in contexts where it has never been employed, as in the case of the 'greengrocer's apostrophe', where the apostrophe is used before a plural -s ending; so-called because it is thought to be particularly prevalent in greengrocers' signs advertising *apple's*, *pear's*, and *orange's*. As Keith Waterhouse notes in his book *English our English* (1991): 'Greengrocers, for some reason, are extremely generous with their apostrophes—*banana's*, *tomatoe's* (or *tom's*), *orange's*, etc. Perhaps these come over in crates of fruit, like exotic spiders' (p. 43).

But, while there may be some confusing and complicated exceptions, is English spelling really so difficult to grasp? Isn't it just a question of mastering a few basic rules and then applying them? Are proposals not to penalize spelling mistakes simply further evidence of the widespread dumbing down of our education system? It is in fact very difficult to generate rules for English spelling, given the large number of exceptions and idiosyncrasies. Most people are aware of the rule 'i before e, except after c', but this actually applies to just eleven of the 10,000 most common English words. While this rule does help with the correct spelling of words like *believe* and *receive*, where the same sound is spelled differently, it doesn't distinguish *siege* and *seize*. Other exceptions include words like *protein*, or *caffeine*, personal names like *Keith* and *Sheila*, as well as words where <ei> represents a different sound, such as *beige*, *eight*, or where it is a plural ending, as in *policies*, or where it represents a diphthong, as in *society*, as well as in words where the <c> represents the sound /ʃ/, as in *ancient* or *efficient*. So, it turns out that the best-known rule for English spelling has a limited application and comes with numerous problematic exceptions.

Similar difficulties bedevil other attempts to devise spelling rules. As an example we might consider the switch from <y> to <i> at the ends of words when suffixes are added. The Penguin Writers' Guide *Improve your Spelling* formulates six separate rules to accommodate the various changes involved here, which depend on whether the <y> is preceded by a vowel or consonant, whether the suffix begins with <a, e, o>, or <i>, or a consonant, and whether the suffix ends with an <s>. This helps to explain why *carry* changes to *carried*, *carrier* but retains the <y> in *carrying*, although it doesn't successfully predict exceptions such as *shyer* (not *shier*), while *dryer* and *drier* are both acceptable. It also explains the preservation of <y> where it is preceded by a vowel in *betray*/*betrayed*, *play*/*played*, but fails with exceptions such as *lay*/*laid*, *pay*/*paid*. Rules formulated to deal with consonants are similarly complex. To predict whether the /s/ sound at the end of a word should be spelled <c> or <s>, the Penguin Guide formulates a series of helpful guidelines to enable certain of these spellings to be successfully predicted, but admits that many such spellings are simply unpredictable. These rules of thumb include the observation that <c> is used after certain vowels, such as *advice*, *choice*, and <s> after others, for example *house*. Nouns related to adjectives ending in <ant/ent> are spelled with a <c>, for example *dominant/dominance*, *evident/evidence*. The Guide also notes that adjectives generally have <s>, for example *dense*, *worse*, although there are exceptions like *nice*, *fierce*, *scarce*. But the majority of such words follow no discernible pattern and must simply be learned individually. There are various aids to learning spellings, such as the mnemonics favoured by schoolchildren. These are valuable methods of enabling children to learn the spelling of frequently misspelled words; as in the mnemonic **B**ig **E**lephants **C**an **A**lways **U**nderstand **S**mall **E**lephants as a cue for the spelling of *because*. But these are not spelling rules and help only in the successful spelling of single words.

But while Mark Twain is right that some people will naturally find spelling easier to learn than others, and while it is true that the present system is full of irregularities making it difficult to formulate rules that can accommodate all the irregularities and exceptions of English spelling, it is at least possible to *explain* these irregularities with reference to the history of the English language. One of the reasons why English spelling is so unpredictable is because its vocabulary consists of many words derived from other languages, which have been adopted with their original spellings intact. Understanding the origins of these words and the languages they have come from will help with spelling them. For example, knowing that *desiccate* and *supersede* derive from the Latin words *siccare* 'to dry', and *sedere* 'to sit' helps to explain the spelling of these two commonly misspelled words. And, rather more simply in the case of Tony Blair's blunder, if you know that *tomorrow* is made up of the preposition *to* and the noun *morrow*, you can work that one out too. One of the goals of this book is to explain the origins and development of our spelling system, to show how a system, which may appear to us as eccentric and irrational, can be understood as the result of a lengthy historical process. This is not a self-help book for bad spellers keen to improve their spelling, but an attempt to explain why English spelling is the way it is. Rather than lamenting the inconsistencies and complexities of English spelling, I want to show how these developed and what they tell us about the fascinating history of our language. Instead of advocating ways of reforming English spelling to make it easier to learn today, I will argue for the importance of retaining it as a testimony to the richness of our linguistic heritage and a connection with our literary past.

Chapter 2
Writing systems

Before we begin to examine the history of English spelling in detail, it will be useful to consider the nature of the alphabetic writing system that it employs. In the previous chapter I noted that the basic principle of the English spelling system is that sounds map onto letters, albeit often in complex ways; in this chapter I want to consider the nature of this sound–letter relationship in more detail. But before I do so, it will be helpful to begin by looking at how this alphabetic principle emerged and its relationship to other kinds of writing system, in which written symbols can stand for different spoken units or even entire words and concepts.

The earliest writing systems were based upon pictograms, in which pictures represented objects such as animals, birds, body-parts, the sun and moon, as in the Egyptian system of hieroglyphics. Egyptian hieroglyphics consist of a large corpus of symbols which represent a series of consonant sounds; they differ from our English alphabet in that there were no symbols to record vowel sounds. There was no single alphabet, in the sense of an established set of signs organized in a standard arrangement; however, modern Egyptologists have assembled a list of some twenty-four symbols that represent those most commonly used for the convenience of modern students of the language. While these twenty-four signs were employed phonetically to represent specific consonant sounds, there were many other signs which functioned as 'ideograms', representing the object itself. In many instances a word may be represented using a combination of these two methods: a series of phonetic signs indicating the pronunciation of a word, followed by a 'determinative', that is a symbol which depicts the object itself, to facilitate decoding.

The Egyptian language and its writing system changed substantially over time; in the seventh century BC a new 'demotic' script emerged which represents a cursive development of the hieroglyphic script. This is one of the three scripts inscribed, along with hieroglyphics and Greek, on the Rosetta Stone, a fragmentary granite tablet whose discovery by French scholars in 1799 led ultimately to the decipherment of hieroglyphics. Although the text that the tablet preserves, a legal decree concerning land ownership, is not of great literary or cultural interest, its linguistic importance is huge. By comparing the representation of the names found in the Greek and Demotic texts, such as that of the king Ptolemaios, with their equivalents in the hieroglyphic text, the French philologist Jean-François Champollion was able to slowly assemble a series of sign–sound correspondences. Building upon this initial breakthrough by extending his findings by comparison with other hieroglyphic inscriptions, Champollion made the major discovery that the symbols could represent both sounds and ideas; he subsequently published his findings in *Précis du système hiéroglyphique* (1824). The Demotic writing system is last recorded in an inscription dated to AD 450; by the second century AD the Egyptian language began to be written using the Greek alphabet, although this needed to be extended to include sounds found in the

Egyptian language not present in Greek. The solution to this problem was to introduce adapted forms of the relevant hieroglyphs; the resultant script came to be known as Coptic.

Our brief survey of Egyptian writing has shown how an initially pictographic writing system changed over time; the symbols which began by depicting a particular object became increasingly stylized so that the picture no longer bore a close resemblance to the object it represented. Another good example of this process is the Cuneiform writing system, which comprises a series of wedges pressed into clay tablets, hence the name, which derives from Latin *cuneus* 'wedge'. The wedge shapes themselves derive from an earlier system of pictograms, which subsequently became stylized to produce an abstract system of wedges. A further development saw individual signs come to represent the name of the object rather than the thing itself, thereby introducing the concept of the phonographic writing system, in which written symbols map onto spoken units. Early examples of phonographic systems were syllabic, in that their constituent signs represented syllables rather than individual sounds. The Linear B writing system, inscribed on a large quantity of clay tablets discovered at Knossos on the island of Crete, is an example of an early syllabary. The writing system itself is a development of an earlier system, known today as Linear A, which was developed to write an unknown language used on the island of Crete in the second millennium BC. Despite attempts to relate this language to other recorded language families, termed 'Minoan', this language appears to have no known extant cognates and texts written in Linear A have yet to be decoded. The Linear B writing system was finally deciphered in the 1950s by an architect called Michael Ventris. Previous attempts to associate Linear B with Greek had failed; Ventris, however, successfully demonstrated that the script was used to write an ancient form of Greek, known as Mycenaean Greek. Rather than representing individual sounds as today, each symbol reflects a syllable constituting a consonant followed by a vowel, *ma*, *da*, *na*, *po*, *to*, *ko*, and so on. Although the Linear B texts themselves represent little more than a random collection of tablets, mostly concerned with accounting, preserved by chance when the temple in which they were stored burned down, thereby baking the clay from which they were made, Ventris's brilliant decipherment allowed new light to be shed on life in Mycenaean Greece.

Although I have characterized the development of writing as a shift from pictograms to alphabetic systems, it is important to emphasize that pictograms are still employed by many of the world's languages, such as Chinese, although Chinese writing is considerably more complex and involves phonetic elements too. In fact, even Modern English relies on logographs (symbols that represent meaningful units rather than sounds) to represent certain concepts, such as our numerals 1, 2, 3, the ampersand & 'and', the pound sign £, and the email sign @ 'at'. These remain convenient shorthand methods of representing ideas which would be much more cumbersome to represent phonetically using the alphabet. Consider how much more long-winded it is to spell out a complex number such as one million, one hundred thousand, two hundred and forty-seven and a half, than to simply represent it numerically: 1,100,247.5. Another advantage a logographic writing system has over a phonetic one is that it can be used by speakers of different languages. Speakers of any language that employs Arabic numerals can understand numbers like 1, 2, 3, irrespective of whether they call them *one*, *two*, *three*, *un*, *deux*, *trois*, *eins*, *zwei*, *drei*, and so on. I remember recognizing the full force of this point when presented with a train timetable in Japan; although

I could make nothing of the names of the various places at which the train would stop, I had no difficulty understanding the times themselves.

In discussing the history of writing systems, I have characterized the English spelling system as phonographic, in that the written symbols map onto sound-segments. However, the relationship between the spoken and written modes is by no means straightforward, and it is important to consider the nature of this relationship before examining the history of English spelling more generally. The principle behind a phonographic writing system is that written symbols, 'graphemes', represent individual spoken units known as 'phonemes'. Thus the graphemes <cot> represent the sounds /kɒt/. If we change the initial sound to /r/, then we change the meaning of the word and thus require a different letter: <r>. It will be apparent from this discussion that angle brackets, <>, are used to indicate graphemes and slash brackets, //, to indicate phonemes. While this might seem fairly straightforward, it is important to be aware that not all spoken distinctions are recorded in the writing system. There are, for instance, various ways of pronouncing, or realizing, the phoneme /r/. Northern dialect speakers, particularly in Leeds and Liverpool, commonly use an alveolar tap, [ɾ], while a different realization is found primarily in the north-east, known as the 'Northumbrian burr', consisting of a voiced uvular fricative, [ʁ]. These different realizations are 'allophones' in that they do not change the meaning of the word. You will have noticed that it is usual for allophones to be indicated using square brackets: []. Because they do not have an impact on meaning, allophonic variations are not encoded in a phonographic writing system. In principle it would be possible to encode such differences, perhaps by using symbols such as [ɾ,ʁ] which belong to the International Phonetic Alphabet, but this would lead to greater complexity for no additional benefit. An important principle of a writing system is that it should be communicatively efficient, that is it should only encode features that are of communicative significance. How an individual speaker pronounces the word *rot* is of little importance when reading a text; the crucial distinction is whether the word is *rot* or *cot*.

The complex nature of the relationship between letters and sounds has been debated for centuries. The medieval understanding of this problem was based upon the writings of the Latin grammarians Donatus and Priscian, who developed the doctrine of the *littera*. Aelius Donatus was a teacher of grammar and rhetoric in the mid-fifth century, whose pupils included St Jerome, author of the Vulgate translation of the Bible. Donatus's grammatical works, *Ars minor* and *Ars maior*, were standard reading in the Middle Ages, to such an extent that any grammatical textbook came to be known as a *donet*. Priscianus Caesariensis, more commonly known simply as Priscian, flourished in the early sixth century AD and wrote a key grammatical work known as the *Institutiones grammaticae*, 'Grammatical foundations', around the year 520. Priscian's work forms the basis for the Anglo-Saxon monk Ælfric's *Excerptiones de arte grammatica anglice*, 'Extracts on grammar in English', a Latin grammar for novice monks, written in Old English. Priscian began by describing what he calls 'voice' (Latin *vox*), dividing this category into four different types: articulate, inarticulate, literate, and illiterate. He then turns to a definition of the letter (*littera*), which he defines as a sound that can be written separately, deriving its name from *legitura*, literally a 'reading-road', the pathway by which people read. In his section on the letter, Donatus listed the letters used to write Latin and their various functions, ending with a definition of their three properties: name (*nomen*), form (*figura*), and force (*potestas*). When translating this doctrine into English in his grammar, Ælfric rendered these concepts *nama*, *hīw* and *miht* and explained them as follows: 'Nama: hū hē gehāten byð (a, b, c); hīw: hū hē gesceapen byð; miht: hwæt hē mæge betwux ōðrum stafum'. This translates as 'Name: how it is called; form: how it is shaped; power: what it has the power to do among other letters'. This is a useful way of thinking about spelling systems in that it enables us to distinguish the uses of a letter, that is the *potestas* 'power', or *potestates* 'powers', with which it is associated, from the form of the letter itself.

The question of how many *potestates* a particular *figura* can be associated with is also one that has been much debated, especially by those who have tried to reform English so that each individual *figura* maps on to a single *potestas*. This medieval understanding of the *littera* is also helpful in enabling us to separate analysis of *litterae* and their *potestates*, from a consideration of the *figurae* themselves, that is the shapes of individual letters, which is technically the province of a distinct scholarly discipline known as palaeography, the study of ancient handwriting, which is concerned with forms of letters and the development of script types. Using this framework, we can distinguish between the *littera* <s>, which can stand for the *potestates* /s/ and /z/ in Modern English, and the various *figurae* with which it has been associated, such as the long-s, <<ʃ>>, found in Early Modern printed texts, or the sigma-shaped form, <<σ>>, used by medieval scribes, or the different fonts of modern computers, such as <<**s**, 𝖘, s>>. As you will have noticed from the above examples, it is conventional to employ angled brackets to signal *litterae*, square brackets for *potestates*, and double angled-brackets to identify *figurae*.

Classical teaching concerning the *littera* implies a blurring of the relationship between letter and sound that would not be appropriate within our modern usage of the word *letter*, which relates solely to a written form. But Classical writers conceived of *litterae* as belonging to a kind of universal alphabet, within which each *littera* refers to a single *potestas*. This is apparent from remarks made by the Roman grammarian Quintilian who complained that Latin is missing the necessary *littera* to represent the /w/ sound in words like *seruus* and *uulgus*, in which the single letter <u> represents both vowel and consonant. This theoretical approach was not followed by medieval scribes, who were much more prodigal than the Romans in their use of *litterae* to represent a variety of *potestates*. However, the concept is revisited in the Early Modern period by certain orthoepists, those concerned with correct pronunciation, and spelling reformers, such as John Hart, who attempted to revise the spelling of English in line with the Classical doctrine whereby each letter should ideally correspond to a single sound, as we shall see in Chapter 4. In attempting to disambiguate *litterae* and speech sounds, Priscian distinguished *litterae* from *elementae*, although the two are often confused by others, and even at times by Priscian himself. As late as 1640, the schoolteacher and author of a spelling book called *Orthoepia Anglicana*, Simon Daines, noted that: 'According to the Etymologie, or strict sense of the term, Letters are but certain Characters, or notes, whereby any word is expressed in writing: and for this cause were they by the antient Latinists distinguished into Letters, as they be Charactericall notes, and Elements, as the first grounds or Principles of speech. But this nicety is confounded in the generall acception, which promiscuously terms them Letters; and this we shall follow' (1640, p. 2). Charles Butler in 1633 refers to 'uncharactered letters', which is clearly a reference to spoken sounds rather than written symbols, while Dr Johnson's definition of *letter* in his *Dictionary* of 1755, 'one of the elements of syllables; a character in the alphabet', appears to combine both spoken and written components. Other ways of distinguishing the speech sound and the written symbol were found, such as John Hart's use of *letter* for writing and *voice* for speech.

Although I have referred to the arbitrary nature of the connection between the letter and the spoken sound to which it refers, or between the *figura* and its *potestas*, attempts have been made by various reformers to devise writing systems in which the two can be linked in some transparent way. One way of rationalizing this relationship has been to deduce some connection between the letter's *nomen* and its *potestas*. While this relationship is apparent in some cases, there are several instances where the connection is decidedly opaque. Thus Alexander Top, in *The Oliue Leafe* (1603), is critical of the 'most improper' names of the letters <h> and <y>, because of the lack of correspondence between the names *aitch* and *wy* and the sounds /h/ and /j/. The desirability of a letter's name being indicative of its sound is apparent from Charles Butler's (1633) criticisms of the name *double-u*, which is 'a name of the forme and not of the force'. This idea continued to find support in a work of 1704, entitled *Right Spelling Very Much Improved*, which states that 'Our Letters should have Names, according to their Sound and Force'. A rather extreme attempt to bring *nomen* and *potestas* into alignment was made by the writer of a work of 1703, entitled *Magazine, Or, Animadversions on the English Spelling*, who goes by the initials G.W., which led him to suggest revising the uses of letters to reflect their names. Thus the letter <h> was given the new sound /tʃ/, thereby reflecting its name *aitch*, while the letter <g> was used to represent /dʒ/, requiring the author to devise new symbols to reflect the sounds /h/ and /g/. The advantages of such names for the purposes of acquiring literacy are apparent in the way that children learning to read today name the letters by their sounds rather than their names: /æ, b, k/; learning the alphabet using the names of the letters, /eɪ/, /bɪː/, /sɪː/, is much less helpful when it comes to learning to read.

Attempts have also been made to devise alphabets to overcome the difficulties caused by this mismatch between *figura* and *potestas*. In the seventeenth century, a number of scholars attempted to create a universal writing system which could be understood by everyone, irrespective of their native language. This determination was motivated in part by a dissatisfaction with the Latin alphabet, which was felt to be unfit for purpose because of its lack of sufficient letters and because of the differing ways it was employed by the various European languages. The seventeenth century also witnessed the loss of Latin as a universal language of scholarship, as scholars increasingly began to write in their native tongues. The result was the creation of linguistic barriers, hindering the dissemination of ideas. This could be overcome by the creation of a universal writing system in which characters represented concepts rather than sounds, thereby enabling scholars to read works composed in any language. This search for a universal writing system was prompted in part by the mistaken belief that the Egyptian system of hieroglyphics was designed to represent the true essence and meaning of an object, rather than the name used to refer to it. Some proponents of a universal system were inspired by the Chinese system of writing, although this was criticized by others on account of the large number of characters required and for the lack of correspondence between the shapes of the characters and the concepts they represent. John Wilkins, in his *Essay towards a Real Character* (1668, p. 375), expressed his view of the desirability of this relationship: ‘there should be some kind of suitableness, or correspondency of the figures to the nature and kind of the Letters which they express', and he set out to devise an alphabet in which the letters express a 'Naturall Character'. Another model that suggested the possibilities of a universal writing system was the emergence of shorthand systems in the sixteenth century, such as Timothy Bright's *Characterie: An Arte of Shorte, Swifte and Secrete Writing by Character* of 1588. Although it was based on an alphabet, users of Bright's Characterie had to learn a symbol or 'charactericall' for every word. Because of the large number of symbols required for such a system, not every word has its own charactericall. Where a symbol is lacking, the user was required to turn to a synonym, supplying the first letter of the intended word alongside. Thus, to represent the word *abandon*, the user must employ the sign for *forsake*, while adding an <a> at the side. If no synonym can be found for the word in question, then an antonym should be selected; a method which was surely fraught with opportunities for serious misunderstanding. Given that many of the scholars engaged in the devising of such systems had mathematical training, the system of algebraic notation and Arabic numerals were also influential factors.

Attempts have also been made to devise writing systems that represent sounds in a more direct way, and which could similarly function as universal systems. In 1867 Alexander Melville Bell, father of Alexander Graham Bell who invented the telephone, published his *Visible Speech*, or 'Self-Interpreting Physiological Letters', which were intended to allow all the world's languages to be written using a single alphabet. This alphabet was based upon a detailed understanding of phonetics, from which an alphabet was devised that would enable all of the possible sounds to be represented in such a way that there would be 'perfect analogy between marks and sounds'. According to Bell, the implications and potential benefits of such a system are on a biblical scale: the illiterate of all countries could be taught to read their own language in 'a few days', the blind could be taught to read, the deaf and dumb to speak, and a universal language could be established so that 'the Linguistic Temple of Human Unity may at some time, however distant the day, be raised upon the earth'. Fundamental to the establishment of Visible Speech is the concept that the significant speech sounds can be represented by a relatively small number of symbols, termed 'radicals', which bear some formal relationship to the sounds they represent. For instance, the sound /h/, made with the throat open, is written O, a sound made with the throat contracted is written 0, while X indicates the closure of the throat. These radicals can then be modified to indicate further phonetic characteristics, such as whether the sound is nasalized, or whether the vocal folds are vibrating and so on. In the case of vowels, various permutations of a single symbol resembling the letter J, written backwards, forwards, upside down, or bisected with a stroke, are used to indicate whether the tongue is high, low, mid, front, back, mixed, and whether the lips are rounded.

This very brief and partial account of a few features of Visible Speech is probably sufficient to indicate its considerable complexity and the high level of phonetic knowledge required to understand how it works. When applying it to the writing of English, Bell recognizes that there will inevitably be disagreements among authorities about the nature of the sounds and their appropriate representation. His solution is to apply the system to 'some approved speakers, and from a comparison of the independent pronunciations of two or three such selected oralists to fix the alphabet for Visible Speech printing'. Bell drew upon his familiarity with the sounds of English to provide a 'Standard English Alphabet', which other 'local' alphabets could follow. Figure 2.1

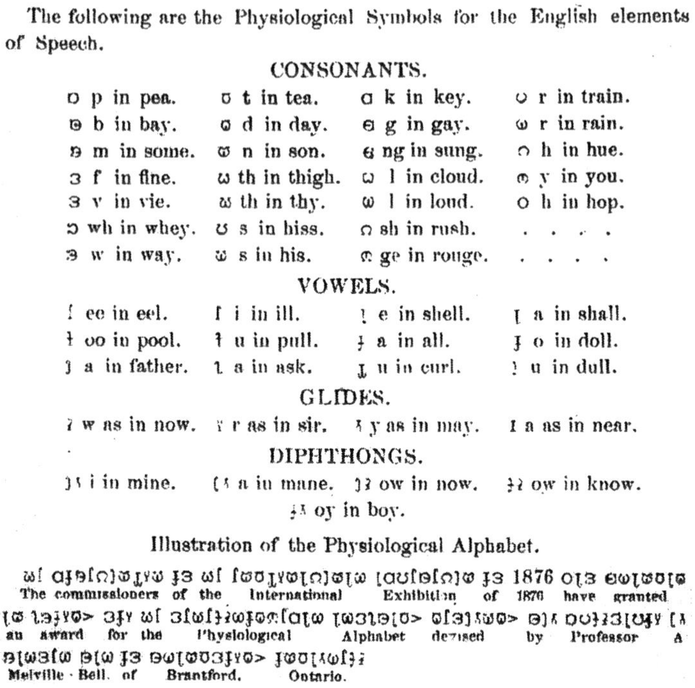

The following are the Physiological Symbols for the English elements of Speech.

CONSONANTS.

p in pea.	t in tea.	k in key.	r in train.
b in bay.	d in day.	g in gay.	r in rain.
m in some.	n in son.	ng in sung.	h in hue.
f in fine.	th in thigh.	l in cloud.	y in you.
v in vie.	th in thy.	l in loud.	h in hop.
wh in whey.	s in hiss.	sh in rush.	. . . .
w in way.	s in his.	ge in rouge.	. . . .

VOWELS.

ee in eel.	i in ill.	e in shell.	a in shall.
oo in pool.	u in pull.	a in all.	o in doll.
a in father.	a in ask.	u in curl.	u in dull.

GLIDES.

w as in now. r as in sir. y as in may. a as in near.

DIPHTHONGS.

i in mine. a in mane. ow in now. ow in know.

oy in boy.

Illustration of the Physiological Alphabet.

The commissioners of the International Exhibition of 1876 have granted an award for the Physiological Alphabet devised by Professor A Melville Bell, of Brantford, Ontario.

Figure 2.1. Bell's Visible Speech applied to English

sets out the results of this attempt to apply the principles of Visible Speech to English. Here you can see how the symbol which resembles an M with rounded sides represents the sound /j/ in *you*; the symbol is designed to indicate the tongue position when the sound is made. The opposite symbol, like a W with rounded sides, indicates the sound /l/, made with the sides of the tongue raised. Following this, Bell supplied a series of lessons which could be adopted to teach the illiterate to read using Visible Speech.

Other examples of artificial alphabets designed to represent phonetic features more directly are the writing systems developed by J. R. R. Tolkien to write the various invented languages in *The*

Lord of the Rings. As we shall see in the next chapter, Tolkien used the Anglo-Saxon runic alphabet in *The Hobbit*; in *The Lord of the Rings* he introduced alphabets of his own invention. Tolkien began devising alphabets early in his life; as early as 1919 he started keeping a diary written in an alphabet that resembled a mixture of Hebrew, Greek, and Pitman's Shorthand. Tolkien returned to the idea in a diary he began writing shortly after the death of his close friend and fellow medievalist, C. S. Lewis. This was a mixture of conventional letters, with different sound values, phonetic characters, and some of the letters he invented for his fantasy fiction. Tolkien referred to it as his 'New English Alphabet', and thought it a considerable improvement on 'the ridiculous alphabet propounded by persons competing for the money of that absurd man Shaw'.

The alphabets used for writing his invented languages in *The Lord of the Rings* are explained in Appendix E to that work. Not only does Tolkien describe these invented scripts and their use, he also gives them a history. Like the Roman alphabet, the scripts and letters of the Third Age have a common origin and are of considerable antiquity. Like the Roman alphabet, they too have their origin in a consonantal script, but by the time of the inscriptions found in *The Lord of the Rings*, they had reached full alphabetic development. There are two main types of alphabets: the Tengwar, meaning 'letters', and the Certar or Cirth, meaning 'runes'. The distinction between these two mirrors the distinction in use between the Anglo-Saxon runic and Roman scripts: the Tengwar was used for writing with a pen or brush, the Cirth were devised specifically for inscribing on hard surfaces. Like runes, the Cirth were originally employed exclusively for scratching names and memorials on wooden or stone surfaces. Over time the Cirth spread and became adopted by a variety of peoples, including

Men, Dwarves, and Orcs, who adapted them to suit the details of their own languages, just as happened with Germanic runes.

An innovative feature of the Tengwar is that the shapes of the individual letters were designed to record phonetic features, meaning that it would be possible to pronounce a word written in the Tengwar without actually knowing what language it was written in. Because of the widespread use of the Roman alphabet for the writing of modern European languages, English speakers are reasonably comfortable with pronouncing words in Spanish, French, German, and so on, without knowing any of those languages. But if we imagine trying to read out a text written in Arabic without any knowledge of its alphabet, then we can appreciate the potential benefits of the Tengwar system. Unlike in the Roman alphabet, the shapes of the Tengwar letters were deliberately designed to highlight links between related sounds. While the arrangement of letters in the Roman alphabet has little bearing on the relationship between the sounds they represent, in the Tengwar the letters were organized according to their sound values. Tolkien was evidently critical of these shortcomings of the Roman alphabet, describing the Tengwar innovations as follows: 'This script was not in origin an "alphabet", that is, a haphazard series of letters, each with an independent value of its own, recited in a traditional order that has no reference either to shapes or to their functions'. He goes on to point out that the only feature of our own alphabet that would have been intelligible to users of the Tengwar is the relationship between <P> and <B>, though the fact that they do not appear next to each other in the alphabet, or near <F>, <M>, <V>, would have seemed 'absurd' to users of the Tengwar. This observation is revealing as to Tolkien's view of the principles of an ideal alphabet. <P> and <B> are closely related letters in form and shape; <P> is constructed from a stem with a single bow, while <B> has an additional bow. This figural relationship mirrors a phonetic connection. The sounds /b/ and /p/ are what are called 'bilabial plosives', sounds made with both lips and a small explosion when the air is exhaled. The difference between them concerns the presence or absence of voicing: /b/ is 'voiced', meaning that the vocal folds are vibrated, while /p/ is voiceless, because the vocal folds are not vibrated. These principles illustrated by <P> and <B> in the Roman alphabet lie behind the construction of the Tengwar. Here, the primary letters are constructed from a single stem and a series of bows, with related shapes appearing close to each other in the alphabet. The names of the letters are based upon their use in Quenya; each name is an actual word, generally beginning with the sound in question, as was the case with the Phoenician letters from which the Roman alphabet derives. So, *tinco* means 'metal', *parma* 'book', *calma* 'lamp', and *quesse* 'feather'. The relationship between letter shapes and sound values was much more haphazard in the Cirth, although, in a later stage in their development, a number of phonetic principles were adopted under the influence of the Tengwar. Thus, the addition of a branch to a letter came to signal the addition of voicing, while the placing of a branch on both sides of the stem indicated both voicing and nasality.

Earlier I suggested that speakers of English, French, German, Spanish, and other European languages have few difficulties pronouncing each others' languages, given their common use of the Roman alphabet. But, while it is certainly true that written English, French, German, and Spanish have considerable similarities in their writing systems, particularly in comparison with a language like Arabic or Chinese, it would be wrong to suggest that they are identical. While these languages all employ the same alphabet, they do not always use it in the same way. For instance, the letter <w> in English is pronounced /v/ in German, while the letter <v> is used to represent the sound /f/. The letter <j> has a particularly wide range of different uses across European languages. While English uses this letter to represent the /dʒ/ sound at the beginning of *jam*, in German it represents /j/ (like the <y> in *yacht*), in French it is sounded /ʒ/ (like the <g> in *beige*), while in Spanish it is pronounced /x/ (a sound not used in English similar to the <ch> in the Scots *loch*).

These examples lead us to another important principle concerning writing systems which we must bear in mind: namely, that the relationship between phonemes and graphemes is conventional. There is no inherent reason why the phoneme /k/ should be represented by the grapheme <k> rather than some other grapheme like <d>. For that matter, it would be equally possible to represent the phoneme /æ/ with the letter <o> and the phoneme /t/ with <g>, with the result that the word /kæt/ would be spelled <dog>. The conventional relationship between letters and sounds becomes more apparent when we recall that in some English words the phoneme /k/ is represented by the grapheme <c>, while <c> can also be used to represent the phoneme /s/, for example *city*, and sometimes even /tʃ/, for example *ciabatta*. The uses of the letters <c> and <k> in English further demonstrate that the principle of individual phonemes mapping onto individual graphemes is an ideal, not necessarily followed in practice. The reason for this is that, throughout its history, English has borrowed words from other languages where different conventions are employed. The word *city* is a French loanword and thus employs the French practice of using <c> to represent the sound /s/, while *ciabatta* is an Italian loanword which preserves the Italian use of <c> for /tʃ/. So one way in which the principle of individual phonemes mapping onto individual graphemes can be disrupted is by borrowing words from other languages and preserving their original spelling forms. Another cause of disruption is sound change. This is best illustrated by a consideration of Modern English spelling, in which a number of words contain graphemes that are not pronounced, such as the <k> and <gh> in the word *knight*. These graphemes were once pronounced but have since ceased to be, so that Modern English spelling contains a number of silent letters. Less commonly, changes in pronunciation can be triggered by the spelling, bringing the two into closer alignment. The introduction of etymological spellings like *adventure* for Middle English *aventure*, or *host* for Middle English *ost*, led to changes in the pronunciation of these words. Changes of this kind are also witnessed by the current pronunciation of the words *waistcoat* and *forehead*, where a pronunciation influenced by the spelling has replaced the older pronunciations /wɛskət/ and /fɒrɪd/ (rhyming with *horrid*). In general, however, while we can describe English spelling as a phonographic writing system in which phonemes map onto graphemes, there are a number of exceptions to this basic principle. One of the tasks of this book will be to examine the reasons why these exceptions have emerged and the various attempts that have been made to remove them.

The above example of the Modern English spelling and pronunciation of the word *knight* raises the question of the relationship between speech and writing. While the spelling system may have been initially designed to reflect speech, it is clearly no longer a faithful record of Modern English pronunciation. But this is the result of standardization of the spelling system, a comparatively recent phenomenon. In earlier periods of the history of English, the relationship between speech and writing was much closer. In medieval texts, for instance, it is common to find different spellings of the same word, like the spellings *ston* and *stan* for the word 'stone', which appear to reflect different pronunciations. But we need to be careful not to assume that all spelling variants are indicative of alternative pronunciations. Spelling variation may have a purely 'graphemic' significance, such as the variation between the Middle English spellings *shall* and *schall*, where it is unlikely that the spelling difference reflects a distinction in pronunciation. While different dialects of Middle English may use different graphemes, or combinations of graphemes, to reflect individual phonemes, they do not necessarily indicate a different pronunciation. A parallel situation exists in the modern distinction between *colour* and *color*, where one spelling is British and another American English, although there is no corresponding pronunciation distinction. A further difficulty with using written language to reconstruct speech is that, even where there are differences in spelling that seem to imply a variant pronunciation, we cannot always determine exactly what the different pronunciation was. In the case of Middle English spellings of the word 'stone', *stan* and *ston*, cited above, we can be fairly sure that these represent different pronunciations of the vowel sound. But there are other spelling differences that seem to imply a spoken difference where the phonetic realization is much less clear. Take the spellings *xal* and *sȝal* for 'shall'. These appear to reflect alternative pronunciations, but what exactly? In many such cases it is simply impossible to reconstruct these pronunciations as we have nothing but the spelling evidence to base our analysis on.

The degree to which a spelling system mirrors pronunciation is known as the principle of 'alphabetic depth'; a shallow orthography is one with a close relationship between spelling and pronunciation, whereas in a deep orthography the relationship is more indirect. It is a feature of all orthographies that they become increasingly deep over time, that is, as pronunciation changes, so the relationship between speech and writing shifts. English has a comparatively deep orthography, whereas Finnish is comparatively shallow. This is in part an accident of history: English spelling began to be fixed in the fifteenth century, whereas Finnish orthography was not standardized until the nineteenth century. This distinction shows that a key reason why English spelling is more irregular than Finnish spelling is to do with the history of their standard varieties. The development of, and attitudes towards, a standard written variety of English are therefore crucial to the question of how English spelling has evolved. Before we look at the development of English spelling in detail, it will be useful to consider what a standard spelling system is and what it is for.

When we talk about English spelling, there is an assumption that we are dealing with a single, fixed, and unified system. This is, of course, largely true of Modern English spelling, although it is important to recognize the artificial nature of this situation. While Modern English is written using a fixed spelling system, in which almost every word has a single correct spelling, this has not always been the case. As we look back over the history of English spelling in the following chapters, we will see the ways in which spelling standards have come and gone, and the lengthy process by which our modern standard has been constructed. Because we are used to having a standard spelling system, it is tempting to believe that it has always been there, and that the individual spellings it supports are somehow inherently correct. But, as we shall see, there is no absolute necessity to have a standard spelling system; indeed, there have been periods in the history of English when there was no standard at all.

Before we consider the processes by which our modern standard spelling system came into existence, it will be useful to consider the function of a standard language. The crucial defining feature of a standard language is its uniformity and resistance to change. This means that the spellings used today should remain identical in the future, thereby ensuring that centuries from now speakers of English will still be able to read works written today. A standard language should also be 'supraregional', that is, it should not be tied to any particular locality. In this way a standard language differs from a dialect, which is a form of language associated with a particular region, such as the Yorkshire dialect. While a child brought up today in Yorkshire may learn to speak with a local accent, and to use features of grammar and vocabulary restricted to that dialect, he or she will learn to write using the same standard spelling system as a child brought up in Devon, Birmingham, or London. A further feature of a standard language is that it is 'elaborated', meaning that it is used for a variety of different linguistic functions. So our standard English spelling system is not just taught to children in schools, it is also used by our government, legal system, and in all printed publications, all of which uses help to reinforce and sustain its continued acceptance as the single acceptable mode of spelling. A further requirement of a standard language is 'codification', that is, its properties should be enshrined in authoritative publications such as grammars and dictionaries.

These features of standard English are all functional: that is to say they are concerned with its use. They are inherently practical, in that having a single system used by everyone, irrespective of upbringing, helps to ensure maximum efficiency of communication. Learning to spell correctly thus becomes a means of ensuring that you are able to participate in this accepted mode of communication: reading texts written using this spelling system, and writing in a way that can be understood by others trained in the same system. However, there is of course another, sociolinguistic, aspect of a standard language, namely its prestige value and its association with correctness and 'proper usage'. The view that standard English is an inherently prestigious variety, better than other regional varieties, is deeply embedded within our society. But this view is a decidedly modern one; as we look back at the history of English spelling we will encounter different attitudes towards standardization, and trace the process by which these ideas of correctness and prestige became enshrined in our attitude towards English spelling.

Despite the dominance of the heavily prescriptive view of language use today, some of the traditional rules of standard English grammar and pronunciation are under threat. The view that it is incorrect to split infinitives by placing an adverb between the 'to' and the verb, as in the famous instance of 'to boldly go', is becoming increasingly obsolete. Despite widespread condemnation of the tendency to omit initial /h/ in the pronunciation of words like *house*, *hand*, and *how*, what is known as 'h-dropping', this is a common feature of nearly every accent of English, and is especially common in colloquial speech. The rules of English spelling differ from those of standard English grammar and pronunciation, in that they are more easily maintained and monitored. In most cases there remains a single correct way of spelling any word, which can be easily identified by reference to a dictionary. By contrast, standard English pronunciation, what is known as Received Pronunciation, or RP, is much less easily defined and maintained. RP is a standard accent in that it is not tied to a particular geographical area, and in that it holds social prestige, but it lacks the fixity and uniformity of standard English spelling. While many people today speak RP, there is considerable variation between their accents, with the result that phoneticians distinguish between Uppercrust, or U-RP, mainstream RP, adoptive RP (spoken by people for whom RP is not their native accent), and near-RP, comprising accents that are similar but not identical to RP. Even within these categories there is considerable fluctuation, dependent upon a range of sociolinguistic and phonetic factors. Another difficulty concerns how to define RP. How do we know when a particular pronunciation is correct or not? There is no equivalent authority to which we can turn, as there is with spelling. Spelling is in fact the most easily defined and regulated aspect of linguistic usage, and therefore the domain that attracts the greatest attention from prescriptivists. While it is comparatively difficult to condemn an individual's pronunciation, incorrect spelling is a much easier target. Yet despite the comparative ease with which spelling can be monitored, incorrect, or alternative, spellings are becoming increasingly common, especially in electronic communications such as email, instant messaging, and text messages. In some cases these spelling variants appear to signal the influence of American spelling, such as *check* instead of *cheque*, *judgment* for *judgement*, while others are abbreviations or shortenings, for example *tho*, *thru*, more suited to the quick-fire exchange of electronic communication. Does this mean that English spelling is being corrupted, and that its hard won and diligently monitored standards are finally in decline? The question of how spelling is used today, and whether the Internet age signals the end of standard English spelling, will be the subject of the final chapter of this book.

Chapter 3
Beginnings

Old English is the term employed to describe the variety of English used by the Anglo-Saxons, which survives in manuscripts and inscriptions written between roughly AD 650 and 1100. During the Old English period there were two writing systems in use: an earlier Germanic system known as runes and the Latin alphabet, which was adopted by the Anglo-Saxons as a result of their conversion to Christianity following the mission of St Augustine in AD 597. The earlier runic writing was subsequently replaced by the Roman alphabet, although there was a period of overlap in which the two systems were used with different functions. If you have read any of J. R. R. Tolkien's works you will be familiar with the runic alphabet. As we saw in the previous chapter, Anglo-Saxon runes formed the basis of Tolkien's invented alphabets in his *Lord of the Rings* trilogy; however, in *The Hobbit* he simply employed the Anglo-Saxon runic system without modification.

Runes are a series of letters comprising short straight lines designed for inscribing on hard materials such as wood, stone, and metals. The runic writing system was first developed in Scandinavia and brought to England by the Germanic tribes who migrated to Britain during the fifth century AD. This

Scandinavian system is known as the 'futhark', a name which is made up of the opening six characters, in the same way as the English word 'alphabet' is based upon the opening two letters of the Greek system from which it ultimately derives: alpha and beta. But while the runic system adopted for use in England was Scandinavian in origin, a number of developments and modifications took place to give the script a distinctively English look. Similar developments are recorded in runic inscriptions found in Frisia (modern Netherlands) and it is unclear whether the Frisian developments were exported to England, or whether the influence went the other way. The adaptations of the older runic system were of two kinds. One kind of development was purely concerned with the appearance of the individual characters and was not connected with the sounds that they represented. A good example of this concerns the *h* rune which in the north Germanic languages is typically represented by ᚺ, with a single bar. Anglo-Saxon runic inscriptions typically show a development of this form with two bars: ᚻ. Another development concerns the *k* rune. In the earliest inscriptions this is represented as ᚲ, which subsequently developed to ᚴ; the equivalent Anglo-Saxon rune ᚳ is evidently a further development of this variant.

The second kind of adaptation is the result of changes in pronunciation that separated the Anglo-Saxon language from its northern Germanic relatives. This is best illustrated by the fate of the Scandinavian *a* rune: ᚨ, given the name *ansuz*, a Germanic word for 'god'. In Old English, and in the related language Old Frisian, the sound represented by this rune underwent various changes. Where it was followed by a nasal consonant, /n/ or /m/, it was pronounced as an 'o' sound; to accommodate this change the shape of the rune was altered to ᚩ. This rune replaced the Scandinavian ᚨ in fourth position, with the result that the

Anglo-Saxon runic alphabet is known as the 'futhork' rather than the Scandinavian 'futhark'. In other positions the 'a' sound, made with the tongue retracted to the back of the mouth, was fronted, made with the tongue at the front of the mouth. To get an idea of the distinction between these two pronunciations compare the *a* sound in *bath* and *cat* in a modern RP accent. This change did not affect all 'a' sounds, so that the back pronunciation was retained in certain contexts, with the result that two different runes were required for the 'a' sound, where Scandinavian had just one. To accommodate these various changes, the Anglo-Saxon runemasters carried out further modifications to the inherited rune. In the Anglo-Saxon futhork the rune ᚨ was used for the fronted vowel, and came to be known as *æsc* 'ash-tree', while a new rune ᚪ was devised for the low back vowel, known as *ac* 'oak'; these two runes were added at the end of the alphabet.

Another development triggered by sound changes that affected the Anglo-Saxons concerns the pronunciation of a group of consonants. These are the sounds /k/ and /g/ found in Modern English words like *king* and *get*. In Old English these sounds developed variant pronunciations depending upon the vowel sound that followed. When they were followed by a back vowel, one made with the tongue positioned at the back of the mouth, they remained unchanged. But when followed by a front vowel, with the tongue at the front of the mouth, they began to be pronounced /tʃ/ and /j/, like the initial sounds in Modern English *church* and *year*. The effects of this sound change can be seen today if we compare the initial sounds in Modern English words like *church* with their continental Germanic cognates, like German *kirche*, or even the Scots word *kirk* (derived from a Scandinavian source: compare Norwegian and Danish *kirke*).

To distinguish between these variant pronunciations the Anglo-Saxon runemasters introduced two new runic characters: ᛣ and ᚸ. These new symbols were used to represent the back pronunciations /k/ and /g/, while the traditional symbols were used for the new front pronunciations /tʃ/ and /j/.

These changes encapsulate several important themes that we will return to throughout this book concerning the ways in which English spelling has developed over the centuries. Some changes are phonetically motivated, that is they are adaptations designed to respond to changes in the pronunciation of the language, while others appear to be little more than changes in fashion: the use of a double-barred *h* rune, rather than one with a single bar. Such changes may appear of little significance in themselves but contribute something important to the appearance of the English spelling system, giving it an identity and setting it apart from related systems used for different languages.

Runic script was employed for a variety of writing tasks, as a simple method of communication, inscribing ownership marks upon everyday objects, writing memorial inscriptions, as well as for recording magical charms. It is the latter aspect of their use that is frequently associated with runes today, though there is little evidence that runes were considered to be primarily associated with pagan religious practices. Their origins were associated with Germanic mythology, as the Norse god Odin learned the secret of runic writing by spending nine days and nights nailed to the world ash-tree called Yggdrasill, as the following lines from the poem *Hávamál* memorably recount:

> Wounded I hung on a wind-swept gallows
> For nine long nights,
> Pierced by a spear, pledged to Odin,

> Offered, myself to myself:
> The wisest know not from whence spring
> The roots of that ancient rood.
> They gave me no bread, they gave me no mead:
> I looked down; with a loud cry
> I took up the runes; from that tree I fell.
>
> (W. H. Auden and P. B. Taylor, *Norse Poems*, Faber, 1983, p. 164)

If runes were exclusively associated with pagan practices, however, we would expect that the advent of Christianity would have signalled their demise, but this is not what happened. Instead there was a period in which the runic script and the Roman alphabet, associated with the new Roman religion, coexisted. Tolerance of runic writing is apparent from a number of significant instances of its use in important Christian contexts. These include runic inscriptions on Christian crosses, such as the famous Ruthwell cross, where a section from the Old English poem *The Dream of the Rood* has been added in runes. Perhaps most striking is the runic inscription that was carved on the coffin of St Cuthbert, which was constructed by the monks of Lindisfarne and carried by them from their monastery to its present resting site in Durham Cathedral. An instance of the use of runes in both pagan and Christian contexts is the small whalebone casket, carved with scenes from the Bible and Germanic myth, known as the Franks Casket (Figure 3.1).

On the left is a depiction of the legend of Weland, the smith, who was hamstrung by king Nithhad to prevent his escape. At the bottom lies the headless body of Nithhad's son, whom Weland has killed. Weland is holding a goblet made from the son's skull and is offering a drink to Nithhad's daughter Bodvild. The drink is drugged; once she drinks it Weland rapes her. He

Figure 3.1. Scenes from the Franks Casket © The Trustees of the British Museum

then escapes by constructing wings from bird feathers; he is depicted collecting the feathers at the right hand edge of this scene. The scene in the right hand panel presents the visit of the Magi; just above their heads you can see the word 'Magi' written in runic letters 'ᛗᚨᚷᛁ'. Around the edges is a lengthy runic inscription which presents a riddle apparently unconnected to the scenes depicted; the inscription is in Old English, and can be translated as follows: 'The flood lifted up the fish on to the cliff-bank; the whale became sad, where he swam on the shingle'. The solution to the riddle is revealed on the left hand edge, which reads 'Whale's bone'. This alludes to the material from which the casket has been crafted, and the surface upon which the runic letters have been cut.

Following the Christianization of the Anglo-Saxons in AD 597 the Roman alphabet was adopted; the letters used for writing Old English are broadly the same as those in use today. As well as being introduced directly from Rome, the Roman alphabet was also imported by Irish missionaries, in the form of the half-uncial script used in Irish monasteries and still used today in Ireland for some road signs and public notices. The half-uncial script is the script of choice for many of the most impressive books produced in the British Isles during this period: it can be seen in the magnificent Lindisfarne Gospels, produced at the monastery on Holy Island in the early eighth century by a monk called Eadfrith, as well as in the Book of Kells. The monks left Lindisfarne in the late tenth century and subsequently settled in Chester-le-Street, where another monk called Aldred added an interlinear gloss in Old English, written in a less impressive, more functional, script known as insular minuscule. In Figure 3.2 you can see the rounded letter forms that are characteristic of the half-uncial script used to copy the Latin text of the gospels themselves.

This was a development of a Roman script devised in the fourth century, its curved lines designed to suit writing in books in preference to the angular letter shapes typical of earlier Roman scripts adapted to carving on stone. Added in a much smaller and scruffier hand above the lines of Latin text you can see the insular minuscule script used by the Anglo-Saxon glossator. I said earlier that the Old English alphabet is similar to that in use today; however, a quick glance at the script used by Aldred for his glosses might seem to contradict this claim. Certainly Aldred's script bears little resemblance to the handwriting used today, but the differences are more to do with the shapes of the individual letters than with the letters themselves. For instance, if we look above the opening word of the Latin text, *plures*, we find the word *monige*, the Old English word for 'many'. Here all the letters are identical to those used today, the major differences being the shape of the letters <e>, and <g>, known as insular <g>, which has a flat top rather than the closed loop we use today. Of course we also spell the word with a <y> rather than a <g>, but this is a differ-

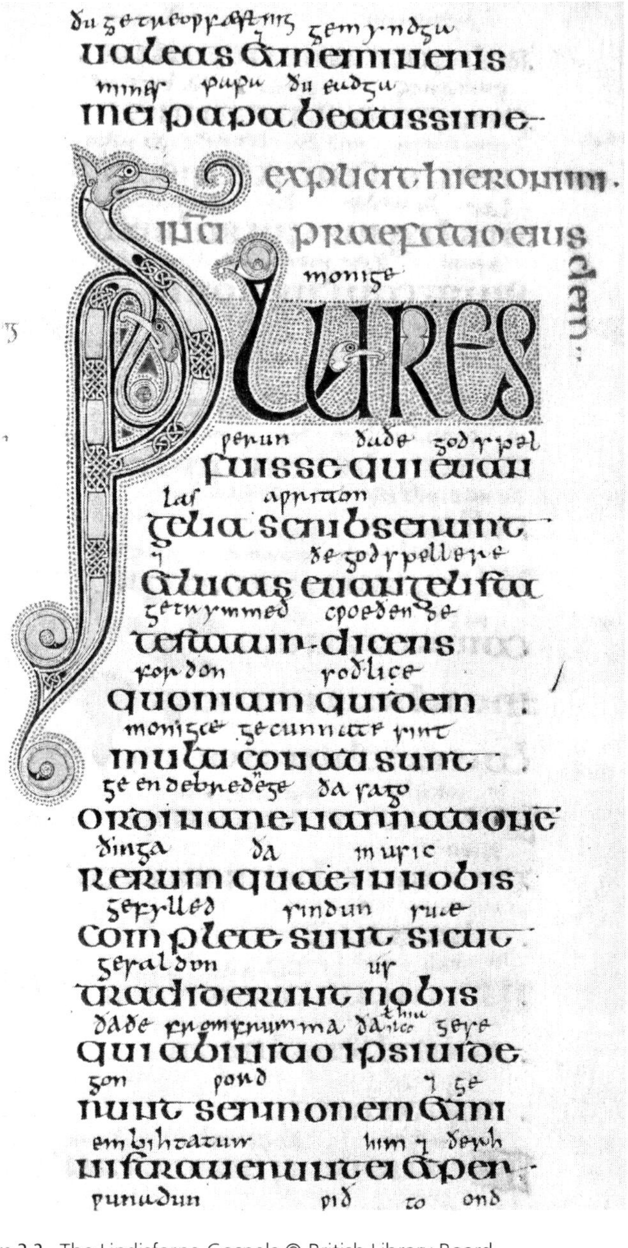

ualeas et memineris
mei papa beatissime.
explicit hieronimi.
incipit praefatio eiusdem.
Plures
fuisse qui euan
gelia scribserunt
et lucas euangelista
testatur dicens
quoniam quidem
multi conati sunt
ordinare narrationem
rerum quae in nobis
completae sunt sicut
tradiderunt nobis
qui ab initio ipsi uide
runt sermonem et mini
strauerunt ei et per

Figure 3.2. The Lindisfarne Gospels © British Library Board

ence of spelling, the use of the letters and the sounds onto which they map, rather than a difference in the actual letters used in the alphabet. Other differences in the shapes of individual letters may be seen in the form of <r>, the third letter of the first word on the following line; similar in shape and easily confused with this long-tailed <r> is a long <s> which is used in the third word of this same line: the word *godspel* (literally 'good news', the ancestor of Modern English *gospel*). The letter <f> has a slightly different shape to our modern letter, as seen at the beginning of the first word of line 6, while the <t> differs in having a flat top.

The differences we have observed so far are all concerned with the shapes of individual letters. However, you will have noticed that there are also several letters in Figure 3.2 that have no obvious equivalent in our modern alphabet. The reason why these letters are unfamiliar is that they are not part of the Roman alphabet, but were used alongside Roman letters by the Anglo-Saxons for writing Old English. The introduction of these letters was driven by the need to represent certain sounds found in Old English that were not part of the Latin language. The Roman alphabet has proved to be a hugely influential writing system and is the most widely used system in the world today, employed for the writing of most European languages as well as many non-European ones too. The spread of the Roman alphabet continues today: following the collapse of the Soviet Union in 1991, Azerbaijan, Uzbekistan, and Moldova have all adopted the Roman alphabet in preference to the Cyrillic alphabet as the official script for writing their different languages. But, while it is clearly a very useful tool for writing a variety of languages, one of the problems involved in adopting a writing system developed for a different language, in this case Latin, is that the sound system that it represents will not necessarily coincide with that of another language. English is a Germanic language and is not directly descended from Latin. Languages descended from Latin are known today as Romance languages, a group which includes French, Italian, and Spanish; these languages are therefore more closely related to Latin than English is. The Germanic and Romance language families do ultimately descend from a single common ancestor, a language known as Proto Indo-European, the hypothetical ancestor of most modern European and Indian languages, but numerous changes took place in the centuries between the emergence of the Romance and Germanic languages. So, while there was a broad overlap between the sounds used in Old English and those used in Latin, there were also some differences. To allow us to consider the ways in which the Roman alphabet was applied to the writing of Old English, we must begin by examining the origins and make-up of the Roman alphabet itself. To do this, we will have to trace its history back beyond the earliest Latin inscriptions, because the Roman alphabet was not originally devised for writing Latin, but was borrowed from the Greek alphabet, with a number of necessary modifications, while the Greek alphabet itself was a development of an even earlier Phoenician writing system.

The origins of the Roman alphabet lie in the script used by Phoenician traders around 1000 BC. This was a system of twenty-two letters which represented the individual consonant sounds, in a similar way to modern consonantal writing systems like Arabic and Hebrew. The Phoenician system was adopted and modified by the Greeks, who referred to them as 'Phoenician letters' and who added further symbols, while also re-purposing existing consonantal symbols not needed in Greek to represent vowel sounds. The result was a revolutionary new system in which both vowels and consonants were represented, although because the letters used to represent the vowel sounds in Greek were limited to the redundant Phoenician consonants, a mismatch between the number of vowels in speech and writing was created which still affects English today. Another innovation made by the Greeks was equally influential. The Phoenicians wrote their letters from right to left; while this direction of writing is perfectly acceptable when carving letters, it is comparatively difficult for a right-handed person writing on papyrus. Early Greek inscriptions abandoned this method of writing in favour of 'boustrophedon', literally meaning 'ox-turning', in which lines are written alternatively from right to left, and then left to right, like an ox ploughing furrows across a field. This sytem was then replaced by the method of writing from left to right, which in turn was passed on to the Romans and down to our Modern English writing system.

The Greek alphabet, including its adaptations, was subsequently adopted by the Etruscans, located in Etruria, north of Rome, who spoke a non-Indo-European language. The Etruscans adopted the Greek alphabet as early as 700 BC; ownership inscriptions using Greek letters can be found on high status objects, such as vases and cups, placed in funeral tombs from this period. The importance of writing and the high value placed on the written word in Etruscan society is apparent from the inscriptions that survive, which frequently draw attention to their status as written documents. An extensive collection of Etruscan inscriptions has survived, carved onto vases and mirrors and painted on walls; estimates vary between nine and thirteen thousand. This is considerably more than survives for other, non-Latin, languages used in Italy, for which only a handful of inscriptions survive. But this large number of extant Etruscan inscriptions represents a fraction of what was originally written down, and provides only a limited insight into the Etruscan language. Despite references to an Etruscan literary tradition and a writer of tragedies, no literary texts have survived. Extant inscriptions are concerned above all with religious and legal practices or are funerary inscriptions, comprising little more than names of the deceased and close family members.

By AD 200 the dominance of Rome had led to the absorption of Etruscan culture and the adoption by the Romans of the Etruscan alphabet for writing Latin. The Etruscans used the western (or 'Euboean') version of the Greek alphabet, which differed in several ways from the eastern ('Ionic') version, the classical Greek alphabet authorized for use in Athens in 403 BC and still used in modern Greece. In this western version, the letter 'eta', <H>, represented the sound /h/ rather than a vowel, while 'chi', <X>, represented the consonant cluster /ks/ rather than the /x/ sound (the fricative sound in Scots *loch* and German *nacht*) of classical Greek. This explains why the same letters, <H> and <X>, represent different sounds in classical Greek and Latin. With the absorption of Etruscan culture, it was this version of the Greek alphabet that was passed on to the Romans, who themselves made further modifications. The Etruscan language did not have the sounds /b d g/ and so dropped the letters <B> 'beta' and <Δ> 'delta'. Because these letters were still known, the Romans simply reinstated them as a means of representing these two sounds in the Latin language. However, since the Etruscans had employed the letter 'gamma', <Γ>, which had now changed in shape to <C>, to represent a /k/ sound, there was no letter available to the Romans to represent /g/. Initially, the Romans used <C> for both /k/ and /g/, which is why the name *Gaius* is sometimes written *Caius*, but later a separate letter was formed by modifying <C>; this is the origin of our modern letter <G>. This letter was added into the alphabet in the seventh position, as a replacement for <Z> 'zeta', which was not used at all by the Romans. The Etruscans had also discarded the letter <K> 'kappa' as it was unnecessary for their system; the Romans too made very little use of <K>, reserving it for a handful of words such as *Kalendæ*, where it was preserved thanks to the widespread use of the abbreviation *Kal.* The Etruscans had a third letter which they used for the /k/ sound, <Q>, although this was only used in the combination <QU>, a restriction which was retained by the Romans and still applies today in English spelling.

Another distinction between the Etruscan alphabet and its Greek ancestor concerned the letter <F>, known in Greek as 'digamma', and used to represent the sound /w/, but in Etruscan used for the /f/ sound. As the /w/ sound fell out of use in Greek, the letter <F> has not survived into the modern Greek language; but because the Roman alphabet is based upon the Etruscan alphabet it continued to employ <F> to represent the sound /f/. The Romans used <V> for the /w/ sound, as well as a vowel sound, while <I> could stand for both /i/ and /j/. Two additional letters were subsequently added by the Romans, to enable them to write Greek loanwords that were adopted into Latin. The first of these was 'upsilon' <Y>, a variant development of the letter <V>, which they used to reflect the /y/ sound, similar to that used in French *tu*, that was found in some Greek words, but not in Latin. The second was the letter 'zeta' <Z>, which had earlier been dropped by the Romans from its original seventh position, and was subsequently added at the end of the alphabet. As the Romans had no need of the Greek letters 'theta', 'chi' and 'phi', <Θ, X, Φ>, these letters were dropped; when writing Greek loan-words containing these letters the Romans used the digraphs

<TH, CH, PH>. The political dominance of the Roman empire means that the Roman alphabet, adapted from the Greek via the Etruscan alphabet for the writing of Latin, is the basic alphabet used by many western European languages, including English.

As we have seen, by the time of its adoption by Anglo-Saxon scribes for the writing of Old English, an established tradition of sound–letter correspondences for the Roman alphabet had been developed, upon which the Anglo-Saxon scribes were able to draw. Because they had been trained to write Latin, Anglo-Saxon scribes were already familiar with these conventions and in many cases the employment of a Roman letter to represent an English sound was quite straightforward. This was especially true for the consonants, which, in most cases, mapped neatly onto the writing of Old English. The Roman alphabet comprised twenty-one letters: <A B C D E F G H I K L M N O P Q R S T V X>, although, as we have seen, the letter <K> had a marginal status. In most cases, Anglo-Saxon scribes simply employed the appropriate Roman letter to represent the equivalent sound in Old English. Just as the Romans made little use of <K>, so this letter was seldom employed by Anglo-Saxon scribes. As we have seen, the Romans used the letter <X> to represent the two sounds /ks/; this practice was followed in Old English, in words like *æx* 'axe', *axian* 'to ask', and in words like *fixas* 'fishes', where an earlier /sk/ has been reversed via a process known as 'metathesis' to give the /ks/ sound. Where the Romans employed a single letter for the vowel and consonant uses of <I>, as can be seen in the spelling of the name *Iulius* 'Julius', this practice was also maintained in Old English; the letter <J>, itself a modification of the letter <I>, was not introduced into English spelling until the seventeenth century. Not all Latin spelling conventions were preserved in Old English. Latin employed the letter <Q>, in combination with the letter <V>, to represent the two sounds /kw/, as in *quattuor* ‘four’ and *equus* ‘horse’. But Anglo-Saxon scribes did not maintain that usage; instead they used the combination <cw> to render these sounds, as in *cwen* ‘queen’, although there are a few instances of the Latin practice being imported in some early texts.

The vowel sounds of Old English were represented by those letters used to reflect the Latin vowels, although Anglo-Saxon scribes required two further symbols to reflect vowel sounds not found in Latin. The first of these is a front a-sound, like the sound in Modern English *apple*. To supply this deficiency in the Latin alphabet, Anglo-Saxon scribes devised a modified form of the letters <a> and <e>, <Æ>, known as *æsc* ‘ash’, meaning ‘ash-tree’, after the rune ᚫ that corresponded to the same sound. Old English also required a symbol for a high front rounded vowel, similar to that in Modern French *tu*, which has not survived into Modern English. To represent this sound, Anglo-Saxon scribes adopted the letter <Y>, derived ultimately from the Greek <Y>, ‘upsilon’, the ultimate source of the three letters <U, V, Y> in the Latin alphabet. As we saw above, Latin <Y> was restricted to Greek loanwords, where it is still found today in Modern English words like *psychology* and *cryptic*. The letter’s Greek origin is signalled by the Anglo-Saxon scholar Ælfric’s reference to it as ‘se grecisca y’, ‘the Greek y’, in the Latin grammar which he wrote in Old English (see pp. 21–2). The name ‘wy’ that we use today is first recorded in 1200, but its origins are unknown. Ælfric’s name for the letter is paralleled in French and Spanish, where it is termed ‘i grec’ and ‘i griega’, while the German and Italian names ‘ipsilon’ are clearly derived from the Greek name for the letter.

Further additional letters were required in Old English to represent consonantal sounds not present in Latin. Old English had several consonant sounds not found in Latin, such as the pair of sounds /ð/ (as in *this*) and /θ/ (as in *thank*). The solution adopted by the earliest Old English scribes was to turn to the runic script and borrow the equivalent symbols. So in Old English the rune <þ> 'thorn' (so-called because it resembles a thorn) was used to represent the sounds /ð/ and /θ/. Where the Romans had used <V> for the /w/ sound, Old English scribes replaced this with another runic letter, 'wynn' <ƿ>, presumably in an attempt to avoid confusion with the other uses of <V>. The runic letter 'wynn' can be seen in use in Figure 3.2. (p. 46) as the first letter of line two; it resembles the letter <p> and the letter <þ> and as a consequence it is often replaced by <w> in modern editions of Old English texts. Alongside <þ>, another letter was used to represent this pair of sounds, a variant form of the letter <d>, known as 'eth', <ð>, a name derived from Icelandic where the letter is still used. This letter appears twice in the second word of line 2 in Figure 3.2. The letters <þ> and <ð> were used to represent both /ð/ and /θ/ indiscriminately and no attempt was made to make an apparently useful distinction between them, associating one letter with one sound. So, while the Latin alphabet mapped closely onto the sound system of Old English, with many letters performing the same functions in both languages, there were a number of misfits and gaps. Where Old English contained sounds not found in Latin, Anglo-Saxon scribes adopted two principal solutions: modifying existing letters, as in the cases of <æ> and <ð>, and importing letters from the runic alphabet in the cases of <þ> and <ƿ>.

So far we have focused on the similarities and overlaps between the use of letters in Latin and Old English. But it is important that we do not assume that the Roman letters used in Old English always represent the same sounds as they did in Latin, or as they do now in Modern English. Many modern languages use the Roman alphabet today, but this does not necessarily mean that they all use it in the same way. Because we have a fixed spelling system today, it is easy to assume that the relationship between letters and sounds is also fixed so that the letter <v>, for example, must always represent the sound /v/. As we saw in the previous chapter, the relationship between a letter and its corresponding sound is conventional; there is no inherent reason why the letter <v> should represent the sound /v/ rather than any other sound.

There are in fact a number of differences in the way the Latin letters were used to represent the sounds of Old English. For instance, the letter <g> was employed to represent two different sounds: the hard /g/ sound found at the beginning of the word *good*, as well as the soft /j/ sound at the beginning of *year*, the sound represented today by the letter <y>. The use of a single letter to represent two different sounds which we today distinguish using two different letters can seem strange to us. How did an Anglo-Saxon reader know which sound was intended? It is customary for modern editors of Old English texts to add a dot above the <g> when it represents the sound /j/ as an aid to a modern reader, but Anglo-Saxon readers had no such helpful marks. In fact the system was not as confusing as it might seem to us today. This is because the sound of the letter was conditioned by its placement in the word. The /j/ sound is generally found before front vowels, ones made with the tongue at the front of the mouth, as in the words *gear* 'year' and *geard* 'yard', while the hard /g/ sound is found before back vowels, made with the tongue retracted at the back of the mouth, like in *god* 'good' and *gold*. Another difference in the use of individual letters between Old English and Modern English concerns the use of the letter <c>, whose use was governed by its placement in the word in a similar way to the letter <g>. The letter <c> was used to represent the hard /k/ sound before back vowels, as in *cuman* 'come', but represents the soft sound /tʃ/, as in the first sound of *church*, when it appears before a front vowel, as in *cild* 'child'. The letter <k> was only used very occasionally at the beginnings of words to represent the hard /k/ sound when it was followed by a front vowel, as in the word *kyning* 'king', an alternative to the spelling *cyning*, which might otherwise have been erroneously pronounced with an initial /tʃ/. Modern editors of Old English typically dot the <c> when it represents the palatal /tʃ/ sound as a guide to modern readers, but, as above, Anglo-Saxon readers were able to make the distinction without aids of this kind.

Earlier we saw two different methods employed by the Anglo-Saxons to extend the Latin alphabet: the importation of runic letters and the combination of two letters to form a single letter, known as a ligature. Another method of representing sounds not present in Latin was to use two separate letters to represent a single sound, what is known as a digraph. This was the solution adopted by Anglo-Saxon scribes for the representation of the sound /ʃ/ which they spelled <sc>. As the sound /sk/ was not found at the beginnings of words in Old English there was no possibility of confusion over its pronunciation. There were instances of its use at the ends of words where confusion was possible, as in *tusc* 'tusk', but this was evidently rare enough not to cause too many difficulties. Another unfamiliar combination of letters used in Old English is <cg>, which represents the sound /dʒ/ in words like *ecg* 'edge'. While this system was generally unproblematic, some Old English scribes were apparently concerned about possible confusion and so added a silent <e> after

<sc> and <cg> to indicate that the consonants should be pronounced as /ʃ/ and /dʒ/, for example *sceolde* 'should', *hycgean* 'think', although these spellings were not widespread.

Despite the need to add letters to its alphabet and to employ digraphs to represent certain sounds, Old English made little use of the letters <v> and <z>, even though they knew both letters from the Roman alphabet. The letter <v> is a variant form of the letter <u>: both originate in a single Roman letter; the angular form <V> was used in inscriptions, while the rounded form was employed for writing on parchment. In Old English, <u> was used for the vowel /u/, while <V> or <U> was used for the upper-case equivalent; neither letter was used to represent the consonant sound /v/. The letter <z> was adopted into Latin from Greek <Z> 'zeta', but it was generally restricted to Greek loanwords. In other instances, the letter <s> was used to represent the voiced as well as the voiceless sound. A similar situation is also found in Old English, where the letter <z> is only found in loanwords, such as *mertze* 'merchandise' (Latin *mercem*), where it represents the sound /ts/. You might assume from this that the two sounds /v/ and /z/ were therefore absent from Old English. However, this is not the case. These sounds did appear in Old English, but their use was conditioned by their placement in the word. The /f/ sound could only appear at the beginnings and ends of words, whereas the /v/ sound was limited to medial position. Because of this, the same letter, <f>, could be used for both sounds. The clue as to whether to pronounce it /f/ or /v/ lay in its position in the word. This may be demonstrated by a comparison of the Old English spelling of the singular and plural forms of the noun *wolf*. In Modern English the singular *wolf* has an /f/ sound at the end but the /v/ sound in the middle; hence *wolf*, *wolves*. The same distribution is found in Old English, but here the letter <f> is used to represent both sounds: *wulf* and *wulfas*. This might seem confusing, but an Anglo-Saxon reader would have known which sound was intended by the position of the <f> in the word. The only disruption of this system was caused by the adoption of Latin loanwords with initial /v/, spelled with a <v>, such as *vannus* 'fan' and *versus* 'verse'. These words were spelled with initial <f> in Old English; in the case of *fann* the pronunciation shifted to /f/, giving us the Modern English pronunciation; *fers* continued to have the /v/ sound which later came to be spelled with a <v>.

The same system used for the representation of /f/ and /v/ applied to the pair of sounds /s/ and /z/. When the letter <s> appeared at the beginning or end of a word it was pronounced as an /s/, when in the middle it was a /z/ sound. Thus the noun *hus* 'house' was pronounced with the unvoiced /s/ sound, while the verb *husian* had the voiced /z/ sound. The same distinction is observed in the pronunciation of the noun *house* and the verb *house* today, showing that such distinctions can be maintained without the need for alternative spellings such as *houze*. As a consequence of this system, based upon placement and use, what is known as 'complementary distribution', Old English had no use for either the letter <v> or <z>.

Another key difference between the Old English and Modern English spelling systems is that in Old English there were very few silent letters; in most cases all the letters were intended to be pronounced. Speakers of Modern English are so inured to the presence of silent letters that they often fail to question why a phonetic spelling system should include letters that are not pronounced. While these silent letters frequently cause problems for children and non-native speakers when learning to read and write English, for philologists they are an invaluable testimony to earlier pronunciations. So, for instance, the word *knot* is spelled with an initial <k> because in Old English it was pronounced that way, as is shown by the Old English spelling of this word: *cnotta*. This same principle also applies to silent letters at the ends of words, such as the now-silent <b> in *lamb* or *comb* (Old English *lamb*, *camb*). Some words which end in a silent <b> do, however, correctly reflect the Old English pronunciation. For example, the word *thumb* is derived from Old English *þuma*, the <b> was not added until the thirteenth century; the same is true of *limb* (Old English *lim*), which gained its <b> in the fifteenth century. The preservation of a mute <b> in these words was probably reinforced in the Middle English period by the introduction of French loanwords like *plumber* and *tomb*, where the <b> derives from their forms in Latin (*plumbarius* and *tumba*) but had ceased to be pronounced. The Old English word *crum* (Modern English *crumb*) began to be spelled with an unsounded <b> from the sixteenth century, although *crum* continued to be used alongside *crumb* until the eighteenth century; both spellings are given by Dr Johnson in his *Dictionary* of 1755. The addition of the <b> is recorded earliest in the word *crumble*, where it is of course still sounded and which probably led to its adoption in *crumb*; the tendency for speakers to add a /b/ after /m/ can be seen in further changes to Old English spellings in words like *þymel* (Modern English *thimble*) and Old English *slumere* (Modern English *slumber*). In the case of Modern English *bramble*, Old English had spellings with and without <b>, *bræmel*/*bræmbel*, suggesting that the pronunciation with an additional /b/ was already current. A similar pronunciation change can be seen by comparing Old English words like *spinel* and *þunor* with Modern English *spindle* and *thunder*, both of which have acquired an intrusive /d/ after the /n/. Interestingly, this same change can be traced in other Germanic languages, such as Dutch and German, while a similar change affected the Modern Swedish word *spindel* 'spider', which is derived from Middle Swedish *spinnil*. An intrusive /p/ was also added to words which had the consonant cluster /mt/ in Old English, giving us the distinction between Old English *æmtig* and Modern English *empty*.

Old English has many instances of consonant clusters that look very unusual to speakers of Modern English, some of which have subsequently been simplified or the words themselves have fallen out of use. For example, Old English had a number of words that began with the consonant cluster <fn>, pronounced with initial /fn/. This combination is no longer found at the beginning of any Modern English word. What has happened to these words? In the case of the verb *fneosan*, the initial /f/ ceased to be pronounced in the Middle English period, giving an alternative spelling *nese*, alongside *fnese*. Because it was no longer pronounced, the <f> began to be confused with the long-s of medieval handwriting and this gave rise to the modern form *sneeze*, which ultimately replaced *fnese* entirely. The *OED* suggests that *sneeze* may have replaced *fnese* because of its 'phonetic appropriateness', that is to say, because it was felt to resemble the sound of sneezing more closely. This is a tempting theory, but one that is hard to substantiate. Another group of Old English words began with <gn> and were pronounced with initial /gn/. Some of these words have fallen out of use, such as *gnorn* 'sorrow', *gnidan* 'to rub', while others like *gnat* (Old English *gnæt*), *gnaw* (Old English *gnagan*), have retained the <gn> in the spelling but have simplified the pronunciation to /n/. Similar changes are recorded in other Germanic languages. As an example we might compare some cognate forms of Old English *gnagan* 'gnaw': modern German has the verb *nagen* for Old High German *gnagan*, modern Icelandic has *naga* for Old Norse *gnaga*. Modern English words spelled with initial <wr>, like *write*, *wreath*, and *wrath* preserve an Old English spelling that reflects a pronunciation with initial /w/: Old English *writan*, *wriða*, *wræððu*. In some combinations, the initial /w/ was dropped very early in the history of English, as in Old English *wlispian* (Modern English *lisp*). The Old English combinations <hl>, <hn>, and <hr> were all pronounced with an initial /h/, *hlaf* 'loaf', *hnutu* 'nut' and *hring* 'ring', but this was lost early in the history of these words and consequently no trace of this pronunciation has remained in our modern spelling system. The early simplification of these pronunciations is probably a reflection of the tendency for speakers of English to drop initial /h/, a phenomenon known today as h-dropping. Although h-dropping is widely condemned as sloppy and ignorant behaviour, it is in fact a very widespread phenomenon, attested in most dialects of English and can be traced back to the Anglo-Saxons. The only group of words that continue to preserve a trace of an earlier pronunciation with initial /h/ that has since been dropped are words spelled today with initial <wh>, *what*, *where*, *whale* (Old English *hwæt*, *hwær*, *hwal*). These words were spelled with initial <hƿ> in Old English, reflecting a pronunciation with initial /hw/. This pronunciation was simplified in a way similar to the other consonant clusters, although it happened comparatively late and consequently has left a trace in our modern spelling system. That the simplification of this consonant group differed from the others we have considered is apparent from the fact that Scots and Irish speakers of English continue to pronounce such words as if they were still spelled with initial <hw>. There are, however, some examples of words spelled today with initial <wh> in which the initial <w> testifies to an Early Modern spelling change rather than an Old English pronunciation. These are the Modern English words *whole* and *whore*, which in Old English were spelled *hal* and *hare*. These two examples are part of a more widespread spelling reform which occurred in the sixteenth century which spelled many such words with initial <wh>, such as *whom* 'home', *wholy* 'holy', *whoord* 'hoard', *whote* 'hot', and *whood* 'hood'; of these reformed spellings only *whole* and *whore* survive. These two unetymological spellings were perhaps retained because they created a useful written distinction from the homophones *hoar* and *hole*.

Modern English words like *folk*, *chalk*, and *half*, which have a silent <l>, are also derived from Old English words in which the <l> was sounded: *folc*, *cealc*, and *healf*. There are other examples where the <l> was dropped from pronunciation considerably earlier and as a consequence this is not reflected in contemporary spelling; examples include *swilc* 'such', *hwilc* 'which'.

Another change which has affected the spelling of some Old English words in Modern English is known as 'metanalysis', a process whereby the form of a word is reinterpreted so as to create a new formation. A good example of this process concerns the Old English word *næddre* which, when preceded by an indefinite article, *a næddre*, was mistakenly interpreted as *an addre*, giving us the Modern English word *adder*. The opposite occurred in the case of the Modern English word *newt*, which is a reinterpretation of *an ewt*, a word ultimately related to Old English *efeta*. Another change that is responsible for the differences in spelling between certain Old English words and their modern equivalents is known as 'metathesis', a term which refers to the process whereby certain sounds are reversed. We encountered this change earlier in the reversal of the /ks/ in the words *axian* 'to ask' and *fixas* 'fishes'; the same process affected the Old English words *brid* 'bird', *þridda* 'third' and *nosterl* 'nostril'.

The majority of Old English words were of Germanic stock, meaning that they were inherited from a hypothetical language which scholars today call 'Proto-Germanic': the variety from which all surviving Germanic languages, including German, Dutch, Norwegian, Danish, and Icelandic, all derive. Where there is a requirement for new words to be formed, Germanic languages tend to favour the use of internal methods of word formation, such as 'compounding', joining two words together, or 'affixation', adding prefixes and suffixes to existing words. This is still a common feature of Modern English, as we can see in recent coinages like *lunchbox, motorway, railway station.* However, Modern English also frequently borrows words from other languages: think of French words like *chic* and *mangetout.* Because these words have been carried over into English with both their French pronunciation and spelling preserved, they have disrupted the pattern of sound–spelling correspondence established in the Old English period. In the two examples mentioned here, you can see that in *chic* <ch> represents the sound /ʃ/ rather than <tʃ>, while in *mangetout* <g> spells /ʒ/ not /dʒ/.

We will examine the effects of borrowing on English spelling more fully in the next chapter, which deals with a period in the history of English in which large numbers of foreign loanwords entered the language. But, while Old English tended to prefer to form new words using existing ones, a number of words of Latin origin did enter the language during this period. The influence of Latin was felt in three distinct stages: the first wave of Latin borrowings occurred during the early period of Anglo-Saxon settlement; the second wave was the direct result of the process of Christianization which took place in the early seventh century;

the third was associated with the Benedictine reform movement which had begun on the continent and which began to influence the English church in the tenth century. The first wave of Latin influence came directly from the Latin spoken by the Romanized Britons who occupied Britain before the Anglo-Saxon invasion. Latin words adopted during this period are generally common, everyday words, although there are also some religious terms, such as *munuc* 'monk' (Latin *monachus*), *mynster* 'monastery' (Latin *monasterium*), and *mæsse* 'mass'. The second period of borrowing from Latin, which followed the Christianization of England, resulted in the adoption of a number of words associated specifically with religion and learning. Most of these words are learned and were probably confined to the written language; these include words like *apostol* 'apostle' (Latin *apostulus*); *abbod* 'abbot' (Latin *abbadem*); *fenix* (Latin *phoenix*); *biscop* (Latin *episcopus*).

What is striking about these loanwords is that their pronunciation and spelling has been assimilated to Old English practices, so that the spelling–sound correspondences were not disturbed. Note, for instance, the spelling *mynster*, which shows the Old English front rounded vowel /y/, the <æ> in Old English *mæsse*, indicating the Old English front vowel. The Old English spelling of *fenix* with initial <f> shows that the word was pronounced the same way as in Latin, but that the spelling was changed to reflect Old English practices. While the spelling of Old English *biscop* appears similar to that of Latin *episcopus*, the pronunciation was altered to the /ʃ/ sound, still heard today in *bishop*, and so the <sc> spelling retained its Old English usage. This process of assimilation can be contrasted with the fate of the later loanword, *episcopal*, borrowed from the same Latin root in the fifteenth century but which has retained its Latin spelling and pronunciation.

However, during the third stage of Latin borrowing in the tenth century, a number of words from Classical Latin were borrowed which were not integrated into the native language in the same way. The foreign status of these technical words was emphasized by the preservation of their latinate spellings and structure, as we can see from a comparison of the tenth-century Old English word *magister* (Latin *magister*), with the earlier loan *mægester*, derived from the same Latin word. Where the earlier borrowing has been respelled according to Old English practices, the later adoption has retained its classical Latin spelling.

From the surviving manuscripts written in Old English it is possible to reconstruct the use of four distinct dialects: West Saxon, Mercian, Northumbrian, and Kentish. However, towards the end of the Old English period, one of those dialects, known today as Late West Saxon, began to be used outside its native area of Wessex. The prominence of this particular dialect was due to the influence of the bishop of the West Saxon see of Winchester, Æthelwold, who, inspired by the Benedictine reform that began with the foundation of monasteries such as Cluny on the continent, put in place a programme of reform of learning and literacy that led to an interest in the vernacular and a flowering of writing in Old English using the West Saxon dialect. One of Æthelwold's star pupils was Ælfric, who became a monk at Cerne Abbas and then abbot of Eynsham, and a prolific author of homilies and saints' lives. Ælfric shared Æthelwold's linguistic interests; he composed a grammar which uses English to teach the Latin language. Ælfric's interest in the Old English language is also apparent from the considerable consistency in spelling, grammar, and vocabulary found in the manuscript copies of his works. Evidence of annotation and correction in these manuscripts shows that Ælfric himself supervised and corrected earlier copies of his works to achieve greater consistency and regularity. Does this mean that Late West Saxon was a 'standard' written variety of Old English, similar in function to standard written English today? To qualify as a standard language a variety must be 'supraregional', meaning that it is no longer tied to a particular region. This seems to have happened in the case of Late West Saxon. While it began life as a south-western dialect, the prominence of the monastery at Winchester led to its adoption in a number of other ecclesiastical centres, including Canterbury, Worcester, and York. A standard written language must also be 'elaborated', meaning that it must be employed for a variety of linguistic functions. Just as modern standard English spelling is employed in all kinds of written texts, Late West Saxon was used to copy a variety of genres of Old English texts, including the *Anglo-Saxon Chronicle*, Old English translations of Latin 'classics', such as Bede's *Historia Ecclesiastica Gentis Anglorum* ('Ecclesiastical History of the English People'), and the *Dialogues* of Pope Gregory the Great. Most strikingly, all four of the major surviving Old English poetic manuscripts are also copied in Late West Saxon, despite the fact that the individual poems were composed much earlier and in different dialects.

So was Late West Saxon a standard variety of Old English? In some ways the answer is clearly 'yes', although it is important to emphasize that it did not achieve full standardization. While they do show considerable internal consistency and regularity, manuscripts copied in Late West Saxon allow a degree of variation that would be unacceptable in modern standard written English. This is perhaps clearest in the large poetic anthologies, which contain traces of a variety of different dialects deriving from their original composition and processes of transmission. Despite imposing the dominant Late West Saxon dialect upon these texts, scribes were evidently willing to tolerate a considerable range of non-West Saxon forms. Another limitation concerns its geographical spread. While Late West Saxon was clearly elaborated and accepted outside the Wessex area, its influence is mostly associated with ecclesiastical centres, some of which show differences in spelling and vocabulary. There is also evidence that Late West Saxon was not viewed as the sole prestigious written variety. Even though there are fewer texts written in the Mercian dialect of the midlands during this period, this dialect appears to have exerted a rival pressure, reinforced by the prestige of the monastery at Lichfield. The Mercian literary language flourished in the early ninth century, as witnessed by the Mercian gloss added to the deluxe copy of the Psalms, known as the Vespasian Psalter, originally copied in Kent in the eighth century. The Mercian literary variety continued to hold prestige in the eleventh century, when the life of St Chad was written in this dialect. As late as the early thirteenth century, works in English show the continuity of conventions associated with the Mercian dialect. But neither of these two varieties achieved full standardization in that they were neither completely fixed nor codified.

To give an idea of what an Old English text looks like, I have included below an extract from the Old English translation of the Gospels, copied in the Late West Saxon dialect. Below this I have supplied the parallel extract from the New International Version to allow a close comparison of the Old and Modern English translations.

Mark 4.1–9: The Parable of the Sower.

And eft hē ongan hī æt þǣre sǣ lǣran; and him wæs mycel menegu tō gegaderod, swā þæt hē on scip ēode, and on þǣre sǣ wæs; and eall sēo menegu ymbe þā sǣ wæs on lande. And hē hī fela on bigspellum lǣrde, and him tō cwæð on his lāre, 'Gehȳrað: ūt ēode sē sǣdere his sǣd to sāwenne. And þā hē sēow, sum fēoll wið þone weg, and fugelas cōmon and hit frǣton. Sum fēoll ofer stānscyligean, þār hit næfde mycele ēorðan, and sōna ūp ēode; and for þām hit næfde eorðan þiccnesse, þā hit ūp ēode, sēo sunne hit forswǣlde, and hit forscranc, for þām hit wyrtruman næfde. And sum fēoll on þornas; þā stigon ðā þornas and forðrysmodon þæt, and hit wæstm ne bær. And sum fēoll on gōd land, and hit sealde ūpp stīgende and wexende wæstm; and ān brōhte þrītigfealdne, sum syxtigfealdne, sum hundfealdne.' And hē cwæð, 'Gehȳre, sē ðe ēaran hæbbe tō gehȳranne.'

(http://archive.org/details/changeliumsecun00brigoog)

New International Version

Again Jesus began to teach by the lake. The crowd that gathered around him was so large that he got into a boat and sat in it out on the lake, while all the people were along the shore at the water's edge. He taught them many things by parables, and in his teaching said: 'Listen! A farmer went out to sow his seed. As he was scattering the seed, some fell along the path, and the birds came and ate it up. Some fell on rocky places, where it did not have much soil. It sprang up quickly, because the soil was shallow. But when the sun came up, the plants were scorched, and they withered because they had no root. Other seed fell among thorns, which grew up and choked the plants, so that they did not bear grain. Still other seed fell on good soil. It came up, grew and produced a crop, some multiplying thirty, some sixty, some a hundred times.' Then Jesus said, 'Whoever has ears to hear, let them hear.'

(http://www.biblica.com/biblichapter/?verse=mark+48version=niv)

There are many differences between the Old English translation of this parable and the New International Version, but here we will focus on differences in spelling. Some words are spelled exactly the same, emphasizing the continuity between Old English and Modern English spelling practices. Consequently we have no difficulty recognizing the Old English words *and*, *he*, *him*, *on*, *up*, *for*.

These are all function words, pronouns, conjunctions, and prepositions, the nuts and bolts of the language, and it is therefore unsurprising that they have not changed over the thousand years that separate these two translations. But many other words are identical with their Modern English equivalents, although they are harder to spot because their spellings have changed. In some cases the Old English letters <þ, ð, æ> mask the similarities between these words; if we replace <þ, ð> with <th> then words like *þornas* 'thorns' and *forð* 'forth' become immediately recognizable. If we replace <æ> with <a> in *wæs*, *æt*, and *þæt* we can easily spot the modern equivalents *was*, *at*, and *that*; if we replace the <ǣ> in *sǣd* with <ee> we get our modern form *seed*, while in the word *sǣ* the <ǣ> is the equivalent of our <ea> in *sea*.

Other correspondences are masked by the differing conventions of representing certain sounds in Old and Modern English. The word *mycel* is the equivalent of the Modern English word *much*, with the main difference being the Old English use of the letter <c> to represent the sound /tʃ/; in Modern English this sound is spelled using the digraph <ch>. As we saw above, in Old English the letter <c> was also used for the sound /k/; Old English made very little use of the letter <k>. Understanding this helps us to recognize the word *þiccnesse* as Modern English *thickness*. In the word *ofer*, the letter <f> stands for the voiced sound /v/; once we change the <f> to <v> the word becomes easily recognizable. The letter <g> could be used to represent both the sounds /g/ and /j/ in Old English; in Modern English the /j/ sound at the ends of words is normally spelled with a <y>. If we apply this rule to the word *weg*, then we can probably recognize it as the equivalent of Modern English *way*. The word *scip* shows the Old English practice of using the digraph <sc> to represent the /ʃ/ sound; in Modern English we spell this sound <sh> and so the equivalent word is *ship*. A slightly more complex case is the word *cwæð*; here we need to make a series of adjustments to identify the equivalent form, although in this instance the word died out in the Early Modern period and is only known in archaic usage today. The <cw> digraph was used in Old English to represent the two sounds /kw/; today these sounds are written <qu>. As we have seen, the <ð> is equivalent to Modern English <th>, while in this word the <æ> is the equivalent to <o>; the word is therefore *quoth*. The correlation between Old English and Modern English vowels is much less stable than consonants, but even here there are still some basic correspondences, representing certain sound changes that have affected particular vowels at various times in the history of English. The long -a sound spelled <ā> in *lare* and *stan* has now become /əʊ/ or /ɔː/, usually spelled with an <o> followed by an <e> at the end of the word. If we apply this rule, we can easily identify these two words as *lore* and *stone*. The long /oː/ sound in Old English, spelled <o>, is generally spelled <oo> in Modern English. This correspondence helps in the identification of *god* as *good* and *sona* as *soon*.

In this sample text I have used an edited text as the basis for comparison and this has masked some of the differences between Old English and Modern English spelling, making it easier for us to identify correspondences between the two versions. To give you an idea of some of the ways in which modern editors update Old English texts to make them easier for a modern audience, the following is a comparison of the opening of the famous Anglo-Saxon epic poem *Beowulf* with a facsimile (Figure 3.3) of the sole surviving manuscript copy.

Edited version of the opening lines of *Beowulf*

Hwæt, wē Gār-Dena   in gēardagum,
þēodcyninga,   þrym gefrūnon,

hū ðā æþelingas   ellen fremedon.

  Oft Scyld Scēfing   sceaþena þrēatum,
monegum mǣgþum,   meodosetla oftēah,
egsode eorlas,   syððan ǣrest wearð
fēasceaft funden,   hē þæs frōfre gebād,
wēox under wolcnum,   weorðmyndum þāh,
oðþæt him ǣghwylc   þǣr ymbsittendra
ofer hronrāde   hӯran scolde,
gomban gyldan.   Þæt wæs gōd cyning!

(G.Jack (ed.), *Beowulf*, Oxford, 1994, ll. 1–11)

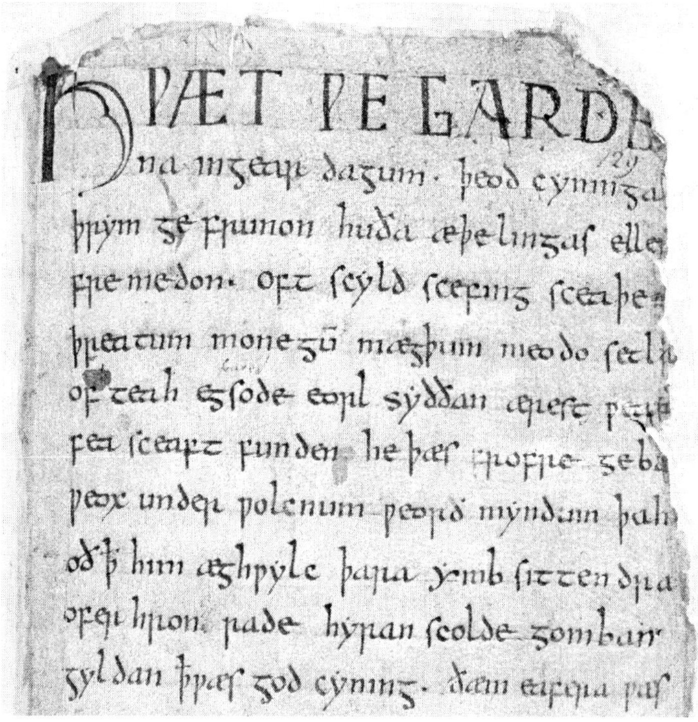

Figure 3.3. The *Beowulf* manuscript © British Library Board

Perhaps the most striking difference that confronts a reader on comparing the edited text with the manuscript concerns their differing layouts. Where the modern edition presents the poem as a series of separate lines, each divided into two half-lines separated by a blank space, the manuscript sets the poem out as if it were prose. Although there are blank spaces between words, the size of these spaces is quite inconsistent, with some words written close together as if a single word, while the two parts of a compound word are often written as if separate words. This is apparent in the second line of the manuscript, where the preposition *in* appears to be joined to the following noun *gēar*, while *gēar* itself is separated from the other element of the compound *dagum*. This rather confusing layout has been tidied up in the modern edition, bringing the word division into conformity with Modern English practice. But if these inconsistencies seem confusing and unhelpful to us, we need to remember that word division and the use of blank spaces between words was a relatively new phenomenon when the *Beowulf* manuscript was written. In Antiquity manuscripts were written using *scriptio continua*, a continuous script without any breaks between words at all. The practice of dividing words in the way we do today was introduced by the Irish monks who brought Christianity to the Northumbrians.

Another obvious difference concerns the relatively sparse punctuation in the manuscript: the commas, semi-colons, capital letters, and exclamation mark are all modern editorial interventions. The only mark used in the manuscript is the 'punctus', the ancestor of our full stop. Another editorial intrusion that concerns spelling directly is found in the opening word: *Hwæt*. The second letter of this word in the manuscript is the runic character wynn, <ƿ>, used in Anglo-Saxon manuscripts to reflect the /w/ sound but commonly replaced by modern editors with the letter <w>. Another editorial convention concerns the addition of the 'macron', the superscript line which is added to mark long vowels. This is a particularly useful aid for distinguishing pairs of words which are otherwise identical in Old English spelling, such as the words *God* and *good*, both spelled *god* in Old English manuscripts. We saw above that the Old English insular script used a number of distinctive letterforms, which can also be seen here in the *Beowulf* manuscript; none of these distinctions are found in the modern edition which simply replaces the insular letters with their modern counterparts. One further difference concerns the treatment of abbreviations. The *Beowulf* manuscript has two of these in the extract reproduced above: the macron over the final letter of *monegu*, signalling an omitted final <m>, and the crossed letter thorn, <ꝥ>, which stands for *that*.

In the two examples of Old English spelling that we have discussed, we have focused on texts written in Late West Saxon, a standardized variety of Old English used widely, but not exclusively, for the writing of Old English. To give a sense of how Old English spelling differed between different dialects, I have included below two copies of the same text, written in different dialects. The text is the short poem in praise of the creation and its creator, recorded by Bede in his *Historia Ecclesiastica Gentis Anglorum*, where he attributes it to the miraculous inspiration of an illiterate cowherd named Cædmon. Bede wrote his account of this miracle in Latin, translating Cædmon's hymn into Latin prose, recording 'the general sense, but not the actual words' used by Cædmon. Bede's work was subsequently translated into Old English and copies survive of this translation written in the Late West Saxon dialect. However, in two of the earliest surviving copies of Bede's Latin original, produced in the early eighth century at Bede's own monastery of Jarrow, an English version of the Hymn has been added in the margin, copied in a Northumbrian dialect. If we compare one of the Northumbrian copies with the Late West Saxon translation, we find a number of spelling differences:

West Saxon version:

Nu sculon herigean heofonrices ƿeard,
Meotodes meahte ond his modgeþanc,
ƿeorc ƿuldorfæder, sƿa he ƿundra gehƿæs
ece Drihten, or onstealde
He ærest sceop eorðan bearnum
heofon to hrofe, halig Scyppend.
Þa middangeard monncynnes ƿeard,
ece Drihten, æfter teode
firum foldan, Frea ælmihtig.

Northumbrian version:

Nu scylun hergan hefaenricaes Uard,
Metudæs maecti end his modgidanc,
uerc Uuldurfadur, sue he uundra gihuaes,
eci Dryctin, or astelidæ.
He aerist scop aelda barnum
heben til hrofe, haleg Scepen.
Tha middungeard moncynnæs Uard,
eci Dryctin, æfter tiadæ
firum foldu, Frea allmectig.

(D. Whitelock (ed.), *Sweet's Anglo-Saxon Reader*,
Oxford, 1984, pp. 467 and 181–2)

Perhaps the most obvious difference is that, where the Late West Saxon copy adopts the runic letter <ƿ> for the /w/ sound, the Northumbrian text carries over the Latin practice of using <u>. The Northumbrian text also lacks the runic letter <þ>, instead employing the digraph <th>. Another spelling difference concerns the Northumbrian use of <b> to represent the /v/ sound in *heben* 'heaven'. Despite this practice appearing in other Northumbrian dialect texts, which include spellings like *ob* instead of *of*, it is only inconsistently employed here; just two words later the word *hrofe*, an inflected form of the Modern English noun *roof*, pronounced with a /v/ sound, is spelled with <f>. Where the Late West Saxon version uses <h> to reflect the velar fricative sound /x/, as in *meahte* and *ælmihtig*, the Northumbrian version spells this sound with <c>: *maecti*, *allmectig*. In its use of the spelling *sceop*, the West Saxon text shows the development of the practice of using a silent <e> to indicate where the preceding consonant was palatalized, in this case distinguishing the /ʃ/ pronunciation of this word from /sk/. The Northumbrian text makes no such disambiguation, simply spelling this word *scop*. Other spelling differences are of a different kind in that they seem to reflect variant pronunciations. The presence of diphthongs in the Late West Saxon *Weard*, *bearnum*, and *heofon*, where the Northumbrian text has *Uard*, *barnum*, and *heben*, reflects a difference in northern and southern accents at this time which is due to much earlier sound changes. Other spelling differences testify to changes affecting the inflexional system of Old English, which underwent a process of decay and loss that was more advanced in the northern dialects. This process of change is apparent in the contrast between the inflected form *foldan* in the West Saxon copy and the Northumbrian spelling *foldu*, showing the loss of the final nasal consonant. Dialectal differences are also apparent in the choice of words; where the Late West Saxon version has the Old English preposition *to*, the Northumbrian version witnesses to a very early adoption of a Scandinavian cognate form *til*.

Although these two texts were copied at different times as well as in different places—the Northumbrian version dates from shortly after Bede's death in 735, while the West Saxon translation was undertaken some time in the ninth century—they do provide valuable traces of the differences between northern and southern dialects of Old English. But, despite these rare and fleeting glimpses into Old English dialect variation, the widespread adoption of the Late West Saxon variety means that our overall picture of Old English is one of considerable stability and uniformity. Such a picture surely masks the considerable diversity that must have been found in the spoken language, of which we sadly have restricted direct evidence. This picture is very different to that of Middle English, an age of astonishing variety and rapid linguistic change, richly and abundantly reflected in its written records, as we will see in the following chapter.

Chapter 4
Invasion and revision

The variety of English used from 1100 to 1500 is known by linguists as Middle English. During this four hundred years the English language changed more radically than in any other period in its history. Many of the characteristic features associated with Old English, the reliance upon special endings to indicate grammatical case and the use of grammatical gender, and the formation of new words by joining two existing words together, known as 'compounding', began to be replaced by different linguistic mechanisms. Much of this was the result of the Norman Conquest of 1066, which saw the English language replaced in its duties as a standard language and relegated to local use. Where the Anglo-Saxons had developed a written language that was used throughout the country for legal, historical, and literary documents, the Normans employed their native French language. The replacement of English with French in these documentary functions had a dramatic effect on the spelling of English. As English was no longer used to communicate on a national level, there was no longer a need for a standard spelling system. As a consequence, writers of English began to spell words in ways that reflected their own local spoken systems more closely. Given that a spelling system is intended to be a guide to pronunciation, this is a logical development, and led to a much closer relationship between spelling and pronunciation in this period. Where individual dialects had ceased to pronounce certain consonant sounds, writers of these dialects were now free to drop these letters from their spellings. In the previous chapter we saw that Old English pronounced words like *which*, *when*, *what*, etc. with an initial aspirate; a pronunciation which has been lost in most varieties of Modern English. This pronunciation, without the initial /h/, first emerged in the Middle English period. We can tell this because spellings of these words began to appear without the <h>: *wich*, *wen*, *wat*. Spellings like these demonstrate how the lack of a standard language enabled writers to adapt their spellings to reflect their pronunciation more closely. As they no longer pronounced these words with an initial /h/, so they ceased to spell them with an <h>. In the northern dialects of Middle English, the Old English /hw/ sound continued to be pronounced and began to be spelled <quh>, so that we find spellings like *quhen* 'when' and *quhy* 'why', perhaps indicative of a different pronunciation to that found in Old English. The northern dialects of Middle English are the ancestors of the Modern Scottish varieties; in Scotland the initial aspirate has been maintained in these words up to the present day. Changes in spelling to reflect local variations in pronunciation like these attested in Middle English would be much more difficult to introduce into a standard spelling system, whose function is to remain fixed and stable.

A similar example of a sound change that took place during this period which has resulted in a major pronunciation distinction between different varieties of English concerns words like *far* and *card*. As the present-day spelling suggests, these words originally had an /r/ sound after the vowel. But, during the

Middle English period, spellings like *cadenall* for *cardinal* appear, indicating that the /r/ sound was no longer being pronounced. But such spellings are not widespread; it is clearly later that the /r/ sound was completely lost from many English accents. In his Grammar of 1640, the playwright Ben Jonson wrote that /r/ was 'sounded firme in the beginning of the words, and more liquid in the middle, and ends' (see Ellis 1874, p. 200). This suggests that the /r/ sound was by this time considerably weakened in these positions, while eighteenth-century phonetician John Walker recorded its complete absence: 'the *r* in lard, bard,... is pronounced so much in the throat as to be little more than the middle or Italian *a*, lengthened into baa, baad' (1791, p. 50). This sound change began comparatively late in the Middle English period, and was only slowly diffused through most of England. As a consequence, by the time it was fully accepted, English spelling had begun to be fixed and spellings with <r> were firmly established.

Another feature of modern English pronunciation which affects the spelling of Middle English is h-dropping, one of the most censured habits of modern pronunciation. Although h-dropping is frequently characterized as a recent phenomenon, typical of lazy, good-for-nothing youths, or the result of the spread of Estuary English, there are instances of words spelled without initial <h> in texts written from as early as the eleventh century, although it is East Anglian texts of the fifteenth century that show the most consistent examples. These include spellings showing loss of initial <h>, and others showing the addition of an <h> where it does not belong, such as *herthe* 'earth', *hoke* 'oak', *herand* 'errand', and *howlde* 'old'.

The history of h-dropping is made more complicated in the Middle English period because of the borrowing of numerous

French loanwords of Latin origin in which the initial /h/ was no longer pronounced. The reason for this was that speakers of post-classical Latin were also h-droppers, which explains why modern Romance languages such as French, Spanish, and Italian are h-less. As an example, we might compare Latin *homo*, with an initial /h/, with its modern equivalents: French *homme*, Spanish *hombre*, and Italian *uomo*. Although both French and Spanish words preserve a written <h>, neither of these languages pronounces these words with initial /h/. There is, in fact, evidence that the omission of initial /h/ was stigmatized by the Romans, and that attempts to avoid such an error led to telling overcompensations. These 'hypercorrections', where initial /h/ is sounded before vowels where it would not normally appear, are mercilessly satirized by the Latin poet Catullus (*ca.*84–*ca.*54 BC) in the following poem:

> 'Hemoluments' said Arrius, meaning to say
> 'Emoluments' and 'hambush' meaning 'ambush',
> Hoping that he had spoken most impressively,
> When he said 'hambush' with great emphasis.
> His mother, her free-born brother and his maternal
> Grandparents, I believe, all spoke like that.
> Posted to Syria he gave the ears of all a rest.
> They heard the same words smoothly and gently spoken
> And had no fear thenceforward of such aspirates,
> When suddenly there came the frightful news
> That after Arrius arrived the Ionian waves,
> Ionian no more, became 'Hionian'.
>
> (Lee (trans.), 2008, LXXXIV)

Because of the loss of initial /h/ in French, numerous French loanwords were borrowed into Middle English without an initial

/h/ sound, and were consequently spelled without an initial <h>; thus we find the Middle English spellings *erbe* 'herb' and *ost* 'host'. But, because writers of Middle English were aware of the Latin origins of these words, they frequently 'corrected' these spellings to reflect their Classical spelling. As a consequence, there is considerable variation and confusion about the spelling of such words, and we regularly find pairs of spellings like *heir, eyr, here, ayre, ost, host*. The importance accorded to etymology in determining the spelling of such words led ultimately to the spellings with initial <h> becoming adopted; in some cases the initial /h/ has subsequently been restored in the pronunciation, so that we now say *hotel* and *history*, while the <h> remains silent in French *hôtel* and *histoire*. The restoration of /h/ in these words is, however, a comparatively recent phenomenon, which explains the tendency in Modern English to write *an historian* and *an hotel*, rather than *a historian* and *a hotel*. But in many examples the <h> is written but not pronounced, as in *honour* and *heir*, while in the case of *herb* it is pronounced in British English, but not in American English. There is a small group of words in which <h> was never restored, such as *able*, borrowed from French *habile* (also spelled *abile*) but descended ultimately from Latin *habilis*.

So, the Middle English period saw a remodelling of the inherited spelling system that reflected regional pronunciations more closely, producing a series of local spelling systems rather than a single national one. This might seem like a useful way of writing English: it must have been considerably easier to learn to use than a standard system based upon a different pronunciation and containing numerous silent letters. While a collection of local spelling systems, rather than a single national standard, may appear to have considerable advantages, it is worth emphasizing the consequent restrictions of such a system. The lack of a standard spelling system meant that there were huge numbers of variant ways of spelling common words, which in turn created huge potential for confusion. There are, for instance, five hundred different spellings of the word *such* recorded in Middle English texts. Some of these would have been relatively easily recognized by users of different dialects, for example *soch*, *swich*, *sech*, *sich*, but others are rather more unusual and would surely have led to confusion if circulated more widely, for example *sik*, *swyche*, *zik*. There was a similarly wide range of variant spellings of *through*, ranging from easily decoded forms like *thurgh*, *thorough*, and *þorowe*, to more opaque spellings such as *drowgȝ*, *trghug*, *trowffe*, and *yhurght*. Variation on this scale can only be accommodated in a system designed purely for local use, where most speakers will only come into contact with spellings with which they are familiar, or which reflect their own pronunciation. As soon as texts written in English are intended for distribution on a national scale, regional variants of this kind become highly inefficient, leading to the potential for considerable confusion and miscomprehension.

This is in fact precisely what happened towards the end of the Middle English period, when English began to be adopted once again as the national language of literature, chronicles, and government records. At first, problems of communication between users of different regional spelling systems were avoided by a system of dialect translation, whereby scribes 'translated' the text they were copying from one dialect to another. A southern scribe who translated the work of the northern writer Richard Rolle into a southern dialect explained that the text had been 'translate oute of Northarn tunge into Sutherne that it schulde the bettir be vnderstondyn of men that be of the selve [same] countre'. But this was an inefficient means of enabling communication between different dialects of English, and relied heavily on the accuracy and consistency of scribes who frequently made mistakes, introducing their own spellings and producing mixed texts that caused even greater problems of interpretation. By the fifteenth century it became increasingly apparent that the solution to the problem was a standard spelling system, and as a consequence such a variety began to emerge.

As we saw in the previous chapter, the Old English standard spelling system had been based upon the West Saxon dialect, associated with the prominent centre of Winchester. By the fourteenth century, the power base had shifted to London, England's largest city and the seat of the royal court, government offices, and the centre of the book trade. As a result, the standard spelling system that emerged was based on the dialect of London. This was a natural development given the prominence of London, but it had major implications for the standard spelling system that emerged. The southern dialects of Middle English were much more conservative than those of the north, which had more immediate and more sustained contact with the Old Norse language of the Viking settlers. Earlier I mentioned that some dialects of Middle English had ceased to pronounce the initial /hw/ of *why*, *which*, *what* and adapted their orthography accordingly. But the more conservative southern dialects continued to reflect the earlier pronunciation in their spelling system and thus maintained the traditional spelling; as a consequence such spellings were adopted into the emerging standard variety. As this new standard formed the basis of the standard spelling system we use today, we continue to spell these words with initial <wh>, despite the fact that the majority of English speakers pronounce them with initial /w/. Another example of a feature of the conservative southern dialects that has been preserved today concerns the spelling of words like *night*, *light*, *sight*, all of which contain the silent letters <gh>. In Old English these words were spelled *niht*, *liht*, *siht*, where the <h> represented a sound rather like that pronounced at the end of the Scottish word *loch*, or in the German equivalent words *nacht*, *licht*, *sicht*. In certain dialects of Middle English this sound ceased to be pronounced and consequently ceased to be written, producing spellings like *nit*, *lit*, *sit*, and *hye* 'high'. Rather than having to learn that the word *knight* is spelled with a silent <k> and <gh>, Middle English children could spell the word *nit*, although its identical appearance to the word *nit*, meaning the egg of a louse, could potentially lead to some unfortunate misunderstandings in accounts of the adventures of King Arthur and the *nits* of the Round Table. The conservative London dialect, however, continued to pronounce this sound, with the result that the spellings *night*, *light*, *sight*, and *high* were adopted in the standard spelling system. The prominence of London in this period made this a logical step, although history has shown it to be an unfortunate development, as the regional pronunciation without the /x/ sound was subsequently adopted more widely. This has had major implications for our modern spelling system, with the result that we continue to spell such words with <gh>, despite the fact that the pronunciation that these letters reflect was dropped some five hundred years ago.

If the modern English standard spelling system had been based upon the northern dialect, the spelling and pronunciation of this group of words would have been very different. In the northern dialect, the /x/ sound developed into /f/, attested by spellings like *thof* 'though', *thruf* 'through' and even the rather unfortunate *dafter* 'daughter'. Some of these dialect pronunciations found their way into the southern dialect and thus survive into modern English, including *laugh*, *enough*, and *rough*; this northern development also explains the origin of the word *duff*, as in *plum-duff*, which originates in the northern pronunciation and spelling of the word *dough*.

The standard spelling system that emerged in fifteenth-century London differed from that of Late West Saxon in a number of key ways, reflecting its fortunes in the intervening centuries. Perhaps the most obvious difference concerns the loss of most of the non-Latin letters that were added to the Latin alphabet employed by Old English scribes. The letter <ð> fell out of use in the twelfth century when it was replaced by the digraph <th>, perhaps because it was considered to be too similar to the letter <d>. However, the rune <þ> survived right up to the fifteenth century, where it is often found in abbreviations like þ^{t} 'that' and in grammatical words like *þe* 'the', and *þis* 'this'. The most decisive factor in the loss of <þ> from the English writing system was the advent of the printing press. Because of restrictions in his printing fonts, which he imported from the continent, William Caxton, England's first printer, tended to use <th> rather than <þ>. Where he did use <þ>, particularly in abbreviations, he employed the similar-looking letter <y>, following a trend that had already been established in the northern dialects. This practice has persisted in faux medieval shop signs such as 'Ye olde tea shoppe', where *ye* represents <þe> and should be pronounced like *the* rather than *ye*. The runic letter <ƿ> continued to be used in England until the end of the thirteenth century, when it was replaced by <u> and by <uu>, the 'double u' from which the letter gets its name.

As well as these straightforward replacements, the Middle English period also saw the tidying up and resolving of certain ambiguities that can be found in the Old English spelling system.

We noted in the previous chapter that Old English scripts drew upon an insular form of <g>, <<ᵹ>> , which could be used to represent two different sounds: the velar stop /g/, and the palatal /j/. Old English scribes who used <g> for both /g/ and /j/ encountered a problem in distinguishing between /g/ and /j/ in initial position. To avoid this confusion some scribes used <i> for /j/, while others adopted the practice of using <e> as a diacritic to indicate that the preceding consonant was a /j/ sound: so we find the word *yoke* being spelled either *ioc* or *geoc*. The Norman Conquest saw the replacement of the insular script with a Carolingian variety, and it is from this script that our modern form of closed <<g>> is derived. This Carolingian <g> was used mainly for /g/, but could also be used for /dʒ/. This Carolingian <g> was known in Anglo-Saxon England, but was reserved mainly for the copying of Latin texts. The insular <g> continued to be used in the Middle English period to represent the /j/ sound, as in Old English, but also the sound /x/; hence its name *yogh*, which is comprised of both the sounds which it represented. Its appearance also changed slightly, so that it came to look more like the number <3> or a French <z>, with a long tail. The letter yogh, <<ȝ>>, was used throughout the Middle English period, but it was replaced in the fifteenth century by the letter <y> in words like *yet*, *you*, and by <gh> in words like *night*, *light*.

Another potential confusion in Old English, which we discussed in the previous chapter, concerned the letter <c>, which was used to represent the /k/ sound before back vowels and /tʃ/ before front vowels. So in Old English we find spellings like *cealc* 'chalk' and *cu* 'cow'. Despite the way that these two different pronunciations were distinguished, this system did result in some potentially confusing word pairs, as in the case of *cinn* 'kin' and *cynn* 'chin'. To avoid such ambiguous spellings the letter <k> was used to represent the sound /k/, especially before front vowels where there was potential for confusion between /k/ and /tʃ/, leaving <c> to represent the same sound before back vowels, for example *can*, *could*. To disambiguate further, a new digraph was added to represent /tʃ/: <ch>. This was adopted from French practice, where it appeared in loanwords such as *champioun* and *chariot*, and then transferred to native words such as *chalk* (OE *cealc*) and *cherl* (OE *ceorl*). Double <cc> was spelled <cch> in Middle English in words like *wacche* (OE *wæcce*). This was subsequently replaced by <tch>, giving the modern practice of spelling such words *watch*, *catch*.

This last example alerts us to another major reason for the differences between Old English and Middle English orthography: the impact of the French language on English. Earlier we noted that the French language used by the Norman invaders replaced English as the standard language used in England during this period. But while the two languages were clearly demarcated in terms of their uses, French for national functions and English for regional use, there was a degree of cross-fertilization. Where Old English had tended not to adopt words from other languages, but rather to rely upon its own resources to form new words, Middle English was very receptive to the introduction of loanwords. When Old English did adopt loanwords, it tended to modify the spelling of the borrowed items to fit Old English conventions. Thus the Latin loanword *phoenix* is spelled *fenix* in Old English, where the Latin use of <ph>, designed to reflect the Greek letter <Φ>, has been replaced by the Old English convention of using <f> for this sound. In the Middle English period, by contrast, French words were often borrowed with their own spelling conventions intact. Thus we find the French words *grace*, *face*, *city* introduced in their French spellings, with the use of the letter <c> to represent the sound /s/. We saw above that the Middle English period witnessed an attempt to regularize the use of the letter <c>, a process which was undermined by the introduction of French words in their native spellings. If the regularization was to be successful, it would have been necessary to respell these loanwords according to English conventions, giving us *grase*, *fase*, *sity*. As it happens, there was an attempt to regularize the spelling system following the introduction of words like these, although this led to the respelling of inherited Old English words according to French conventions, giving us our modern English spelling *ice*, where the Old English word was *is*, as well as *nice* for Old English *nys*, and *once* for Old English *ænes*. As is often the case with such revisions, they were not implemented consistently, in some cases leading to greater confusion. This is apparent from the modern English spelling of the word *mouse*, which is spelled with <s>, as in Old English *mūs*, while its plural form *mice* has been respelled according to French conventions. In some cases, words of French etymology which had <s> for /s/ were respelled according to this convention, giving us spellings like modern English *device* (French *devys*) and *defence* (French *defens*).

A further result of the impact of French conventions on English spelling can be seen in the replacement of the Old English practice of using <cw> to represent the sound /kw/, with the digraph <qu> following the introduction of French loanwords such as *quality* and *quiet*. The adoption of this French convention added an unnecessary complexity to English spelling; the <cw> spelling of Old English was a more logical means of representing these sounds. It is ironic that, having inherited this digraph from Latin, French had retained it, despite the fact that it no longer pronounced the /w/ sound, as is apparent from a comparison of cognates such as Latin *quattuor* 'four' and French *quatre*. So, rather than adjust the spelling of the French loans to fit the native pattern, existing English words were respelled according to the French practices, leading to the respelling of words like Old English *cwen*, *cwic* as *queen* and *quick*. Unlike the previous example we considered, this process was at least carried out consistently, with the result that there are no words today that begin with the sound /kw/ that are spelled <cw>. Contact with French also led to the introduction of the digraph <ou> as an alternative way of indicating /u:/, removing potential confusion with /v/ and /ʊ/. The <ou> digraph appeared first in French loanwords, such as *doute*, and was subsequently transferred to words inherited from Old English, for example *house*, *toun* 'town'.

Borrowing of words from French also led to the introduction of two new letters into the English alphabet, namely <v> and <z>. In Old English there were no words that began with the sounds /v/ or /z/; these sounds only appeared within words. Because of this clear distinction in distribution, the letters <f> and <s>, representing the voiceless equivalents, could be used for both pairs of sounds. So, where <f> and <s> appeared at the beginning of an Old English word they represented the sounds /f/ and /s/, as in *folc* 'folk' and *sunu* 'son'. Where the same letters appeared within a word, it was the voiced sounds, /v/ and /z/, that were intended, as in the words *lufu* 'love' and *wise* 'wise'. The introduction of French words which began with the sounds /v/ and /z/ upset this system, leading to the potential for considerable confusion. The French loanword *vine*, for instance, could not be spelled according to the Old English system, as it would be identical with the word *fine*. How would a reader know which sound, and thus which word, was intended? To remedy this problem the letter <v> was adopted, although initially its use differed from that of modern English. In origin the letter <v> is a variant of the letter <u>, and this is how it was used throughout the entire Middle English and much of the Early Modern periods. Thus the letters <v> and <u> were used to indicate both the vowel and consonant sounds, with the choice of letter depending on its position in the word. When it appeared at the beginning of a word it was written as a <v>, when it appeared within the word it was written <u>, giving rise to spellings such as *vntil* 'until' and *loue* 'love'.

The need for the letter <z> was much less pressing as there were fewer words that began with the sound /z/ that could cause confusion. One example is the word *zeal*, introduced in the Middle English period, and potentially confused with *seal*. But there are few examples of pairs of words of this kind. This small number of words beginning with <z> is still true of Modern English, in which the majority of such words are highly specialized and technical loanwords, such as the Arabic loan *zedoary*, an aromatic plant, the Greek loanword *zygoma*, a part of the skull, and the Italian *zucchetto*, an ecclesiastical skull cap. Nevertheless, the letter <z> was introduced into English spelling during this period and is found in words like *gaze* and *maze*, although the letter <s> continued to function as an alternative to <z>, as it had done in Old English. The marginal status of the letter <z> in English spelling is still apparent today in the use of the letter <s> to represent the /z/ sound in verbs like *realise*, *memorise*, although the American spelling of these verbs with <z> is becoming more widespread, for example *digitize*, *standardize*. The liminality of the letter <z> in English spelling is alluded to by Shakespeare in a memorable insult in *King Lear*, in which Kent calls Oswald 'Thou whoreson zed! thou unnecessary letter!' (Act II scene ii).

The shape of the letter <z>, written originally with a tail as it still is in France, meant that it resembled the letter yogh, as noted in a contemporary description: 'Þe carect [character] yogh, Þat is to seie . ȝ. is figurid lijk a zed'. This similarity provoked confusion for Scottish printers, who often used <z> in place of the unfamiliar yogh, leading to the introduction of spellings such as *zeir* 'year', *ze* 'ye' and *capercailzie* 'capercailye'. This confusion of <z> and <ȝ> also explains the spelling of Scottish proper names such as *Dalziel* and *Menzies*, often incorrectly pronounced with a /z/ sound.

The Middle English period also saw the introduction of the digraph <ea>, which was imported from the Anglo-Norman dialect of French spoken by the Norman conquerors, where it represented the vowel /ɛː/. It appeared initially in loanwords such as *ease*, *reason* and was subsequently transferred to native words like *meat* and *heat*. This allowed a distinction in the writing system between the two sounds /ɛː/ and /eː/, a distinction not found in Chaucer's English. Chaucer spelled the verb 'to meet' *meten* and the noun 'meat' as *mete*, despite pronouncing them differently. But the distinction introduced in the spelling of *meat* and *meet* is less useful today, as both groups of words have merged on /iː/, so that *meet* and *meat* are spelled differently but pronounced the same (for a more detailed discussion of this sound change see pp. 157–9). Subsequent sound changes have meant that the <ea> graph now represents a range of various sounds; compare for instance the vowel sounds in the following group: *pleasure*, *break*, *earth*, *heart*.

Another French practice borrowed in this period was the <ie> graph, which represented /eː/ and now survives in both native and borrowed vocabulary, like *piece* and *friend*. But not all vowel digraphs have their origins in French. Some were adopted from

Old English, such as the <eo> digraph, which in Old English had represented a genuine diphthong, a combination of both vowel sounds, but which in Middle English was employed to represent the mid rounded vowel /ø/ in French loanwords like *people*. French *peuple* still preserves that rounded vowel sound, while present-day English *people* has an unrounded vowel, but has preserved the medieval spelling.

Because the influence of the French language was felt in two distinct phases from two separate sources, there are a number of differences in the French words adopted during this period. The initial wave of French borrowings occurred directly following the Norman Conquest in 1066; words adopted at this time were derived from the Norman French dialect. The second stage of French influence, which was felt in the fourteenth century, comprised words taken from the Central French dialect. This dialect distinction explains why the modern English borrowing *war* differs from its modern French form *guerre*. The difference is due to a pronunciation distinction between the Norman and Central French dialects when these two words were borrowed. In some cases, the same word was borrowed in both its Norman and Central French spellings, giving us doublets such as *warranty* and *guarantee*, *wile* and *guile*, *warden* and *guardian*, *reward* and *regard*. Another difference in pronunciation between these two dialects is preserved in pairs such as Modern English *catch* and *chase*, *cattle* and *chattel*, which attest to a distinction between Norman French /k/ and Central French /tʃ/. Also of Norman French origin are the Modern English words *garden* and *gammon* (compare modern French *jardin* and *jambon*); this alternative pronunciation also led to the doublet *gaol* and *jail*. As you can see from these few examples, where the same word has been borrowed twice, the meaning has often changed to create a distinction in usage. While many English words do have similar meanings, what is known as 'synonymy', it is very unusual for two words to have precisely the same meanings. Where a pair of words borrowed in this way do retain similar meanings, they tend to differ in connotation or register. While *warranty* and *guarantee* do have closely overlapping meanings, the former is generally restricted to specialized legal contexts. In the case of *gaol* and *jail* a dialectal distinction was introduced. The former was initially the preferred spelling in British English, while *jail* was generally adopted in American English; although, as we will see in Chapter 7, this distinction has since changed considerably.

As well as numerous words derived from French, Middle English also adopted a group of loanwords from Latin. The number of Latin loans, however, is comparatively small: Latin loanwords tend to be associated with specialized areas such as religion and learning, including words like *scripture*, *history*, and *allegory*. The use of Latin alongside French in the English administration also led to the introduction of Latin words in this domain; examples include *client*, *conviction*, and *executor*. Determining the extent of Latin borrowing in this period is complicated by the fact that many words which are ultimately of Latin origin entered English via French, so that it is often difficult to determine whether a word was borrowed from Latin or French. Sometimes it is possible to tell whether a word has been borrowed from French or Latin by its spelling, although the unpredictability of spelling in this period means that it is not always a reliable guide. The verb *enclinen* may be spelled either *enclinen* or *inclinen* in Middle English. This variation means that it is very difficult to determine whether it derives from French *encliner* or from Latin *inclinare*. The verb *embrace* can be spelled both as *imbrace* or *embrace* in Middle English, suggesting derivation from either

French *embracer* or Latin *in+bracchium*. In the case of *enquire /inquire*, both the French and Latin derived spellings have survived into modern English as alternative spellings. The same is true of the pair of spellings *ensure/insure*, although in this case a semantic distinction has arisen. Where both spellings were initially used in all senses, *insure* has now become restricted to legal usage. Similar difficulties can be found in determining the origin of certain suffixes. For example, words derived from Latin present participle endings tend to preserve the Latin stem vowel in their unstressed syllables. Thus Modern English *opponent* derives from Latin *opponere* 'set against' and thus has <-ent>, while *ignorant* derives from Latin *ignorare* 'do not know' and so has <-ant>. But this pattern was disrupted by the introduction of French loanwords in which these words were all spelled <-ant>, irrespective of their Latin origin. Thus we also find modern English words like *repentant*, from French *repentant*, but which is ultimately derived from Latin *repaenitere*. Words which end with the suffix <-cioun> in Middle English present similar difficulties. This suffix was added to verbs to form a noun of action, an equivalent to the native ending <-ing>; thus in Middle English we find the earliest attestation of words like *attencioun* 'attention', *attracioun* 'attraction', *confirmacioun* 'confirmation'. This suffix is found in Old French in the form <-cion>, giving rise to the Middle English spellings. However, it is ultimately derived from Latin <-tion>, which has since influenced our modern spelling of such words, so that it is often difficult to tell which language is the source of a particular form.

French and Latin were not the only languages which affected the spelling of English during this period. During the Middle English period we also find large numbers of loanwords from the Scandinavian languages. The appearance of these words in

English for the first time during the Middle English period may seem odd, given that the Viking invasions had taken place as long ago as the eighth century. However, Old English tended to resist the adoption of words from other languages, preferring to form new words by drawing on its own resources, so it is unsurprising that Scandinavian words do not appear in writings from that period.

But if these words were not adopted into Old English following the Viking raids and settlements, how did they survive in England until the Middle English period? The answer is that these words must have been adopted into the spoken language during the Old English period, but were not felt appropriate for use in the more formal written register. Another reason for the invisibility of such words in the written register is the dominance of the West Saxon dialect during this period. As we saw in the previous chapter, this variety, which was based upon the dialects of the south-west, was used for written communication throughout the country. The Viking settlements were located in the east midland and the northern counties of England; as a consequence there was little interaction between Scandinavian speakers and the West Saxons. While English speakers in areas of dense Norse settlement adopted Norse words into their spoken language, their use of a standard language based upon the south-western dialect obscured such adoptions. The loss of the Old English standard language, and the subsequent adoption of regional spelling systems in the Middle English period, led to the introduction of large numbers of Scandinavian loanwords into English. Many such words began with the sounds /sk/, like *skin* and *sky*; a sound which did not appear in Old English as a result of an earlier sound change, which predates the earliest written records, in which this sound had become /ʃ/. In Old English, the sound /ʃ/

was spelled <sc>, which created the potential for confusion with words pronounced with /sk/, as in the pair of words *scyrte* 'shirt' and *skyrta* 'skirt'. The problem was compounded after the Norman Conquest as words were introduced from French with initial /sk/ and spelled with <sc>, such as the word *scale*, as well as words spelled <sc> but pronounced with initial /s/, for example *science*. Various remedies were implemented to rectify this problem. The simplest was to replace the <sc> spelling of /ʃ/ with <sch> or <sh>, in which the additional <h> serves as a diacritic to indicate that the consonant represents the palatal sound, just as it does in the combination <ch>. The other remedy was to distribute the two spellings for /sk/ according to etymology. As a result of this process, words in modern English spelled with <sk> tend to be Germanic loans, from Old Norse, for example *skin*, *sky*; or of Dutch origin, for example *skipper*, *skate*. Those with <sc> are drawn from Old French, for example *scare*; although some are also of Greek extraction, for example *scope* (Greek *σκοπος*). In some words there is variation between <sk> and <sc>, as in *sceptic/skeptic* (the *OED* gives both spellings in its headword). This word was introduced into English from French *sceptique*, which was pronounced with initial /s/ rather than /sk/. The /sk/ pronunciation adopted in English is due to influence from its Greek etymology, and this led to early spellings with initial <sk>. However the <sc> spelling subsequently replaced <sk> and has become the more common in Britain, although *skeptic* is now widely accepted in the USA.

Another important change that took place in the Middle English period concerns the use of the letters <i> and <y>. In Old English the letter <y> was used to represent a vowel sound similar to that found in the French word *tu*; in Middle English this sound became identical with the sound /ɪ/ and so could be spelled with the letter <i>. As a result of this, the letters <i> and <y> became interchangeable, and remained so throughout the whole of the Middle English period. There were, however, some constraints upon the use of these letters. In medieval handwriting the letters <i, n, m, u> were made up of identical individual strokes known as 'minims', with the result that a word like *sin* was written as a letter <s> followed by three identical minim strokes. Combinations of minim strokes were often difficult to decipher: should four minims be read as <ini>, <un>, <nu>, <im>, <mi>? One way of reducing this potential for confusion was to employ the letter <y> in such positions, as in the Middle English spelling *synne* 'sin'. In Modern English spelling <y> has been replaced by <i> in such cases, although it has become conventional to use <y> to represent /ɪ/ at the ends of words, for example *city*, *family*, although <i> is employed when such words take a plural ending, as in *cities*, *families*. One group of exceptions to this rule is those words where final <y> follows a vowel, for example *toy*, *play*, which preserve the <y> in their plural forms: *toys*, *plays*. The use of <y> for /ɪ/ in other contexts is restricted to loanwords, as in its use to represent the Greek letter upsilon, <Y>, in *physics*, *psychology*. An alternative way of avoiding the confusion caused by multiple minims was to use <o>, which gives rise to modern spellings such as *son* (OE *sunu*) and *come* (OE *cuman*).

These examples provide further evidence of the practical considerations that have contributed to the English spelling system, often at the expense of phonetic factors. Here, a constraint derived from medieval handwriting practices, and thus of no relevance to our post-print era, continues to be observed in our spellings of common words like *come*, *son*, and *love*, which seem to indicate a completely different vowel sound. One further method of distinguishing between minim strokes employed by medieval scribes was to add an accent above the letter <i>; this accent subsequently developed into a curved flourish and from there to the round dot employed up to the present day.

We began this chapter by considering the effects of the Norman Conquest and the break-up of the Late West Saxon standard written language on the spelling of Middle English. We saw that during this period there was no supraregional standard, but rather a series of local and individual attempts to devise consistent ways of representing sounds with spellings. One well-known instance of this is what has come to be known as 'AB language': so-called because it is represented by two manuscripts, one containing an instructional treatise for women intending to become anchoresses (female recluses), known as *Ancrene Wisse* 'Guide for Anchoresses' (A), and the other containing various related lives of female saints and religious tracts (B). Where most early Middle English texts are relatively prodigal and idiosyncratic in their spelling habits, these two manuscripts are very closely related. This close linguistic relationship was first spotted by J. R. R. Tolkien, who coined the uncharacteristically unimaginative term AB language in an influential essay on the topic. Because these texts were copied by different scribes using a very similar spelling system, Tolkien assumed that they represent an early attempt to devise and enforce a standard language, presumably within a monastic school or scriptorium. A single example of the orthography of these two manuscripts will give a flavour of the degree of detailed attention that was paid to standardizing this aspect of their orthography. In both of these manuscripts the scribes adopted the well-established convention of using <þ> only at the beginnings of words and <ð> only in the middle and at the ends. But there is a single word that consistently disrupts this pattern, using <þ> in medial position. This is the adjective *oþer* 'other', where the letter thorn was consistently employed to enable it to be distinguished from the homophone *oðer*, the conjunction 'or'. But while AB language was indeed a highly consistent and meticulously observed system, devised by someone with considerable interest in orthography, it would be wrong to think of it as a 'standard language'. While AB language was used by more than one scribe, it is only found in two closely related manuscripts written at the same time and in a similar location. To be considered a standard language, AB would need to have been employed on a wider geographical scale, by many more scribes, and in a much greater range of types of text.

A more extreme example of an individual attempt to revise and reform English spelling during the early Middle English period was carried out by an Augustinian canon from Bourne in Lincolnshire, named Orm. Orm wrote an extensive metrical paraphrase of the Bible, which he modestly called *Ormulum*, after himself. The surviving work is 20,000 lines long, although the table of contents reveals that the original work was closer to 160,000 lines. The *Ormulum* survives in a single manuscript, whose shape and appearance, including the presence of numerous corrections and revisions inscribed on scraps of parchment pasted to the main text, suggest that it was the author's own working draft. Like other early Middle English scribes, Orm was concerned with devising a spelling system that would accurately reflect the way the text was to be pronounced, especially important given that the work was written in verse and intended for reading aloud. Where Orm differed from other contemporary copyists was his greater concern with absolute regularity and consistency.

Examples of Orm's idiosyncratic spelling system include his use of the insular <g>, <<ᵹ>>, for the /j/ sound in words like *year*, as well as an adapted version of the Caroline <g>, <<ḡ>>, with a thick bar at the top for the hard /g/ in words like *god*; he also distinguished between the use of <ȝh> for the fricative sound /x/ in words like *leȝhenn* 'to tell a lie', where the <h> is raised above the line, and the use of <ȝh> in the single word *ȝho* 'she', where the <h> is written on the line. The consistency with which this distinction was observed clearly indicates that the alternative positions of <h> were intended to signal a phonetic difference. It is thought that the combination <ȝh> in the single word *ȝho* was intended to represent the sound /hj/, and that this form of the feminine pronoun *she* is important evidence of an otherwise unattested stage in the development of present-day *she* from the Old English equivalent *heo*.

In addition to these spellings which attest to revisions of Old English practices, the *Ormulum* is also the first Middle English text to use <wh> in words like *which*, *when*, *where*, and to employ the <sh> digraph in place of Old English <sc> in *shall* and *should*. But, alongside this tendency to revise and replace Old English spelling practices, Orm continued to use many of the characteristic Old English letterforms, such as long-r, long-s, <ƿ>, and <æ>. In addition to implementing a number of changes, which are common to many early Middle English texts, Orm also imposed a number of further revisions of his own devising. Probably the most conspicuously unusual feature of Orm's spelling system is the prodigal use of double consonants. The system of doubling consonants to indicate where a preceding vowel is short is still used today: it is how we distinguish the long and short vowels in the words *later* and *latter*. But Orm took this to its logical extreme, and added double consonants after all such examples, even in the case of common words like *iss*, *wass*, *itt*, and in endings like <-ess> for noun plurals.

Orm's system was evidently meticulously planned in advance and executed with considerable care and effort. Nevertheless, a number of mistakes crept in which were subsequently corrected, while a number of further corrections show Orm changing his mind during the process of composition. For instance, Orm began using the Old English derived digraph <eo> in a large number of words, for example *þreo* 'three', *steorre* 'star', *deope* 'deep', but he subsequently switched to the consistent use of <e>, *þre, sterre, depe*. This change shows Orm, perhaps reluctantly, acknowledging a change in pronunciation that had already taken place in the spoken language. Being Orm, it was not enough to simply make the change from the point at which the decision was made: he returned to the beginning of the manuscript and changed every instance of <eo> to <e>. Of the several hundred instances of <eo> corrected to <e> by Orm, just three instances were missed, a testimony to his dedication to orthographic consistency. But not all such changes provoked Orm to return and correct earlier instances. For the first two-thirds of the work Orm used both <þ> and <ð>, although the latter was much rarer, and used mostly at the beginnings of lines. But in the final third of the manuscript Orm abandoned <ð> and relied solely upon <þ>. At a similar point in the copying process, Orm ceased using the <ph> digraph in Latin words, from that point preferring spellings with the native <f>, for example *profete* 'prophet'. However, despite this clear change in preference, Orm did not return to previous examples and revise their spelling. Perhaps he intended to do so but never completed the process of checking and correction, or, less likely given his orthographic obsessiveness, perhaps he deemed these variations of less consequence.

A final example of an early Middle English writer with demonstrable interests in spelling is a Worcestershire priest named

Laȝamon who wrote a verse chronicle in English, translated from an Anglo-Norman version, in the early thirteenth century. In addition to providing a history of Britain, Laȝamon's *Brut* contains the first account of the deeds of King Arthur in English. It is written in a verse form that revives alliterative practices of Old English poetry and which uses many distinctive poetic words associated with that verse. As part of that nostalgic evocation of the past Laȝamon employed a number of archaistic spelling forms. As we have seen, the Middle English period witnessed the demise of the letter <æ>, which was no longer required. However, for Laȝamon this letter evidently carried archaic connotations, and he employed it liberally in words where it had never appeared in Old English. Although we cannot be sure of Laȝamon's intentions in introducing these unusual spellings, it seems likely that they are consciously intended to evoke a nostalgic and antiquarian flavour rather like 'ye olde' spellings today.

Towards the end of the Middle English period, varieties of written English appeared which began to be adopted more widely. These were focused on the dialect of London; it was this variety that became the basis of the written standard that emerged in the fifteenth century and which formed the basis of our modern standard variety. This London dialect is close to that used by the fourteenth-century poet Geoffrey Chaucer, although there are a number of differences which appeared in the fifteenth century as the result of contact with more northerly dialects, the result of immigration into the capital from the northern and midland counties. The influence of the northern dialects during the fifteenth century accounts for the differences between the spelling of common words like Chaucer's *her* and *hem* (derived from Old English) and modern English *their* and *them* (derived from Old Norse); Chaucer's *yeve* (from Old English) and modern English *give* (from Old Norse). The variety of London English that emerged in the fifteenth century is sometimes referred to as 'Chancery Standard' because of its association with the offices of state, the Chancery, Privy Seal, and Signet Offices. This bureaucratic association helped to lend Chancery Standard status and authority, which in turn led to its gradual adoption as the basis of the standard written variety of English. However, the process by which this variety was adopted was by no means immediate or straightforward. By the end of the Middle English period, a standardized variety of written English had emerged, one which is recognizable as the ancestor of our modern written standard. But, while this process of standardization was to be maintained and accelerated within the Early Modern period, other factors intervened that led to the introduction of further reforms and revisions.

To give a glimpse of what the Middle English of this period looks like, I include here an extract from Chaucer's *Canterbury Tales*. This excerpt is the opening of the General Prologue, taken from the standard modern edition of the poem, the *Riverside Chaucer*, used in University English departments to introduce undergraduate students to Chaucer's works.

Whan that Aprill with his shoures soote	showers; sweet
The droghte of March hath perced to the roote,	drought; pierced
And bathed every veyne in swich licour	such liquid
Of which vertu engendred is the flour;	power; flower
Whan Zephirus eek with his sweete breeth	the west wind
Inspired hath in every holt and heeth	wood
The tendre croppes, and the yonge sonne	shoots
Hath in the Ram his half cours yronne,	the sign of Aries; run

And smale foweles maken melodye,   birds
That slepen al the nyght with open ye   eye
(So priketh hem nature in hir corages),   incites; hearts
Thanne longen folk to goon on pilgrimages,   desire
And palmeres for to seken straunge strondes,   pilgrims; foreign shores
To ferne halwes, kowthe in sondry londes;   distant shrines; known
And specially from every shires ende   particularly
Of Engelond to Caunterbury they wende,   travel
The hooly blisful martir for to seke,   seek
That hem hath holpen whan that they were seeke.   helped; sick

While there is a fair amount of spelling variation that is foreign to modern spelling practices in the above extract, you may be struck by how many similarities there are. Many words are spelled just as they are today, not just small function words like *the*, *in*, *with*, *they*, *to*, but longer words too, such as *specially*, *every*, *april*, *march*, *bathed*, *nature*, *pilgrimages*, *shires*. Although these words were pronounced differently by Chaucer, their spelling is the same today. While there are many differences in the spelling of individual words in this extract, these differences are fairly minor. The most common differences include the tendency to add an extra <e> on the ends of words, as found in *roote*, *melodye*, *ende*, and the use of <y> where we would use <i>, *nyght*, *veyne*. Perhaps the most striking differences concern the representation of the vowel sounds. In some cases, these are simply different conventions of representing similar pronunciations, as with *yonge*, *sonne*, and *sondry*, with <o> instead of <u> to avoid confusion caused by consecutive minim strokes, or with the double <o> to indicate the long vowel in *hooly* and *goon*. Different conventions in the use of consonants are also apparent, in the use of <k> before back vowels (compare *kowthe* with its Modern English equivalent *couth*), and in the use of <c> instead of <qu> in *licor*. But in other cases these spelling differences represent genuine differences in pronunciation, as in *Engelond* and *londes*, for modern spellings with <a>: *England* and *lands*. Some of the differences in pronunciation are masked by the similarities in spelling; Chaucer would have pronounced the <l> in *folk* and would have pronounced *whan* with a /hw/ sound at the beginning. Because we still spell these words the same but no longer use these pronunciations, it is easy to overlook such differences. However, the familiarity of much of the spelling of this extract is in part due to modern editorial practice, which tends to modernize the spelling of a Middle English writer like Chaucer. If we compare this extract with the equivalent lines as they appear in the earliest surviving manuscript copy of the *Canterbury Tales*, then we find considerably more differences.

> Whan that Aueryll w^t his shoures soote
> The droghte of March / hath p*er*ced to the roote
> And bathed euery veyne in swich lycour
> Of which v*er*tu engendred is the flour
> Whan zephirus eek w^t his sweete breeth
> Inspired hath in euery holt and heeth
> The tendre croppes / and the yonge sonne
> Hath in the Ram / his half cours yronne
> And smale foweles / maken melodye
> That slepen al the nyght with open Iye
> So priketh hem nature / in hir corages
> Thanne longen folk to goon on pilgrymages
> And Palmer*es* for to seeken straunge strondes
> To ferne halwes / kouthe in sondry londes

And specially / from euery shyres ende
Of Engelond to Caunterbury they wende
The holy blisful martir for to seke
That hem hath holpen whan þ[t] they weere seeke

(Taken from the 'Hengwrt' manuscript;
National Library of Wales, MS Peniarth 392D)

Reading the same passage in its manuscript spelling reveals a number of additional differences that have been masked by the process of editorial modernization. Perhaps most striking is the use of abbreviations in the manuscript, *w[t]* for 'with' and *þ[t]* for 'that'. The latter is particularly interesting, as it shows how the runic letter thorn was still in use in the fifteenth century, even in the works of Chaucer. The tendency for editors to remove the letter thorn in modern editions of Chaucer is quite different from the way other Middle English texts are treated. The poem *Sir Gawain and the Green Knight*, for instance, written by a contemporary of Chaucer's, is edited with its letters thorn and yogh left intact. This difference in editorial policy is perhaps a reflection of different attitudes to the two authors. Because Chaucer is seen as central to the English literary canon, there is a tendency to present his works as more 'modern', thereby accentuating the myth of an unbroken, linear tradition. The anonymous poet who wrote *Sir Gawain and the Green Knight*, using a western dialect and the old-fashioned alliterative metrical form, is further cut off from this literary canon by presenting his text in authentic Middle English spelling.

Chaucer exploited the possibilities provided by variant spelling on one memorable occasion in the *Canterbury Tales*, where he employs alternative spellings to represent a northern pronunciation distinct from his own southern accent. *The Reeve's*

Tale concerns two Cambridge students, called Aleyn and John, who come from 'fer in the North'. As well as giving the students this northern pedigree, Chaucer also gives them northern accents. But, far from being a consistent representation of northern speech patterns, the students' northern accents are characterized by a few distinctive features. For example, the students' speech includes numerous words with an <a> which Chaucer would usually spell with an <o>, words like *na* 'no', *banes* 'bones'. This is not simply spelling variation; the <a> spellings reflect a distinctively northern pronunciation in which the Old English long-a sound in words like *ham* 'home' and *gan* 'gone' was preserved, where southerners had begun to pronounce these words with a long-o sound. It is tempting to assume that this depiction of a northern accent is an early instance of a north–south prejudice that is apparent in modern English society. But the facts are more complex than this simple solution allows, and scholars continue to debate the possible reasons for this early instance of northern dialect depiction. For while Chaucer may have been a southerner writing for a largely southern audience, the two students with northern accents are Cambridge undergraduates and of a much higher class than the Miller in the tale, who speaks with a southern accent. It is also notable that, despite being tricked by the Miller, they do in fact get their revenge, and they are the ones who come out with the advantage at the end of the tale.

So the Middle English period is characterized by orthographic variety and a wealth of individual spelling systems, rather than a single supraregional standard. As such, Middle English spelling challenges our modern acceptance of the need for a single spelling standard and the unquestioned assumption that there is only one correct way of spelling every English word. Nevertheless, shifting social conditions meant that this situation began to change towards the end of this period, with the result that the London dialect was selected as the foundation for a new spelling standard which would continue to be adapted and codified in the early Modern period that followed.

Chapter 5
Renaissance and reform

The spelling of Early Modern English (1500–1700) is much more familiar to modern English readers than that of Old or Middle English. It shows considerably more consistency, and the relationship between spelling and pronunciation is closer to that of our contemporary spelling system. It might therefore seem as if there is little to discuss in this chapter dealing with the relationship between spelling and sound in the Early Modern English period. But in fact this is a period in which the spelling system and its effectiveness as a means of representing the spoken system came under widespread scrutiny, leading to a wealth of publications on the subject. The issue was taken up by a group of spelling-reformers, concerned with adapting the spelling system in order to create a closer relationship between speech and writing. Another group of reformers had a quite different goal: they wished to see the spelling system revised so as to reflect etymology, that is the spelling of the roots from which the words were derived, thereby widening the gap between spelling and pronunciation. An increase in literacy in this period led to a concern with the teaching of spelling to children, and consequently a debate was initiated concerning the best way of educating children in a system that is rife with inconsistency and complexity. The result was a series of publications concerned with explaining the rules that lie behind the spelling system, and the means by which they may be learned. Inevitably, such concerns also led to a desire to see the spelling system fixed, so that children could be taught the single acceptable way of spelling a word, rather than one of several optional variants.

The desire to establish a fixed spelling system was sparked by the process of standardization that had begun to emerge at the close of the Middle English period, but which became fully enshrined within English society during the Early Modern period. The process of standardizing English also led to the establishment of an ideology of standardization, a notion that there are right and wrong ways of using language and that variation should be suppressed. In establishing and fixing a standard spelling system, reformers were also driven by a desire to give English spelling greater prestige, so that it might stand as a worthy rival to the classical languages. By comparison with their fixed spelling systems, English spelling appeared chaotic and unstable; greater stability and fixity was therefore seen as necessary to ensuring its enhanced status. This was, of course, a false comparison in that classical Latin and Greek were not living languages and thus no longer subject to the constant variation and modification that are features of all languages in use.

The reform of English spelling was addressed by a number of books on spelling which were published between 1540 and 1640, each proposing various alterations of the spelling system, some more radical than others. The major concern of these spelling reformers was the lack of correlation between spellings and sounds. As we saw in the previous chapter, this was in part a problem inherited from the Middle English period, when Norman French conventions were applied to the spelling of English words. But it was further complicated in the Early Modern English period, as spelling became standardized and pronunciation changed, leaving the gap between spelling and speech even wider.

As part of an attempt to accord greater status to English, the spellings of certain English words were adjusted to align them with their presumed Latin etymons, the words from which they were thought to derive, creating a further discrepancy between spelling and sound. So the letter <b> was added to the Middle English spellings *dette* 'debt' and *doute* 'doubt' to align them with the spelling of the corresponding Latin words (*debitum* and *dubitare*), even though they were in fact borrowed directly from French words without the <b>. The letter <c> was added to Middle English *vitailes* to give *victuals* (Latin *victualia*); it was also added to Middle English *sisours* to give *scissours* (Latin *scissor*; compare modern English *scythe*); <u> was added to Middle English *langage* producing *language* (Latin *lingua*); a <p> in Middle English *receite* (Latin *receptum*); <l> in Middle English *samon* to give *salmon* (Latin *salmo*). The Middle English word *quire*, modern English *choir*, was respelled to reflect the Latin word *chorus* more closely, although, as with all these examples, the pronunciation remained unchanged. The word *nephew* has a particularly complex history. It was borrowed from Old French *neveu* in the Middle English period and retained the French pronunciation with /v/. Attempts to align its spelling more closely with its Latin etymon *nepos/nepot-* (compare modern English *nepotism*) led to the insertion of a silent <p>, giving the spelling *nepveu*. This change opened the door to a wealth of different spellings, including ones with <f> and <ph>: *nefeu*, *nepheu*. It was the <ph> spelling which finally became adopted; the use of this spelling ultimately led to a pronunciation with /f/ rather than /v/.

In some cases these adjustments led to a corresponding change in pronunciation. Middle English *aventure* began to be spelled *adventure*, to reflect the Latin form *adventura*; the new spelling led to a change in pronunciation. A similar situation affected Middle English *avis*, which became *advice* by comparison with Latin *advisum*. Middle English *perfeit* shows the spelling of the French word from which it was borrowed. It was only after 1500 that a spelling with a <c>, based upon the Latin *perfectus*, started to appear, which subsequently influenced the pronunciation of this word. The reintroduction of the <l> in *fault* (Middle English *faute*), has also led to its pronunciation in Modern English. The Middle English spelling *cors* began to be spelled with a <p>, by analogy with Latin *corpus*, which subsequently began to be pronounced. Interestingly, the addition of the <p> has also occurred in French, although this has never been pronounced; in fact French has since ceased to pronounce the final <s> as well. A subsequent change in the English spelling of this word was the nineteenth-century addition of a final <e> to distinguish *corpse* from the military word *corps*, meaning an army unit. But, while these instances show spelling and pronunciation changing in tandem, in many cases these changes simply widened the gap between spelling and pronunciation. The drive towards etymological spelling also led to the introduction of the Latin digraphs <æ> and <œ> in words of Latin origin, producing spellings like *æstimate* 'estimate' (Latin *æstimare*), *æqual* 'equal' (Latin *æqualis*) and *œconomy* 'economy' (Latin *oeconomia*), although these have not survived into Modern English.

In some cases the respellings were based upon entirely false etymologies. For instance, the Middle English word *iland* was respelled *island* on the basis of a comparison with the French word *isle*, derived ultimately from Latin *insula*. But Middle

English *iland* was not a French borrowing at all; rather it derives from Old English *ieȝland*. Confusion between *isle* and the unrelated Middle English word *eile* led to the parallel insertion of an <s> into the latter, leading to our modern English spelling *aisle*, derived from French *ele*, and ultimately from Latin *ala* 'wing' (an aisle is metaphorically a wing of a church). Another example is the Middle English word *amyrel*, which was respelled as *admiral* on the assumption that it was derived from a form with the Latin <ad-> prefix, as was the case with the words *adventure* and *advice*. The word did indeed enter Middle English from Old French, where it is spelled *amiral*, but its ultimate origin is the Arabic word *amir*. Etymological respellings were not just influenced by supposed Latin etymons; there were also attempts to reform spelling to reflect supposed Greek etymologies. The Middle English words *autor* and *anteme* were respelled as *author* and *anthem* because they were thought to derive from Greek etymons spelled with the letter <θ> 'theta'.

The contemporary view that spelling should reflect etymology was satirized by Shakespeare in his play *Love's Labour's Lost*, which draws on many of the debates concerning the status of English in this period. His character Holofernes is a pedant who insists on correct usage and who favours etymological spellings, arguing that etymologically respelled words should be pronounced as they are written. This leads him to propose that the unhistorical <b> in words such *doubt* and *debt* should be pronounced, along with the <l> in *calf* and *half*, as he explains in the following speech:

PEDANT

He draweth out the thred of his verbositie, finer then the staple of his argument. I abhore such phanaticall phantasims, such insociable and poynt deuise

> companions, such rackers of ortagriphie, as to speake dout *sine* b, when he should say doubt; det, when he shold pronounce debt; d e b t, not d e t: he clepeth a Calfe, Caufe: halfe, haufe: neighbour *vocatur* nebour; neigh abreuiated ne: this is abhominable, which he would call abbominable, it insinuateth me of *infamie*: *ne intelligis domine*, to make frantick lunatick?
>
> (*Love's Labour's Lost*, V.i.17–23)

Holofernes' desire that spelling should reflect pronunciation leads him to propose changes to pronunciation rather than changes to the orthography. A more common solution proposed by reformers at the time was to reform spelling to reflect pronunciation, although in some cases analogy with other English words led to further complications in the relationship between writing and speech. As we saw in the previous chapter, the words *light* and *night* were once pronounced with a medial fricative consonant, hence their spelling with medial <gh>. But in the Middle English period this sound was lost in such words, although because of standardization the spelling remained unchanged. Thus *light* and *night* were now pronounced as in the second syllable of the French loanword *delit* 'delight'. But to tidy up the inconsistency of having two words with the same pronunciation spelled differently, the word *delite* began to be spelled *delight*. Superficially, this might seem like a logical step, although it would surely have been more sensible to remove the <gh> from *light* and *night*, thereby bringing their spelling closer to their actual pronunciation. In the cases of *despite* and *spite*, both words began to be spelled *despight* and *spight* in the sixteenth century, but these analogical spellings were soon replaced by earlier spellings without <gh>. A more extreme instance of this kind of respelling of a Latin borrowing according to Germanic spelling practices can be seen in the history of *haughty*, which is derived from Latin *altus*, but which acquired an initial <h>, <gh>, and final <y> in an attempt to nativize it.

It was a desire to restore the link between spelling and sound that motivated the spelling reformers of the sixteenth century. One of the earliest of these reformers was Sir John Cheke, the Regius Professor of Greek at Cambridge University, whose interest in spelling stemmed from his involvement in the Cambridge controversy over the correct pronunciation of classical Greek which raged throughout the 1530s. This controversy was sparked by Erasmus, who proposed that the pronunciation of classical Greek should be based upon a reconstruction of its sound system using the written language, rather than on the contemporary pronunciation of Greek. Cheke followed Erasmus's arguments and adopted this reconstructed pronunciation in his teaching at Cambridge. This seems to have been a popular move with students, but sparked opposition from more conservative members of the university, in particular its Chancellor, Stephen Gardiner, who issued a decree in 1542 prohibiting its use; anyone found using this pronunciation faced expulsion and caning. Gardiner argued that the reconstructed pronunciation was based upon dubious evidence and was little more than an attempt to revive a pronunciation that had been superseded by a better one. Cheke countered by setting out the principles upon which the reconstructed pronunciation had been built, discussing the individual sounds in detail. This debate led to a reconsideration of the nature of the English spelling system and its adequacy as a means of representing the English sound system, and the devising of a reformed spelling system which better reflected contemporary speech. Although Cheke did not produce any defence or explanation of his proposed spelling reforms, he employed his reformed system in a translation of the Gospel of St Matthew, and in a partial translation of the first chapter of St Mark's Gospel, dated *ca.*1550. In his desire to tidy up the problems presented by final

<e>, which often appeared at the ends of words but was not sounded, Cheke used doubled vowels to indicate long vowel sounds, for example *taak*, *haat*, *maad*, *mijn*; employed doubled consonants to indicate where a preceding vowel was short, *Godd*; and omitted final <-e> in words where it had no function, for example *giv*, *belev*. Where both <y> and <i> were used for /ɪ/ Cheke used only <i>, for example *mighti*, *dai*. Cheke's reforms were fairly systematic, although there were inconsistencies, such as the spelling *ned* not *need*, and alongside the spelling *fruut* we find *fruit* and *frute*. Cheke also introduced special letterforms, including a form of <g> with a curling headstroke intended to represent /dʒ/, and a <y> with an acute accent above it for the sounds /ð, θ/, presumably an attempt to revive a version of the Old English letter <þ>. Above all, Cheke's system was idiosyncratic, representing little more than his own preferences. The following is an extract from his translation of the Gospel of St Matthew (chapter 14):

> Jesus heering y^{is} went from ýens in a boot himself aloon, into á wildernes. ye pepil heering y^{is} cām folowed him out of ye citees on foot. Jesus cōming forth and seing great resort ýeer piteed ýem and healed ýeer diseased. And whan it was som thing laat, his discipils cam vnto him and said, This is á wild place, and ye tijm is wel goon, let ýis resort go now, y^{t} y^{ei} maí go into villages and bi ýemselves sōm meat. ýei have no need said Christ to ýem to go awaí. Giue yow ýem sōm meat.
>
> (Goodwin (ed.), 1843, p. 61)

Cheke's biblical translations are also of interest for the author's determination to employ native words rather than borrowed vocabulary. Where there were no native equivalents, Cheke resorted to the introduction of 'calques', or loan-translations, where the concept is translated into English, as in the case of modern English *superman*, a literal translation of the German *Übermensch*. Examples of calques, termed 'trutorns' by Cheke, in his translation of St Matthew's Gospel include *gainbirth* 'regeneration', *gainrising* 'resurrection', *onwriting* 'superscription', *moond* 'lunatic', *biwordes* 'parables', *hundreder* 'centurion', *washing* 'baptism', *forschewers* 'prophets', *frosent* 'apostle'. Unsurprisingly, such ungainly and idiosyncratic terms did not catch on, although they do prefigure the purist vocabulary of later Protestant writers such as Edmund Spenser.

Another Cambridge scholar who supported Erasmus's reformed pronunciation of classical Greek was Sir Thomas Smith. Smith was appointed Regius Professor of Civil Law at Cambridge in 1543, and in the same year he assumed the role of Vice Chancellor. Smith's support for Cheke's position came with the publication of a tract on the pronunciation of Greek, published in 1568. In this work Smith argued that the correct pronunciation of a classical language should be that used by its greatest writers; in the case of Latin this was the pronunciation used by Cicero and his age. Subsequent pronunciations represent a corruption of the correct usage and so have no authority for contemporary scholars. Just as no Frenchman would recognize French spoken with an English accent as his own language, so the classical Greek language should not be spoken according to contemporary standards of pronunciation. Smith's work recounts the process by which he and Cheke came to introduce their pronunciation, first practising it together in private and only then introducing the reforms in their lectures.

Alongside his work on the pronunciation of Greek, Smith also wrote a companion tract on the pronunciation and spelling of English, the first tract advocating reform of the English spelling system: *De recta et emendata linguæ anglicæ scriptione dialogus* (1568). Smith argued that the number of letters in the English spelling system should be the same as the number of ‘voyces or breathes in speaking and no more’. Where sounds exist in English for which the Roman alphabet does not provide an equivalent letter, a new letter should be introduced, rather than ‘abusing’ existing letters by making them represent more than one sound.

Smith thus proposed introducing a number of new letters into the alphabet, some of his own devising, while others were introduced from Greek and from earlier varieties of English. So he proposed that the letter <g> should only be used for the hard /g/ sound, while the sound /dʒ/ should be represented by the reintroduction of the Old English insular <g>: <<ȝ>>. Smith appealed to Old English usage in his proposal to use <c> for /tʃ/, reserving the letter <k> for the sound /k/. He objected to the use of <h> as a diacritic to ‘soften’ other letters, as it does in <sh>, <ch>, and <th>; its true use should be limited to the representation of an aspirate. For the sound /ð/ Smith employed two different letters, the Greek delta <Δ> and what he called ‘thorn d’, that is the Anglo-Saxon letter ‘eth’: <ð>. For the /θ/ sound he proposed a further two letters, one Greek and the other derived from Old English: the Greek theta <θ> and the Old English thorn <þ>, giving spellings like *θin* ‘thin’ and *þik* ‘thick’ and *Δöu* ‘thou’ and *ðër* ‘there’. The use of two different symbols to represent each of these two sounds runs contrary to Smith’s view of an exact correspondence between sounds and letters, and it is unclear why he felt it necessary to propose alternatives for these sounds in particular. The explanation presumably lies in the fact that these reforms were merely offered as alternative suggestions from which a single solution could be chosen. A further anomaly in Smith's proposals is that he retained the letter <x>, which represents two sounds, and which therefore violates his strictly phonetic principles. The appearance of the spellings <ks> alongside this letter again suggests that these reforms were intended to be proposals rather than final solutions. For the representation of vowel sounds, Smith did not favour the use of silent letters to indicate the length of vowels, preferring to use a system of accents, such as the diaeresis or circumflex, giving his system an unfamiliar and offputting appearance.

In 1569 John Hart, a Chester Herald, published his *Orthographie*, subsequently elaborated as *A Method or Comfortable Beginning for All Unlearned, Whereby They May Bee Taught to Read English* (1570). He also wrote a third work, *The opening of the unreasonable writing of our inglish toung*, although this was not published in Hart's lifetime, and survives only in a manuscript dated to 1551. Hart was probably a Londoner by birth and, despite not having a university education, he read Latin and Greek, and knew the work of Smith and Cheke and the debate concerning the pronunciation of Greek. While he was familiar with their work, Hart differed from Smith and Cheke in his preferred pronunciation: in the case of Latin, Hart advocated using the pronunciation of modern Italian. In his work on English spelling, Hart began by listing the sounds of English: what he called 'voices', claiming that a writing system should contain the same number of letters as there are voices in the language. The problem with English, according to Hart, was that there were not as many letters as sounds, creating a writing system which was 'corrupt'. Hart claimed that English suffered from 'superfluity', in that many words contained superfluous letters (letters which are not pronounced), such as the <b> in *doubt*, the <gh>

in *eight*, the <h> in *authority*. Hart believed that a spelling system should reflect pronunciation not etymology, and that silent letters were an unnecessary feature of English spelling. Their only useful function was to indicate where a preceding vowel is long, as in *hope*, although Hart proposed using accents for this purpose. A further problem was what he called 'usurpation': the use of a letter to represent more than one sound, as in the <g> of *gentle* and *game*. A related problem is that of 'diminution', or the lack of sufficient letters to represent all of the sounds in English. So Hart's view of spelling was that each phoneme should map onto a single grapheme, and that no grapheme should represent more than one phoneme. Hart considered etymological spellings pointless, since they are of no help as a guide to pronunciation for a foreigner trying to learn English. He also rejected the view that spelling variants were useful as a way of distinguishing homophones, as in the modern English spelling of *bear* and *bare*, since these can be distinguished by context, just as they are in speech. Hart's response to the perceived criticism that a strictly phonetic system is impossible, given the various regional accents of English, was to advocate alternative systems for different pronunciations.

In many ways Hart's is a sensible, if slightly over-idealized, view of the spelling system. Some of the reforms introduced by him have in fact been adopted. Before Hart the letter <j> was not a separate letter in its own right; in origin it is simply a variant form of the Roman letter <I>. In Middle English it was used exclusively as a variant of the letter <i> where instances appear written together, as in *lijf* 'life', and is common in Roman numerals, such as *iiij* for the number 4. Its Middle English use is thus similar to that of <o> in spellings like *son*, *love*, where a series of minim strokes could lead to confusion. Hart advocated using the letter <j> as a separate consonant to represent the sound /dʒ/, as we still do today. A similar situation surrounds the letters <u> and <v>, which have their origins in the single Roman letter <V>. In the Middle English period they were positional variants, so that <v> appeared at the beginning of words and <u> in the middle of words, irrespective of whether they represented the vowel or consonant sound. Thus we find Middle English spellings like *vntil*, *very*, *loue*, and *much*. Hart's innovation was to make <u> the vowel and <v> the consonant, so that their appearance was dependent upon use, rather than position in the word. The use of <j> for /dʒ/ meant that <g> no longer had this value and so could be used unambiguously for /g/, while he also reorganized the use of <c> and <k> so that <k> was always used for /k/, leaving <c> for /tʃ/, thereby returning to an Old English spelling practice. All silent letters were abolished in Hart's system, as were digraphs such as <th> and <sh>, which he represented using specially constructed characters, although he proposed <þ> as an alternative for /θ/ and <dh> for /ð/. Hart objected to the use of final <e> as a length marker, preferring to use accents or vowel doubling instead. But while Hart's proposals did have some impact, in their reorganization of the letters <v/u> and <i, j>, the majority of his proposed changes were too radical to be adopted.

Another spelling reformer of this period was William Bullokar, a teacher who had given up his profession to focus on finding a solution to the problems posed by English spelling. In his *Book at Large, for the Amendment of Orthographie for English Speech* (1580) Bullokar avoided inventing new characters, which he considered the reason for the failure of his predecessors, and rejected the revival of ancient letters, 'many yeeres forgotten' and 'known but to a few in corners'. But, despite his rejection of new letterforms, Bullokar did introduce accents and other diacritics, what he calls 'strikes', which proved just as offputting to the general public. In an effort to bring the names of letters closer to their sounds, Bullokar proposed renaming the letters <w> and <z> 'wee' and 'zee'. The latter of these has become the standard name in America for the letter the British call *zed*; the former has, perhaps unsurprisingly, not caught on. Bullokar proposed the use of ligatured versions of digraphs such as <sh>, <th>, and <ch> so as to make them single letters; to distinguish between /θ/ and /ð/ he used a diacritic. Despite his extensive output and years of dedication to the cause, at considerable personal expense, Bullokar added little new to the theory or practice of spelling reform, although his zeal as a committed teacher to promote an improved system for the benefit of his pupils led to the publication of many pamphlets and translations employing his revisions. While Bullokar's proposed revisions have had little influence upon our modern spelling system, the copious amount of material he published in his revised spelling system is of considerable value to modern study of sixteenth-century pronunciation.

Perhaps the most effective and influential of the spelling reformers of the sixteenth century was Richard Mulcaster, whose *Elementarie* (1582) presented a more balanced solution to the problem of English spelling. Mulcaster was educated at Eton, Cambridge, and Oxford, taking his MA at Christ Church in 1556. He then went into schoolteaching, and in 1561 he was appointed headmaster of the Merchant Taylors' School. He resigned this post in 1586 and then in 1596 became headmaster of St Paul's School in London. Mulcaster's *Elementarie*, intended to cover the rudiments of an elementary education, was published in 1582, during his period as head of Merchant Taylors', and the work was designed to function as a textbook to be used in schools. Mulcaster attempted to negotiate a compromise between the ideals of his predecessors and the actual system in place. He understood that no writing system could ever be truly phonetic, and that the use of a single letter to represent two sounds was perfectly acceptable. This is, after all, a feature of many other languages which seem to manage perfectly well. Similarly, the current alphabet is used successfully in other languages, as it currently is for English, while new letterforms are considered ugly and 'unreadie for a penman'. These arguments suggest a certain amount of confusion in Mulcaster's understanding of the proposed reforms and the history of English spelling. To appeal to the use of the Roman alphabet for writing Latin is to fail to recognize that a fundamental problem for English spelling concerned those sounds not found in Latin, for which there were therefore no corresponding letters. Nor does Mulcaster appear to have been aware of the different uses of the Roman alphabet in other languages, such as German, Spanish, French, and Italian. Mulcaster recognized the inevitability of the discrepancy between spelling and sound, given that pronunciation was constantly changing, although he confused the concept of a phonetic alphabet with one that could represent all the different accents of English.

Mulcaster's proposals were based upon sound, reason, and custom; his concern was more with explaining and rationalizing the current system than attempting to impose innovations of his own design. He did note some of the confusions in the English spelling system that others attempted to remove, such as the 'weak' and 'strong' pronunciations of <g>, which 'semeth to giue som matter to confusion in our writing'. Mulcaster was also influential in reforming and regularizing the use of a final <e> to indicate the quality of the preceding vowel: a feature of the modern spelling of words like *like, wise, life*. In Middle English long vowels tended to be represented by double vowels, although this was often very erratic, so that there was uncertainty about the length of the vowels in words like *hom, hoom* 'home'. In Early Modern English final <e> was frequently added to words which end in a consonant. Mulcaster introduced a system of using the final <e> as a marker of a preceding long vowel, although the tendency to add it after short vowels followed by a consonant survives in some examples, such as *live, done, gone*. Another group of spellings which do not conform to Mulcaster's rule are words like *sneeze, groove, seethe*, where the vowel length is indicated both by double letters and final <e>.

Mulcaster was less interested in reforming the spelling system to remove such inconsistencies and to promote phonetically consistent spellings, and more concerned with the promotion of a single spelling for each word. At the end of his book he provided a General Table of the spellings of the 7,000 commonest words, many of which, though not all, are identical to those used in Modern English. The list includes many high-frequency grammatical words that are in their modern spelling: *through, such, after, again, against*; although there are also differences from modern usage, such as the use of <ie> rather than <y> in words like *anie, verie*, and of <k> instead of <ck> in words like *quik, stik, pik*. While Mulcaster followed Hart in his explanation that <v> should always be used to represent the consonant and <u> the vowel, he did not observe this in practice. Spellings in his General Table like *auenge* 'avenge' and *vpon* 'upon' show that Mulcaster continued to employ these letters as positional variants. Despite the objective and descriptive attitude shown by Mulcaster in this work, his list of spellings does include some of his own innovations. For instance, he objected to the inclusion of a <u> after ‘strong’ <g> in words like *guest* and *guess* and advocated its omission. In the list of words at the end of the work these words are spelled *gest* and *gess*, reflecting Mulcaster’s prejudice rather than current practice. Mulcaster also felt that <ph> for /f/ in Greek loanwords could easily be replaced by <f>, making such words easier to pronounce by speakers illiterate in Greek or Latin. He included such words under <ph> in his list, but added a marginal note questioning the necessity for such spellings: ‘Why not all these with f?’. In fact the word *pheasant* appears under both <f>, *feasant*, and <ph>, *pheasant*. Mulcaster’s list of spellings was influential in setting the model for spelling books of the following century; in addition he highlighted the need for a monolingual English dictionary, paving the way for the first English dictionaries of the seventeenth century: ‘It were a thing verie praiseworthie in my opinion and no lesse profitable then praise worthie, if som one well learned and as laborious a man, wold gather all the words we vse in our English tung . . . into one dictionarie’.

In addition to Mulcaster’s list of common words, the spelling system employed by the early printers provided a model for private spelling habits. Whether Mulcaster’s list of spellings influenced printers, perhaps indirectly as a result of his influence among other teachers, or whether he was simply proposing spellings already commonly in use, his spellings bear a close resemblance to those employed in printed books of the late sixteenth century. The practical difficulties encountered by compositors responsible for setting the type, and thus for the spelling system employed, led to a tolerance of variation in spelling. This was partly driven by the need to justify the ends of lines of texts, for which variable spelling was a distinct advantage. Additional letters, such as double consonants or a final <e>, could be added or omitted to aid a compositor to fit the line of text neatly on the page. Variation between the size of individual pieces of type could also lead to variant spellings; for example, the variation in width between <ee> and <ei> could have influenced the choice between these two vowel combinations. However, a study of the spelling of two plays in the First Folio collection of Shakespeare's plays (published in 1623) identified a remarkably small number of spellings which could confidently be attributed to the need to justify the line of type, suggesting that justification of lines was not a major factor in the inconsistency of the spelling of printed books. Furthermore, the same study argues that, for the early printers, a fixed spelling system had its own benefits. Once a page of type had been printed, the junior apprentices were tasked with the tedious job of putting the individual pieces of type back in the correct compartments in the cases. Because the earliest printing types were designed to closely resemble handwriting, they included a large number of ligatures, combinations of letters such as *ff*, *fl*, *ffl*, as well as various vowel combinations.

It has been estimated that a single type case in William Caxton's print shop would have had around 250 compartments, each containing a different piece of type. The job of distributing type correctly was thus laborious and tedious, but crucial in ensuring that the compositor, who set the type to be printed, could perform his task accurately and efficiently. It would clearly be advantageous to the apprentice faced with this task to learn the spellings he encountered by heart, enabling him to perform his task more quickly and accurately. In this way, the spellings of the more experienced compositors were passed on to the more junior apprentices. Such a process has been shown to have occurred in the printing of Shakespeare's First Folio, where a novice compositor who began by employing his own non-standard spellings, as well as those of his copytext, switched to the standard forms employed by a more experienced colleague with whom he collaborated.

So where orthoepists and spelling reformers considered spelling from a theoretical standpoint, for the printers and compositors it was a more pragmatic concern. To make matters more complicated, many of these compositors were recruited from the continent, leading to the introduction of a small number of their own spelling conventions, such as the Dutch convention of using <gh> for /g/ in words like *ghost* and *ghest* 'guest'; or the use of <oe> in *goed* 'good', which appear from time to time in some of William Caxton's printed editions, although of these only *ghost* has survived into Modern English.

The influence compositors and printers had over the spelling systems employed in the works they printed is apparent from Mulcaster's *Elementarie*. As we have seen, this work ended with a list of recommended spelling forms for numerous common English words. It is striking, however, that not all of these recommended forms appear in the printed text which accompanied them. For example, the word *through* appears in Mulcaster's list in its Modern English spelling, but in the text both *through* and *thorough* appear. Similarly, the spelling of *though* recommended by Mulcaster in his list is the Modern English spelling, whereas this spelling is never used in the text; here we see the spelling *tho*, without the <gh>. While Mulcaster preferred to use a single <k> at the ends of words, as we have seen, there are instances of the <ck> spelling in the text itself, for example *musick*.

By the end of the sixteenth century, a common core of acceptable forms had been established; in most cases these were the basis of our Modern English standard spelling system. Once this core of spellings had been established and the battle for spelling reform appeared to be lost, the reformers redirected their efforts to the design of transitional alphabets. These were intended to function as a compromise between reformed and traditional alphabets, to enable pupils to master the complexities of English spelling in a more gradual and accessible way. Once again it was the school teachers that led the way, as demonstrated by Edmund Coote's *The English Schoole-maister* of 1596. This work was not intended to offer a reformed orthography, but rather presented a more phonemic system as a bridge to acquiring the traditional system.

The most successful of this variety of scripts was that formulated by another schoolmaster, Richard Hodges, in his *The English Primrose* of 1644. This publication set out in detail his transitional system, which employed special accents, known as 'diacritics', to indicate the length and quality of vowels, as well as a system of underlining to indicate letters that were not to be pronounced. *The English Primrose* built upon an earlier publication, *A Special Help to Orthographie: or, The True Writing* (1643), slightly revised and reprinted as *The Plainest Directions for the True-Writing of English* in 1649. These earlier works consist of lists of homophones: words pronounced identically but with different spellings and meanings, and words of similar pronunciation which are often confused. While such lists may not make the most stimulating reading material, they are of considerable interest for philologists interested in reconstructing contemporary pronunciation. Of particular interest in Hodges's list is the treatment of the <eth> ending of the third person singular of the present indicative: *maketh*, *leadeth*. This ending derives from an Old English ending in <-eþ>, but during this period it was replaced by a northern alternative ending <s>, the ending we use today: *makes*, *leads*. Hodges grouped words with this ending with other words ending in /s/, indicating that, even though the conventional spelling was still in use, it was by this time being pronounced as if written <s>. Thus Hodges's list includes: *cox*, *cocks*, *cocketh*; *clause*, *claweth*, *claws*; *Mr Knox*, *he knocketh*, *many knocks*. The significance of this was noted by Hodges himself, who explained that 'howsoever wee use to write thus, *leadeth it*, *maketh it*... yet in our ordinarie speech wee say *leads it*, *makes it*'. This example reinforces the way conventional spellings can be tolerated after a change has taken place in the spoken language. Hodges followed these lists with recommendations for the spelling of certain words, most of which correspond closely with the spellings current today. He preferred not to employ final <e> as a length marker, especially in words where the vowel or diphthong is already apparent from its spelling, as in *lead* and *seed*, which should never be spelled *leade* or *seede*. He did, however, suggest that final <e> be used in the case of the present tense form of the verb *reade*, to distinguish it from the past tense form *read*. Although concerned with promoting consistency, Hodges was willing to make concessions to etymology and custom. So while he advised the use of <i> in all instances, he permitted the use of <y> in Greek words, and in words like *my* and *by*, where it had been long established.

The English Primrose aimed to help students learn to read English, explaining the faults of English spelling and setting out a system to enable students to cope with these difficulties using a series of special marks. To achieve this goal, Hodges used a system of underlining silent letters as an indication that they should not be pronounced, adding diacritics to vowels to indicate their different sound values, or to consonants like <g> to distinguish between /g/ and /dʒ/. Hodges was an accomplished phonetician who grouped consonants into voiced and voiceless pairs, using a system of diacritics to indicate whether <s> represents the voiceless sound /s/, or the voiced equivalent /z/. Following this explanation, *The English Primrose* proceeds by setting out lists of syllables and words composed using them, as a way of helping learners understand the structure of English spelling.

The fifteenth and sixteenth centuries were a period of considerable interest and engagement with English spelling, with many writers proposing alternative methods to refine and improve this system. But in most cases their proposals were too extreme and too complex to be acceptable, and most had no effect upon the spelling employed in printed books. There were exceptions: Hart's proposals concerning the regulation of <i/j> and <u/v>, for example; but these are few, and the implementation of these proposals was only very gradual. The most influential intervention was the most modest: by supplying a list of common spellings which could function as a model for contemporary usage, William Bullokar did far more to influence standard spelling practices than any of his more qualified fellow reformers. Bullokar's list of preferred spellings, along with his call for the compilation of a dictionary, in which every word might have a single preferred form, established the framework for the developments of the eighteenth century during which fixing English spelling, by establishing a single acceptable spelling for every English word, became the chief concern.

The Early Modern period also saw the introduction of the neuter possessive pronoun *its*, and with it considerable confusion for future spellers who would muddle it with the abbreviation for *it is*: *it's*. The older form found in Old and Middle English was *his*, but this was increasingly considered unhelpful given that it was identical with the masculine pronoun. I quoted, in

Chapter 1, Lynne Truss's scornful attack on people who confuse the two spellings, *its* and *it's*, and I expressed some sympathy for people who find the situation confusing. Given that the apostrophe is used to signal possession in nouns, like *Lucy's picture*, it is unsurprising that the apostrophe often crops up in sentences like *the dog wagged it's tail*. In fact the origin of the *its* pronoun lends further support to such errors, given that it was formed by adding the genitive inflexion <-s> to the subject form *it*. At first the neuter pronoun *its* was considered colloquial and inappropriate for serious writing; the 'King James' Bible (1611) ignored it completely in favour of the more formal *his*: 'Ye are the salt of the earth: but if the salt have lost his savour, wherewith shall it be salted?' (Matthew 5:13) The Shakespeare First Folio includes only a handful of instances of *its*, and these are limited to late plays, suggesting Shakespeare may have adopted the form towards the end of his career. But what is striking about these instances is that they are spelled with the apostrophe, as in the case of this example from *The Tempest*:

> This Musicke crept by me vpon the waters,
> Allaying both their fury, and my passion
> With it's sweet ayre
>
> (*The Tempest* I.ii.392–4)

The spelling of *its* with an apostrophe was in fact used right up to the beginning of the nineteenth century, and thus our modern use of *it's* exclusively as an abbreviation for *it is* is a comparatively recent phenomenon. While the decision to create a possessive neuter pronoun using the genitive inflexion <-'s> may seem unnecessarily confusing, we need to remember that the difficulty that such a spelling causes for us today was not an issue in the Early Modern period, where the form *'tis* was used as an abbreviation for *it is*. Furthermore, there was a logic to the use of the apostrophe in cases like *the boy's book*, on which *it's* was modelled. The apostrophe was initially introduced in the Early Modern period to mark omissions, as it does today in *won't*, *I'll*, and in poetic forms like *yawn'd*. Given this, it seems illogical to use it to mark possession in nouns. But the reason the apostrophe was introduced in possessives was because it was thought that the -s inflexion was an abbreviation for the masculine possessive pronoun *his*. Thus a phrase like *the boy's book* was interpreted as an abbreviation of the phrase *the boy his book*, and so the apostrophe was added to signal the omission. The apostrophe commonly appeared in plurals too, such as *box's*, now viewed as a major grammatical solecism. But such forms were also understood as abbreviations, in which an <e> had been omitted.

The obsession with the classical languages of Latin and Greek reflected in the spelling reforms discussed above also led to the introduction of large numbers of loanwords from these sources during this period. While a number of Latin loanwords were adopted during the Middle English period, many of these entered the language via French rather than directly from classical Latin. During the Early Modern period these loans came directly from Latin, creating doublets where the same Latin word appears twice: once in a form borrowed from French and subsequently in a spelling reflecting its Latin origins. The verb *count* was borrowed from French *conter* in the fourteenth century, although its ultimate root is the Latin verb *computare*. The same verb was reborrowed during the sixteenth century, this time directly from Latin in the form *compute*. Other examples of this process include the Middle English *poor* (Old French *povre*) and Early Modern English *pauper* (Latin *pauper*), and Middle English *sure* (Old French *seur*) and Early Modern English *secure* (Latin *secu-*

rus). The desire to reflect Latin spelling led to the adoption of many Latin loanwords with their Latin endings still intact, as in the cases of *folio* and *proviso*, in which the <o> ending indicates the ablative case. In other examples we find verb endings preserved, as in *exit* 'he goes out', *ignoramus* 'we do not know', as well as entire verb phrases in *facsimile* (Latin *fac* 'make' + *simile* 'like') and *factotum* (Latin *fac* 'do' + *totum* 'the whole', i.e. someone who does everything). Despite this desire to preserve the Latin appearance of loanwords, some words did lose their endings in order to make them fit better with English morphological patterns: for example *immature* from Latin *immaturus*, *exotic* from *exoticus*.

While Latin was by far the largest source of new words, a number of borrowings came from French, although, as was also the case in the Middle English period, it is not always possible to determine whether a Latin loan came directly from Latin or indirectly via French. In some cases the spelling of the word reveals the direction of the borrowing. So, for instance, in the mid sixteenth century the two verbs *prejudge* and *prejudicate* were borrowed. *Prejudge* is clearly from the French *prejuger*, while *prejudicate* was borrowed directly from the Latin *prejudicare*. Presumably the similarity in the meanings of these two verbs meant that only one was required, and it was the French loan that survived. But in other cases it is simply not possible to determine from the form of the word whether it is derived from Latin or French. The noun *proclivity*, meaning 'inclination', is first recorded in 1591. The *OED* gives it a Latin etymon, deriving it from Latin *proclivitas* 'tendency, propensity', although it also compares the similar French form *proclivité*. French loanwords adopted towards the end of the seventeenth and into the eighteenth centuries are much easier to identify, as these often remained deliberately unassimilated. During this period French manners and culture were very fashionable, leading to the adoption of French words with their native spelling and even pronunciation intact. In most cases the words that were borrowed during this period still retain their French spelling and pronunciation today, such as *liaison*, *faux pas*, and *beau*. The vogue for borrowing French words with their native spellings led to the introduction of some new correspondences between letters and sounds, as well as the adoption of French accents, like the e-acute in *café*, first recorded in the late eighteenth century. Although the accent is still commonly used today, it is frequently dropped; a fully assimilated spelling *caff*, reflecting an anglicized pronunciation, is attested from the 1930s. Meanwhile the introduction of French words and expressions, with their French pronunciations still intact, also led to etymological respellings. The word *bisket* was replaced by the French spelling *biscuit*, and the adjective *blew* took on the French spelling *blue*. The fashion for French words among the nobility was condemned by eighteenth-century writers like Joseph Addison and Dr Johnson; it subsequently died out following the rise in prominence of the middle class and the corresponding fall in prestige of the nobility following the French Revolution.

A much smaller group of loanwords entered English during this period from other European languages, such as Italian, Spanish, and Dutch, frequently causing further disruption to the English sound-spelling patterns. The majority of these borrowings are the result of trade between Britain and these countries during this period; the words themselves relate particularly to commodities and foodstuffs. A feature of these borrowings is that unusual spellings of their donor languages were frequently modified to reflect English practices; thus we find the Italian word *articiocco* rendered as *artichoke*, with the medial /tʃ/ sound spelled <ch> rather than <c>. The spelling of the second element was influenced by a folk etymological association with the word *choke*, on the basis that the plant was thought to have an inedible centre which would choke anyone foolish enough to attempt to eat it, while an alternative spelling *hartichoke* reflects a popular association of the first element with *heart*. It is striking to contrast this kind of orthographic assimilation with the preservation of the Italian use of <c> for /tʃ/ in more recent loans such as *ciao* (first recorded in 1929) and *ciabatta* (first recorded in 1985). The Portuguese loanword *molasses* (Portuguese *melaços*) shows the replacement of the foreign <ç> with a native equivalent <ss>.

Confusion over the correct pronunciation of unstressed syllables in foreign loans like the Spanish *tomate* and *patata*, led them to be remodelled, perhaps by comparison with words like *fellow* and *pillow*, and ultimately to their respelling as *tomato* and *potato*. Different patterns of assimilation are reflected in regional spellings like *pottatie* (subsequently abbreviated to *tattie*) and *potater* (now frequently shortened to *tater*). Non-European loans reflect travel further afield and include words of Persian or Arabic origin, frequently borrowed via Turkish and assimilated to English spelling practices. The word *sherbet* was borrowed from a Turkish word, itself derived from the Arabic verb *shariba* ‘to drink’, and was initially used to refer to a cooling drink of fruit juice mixed with water. Uncertainty over the representation of its initial consonant is apparent from early spellings such as *zerbet*, *cerbet*, *sarbet*; these were subsequently replaced by our modern spelling with initial <sh>. Contact with Indian languages led to the adoption of a number of words into English, such as *shampoo*, derived from the Hindi verb *čāmpo* ‘press’ and originally referring to a kind of massage. The earliest instances of this word in English vary in their spelling between *champo* and *shampo*. The spelling of words of Indian origin is especially complex in the case of words borrowed during the period of the British Raj, where Anglo-Indian elements combine in surprising ways. An example is *gymkhana*, a word with several non-native spelling features, not all of which are due to its Indian origins. The word is derived from the Hindi word *gend-khāna* 'ball-house', a court used for playing racquet sports; it was initially used to refer to any kind of sporting resort and only later specialized to refer solely to horse riding contests. The pronunciation and spelling of the first syllable of the English word is due to a false association with the English word *gymnasium*, a word ultimately of Greek origin, whose first element means 'naked', a reminder that Greek athletes competed in the nude. From Afrikaans are derived a handful of words which retain their distinctive <aa> spelling, such as *aardvark* (from *aarde* 'earth' + *varken* 'pig'), first recorded in 1785 and consequently too late for Dr Johnson, despite Blackadder's famous criticism, in the BBC TV comedy series, that he 'left out aardvark'.

Having discussed the various trends in Early Modern spelling habits, it will be useful to examine a sample text in some detail. The text I have chosen is the famous soliloquy from *Hamlet*, reproduced here as it is found in the recent third edition by Arden Shakespeare:

> O that this too too solid flesh would melt,
> Thaw, and resolve itself into a dew,
> Or that the Everlasting had not fixed
> His canon 'gainst self-slaughter. O God, God!
> How weary, stale, flat and unprofitable
> Seems to me all the uses of this world!
> Fie on't, O fie, fie, 'tis an unweeded garden

That grows to seed: things rank and gross in nature
Possess it merely. That it should come to this:
But two months dead—nay not so much, not two—
So excellent a king, that was to this
Hyperion to a satyr, so loving to my mother
That he might not beteem the winds of heaven
Visit her face too roughly. Heaven and earth,
Must I remember? Why, she would hang on him
As if increase of appetite had grown
By what it fed on. And yet within a month!
(Let me not think on't—Frailty, thy name is Woman),
A little month, or e'er those shoes were old
With which she followed my poor father's body,
Like Niobe, all tears. Why, she, even she—
O heaven, a beast that wants discourse of reason
Would have mourned longer—married with mine uncle,
My father's brother (but no more like my father
Than I to Hercules). Within a month!
Ere yet the salt of most unrighteous tears
Had left the flushing of her galled eyes,
She married. O most wicked speed, to post
With such dexterity to incestuous sheets,
It is not, nor it cannot come to good;
But break, my heart, for I must hold my tongue.

Hamlet I.ii.127–57

The spelling, punctuation, and capitalization used in this passage are essentially those of Modern English. The reason for this is that the editor has modernized the spelling of the original text of *Hamlet* from which the play has been edited, to make it more accessible to a modern audience. But, while making Shakespeare's language appear more familiar and more modern, this procedure simultaneously has the contrary effect of implicitly making the language appear more distant, shielding readers from its true form. But, if we compare the above extract from *Hamlet* with the equivalent passage in the First Folio, we find that the spelling is actually very similar:

> Oh that this too too solid Flesh, would melt,
> Thaw, and resolue it selfe into a Dew:
> Or that the Euerlasting had not fixt
> His Cannon 'gainst Selfe-slaughter. O God, O God!
> How weary, stale, flat, and vnprofitable
> Seemes to me all the vses of this world?
> Fie on't? Oh fie, fie, 'tis an vnweeded Garden
> That growes to Seed: Things rank, and grosse in Nature
> Possesse it meerely. That it should come to this:
> But two months dead: Nay, not so much; not two,
> So excellent a King, that was to this
> *Hiperion* to a Satyre: so louing to my Mother,
> That he might not beteene the windes of heauen
> Visit her face too roughly. Heauen and Earth
> Must I remember: why she would hang on him,
> As if encrease of Appetite had growne
> By what it fed on; and yet within a month?
> Let me not thinke on't: Frailty, thy name is woman.
> A little Month, or ere those shooes were old,
> With which she followed my poore Fathers body
> Like *Niobe*, all teares. Why she, euen she.
> (O Heauen! A beast that wants discourse of Reason
> Would haue mourn'd longer) married with mine Vnkle,
> My Fathers Brother: but no more like my Father,

Then I to *Hercules*. Within a Moneth?
Ere yet the salt of most vnrighteous Teares
Had left the flushing of her gauled eyes,
She married. O most wicked speed, to post
With such dexterity to Incestuous sheets:
It is not, nor it cannot come to good.
But breake my heart, for I must hold my tongue.

Compared with the example of Chaucer's spelling in the previous chapter, there are a very large number of words spelled exactly as they are today. Perhaps the most significant difference concerns the distribution of <u> and <v>, which are used according to their position in the word rather than to distinguish vowel and consonant, as in *resolue* and *vnrighteous*. Another difference concerns the more erratic word division and the more liberal use of capital letters: the Folio text uses *it selfe* rather than *itself*, and capitalizes a number of words not capitalized in Modern English. Capitalization in this period was designed to give emphasis to important words, as is clearly true of words such as *Euerlasting*, *Cannon*, and *King*, and to highlight links between words, as in the case of *Flesh* and *Dew*, *Heauen* and *Earth*. But the practice was often implemented more randomly; it is unclear why *Month* should be capitalized and not *months*, for instance. The one example in the modern edition of capitalization within the sentence is the word *Woman*, indicating that it is the female sex that is being referred to rather than an individual woman; interestingly, this word is not capitalized in the First Folio. Another striking difference from modern practice is the large number of abbreviated forms, where omitted letters are marked by an apostrophe: *'gainst*, *on't*, *'tis*. Because these contracted forms are important for the metre, they are preserved in the modern edition. As we still use the apostrophe to signal a contraction, as in *can't*, *won't*, *isn't*, these forms pose little difficulty for modern readers. Where such forms appear in prose, and therefore carry no metrical significance, the editorial decision whether to include or omit an apostrophe to indicate a contraction becomes more problematic. Leaving out the apostrophe in such cases may give the false impression that *'gainst* and *gainst* are different words, while including it suggests a colloquial tone that may be inappropriate. Perhaps more confusing is the lack of an apostrophe where we would expect it, namely in the case of possessive nouns like *Fathers body*. As we saw above, this use of the apostrophe was a later development, based upon a misunderstanding of the origins of the <-s> inflexion, which was taken to be a contracted form of *his*. Where individual words are spelled differently from modern practices, these differences are generally comparatively trivial. Often the differences concern little more than the addition of a final <e>, as in *possesse*, *satyre*, *breake*, *grosse*, *selfe* and in plural forms, as in *windes* and *Teares*. Although editions of Shakespeare's plays are intended for a wide audience, it is surprising that editors consider this level of spelling variation to be beyond the abilities of their readership and so present his works in modern spelling.

If we look beyond this single extract from *Hamlet* at the spelling of the First Folio as a whole, then we can see that this passage is fairly representative of the volume. For, while there are spelling variants in the First Folio, it is striking how uncommon the older forms are. For instance, a common variation concerns the spelling of the final unstressed syllable in words like *truly/trulie*, *many/manie*, *very/verie*, *bury/burie*. But while both spellings are found throughout the First Folio, the <ie> spellings are very much less common than those with <y>. Another common area of variation concerns the tendency to double the final consonant and add an additional <e> in words like *had/hadde*, *bad/badde*, *sad/sadde* and so on. But, once again, the irregular spellings turn out to be very much less common than we might have thought. What these examples show is that, by the early seventeenth century, spelling variation was considerably less tolerated; the preferences associated with modern standard spelling were well on their way to being recognized as the single acceptable spelling. One area where there does appear to be continued uncertainty concerns the representation of the vowel sound in words like *believe*, *receive*, *grieve*. While the spelling with <ie> is more common in *chiefe*, there is only one instance of *belieue* alongside numerous occurrences of *beleeue*. Interestingly, the spelling *receiue* is very much more common than *recieue* (which appears just twice), while *deceiue* and *perceiue* are the only recorded spellings of these words; this suggests that the modern tendency to prefer <ei> after <c> was fairly well established by Shakespeare's day.

Because of this relatively high level of uniformity in Shakespeare's spelling, it was possible to employ non-standard spelling features to signal dialect usage, in a similar way as Chaucer did in the Reeve's Tale. Although dialect usage is not a particularly marked aspect of Shakespeare's works, he did draw upon non-standard spelling features in several of his plays. A good example of this appears in Act IV Scene vi of *King Lear*, where Edgar, disguised as a peasant, leads his blind father, the Duke of Gloucester, towards the cliffs of Dover. When accosted by a steward, Edgar affects a provincial accent to add greater authenticity to his rustic disguise:

Stew. Wherefore, bold Pezant,
Dar'st thou support a publish'd Traitor? Hence,
Least that th'infection of his fortune take
Like hold on thee. Let go his arme.
Edg. Chill not let go Zir,
Without vurther 'casion.
Stew. Let go, slave, or thou diest!
Edg. Good Gentleman, goe your gait, and let poore volke passe: and 'chud ha'bin zwaggerd out of my life, 'twould not ha'bin zo long as 'tis, by a vortnight. Nay, come not neere th'old man: keepe out, che vor'ye, or ice try whither your Costard, or my Ballow be the harder; chill be plaine with you.
Stew. Out Dunghill.
Edg. Chill picke your teeth Zir: come, no matter vor your foynes.

King Lear IV.vi.227–41

Rather than attempting to portray a genuine dialect with consistency and authenticity, the spellings in this exchange show Shakespeare rendering the flavour of rustic pronunciation through a few marked spellings. The most obvious of these is the use of the letter <v> instead of <f>, and <z> where we would expect <s>, as in the words *vor* 'for', *vortnight* 'fortnight', *vurther* 'further', *zir* 'sir', *zo* 'so', and *zwaggered* 'swaggered', representing an alternative pronunciation in which initial fricatives have been voiced. This is a feature of southern dialects of this period, which would fit with the Kentish location in which the action is set; today voiced fricatives are typical of the south-west and are witnessed in local pronunciations of *Dorset* as 'Dorzet' and *Somerset* as 'Zumerzet'. Another aspect of the dialect is the use of the pronoun *Ich* instead of *I*; this pronoun appears in *Chill* (Ich + will) and *'chud* (Ich + should) and in the expression *che vor'ye*

'I warrant you'. This form of the pronoun, an alternative development of the Old English pronoun *Ic*, is particularly associated with southern dialects. But, even though both these dialect features coincide broadly with southern dialect usage of this period, Shakespeare is not attempting a realistic depiction of a local accent. Spellings showing voiced fricatives, *chud* and *chill*, and even the expression *che vor'ye*, form part of a stereotyped stage dialect that was commonly employed on the Elizabethan stage as an indicator of rustic usage. Given our modern sociolinguistic prejudices, it is tempting to interpret this depiction of dialect use as evidence that non-standard speech had begun to be linked with social status. But, even though Edgar is evidently adopting a lowly social persona, the dialect forms are indicative of rusticity rather than social class. Despite this example, Shakespeare's works contain few instances of dialect usage; the main exceptions all concern the depiction of national, rather than regional, accents, such as the Irish, Scottish, and Welsh speech found in *Henry V*. This suggests that, for Shakespeare and his audience, accents were principally badges of national identity, as well as a means of distinguishing courtly and rustic speakers. The association of accents with social class was a development of the eighteenth and nineteenth centuries, as we shall see in the following chapter.

Chapter 6
Fixing spelling

In the eighteenth century the focus was on enshrining English spelling to prevent further corruption, rather than on reforming it. The great linguistic authority Dr Johnson was dismissive of attempts to reform spelling and their lack of success, arguing that no spelling system should be adapted to conform to current fashions of pronunciation which will inevitably continue to change. Johnson's view of orthography was that a lexicographer should aim to correct mistakes that were of recent introduction, while tolerating those inconsistencies that have the authority of age and tradition:

> In adjusting the ORTHOGRAPHY, which has been to this time unsettled and fortuitous, I found it necessary to distinguish those irregularities that are inherent in our tongue, and perhaps coeval with it, from others which the ignorance or negligence of later writers has produced. Every language has its anomalies, which, though inconvenient, and in themselves once unnecessary, must be tolerated among the imperfections of human things, and which require only to be registered, that they may not be increased, and ascertained, that they may not be confounded: but every language has likewise its improprieties and absurdities, which it is the duty of the lexicographer to correct or proscribe.
>
> (Preface to *A Dictionary of the English Language* 1755)

Johnson saw the inconsistencies of English spelling as the inevitable result of a process by which language was transferred from speech to writing, explaining the variety of spelling as the direct result of a diversity of accents. From such variety emerged pairs of related words with quite different spellings, such as *strong* and *strength*, *dry* and *drought*, *high* and *height*. Johnson is dismissive of Milton's spelling of *height* as *highth*, attributing it to a 'zeal for analogy', quoting the Roman poet Horace 'Quid te exempta juvat spinis de pluribus una', or 'Out of so many thorns, how does one extracted help you?'. As Johnson himself went on to put it: 'to change all would be too much, and to change one is nothing'. While Johnson tolerated orthographic oddities, what he labelled 'spots of barbarity', which have built up over time, he aimed to correct more recent errors by reference to what he called their 'true spelling', essentially that implied by their etymology. So, Johnson prefers the spellings *enchant* and *enchantment*, following the French spelling, but *incantation* after the Latin. Similarly, he adopted *entire* rather than *intire*, because the word is derived from French *entier* rather than directly from its Latin ancestor *integer*, the source of our English word of the same spelling. Although he acknowledged the frequent difficulties in distinguishing whether a Romance borrowing is from Latin or French, Johnson argued that French is generally the more likely given its greater influence upon contemporary usage.

Johnson preferred to stick with established usage, to 'sacrifice uniformity to custom', even when this led to contradictory pairs of spellings where the same sounds were spelled differently, such as *deceit* and *receipt*; *fancy* and *phantom*. In other instances he noted that some combinations of letters which represent the same sound are used interchangeably without any obvious rationale, giving examples such as *choak*, *choke*; *soap*, *sope*; *jewel*, *fuel*. In such cases Johnson preserved the variation, even to the extent of inserting some instances twice, so that 'those who search for them under either form, may not search in vain'. Johnson's acceptance of current spelling habits in the *Dictionary* is further apparent from his tolerance of variant spellings in the headwords: the headword *complete* may be found under both spellings *complete* and *compleet*. There are other inconsistencies in his *Dictionary* that he made no attempt to clear up, such as the use of single or double consonants in final position, as in *downhil*, *uphill*, *distil*, *instill* and the use of <ou> or <o> in words like *anteriour* and *exterior.* Ultimately, the spelling form used for the headword is to be considered Johnson's preferred form, although he observed that he left unchanged the spellings of the quotations, allowing the reader to judge for himself which spelling is preferable. In general, Johnson's preference is always for a spelling that is closest to etymology, and he had little time for those who wished to tinker with spelling to reflect contemporary pronunciation more closely: 'Much less ought our written language to comply with the corruptions of oral utterance, or copy that which every variation of time or place makes different from itself, and imitate those changes, which will again be changed, while imitation is employed in observing them'.

For Johnson, spelling is above all about etymology and tradition; his guiding principle in handling this aspect of his dictionary is 'to proceed with a scholar's reverence for antiquity, and a grammarian's regard to the genius of our tongue'. Johnson admits that the spellings he has chosen are not incontrovertible, while his etymologies remain uncertain and in some cases possibly wrong. Johnson was candid about the uncertainty he felt towards some of the etymologies he provided, admitting that he knew of no satisfactory etymology for the word *gun*, and that the etymology of *smell* is 'very obscure'. Other efforts to propose plausible etymologies are rather misguided, such as his suggestion that spider is derived from '*spy dor*, the insect that watches the dor'. In some cases, Johnson's mistaken etymology led to the adoption of a spelling that is still current today, as in the case of the verb *to ache*, which he believed was derived from Greek *achos* 'pain, distress' rather than from Old English *acan*. Even though he used the form *ake* for the headword, Johnson commented that it should be 'more grammatically' written *ache*. In another case, the misreading of the word *adventive* (Latin *adventivus* 'of foreign origin'), in a quotation from Francis Bacon, as *adventine*, led to the introduction of a ghost word into Johnson's *Dictionary*. Johnson glossed the word as 'adventitious, that which is extrinsically added', recognizing at least that it is 'a word scarcely in use'. The subsequent influence of Johnson's *Dictionary* is apparent from the word's appearance in later dictionaries, such as Thomas Sheridan's *Complete Dictionary of the English Language* (1789), which includes the entry 'Adventine: adventitious, that which is extrinsically added', and John Ogilvie's *Imperial Dictionary* (1850), a work specializing in technical scientific vocabulary, which includes '*Adventine*, adventitious'.

In some instances Johnson's bogus etymologizing has led to the selection of a variant spelling which conceals a word's actual etymology. For example, the word *bonfire* is a compound of the words *bone* and *fire*, so-called because of the tradition of burning bones, an etymology reflected in its earliest spellings: *banefire* and *bonefire*. The tendency for the first syllable to be shortened in speech led to the use of the alternative spelling *bonfire*; this spelling was subsequently adopted by Johnson on the false assumption that the first element was derived from

French *bon* 'good'. As well as being guided by etymology, Johnson also retained spellings that maintained visual links with derived words. Under the headword for *beggar*, Johnson stated that he adopted this spelling, even though the word should 'more properly' be written *begger*, because its derivatives are also spelled with an <a>. Although both spellings are attested in Middle English, the <-er> spelling was more common in Johnson's day, demonstrating his willingness to ignore popular usage in favour of other factors. In other instances, Johnson gives a spelling preference but does not adopt it for the headword, as in the case of *devil*, which he noted 'is more properly written divel', although he does not explain why. In the case of *grocer* Johnson preserved the traditional spelling rather than adopt his preferred spelling *grosser*, which highlights its derivation from *gross* 'large quantity': 'a grocer originally being one who dealt by wholesale'.

Although Middle English used both spellings interchangeably, the distinction between *practice* (noun) and *practise* (verb) was established in English usage by the publication of Johnson's *Dictionary*, which observes this distinction in spelling the headword for the noun *practice* and the verb *practise*. But, while Johnson observed this distinction in the spelling of the headwords, he did not follow it elsewhere in the *Dictionary*, preferring to use the spelling *practice* for both noun and verb. Thus the definition of *To Cipher* is given as 'To practice arithmetick', while *Coinage* is defined as 'The act or practice of coining money'. In the case of the pair of words now spelled *licence* (noun) and *license* (verb), Johnson did not make any distinction in spelling. Both headwords are spelled *license*, an odd choice given that, as Johnson himself noted, the noun is derived from French *licence* (originally Latin *licentia*) and the verb from French *licencier*; the spelling with <s> therefore has no etymological justification. In the case of the noun *defence* Johnson adopted the spelling with <c> rather than <s>, even though he correctly identified its origin in Latin *defensio*. The spelling *defence* arose during the Middle English period as a respelling according to French practices, parallel to the change from *is* to *ice*, *mys* to *mice* (see pp. 87–8), and has now become the more common spelling in Modern English; the *OED*, however, gives both spellings as alternatives, while noting that *defense* is usual in American English. Johnson introduced a distinction between the two words *council* and *counsel*; these words derive from two separate Latin words with distinct, but related, meanings: *concilium* 'assembly' and *consilium* 'advice'. In Middle English the two spellings were interchangeable; Johnson, however, distinguished between *council* (Latin *concilium*), 'an assembly of persons met together in consultation', and *counsel* (Latin *consilium*), 'advice, direction', a distinction which has survived into Modern English.

Another modern distinction codified by Johnson concerns the spelling of the two adjectives *discreet* and *discrete*. In Middle English *discrete* is the more frequent spelling, reflecting its immediate source in French *discret*. By the sixteenth century, the spelling *discreet* became much more commonly used, particularly in the sense of 'judicious, prudent'; the spelling *discrete* subsequently began to be employed solely in the technical sense of 'separate, distinct'. This sense is derived from the classical Latin *discretus*; it may be that the spelling *discrete* was preferred for this sense as it reflects that etymology more closely. Johnson did not introduce this distinction, but by assigning the two spellings to two separate headwords he did help to ensure its widespread adoption. Other modern distinctions which were not adopted by Johnson include the separation of *flower* and *flour*. The these two words are from the same root, ultimately Latin *flos/florem*; the sense 'finest quality of meal' now represented by the distinct word *flour*, was originally a sub-sense of the word *flower*. In Johnson's *Dictionary*, there is a single entry for the word *flower* which incorporates both senses. But, despite this, Johnson does use the spelling *flour* in the *Dictionary* itself, as in the definition of *biscotin*: 'a confection made of flour, sugar, marmalade, eggs, &c.'

The emphasis upon fixing, or 'ascertaining', the English language, so as to protect it from further corruption, is also apparent in the works of writers like Jonathan Swift and Joseph Addison. This desire to standardize the language inevitably focused on the spelling system, as this is the aspect of the language that is most easily regulated. A further objection to spelling reform arose from changing attitudes to pronunciation. One of the principles of spelling reform had been to bring the spelling system in line with current pronunciation. Eighteenth-century attitudes to pronunciation, however, were highly prescriptive. As a consequence, there was no suggestion that spelling should be reformed to accommodate the corrupt and ignorant pronunciation that was in common use. Thomas Sheridan, in his 1763 Course of Lectures on Elocution, was equally censorious of incorrect spellings as he was of erroneous pronunciation: 'It is a disgrace to a gentleman, to be guilty of false spelling, either by omitting, changing, or adding letters contrary to custom'. Swift was particularly dismissive of the notion that spelling should reflect pronunciation, and of what he called the 'barbarous Custom of abbreviating Words', ridiculing spellings such as *tho* 'though', *agen* 'again', *thot* 'thought', *brout* 'brought'. In a letter to the *Tatler*, published in 1710, Swift castigated poor spellers and ridiculed the notion that the spelling system should be subject to their erroneous pronunciations:

> These are the false refinements in our style which you ought to correct: First, by argument and fair means; but if those fail, I think you are to make use of your authority as Censor, and by an annual *index expurgatorius* expunge all words and phrases that are offensive to good sense, and condemn those barbarous mutilations of vowels and syllables. In this last point the usual pretence is, that they spell as they speak; a noble standard for language! to depend upon the caprice of every coxcomb, who, because words are the clothing of our thoughts, cuts them out, and shapes them as he pleases, and changes them oftener than his dress. I believe, all reasonable people would be content that such refiners were more sparing in their words, and liberal in their syllables.
>
> (*Tatler*, 1710, no 230)

But, despite his stringent objections to such spellings, Swift himself used *tho* and *thro* frequently in his published works, while the spelling *enuff* 'enough' appears as an eye-rhyme in his *Epistle to a Lady*:

> Conversation is but *carving*,
> Carve for all, yourself is starving.
> Give no more to ev'ry Guest,
> Than he's able to digest:
> Give him always of the Prime,
> And, but little at a Time.
> *Carve* to all but just enuff,
> Let them neither starve, nor stuff:
> And, that you may have your Due,
> Let your Neighbours *carve* for you.
>
> (ll. 132–41)

Swift satirized what he termed 'beau spelling' in his poem *Verses wrote in a Lady's Ivory Table-Book* (1698), where the lady's notebook is annotated with poorly spelled heartfelt expressions of love:

> Peruse my Leaves thro' ev'ry Part,
> And think thou seest my owners Heart,
> Scrawl'd o'er with Trifles thus, and quite
> As hard, as senseless, and as light:
> Expos'd to every Coxcomb's Eyes,
> But hid with Caution from the Wise.
> Here you may read (*Dear Charming Saint*)
> Beneath (*A new Receipt for Paint*)
> Here in Beau-spelling (*tru tel deth*)
> There in her own (*far an el breth*)
> Here (*lovely Nymph pronounce my doom*)
> There (*A safe way to use Perfume*)
> Here, a Page fill'd with Billet Doux;
> On t'other side (*laid out for Shoes*)
> (*Madam, I dye without your Grace*)
> (Item, *for half a Yard of Lace*.)
> Who that had Wit would place it here,
> For every peeping Fop to Jear.

In his *A Proposal for Correcting, Improving, and Ascertaining the English Tongue* (1712) Swift argued that a phonetic spelling would destroy the link with etymology, as well as requiring continual updating. Because of the great diversity of pronunciations, both across different dialects and within individual cities, such processes of reform would result in 'confounding orthography':

> Another Cause (and perhaps borrowed from the former) which hath contributed not a little to the maiming of our Language, is a foolish Opinion,

> advanced of late Years, that we ought to spell exactly as we speak; which beside the obvious Inconvenience of utterly destroying our Etymology, would be a thing we should never see an End of. Not only the several Towns and Countries of *England*, have a different way of Pronouncing, but even here in *London*, they clip their Words after one Manner about the Court, another in the City, and a third in the Suburbs; and in a few Years, it is probable, will all differ from themselves, as Fancy or Fashion shall direct: All which, reduced to Writing would entirely confound Orthography. Yet many People are so fond of this Conceit, that it is sometimes a difficult matter to read modern Books and Pamphlets, where the Words are so curtailed, and varied from their original Spelling, that whoever hath been used to plain *English*, will hardly know them by sight.

Swift's concern was above all with fixing the spelling of the language to prevent it from changing further. In his proposal he called upon the Earl of Oxford to set up an academy, on the model of the French Académie Française, which would be charged with establishing rules of correct usage and presiding over any further changes. Swift lamented the alterations to which the language had been subjected during the preceding century, and the problems these had caused for the reading of earlier works. Clearly Swift had an investment in ensuring that the language would not continue to change at such a rate, given the impact that could have upon the fortunes of his own works.

During the eighteenth century, female spelling was a particular target of censure. In a letter to Mrs Whiteway Swift praised her spelling, reporting that, on receiving her letter, he thought it came from a man, 'for you have neither the scrawl nor the spelling of your sex'. Swift's correspondence with Stella reveals several attempts to correct her poor spelling, as in the following example:

REdiculous, madam? I suppose you mean rIdiculous: let me have no more of that; 'tis the author of the Atalantis's spelling. I have mended it in your letter.

(14 December 1710)

In a letter of 1711, Swift listed a number of spelling errors made by Stella, asking how many of these are slips of the pen and how many due to ignorance:

Plaguely, Plaguily.
Dineing, Dining.
Straingers, Strangers.
Chais, Chase.
Waist, Wast.
Houer, Hour.
Immagin, Imagine.
A bout, About.
Intellegence, Intelligence.
Merrit, Merit.
Aboundance, Abundance.
Secreet, Secret.
Phamphlets, Pamphlets.
Bussiness, Business.
Tell me truly, sirrah, how many of these are mistakes of the pen, and how many are you to answer for as real ill spelling? There are but fourteen; I said twenty by guess. You must not be angry, for I will have you spell right, let the world go how it will. Though, after all, there is but a mistake of one letter in any of these words. I allow you henceforth but six false spellings in every letter you send me.

(23 October 1711)

Ironically, just a few weeks later, Swift found himself unable to remember the correct spelling of *business*, although he showed no compunction in blaming his uncertainty upon his exposure to Stella's erroneous spellings of the word:

> Pray let us have no more bussiness, but busyness: the deuce take me if I know how to spell it; your wrong spelling, Madam Stella, has put me out: it does not look right; let me see, bussiness, busyness, business, bisyness, bisness, bysness; faith, I know not which is right, I think the second; I believe I never writ the word in my life before; yes, sure I must, though; business, busyness, bisyness.–I have perplexed myself, and can't do it. Prithee ask Walls. Business, I fancy that's right. Yes it is; I looked in my own pamphlet, and found it twice in ten lines, to convince you that I never writ it before. Oh, now I see it as plain as can be; so yours is only an s too much.
>
> (1 December 1711)

To remedy the deficiencies in female spelling, spelling books were written specifically to instruct young ladies in correct spelling. In Edinburgh a ladies' club, The Fair Intellectual Club, formed in response to the all-male Athenian Club, met regularly to receive instruction in spelling from James Robertson, author of *The Ladies Help to Spelling* of 1722. *The Ladies Help* is a dialogue between the lady and her master, designed to make good this perceived omission in female instruction. That spelling did not otherwise form part of a young lady's education is made explicit in the following exchange:

> Master: There are certain rules which may be taught, as the grounds of the English tongue; the observation of which will discover the various sounds of the letters, and make Spelling a work of no difficulty.
>
> Lady: I know nothing of these Rules, for my Education was too like to that bestowed on most of my Sex; viz. Sewing, Dancing, Musick, Paistry, &c, mean

> while, ignorant of our Mother tongue, not daring to speak in company lest we blunder; much less to communicate our thoughts to an absent Friend, or Commerade, lest our bad Spelling afford matter of Laughter to the Reader.
>
> (Robertson, 1722, pp. 2–3)

Despite this desire to standardize the spelling system and to ensure that it was correctly employed by members of both sexes, private spelling habits continued to tolerate considerable variation: eighteenth-century diaries and personal letters reveal a variety of different non-standard spellings. While most writers adopted the conventions of the standard in their public writings, they continued to use their own personal spellings in private documents. This is perhaps most striking in the case of Samuel Johnson, whose *Dictionary* provided the opportunity for a single system to be codified and adopted as a national standard. We saw earlier that Johnson based his spelling of headwords upon custom, and that he was unsympathetic to the view that spelling should reflect individual preference and pronunciation. But the spellings he employed for his private writings show considerable variation, such as *complete/compleet*, *pamphlet/pamflet*, *stiched/dutchess*, and *dos/do's/does*, suggesting that Johnson felt no constraint to employ the spellings of his own dictionary in private correspondence.

The tolerance of spelling variation in private writings was quite common in the seventeenth and eighteenth centuries. Writers like Addison, Dryden, and Swift used a large number of variant spelling forms in their private writings, such as letters and diaries. While he chastized Stella for her spelling mistakes, Swift himself used numerous erroneous spellings in his letters. Examples include *gail* 'jail/gaol', *belive* 'believe', *hear*, *here*, *heer* 'hear', *college/colledge* 'college', *reach* 'rich', *their* 'there', *legnth* 'length', *scheem* 'shame', *mak* 'make', *malic* 'malice', *hom* 'home'. Rather being a marker of literacy or education, non-standard spellings in private letters seem to have been considered a marker of relative formality. These private spellings do not appear in printed works, which follow the accepted standard spelling conventions. The convention seems to have been for a writer to leave the business of spelling to the compositors who were responsible for setting the type for printed texts. This practice led to the introduction of a distinctive feature of punctuation found in this period: the capitalization of nouns. This practice has its origins in an author's wish to stress certain important nouns within a piece of writing. Because ultimate authority for spelling and punctuation lay with compositors, who were often unable to distinguish capital letters from regular ones in current handwriting, they adopted a policy of capitalization of nouns by default.

While much of the debate concerning spelling during this period centred upon notions of correctness, several changes to the spelling system were introduced that have left their mark upon English spelling. During the Middle English period the letter <g> was no longer used for the sound /j/, as it had been in Old English, but it was used for two separate sounds: /g/ or /dʒ/. The potential for confusion that this created was resolved in this period by the introduction of <u> as a diacritic to indicate when the <g> represented the hard /g/ sound, thereby creating a distinction between *guest* and *geste* 'feat'. There was, however, another method of representing the sound /dʒ/ initially, namely the use of the letter <i>, as found in French. As we saw in the previous chapter, it was the spelling reformer John Hart who first suggested that the letter <i> be replaced in these contexts by the letter <j>, a use that has survived into modern English: compare our spellings of *guest* and *jest*.

The spelling of vowel sounds was disrupted in the period 1500–1700 by the most dramatic sound change in the history of English, what is now known as the Great Vowel Shift. The Great Vowel Shift is comprised of a series of changes whereby each long vowel was raised one place in the vowel space, while the high vowels, those made with the tongue pressed on the roof of the mouth, diphthongized. The results of this change can be represented by a comparison of the Middle and Early Modern pronunciations of the following words:

/liːf/ > /ləɪf/	'life'
/huːs/ > /həʊs/	'house'
/deːd/ > /diːd/	'deed'
/foːd/ > /fuːd/	'food'
/dɛːl/ > /deːl/	'deal'
/bɔːt/ > /boːt/	'boat'
/naːmə/ > /nɛːm/	'name'

The Great Vowel Shift is the reason why the Middle English vowel /eː/ is now pronounced /iː/ in words like *been*, *seen*, and why the Middle English vowel /iː/ is now a diphthong in words like *wife* and *life*. This change had far-reaching consequences for the spelling system too, as traditional spellings began to represent new sounds. Where previously the letter <i> represented the long and short i-sound, it could now also signal the diphthong in *life*. Where the letter <a> was used to signal the long and short a-sound, it could now represent the diphthong in *name* as well. Where Middle English <oo> had represented a long-o sound in words like *mood* and *food*, following the Great

Vowel Shift this digraph had come to represent a u-sound, as it does still today in those words. So where Middle English had a more straightforward system whereby a letter represented long and short values of a single vowel sound, this pattern was disrupted and replaced with a much less predictable system. Only in the case of words spelled with <ou> or <ow> did the Great Vowel Shift have the effect of creating a more predictable pronunciation. Words like *house* and *town* derive from Old English words with /u:/, *hus* and *tun*; the <ou> and <ow> digraphs were introduced by the Normans in the Middle English period (see p. 89). While <ou> and <ow> are not especially logical ways of representing the /u:/ sound in English, they are comparatively efficient ways of representing the diphthong with which these words came to be pronounced following the Great Vowel Shift.

After the Great Vowel Shift, another change affected the long vowel system, bringing it closer to the system of pronunciation we use today. Throughout much of the Middle English period words spelled with <ee> could be pronounced with either /ɛ:/ or /e:/ (see p. 91). Following the Great Vowel Shift this distinction was maintained, so that a word like *see* was now pronounced /si:/ while *sea* was pronounced /se:/; *meet* was pronounced /mi:t/ and *meat* was pronounced /me:t/. But in the eighteenth century these two sounds merged in the East Anglian dialect, giving a single pronunciation /i:/. Because of immigration into London from this area, this merger came to be adopted in the London dialect, giving us the Modern English situation where *sea* and *see* are pronounced identically. The process by which this merger was adopted was evidently gradual; prestigious writers like Swift and Pope continued to rhyme pairs like *speak* and *awake*; *speak* and *take*. This merger led to a further confusion of the sound–spelling relationship in Modern English, resulting in a situation where words with different vowel digraphs, like *meet/meat*, *sea/see* are in fact homophones. There are one or two exceptions to this: in the case of *break*, *great*, *steak*, *yea*, all spelled with <ea>, modern English still preserves the pronunciation that predates the merger.

The nineteenth century witnessed a huge increase in literacy, especially in the second half of the century. In 1850 30 per cent of men and 45 per cent of women were unable to sign their own names; by 1900 that figure had shrunk to just 1 per cent for both sexes. It is often assumed that the language used in the nineteenth century remained stable and was largely identical to that used today. But there were changes in spelling during this period, and non-standard forms that survived the eighteenth century continued to be used, particularly in private writings. Thus the distinction between private and public spelling practices that was maintained throughout the eighteenth century persisted during this period, so that writers continued to use non-standard spellings in their letters and diaries, while printers and compositors functioned as the guardians of correct usage in printed works.

A writer's reliance upon printers for their spelling and punctuation is apparent from comments made by Caleb Stowe in his *Printer's Grammar* of 1808: 'Most Authors expect the Printer to spell, point [punctuate], and digest their Copy, that it may be intelligible and significant to the reader'. One of the reasons why we view the nineteenth century as a period of stable and fixed spelling is that modern editors generally emend non-standard spellings found in nineteenth-century manuscripts. The manuscripts of Charles Dickens's novels, for instance, contain persistent non-standard spellings, such as *recal*, *pannel*, *poney, trowsers*. Described by modern editors as 'life-long misspellings', such irregularities are generally replaced with correct forms for the modern reader. Some unusual spellings have been retained by editors as appearing to have been authorized by the author. The World Classics edition of *Bleak House* preserves the use of <-or> spellings of *favored*, *parlor*, and *humor*, as well as more unusual spellings like *secresy*, *gypsey*, *chimnies*; in other unusual forms, such as *villanous*, editors are uncertain as to whether these are errors or genuine Dickensian spellings. The recent publication of a facsimile edition of Dickens's manuscript of *Great Expectations*, complete with authorial revisions, additions, and non-standard spellings, led one journalist to marvel at Dickens's 'terrible handwriting' and use of 'text-speak' (*The Guardian*, 8 December 2011). Despite his use of non-standard spellings, Dickens's portrayal of Pip's initial efforts at literacy reveal his clear sociolinguistic prejudices. Pip's efforts do not just reveal his lack of education; they also give away his poor pronunciation, including a propensity for h-dropping, seen in the spelling *ope* for *hope*, and hypercorrection, in his addition of an <h> in *habell* 'able'.

"MI DEER JO i OPE U R KR WITE WELL i OPE i SHAL SON B HABELL 4 2 TEEDGE U JO AN THEN WE SHORL B SO GLODD AN WEN i M PRENGTD 2 U JO WOT LARX AN BLEVE ME INF XN PIP."

(*Great Expectations*, chapter 7)

Examples like this show how the establishment of a standard spelling system and its association with education and other social accomplishments enabled writers to employ non-standard spellings as a badge of provincialism and social inferiority. The stage dialect employed sparingly by Shakespeare is employed much more liberally in the eighteenth- and nineteenth-century novel. As an example we may consider the character of Squire Western in Fielding's novel *Tom Jones*, whose south-western origins are flagged by non-standard spellings which serve to indicate a rustic accent:

> 'Come, my lad,' says Western, 'd'off thy quoat and wash thy feace; for att in a devilish pickle, I promise thee. Come, come, wash thyself, and shat go huome with me; and we'l zee to vind thee another quoat.'
>
> (*Tom Jones*, V, 12)

A marked feature of this literary dialect is the voicing of initial fricatives, that is, words pronounced with /f/ and /s/ in the standard accent are here pronounced with /v/ and /z/; a feature which also marked the rustic dialect used by Edgar in *King Lear*. This distinctive aspect of south-western accents is found in Middle English texts written in southern dialects too, but the lack of a standard spelling system meant that such spellings were merely indications of an alternative pronunciation. By the eighteenth century such deviant pronunciations were socially stigmatized; Fielding's depiction of Squire Western is clearly intended to cast a smear on his manners rather than to add verisimilitude to his characterization. That this is the case is further supported by the lack of consistency in its depiction: if it were intended to be a realistic portrayal of an authentic accent we would expect such spellings to appear throughout the Squire's dialogue. But the accent is used unpredictably, implying that it is being employed as a social marker. It is also striking that the Squire's daughter, Sophia, does not speak with a rustic accent, despite being brought up in that part of the country, perhaps because its lowly and provincial connotations were felt inappropriate for the novel's heroine.

In Chapter 1, I described the media frenzy provoked by comments about Jane Austen's spelling and punctuation habits in her unpublished manuscripts. But when we consider these alongside similar misspellings in the letters and manuscripts of other nineteenth-century writers, we recognize that such spellings would have been entirely acceptable at the time. Austen's manuscripts include such misspellings as the Dickensian *poney*; uncertainty over double or single <l>, for example *fearfull*; and the use of contractions such as *tho'*, *thro'*, spellings condemned by Swift but ones which he himself regularly employed. Jane Austen's consistent use of <ei> for the /iː/ sound—*teizing*, *teized* 'teasing, teased' and *beleive*—might appear to be less easily condoned, given that it shows a failure to apply the best known of all spelling rules: 'i before e, except after c'. But Austen was not alone in using this spelling; the *OED* records several instances under its entry for *believe*. In a letter of 1716 quoted by the *OED*, Lady M. W. Montagu writes, 'I find that I have a strong disposition to beleive in miracles'. Such examples indicate that it would be anachronistic to view Austen's spellings as errors; instead we should view these spellings as alternatives that would have been considered perfectly acceptable in unpublished writings.

Another writer of the nineteenth century whose diaries and notebooks attest to considerable variation in spelling habits, and a lack of concern with consistency and standardization, is Charles Darwin. Darwin compiled notebooks of some 3,000 manuscript pages, collecting together various observations made during his voyage on the *Beagle*. Scholars have observed changes in his spelling practices throughout the notebooks, and have used these to date his various entries in this work. But while there do seem to be datable patterns in changes of preference, it is not a straightforward case of a non-standard spelling being replaced with the standard equivalent. Darwin used the standard spelling *occasion* for brief periods, but the non-standard spelling *occassion* remained his preferred form throughout much of the period 1831–6. In the case of *coral*, spellings with single <l> and double <ll> are found during the first two years of the voyage, to be replaced exclusively by the form *corall*. In 1835 the *coral* spelling then replaced *corall*, in Darwin's essay on coral islands. *Pacific* was the only spelling of this word up to 1834, when it was replaced by the spelling *Pacifick*, which was used until 1836, at which point the earlier form reappeared. Other non-standard spellings recorded in Darwin's notebooks, none of which was ever corrected or replaced, include *neighbourhead*, *thoroughily*, *yatch*, *broard*, *mœneuvre*, *Portugeese*.

Darwin's tendency to misspell words was highlighted by his sister Susan in a letter of 1834, in which she drew attention to the following erroneous spelling forms which appear in his diary: *lose* 'loose', *lanscape* 'landscape', *higest* 'highest', *profil* 'profile', *cannabal* 'cannibal', *peacible* 'peaceable', *quarrell* 'quarrel', *berrys* 'berries', *barrell* 'barrel', *epock* 'epoch', *untill* 'until', *priviledge* 'privilege'. In commenting on these mistakes, Susan adds, 'I daresay these errors are the effect of haste, but as your Granny [her nickname] it is my duty to point them out'.

A casual attitude towards spelling, combined with the frequent use of abbreviations, is also found in private letter collections of this period. To demonstrate this, I have taken some examples from the Corpus of late Modern English Prose, assembled by David Denison, a linguistics professor at the University of Manchester. This electronic corpus mainly consists of 20,000-word blocks of extracts from the letters and diaries of Lord and Lady Amberley, the letters of Gertrude Bell, Ernest Dowson, John Richard Green, and Sidney and Beatrice Webb. A survey of the spelling practices recorded in these letter and diary collections reveals considerable reliance upon abbreviations, including such common instances as *altho'*, *tho'*, *thro'*, *shd*, *cd*, *wch*, *wd*, *yr*. The writers reveal a casual attitude to the apostrophe, frequently omitting it in abbreviations, such as *cant* 'can't', *wont* 'won't', *dont* 'don't' *oclock* 'o'clock', and adding it where it is not required: *her's* 'hers'. The letters also show a tendency to confuse *its* and *it's*, a solecism which engenders considerable scorn today, but appears to have been reasonably common then. Alongside these abbreviations and contractions we find a host of spelling errors, some of which are commonly made today, such as *embarassed*, *installment*, and *accomodation* (a familiar feature of many Bed and Breakfast signs), deriving from uncertainity over number of double consonants. Other misspellings are more idiosyncratic, revealing difficulties with some of the complexities of English spelling. The spelling *doutbless* 'doubtless' shows an awareness of the silent <b>, but an uncertainty about where to place it, similar to the confusion over the placement of the <u> in *guages* 'gauges'. Other misspellings found in this collection of letters reveal difficulties with foreign spellings, such as *buscuits* 'biscuits', *croqueses* 'crocuses', *maccaronni* 'macaroni', while others suggest attempts to spell phonetically, for example *umberella* 'umbrella', *subtilty* 'subtlety', *tempemomy* 'temporary', *sens* 'sends', and perhaps *werry* 'very'.

Yet, despite this more casual attitude towards it in private writings, spelling had clearly become institutionalized as a badge of social status. Despite acquiring his new identity as Lord Fauntleroy in Frances Hodgson Burnett's novel of 1886, seven-year old Cedric Errol's humble upbringing is still apparent from his unreliable spelling and lack of awareness of the conventions of letter writing: 'Dear mr. Newik if you pleas mr. higins is not to be interfeared with for the present and oblige, Yours rispecferly, FAUNTLEROY'. When he asks his grandfather if this is the correct spelling of *interfered*, the Earl of Dorincourt replies: 'It is not exactly the way it is spelled in the dictionary', a response which highlights the status of dictionaries as the ultimate authority in determining correct orthography. But, while dictionaries are commonly cited as authorities on correct spelling, we should remember that lexicographers often have to choose between a number of variant forms for the spelling of each headword and the final choice can be arbitrary and provisional. Lynda Mugglestone records how James Murray, chief editor of the *New English Dictionary* (later the *Oxford English Dictionary*), received frequent requests from members of the public for an authoritative judgement as to the correct spelling of a particular word. Murray, however, often found himself unable to provide a definitive response, responding to one enquiry concerning the correct spelling of *whisky* with the suggestion 'when in a hurry you may save a fraction of time by writing whisky, and when lingering over it you may prolong it to whiskey…in matters of taste there is no "correct" or "incorrect"; there is the liberty of the subject'. Such a response must have seemed disconcertingly vague and permissive to a correspondent in search of certainty and uniformity. The *OED* itself refused to be committed over the spelling of this particular word, providing both spellings as alternatives in the headword. The *OED* also notes that the variation reflects a modern trade distinction between Scotch *whisky* and Irish *whiskey*; in general use the *OED* suggests that *whisky* is the usual spelling in Britain and *whiskey* that in the USA.

In the preface to the first volume of the dictionary, published in 1888, Murray explained that the choice between alternative spellings was often based upon the 'preponderance of modern usage'; where there was no clear preference, etymology, pronunciation, and practical convenience were taken into account. He concludes with the warning that 'in many cases, it is not implied that the form actually chosen is intrinsically better than others which are appended to it'. The various factors involved in the adoption of a particular spelling as the *OED* headword is apparent from a comparison of the treatment of *axe* in the first and second editions. The first edition (published in 1885) employed the spelling *ax* for its headword, noting that this spelling is 'better on every ground, of etymology, phonology, and analogy, than axe, which has of late become prevalent'. But, despite its continued support for the spelling *ax*, the second edition of 1989 adopted *axe* for its headword, noting that, despite its many advantages, *ax* had by this time fallen out of use entirely.

In the 1860s, the question of spelling reform was revisited by the learned society of linguists known as the Philological Society. In 1871 A. J. Ellis produced 'Glossic', a more phonetic spelling system based upon the existing values of the English letters, intended to be used concurrently with the traditional spelling when teaching children to read and write, as a means of mitigating some of the difficulties involved in the acquisition of standard English spelling. Ellis's desire to aid learners while introducing minimum adjustments to standard orthography had an important effect upon subsequent proposals for reform. The interest provoked by this foray into the creation of a reformed alphabet led to the founding of the British Spelling Reform Association in 1879, whose membership included Lord Tennyson and, perhaps unsurprisingly given his evident difficulties with spelling, Charles Darwin. This Association produced several new schemes, finally settling on a proposal to eradicate the most severe anomalies, such as the plethora of superfluous letters. But this remained little more than a purely academic activity; its proposals were never adopted outside philological circles.

A society founded specifically with the aim of reforming spelling was The British Simplified Spelling Society, founded in the early twentieth century, and initially supported by respected members of the Philological Society, including the etymologist Walter Skeat and James Murray. But while these professional linguists were generally in support of spelling reform, they expressed some concern about the easy assumption that writing was simply intended to reflect speech. While Henry Bradley, another of the editors of the *New English Dictionary*, agreed that the lack of correspondence between spelling and pronunciation is a 'serious evil', he argued for the 'ideographic' value of writing (Bradley, 1913, p.13), by which printed symbols were directly translated into thoughts. Bradley accepted that spelling need not reflect etymology in common words, whose meanings had long since diverged, but he felt that preserving this link was important in 'literary' words, especially those derived from classical sources where spelling's link with etymology is a valuable guide to meaning. The benefits of etymology extended to the pronunciation of such words, which Bradley argued should be closely based on spelling, rather than the other way around. Bradley recognized that a spelling system's capacity for communication does not depend on accurate reflection of speech, and may be impaired by homonymic clash. He stressed the potential problems in a spelling system in which pronunciation is the sole guide to spelling, and the potential for confusion it could cause: 'If I receive a letter beginning "My *deer* Sir", I have seen a vision of a horned animal before my mental ear has perceived the sound of the word' (1913, p. 5). Without spelling distinctions, imagine the confusion that could occur in a statement such as: 'We must consider Oxford as a whole, and what a whole it is' (p. 15).

For Bradley, the true purpose of written language is the 'expression of meaning'; he saw words as 'direct symbols of their meanings' (p. 5), which are conveyed directly to the brain, bypassing the speech organs entirely. Bradley's conservative attitude towards spelling is guided by a view of spelling as a means to reading for those who have already mastered the basic system; a more phonetic system is much more useful for those learning to read and write. The one area where Bradley was in favour of a more phonetic spelling concerned foreign loanwords. He argued that the adoption of a phonetic spelling system would reduce the ease with which foreign words were being borrowed, enabling us 'to free our language from its unnatural bondage to the alien, to compel the development of its native resources, and to revive its decayed powers of composition and derivation' (p. 16). Bradley did also recognize that a more phonetic spelling system would contribute to what he called the 'democratization' of the language, making it more accessible to a readership not educated in Latin and Greek. While it may be true that a more phonetic spelling system would help with the pronunciation of classical loanwords, being able to pronounce a word is of little help to being able to understand it.

The one other area where Bradley favoured reform of spelling was in proper nouns. But this suggestion seems even more likely to fail than many of the more radical changes proposed by spelling reformers. As well as being a way of differentiating between different families, such as between *Wild* and *Wilde*, the traditional spelling of certain names holds considerable cultural cachet. Names such as *Beaulieu* and *Beauchamp*, which preserve French spellings that vary widely from their pronunciation, function as badges of prestige, signalling an ancestry traceable to the Norman aristocracy. Names like *Cholmondeley*, *Colquhoun*, and *Mainwaring*, pronounced as if they were spelled 'Chumley', 'Cahoon', and 'Mannering', also serve to mark out the uninitiated, who incorrectly assume that they are pronounced according to their spelling. The social desirability of such phonetically unpredictable surnames is the rationale behind the many exotic titles given to his characters by P. G. Wodehouse, such as Stanley Featherstonehaugh Ukridge, pronounced 'Fanshaw Ewkridge'. The social distinction supplied by such names lies behind the name of Wodehouse's creation Psmith, who adopted the <p> to distinguish himself from the many other people with the same name, as he explains in the following exchange with a bemused maid:

> 'Will you inform her that I called. The name is Psmith. P-smith'.
>
> 'Peasmith, sir?'
>
> 'No, no. P-s-m-i-t-h. I should explain to you that I started life without the initial letter, and my father always clung ruggedly to the plain Smith. But it seemed to me that there were so many Smiths in the world that a little variety might well be introduced. Smythe I look on as a cowardly evasion, nor do I approve of the too prevalent custom of tacking another name on in front by means of a hyphen. So I decided to adopt the Psmith. The p, I should add for your guidance, is silent, as in phthisis, psychic, and ptarmigan.'
>
> (*Leave it to Psmith*, chapter 2)

Despite the initial <p> being silent, Psmith can somehow always detect when someone calls him plain Smith, and is quick to correct them. Interestingly, James Murray actually proposed that the <p> in scientific words like *psychic*, *psychology*, which ultimately derive from Greek roots beginning with the letter <Ψ> 'psi', should be pronounced, although he felt this should not be extended to more common words like *psalm*.

Quite what the advantages of this change were felt to be is not clear; technical terms used by a restricted and specialized group are unlikely to cause much difficulty. It is surely in instances of such words in more widespread use that there would be real benefits from either dropping the <p>, or pronouncing it.

The twentieth century saw a continued appetite for spelling reform, largely based upon economic principles. Traditional spelling, with its prodigal use of silent letters and digraphs, increased the costs associated with printing books, requiring more time and paper to print. A leaner spelling system, in which silent letters and digraphs were reduced, would therefore represent a major economic saving for both printer and reader. This is a guiding principle of 'Cut Spelling', the brainchild of the English Spelling Society member Christopher Upward, in which redundant letters are simply cut out. In Cut Spelling the <b> in *doubt* and in *lamb* are removed, as is the <gh> in *night*, as are vowels in unstressed syllables, so that *given* becomes *givn*, *helped* becomes *helpd*. While such changes appear to introduce greater economy into the spelling system, without any loss to comprehension, it is important to recall that spelling systems operate with a degree of redundancy which helps to limit misunderstandings. While such cut spellings may avoid homographs, they do introduce considerable unfamiliarities which will undoubtedly require longer and more careful processing by readers. The most significant objection, however, lies in the difficulties inherent in any attempt to implement a reform of this nature. Its designers are not afraid to confront the practical difficulties associated with implementation, but recognize that their ideal scenario, in which Cut Spelling is simultaneously adopted by all worldwide users of English, from 'lexicographers in the great metropolitan centres of the English-speaking world, to peasant farmers and market traders in countries that have at best indirect contact with the English language' is a utopian dream.

More moderate proposals were set out by the Swedish linguist Axel Wijk, in a system published as *Regularized Inglish: An Investigation into the English Spelling Reform Problem with a New, Detailed Plan for a Possible Solution* (1959). This was not so much a proposal for reforming spelling, but rather a plan to replace its inconsistencies with more regular equivalents. For Wijk, the aim of regularizing English spelling was systematically to preserve, as far as was practical and possible, the various symbols of the current spelling system in their most frequent usages. Wijk saw the goal of a strictly phonetic spelling system as impossible, and was concerned more with limiting the number of sounds which could be associated with a single letter, or the number of ways of writing a single sound. Thus, because established by centuries of use, and because they cause little difficulty, the two uses of <g> (*gentle*, *get*) are retained, as are the three letters employed to represent /k/ (<c, k, qu>), and the use of <s> and <c> for /s/. Wijk's respect for established precedent extends even to the maintenance of the <gh> digraph where it is silent, as in *high*, *sight*. The only regular symbol that he proposed dropping was the <ph> digraph, which serves no other purpose than marking words of Greek etymology, and which could just as easily be spelled with <f>. Wijk proposed the introduction of just two new sound symbols: <aa> for the long-a sound in *father*, and a <dh> ligature to allow a distinction between the voiced /ð/ and its voiceless equivalent /θ/, both spelled <th> in contemporary orthography. Oddly, the second of these was only to be used within words; initially <Th> would continue to be used for both sounds. Consistently marking contrasts between voiced and voiceless fricatives led to the suggestion of extending the use of <z>, so that it would regularly mark the voiced /z/ sound in words like *abzolve*, *theze*, *vizion*, and *pleazure*. Wijk's attempt to rationalize the variety of ways of representing individual vowel sounds led to a number of proposed revisions. For example, having determined that the predominant use of the symbols <ai, ay> is for the long-a sound, for example *maid*, *day*, Wijk proposed that all other instances of this sound be spelled accordingly and that words like *says*, *plait*, *quay*, *aisle*, in which the <ay, ai> digraph has a different pronunciation, be changed to *sez*, *plat*, *kea*, *yle*.

Regularized Inglish agrees with between 90 and 95 per cent of the words of ordinary English, as well as maintaining the principal regular features of that orthography, and was seen by Wijk as a more moderate stance that was more likely to win support where other schemes had failed. Wijk viewed the older generation as beyond conversion, and consequently advocated introducing his reforms gradually among children, with a sufficiently extensive period of transition. Wijk claimed that children, of six years of age and 'normal intelligence', should be able to acquire the system so that they can read tolerably well in a year. One obvious difficulty of phasing in the new system was that children would be required to learn both systems. To enable a gradual introduction to the current system, with all its irregularities, Wijk recommended children should spend two years learning nothing but the Regularized system. Having mastered this, they would be ready to engage with the complexities of the traditional system, which they would meet in all forms of adult publications and which would continue to be required by employers. Despite *Regularized Inglish* being principally a system for reading English, Wijk advocated teaching children to write using this system as well, so as to ensure its acceptance was not unduly delayed.

Despite Wijk's assurances that learning to use two spelling systems rather than one would not cause any additional difficulties, it is hard to understand how this could be seen as an advantage for those children who would function as the guinea-pigs in this linguistic experiment. For Wijk the truth of this claim was self-evident, but sceptics could easily be convinced by practical testing. Although he did not actually carry out any such tests, Wijk confidently anticipated a successful outcome: 'It is not hard to predict what the outcome of these tests will be. They will show beyond all doubt that in this particular case it is far easier to learn two systems than one'. While he conceded that it may seem unfair on those children required to learn both systems during this period of transition, they are the generation who will benefit most from the reform and 'it is perhaps only fair that they themselves should have to contribute the main part of the "cost" '. By suggesting that only those children who were likely to go on to careers that would require them to write the traditional orthography need to learn it, Wijk advocated introducing a form of linguistic discrimination that would be particularly divisive. He also tackled the potential objection that the new spellings appear 'ugly, grotesque and generally detestable', arguing that such aesthetic considerations are little more than prejudices stemming from a lack of familiarity with the new spellings. Rather than denouncing the reformed spellings on these grounds, He advocated reading a brief outline of the historical development of the English language and a handbook on English phonetics, after which anyone would be able to appreciate the greater value of the new spellings. It will be apparent that Wijk's proposed reforms, although moderate and reasonable in their attitude to the traditional orthography, demanded considerable tolerance and investment from those teachers and children required to learn it.

Two final objections that are confronted by Wijk concern the difficulties presented by different pronunciations and accents, and the question of the spelling of loanwords. In the case of American English Wijk accepted that different spellings would be needed to accommodate differences in accent, so that where British English has *baath*, *caar*, American English would have a single <a>, reflecting the short vowel. Introducing additional differences between British and American English seems rather unfortunate, although for Wijk this is easier to justify than the preservation of purely written differences such as *color/colour*, *centre/center*. But while he accepts different spellings for American English, he is not prepared to allow regional English accents to be accommodated in this way. It is self-evident for Wijk that the model for American and British English spelling should be 'the dialect of its cultivated circles'. On the question of the spelling of loanwords using the new system, Wijk concluded that loanwords that have been fully assimilated should be respelled using the new system, while those which 'give the impression of being foreign' should retain their foreign spelling until they are fully assimilated. While this seems a sensible policy in theory, in practice it would of course be extremely difficult to draw a distinction between these two groups.

A much more radical approach to spelling reform was proposed by George Bernard Shaw, who advocated the replacement of the Roman alphabet with a completely revised, purpose-built system, in which each of the sounds had its own letter, and related sounds were represented by related letterforms. Shaw was opposed to English spelling because of its considerable wastage:

the large number of digraphs, in which two letters represented a single sound, and silent letters which unnecessarily wasted paper and ink, as well as writers' and printers' time.

'To any others the inadequacies of our 26 letter alphabet seem trifling, and the cost of a change quite prohibitive. My view is that a change, far from being an economic impossibility, is an economic necessity. The figures in its favor, hitherto uncalculated and unconsidered, are astronomical.... To spell "Shaw" with four letters instead of two, and "though" with six, means to them only a fraction of a second in wasted time. But multiply that fraction by the number of "thoughs" that are printed every day... and the fractions of a second suddenly swell into integers of years, of decades, of centuries, costing thousands, tens of thousands and millions.... Shakespear might have written two or three more plays in the time it took him to spell his name with eleven letters instead of seven'.

(From a letter in *The Author,* 1944; quoted from Carney, 1994, pp. 483–4)

Shaw also deplored the time wasted by schoolchildren who were forced to learn this system, which he dismissively referred to as 'Johnsonese', blaming its current form on Dr Johnson. Shaw's campaign against the English spelling system occupied much of his later life; he wrote lengthy letters to *The Times* drawing the public's attention to its shortcomings. But as early as 1916, in the preface to his play *Pygmalion*, in which a Cockney flower girl is taught to speak correctly by a phonetics professor, Shaw complained about the flaws in English spelling: 'The English have no respect for their language, and will not teach their children to speak it. They spell it so abominably that no man can teach himself what it sounds like'. Shaw himself wrote using the shorthand system devised by Isaac Pitman, although he recognized that a shorthand system, designed principally to facilitate writing by hand and therefore unsuitable for printing, could never function as a national spelling system. This was made particularly apparent to him from the postcards he received from the Oxford phonetics professor Henry Sweet, on whom the character of Professor Henry Higgins was partially modelled. These were written in Sweet's own 'Current Shorthand', intended to function as an alphabet in which each sound had its own symbol, although his use of the script as a form of shorthand reduced it to 'the most inscrutable of cryptograms'. In Shaw's *Pygmalion*, Higgins's mother complains that she is unable to read the pretty postcards her son sends her in his patent shorthand, forcing her to rely upon the copies in 'ordinary writing' (Act III).

Recognizing his own lack of qualifications for designing a new alphabet, Shaw left money in his estate for the development of a 'British' alphabet which would comprise at least forty letters, with related sounds being represented by related letter-forms, so as to enable a more efficient mode of writing English. Various legal difficulties concerning the terms of his will meant that it was not until seven years after his death that a competition was announced, in which the designer of the best alphabet following the stipulations of Shaw's will would receive £500. A total of 467 entries was received and the judges decided to split the prize money between four of these, none of which was deemed to be entirely acceptable. Because another of the goals of the competition was the publication of an edition of Shaw's play *Androcles and the Lion* in the new alphabet, it was proposed that the four winning entries be asked to continue to refine their proposals with advice from the panel of expert judges.

The result of this process was the selection of the revised version of the alphabet devised by Kingsley Read as the 'Bernard Shaw', or 'Shavian' alphabet. The Shavian alphabet (Figure 6.1)

The Shaw Alphabet for Writers

Double lines : between pairs show the relative height of Talls, Deeps, and Shorts. Wherever possible, finish letters rightwards; those starred * will be written upwards. Also see heading and footnotes overleaf.

	Tall		Deep				Short		Short	
peep	𐑐	:	𐑚	bib		if	𐑦	:	𐑰	eat
tot	𐑑	:	𐑛	dead		egg	𐑧	:	𐑱	age
kick	𐑒	:	𐑜	gag		ash*	𐑨	:	𐑲	ice
fee	𐑓	:	𐑝	vow		ado*	𐑩	:	𐑳	up
thigh	𐑔	:	𐑞	they		on	𐑪	:	𐑴	oak
so	𐑕	:	𐑟	zoo		wool	𐑫	:	𐑵	ooze
sure	𐑖	:	𐑠	meaSure		out	𐑬	:	𐑶	oil
church	𐑗	:	𐑡	judge		ah*	𐑭	:	𐑷	awe
yea	𐑘	:	𐑢	*woe		are	𐑸	:	𐑹	or
hung	𐑙	:	𐑣	ha-ha		air	𐑺	:	𐑻	err
	Short		Short			array	𐑼	:	𐑽	ear
loll	𐑤	:	𐑮	roar					Tall	
mime*	𐑥	:	𐑯	nun		Ian	𐑾	:	𐑿	yew

Figure 6.1. The Shavian alphabet

is a radical departure from the Roman alphabet, employing a series of letter shapes which resemble the current shorthand systems, such as that devised by Henry Sweet, favoured by Shaw himself. There are continuities with the traditional system: letters are written from left to right and words are separated by blank spaces. But there are departures too, such as the lack of distinction between upper and lower case letters, which were seen as an unnecessary complication. Where a capital letter is required, the Shavian alphabet employs a 'namer' dot which is placed above the letter. The individual letterforms were devised so as to discourage cursive writing, that is the addition of unnecessary linking strokes which add to uncertainty and confusion when written quickly. Much time and effort was expended upon ensuring that the letters were sufficiently distinct so as to avoid the possibility of confusion when being read at speed. At the same time, care was given to creating letters that could be formed with the fewest possible number of strokes so as to facilitate speedy writing; many Shavian letters consist of single strokes. Shaw's requirement that related sounds be accorded related shapes is satisfied to a certain extent. All vowels are represented by 'shorts', letters with no ascenders or descenders, while voiced and voiceless pairs of consonants are given related shapes.

If we read down the first column of the alphabet set out in Figure 6.1, we can see that the left-hand side gives the voiceless consonants, while the right-hand side contains the voiced equivalents. In each case the two letters are simply reversed, so that the symbol for the voiceless /p/ is simply written upside down to give its voiced partner /b/. In providing for a much larger number of sounds and linking related sounds in this way, the Shavian system comes closer to achieving certain key features of the ideal alphabet discussed in Chapter 2. But there are obvious shortcomings of the Shavian alphabet. Most evident are the practical problems concerning the economic and social costs of introducing such a radical departure from the traditional system. Given that Shaw's major objection to the Roman alphabet was the amount of time and money that were wasted unnecessarily, it seems ironic that the system that bears his name would have required huge financial outlay to ensure its adoption. Another obvious shortcoming concerns the similarity in some of the lettershapes. Even though this was a desideratum of any proposed new alphabet, the letters employed in the Shavian system are insufficiently distinctive. The three symbols used for the sounds in **z**oo, mea**s**ure, **j**udge have only minor differences in the formation of the curl at the foot of the letter, which would surely be blurred when written by hand at speed. Read himself accepted that the shapes, though designed to be easy to write, were not always sufficiently distinctive, advising his readers to be careful to distinguish between similar short letters. Yet, if reading this alphabet takes greater care and therefore time, the principle of economy on which it was built has been flouted. The same problem also applies to the writer, who must take greater care in ensuring that the letters are written sufficiently distinctively. One begins to wonder if such a system would really have resulted in additional plays by Shakespeare, as claimed by Shaw. The Shavian system devised and refined by Kingsley Read was duly employed for the publication of Shaw's play *Androcles and the Lion* in 1962, thereby fulfilling the terms of Shaw's will. The publication was accompanied by a preface written by James Pitman, in which he outlines the merits of the proposed system. Pitman argues that the Shavian letters are more legible, one-third more economical in terms of space, and some 80–100 per cent quicker to write. Pitman devised an unusual way of demonstrating that the Shavian script could be read at a greater distance than the traditional letters, requesting his readers to hold the book upside down in front of a mirror, so that both scripts are illegible. Keeping the book pressed against his lips, the reader is urged to advance towards the mirror, whereupon he will notice that the Shavian script becomes legibile before the traditional letters. According to Pitman, mastering the unusual letters would take just three or four hours, and he urges the reader to learn the system and encourage others to do the same: 'This book costs very little. Get your friends to buy one and to learn the alphabet so that you can write to one another—or, if you become so skilled that you no longer need to "keep your eye in", give it away' (Pitman, 1962, p. 15). But, despite Pitman's evangelical injunction, and the publication of *Shaw-Script*, a quarterly periodical printed using the Shavian system, the Shavian system was yet another dead end in the search for a new improved English spelling system.

One final attempt to reform spelling which we will consider here is the Initial Teaching Alphabet, devised by James Pitman himself. Sir James Pitman was the grandson of Isaac Pitman, inventor of the highly successful shorthand system, and was one of the trustees of Bernard Shaw's will. He was born in London in 1901. As well as serving as MP for Bath between 1945 and 1964, and acting as Pro-Chancellor of Bath University from 1972 to 1981, he was a vigorous spelling reformer. Pitman's major contribution to spelling reform was his Initial Teaching Alphabet, or ITA, a revised version of the Roman alphabet intended to simplify the process of learning to read (Figure 6.2). The ITA worked by introducing children to an easier, more phonetic system, before they made the transition to the more complex standard spelling system at the age of seven. The Initial Teaching Alphabet consists of forty-two letters, comprising the standard Roman letters with

THE INITIAL TEACHING ALPHABET

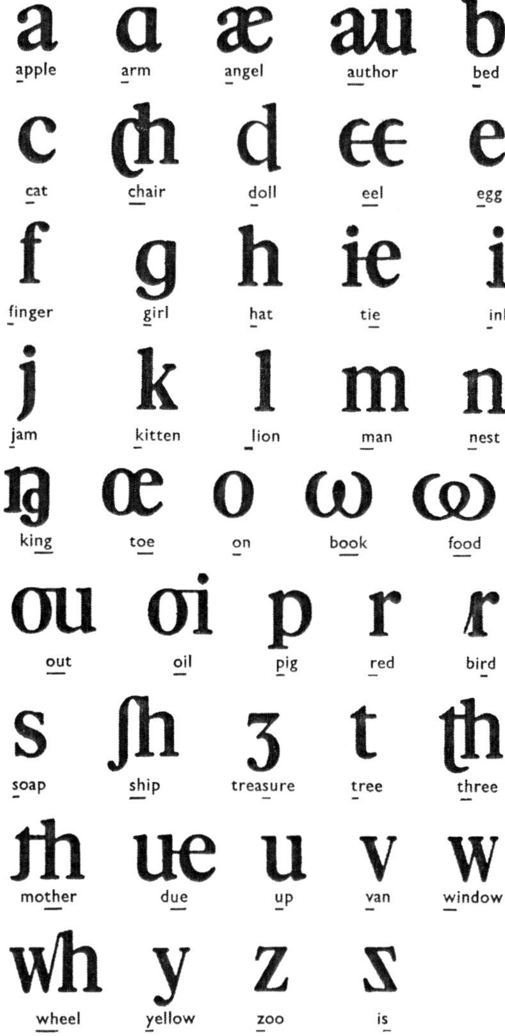

Printers use the 'ɑ' in words like path where regional pronunciation differs. Teachers and children use a or ɑ, whichever corresponds to their own speech.

Figure 6.2. The Initial Teaching Alphabet © The Pitman Collection, University of Bath

an additional set of Roman letters with slight modifications designed to help children determine the correct sound value. So the letter <t> has a slight turn at the foot when used in the digraph <th>, the long <s> of the eighteenth century is revived in the digraph <sh>, while a reversed <z> is used instead of <s> when it reflects the voiced sound /z/.

Unlike many of the reformed spelling systems we have examined in the course of this book, Pitman's system was actually implemented, albeit in an exploratory manner. The Initial Teaching Alphabet was introduced in British schools in the early 1960s; the comparatively minor reforms it required, combined with its use as a training alphabet rather than a replacement for the standard, meant that some educationalists considered it a useful teaching tool. But there were problems too. In being based upon the standard southern accent, Received Pronunciation, the system was much less straightforward for children who spoke with regional accents. More problematically, many children who got to grips with the Initial Teaching Alphabet encountered difficulties in making the transition to the standard system. Despite being used in some progressive schools throughout the 1960s and 1970s, and spreading as far afield as Australia and North America, Pitman's ITA was never accepted as a mainstream teaching tool. A BBC report on Pitman's ITA in 2001, marking its fortieth anniversary, provoked a number of adverse memories from people who were first taught to read using this system. One former pupil of ITA recalls: 'I suffered ITA for my first few years at school, with the consequence that at the age of seven I could barely read or write', while another reports difficulties in corresponding with older family members not familiar with the system. The most extreme cases are those former pupils who blamed a lifetime of poor spelling on their early encounter with Pitman's ITA. But, despite its lack of widespread success, the ITA continues to be promoted by the Initial Teaching Alphabet Foundation, whose mission is to promote, maintain and advance the use of the ITA. To achieve this goal the Foundation provides funding and technical assistance to nonprofit organizations who wish to employ the ITA in order to promote literacy. While the Foundation offers support for training and programme development, as well as publishing manuals for those wishing to implement ITA and reading materials for children learning to read using this system, nowhere on their website is there any evidence to justify their bold claim that children trained to use ITA 'become proficient readers and writers without struggling with complex spelling patterns in the beginning stages of literacy development'.

Despite its lack of success, ITA advanced further than most other schemes for a reformed spelling in actually being trialled in British classrooms. The many other various proposals for spelling reform that we have examined in this and the previous chapter have had little or no effect upon the way English is spelled today. In the next chapter we will turn to a consideration of the development of American spelling, where we will encounter one of the few spelling reformers whose proposals have had a long-lasting effect upon the way English is spelled.

Chapter 7

American spelling

American English has its roots in the Early Modern English variety that was used by the Pilgrim Fathers and first settlers; thus the earliest American documents were written using the emerging English standard spelling system. This practice continued to be reinforced by the importing of books from Britain, with the result that authoritative English models, the works of Dryden, Pope, and Addison, continued to exert an influence in the New World. The earliest colonial authors sent their books to Britain to be printed, thereby further ensuring the continued support of the English standard spelling system. The comparatively slow establishment of higher education colleges in the colonies meant that the sons of rich colonialists were frequently sent to Oxford and Cambridge for their education. The maintenance of this link with English spelling practices meant that American spelling underwent a process of standardization similar to Early Modern English, including such developments as the regularization of the use of <u> and <v> and the introduction of <j> as an alternative to <i>. The influence of the Augustan English writers, such as Swift and Addison, can also be seen in calls for the establishment of an academy to oversee the fixing and preservation of a standard American usage.

The direct influence of the English writers in such proposals can be seen in the suggested title for a proposed institution: 'The American Academy for Refining, Improving, and Ascertaining the English Language'. Although this proposal of 1780 was never acted upon, an American Academy of Language and Belles Lettres was founded in 1820, under the presidency of John Quincy Adams. But neither this, nor subsequent academies of this kind, have had any effect upon American spelling.

One area where Government attempts to regulate spelling did meet with success concerned the transliteration of foreign words into American spelling. This was particularly pressing in the case of geographical names, where the original forms were often drawn from foreign languages and the adoption of a fixed spelling was a practical necessity. In 1890 the then president, Benjamin Harrison, established the United States Board on Geographic Names, whose remit was to ensure uniformity in the spelling of place names throughout Government offices and on maps and charts. An important function of this Board was to ensure that the spellings of place names were suitable guides to their pronunciation for American citizens. The problems of variant spellings, especially ones using foreign spelling practices, had been highlighted by Noah Webster in the preface to his 1803 spelling book, where he argued against the maintenance of French spellings in place names like *Ouisconsin* and *Ouabasche*:

> How does an unlettered American know the pronunciation of the name Ouisconsin or Ouabasche, in this French dress? Would he suspect the pronunciation to be Wisconsin or Waubash? Our citizens ought not to be thus perplexed with an orthography to which they are strangers.

But, despite the adoption of the spelling *Wisconsin* in a map of 1822 and a resolution of 1845 declaring this spelling the correct one, a report by the US Board on Geographic Names published in 1901 states of the Wisconsin river: 'Wisconsin; river in Wisconsin. (Not Ouisconsin, nor Wiskonsin).' Names taken from indigenous American languages also underwent a process of Americanization, bringing them into line with more accustomed spelling conventions. Thus *Unéaukara* is now *Niagara*, *Mauwauwaming* became *Wyoming*, while *Machihiganing* became *Michigan*. Not all such names were Americanized in this way; many Native American names survive, preserving rich and fiendishly difficult combinations of consonants ready to trap the unwary: *Tallahassee*, *Susquehanna*, *Mississippi*, *Allegheny*, *Massachusetts*.

The Board was also responsible for introducing other reforms whose influence can still be seen in US place names today. For instance, the word *borough* was abbreviated to *boro*, introducing the contrast between the English and American place names *Marlborough* and *Marlboro*. A further change involved the dropping of a final <h> in *burgh*, producing names like *Gettysburg* and *Williamsburg*. *Pittsburgh* is, however, an exception and should be spelled with its final <h>. This is because the citizens of *Pittsburgh* refused to change the spelling of their home town; the Board finally relented in 1911, allowing *Pittsburgh* to retain its traditional spelling and simultaneously making it the most commonly misspelled place name in the USA.

The problems encountered in the Americanization of indigenous place names were paralleled in the establishment of spellings of other words borrowed from Native American languages. The word *raccoon*, derived from the Algonquian word *aroughcun*, is attested in a rich range of variant forms, including *rahaugcum*, *raugroughcum*, and *aroughcoune*, before the standard spelling was established. Another native animal whose Algonquian name was adopted by the settlers is the *opossum*, derived from the Algonquian word *op* 'white' + *assom* 'dog'. The earliest forms of this word spell the initial vowel sound with an <a>: *aposoum*, *apossoun*, while an 'aphetic' form *possown*—dropping the initial vowel entirely—is the ancestor of modern English *possum*. The Powhatan borrowing *mockasins*, referring to the soft-soled leather shoes worn by Native Americans, is attested in a similarly wide range of spellings, including *maccaseene*, *mogginson*, *mawkisin*, *mogasheen*, and *mognesan*, before the modern English spelling *moccasin* was settled upon. The word *wigwam* derives from the Ojibwa word *wigwaum*, a variant of the Algonquian word *weekuwom*, *wikiwam* (literally meaning 'their house'). The Narragansett term for a shaman or healer, *powwaw*, ultimately derived from an Algonquian root meaning 'he who dreams', is first attested in the spelling *powahe*. Different spellings recorded in the sixteenth and seventeenth centuries appear to reflect alternative pronunciations. In the case of *pauwau* the second syllable has been assimilated to the first; the opposite is true of *powwow*, the spelling from which our modern English form is derived. The word *toboggan* is derived from the Canadian Native American language Micmac, *tobâkun*. The word was first adopted into French as *tabaganne*, from which the earliest English spellings, *tabagane*, *toboggnay*, *tarboggin*, *treboggin*, are derived.

The major break between American and British spelling took place in the late eighteenth and early nineteenth centuries, when several notable American linguists attempted to reform American spelling. One proponent of spelling reform was Benjamin Franklin, who created a revised alphabet which was intended to bring English spelling closer to its pronunciation. Franklin thought that spelling should reflect pronunciation rather than etymology, as the latter was not a reliable guide to the meaning of a word. While he recognized that a purely phonetic system would introduce a larger number of homophones, he believed that any potential confusion could be easily avoided by simple re-reading. Franklin advocated a phonemic system in which each letter should represent a single sound, and in which there should be no superfluous, that is silent, letters. The result was an alphabet which dropped the Roman letters <c, j, q, w, x, y>, and which employed six new letters, which were modifications of existing ones (Figure 7.1). He began by reorganizing the alphabet into a more 'natural' order, using an arrangement that began with 'simple sounds', made without any of the vocal organs, and then followed the tongue as it moved forward in the mouth. In this arrangement <g> appears early, because it is made with the

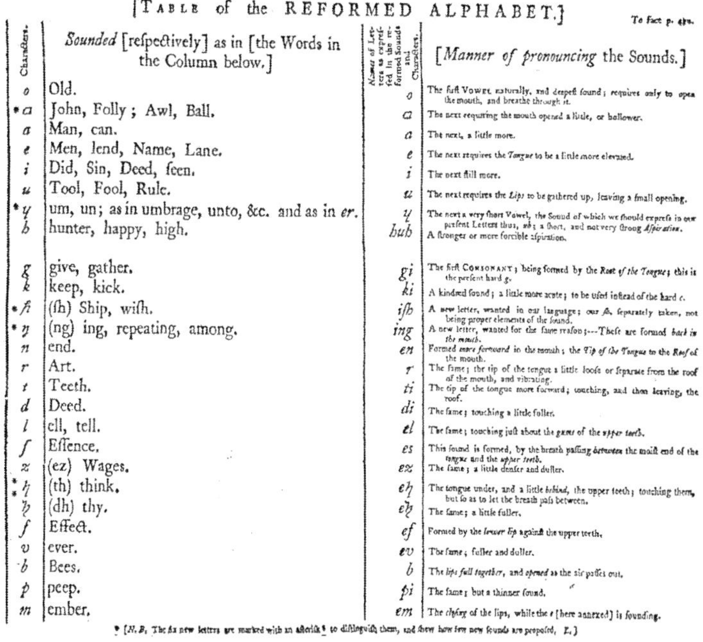

[TABLE of the REFORMED ALPHABET.] *To face p. 470.*

Characters.	*Sounded* [respectively] as in [the Words in the Column below.]	Names of Letters as expressed in the reformed Sounds and Characters.	[*Manner of pronouncing* the Sounds.]
o	Old.	o	The first VOWEL naturally, and deepest sound; requires only to open the mouth, and breathe through it.
* ɑ	John, Folly; Awl, Ball.	ɑ	The next requiring the mouth opened a little, or hollower.
a	Man, can.	a	The next, a little more.
e	Men, lend, Name, Lane.	e	The next requires the *Tongue* to be a little more elevated.
i	Did, Sin, Deed, seen.	i	The next still more.
u	Tool, Fool, Rule.	u	The next requires the *Lips* to be gathered up, leaving a small opening.
* ʏ	um, un; as in umbrage, unto, &c. and as in *er*.	ʏ	The next a very short Vowel, the Sound of which we should express in our present Letters thus, *uh*; a short, and not very strong *Aspiration*.
h	hunter, happy, high.	huh	A stronger or more forcible aspiration.
g	give, gather.	gi	The first CONSONANT; being formed by the *Root of the Tongue*; this is the present hard *g*.
k	keep, kick.	ki	A kindred sound; a little more acute; to be used instead of the hard *c*.
* ʃh	(sh) Ship, wish.	iʃh	A new letter, wanted in our language; our *sh*, separately taken, not being proper elements of the sound.
* ŋ	(ng) ing, repeating, among.	iŋ	A new letter, wanted for the same reason;---These are formed *back in the mouth*.
n	end.	en	Formed *more forward* in the mouth; the *Tip of the Tongue* to the *Roof* of the mouth.
r	Art.	r	The same; the tip of the tongue a little loose or separate from the roof of the mouth, and vibrating.
t	Teeth.	ti	The tip of the tongue more forward; touching, and then leaving, the roof.
d	Deed.	di	The same; touching a little fuller.
l	ell, tell.	el	The same; touching just about the *gums* of the *upper teeth*.
ſ	Effence.	es	This sound is formed, by the breath passing *between* the moist end of the tongue and the *upper teeth*.
z	(ez) Wages.	ez	The same; a little denser and duller.
* ħ	(th) think.	eħ	The tongue under, and a little *behind*, the upper teeth; touching them, but so as to let the breath pass between.
* ð	(dh) thy.	eð	The same; a little fuller.
f	Effect.	ef	Formed by the *lower lip* against the upper teeth.
v	ever.	ev	The same; fuller and duller.
b	Bees.	b	The *lips full together*, and *opened* as the air passes out.
p	peep.	pi	The same; but a thinner sound.
m	ember.	em	The *closing* of the lips, while the *e* [here annexed] is sounding.

* [N.B. The six new letters are marked with an asterisk* to distinguish them, and shew how few new sounds are proposed. E.]

Figure 7.1. Benjamin Franklin's reformed alphabet

tongue in contact with the 'velum', or hard palate, then come the dental consonants, such as <dh> and <th>, where the tongue comes into contact with the teeth, while <m>, in which the lips are closed, is the final letter. Franklin omitted the letter <c> on the basis that it was unnecessary; he used <k> for /k/ and <s> for /s/. While this may seem a logical rationalization of the alphabet, Franklin's decision to drop <j> and then add a new letter, a kind of modified <h>, in its place is hard to explain. He claimed that this letter is particularly useful in combination, so that *James* can be written *dhames*, *cherry* as *theri*, and the French word *jamais* can now be rendered *zhamais*. Why these new spellings were considered to be improvements is hard to fathom, although it did have the side benefit that the letter <g> could be reserved for the hard /g/ sound. Long and short vowels were distinguished by using single and double letters, as in the examples of *mend* 'mend' and *remeend* 'remained'; *did* 'did' and *diid* 'deed'. Franklin's system used a ligature of <t> and <h> for /θ/ and a ligature of <d> and <h> for /ð/, <n> with the tail of <g> for /ŋ/ and <s> with a bow in the vertical stroke for /ʃ/.

Franklin explained his reformed alphabet in a letter of 1768 to Mary Stevenson, to which she responded in a letter of the same year. Stevenson's reply was written using the reformed alphabet, which she had evidently studied in some detail, and to which she made a number of cogent objections. Stevenson noted that there could well be benefits in enabling readers to determine the correct pronunciation of a particular word, if such a thing could ever be finally determined, but she drew attention to the problem of homophones having identical spellings, as well as the break with etymology and the corresponding difficulties in determining the meaning of words. Stevenson concluded by advising Franklin to leave people to spell in the traditional way, hammering home the point by switching back to spelling in the 'old way' to sign off: 'With ease and with sincerity I can, in the old way, subscribe myself, Dear Sir, Your faithful and affectionate Servant'. Franklin responded by claiming that the reformed alphabet would be of most benefit in the promotion of literacy, enabling better spelling among those bad spellers whose difficulties are the fault of the spelling system. Franklin also pointed out that etymology was not a sound guide to current meaning, giving the example of *Neev* 'knave' and *Vilen* 'villain', which would be much less insulting if understood according to their older meanings of 'lad' and 'inhabitant of a village'. He subsequently published his reformed spelling system as 'Scheme for a New Alphabet and Reformed Mode of Spelling', in his *Political, Miscellaneous, and Philosophical Pieces* (1779), but it had no effect upon American spelling. Franklin's influence on the spelling of American English was destined to be an indirect one, through his encouragement and inspiration of the most important American linguist of this period, Noah Webster.

Noah Webster trained as a lawyer but turned to school teaching, setting up his own school in his home state of Connecticut. When this closed in 1782, Webster turned to writing elementary textbooks, the first of which was a spelling book designed to replace the standard work, *A New Guide to the English Tongue*, by the British author Thomas Dilworth, which by 1785 had appeared in over forty editions. While Webster objected to the use of Dilworth on academic grounds, his desire to replace this work was also partly nationalistic: 'He [Dilworth] is not only out of date; but is really faulty & defective in every part of his work. America must be as independent in *literature* as she is in *politics*—as famous for *arts* as for *arms*' (see Monaghan, 1983, p. 28).

Webster had to borrow money to enable publication of his work, which appeared as *A Grammatical Institute of the English Language, Part I* in 1783; the punchier original title *American Instructor* was ditched in favour of the pompous title at the suggestion of the president of Yale. This was followed in 1784 by the second part, a grammar, and a reader in 1785. But it was the speller that was the success of this publishing venture, becoming a 'riproaring bestseller', with publishers in the states of Pennsylvania, New York, Connecticut, Vermont, and Massachusetts.

Webster's spelling book was designed to teach children to read by helping them to understand how to pronounce hundreds of words. For Webster there was a close correlation between spelling and pronunciation; consequently he conceived of his speller as setting-out a method of speaking as well as of writing English. Webster lamented the lack of care that had been shown to pronunciation by previous grammarians, whom he blamed for the great variety of dialects. Webster's speller differed from that of Dilworth in the way he divided up syllables: where a word like *salvation* had been divided up as *sal-va-ti-on*, Webster divided it as *sal-va-tion*. Webster also sought to provide an exhaustive account of the sounds of each individual letter, such as that for G: '*G* has two sounds: one as in *go*, the other like *j*, as in *gentle*. It has its first or hard sound before *a*, *o*, *u;* in general its second or soft sound before *e* and *y*, and is either hard or soft before *i*' (1793, p. 16). In explaining the various uses of each letter, Webster distinguished between a proper use and secondary ones, employing the concept of usurpation in cases where a letter takes on the role of another, as in the description of Z: '*Z* has two sounds; its proper sound, as in *zeal;* and that of *zh*, as in *azure*. Its place is commonly usurped by an *s*, as in *wisdom*, *reason*' (1793, p. 16). Another of Webster's innovations was the introduction of a numerical index as a guide to identifying vowel length and quality, so that a superscript 1 indicates that a vowel is long, a superscript 2 that a vowel is short, number 3 the 'broad a' in *hall*, number 4 the vowel sound in *not* and *what*. At the end of the work there are a number of lessons comprising easy words, which were designed to teach children to read and to 'know their duty'. These are overtly moralistic and Christian in their teaching, urging the young learner to avoid sin, trust in the Lord, and to work hard at school. The following example gives a flavour: 'A good child will not lie, swear nor steal. He will be good at home, and ask to read his book, when he gets up, he will wash his hands and face clean; he will comb his hair, and make haste to school; he will not play by the way, as bad boys do' (1793, p. 56).

Despite its huge popularity, Webster's speller was criticized by some, including an anonymous writer calling himself 'Dilworth's Ghost', who accused Webster of plagiarizing Dilworth, and who ridiculed his attempt to fix pronunciation. Webster responded by defending the originality of his work, while conceding that all such works were necessarily compilations, and arguing that a language could become fixed once it had attained a standard of perfection. Webster followed the first edition of 1783 with a revised edition in 1787, under the new title *The American Spelling Book*. A revised version of the second edition appeared in 1804 with the same title, and again with a handful of further corrections in 1818. The major difference between the first and second editions concerns the additions of proverbs and fables, symptomatic of Webster's deeply moral view of elementary education. Thus we find lists of maxims such as: 'A wise man hath his tongue in his heart, but a fool hath his heart on his tongue. Be more apt to hear than to speak, and to learn than to teach. Youth, like the spring will soon be past. All is not gold that shines'. As well as these gnomic sayings, we find moral precepts such as 'He that speaks loud in school will not learn his own book well, nor let the rest learn theirs; but those that make no noise will soon be wise, and gain much love and good will'.

While Webster began his speller with a complaint about the difficulties posed by English spelling, with its numerous ways of writing a single sound, he adopted the spellings used by Johnson's *Dictionary* as his model, and made no attempt to suggest alternatives. At this stage in his career Webster remained staunchly conservative in his attitude to spelling, ridiculing those who proposed alterations. He was especially critical of those who advocated removing the <u> in *favour* and *honour*, a reform that Webster himself would later go on to promote, claiming that such reforms drop the wrong letter; instead Webster proposed spelling these words *onur* and *favur*. Webster also opposed dropping the <e> in *judgement*, another of his subsequent reforms, arguing that its use to indicate a soft <g> makes it 'the most necessary letter in the word'. For Webster, the fact that spelling and pronunciation are not closely related 'is an inconvenience we regret, but cannot remedy'.

It was Webster's meeting with Benjamin Franklin in 1786 in Philadelphia, and their subsequent friendship, that was to stimulate him to consider seriously the possibilities of spelling reform. Shortly after that meeting, Webster wrote to General George Washington, informing him of his desire to serve his country by refining the language so as to reduce it to perfect regularity, and thereby to improve education. Webster's proposal attempted to modify that of Franklin by reducing the number of innovative characters, and he attempted to seize on the nationalistic political climate to encourage its adoption. Of the various reasons he offered in support of a reformed alphabet, the most striking was its nationalistic advantages: 'A national language is a national tie, and what country wants it more than America?' (see Lepore, 2003, p. 5). Webster saw a reformed alphabet as a means of bestowing a national identity on the American use of English, and as a way of ensuring linguistic uniformity across the country. But, in viewing a standard spelling system as a means of fixing pronunciation, Webster misunderstood the relationship between spellings and sounds. Webster's reforms were set out in an appendix to his *Dissertations*, in which he proposed removing silent letters, making the relationship between letters and sounds more predictable, as well as tinkering slightly with the forms of letters. In 1790 he published *A Collection of Essays and Fugitiv Writings*, which collected together various of his writings. The more recent essays were written in the reformed orthography, which was not received favourably by the reading public. When Webster proposed producing a dictionary in 1793, there was considerable hostility to the idea that such a work should incorporate his reformed spelling system. The editor of the *Philadelphia Gazette of the United States* responded by satirizing a dictionary in which the spelling privileges the whims of an individual, rather than the accepted norms of the standard language (see Monaghan, 1983, pp. 118–19):

To Mr. Noab Wabstur

Sur,

by rading all ovur the nusspaper I find you are after meaking a nue Merrykin Dikshunary; your rite, Sir; for ofter lookin all over the anglish Books, you wont find a bit of Shillaly big enuf to beat a dog wid. so I hope you'll take a hint, and put enuff of rem in yours, for Och 'tis a nate little bit of furniture for any Man's house so it 'tis.

PAT O'DOGERTY

As I find der ish no DONDER and BLIXSUM in de English Dikshonere I hope you put both in yours to oblige a Subscrybur

HANS BUBBLEBLOWER

Unperturbed by such scornful criticism, Webster began the dictionary project in earnest in 1800, although it was not until 1806 that he published *A Compendious Dictionary*. In the introduction Webster distanced himself from Franklin's proposed reforms, with their more extreme changes and new characters, adopting a more moderate stance. He advocated reform of spelling to reflect changing pronunciation, no longer subscribing to the view that a language should be perfected and fixed to prevent further changes. The spelling of his *Compendious Dictionary* included some of his preferred spellings, such as *tung*, *determin*, *melasses* 'molasses', *honor*, *music*, *theater*, but avoided others, such as *hed*, *giv*, *nabor* 'neighbour'. Webster's shrewd business sense, sharpened by his financial dependence upon the economic success of the speller, meant that he sensibly avoided incorporating his reformed spellings into that work, sticking carefully to traditional orthography. He did however make another attempt to sneak some of his reformed spellings into print, in his *The Little Reader's Assistant*, published in 1790. This book was intended to function as an intermediate reader, providing students with the rudiments of English grammar, as well as accounts of American history and government. It is particularly remarkable for its early attack on the practice of slavery; Webster himself was heavily committed to the abolitionist movement. Traces of Webster's early spelling reforms can be seen in its use of spellings such as *nabors*, *yung*, *hed*, *giv*, and *cathecizm*.

It was with the publication of the version of his dictionary for schools in 1830 that Webster most clearly set out his final views on a reformed English orthography. Here he proposed many of the revised spellings that are now familiar as accepted features of American spelling, such as the use of <-or> rather than <-our>, *honor, favor*, <-er> instead of <-re> in *meter* and *theater*, and single consonants before suffixes beginning with vowels, for example *traveled*. As well as creating a distinction between British and American spelling that served contemporary nationalistic interests, Webster's spelling reforms brought American spelling more closely into line with its pronunciation. The <u> in *colour, honour* was not sounded and so redundant, while the <re> spelling in *theatre* and *metre* is an unhelpful guide to its pronunciation. Webster also argued that a spelling like *labor* established a link with derived forms, like *laborious*, although in the case of *meter* and *theater*, the <re> spelling would be preferable, given related forms such as *metrical* and *theatrical*. Despite Webster's claims, there are benefits to the British spellings of these words that are lost in his reforms. The <our> spelling of *colour, favour* helps to distinguish these words from agent nouns in <or>, such as *actor* and *author*, while the <re> ending in *theatre, centre*, enables such nouns to be distinguished from agent nouns in <er>, such as *helper* and *driver*. This is the basis of the distinction between the British spellings *metre* 'unit of length' and *meter* 'a measuring instrument'.

Webster also removed the British English distinction between *practise* (verb) from *practice* (noun) by spelling both words *practise*. Webster's rationale for this change was that words like *defence* were spelled with an <s> in Latin (Latin *defensum* means something that is forbidden), as were derived words like *defensive* and *offensive*. But why modern English spelling should reflect Latin usage is neither explained nor defended; while the proposed change does bring spelling and sound into alignment, it also removes a useful visual distinction between different grammatical classes of the same word. Webster also adopted the letter <z> in verbs like *patronize*, although this reform was not implemented systematically: his dictionary includes spellings like *criticise, circumcise*. Webster also implemented the rule that suffixes beginning with vowels, like <-ed>, should follow single consonants in words of two or more syllables: compare American *traveled* with the British spelling *travelled*. In words like *distil* and *fulfil*, Webster advocated employing double <l>, but proposed dropping the silent <e> in *movable, debatable*. Some of the changes proposed by Webster corresponded with changes adopted in British spelling, such as the simplifying of the spellings of *logick* and *masque* to *logic* and *mask*, although this was not carried through systematically, so that *traffick, almanack, frolick* continued to appear. Webster advocated spelling words like *connection, inflection*, and *reflection* with <ct> rather than <x>, by analogy with the verbs *connect, inflect*, and *reflect*. He also attempted to regularize the spelling of words ending <ise/ize>. Words derived from Latin and Greek were to be spelled <ize>, *moralize, legalize*, while those from French should be spelled <ise>, *surprise, enterprise*. Other changes did not catch on, such as Webster's proposal to omit silent letters in *determin* 'determine', *altho* 'although', *crum* 'crumb', *ile* 'isle', *fashon* 'fashion'.

Despite the influence of the 1806 *Dictionary* it was the 1828 edition that established the majority of American spellings; these were further promoted by Webster's revised speller, *The Elementary Spelling Book* (1829). This version of Webster's *Dictionary* saw him drop some of the more radical reforms offered by his 1806 publication, especially the proposed removal of silent letters in words like *ake, aker, bild, nightmar, tung, wo, iland*. In the mid-nineteenth century the speller continued to sell huge numbers of copies, in excess of 1,500,000 a year. When, in 1880, William H. Appleton, director of the printing house contracted to publish Webster's speller, was asked which was his bestselling title, he replied: 'Webster's *Speller*... it has the largest sale of any book in the world except the Bible'. While Webster's speller came to be replaced as the main instructional text in American classrooms, it retained its status as an authority on spelling, as can be seen by its frequent adoption as the rulebook for that most American of entertainments, the spelling bee.

The spelling bee began in the 1700s, forming a key part of a child's education. By 1800 spelling bees came to be viewed more as social entertainment; such frivolous pleasures were frowned upon by the Puritans who rebranded them as spelling 'schools' in an attempt to re-assert their educational focus. Spelling bees gradually fell out of favour in New England society, but were preserved within a frontier culture for whom correct spelling was still considered an index of education and social status. The publication of *The Hoosier Schoolmaster* (1871) led to a craze in spelling contests, following which the term *spelling bee* was introduced. The use of the word *bee* to refer to a social gathering of this kind, often with a specific purpose in mind, such as *apple-bee*, *husking bee*, or the more sinister *lynching-bee*, is based upon an allusion to the social character of bees themselves. But while these gatherings were social, they were also highly competitive, as the alternative name 'spelling fight' makes clear. The first national spelling bee was introduced in 1925, subsequently rebranded the Scripps bee in 1941, and is still going strong today. As I noted in Chapter 1, the contest has become progressively harder over the years, with recent winners successfully spelling rare, technical words of foreign extraction in order to claim their prize. The cash value of the prize, $20,000, a considerable sum for a young child, is relatively small compared to the social prestige and national acclaim attached to winning the Scripps Spelling Bee. The popularity of the spelling bee is far greater in America than it is in Britain. The British equivalent, the BBC's *Hard Spell*, was a short-lived innovation, while British schools rarely engage in the spelling bees that are such an integral part of American society.

The distinctions between American and British spelling are gradually being eroded, as some American spellings introduced by Webster have become more prevalent in British usage. One such simplification was the removal of the silent <e> after <dg> when it is followed by the suffix <ment>, as in *abridgment*, *judgment*. British spelling retained this silent <e>, which functions as a diacritic, a marker to indicate that the preceding <g> should be pronounced soft rather than as a hard /g/ sound, as in the initial sound of *get*. So, while the silent <e> may appear entirely redundant, it does serve a useful function. James Murray, editor of the *Oxford English Dictionary*, condemned the omission of <e> in such cases, arguing that it goes against 'all analogy, etymology, and orthoepy, since elsewhere g is hard in English when not followed by e or i'. But, while this may be true, since the combination <dg> is always soft, it seems less necessary to employ a diacritic in words like *badge*, *bridge*, and *abridgement*, *judgement*. The use of <e> in such words derives from the Middle English period, when these words were spelled with <gg>, with consequently greater potential for confusion between hard and soft pronunciations. Despite Murray's objections, it is clear that the American spelling of *judgment* is becoming widely adopted in British usage. The spelling *judg(e)ment* was discussed by H. W. Fowler in his classic *Dictionary of Modern English*

Usage (1926): despite his acknowledgement that 'modern usage' favours *judgment*, Fowler recommended 'the older and more reasonable spelling' *judgement*. This judg(e)ment had changed by the publication of the third edition of Fowler, edited by R. W. Burchfield (1996), who stated that 'The presence or absence of-e-is not a matter of correctness or the reverse, but just one of convention in various publishing houses' (p. 429).

The use of the ligatures <æ> and <œ> in classical borrowings, for example *gynæcology*, *œcology*, was common to both American and British spelling in the nineteenth century. But by the late 1950s these had been simplified to <ae>, or more commonly replaced by <e>: *gynaecology*, *gynecology*. The <æ>/<ae> and <œ>/<oe> spellings were initially introduced in Latin as ways of representing the diphthongs <AI> and <OI> in Greek loanwords. British usage has tended to be more conservative in its retention of <æ> or <ae> in such words, although the American spellings with <e> have begun to be more widely adopted, as in *encyclopedia*, *medieval*, *paleography*; however, spellings of these words with <ae> are still found. Despite this change, the <ae> spelling is still favoured in words like *archaeology*, *Caesar* and in the personal name *Michael*; the <oe> spelling tends to have been maintained in medical and scientific words like *amoeba*, *diarrhoea* (compare the American spellings of these words: *archeology*, *Cesar*, *ameba*, *diarrhea*). In some instances, variant spellings have developed their own specific uses, as in the case of *dæmon* (Greek *daimon*). This word was used in Greek mythology to refer to an inferior divinity, intermediate between gods and men; the spelling of this word with <æ> was a deliberate means of maintaining a distinction from the specifically Judaeo-Christian sense of 'evil spirit' associated with the spelling *demon*. More recently, the spelling *dæmon* has been revived by Philip Pullman in his *Dark Materials* trilogy, to refer to the animal manifestations of the humans which occupy his fictional worlds. Although the word is consistently spelled *dæmon*, the author adds a note indicating that the intended pronunciation is identical with our modern English word *demon*; presumably the adoption of this variant spelling was an attempt to allow a similar kind of distinction from the modern English sense of *demon* as attested in earlier usage.

The choice between the verb endings <ise/yse > and <ize/yze> is rather more complex. The <ize/yze> suffix originated in Greek, and British practice is to reflect words that derive ultimately from Greek by spelling them <ize/yze>. We saw above that Webster also advocated following this etymological principle in these words. There remain, however, a number of exceptions. A small group of words are of French or Latin, not Greek, origin and consequently spelled with <ise>. These include *advertise*, *advise*, *apprise*, *arise*, *chastise*, *circumcise*, *compromise*, *despise*, *disguise*, *improvise*, *supervise*, *surprise*. But the widespread use of <ize> in words pronounced with the identical ending, /aɪz/, means that there is considerable potential for confusion. American usage today tends to spell all such words <ize/yze>; more recently, the <ize/yze> spelling has become increasingly common in British usage, suggesting the influence of American spelling habits. However, the explanation is not so straightforward; the <ize/yze> spelling is not simply an Americanization, it is also the preferred spelling of the *OED*. As the dictionary explains: ‘Some have used the spelling *-ise* in English, as in French, for all these words, and some prefer *-ise* in words formed in French or English from Latin elements, retaining *-ize* for those formed from Greek elements. But the suffix itself, whatever the element to which it is added, is in its origin the Greek *-ιζειν*, Latin *-izāre*;

and, as the pronunciation is also with *z*, there is no reason why in English the special French spelling should be followed, in opposition to that which is at once etymological and phonetic. In this Dictionary the termination is uniformly written *-ize*' (*OED* 2nd edn, 1989: s.v. -ize suffix). So, while it may seem as if the spread of spellings with <ize> represents the Americanization of British spelling, it may in fact reflect a return to its Greek origins.

After Webster's death, the question of spelling reform was revisited by The American Committee for spelling reform, known as the Simplified Spelling Board, established in 1906 with funding from Andrew Carnegie. The committee comprised respected academics and lexicographers, including Melville Dewey, inventor of the Dewey Decimal system of library classification, and the author Mark Twain; later members included the prominent British lexicographers W. W. Skeat, Henry Bradley, and James Murray. The Board began by publishing a list of three hundred revised spellings, many of which were already in use by major dictionaries like that of Webster, and which received the endorsement of the President, Theodore Roosevelt, who ordered its adoption by the Government Printing Office. Roosevelt's attempts to enforce this revised spelling met with derision among British journalists, who mocked his attempts to tinker with the language of Shakespeare. *The Sun* newspaper provided a typically dismissive response: 'Mr Andru Karnegi (or should it be Karnege?) and President Rusvelt (or is it Ruzvelt?) are doing their (or ther) best to ad to the gaiety of nations (or nashuns) by atempting to reform the spelling of the English langwidge. No dowt their (or ther) intentions (or intenshuns) are orl rite, but their (or ther) objekt is orl rong, not to say silly (or sily)'. By the end of the year, Congress resolved that the former spelling system be reinstated and Roosevelt gave up the fight for spelling reform. Carnegie continued to fund the Simplified Spelling Board, but grew increasingly dissatisfied with its lack of results. In 1915 Carnegie wrote to the chairman of the Board attacking its members: 'A more useless body of men never came into association, judging from the effects they produce. Instead of taking twelve words and urging their adoption, they undertook radical changes from the start and these they can never make.... I think I hav been patient long enuf.... I hav much better use for twenty thousand dollars a year' (see Anderson, 1999). No provision was made for spelling reform in Carnegie's will, and so the Simplified Spelling Board was disbanded in 1920 following Carnegie's death a year earlier. This led to the publication of a *Handbook of Simplified Spelling*.

The *Handbook* begins by offering an account of English spelling, its 'true function', its anomalies, and its 'fonetic' origins, which were disrupted by the importation of different spelling systems, attempts to reform it on 'etimological' grounds, what is quaintly termed the 'etimological bugaboo', and the 'bungling' attempts of the earliest printers. While the historical account is largely accurate, if rather crude, the impact of the printers, 'nativs of Holland, who, with far too little knowledge of English or of its proper pronunciation to fit them to be arbiters of English spelling, nevertheless changed the forms of many words to conform with their Dutch habits of orthography' (Grandgent *et al.*, 1920, p. 3), is hugely overstated. We saw in Chapter 5 that only a few spellings could be attributed to such influence, and that of these only *ghost* survives into Modern English. The *Handbook* cites this instance, alongside three others: *aghast* and *ghastly*, which derive from the same word as *ghost*, and *gherkin*. The <gh> spelling of *gherkin* cannot be attributed to the early Dutch printers, as the <gh>

spelling is first recorded by the *OED* in 1834. The other examples are all spellings which do not survive today, including *ghospel*, *ghizzard*, *ghossip*, *ghest*, *ghittar*, and so hardly provide strong support for further reform. Printers are further condemned for their tendency to vary their spelling habits, which could even lead to inconsistencies in spelling as late as the eighteenth century. Such a situation is a case for standardizing spelling rather than reforming it; indeed, further changes might well exacerbate the problem of inconsistency among printing houses, rather than improve it. The difficulties caused by attempts to reform spelling on etymological grounds are also overstated, with the claim that current spelling is misleading as to the 'true derivation of many words'; erroneous spellings of this kind represent but a small proportion of the English vocabulary, as we saw in Chapter 5.

While urging the need for reform, the *Handbook* criticized radical revisions, like the stenographic system of Isaac Pitman, which introduced sixteen new letters into the alphabet. The *Handbook* considered such 'violent' changes to be unacceptable to the general public, as well as too costly and time-consuming to be adopted by printing houses. Interestingly, the *Handbook* also complains about the aesthetic quality of Pitman's new letters, which were not 'tipographically good'. The *Handbook* proposed a system that was neither radical nor revolutionary, but which would produce greater regularity and consistency within the existing system. However, the Simplified Spelling Board made clear that it had no particular authority in proposing such a system, and its aims were somewhat undermined by its admission that 'there is, in fact, no final standard of orthografy. Nowhere is there any authority to set up such a standard. Spelling is never stable. All that the accepted dictionaries can legitimately do is to record the varying usages' (Grandgent *et al.*, 1920, p. 17). Despite this admission, the Board appealed to the public for support on the basis of the purity of its ambition and the reasonableness of its intentions: 'The Simplified Spelling Board, however, as an independent body of men, who hav at heart only the interests of civilization, makes its appeal to the reason of mankind'. The guiding principles of the Board were to choose the shortest and simplest available spelling, thus *honor* not *honour*, *tho* rather than *though*, *catalog* instead of *catalogue*. Whenever possible, silent letters should be removed, giving *frend*, *helth*, *scool*, *crum*, *thru*, and *yu* 'you', while more complex spellings should be simplified, thus *enuf* should replace *enough*, with *wize* instead of *wise*, *rime* instead of *rhyme*, and *tung* replaces *tongue*. Final <e> is dropped in cases like *activ*, *comparativ*, *favorit*, but is preserved where it serves to mark the length of the preceding vowel, as in *arrive*, *care*, *confuse*. Where a verb ending is pronounced /t/, the <ed> spelling is replaced by <t>, as in *stopt*, *kist*, *notist*.

As justification for these changes, the *Handbook* claimed that they would make it easier for children to acquire literacy, cutting a year from the amount of time taken to teach a child to learn to spell, and thereby shaving around $100,000,000 from the education budget, although no evidence is offered to substantiate this bold claim. The Board also claimed that its reforms would have an incalculable effect on the health and 'nervous energy' of students and teachers, reducing the friction and 'nervous temperature' in the classroom, which can be so 'destructiv of temper and material'. The claim that a truly phonetic system would enable English speakers to spell a word as they pronounce it is hardly relevant, since the proposed system is far from being entirely phonetic. The Board accepted that there would be difficulties with such a system, given the variety of modern accents, but considered the variations in spelling that would be needed to accommodate such a system to be a minor issue. 'Since the spelling would correctly represent the speech of the writer, it would present no more difficulty to the eye of the ordinary reader than the current variations in English pronunciation present to the ear of the ordinary listener'. This is a bizarre statement, given that it flies in the face of all previous statements concerning the difficulties caused by an inconsistent and unstandardized system. The claim that variation of this kind would be of great benefit as a record of the variety of English speech for future generations seems completely at odds with the aims of the Board, while the suggestion that a truly phonetic spelling system would ultimately lead to the loss of different accents shows a total linguistic naivety. Such discussions begin to make clear that, lurking behind such appeals to science, education, and literacy, is a conservative view of correct pronunciation and a desire to regulate regional usage: 'Every step taken now to simplify English spelling, to make it represent more accurately the spoken word, is a step toward restoring the purity and precision of English speech'.

I noted above that the committee of the Simplified Spelling Board included the author Mark Twain. Twain's interest in spelling, and dissatisfaction with the contemporary system, are apparent from his essay 'A Simplified Alphabet', written in 1899. But while Twain voiced familiar complaints against the irregularities of English spelling, his solution was not a reformed spelling system, which he thought would at best deliver only a partial solution to the problem:

> I myself am a Simplified Speller; I belong to that unhappy guild that is patiently and hopefully trying to reform our drunken old alphabet by reducing his whiskey. Well, it will improve him. When they get through and have reformed him all they can by their system he will be only HALF drunk. Above that condition

their system can never lift him. There is no competent, and lasting, and real reform for him but to take away his whiskey entirely, and fill up his jug with Pitman's wholesome and undiseased alphabet.

(Twain, 2010, p. 195–6)

Another of Twain's oppositions to a reformed spelling based upon the current system stemmed from a distaste for the ugly appearance of reformed spellings, which detract from what he calls the 'expression' of the words. Quoting a line from Shakespeare's *Macbeth* in reformed spelling: 'La on, Makduf, and damd be he hoo furst krys hold, enuf!', Twain comments: 'It doesn't thrill you as it used to do. The simplifications have sucked the thrill all out of it'. This is an interesting objection in that it highlights the aesthetic value of spelling, suggesting that traditional spellings like *enough* have an aesthetic appeal which would be lost if they were replaced with *enuf*. Twain's disdain for reformed spellings which interfere with the conventional spellings is apparent in a satirical proposal for the improvement of English spelling frequently attributed to him:

For example, in Year 1 that useless letter *c* would be dropped to be replased either by *k* or *s*, and likewise *x* would no longer be part of the alphabet. The only kase in which *c* would be retained would be the *ch* formation, which will be dealt with later. Year 2 might reform *w* spelling, so that *which* and *one* would take the same konsonant, wile Year 3 might well abolish *y* replasing it with *i* and Iear 4 might fiks the *g/j* anomali wonse and for all.

Jenerally, then, the improvement would kontinue iear bai iear with Iear 5 doing awai with useless double konsonants, and Iears 6-12 or so modifaiing vowlz and the rimeining voist and unvoist konsonants.

Bai Iear 15 or sou, it wud fainali bi posibl tu meik ius ov thi ridandant letez *c*, *y* and *x*— bai now jast a memori in the maindz ov ould doderez—tu riplais *ch*, *sh*, and *th* rispektivli.

> Fainali, xen, aafte sam 20 iers ov orxogrefkl riform, wi wud hev a lojikl, kohirnt speling in ius xrewawt xe Ingliy-spiking werld.
>
> (Kimball, 2002)

Twain's solution for English spelling was not a reformed version of the current system, but an entirely new alphabet. His preference was for the use of a 'phonographic alphabet', 'that admirable alphabet, that brilliant alphabet, that inspired alphabet', to write English words, thereby enabling any word to be correctly spelled and pronounced. The great advantage of such a system would be that English spelling would become entirely predictable. Twain gave the example of the 'simple, every-day word' *phthisis*, which, if spelled phonetically would be rendered *tysis*, thereby causing the speller to be 'laughed at by every educated person'. Twain also valued the phonographic alphabet for its economy: it required fewer letters to be printed, and fewer strokes of the hand to write, although curiously he advocated that this shorthand alphabet should be used to write words in full. Twain argued that, while a simplified spelling system would take considerable time to be learned, a phonographic system is quick to grasp and, once mastered, every word can be spelled according to its pronunciation. Whereas simplified spellings are individual and subject to further modifications, once introduced, the phonographic alphabet would remain fixed. He writes: 'It has taken five hundred years to simplify some of Chaucer's rotten spelling…and it will take five hundred years more to get our exasperating new Simplified Corruptions accepted and running smoothly. And we sha'n't be any better off then than we are now; for in that day we shall still have the privilege the Simplifiers are exercising now: ANYBODY can change the spelling that wants to'. This is, of course, a rather optimistic view. Twain appears not to have considered the fact that pronunciations can and do change, so that the spellings of individual words using the phonographic alphabet would also have to change if they are to continue to be phonetic. In fact, a phonographic alphabet would be liable to considerably greater changes than a spelling system that is less phonetic. What is more, one wonders how Twain would feel about reading Shakespeare in a phonographic system; surely this would be equally likely to detract from the grace and grandeur of the language?

The case for spelling reform in America was carried forward by the editor of the newspaper *The Chicago Tribune*, Robert R. McCormick, who employed reformed spellings in an attempt to promote their more widespread use in the 1930s. McCormick's preferred spellings included *frate* 'freight', *grafic* 'graphic', *tarif* 'tariff', *soder* 'solder' and *sofisticated* 'sophisticated'. But, following the editor's death, the paper gave up the reformed spellings in 1955. An article in *Time* magazine reporting the change of policy quoted the newspaper's new Managing Editor Don Maxwell, who attributed the decision to the lobbying of frustrated Chicago schoolteachers fed up with having the Tribune's spellings undermining their own attempts to teach children to spell correctly (*Time*, 29 August 1955).

The history of spelling reform in America thus differs from that in Britain as it did meet with some success, in that many of the reformed spellings proposed by Noah Webster are still in use today. So why was Webster successful where all others have failed, including many of his countrymen? Part of the reason for the success of Webster's reforms was their relative modesty: his more moderate stance, which led him to drop some of the more radical reformed spellings, undoubtedly aided their acceptance. Clearly the most important factor in the success of his enterprise was that he was able to enshrine his reformed spellings in his hugely successful and authoritative spelling-books and dictionaries, thereby ensuring their dissemination and adoption throughout the United States. If Dr Johnson had taken a different view of orthography, might his *Dictionary* have been a similarly authoritative source for promoting reformed spellings? This is possible, although Johnson had little sympathy for spelling reform. However, there is another factor behind the success of Webster's spelling reforms. The motivation behind these reforms, and the reason for their success, was not solely educational. Webster wanted to set American English apart from British English, to create a national language which he believed would help forge a sense of patriotism and national identity:

> A national language is a band of national union. Every engine should be employed to render the people of this country national; to call their attachments home to their own country; and to inspire them with the pride of national character. However they may boast of Independence, and the freedom of their government, yet their opinions are not sufficiently independent; an astonishing respect for the arts and literature of their parent country, and a blind imitation of its manners, are still prevalent among the Americans.
>
> (Webster, 1789, pp. 397–8)

George Bernard Shaw famously described England and America as 'two countries separated by a common language'; the introduction and promotion of Webster's spelling reforms were a conscious and deliberate attempt to enshrine that separation in the spelling system.

Chapter 8

Spelling today and tomorrow

The major factor affecting English spelling today, which may have implications for the future of our spelling system, is the influence of electronic modes of communication, such as texting, email, and Twitter. The electronic medium is a much more immediate means of communication, and messages are written quickly, often without careful attention to details of grammar, spelling, and punctuation. Just as medieval scribes tended to employ numerous abbreviations when writing cursively to speed up the process of copying books by hand, so electronic communication draws upon shortened forms of common words and an established set of abbreviations. Common methods of abbreviation include the use of numbers to replace syllables, as in *2moro* 'tomorrow', *4ever* 'forever' and *gr8* 'great', or the names of letters, as in *cu* 'see you'. Another typical mode of abbreviation is the reduction of an expression to its initial letters, for example *btw* 'by the way', *imho* 'in my humble opinion', *omg* 'oh my God'. Less commonly, we find the dropping of unnecessary letters, especially vowels, as in *pls* 'please' and *txt* 'text'; in a handful of instances the letter <x> is substituted for the digraph <ks>, for example *thx* 'thanks'. It is striking how few of the commonest abbreviations used in textspeak could be misunderstood as genuine attempts at a standard spelling; in most cases it is difficult to imagine someone mistakenly transferring such abbreviations into more formal modes of writing.

Nevertheless, this tendency to flout the rules of English spelling and to employ abbreviations has provoked considerable outrage among traditionalists, who view such practices as corrupting the English language. According to John Sutherland of University College London, writing in *The Guardian* in 2002, textspeak is 'bleak, bald, sad shorthand. Drab shrinktalk.... Linguistically it's all pig's ear... it masks dyslexia, poor spelling and mental laziness. Texting is penmanship for illiterates'. In an article published in the *Daily Mail* in 2007, the broadcaster John Humphrys accused the texting generation of wrecking the English language, describing them as 'vandals who are doing to our language what Genghis Khan did to his neighbours eight hundred years ago'. Humphrys was responding to the *Oxford English Dictionary*'s decision to remove the hyphen in some 16,000 words for the publication of the sixth edition. The editor of the *OED* explains this change in policy as a response to the fact that people no longer use the hyphen in such words, because of the additional time it takes to type this additional character. Humphrys suspects that the real culprits leading this change are text-messagers, with their frequent use of abbreviations. But the use of hyphens to mark compound words is an area of considerable linguistic uncertainty, so that it is unsurprising to find that many people have resorted to omitting them. The purpose of the hyphen is to indicate when a combination of two words is considered to be a single compound, and so it is often added to newly formed compounds. Once these compounds have become widely accepted it is common for the hyphen to be dropped. Many modern English words have their origins in compounds that were originally separated by a hyphen, but which today have come to be written as single words. Words like *instead* and *tomorrow* began life as phrases comprising a preposition followed by a noun, *in stead*, literally 'in place of', *to morrow*, literally 'for the next day'. As they came to be considered compound words rather than phrases, so they began to be written with hyphens, *in-stead*, *to-morrow*, and subsequently as single words. Determining the point at which a phrase can be considered to have become an established compound and should thus lose its hyphen is necessarily a rather speculative and subjective activity; in deciding to remove hyphens in these words dictionaries are merely responding to established practice. In Old and Middle English, there was a tendency to separate prefixes from their stems, so that we find spellings like *mis deed* and *mis fortune*. In the Middle English period a hyphen began to be added to such formations; many of these have since come to be written as single words, although there is still confusion concerning words where the prefix *mis* is added to a base that begins with the letter *s*. Should we for example write *misspell* or *mis-spell*, or even *mispell*?

Often dictionaries are conservative in their treatment of features such as hyphenation, tending to preserve these long after they have been dropped in contemporary usage. For instance, the *OED* spells *lap-top* and *desk-top* with hyphens, whereas these words almost always now in fact appear as *laptop* and *desktop*. In his article Humphrys complains that 'in the future we are required to spell *pigeon-hole*... as *pigeonhole* and *leap-frog* as *leapfrog*'. The spelling *pigeonhole*, without the hyphen, has in fact been in widespread use for some time; the *OED* contains a quotation containing that spelling from the *Daily Mail*, the very newspaper in which Humphrys' article was published, dated to 1994. What this shows is that dictionaries do not make decisions about what is correct usage, but instead reflect the ways in which the language is changing. The publication of a new edition of a dictionary draws our attention to particular linguistic changes, but for these to warrant inclusion in a dictionary they will already have been in widespread use for some time.

As we saw above, a distinctive feature of textspeak that is often singled out for censure is its use of abbreviations and logograms, such as the letter <u> to represent the second person pronoun *you*, or the numerals <2> and <4> in place of *to* and *for*. But the linguist David Crystal, in his book on texting, has argued that it is only a small percentage of text messages that employ abbreviations, and these are often limited to relatively transparent examples like those given above. While this practice may have been prevalent in the early days of text-messaging, the advent of predictive texting has made such abbreviations much less useful. Naomi S. Baron, in her book *Always On: Language in an Online and Mobile World*, carried out a study of the linguistic habits of her students when using Instant Messaging. Her findings agree closely with David Crystal's conclusions, in that the use of abbreviations and acronyms was found to be quite rare, while contractions were employed much less commonly than in spoken language. Her study also found that the number of spelling mistakes and instances of omitted punctuation was strikingly low: only one spelling mistake appeared for every 12.8 transmissions. Furthermore, 9 per cent of these errors were self-corrected, suggesting that these students retain a strong sense of Instant Messaging as belonging to the formal written mode rather than to the less formal mode of spoken discourse. But, while the treatment of spelling and punctuation remained relatively formal, there was considerably less concern for the correct use of the apostrophe. More than a third of the punctuation and spelling mistakes observed by Baron involved the omission of an apostrophe, none of which was subsequently remedied by self-correction. Baron concludes from this study that, despite its more casual and informal style, Instant Messaging remains strongly associated with the formal domain of written usage; students tend to observe the same rules of spelling and punctuation as they would in more formal contexts like composing an essay or a job application.

When assessing the potential impact of text-messaging, we should also bear in mind that the phenomenon of using abbreviations in writing is not new; spellings such as *every1* 'everyone', *4eva* 'forever' were in widespread use in adolescent graffiti long before the technology of texting was developed. Latin texts transmitted throughout Antiquity and the Middle Ages also employed large numbers of abbreviations. One type of abbreviation employed by medieval scribes is the use of modified letters to represent a prefix or suffix, such as a letter resembling the number 9 to abbreviate the prefix *con-* in words like *9tra* 'contra', or *p* with a line through the tail as an abbreviation for the prefix *per-*. Another type of abbreviation is the use of a single symbol to stand for a complete word, such as the crossed <ł> for *vel* 'or', and <ē> or <÷> for *est* 'is'. Old English scribes adopted many of these same methods of abbreviation, while introducing others of their own invention, such as the crossed thorn, <Ꝥ>, as an abbreviation for 'that'. Another Latin logogram introduced during this period has survived as our modern symbol for 'and', <&>, which is based upon a conjunction of the Latin letters <e> and <t> that make up the Latin word *et* meaning 'and'. This symbol is known as the 'ampersand', literally 'and, per se = and', meaning 'and, by itself = and'. The origin of its name lies in its appearance at the end of the medieval alphabet, so that when the full alphabet was recited it would end with 'and, by itself, and'. Another abbreviation for 'and', used throughout the Middle Ages, is the ancient symbol <⁊>, known as the 'Tironian nota' because it was first introduced by Cicero's secretary Tiro as part of a system of shorthand developed for the speedy recording of dictation. Although this particular abbreviation has dropped out of use in modern English, another of Tiro's abbreviation marks survives in the word *viz*. The letter <z> here descends ultimately from the use of the semi-colon as an abbreviation for a sequence of letters ending in <et>. Once this semi-colon began to be written as a single stroke, it started to resemble the numeral 3. So, in Latin texts we commonly find abbreviations such as *l3*, for *licet* 'it is allowed', and *o3* for *oportet* 'it is proper'; the modern word *viz* derives from *vi3*, an abbreviation of *videlicet* 'it may be seen'. Once the origins of this abbreviation had been lost, early printers mistook the <3> for the letter <z> and consequently the word *vi3* began to be printed as *viz*.

Just as textspeak uses logograms as part of larger words, such as *2nite* 'tonight', Old English scribes used the Tironian nota to abbreviate longer words, for example *h⁊* 'hand'. Anglo-Saxon scribes writing with the Roman alphabet occasionally drew upon the names of individual runes as logograms, particularly the runes ᛗ *mon* 'man', ᛞ *dæg* 'day', ᚹ *wynn* 'joy', ᛟ *eþel* 'homeland', whose names correspond to common words. These substitutions are most commonly associated with texts added as interlinear glosses, such as in the Old English gloss added to the *Lindisfarne Gospels*, where they are an obviously helpful way of conserving space (see Figure 3.2 on p. 46). But runes were also used in poems, including the famous Anglo-Saxon epic poem *Beowulf*, and in riddles and prose, either alone, or as a contraction of a longer word like Soloᛗ 'Solomon'. The poet Cynewulf employed runes as an authorial signature, building into his poems the words which are the names of the runes that spell out his name; the runic symbols themselves were used in place of the words, thereby highlighting their status as both logograms and individual letters spelling out the name of the poem's author.

Other kinds of abbreviation found in medieval Latin manuscripts which are still commonly used today include *e.g.* 'exempli gratia', *i.e.* 'id est', *etc.* 'et cetera', *et al.* 'et alii' and so on. Despite being commonly followed by a full-stop, the word *re*, meaning 'concerning', is not an abbreviation of the word *regarding*; it is in fact the ablative form of the Latin word *res* 'thing'. A slightly different kind of abbreviation consists of the initial letter of each syllable of the abbreviated word, as in *cf.* for *confer* 'compare', or *lb.* for *libra* 'pound'. The Latin word *libra* is also the basis of the abbreviation <£>, which is made up of a capital <L> with a line through it; the crossed line is based upon an alternative medieval method of marking abbreviations which was subsequently replaced by the period, now employed in more recent abbreviations, such as *Mr.*, *Dr.*, *Ms.*. Another instance of an abbreviation formed according to the older method which has experienced a recent resurgence of use thanks to Twitter is the 'hash tag' <#>, ultimately based upon the letter <N>, standing for Latin *numerus* 'number', with a line through it. A further abbreviation which has become particularly prominent in the electronic age but which has a similarly ancient pedigree is <@>, the 'at' sign used in email addresses. This symbol is in origin the Latin preposition *ad* 'to, towards', made up of the ligature <ad>, with an exaggerated upstroke which loops round the letter <a>. Abbreviations were thus widespread in medieval manuscripts and not limited to books that were hastily written and carelessly produced. Another group of words that was commonly abbreviated consists of the sacred names commonly found in biblical manuscripts, DS (*Deus* 'God') and SPS (*spiritus* 'Holy Spirit'). The Latin abbreviation IHS for 'Jesus' is a partial adaptation of the Greek abbreviation IHC; the original Greek letter 'eta', <H>, remained unchanged because it was mistakenly held to be the Roman letter; this confusion subsequently led to the widespread use of the spelling *Ihesus* for *Jesus*. This confusion subsequently spawned various ingenious attempts to interpret this abbreviation, such as *Iesus hominum salvator* 'Jesus, saviour of men'. The regular appearance of abbreviations of sacred names in deluxe biblical manuscripts serves to reiterate the point that abbreviated forms were certainly not considered sloppy or informal.

But while scribes throughout Antiquity and the Middle Ages could use abbreviations without attracting censure, this permissive attitude changed in the eighteenth century. During this period, prescriptivists like Jonathan Swift and Joseph Addison opposed abbreviated spellings like *tho* 'though', *agen* 'again', *thot* 'thought', *brout* 'brought', casting scorn upon what Swift called 'that barbarous Custom of abbreviating Words'. Such condemnation anticipates the hostility which modern commentators have directed towards the use of abbreviations in text-messaging. Alongside the outspoken opponents of the practice of abbreviation, we find others resisting its acceptance by recourse to silent protest. The broadcaster and writer Lynne Truss, whose 'zero-tolerance' towards incorrect punctuation is well-known from her bestselling book *Eats, Shoots and Leaves*, wrote the following account of her texting practice in *The Guardian*:

> As someone who sends text messages more or less non-stop, I enjoy one particular aspect of texting more than anything else: that it is possible to sit in a crowded railway carriage laboriously spelling out quite long words in full, and using an enormous amount of punctuation, without anyone being aware of how outrageously subversive I am being. My texts are of epic length.
>
> (*The Guardian*, 5 July 2008).

Another factor that is often cited by those opposed to abbreviations in text-messaging is the detrimental effect it could have on literacy levels in children. A study by researchers at Coventry University, however, found that there were no ill effects from the use of abbreviations. In fact, the findings suggested that playful experimentation of this kind, in a context where standard English is not required, could have a beneficial effect in the acquisition of literacy. Texting increases a child's exposure to text at a critical stage in the development of their literacy, while the use of abbreviations helps in the development of 'phonological awareness', that is an awareness of the sound structures of words. The concern that children will import these informal conventions into more formal modes of writing also appears unfounded, at least with respect to children taking A-levels. David Crystal reports that interviews with A-level students revealed that most expressed incredulity and disbelief when asked if they would ever use text abbreviations in their schoolwork: 'They were perfectly clear in their minds that texting was for mobile phones and not for other purposes' (Crystal, 2008, p. 152).

There are, perhaps, more reasons for concern with younger children, at GCSE level. Examiners for the Scottish Standard Grade exams in 2006 reported some problems attributable to text-messaging. Students encountered problems with 'observance of the conventions of written expression and, in a few cases, the avoidance of the inappropriate use of the informalities of talk and, occasionally, "text language"'. The solution is not to ban the use of abbreviations but rather to educate the children about when to use them, and when only standard English is acceptable. As such, texting is no different from other kinds of informal usage, such as slang words, dialect words, and non-standard grammar that children use freely in the playground, but must learn not to use in the classroom.

Similar issues concern the prevalence of non-standard spelling often used in emails. Email is generally considered an informal medium, in which it is acceptable to relax the rules of standard English grammar, spelling, and punctuation. As the style guide for the electronic age, *Wired Style*, advises: 'Think blunt bursts and sentence fragments.... Spelling and punctuation are loose and playful. (No one reads email with red pen in hand)' (Hale, 1997, p. 3). There are of course difficulties in knowing exactly how loosely one can apply spelling conventions, especially given that email is used for a greater range of linguistic functions than texting. While it may be acceptable to scatter emails to friends with non-standard spellings, it would be inappropriate and unhelpful when writing to a prospective employer. While emails share the ephemeral status of text messages, they are often preserved much longer and can be printed off and filed. Even though style guides are available, like that quoted above, there remains considerable uncertainty about the correct register and tone to adopt in an email. While it is more informal than sending a letter through the post, it is still a written mode of communication, and is often used in parallel situations. What then is the correct mode of address? Should one begin an email 'Dear X', or is that overly formal? But the alternatives, 'Hi X, or just 'X', often seem too casual and brusque. Signing off is another area of uncertainty. Where conventional letter-writing requires 'Yours sincerely', this seems inappropriate in an email. Various alternatives are available, 'Best wishes', 'Cheers', 'Yours', making it difficult for the sender to gauge the appropriate level of formality.

Another aspect of the informal spelling of emails, what we might call 'e-spelling', is that it often plays around with the conventions of English spelling, rather than simply breaking them. While spellings like *nite* 'night', *kool* 'cool', and *coz* 'because' might seem to reflect illiterate attempts at correct spelling, other e-spellings, like *phat*, a backspelling of *fat*, meaning 'cool', 'terrific', *phishing* 'sending out hoax emails', and *phreak*, 'a hacker', show an understanding of the different conventions of English spelling. Here the <ph> digraph, used to represent the sound /f/ in learned words of Greek origin like *philosophy*, has been applied to slang words peculiar to the electronic medium. Substituting <ph> for <f> requires an additional letter, demonstrating that such substitutions are not always driven by economies of speed. In spite of fears that text-messaging is driving a complete loss of literacy in the younger generation, it is striking how 'traditional' non-standard spellings most commonly employed in text messages are: spellings like *gonna*, *coz*, *skool*, *goin*, and so on were in widespread use long before the dawn of electronic communication.

Other deviant spelling practices, identified by David Crystal in his book *Language and the Internet*, are more unconventional, but rather more specialized, belonging to a teenage jargon that has been termed 'Leeguage'. Such spellings include the replacement of the letter <o> with the number <0>, as in *d00dz* 'dudes', or with the percentage sign, *c%l* 'cool'. Another feature of this jargon is the use of the letter <k> as an emphatic prefix, as in words like *k-kool*. While such a usage is unprecedented in earlier non-standard spellings, it is striking how frequently it is the letter <k> that is used in such novel forms.

Further evidence in response to the prophets of doom is the sophisticated way in which young people have been shown to vary their spelling according to the recipient of the text-message. In the study of the texting habits of Coventry University undergraduates mentioned earlier, it was found that non-standard spellings were used to establish an informal register, within which more intimate small talk could be exchanged. Far from being an exclusive and incomprehensible medium, the text messages of this group were readily understood by outsiders; only 20 per cent of the message content was in the form of abbreviations. There was a relatively infrequent use of typographic symbols; the main use of non-alphabetic symbols was limited to kisses <xxx> and exclamation marks <!!!>. Emoticons were seldom used, and in some cases only ironically, while abbreviations using numbers and letters, such as *GR8* and *RU*, were only rarely attested. Much more common were exclamations, such as *haha!*, *arrrgh*, and dialect spellings, which presumably were felt to contribute towards the personal and informal register.

This playful and subversive attitude to standard English spelling draws on a history of deliberate misspelling, such as the frequent appearance of phonetic spellings, such as *woz*, *wot*, *skool* in graffiti. Graffiti is a subversive and countercultural medium; spellings like these are not evidence of illiteracy, but of deliberate opposition to the norms of standard English spelling. By deliberately flouting such rules, those responsible make clear their unwillingness to conform to society's conventions. Something similar lies behind the spelling used by Nigel Molesworth, 'the Curse of St Custard's', in his accounts of his 'skool' days. Molesworth's prose is littered with spelling mistakes, partly a reflection of his basic illiteracy, but also a reflection of his refusal to conform to standards, as witnessed by his description of English masters: 'They teach english e.g. migod you didn't ort to write a sentence like that molesworth' (see Willans and Searle, 1985, p. 35). Modern sociolinguistic research into the use of non-standard orthography suggests that such spellings are often employed for their unique capacity to capture the 'authenticity' and ‘flavour’ of speech, as well as for their potential to challenge linguistic hierarchies. Research into the use of non-standard spelling systems has shown that, once people step outside the standard orthographical framework, they perceive a desire to use their spelling to express their linguistic identity more closely. This can be seen in the ways in which British Creole authors frequently employ non-standard spellings, even when the sounds being represented are identical to those of standard English. In such cases, the use of a non-standard spelling represents an awareness and rejection of the standard orthography, rather than a lack of understanding.

The idea that variant spelling could be used as a means of expressing identity was proposed by H. G. Wells in his essay ‘For Freedom of Spelling: The Discovery of an Art’ (1901). Wells opens with an attack on the need for a standard orthography, questioning the value of the hours spent teaching children to acquire a system of dubious authority and for which there was no ‘grammatical Sinai, with a dictionary instead of tables of stone’ (p. 99). Wells goes on to suggest that, rather than being a question of right and wrong, spelling might better be considered an art form, allowing users to employ variant spellings to encode subtle nuances of meaning. Rather than addressing a loved one in the ‘clean, cold, orderly’ phrase ‘my very dear wife’, Wells proposes using the ‘infinitely more soft and tender’ endearment ‘Migh verrie deare Wife’, asking ‘Is there not something exquisitely pleasant in lingering over those redundant letters, leaving each word, as it were, with a reluctant caress?’ (p. 101). While Wells accepts that some people already employ a personal orthography in letters and other informal writing, his vision is for non-standard spelling to be licensed for use in printed books, thereby overcoming the ‘tyranny of orthography’ imposed by editors and printers, whom he dismisses as 'a mere orthodox spelling police' (p. 102). Similar views were expressed by Mark Twain, who was scornful of people who could only spell words one way. As he wrote in his *Autobiography*: 'I never had any large respect for good spelling.... Before the spelling-book came with its arbitrary forms, men unconsciously revealed shades of their characters and also added enlightening shades of expression to what they wrote by their spelling'.

The idea of a personalized orthography is also apparent in the recent phenomenon of giving children names with non-standard spellings. Often such names remain phonetically identical to their traditional spellings, but changes in spelling are used to endow them with a unique identity. While many popular names already have variant spellings, such as *Catherine*, *Katherine*, *Kathryn*, others have only recently developed alternative forms. Parenting websites are littered with helpful suggestions of ways of respelling traditional names to give 'funky' alternatives, such as *Katelyn*, *Rachaell*, *Melanee*, *Stefani*. Discussion boards record lively debates between those in favour of such personalizations, and those for whom such 'incorrect' spellings are complete anathema.

Non-standard spellings are frequently employed in trade names to give a sense of novelty to the brand name and consequently make it stand out in the market. As with personal names, these are often based upon alternative ways of spelling particular sounds using the conventional orthography, as in the numerous hotels, bars, and clubs which adopt the spelling *nite* in their names. Despite the desire for novelty implied by such spellings, it is striking how conventional they in fact are. Although it would be an equally permissible alternative spelling, there are no examples of *nayt clubs*, or *niit spots*. Another popular example is the spelling *kwik*, best-known from *Kwik-Fit* and the *Kwik-E-Mart* in

The Simpsons, but used by a large number of service providers. This spelling has become an accepted non-standard alternative spelling, specifically used in trade names, even though plenty of alternatives—*cwik*, *cwic*, *quik*, *cwiq*—are equally possible, and potentially even more eye-catching. An important feature of many such novel spellings is that they should be easily understood as an alternative spelling of a particular standard English word, so that the product or service to which it relates is clear.

In some cases the motivation for the alternative spelling appears to be linked to a desire to emphasize an alliteration, which might not be so clear from the conventional orthography. Thus a trade name like *Kwik Kopy* highlights the alliteration in a way that is less apparent from the conventional spelling *Quick Copy*. A further type of such spellings is the deliberate archaism, found especially in numerous quaint antique and tea *shoppes*, and in the traditional country *fayre*. Another common abbreviation is the use of *'n'* for *and*, as in *fish 'n' chips*, lending the company a more friendly and informal association.

An interesting aspect of these alternative spellings is that they rarely signal a non-standard pronunciation. Just as with personal names, company bosses seldom wish to identify their product or service with a dialect pronunciation. Even popular music bands are only moderately more alternative in their spellings, with some being willing to employ spellings that reflect non-standard pronunciations. The omission of the <g> in *Fun Lovin' Criminals* reflects a Cockney pronunciation, although the provision of the apostrophe to signal the omission is a striking concession to standard English conventions. Otherwise pop groups employ similar kinds of replacements as we saw above with trade names, as in the replacement of <k> for <c>, although *Mis-Teeq* has a final <q>, and the use of *'n'*, as in *Guns 'n' Roses*, *Salt 'N' Pepa*.

Another electronic communication medium where the conventions of spelling and punctuation are relaxed is Twitter. A list of eight spelling and grammar rules you can break on Twitter includes leaving out punctuation, including additional punctuation for emphasis, and using abbreviations such as *shuld*, *b*, *2*, *u*, and *ya*. Interestingly, these abbreviations are only permitted when they are necessary in ensuring that the message remains within the 140 character limit. But, while many feel licensed to spell and punctuate as they wish, others adopt a much more prescriptive attitude. Jamie Oliver was recently criticized on Twitter for his poor spelling by a follower who felt that as a role model to young children he should spell properly. Rather than defending his spelling as appropriate for the medium, Oliver responded by attributing his poor spelling to his dyslexia: 'Get lost you idiot im dislexic and i cant spell so stick that in your pipe! its better than being smug'. The American actor John Cusack has come in for similar vitriol in response to errors in spelling and grammar in his tweets. He describes his relaxed attitude to spelling when tweeting as follows: 'I basically get in the general ballpark and tweet it'. His use of spellings such as *breakfasy* and *hippocrite* have earned him a barrage of abusive responses; as one supposed fan tweeted: 'If you're going to be political, maybe learn how to spell Pakistan, and all words in general' (*New York Times*, 28 April 2010).

Far from showing a relaxed attitude to spelling and grammar on Twitter posts, a number of individuals have taken it upon themselves to police linguistic usage and draw attention to errors. These include YourorYou're, whose self-appointed role is to monitor the frequent confusion between these two homophones, and Grammar Hero, who retweets messages with their spelling and grammar corrected. Evidently such users consider that standard English grammar and orthography should be preserved on Twitter, although these views are not always shared by those whose usage is subject to correction. One such correction was met with the response: 'yu can miss me w. dem corrections . dis is twitter why wuld i spell erythang out . time i dnt have'. Undaunted by this lack of gratitude and unwillingness to bend to convention, Grammar Hero retweeted the response, correcting the grammar and spelling: 'Those corrections aren't for me. This is twitter, why would I spell everything out? Time I don't have!' The Twenglish Police, whose goal is to prosecute crimes against the English language carried out on Twitter, corrects misspellings such as *definately*, *tendancies*, *dissapointed*, the confusion of *its* and *it's* and other grammatical errors, signing off cheerily, and somewhat ironically, with 'You're Welcome!'

Many readers of this book will find the acceptance of erratic spelling as impossible to support and will consider the maintenance of our standard English spelling system a highly desirable and necessary activity, more than compensating for the hours of classroom time dedicated to learning its numerous complexities and exceptions. But should we really be so concerned about the casual attitude to spelling employed in texts, tweets, and emails? Why is correct spelling so important? Throughout this book I have tried to show that standard English spelling comprises a variety of different forms that have developed in an erratic and inconsistent manner over a substantial period of time. We must therefore discard any attempt to impose a teleological narrative upon the development of English spelling which would argue that our standard spelling system is the result of a kind of Darwinian survival of the fittest, producing a system that has been refined over centuries. We have also seen that the concept of a standard spelling system is a relatively modern one; earlier periods in the history of English were able to manage without a rigidly imposed standard perfectly well. What is more, informal writing, as in the letters and diaries of earlier periods, has often allowed more relaxed rules for spelling and punctuation. So why should it be different for us? I suspect that one reason is that spelling is the area of language use that is easiest to regulate and monitor, and thus the area where users come under the strongest pressure to conform to a standard. Older generations of spellers have considerable personal investment in maintaining these standards, given that they themselves were compelled to learn them. As David Mitchell confesses in his *Soapbox* rant against poor spellers, 'I'm certainly happy to admit that I do have a huge vested interest in upholding these rules because I did take the trouble to learn them and, having put that effort in, I am abundantly incentivized to make sure that everyone else follows suit. The very last thing I want is for us to return to a society where some other arbitrary code is taken as the measure of a man, like how many press-ups you can do or what's the largest mammal you can kill'. But is correct spelling more than simply a way of providing puny men with a means of asserting themselves over their physically more developed peers?

One argument often made is that spelling mistakes are indicative of someone too lazy to bother consulting a dictionary. It is certainly true that it is relatively easy to find out how to spell a word correctly; with the advent of electronic spell-checkers it has become even easier to do so. But this argument does little more than reinforce the assumption that correct spelling is important; it does not attempt to justify why it is important that a word always be spelled in the same way. One possible answer to this question is that consistency is important for clarity of communication: tolerating variant spellings would lead to all kinds of confusion. But while this was evidently true of some of the inventive spellings employed in the Middle English period, it is hardly a risk in the cases of commonly misspelled words like *accomodation*, *ocurring*, *recieve*, *supercede*, *definately*, and so on. Another possible explanation lies with professional copyeditors, who have an even stronger incentive for maintaining standards of spelling, punctuation, and grammar, since their livelihoods depend upon them. In an attempt to investigate the rationale behind such practices of 'verbal hygiene', the linguist Deborah Cameron interviewed several copyeditors and asked them why imposing certain linguistic conventions, so-called 'house style', upon different authors' works was so important. 'My informants conceded that in some cases—not all—one variant might have no intrinsic superiority over another. Their cardinal principle, however, was that text must be *consistent*' (2012, p. 37).

According to the copyeditors whose job it is to regulate spelling in printed books, therefore, a standard spelling is an important aspect of the consistency of a text; errors in spelling affect the clarity of a text by distracting a reader's attention from its content. But does variation of this kind really distract from the text's message in such a way as to run the risk of miscomprehension? If we look at a modern style guide, such as that observed by copyeditors for *The Guardian* newspaper, we find a series of clear and unambiguous statements about spelling preferences. Yet, despite the authoritative tone that is adopted, the rationale behind these simple instructions is not beyond question. For example, journalists and editors are instructed to use the spelling *admissible* rather than *admissable*. The *OED*, however, has separate headwords for both spellings, suggesting that either is acceptable; of the two, *admissable* is in fact recorded earlier. The preference for *dispatch* over *despatch* is similarly debatable, given that *despatch* is recorded as an alternative spelling for *dispatch* by the *OED*. The spelling *dispatch* is certainly to be preferred on etymological grounds, since its prefix derives from Latin *dis-*; the *despatch* spelling first appeared by mistake in Johnson's *Dictionary*. However, since the nineteenth century *despatch* has been in regular usage and continues to be widely employed today. Given this, why should *despatch* not be used in *The Guardian*? In fact, a search of this newspaper's online publication shows that occurrences of *despatch* do sneak into the paper: while *dispatch* is clearly the more frequent spelling (almost 14,000 occurrences when I carried out the search), there were almost 2,000 instances of the proscribed spelling *despatch*. Searches of the online newspaper for other spellings outlawed by its style guide reveal similar patterns of distribution; there are, for instance, 590 instances of *admissible* against 45 of *admissable*.

In the case of less frequent words, like *foetus* and *gobbledegook*, the preferred spellings of the style guide run contrary to those employed by the *OED*, in which the headwords are spelled *fetus* and *gobbledygook*. In the case of *fetus*, the spelling with <e> only is etymologically preferable, reflecting its Latin etymon. But how do we begin to decide whether *gobbledygook* or *gobbledegook* is the correct spelling of that word? The word is probably onomatopoeic in origin, deriving from an attempt to represent a turkey's gobble; to engage in a discussion about which spelling was closer to the turkey noise would clearly be absurd. Once again, despite the clear proscription of these spellings, both *fetus* and *gobbledygook* appear in *The Guardian*'s online publication. This kind of inconsistency is not a recent phenomenon. Searching an electronic database containing the full-text of *The Guardian* and *The Observer* newspapers for the period 1791–2003 reveals 212 instances of *admissable*, an impressive 93,726 occurrences of *despatch*, 3,454 of *fetus* and 243 of *gobbledygook*. That these instances of spelling variation have gone unnoticed, both by copyeditors and by readers, undermines the argument that a lack of consistency disrupts the reading experience.

Along with this conservative backlash that attempts to impose notions of correctness and propriety on the spelling of online communication, there are movements which continue to call on the government to reform English spelling. A primary driving factor behind such organizations is a desire to make spelling more phonetic as a means of promoting child literacy. A more phonetic spelling system is also deemed desirable in the promotion of English as a global language, while an economic rationale argues that removing superfluous letters will mean savings for the publishing industry by lowering costs of production and materials. Opposition to spelling reform is often based upon concern that older literature will become inaccessible, the presumed suppression of regional accents, or simple conservatism based upon concern over unforeseen consequences. Reforming spelling could also lead to the loss of the morphological transparency which allows us to identify correspondences between words which are not apparent from their pronunciation, such as *sign* and *signature*. Attempting to make spelling agree more closely with pronunciation raises the further practical and sociolinguistic problem of whose pronunciation should be represented. This issue is even thornier today than it was in earlier periods in the history of English. Now that English has become a global language, changing its spelling system would have major repercussions for users of English throughout the world.

Despite these many legitimate objections, it is interesting to observe that such reforms have been successfully implemented in other languages. Perhaps the most drastic orthographic reforms have affected the peoples of those states of the former

Soviet Union whose languages have been switched from the Roman alphabet to the Cyrillic alphabet and back again, notably Moldovan, a variety of Romanian, a Romance language written using the Cyrillic alphabet for four centuries before it switched to the Roman alphabet in the late nineteenth century. The switch to a Roman alphabet was accompanied by a move to phonetic spelling, which severed etymological links between related words, thereby removing what was considered to be an important aspect of the language.

Attempts to reform the spelling of other languages, however, have met with considerable objections. In Germany, discussion of spelling reform began in the 1960s, when it was felt that tidying up certain ambiguities in the current system would make it easier for children to learn to write. This reform was concerned with introducing a closer relationship between symbol and sound, clarifying the principles of word division and the use of capital letters, as well as rationalizing the use of the letter <ß>. A revised system was introduced in 1998, which was intended to be phased in over a seven-year period. But the process was far from smooth: several individuals, including Rolf Gröschner, professor of Law at the University of Jena, and his fourteen-year old daughter, challenged the legality of the reform in the law courts. The state of Schleswig-Holstein voted in a referendum to opt out of its implementation, while in 2000 the influential newspaper *Frankfurter Allgemeine Zeitung* decided to abandon the reforms and revert to the old orthography. Only in 2007 did it reverse this decision and finally adopt the reformed spelling.

In Germany, spelling reform is treated with suspicion and hostility; attempts to enforce a revised system are considered repressive and authoritarian. Such an attitude is partly historical, a hangover from the Nazi Party's plans to introduce a simplified writing system to facilitate the acquisition and use of German among nations of their newly occupied territories. The reforms proposed by the National Socialist Minister for Education, Bernhard Rust, included the replacement of digraphs <ph, th, rh> in foreign loans with <f, t, r>, the reduction of consonant clusters, and the abolition of the comma before the co-ordinating conjunctions *und* 'and' and *oder* 'or'. The status of these reforms became the basis of the legal challenges to the 1998 proposals, although Hitler's determination that all debate over reform be deferred until after the war indicates that they were never officially implemented. The recognition of the importance of a writing system as a means of communication lay behind the Nazi Party's successful reform, the replacement of the traditional Gothic script (known as 'Fraktur') with the widespread Roman script. Such a change is striking, given that one might have expected the Nazi Party to have wished to exploit the ideological and nationalistic associations of the traditional Gothic script and its Germanic origins.

The 1998 reforms include the replacement of <ß> with <ss> after short vowels: so that the noun *Haß* 'hate' is now spelled *Hass*; *Kuß* 'kiss' is now written *Kuss*. This has the advantage of identifying connections between related words, so that *Hass* and *Kuss* now resemble the verbs *hassen* 'to hate' and *küssen* 'to kiss'. The letter <ß> continues to be used after long vowels, in words like *Straße* 'street'. A desire to reduce the complexities caused by non-native spelling patterns found in words adopted from foreign languages led to the respelling of certain common loanwords, although in some cases variant spellings continue to be tolerated. As a result of this reform the following respellings were produced: *Geografie* (formerly *Geographie*), *Fotometrie* (formerly *Photometrie*), *Jogurt* (formerly *Joghurt*), *Spagetti* (formerly *Spaghetti*), *Exposee* (formerly *Exposé*), *Varietee* (formerly *Varieté*), *Kommunikee* (formerly *Kommuniqué*), *Ketschup* (formerly *Ketchup*), *Katarr* (formerly *Katarrh*), *Nessessär* (formerly *Necessaire*), *Panter* (formerly *Panther*). Other changes included the introduction of greater consistency in the use of hyphens in compounds derived from English: *Hair-Stylist* has thus become *Hairstylist*, and *Midlifecrisis* replaces *Midlife-Crisis*. The most vehement public outrage was directed at the attempt to drop the practice of capitalizing all nouns. As we have seen, this was a feature of Early Modern English but, as with every other modern language, English no longer capitalizes nouns. The public opposition to this reform led to a much more moderate proposal which aimed merely to remove inconsistencies in the use of capital letters.

Even these more moderate proposals have been criticized for their lack of logic and consistency, and many of the criticisms recall those which have been levelled at proposals for the reform of English spelling. The question of which accent of German should be the basis of the reform is a key issue: the replacement of <ß> with <ss> after short vowels is based upon the standard German accent; other accents have a different distribution of long and short vowels. The attempt to respell words to allow consistency between forms of the same word is undermined by counter-examples. For instance, the proposal to respell *numerieren* 'to number' as *nummerieren*, thereby aligning it with the noun *Nummer*, ignores other semantically related forms with a single <m>, such as *Numerale* 'numeral' and *numerisch* 'numerical'. The criterion for respelling foreign loanwords was the degree to which a word was considered to have been assimilated into the language, a subjective assessment which led to considerable uncertainty and difference of opinion. Thus *Phonologie* was proposed as the primary spelling of this word, alongside the secondary spelling *Fonologie*, while *Phonetik* remained the only permitted spelling. The capitalization of pronouns of address remains inconsistent, in that the familar forms *du* and *ihr* are written without capitals, while the polite forms *Sie* and *Ihr* are capitalized. But, although the reforms can be criticized on these largely technical, linguistic grounds, public opinion was more focused on ideological issues. For the German public these reforms represented an attack on the German language and its link with national and cultural history and identity.

It is apparent from this discussion that proposed changes to a spelling system can meet with vociferous objections, even when the resulting reforms will introduce only minor differences. The German reform proposals of the 1990s affected a mere 0.5 per cent of the total lexicon of the language, but nevertheless generated considerable outrage. A similar set of minor proposals to reform French spelling, known as the Druon reforms after Maurice Druon, permanent secretary of the Académie Française, generated similar levels of public hostility, despite affecting a total of just over 2,000 words, none of which included the 500 most commonly used. Despite the relatively trivial impact of the reforms, which were mostly concerned with regularizing the use of hyphens and removing the circumflex accent over <u> and <i> where it serves no function, and despite attempts to present these reforms using a discourse of continuity and permanence, the proposals incurred a volley of public hostility. The public attitude towards French spelling is based upon a mythical view that modern French orthography represents an unbroken continuity with the great French writers Molière and Racine, failing to recognize that their works are read today in modernized spelling editions. For many teachers, the spelling reforms threatened a lowering of educational standards, while for others the difficulty of French spelling was considered to be part of its charm. The national backlash which these proposals caused was reported by *The New York Times* as follows: 'Several members of the Academy have accused its permanent secretary of acting in unseemly haste to alter the hallowed language of Voltaire, Molière, Flaubert, and Proust. The official, Maurice Druon, has been accused of lying and trickery, and has threatened to sue his detractors on charges of defamation' (*The New York Times*, 2 January 1991).

Proposals to modify Dutch spelling in 1969 met with similar objections. These proposals were mostly limited to the spelling of new loanwords entering the language, but nevertheless prompted fierce objections. For traditionalists, the proposals were far too radical; for advocates of change, they were too conservative. Dutch education ministers favoured the proposals, arguing that they would make learning to read and write easier, freeing up more time for other aspects of language learning. Dutch educationalists were broadly supportive, although many favoured more radical simplification. Dutch writers, irrespective of their political leanings, were unanimous in their opposition to the proposed simplifications, which were seen as an unlicensed jettisoning of tradition. As one writer wrote: 'I am against a world which only considers the current moment important. Against a world in which roots are eradicated, origins made unrecognizable, traditions thrown away as useless rubbish' (see Geerts *et al.*, 1977, p. 223). In the conservative press the reforms were ridiculed, caricatured as the 'odeklonje spelling', based upon a respelling of the loanword *eau de Cologne*. Reactions to these proposals demonstrate how tightly spelling is bound up with national identity; even a single letter can stir up strong nationalistic sentiments. In Belgium there was an acceptance that maintaining unity with Dutch spelling was important, although Dutch speaking Belgians objected to the use of the letter <c> in words like *kultuur* because of its similarity to French. Many of the Dutch, however, prefer to write *cultuur* to distinguish their spelling from that of German.

The sociolinguistic implications of the choice between the letters <c> and <k> are not limited to Dutch speakers; the choice between <c> and <k> has been deliberately exploited by various subcultural groups in countries where the Roman alphabet is in use. Because both letters can be used to represent the same sound, it is possible to substitute one for the other as a statement of subversion and opposition to societal norms. In Spain, where <k> is only found in a handful of foreign words, graffiti writers often employ it where the Castilian standard requires <c> or <qu>. The fact that the letter <k> is used to represent /k/ in Basque, a non-Indo-European language spoken by inhabitants of the Basque Country, a region occupying part of north-eastern Spain and south-western France, adds to the charged political connotations the letter holds in Spanish society.

A subsequent attempt was made to reform Dutch spelling in the early 1990s, when the Dutch government established a committee of professional linguists and scholars who were tasked with devising proposals for reform. Their work was carried out in secret in order to avoid adverse coverage in the national press. But in 1994 their proposals were leaked to the media, who reported them in a wholly negative light, leading to their complete rejection by the politicians who had initially commissioned the proposals.

Calls for reform of English spelling today are generally focused on the lack of phonetic consistency, which reformers cite as the cause of poor literacy standards in Britain. In an article in *The*

Times Education Supplement in 2010, a teacher claimed that the English spelling system is putting schoolchildren at a disadvantage compared to their Finnish, Greek, and Spanish contemporaries, whose native languages have a more straightforward correlation between letters and sounds. Identifying twenty-five words she believes are holding children back, she suggested reformed spellings which would make literacy easier to acquire.

beautiful	butiful	any	enny
head	hed	many	menny
learn	lern	are	ar
read	reed/red	gone	gon
great	grate	give	giv
there	thair	have	hav
were	wer	live	liv
where	whair	you	u/yoo
friend	frend	your	yor
believe	beleev	rough	ruf
pretty	pritty	tough	tuf
said	sed	bought	bawt
		thought	thawt

The evidence in support of the claim that languages with a close relationship between spelling and sound are easier to learn than those with a less direct relationship is ambiguous at best. Such claims are frequently made without any reference to research, relying on a simplistic view of the relationship between a reformed spelling and improved literacy, without considering the complex socioeconomic factors which lie behind British child literacy figures. Complaints about the problems inherent in English spelling by spelling reformers are often based upon exaggerated and mistaken views of the history of the spelling system. In an interview on BBC Radio 4's *Learning Curve*, Masha Bell of the English Spelling Society, formerly the Simplified Spelling Society, claimed that the messy nature of English spelling is the result of its being corrupted by the earliest printers, who couldn't speak English:

> It all happened when the printing press first came to England, in 1476. When Caxton came back from the Continent, where he had lived for 30 years and learned to print, he brought with him assistants who none of them spoke English. So they were type-setting English books and making a right hash of it. And it got far worse when the Bible was first translated into English, because bible-printing wasn't even allowed in England, so all the different editions of the New Testament of 1525, the 40 different editions, were all printed on the Continent, and that made it worse. That put the spelling in a mess.
>
> (<http://www.englishspellingsociety.org/news/media/bell.php>)

This is, as we have seen, a completely inaccurate claim about the origins of our modern spelling system, designed to give credence to an extreme view of its inadequacy so as to bolster calls for its reform. While early printers were often recruited from the continent, it is nonsensical to suggest that the books they printed completely corrupted English spelling. As we saw in Chapter 5, the influence of continental spelling habits on English as a result of early printers was extremely small. What's more, the advent of the printing press was one of the decisive factors in bringing about a standard spelling system.

In her book, *Understanding English Spelling* (2004), Bell includes a chapter outlining the history of English spelling, with the subtitle 'How we ended up with such a mess on our hands'. Here she claims that the Chancery clerks of the fifteenth century were tasked with devising an entirely new system of English spelling out of nothing and at speed: 'The Chancery clerks did not merely have to abandon their old writing habits. Hurriedly, they also had to invent a system for spelling English; and they had to do it almost from scratch'. This account ignores the obvious continuities between the spelling of Chancery documents and earlier Middle and Old English texts. The spelling system 'hurriedly cobbled together' by Chancery clerks was then adopted by Caxton for his printed books, which had a huge impact upon the development of standard English. Because Caxton had lived abroad for thirty years his grasp of English was unsound, while those compositors he brought from the continent were 'even less proficient in English than he was'. This is another hugely exaggerated claim; Caxton translated a number of the works he printed into a stylistically elevated literary English, adding prefaces and epilogues to many of his publications. The claim that he had a poor grasp of English is nonsense, as is the implicit suggestion that the spelling of his books was a complete mess.

To question Caxton's proficiency in English is merely to take seriously the printer's own comments about his lack of suitability to the task of translating the *History of Troy* from French into English on account of the 'symplenes and unperfightnes' of his grasp of both languages. Caxton attributes his poor linguistic skills to his upbringing in the Weald of Kent, 'where I doubte not is spoken as brode and rude Englissh as is in ony place of Englond', and the thirty years he has spent on the continent. But these are nothing more than literary conventions, the so-called 'humility topos' made famous by Geoffrey Chaucer, who consistently apologized for his literary and linguistic failings, presenting himself as the innocent pilgrim narrator whose contribution to the *Canterbury Tales* is so awful it has to be cut short by the host. Caxton is here invoking this literary device as a means of flattering Margaret of Burgundy, to whom he dedicated his translation, committing himself to her service and flattering her in the highest terms: 'the ryght hyghe excellent and right vertuous prynces, my ryght redoughted lady, my Lady Margarete by the grace of God suster unto the Kynge of Englond and of France'. By imputing the current state of English spelling to Caxton's illiteracy, Bell has taken Caxton's comments at face value and stubbornly ignored the considerable linguistic and literary sophistication of Caxton's writing. Such a deliberate misreading is typical of this book, whose tone is alarmist, making exaggerated and unfounded claims about the difficulties of English spelling. Bell goes on to suggest that the difficulties English spelling poses are the cause of youth illiteracy and unemployment, even implying that they are the reason that, in 2000, 25 per cent of British women failed to take up an offer of free breast screening.

While it may be true that a transparently phonetic spelling system is beneficial for children in the early stages of learning to read, more advanced readers are primarily concerned with the identification of meanings rather than sounds. For them a deep orthography, where individual words have distinctive spellings and in which homophones, like *too*, *to*, and *two*, are distinguished, is a more useful tool. Most studies of the relative benefits of shallow and deep spelling systems have focused on learning to read, rather than learning to spell, which is a very different kind of task. One such study, examining the spelling habits of children in Tanzania learning to write Kiswahili, a language with a relatively shallow orthography, found that the various factors involved were numerous and complex: 'multiple facets of language knowledge are simultaneously necessary to achieve good spelling in even this regularly spelled language with a relatively simple syllabic structure' (Alcock and Ngorosho, 2003, p. 657). The various practical issues involved in making cross-linguistic comparisons in the acquisition of literacy, such as the different levels of pre-school literacy, mean that any straightforward claims that English spelling is retarding children's literacy are hard to support. As Richard Venezky has observed: 'For the beginner, the orthography is needed as an indicator for the sounds of words, but for the advanced reader, meanings, not sounds, are needed. This conflict between the needs of beginning and advanced readers forces certain compromises upon the design of a practical writing system, depending upon the intended function of the system' (1999, p. 42). The social context within which literacy is acquired is clearly an important aspect of this debate, but one which is rarely taken into account in simplistic claims that poor spelling leads to poor education and further social ills. In *Understanding English Spelling*, Masha Bell highlights the low reading ages of many of the UK's prisoners and suggests that the large prison population may be due to the 'fiendish' difficulty of English spelling.

Another practical problem associated with spelling reform along phonetic principles, which we have met several times in this book, is the issue of diversity of pronunciation. While some reforms could be implemented uncontroversially, such as the removal of consonants that are no longer pronounced in any dialect of English, like the <k> in *knight*, others are more problematic. We have seen how the <wh> in *which*, *when*, *what*, derives from an Old English pronunciation with an initial aspirate that is no longer sounded in southern dialects of English. But this sound is preserved in Scottish accents and so different spellings would be required for the Scottish and English pronunciations. Attaining a one-to-one correspondence between vowel sounds and letters is considerably more complex; in earlier attempts at reform this has led to the adoption of complex systems of diacritics and accents that can hardly be said to make spelling easier. Dialect variation is the norm, and would pose considerable problems for any spelling reform designed to bring spelling and pronunciation into harmony. For instance, a major difference between the northern and southern accents of English concerns what linguists term the BATH vowel, that is the vowel used in the pronunciation of that word and others like it: *grass*, *fast*, *glass*. For northern speakers, this vowel is identical with the TRAP vowel, that is, the vowel sound used in *trap*, *map*, *clap*; for southern speakers there are two separate vowels. At present, a single spelling accommodates both southern pronunciations, but a one-to-one spelling system would need different spellings for the southern realization of the TRAP and BATH groups of words. One way to do this would be to reintroduce the Old English ligature <æ> for the TRAP vowel, thereby enabling a distinction between *træp* and *bath*. But this adjustment would only make sense for southern speakers; for northerners the word *bath* would need to be spelled *bæth*.

Many American speakers and other international users of English would require further alternative spellings. An unstated assumption that is often built into such discussions is that the model for any phonetic spelling reform should be the standard reference accent of English, what is called Received Pronunciation (RP). But this assumption derives from a sociolinguistic prejudice that accords social prestige and dominance to one particular accent over another, completely lacking in linguistic justification. Reforming spelling to bring it closer to an accent that is not their own will hardly help children who do not speak with

RP accents to learn how to spell. While it would be possible for northerners to use the <æ> as their spelling of the BATH vowel, spelling such words as *bæth*, *fæst*, *græs*, introducing different spelling systems for different accents would introduce a whole new set of communicative barriers.

While removing silent letters no longer pronounced might seem uncontroversial enough, we have seen that silent letters have a secondary function: to distinguish identically-sounding words, what are called 'homophones', preventing them from becoming 'homonyms', words which are identical in both pronunciation and spelling. In the case of *knot*, the preservation of the silent <k>, derived from the Old English word *cnotta*, helps to distinguish it from the word *not*. This is not quite as straightforward as may at first appear, because there are a number of words whose spellings have changed to reflect adjustments in pronunciation from Old to Modern English. For example, the word *ring* derives from the Old English word *hring*, in which the initial /h/ was pronounced. During the Middle English period this /h/ was dropped from the pronunciation, and the spelling changed to reflect this. But if there is a need for written distinctions to prevent homophones becoming homonyms, why has the <h> in *hring* not been retained? In modern English there are two words *ring* that have the same spelling and pronunciation. The noun *ring*, derived from Old English *hring*, refers to a small circular band, while the verb *ring*, derived from Old English *hringan*, describes the sound a bell makes. Although both of these words derive from Old English words spelled with an initial <h>, it would have been perfectly possible to have preserved the <h> in one case while dropping it in the other. But, instead of observing such a distinction, modern English spells both words identically. This does mean that there is some potential for confusion:

take a sentence like 'Give me a ring', for example, where it is unclear whether the speaker is requesting a metal band or a phone call. But this is only unclear because it is decontextualized: when heard or read within context there will be numerous other clues to enable the hearer to determine which meaning is correct. If we return to the case of *knot* and *not*, where different spellings have been preserved, then we can see that here the potential for confusion is considerably less, given that the two belong to different word classes. So where it is possible to say 'Give me a ring' and for either word to be possible, there are no parallel situations with regards to *not* and *knot*. Functionally, therefore, there is comparatively less requirement for a written distinction between *not* and *knot* than there is between *ring* and *ring*, but yet our spelling system has different spellings for the first pair and not (not knot) for the second pair.

Another problem caused by spelling reform along phonetic lines concerns the resulting loss of etymological spellings. We saw how the principle of etymology was a determining factor in a number of spelling reforms implemented in the Early Modern period; however, this is often ignored in modern debates. It is easy to write off etymology as a significant factor in spelling, drawing attention to the introduction of silent letters in *doubt* and *debt* as misguided attempts to give English a Classical pedigree. But etymological spellings preserve visual links between words that are semantically related, which have been lost in the spoken language. Although we no longer pronounce the <g> in *sign*, its presence alerts us to the link it shares with *signal* and *signature*, where the <g> is still preserved. Although the <b> in *doubt* has never been pronounced, it signals a link with *indubitable* 'not to be doubted', and thus may well help a reader to discern the meaning of the more unfamiliar word. The same is true of the pair of words *debt* and *debit*. Such spellings recall our earlier comments about the way that English spelling is concerned with consistent spelling of morphemes, sometimes at the expense of phonetic consistency. The spellings of *electrical* and *electricity* are phonetically inconsistent in that they both use the letter <c>, despite the fact that one has a /k/ sound and the other /s/. However, the use of the single letter <c> in both preserves the link with the base form *electric*. A similar example can be seen in *medical* and *medicine*, both of which preserve an orthographic link with *medic*. The spellings *allege* and *allegation* have different pronunciations of the letter <g>, but by preserving the same letter they maintain a visual link between verb and noun. Similar links are preserved in other phonetically irregular pairings, as in the case of *sane* and *sanity*. The first of these has a diphthong and the second has a short monophthong, but they are both spelled <a>. While this is phonetically inconsistent, it preserves a visual link between two semantically-related words.

There are, however, some exceptions to this practice, such as *pronounce* and *pronunciation*, commonly misspelled *pronounciation* because speakers tend to make a semantic link with the verb and assume a similar spelling despite the different pronunciation. While it may seem unnecessarily complicated to retain silent letters in such words, these etymological links can be a useful guide to spelling. So, while the silent final <n> in *autumn* may cause people difficulties, the related adjective *autumnal* provides a helpful prompt. Such links can also be useful ways of correctly spelling unstressed vowels in words like *grammar* or *parent*. Because these vowels are not stressed, they are pronounced identically, using the sound known as 'schwa'. In some words this sound is spelled <e> and in others <a>, making it a common area of uncertainty. But the link between *grammar* and its adjective, *grammat-*

ical, where the stress falls on the second syllable, provides a useful hint as to the correct spelling of the noun. In the case of *parent*, the unstressed schwa sound is spelled with an <e>, as can be gleaned from comparison with the adjective *parental*.

In other cases the preservation of a traditional spelling helps to signal the word's etymology, which can also help with the understanding of unfamiliar words. Words that begin with the /f/ sound spelled with <ph> are generally of Greek origin: preserving that <ph> digraph signals that etymology; while words that begin with /s/ spelled <c> are generally of French origin. For those who can refer to a knowledge of those languages, these etymological triggers are helpful in allowing the decoding of unfamiliar words. For example, the word *century* belongs to a group of words containing the French numeral *cent* '100', derived from Latin *centum*. Being able to identify this etymological link is extremely helpful when encountering similar formations, such as *centipede*, *centenary*, and *centennial*. But if the spelling of these words was reformed on a phonetic basis, producing a form with initial <s>, *sentury*, *sentipede*, *sentenary*, and *sentennial*, then this origin would be obscured, allowing the potential for confusion with words derived from Latin *sentientem* 'feeling': *sentient*, *sentiment*, *sententious*, and so on. Etymology also helps in the correct pronunciation of words with similar spellings but different pronunciations, such as the different uses of the digraph <ch> in English words. In native English words like *cheese*, *choose*, <ch> is pronounced /tʃ/, in words of French origin it represents /ʃ/, as in *champagne*, *chef*, while in Greek derivations it is sounded /k/, for example *chord*, *chorus*. Etymology may seem an arcane and purely scholarly criterion on which to base a spelling system, but it has a valuable role to play in helping readers to understand and pronounce the words they are reading.

Etymological relicts like these have a further function in making visible the linguistic richness of the language and its varied history, thereby maintaining a closer relationship with texts of the past. While silent letters like those in *knight* are unhelpful for modern learners of English, they do enable us to retain a link with the language used by Chaucer, and before him by the Anglo-Saxons. When English speakers turn to Chaucer they may find much unfamiliar, but they will not be thrown by the line 'A knyght ther was, and that a worthy man' thanks to our preservation of the medieval spelling. Historical spellings like this preserve that link with the past, as well as standing as a monument to the language's history.

The English language has come into contact with numerous other languages throughout its history, all of which have left their mark, to a greater or lesser degree, on the language's structure, spelling, and vocabulary. To remove the idiosyncrasies that they have contributed to our spelling system would be to erase the evidence of that history. Our spelling system could be likened to a cathedral church, whose origins lie in the Anglo-Saxon period, but whose structure now includes a Gothic portico added in the Middle Ages, a domed tower added in the Early Modern period, and a gift shop and café introduced in the 1960s. The end result is an awkward mixture of architectural styles which no longer reflects the builders' original plan, nor is it the ideal building for the bishop and his clergy to carry out their diocesan duties. But, in spite of these practical limitations, it would be hard to imagine anyone suggesting that the cathedral be demolished to allow the rebuilding of a more functional and architecturally harmonious modern construction. Quite apart from the practical and financial costs of such a project, the demolition and reconstruction of the cathedral would erase the rich historical record that such a building represents.

One thing that is clear from this historical overview of spelling is that spelling matters. People have very strongly held views about spelling; this was just as true in the eighteenth century as it is today. Variation and change in any area of language use typically incites outrage in today's media; incorrect spelling, h-dropping, and split infinitives provoke strongly held linguistic prejudices. Rather than being seen simply as mistakes, incorrect spellings are often viewed as a reflection of a person's intelligence, social class, and even morality. Many people are highly conservative when it comes to language, and are very reluctant to allow the linguistic system that they have grown up with to be altered. Attempts to introduce change are therefore treated with considerable scepticism; innovators are dismissed as trendy and permissive thinkers without proper standards. Learning how to spell correctly is seen as character-forming, a rite of passage, a test of character and moral fibre, similar to cold showers or early morning runs. Such views can also have a nationalistic impulse, so that mastering English spelling is considered part of the process of becoming assimilated into English, or British culture. As one scholar has written of French attitudes to their spelling system: 'Though it may sometimes be illogical, that is its beauty, and to learn its rules imparts a kind of discipline that is good for people. One submits oneself to it for the greater glory of France'. Any attempt to overturn such views, and to introduce a reformed spelling system of any kind seems doomed to failure.

A lesson that is clear from the German spelling reforms is that once a tradition of spelling has been established it is highly resistant to change, even if such reforms are relatively trivial. In Germany, the reform of the 1990s only affected an estimated 0.5 per cent of the vocabulary, yet this still resulted in public uproar. Other countries, such as France and Holland, have experienced similar problems when politicians and linguists have attempted to introduce minor reforms affecting only a small number of words. Where there is a clear political or ideological agenda that has public support, orthographic change has been successful. This is most evident in the former Soviet states, such as Tatarstan, which introduced spelling reform despite opposition from Moscow. But such changes come at a cost, especially to the publishing industry, while also creating difficulties in preserving access to historical texts written in a spelling system from which modern speakers will become increasingly distant.

Advocates of spelling reform predict a dire future for English spelling, arguing that reform is a necessary protection against problems with declining literacy and linguistic standards. To strengthen their case, such commentators often exaggerate the extent of the problem. Just as Masha Bell claimed that the printing revolution produced a spelling system that was an unintelligible mess, so Naomi Baron suggests that a possible future for English spelling in the digital age will be to 'revert to an attitude toward spelling and punctuation conventions redolent of the quasi-anarchy of medieval and even Renaissance England'. As we have seen, the spelling of Middle and Early Modern English was far from anarchic; many words were spelled identically to Modern English spellings. While it is true that there was greater tolerance of variation, this rarely caused problems for comprehension. Similarly, the tolerance of non-standard spelling, punctuation and the use of abbreviations in email, texting, and other forms of electronic communication, do little to hinder communication.

Finally, there seems to me another reason for resisting any attempts to reform English spelling, and retaining traditional spellings, silent letters and all: such spellings are a testimony to the richness of our language and its history. This argument is harder to defend as it has no practical purpose, although it does help to maintain a connection between our present-day language and that of the past. Modern English speakers would surely find it more difficult to read the works of Chaucer and Shakespeare if our spelling system was to be radically reformed. Silent letters are silent witnesses to pronunciations that have since been lost, but which continue to be preserved in a spelling system that boasts a long and rich heritage.

Further reading

Chapter 2

Excellent introductions to writing systems can be found in Andrew Robinson, *Writing and Script: A Very Short Introduction* (Oxford: OUP, 2009) and Michelle Brown, *The British Library Guide to Writing and Scripts* (London: British Library, 1998). For a linguistically more sophisticated discussion see Geoffrey Sampson, *Writing Systems: A Linguistic Introduction* (London: Hutchinson, 1985). John Chadwick, *The Decipherment of Linear B* (Cambridge: CUP, 1990) is a very readable account of the fascinating journey that led to the successful cracking of Linear B. The classical doctrine of the *littera* and its later influence are discussed in David Abercrombie, 'What is a "letter"?', in D. Abercrombie (ed.), *Studies in Phonetics and Linguistics* (London: OUP, 1965), 76–85; the medieval applications of this theory are assessed by Margaret Laing and Roger Lass in the Introduction to *A Linguistic Atlas of Early Middle English, 1150–1325*, http://www.lel.ed.ac.uk/ihd/laeme1/laeme1.html. On Tolkien's invented languages see Elizabeth Solopova, *Languages, Myths and History: An Introduction to the Linguistic and Literary Background of J.R.R. Tolkien's Fiction* (New York: North Landing Books, 2009). The nature of a standard language and its role in modern English society is debated by J. Milroy and L. Milroy, *Authority in Language: Investigating Standard English*, 3rd edn (London: Routledge, 1999).

Chapter 3

The best introduction to runes is that by R. I. Page, *An Introduction to English Runes* (Woodbridge: Boydell Press, 2006). If you want to extend your knowledge and experience of Old English, there are many introductions to Old English with sample texts. A good place to begin is Bruce Mitchell and Fred C. Robinson (eds), *A Guide to Old English*, 7th edn (Oxford: Blackwell, 2006). For a more linguistically-informed account, including some discussion of Old English dialect variation, try Jeremy J. Smith, *Old English: A Linguistic Introduction* (Cambridge: CUP, 2009). If the opening lines of *Beowulf* have inspired you to read more of the poem, there is an excellent translation by Seamus Heaney, *Beowulf: A New Translation* (London: Faber, 1999).

Chapter 4

For an overview of Middle English see Simon Horobin and Jeremy Smith, *An Introduction to Middle English* (Edinburgh: Edinburgh University Press, 2002); Chaucer's language, including his use of dialect in *The Reeve's Tale*, is the focus of Simon Horobin, *Chaucer's Language* (Basingstoke: Palgrave Macmillan, 2006; 2nd edn 2012). There is an interesting sociolinguistic account of the status of English during the Middle Ages in Tim William Machan, *English in the Middle Ages* (Oxford: OUP, 2003). For an assessment of the process of standardization in Middle English see Jeremy J. Smith, *An Historical Study of English: Function, Form and Change* (London: Routledge, 1996). J. R. R. Tolkien's classic study of AB Language is '*Ancrene Wisse* and *Hali Meiðhad*', *Essays and Studies by Members of the English Association* 14 (1929), 104–26.

Chapter 5

Terttu Nevalainen, *An Introduction to Early Modern English* provides a very readable account of Early Modern English. There are several books that focus specifically on Shakespeare's language; the best introduction available is David Crystal, *Think on my Words: Exploring Shakespeare's Language* (Cambridge: CUP, 2008). For an account of the various spelling reformers and their views on language, see E. J. Dobson, *English Pronunciation 1500–1700* (Oxford: Clarendon Press, 1957), volume 1.

Chapter 6

For a more detailed overview of the development of English in this period see Joan C. Beal, *English in Modern Times* (London: Hodder, 2004). Darwin's spelling is discussed in two articles by Frank J. Sulloway, 'Darwin's conversion: the *Beagle* voyage and its aftermath', *Journal of the History of Biology* 15 (1982), 325–96 and 'Further remarks on Darwin's spelling habits and the dating of *Beagle* voyage manuscripts', *Journal of the History of Biology* 16 (1983), 361–90. Lynda Mugglestone quotes James Murray's views on the spelling of *whisky* in *Dictionaries: A Very Short Introduction* (Oxford: OUP, 2011). Henry Bradley's views on spelling can be found in his essay *On the Relations between Spoken and Written Language, with Special Reference to English* (London: Oxford University Press, 1913). The Cut Spelling handbook setting out its various proposals can be found on the website of the English Spelling Society: <http://www.spellingsociety.org/>. For an account of the 'Shavian' alphabet see P. A. D. MacCarthy, 'The Bernard Shaw Alphabet', in W. Haas (ed.), *Alphabets for English* (Manchester: Manchester University Press, 1969), 105–17.

Chapter 7

For an overview of American English see R. W. Bailey, *Speaking American: A History of English in the United States* (Oxford: OUP, 2012) and the account of orthography in R. Venezky, *The American Way of Spelling* (New York: The Guilford Press, 1999). The life and career of Noah Webster, and the phenomenal success of his Blue-Back speller, are assessed in Jennifer E. Monaghan, *A Common Heritage: Noah Webster's Blue-Back Speller* (Hamden, CT: Archon Books, 1983). For a cultural history of the American spelling bee see James Maguire, *American Bee: The National Spelling Bee and the Culture of Word Nerds* (New York: Rodale, 2006).

Chapter 8

The Coventry University study of young people's texting habits is published as Crispin Thurlow, 'Generation Txt: Exposing the sociolinguistics of young people's text-messaging', *Discourse Analysis Online* 1:1 (2003) <http://extra.shu.ac.uk/daol/previous/v1_n1.html> David Crystal discusses these findings and the potential impact of text-messaging upon literacy levels in *Txtng: The Gr8 Db8* (Oxford: OUP, 2008). For a wide-ranging discussion of spelling reform see Mark Sebba, *Spelling and Society: The Culture and Politics of Orthography Around the World* (Cambridge: CUP, 2007), ch. 6. The German spelling reform is discussed by S. Johnson, *Spelling Trouble? Language, Ideology and the Reform of German Orthography* (Clevedon: Multilingual Matters, 2005). On the proposals for Dutch spelling reform see D. Jacobs, 'Alliance and betrayal in the Dutch orthography debate', *Language Problems and Language Planning* 21 (1997), 103–18. Masha Bell's interview with Libby Purves is quoted from a transcript available at <http://www.englishspellingsociety.org/news/media/bell1.php>; her views on English spelling are set out at greater length in *Understanding English Spelling* (Cambridge: Pegasus, 2004). The study of children learning to spell Kiswahili is found in K. J. Alcock and D. Ngorosho, 'Learning to spell a regularly spelled language is not a trivial task—patterns of errors in Kiswahili', *Reading and Writing* 16 (7) (2003), 635–66.

Bibliography

Abercrombie, David, 'What is a "Letter"?', in D. Abercrombie, *Studies in Phonetics and Linguistics* (London: Oxford University Press, 1965), 76–85.

Ager, Dennis E., *Language Policy in Britain and France: The Processes of Policy* (London: Cassell, 1996).

Alcock, K. J. and D. Ngorosho, 'Learning to spell a regularly spelled language is not a trivial task—patterns of errors in Kiswahili', *Reading and Writing* 16 (7) (2003), 635–66; quoted from Mark Sebba, *Spelling and Society: The Culture and Politics of Orthography Around the World* (Cambridge: Cambridge University Press, 2007).

Anderson, George B., 'The forgotten crusader: Andrew Carnegie and the Simplified Spelling movement', *Journal of the Simplified Spelling Society* J26 (1999), 11–15; <http://www.spellingsociety.org/journals/j26/carnegie.php> last accessed 28 August 2012.

Bailey, R. W., *Speaking American: A History of English in the United States* (New York: Oxford University Press, 2012).

Baron, Naomi S., *Always On: Language in an Online and Mobile World* (Oxford: Oxford University Press, 2008).

Bell, Masha, *Understanding English Spelling* (Cambridge: Pegasus, 2004), quoted from a transcript available at <http://www.englishspellingsociety.org/news/media/bell1.php> last accessed 28 August 2012.

Bradley, Henry, *On the Relations between Spoken and Written Language, with Special Reference to English* (London: Oxford University Press, 1913).

Brown, Michelle, *The British Library Guide to Writing and Scripts* (London: British Library, 1998).

Burchfield, R. W. (ed.), *The New Fowler's Modern English Usage* 3rd edition (Oxford: Clarendon Press, 1996).

Butler, E. H., *The Story of British Shorthand* (London: Isaac Pitman & Sons, 1951).

Cameron, Deborah, *Verbal Hygiene* 2nd edn (London: Routledge, 2012).

Carney, E., *A Survey of English Spelling* (London: Routledge, 1994).

Chadwick, John, *The Decipherment of Linear B* (Cambridge: Cambridge University Press, 1990).

Cook, Vivian, *Accomodating Brocolli in the Cemetary* (London: Profile Books, 2004).

Cook, Vivian, *The English Writing System* (London: Hodder, 2004).

Crystal, David, *Language and the Internet* 2nd edn (Cambridge: Cambridge University Press, 2006).

Crystal, David, *Txtng: The Gr8 Db8* (Oxford: Oxford University Press, 2008).

Daines, Simon, *Orthoepia Anglicana* (London, 1640).

Davidson, George, *Penguin Writers' Guide: Improve Your Spelling* (London: Penguin, 2005).

Davies, Eirlys E., 'Eyeplay: on some uses of nonstandard spelling', *Language and Communication* 7 (1987), 47–58.

Dobson, E. J., *English Pronunciation 1500–1700*, 2 volumes (Oxford: Clarendon Press, 1957).

Ellis, A. J., *On Early English Pronunciation* Early English Text Society ES 2 (London: Asher, 1874).

Essinger, James, *Spellbound: The Surprising Origins and Astonishing Secrets of English Spelling* (New York: Bantam Dell, 2006).

Franklin, Benjamin, 'Scheme for a New Alphabet and Reformed Mode of Spelling', in Benjamin Franklin, *Political, Miscellaneous, and Philosophical Pieces* (London, 1779), 475–87.

Geerts, G., J. Van Den Broeck, and A. Verdoodt, 'Successes and Failures in Dutch Spelling Reform', in J. A. Fishman (ed.), *Advances in the Creation and Revision of Writing Systems* (The Hague: Mouton, 1977), 179–245.

Goodwin, James (ed.), *The Gospel According to Saint Matthew Translated into English from the Greek by Sir John Cheke* (Cambridge: J. and J.J. Deighton, 1843).

Grandgent, Charles H., Calvin Thomas, and Henry Gallup Paine, *Handbook of Simplified Spelling* (New York: Simplified Spelling Board, 1920).

Hale, Constance, *Wired Style: Principles of English Usage in the Digital Age* (San Francisco: Wired Books, 1997).

Heaney, Seamus, *Beowulf: A New Translation* (London: Faber, 1999).

Horobin, Simon, *Chaucer's Language* (Basingstoke: Palgrave Macmillan, 2006; 2nd edn 2012).

Horobin, Simon and Jeremy Smith, *An Introduction to Middle English* (Edinburgh: Edinburgh University Press, 2002).

Jacobs, D., 'Alliance and betrayal in the Dutch orthography debate', *Language Problems and Language Planning* 21 (1997), 103–18.

Joffe, Alexander, 'Introduction: non-standard orthography and non-standard speech', *Journal of Sociolinguistics* 4 (2000), 497–513.

Johnson, S., *Spelling Trouble? Language, Ideology and the Reform of German Orthography* (Clevedon: Multilingual Matters, 2005).

Kimball, Cornell, 'Investigating spelling reform satire', *Journal of the Simplified Spelling Society*, J31 (2002), 20–2. <http://www.spellingsociety.org/journals/j31/satires.php> last accessed 28 August 2012.

Laing, Margaret and Roger Lass, *A Linguistic Atlas of Early Middle English, 1150–1325* (Edinburgh: The University of Edinburgh, 2007) <http://www.lel.ed.ac.uk/ihd/laeme1/laeme1.html> last accessed 12 August 2012.

Lee, Guy (trans. and intro.), *Catullus: The Complete Poems* (Oxford: Oxford University Press, 2008).

Lepore, Jill, *A is for American: Letters and Other Characters in the Newly United States* (New York: Vintage Books, 2003).

Machan, Tim William, *English in the Middle Ages* (Oxford: Oxford University Press, 2003).

Maguire, James, *American Bee: The National Spelling Bee and the Culture of Word Nerds* (New York: Rodale, 2006).

Milroy, J. and L. Milroy, *Authority in Language: Investigating Standard English*, 3rd edn (London: Routledge, 1999).

Mitchell, Bruce and Fred C. Robinson (eds), *A Guide to Old English*, 7th edn (Oxford: Blackwell, 2006).

Mitchell, David, 'Standards of spelling', <http://www.youtube.com/user/davidmitchellsoapbox> last accessed 28 August 2012.

Monaghan, Jennifer E., *A Common Heritage: Noah Webster's Blue-Back Speller* (Hamden, CT: Archon Books, 1983).

Mugglestone, Lynda, *Dictionaries: A Very Short Introduction* (Oxford: Oxford University Press, 2011).

Page, R. I., *An Introduction to English Runes* (Woodbridge: Boydell Press, 2006).

Pitman, James, *The Shaw Alphabet Edition of Androcles and the Lion* (Harmondsworth: Penguin, 1962).

Robertson, James, *The Ladies Help to Spelling* (Glasgow, 1722).

Robinson, Andrew, *Writing and Script: A Very Short Introduction* (Oxford: Oxford University Press, 2009).

Sampson, Geoffrey, *Writing Systems: A Linguistic Introduction* (London: Hutchinson, 1985).

Sauer, W. W. and H. Glück, 'Norms and Reforms: Fixing the Form of the Language', in P. Stevenson (ed.), *The German Language and the Real World* (Oxford: Clarendon Press, 1995).

Schiffman, H. F., *Linguistic Culture and Language Policy* (London: Routledge, 1996).

Scragg, D. G., *A History of English Spelling* (Manchester: Manchester University Press, 1974).

Sebba, Mark, *Spelling and Society: The Culture and Politics of Orthography Around the World* (Cambridge: Cambridge University Press, 2007).

Simpson, J. A. and E. Weiner (eds), *Oxford English Dictionary* 2nd edn (Oxford: Oxford University Press, 1989).

Smith, Jeremy J., *An Historical Study of English: Function, Form and Change* (London: Routledge, 1996).

Smith, Jeremy J., *Old English: A Linguistic Introduction* (Cambridge: Cambridge University Press, 2009).

Solopova, Elizabeth, *Languages, Myths and History: An Introduction to the Linguistic and Literary Background of J.R.R. Tolkien's Fiction* (New York: North Landing Books, 2009).

Swift, Jonathan, Letters quoted from <http://www.swiftiana.com/stella> last accessed 28 August 2012.

Thurlow, Crispin, 'Generation txt: exposing the sociolinguistics of young people's text-messaging', *Discourse Analysis Online* 1:1 (2003) <http://extra.shu.ac.uk/daol/previous/v1_n1.html> last accessed 12 August 2012.

Tolkien, J. R. R., '*Ancrene Wisse* and *Hali Meiðhad*', *Essays and Studies by Members of the English Association* 14 (1929), 104–26.

Truss, Lynne, *Eats, Shoots and Leaves: The Zero-Tolerance Approach to Punctuation* (London: Profile, 2003).

Twain, Mark, 'A Simplified Alphabet', in *What is Man? And Other Essays* (Bremen: Europäischer Hochschulverlag GmbH & Co. KG, 2010).

Upward, Christopher and George Davidson, *The History of English Spelling* (Oxford: Wiley-Blackwell, 2011).

Venezky, Richard, 'Principles for the Design of Practical Writing Systems', in Joshua A. Fishman (ed.), *Advances in the Creation and Revision of Writing Systems* (Berlin: Walter de Gruyter, 1977), 37–54.

Venezky, Richard, *The American Way of Spelling* (New York: The Guilford Press, 1999).

Walker, John, *A Critical Pronouncing Dictionary* (London, 1791).

Waterhouse, Keith, *English our English* (London: Viking, 1991).

Webster, Noah, *Dissertations on the English Language* (Boston, 1789).

Webster, Noah, *The American Spelling Book* (Boston, 1793).

Wells, H. G., 'For Freedom of Spelling: The Discovery of an Art', in *Certain Personal Matters* (London: T. Fisher Unwin, 1901).

Wilkins, John, *Essay towards a Real Character* (London, 1668).

Willans, Geoffrey and Ronald Searle, *The Compleet Molesworth* (London: Pavilion Books, 1985).

Word Index

aardvark 136
abandon 26
abbot 64
able 81, 161
abridgement/abridgment 200
absolve 173
accommodate 8
accommodation 165
ache 147, 198
active 206
adder 62
admiral 113
admissible/admissable 230–1
admit 9
adventine 147
adventure 33, 112, 113
advertise 202
advice 112, 113
advise 202
after 124
again 124, 219
against 124
aghast 204
air 81
aisle 113, 173
allegation 247
allege 247
allegory 93
almanac 198
almighty 75
although 198
amoeba 201
ampersand 216
ancient 13
anthem 113
any 124
apostle 64
apple 9
appoggiatura 6, 7
apprise 202
archaeology 201
arise 202
arrive 206
artichoke 135
ask 52, 62
at 69
attention 94
attraction 94
author 113
autochthonous 6
autumn 247
autumnal 247
avenge 124
axe 52, 167

bad 141
badge 200
bath 175
beagle 6
beau 134
because 14, 222
been 158
beggar 148
beige 13, 32
believe 13, 141, 156, 163
big 9
bird 63
biscuit 134, 165
bishop 64
blue 134
bonfire 147–8
bramble 59
break 91, 160
bridge 200

brought 219
build 198
bury 140
business 155

Caesar 201
café 134
caffeine 13
can 87
capercailye 91
car 9, 10, 175
card 10
cardinal 79
care 206
castle 9
cat 9
catalogue/catalog 206
catch 87, 92
catechism 196
cattle 92
centenary 248
centennial 248
center/centre 175, 197
centipede 248
century 248
chalk 62, 86, 87
champion 87
chariot 87
chase 92
chastise 202
chattel 92
check 38
cheese 9
cheque 38
cherry 190
chic 63
chief 141
child 56
chimneys 161
chin 86
chips 9
choke 145
church 56
churl 87
ciabatta 32, 135
ciao 135
circumcise 198, 202
city 9, 32, 87, 97
client 93
colo(u)r 34, 175, 197
comb 59
come 56, 97
comparative 206
complete 146, 156
compromise 202
compute 132
confirmation 94
confuse 206
connection 198
conviction 93
cool 222
corps 112
corpse 112
cot 20, 21
could 87
council/counsel 149
count 132
cow 86
criticise 198
crocuses 165
crumb 59, 198, 206
cryptic 53
cut 10
cute 10

dachshund 7
Dalziel 91
datable 198
daughter 84
day 173
debatable 198
debt 111, 246
deceit 145
deceive 141
deed 190
deep 101
defence/defense 88, 149, 197
delight 114
demon/dæmon 201–2
desiccate 8, 14

desktop 214
despise 202
despite 114
determine 196, 198
device 88
devil 148
diarrhoea 201
did 190
digitize 90
discreet/discrete 149
disequilibrium 7
disguise 202
dispatch/despatch 230–1
distil 198
does 156
dog 9, 32
done 124
doubt 111, 171, 246
doubtless 165
dough 85
drought 145
dry 145

earth 79, 81
ease 91
ecology 201
economy 112
edge 56
efficient 13
eight 13
electric/electrical/electricity 247
embarrassment 8, 165
embrace 93–4
empty 60
enchant 145
enchantment 145
encyclopedia 201
enough 85, 206, 208
enquire 94
ensure 94
enterprise 198
entire 145
episcopal 64
equal 112
errand 79
estimate 112
everyone 216
executor 93
exit 133
exotic 133

face 87
facsimile 133
factotum 133
family 97
fan 58
fancy 145
fashion 198
fault 112
faux pas 134
favo(u)r 161, 194, 197
favourite 206
fetus 231
fine 89
fish 10, 52, 62
flour 149–50
flower 149–50
fnese 60
folio 133
folk 62, 89
food 158
forehead 33
forever 212, 216
forsake 26
forth 69
fracas 6
freight 210
friend 91, 206
frolic 198
fuel 145
fulfil 198

gammon 92
gaol 92, 93, 156
garden 92
gaze 90
gear 9
gentle 9
ghastly 204
gherkin 204

ghost 127, 204
give 103, 196
given 171
gladiolus 6
gnat 60
gnaw 60
gnidan 60
gnorn 60
goat 9
gobbledegook 231
god 73, 100
gone 124
good 70, 73, 127
gospel 47
grace 87
grammar 247–8
grammatical 247–8
graphic 210
great 160, 212
grieve 141
grocer 148
groove 124
guages 165
guarantee 92, 93
guardian 92
guerdon 6
guess 125
guest 125, 127
guile 92
gymkhana 136
gynaecology/gynecology 201
gypsy 161

had 141
half 62
hand 217
hate 9
haughty 114
have 10
head 196
health 206
hear 9, 156
heart 91
heat 91
heaven 75
height 145
heir 81
helped 171
herb 81
high 84, 145
history 81, 93
historian 81
home 124
honour 81, 194, 196, 197, 206
hope 10, 161
host 33, 81
hotel 81
house 58, 89, 159
humor 161

ice 88, 149
ignoramus 133
ignorant 94
immature 133
improvise 202
incantation 145
incline 93
indubitable 246
inflection 198
inquire 94
instalment 165
instead 214
insure 94
intelligible 7
interfered 166
island 112–13, 198
isle 112–13, 198
its/it's 11–12, 130–2, 165

jail 92, 93, 156
jam 32
jamais 190
James 190
jewel 145
judgement 38, 194, 200
Julius 52

Keith 13
kin 86
king 56
kissed 206
knack 6

knave 191
knight 10, 33, 243, 249
knot 59, 245–6

labour/labor 197
lamb 10, 59, 171
language 111
Laodicean 6
laptop 214
later 100
latter 100
laugh 85
lead 129
leapfrog 214
legalize 198
liaison 134
licence/license 148
lie 100
life 124, 158
light 84, 86, 114
like 124
limb 59
lisp 61
live 124
loaf 61
loch 32
logic 198
lore 70
love 89, 90, 97, 120, 121

macaroni 165
magister 65
maid 173
mangetout 63
many 140
mask 198
mass 64
maze 90
meat 9, 91, 159–60
medic/medical/medicine 247
medieval 201
meet 91, 159–60
memorise 90
mend 190
Menzies 91
metre/meter 197
mice 88, 149
Michael 201
misdeed 214
misfortune 214
misspell 214
moccasin 188
molasses 135, 196
monk 64
mood 158
moralize 198
mouse 88
movable 198
much 69, 121
music 127, 196
mynster 64

name 158
neighbour 196
nephew 111
nese 60
newt 62
nice 88
night 84, 86, 114, 171, 222, 225
nightmare 198
nostril 63
not 245–6
noticed 206
nut 61
nyctophobia 7

oak 79
occasionally 8
old 79
olde 85, 102
once 88
opossum 187–8
opponent 94
other 99
over 69

paleography 201
pamphlet 156
panel 160
parent/parental 247–8
parlor 161
patronize 198

pauper 132
peace 91
people 92
perceive 141
perfect 112
phantom 145
phat 222
pheasant 125
phishing 222
phoenix 64, 87
phreak 222
phthisis 209
physics 97
pick 124
pigeonhole 214
plait 173
play 97
please 212
pleasure 91, 173
plumber 59
plum-duff 85
policies 13
pony 161, 163
poor 132
possum 188
potato 2, 135
powwow 188
practice/practise 148, 197
prejudge 133
prejudicate 133
proclivity 133
pronounce/pronunciation 247
prophet 101
protein 13
proviso 133
psalm 170
Psmith 170
psychic 170
psychology 53, 97, 170

quality 88
quay 173
queen 53, 89
quick 89, 124, 225–6
quiet 88
quire 111
quoth 70

raccoon 187
read 9, 129
realise 90
reason 91
recall 160
receipt 111, 145
receive 13, 141
reflection 198
regard 92
remained 190
repentant 94
reward 92
rhyme 206
ring 61, 245–6
roof 75
rot 21
rough 85

sad 141
salmon 111
sane 247
sanity 247
says 173
scale 96
scare 96
sceptic 96
school 206, 223
science 96
scissors 111
scope 96
scripture 93
scythe 111
sea 69, 159–60
seal 90
secrecy 161
secure 132
see 159–60
seed 69, 129
seen 158
seethe 124
seize 13
sends 165

separate 8
serrefine 6, 7
shall 34, 100
shampoo 135
she 100
Sheila 13
sherbet 135
ship 9, 69
shirt 96
should 100, 165
siege 13
sight 84
sign/signal/signature 246
sin 97
sing 10
skate 96
skeptic 96
skin 95, 96
skipper 96
skirt 96
sky 95, 96
slumber 59
sneeze 60, 124
soap 145
society 13
solder 210
son 89, 97, 120
soon 70
sophisticated 210
speak 159
spindle 59
spite 114
standardize 90
star 101
steak 160
stick 124
stone 33–4, 70
stopped 206
strength 145
stromuhr 6, 7
strong 145
subpoena 7
subtlety 165
such 62, 82, 124
supersede 8, 14
supervise 202
sure 132
surprise 198, 202

take 159
tariff 210
tear 9
tease 163
temporary 165
text 212
thanks 9, 212
that 69
theatre/theater 196, 197
their 102
them 102
these 173
thickness 69
thimble 59
third 63
thorn 69
though 9, 38, 84, 127, 163, 165, 206, 219
thought 219
three 101
through 38, 82, 84, 124, 127, 163, 165, 206
thumb 59
thunder 59
toboggan 188
tomato 135
tomb 59
tomorrow 3, 15, 212, 214
tongue 196, 198, 206
tonight 217
torsion 7
town 89, 159
toy 97
traveled/travelled 197, 198
troglodyte 7
trousers 161
truly 140
tusk 56

umbrella 165
until 90, 121
upon 124
ursprache 6, 7

van 9
verse 58
very 121, 124, 140, 165
victuals 111
villain 191
villainous 161
vine 89
vision 173

waistcoat
war 92
warden 92
warranty 92, 93
was 69, 223
watch 87
way 69
whale 61
what 61, 78, 83, 223, 243
when 78, 100, 243
where 61, 100
which 62, 78, 83, 100, 165, 243
whisky 166
whole 62
whore 62
why 78, 83
wife 10, 158
wigwam 188
wile 92
wise 89, 124, 206
woe 198
wolf 57–8
wrath 61
write 61
wreath 61

yacht 32
yard 55
ye 85, 91, 102
yea 160
year 55, 91, 99
yet 86
yoke 86
you 86, 206
young 196
your 165

zeal 90
zedoary 90
zucchetto 90
zygoma 90

Subject Index

AB language 98–9
abbreviations 38, 73, 85, 106, 132, 139, 150, 163, 165, 212–13, 215–20
Ælfric 21–2, 53, 65
affixation 63
alphabetic depth 34–5
apostrophe 11–12, 131–2, 139–40, 165, 215–16, 226
Austen, Jane 4, 162–3

Baron, Naomi 215–16, 251
beau spelling 152
Bell, Alexander Melville, *see* Visible Speech
Bell, Masha 240–3, 251
Beowulf 70–3, 217
Blackadder 136
borrowing 33, 63–5, 79–80, 89, 94–6, 133–4, 145, 188, 201
boustrophedon 49
Bradley, Henry 168–9
Brown, Charlie 6
Bullokar, William 121–2, 130

calques 117–18
Cameron, Deborah 230
capitalization 139, 157, 236
Carolingian minuscule 86
Caxton, William 85, 126–7, 240–2
Chaucer, Geoffrey 91, 102–7, 209, 241, 249, 252
Cheke, Sir John 115–17, 119
complementary distribution 58
compounding 63, 77
Coote, Edmund 128
Crystal, David 215, 220, 222
cuneiform 18
cut spelling 171–2

Daines, Simon 23
Darwin, Charles 163–4, 167
Denison, David 164
dialect translation 82–3
dialects 65, 73–6, 78, 102–3, 107, 141–3
Dickens, Charles 160–1, 163
Donatus 21
doublets 92, 132

Etruscan 49–52
etymological respelling 33, 62, 111–14, 134, 246

female spelling 153–6
Fielding, Henry 162
folk etymology 135
Fowler, H.W. 200–1
Franklin, Benjamin 188–91, 194, 196
Franks Casket 43–4
French dialects 92

glossic 167
great vowel shift 158–9

h-dropping 10, 37, 61, 79–81, 161, 250
half-uncial 44–5
Hart, John 23, 24, 119–21, 124, 130, 157–8
hieroglyphics 16–8, 25
Hodges, Richard 128–30
homonymic clash 168
homophones 120, 128–9, 160, 189, 190, 227, 242, 245

Humphrys, John 213–14
hypercorrection 80, 161
hyphens 213–14, 235

Initial Teaching Alphabet 180–4
insular minuscule 45–7, 73, 86

Johnson, Dr Samuel 1, 23, 59, 134, 136, 144–50, 156, 176, 194, 211, 231
Jonson, Ben 79

Laȝamon 101–2
Late West Saxon 65–7, 73–6, 83, 85, 95, 98
lindisfarne gospels 45–6, 217
linear B 18–19
logographs 19, 215, 216

metanalysis 62
metathesis 52, 62
minim confusion 97, 104, 120–1
mnemonics 14
Mugglestone, Lynda 166
Mulcaster, Richard 122–5, 127
Murray, James, 166–7, 168, 170, 200, 203

Orm 99–101

pictograms 16, 18, 19
Pitman, Isaac 176, 205, 208
Pitman, James 180–3
place names 186–7
printing 85, 97, 125–7
Priscian 21
proper nouns 169–70, 225
Pullman, Philip, 201–2
punctuation 4, 11, 72, 137, 157, 160, 215–6, 219, 221, 227, 251

Received Pronunciation 37–8, 183, 244
Robertson, James 155
runes 29–30, 39–44, 54, 85, 217

scriptio continua 72
Shakespeare, William 90, 113–14, 126–7, 131, 136–43, 161, 180, 203, 208, 210, 252
Shavian alphabet 177–81
Shaw, George Bernard 10, 175–7, 179, 180, 211
Sheridan, Thomas 147, 150
shorthand 26, 29, 176–7, 181, 209, 217
Simplified Spelling Board 203–7
Sir Gawain and the Green Knight 106
Smith, Sir Thomas 117–19
spelling bee 2, 5–8, 199–200
spelling pronunciations 33
spelling reform:
  Dutch 237–8
  English 238–48
  French 236–7, 250–1
  German 233–6, 250
spelling rules 12–14, 163
standardization 33, 35–8, 65–7, 103, 110, 114, 163, 185
synonymy 93, 133
Swift, Jonathan 150–5, 156–7, 159, 163, 185, 219

Tennyson, Lord 167
Tironian nota 217
Tolkien, J.R.R. 28–31, 39, 98
trade names 225–6
transitional alphabets 128
Truss, Lynne 11, 131, 219
Twain, Mark 8–9, 14, 203, 207–10, 225
Twitter 218, 227–8

United States Board on Geographic Names 186–7

Visible Speech 26–8

Walker, John 79
Webster, Noah 186, 191–9, 203, 210–11
Wells, H.G. 224–5
Wijk, Axel 172–5
Wilkins, John 25–6
Wodehouse, P.G. 170

yogh 86, 91